Lecture Notes in Electrical Engineering

Volume 1221

The book series *Lecture Notes in Electrical Engineering* (LNEE) publishes the latest developments in Electrical Engineering—quickly, informally and in high quality. While original research reported in proceedings and monographs has traditionally formed the core of LNEE, we also encourage authors to submit books devoted to supporting student education and professional training in the various fields and applications areas of electrical engineering. The series cover classical and emerging topics concerning:

- Communication Engineering, Information Theory and Networks
- Electronics Engineering and Microelectronics
- Signal, Image and Speech Processing
- Wireless and Mobile Communication
- Circuits and Systems
- Energy Systems, Power Electronics and Electrical Machines
- Electro-optical Engineering
- Instrumentation Engineering
- Avionics Engineering
- Control Systems
- Internet-of-Things and Cybersecurity
- Biomedical Devices, MEMS and NEMS

For general information about this book series, comments or suggestions, please contact leontina.dicecco@springer.com.

To submit a proposal or request further information, please contact the Publishing Editor in your country:

China

Jasmine Dou, Editor (jasmine.dou@springer.com)

India, Japan, Rest of Asia

Swati Meherishi, Editorial Director (Swati.Meherishi@springer.com)

Southeast Asia, Australia, New Zealand

Ramesh Nath Premnath, Editor (ramesh.premnath@springernature.com)

USA, Canada

Michael Luby, Senior Editor (michael.luby@springer.com)

All other Countries

Leontina Di Cecco, Senior Editor (leontina.dicecco@springer.com)

**** This series is indexed by EI Compendex and Scopus databases. ****

Sabu M. Thampi · Vipin Chaudhary ·
Al-Sakib Khan Pathan · Kuan Ching Li ·
Dilip Krishnaswamy
Editors

Fifth International Conference on Computing and Network Communications

Proceedings of CoCoNet 2023, Volume 2

Springer

Editors
Sabu M. Thampi
Department of Computer Science and Engineering
Digital University Kerala
Trivandrum, Kerala, India

Al-Sakib Khan Pathan
Department of Computer Science and Engineering
United International University
Dhaka, Bangladesh

Dilip Krishnaswamy
Quantum Walks Technologies
Cladwell, NJ, USA

Vipin Chaudhary
Department of Computer and Data Sciences
Case Western Reserve
Cleveland, OH, USA

Kuan Ching Li
Department of Computer Science and Information Engineering
Providence University
Shalu, Taiwan

ISSN 1876-1100 ISSN 1876-1119 (electronic)
Lecture Notes in Electrical Engineering
ISBN 978-981-97-4713-9 ISBN 978-981-97-4711-5 (eBook)
https://doi.org/10.1007/978-981-97-4711-5

This Springer imprint is published by the registered company Springer Nature Singapore Pte Ltd.
The registered company address is: 152 Beach Road, #21-01/04 Gateway East, Singapore 189721, Singapore

Organized By

PES University, Bengaluru, India

Conference Organization

Chief Patron

M. R. Doreswamy, Chancellor, PES University

Patrons

D. Jawahar, Pro-Chancellor, PES University
Ajoy Kumar, COO, PES Institutions
J. Surya Prasad, Vice Chancellor, PES University
K. S. Sridhar, Registrar, PES University

General Chairs

Ljiljana Trajkovic, Simon Fraser University, Canada
Jaime Lloret, Universitat Politecnica de Valencia, Spain
Shikha Tripathi, PES University, Bengaluru

General Executive Chair

Sabu M. Thampi, Digital University Kerala, India

Steering Committee

Sudarshan T. S. B., Dean of Research, PES University (Chair)
San Murugesan, BRITE Professional Services and University of Western Sydney, Australia
Kuan Ching Li, Providence University, Taiwan

TPC Chairs

Patrick Siarry, University Paris-Est Créteil Val de Marne, France
Mohammed Atiquzzaman, School of Computer Science, University of Oklahoma, USA
Dilip Krishnaswamy, Quantum Walks Technologies, USA
Vipin Chaudhary, Case Western Reserve University, USA

Workshop and Symposium Chairs

Peter Mueller, IBM Zurich Research Laboratory, Switzerland
Axel Sikora, University of Applied Sciences Offenburg, Germany

Tutorial Chairs

Sameep Mehta, IBM Research AI, India
Al-Sakib Khan Pathan, United International University (UIU), Dhaka, Bangladesh

Organizing Chair

Anuradha M., PES University

Organizing Secretaries

M. S. Sunitha, PES University

Event Management Chair

M. Rajasekar, PES University

TPC Members

https://coconet-conference.org/2023/?q=committee

Preface

The Fifth International Conference on Computing and Network Communications (CoCoNet'23) took place in Bengaluru (Bangalore), India, during December 18–20, 2023. The conference provided a platform for experts in the emerging areas of computing and communications to share original research outcomes and practical development experiences. The event was organized by PES University, Bengaluru, with technical support from the IEEE Signal Processing Society Bangalore Chapter and the IEEE Communications Society Bangalore Chapter.

The program included keynote talks, a Start-up Expo, Panel Discussions, lightning talks, tutorials, symposiums, parallel sessions, workshops, Hands-on Sessions, and Hot off the Press. CoCoNet'23 convened a tailored and selected collection of eminent speakers from renowned universities and industries in and outside India. The keynote speakers included Dr. C. Mohan, Dr. Ranga Rao Venkatesha Prasad, Dr. Sanjay K. Madria, Dr. S. R. Mahadeva Prasanna, and Dr. Sri Krishnan. An industrial keynote was delivered by Dr. Vishwas Lakkundi. The tutorial speakers were Dr. Abdullah Uz Tansel, Dr. Nithin Nagaraj, Dr. Amitava Mukherjee, Dr. Vishnu Srivastava, Dr. H. L. Shashirekha, Asha Hegde, Karthik Sivarama Krishnan, and Dr. Jaya Dofe.

The conference received 212 submissions, with 91 papers accepted after a rigorous review process evaluating their significance, novelty, and technical quality. The accepted papers were presented in regular paper sessions, symposiums, workshops, and short paper sessions.

The proceedings of the conference are organized into two volumes. This volume comprises 47 papers. The topical sections include Artificial Intelligence and Machine Learning, Healthcare and Medical Applications, Computer Vision, Natural Language Processing, and Education Informatics.

The success of the conference ultimately depends on the numerous individuals to plan and organize the technical program and local arrangements. In particular, we extend our gratitude for the insightful advice and innovative ideas provided by the Program Chairs, Workshop and Symposium Chairs, and Tutorial Chairs. Our heartfelt thanks also go to the organizing committee and steering committee members.

The accomplishment of the conference is attributed to the tireless efforts of many individuals. We extend our sincere gratitude to the TPC members and reviewers who generously shared their technical expertise and assisted us in the thorough review of all submitted papers. Heartfelt thanks go to the general chairs, organizing committee members, steering committee, keynote speakers, tutorial speakers, session chairs, and conference attendees. We appreciate the authors for choosing this conference as the platform to present their research works. Our wholehearted appreciation goes to all those who dedicated their valuable time to ensure the success of CoCoNet'23.

We are thankful to PES University for hosting the conference. Recognition is due to the Local Organizing Committee members who worked diligently on the crucial aspects of the conference programs. We acknowledge the contributions of the faculty and staff of PES University, as well as the student volunteers who selflessly dedicated their time to contribute to the success of the conference. We would like to thank Kamiya Khatter, Associate Editor at Springer Nature, for her valuable assistance and cooperation.

We sincerely believe that CoCoNet'23 served as a platform for excellent discussions that facilitated the emergence of new ideas and the promotion of collaborative research. We are confident that this proceedings will serve as a significant source of research references and knowledge, contributing to scientific and engineering findings as well as the development of new products and technologies.

Trivandrum, India
Cleveland, USA
Dhaka, Bangladesh
Shalu, Taiwan
Cladwell, USA
December 2023

Sabu M. Thampi
Vipin Chaudhary
Al-Sakib Khan Pathan
Kuan Ching Li
Dilip Krishnaswamy

Contents

Healthcare and Medical Applications

Computer Vision

Natural Language Processing

About the Editors

Sabu M. Thampi is a Professor at the School of Computer Science and Engineering. His research interests include the Internet of Things (IoT), Cognitive Security, Endpoint Security, Cognitive Surveillance, and Smart Cyber-Physical Systems. He also coordinates the Connected Systems and Intelligence (CSI) Lab. With a Ph.D. in Computer Engineering from the National Institute of Technology Karnataka, Sabu has actively contributed to funded research projects and has a notable record of publications in books, journals, and conferences. He has authored and edited books and conference proceedings published by reputed publishers. Dr. Sabu's contributions extend to serving as a Guest Editor for special journal issues and participating as a program committee member in international conferences and workshops. He has taken on leadership roles as a co-chair for international workshops and conferences and has initiated and organized multiple annual conferences and symposiums.

Vipin Chaudhary is the Kevin J. Kranzusch Professor and Chair of the Department of Computer and Data Sciences at Case Western Reserve University. Most recently, he was a Program Director at the National Science Foundation where he was involved in many national initiatives and the Empire Innovation Professor of Computer Science and Engineering at SUNY Buffalo. He cofounded Scalable Informatics, a leading provider of pragmatic, high-performance software-defined storage and compute solutions to a wide range of markets, from financial and scientific computing to research and big data analytics. Dr. Chaudhary was awarded the prestigious President of India Gold Medal in 1986 at the Indian Institute of Technology (IIT) Kharagpur where he received the B.Tech. (Hons.) degree in Computer Science and Engineering and a Ph.D. degree from The University of Texas at Austin.

Al-Sakib Khan Pathan is a Professor at the Computer Science and Engineering department at United International University (UIU), Bangladesh, and serves as a Ph.D. Co-supervisor (external) at the Computer Sciences Department, University Ferhat Abbas Setif 1, Algeria. He earned his Ph.D. in Computer Engineering from Kyung Hee University, South Korea, in 2009, and a B.Sc. in Computer Science and Information Technology from the Islamic University of Technology (IUT),

Bangladesh, in 2003. With a diverse academic background, he has held faculty positions at various institutions, including Independent University, Bangladesh (IUB), Southeast University, Bangladesh, and International Islamic University Malaysia (IIUM). His research interests encompass wireless sensor networks, network security, cloud computing, and e-services technologies. He has delivered 33 keynote and invited speeches at international conferences, received several awards and best paper accolades, and actively contributed to top-ranked journals and conferences.

Prof. Kuan Ching Li is currently a Professor in the Department of Computer Science and Information Engineering at the Providence University, Taiwan. He was a department chair in 2009, the Special Assistant to the University President since 2010, and appointed as Vice-Dean for Office of International and Cross-Strait Affairs (OIA) in this same university since 2014. Dr. Li is recipient of awards from Nvidia, Ministry of Education (MOE)/Taiwan and Ministry of Science and Technology (MOST)/Taiwan, as also guest professorship from universities in China, that includes Xiamen University (XMU), Huazhong University of Science and Technology (HUST), Lanzhou University (LZU), Shanghai University (SHU), Anhui University of Science and Technology (AUST), and Lanzhou Jiaotong University (LZJTU). He has been involved actively in conferences and workshops as a program/general/steering conference chairman positions, numerous conferences and workshops as a program committee member, and has organized numerous conferences related to high-performance computing and computational science and engineering.

Dr. Dilip Krishnaswamy currently the Founder of Quantum Walks Technologies (QWalks) based in New Jersey, USA since 2022, is a visionary technologist with a profound interest in leveraging cutting-edge technologies to address contemporary challenges. His expertise spans Quantum and AI Edge Acceleration, Zero Trust Processing, Trustworthy Distributed AI, Quantum Key Distribution Protocols, Digital Twin Modeling, Risk-aware Generative AI, Sustainable Systems, and System Optimization. With a remarkable track record boasting 70+ granted US patents and 70+ research publications, Dr. Krishnaswamy has received accolades such as the "Technologist of the Year 2020" award from IEEE Bangalore. Dr. Krishnaswamy has held key roles in renowned organizations, including Senior Principal Architect at STL Edge Networks, Vice President (New and Emerging Tech R&D) at Reliance Jio Platforms Ltd., Senior Scientist at IBM Research, and Senior Staff Researcher at Qualcomm. His educational background includes a Ph.D. in Electrical Engineering from the University of Illinois at Urbana-Champaign, where his research focused on Parallel Processing Applications in VLSI-CAD, Signal and Image Processing, and Numerical Methods. Dr. Krishnaswamy has been an influential figure in the field, contributing to various conferences, events, and initiatives, demonstrating his commitment to pushing the boundaries of technology.

Artificial Intelligence and Machine Learning

Review of Optimization Techniques and Barren Plateaus in Training of Quantum Machine Learning Problems

Mandaar B. Pande

Abstract Variational quantum circuits (VQC) are being used quite efficiently for Noisy Intermediate Scale Quantum Computers (which are not yet robust and fault tolerant), which have small number of qubits and shallow circuit depth. VQCs are a category of algorithms used in solving several categories of quantum machine learning problems. However, it has been discovered recently that during training these algorithms, the gradient vanishes exponentially as the qubit numbers and the circuit depth increase. This is the barren plateau problem which has a serious impact on many of the VQCs including the variational eigensolver, quantum approximate optimization algorithm, and quantum neural networks. Barren plateaus thus lead to significant challenges in scaling VQCs which are necessary to address problems involving higher qubit number and larger depth quantum circuits. In this paper, we focus on the impact of barren plateaus during training of a VQC model. We review the various algorithms which are used for training and optimizing VQCs for learning. Next, we present the various causes leading to the barren plateau problems. Finally, we review the solutions available to mitigate this problem.

Keywords Quantum machine learning · Variational quantum circuits · Barren plateau · Trainability · Gradient descent

1 Introduction

The computers that exist today are termed as classical computers. The fundamental building block of today's classical computers is the bit or binary digit. All computing devices which exist today including laptops, smart phones, tablets, and so on are based on technology developed using the bit as their fundamental building block. Unlike classical computing, quantum computing is a new technology which uses the

---

M. B. Pande (✉)
Symbiosis Centre for Information Technology, Symbiosis International (Deemed University), Plot No. P-15, MIDC, Rajiv Gandhi Infotech Park, Hinjawadi, Pune 411057, India
e-mail: mandaar@scit.edu

S. M. Thampi et al. (eds.), *Fifth International Conference on Computing and Network Communications*, Lecture Notes in Electrical Engineering 1221,
https://doi.org/10.1007/978-981-97-4711-5_1

fundamental laws of quantum mechanics for developing quantum computers. The first scientist to consider using the principles of quantum mechanics for simulating physics was Feynman [1]. The principles which are primarily used include superposition, entanglement, and interference [2]. The qubit is the fundamental component of constructing a quantum computer. The qubit (short form of quantum bit) is the general case of a classical bit (binary digit). Using the terminology of quantum mechanics, a qubit is represented mathematically as $|\psi\rangle = \alpha|0\rangle + \beta|1\rangle$. $|\psi\rangle$ is the general state of any quantum system. $|0\rangle$ and $|1\rangle$ are basis states, and α, β are complex numbers. We say that $|\psi\rangle$ is a superposition of $|0\rangle$ and $|1\rangle$. From the perspective of a physical/operational realization of a qubit, it is created from various materials and approximated as a two-level quantum system. There are many technologies using which one can create qubits including, superconductors, ion traps, photonic, neutral atom, Rydberg atom, quantum annealing, and so on. Entanglement is a property which does not exist in classical systems. From a mathematical perspective, we say that a pair of qubits are entangled if both can be expressed through one quantum state. If the state of one qubit is changed, the state of the other qubit also changes in a very predictable manner, even at very far distances. An example of a maximally entangled pair is a Bell state, which is expressed as $\frac{1}{\sqrt{2}}(|00\rangle + |11\rangle)$. The quantum state $|00\rangle$ refers to two qubits, each existing in state $|0\rangle$, while the quantum state $|11\rangle$ refers to the two qubits, each existing in state $|1\rangle$. Interference is understood through well-known Young's double-slit experiment of light and more details are given here [2].

2 Brief Overview of Quantum Machine Learning

Quantum machine learning (QML) is a cross-disciplinary research field of quantum computing. In recent years, QML has been addressing some of the NP complexity class of decision problems which are difficult for classical computers to solve. The equivalent quantum complexity class for NP in quantum computing is called as bounded-error quantum polynomial time (BQP). As the name suggests, it combines two rich areas, namely quantum mechanics from physics and machine learning from computer science. Quantum computers are still not yet mature or robust enough to replace classical computers in most of the tasks. The current class of quantum computers is termed as noisy intermediate scale quantum (NISQ) computers [3]. An excellent overview of QML, current challenges, and opportunities can be found here [4–6].

In this paper, the general term "quantum machine learning" encompasses all the variants including quantum machine learning, quantum deep learning, and quantum AI-based systems.

There are four steps for computationally solving quantum machine learning problems. The **first step** is to encode data which may be either classical or quantum, into a quantum computer using various data encoding techniques. The **second step** is to create a model of the system being addressed which is essentially the development of a quantum circuit consisting of 1 and 2 qubit quantum gates. Examples of 1 qubit quantum gates are the qubit flip (X) gate, the phase flip (Z) gate, rotation gate (R), and Hadamard (H) gate which creates a superposition of the $|0\rangle$ and $|1\rangle$ states. A prime example of a 2 qubit gates is the *CNOT* (Controlled X) gate. This gate creates the entanglement between two qubits. The **third step** is to train the model using gradient-based or evolutionary/gradient-free optimization techniques. The **fourth step** is measurement and analysis of classical outputs of the model and improves various metrics of the model to make it robust for providing accurate results. All the four steps are crucial for addressing complex problems using QML. There are multiple ways in which classical data are encoded so that quantum computers can appropriately process them. A summary of various encoding techniques is given here [7]. The model is essentially the development of a VQC. This is done using a combination of 1 and 2 qubit gates. The 1 qubit gates are parameterized rotation gates, and 2 qubit ones are gates which create entanglement between pairs of qubits which form part of the VQC. As mentioned earlier, examples of such entangling gates are Controlled X, Y, Z gates. The final step of measurement is essentially to get information from the quantum computational system. It is important to note that in classical systems which are in general macroscopic, measurement usually has no effect on the system; in quantum systems, it has a profound effect in the sense that a quantum state is altered irreversibly after a measurement is made.

The third step which trains the model is the topic of interest in this paper. Just as in classical machine learning models, VQCs are used to carry out data-driven tasks. The purpose of training a model is the minimization of a cost function $f(\vec{\boldsymbol{\theta}})$ with respect to the parameter vector $\vec{\boldsymbol{\theta}}$. This is the task of making an arbitrary function to learn from data. In QML when one is training the VQC model, the function that is being optimized (usually it is being minimized) is the expectation value $\left\langle \boldsymbol{\Psi}(\vec{\boldsymbol{\theta}}) \middle| \hat{\boldsymbol{H}} \middle| \boldsymbol{\Psi}(\vec{\boldsymbol{\theta}}) \right\rangle$. Here, $\hat{\boldsymbol{H}}$ is the Hamiltonian operator and $\left| \Psi(\vec{\boldsymbol{\theta}}) \right\rangle$ is the parameterized quantum state of the system being optimized. There are various types of algorithms which are used to optimize the parameters of a VQC, including gradient-based, evolutionary, and gradient-free. Specifically with respect to gradient-based methods, when the qubits and the layer depth in the VQC model are increased, and the gradient is evaluated, there can be a situation where the gradient vanishes. This feature of the vanishing gradient is the ***barren plateau*** problem and is usually found in NISQ quantum machine learning problems.

In this paper, the following research questions are considered:

1. *Which gradient-based algorithms and natural evolution strategies are used for training and optimization of variational quantum circuits?*
2. *What are the causes leading to barren plateaus in variational quantum circuits?*
3. *What are the mitigation strategies to avoid barren plateaus for minimizing cost functions for QML problems?*

Thus, in Sect. 3, we address the first research question of optimization techniques for VQCs. Section 4 discusses the causes leading to barren plateaus. In Sect. 5, we discuss the mitigation strategies for avoiding barren plateaus in VQCs. In Sect. 6, we conclude the paper and suggest related areas of research.

3 Optimization Algorithms for Variational Quantum Circuits

There are two categories of algorithms which can be used for training the VQC model in quantum machine learning problems, namely

1. Gradient-based optimization algorithms.
2. Natural evolution strategies.

In the subsequent paragraphs, we summarize the various algorithms in each category.

3.1 Gradient-Based Optimization Algorithms

In this section, we briefly summarize the various gradient-based optimization algorithms. Some of the discussed algorithms are well-known in classical machine and deep learning problems and have been used for almost all optimization problems. Other algorithms presented are versions that have been modified for quantum machine learning problems. We consider a function $f(\vec{\boldsymbol{\theta}})$ which is a function of $\vec{\theta}$ which is the vector of parameters. $\vec{\nabla}$ is a gradient function denoted by the Greek symbol nabla. Thus, $\vec{\nabla}$ operates on $f(\vec{\boldsymbol{\theta}})$ for obtaining the gradient, which is written mathematically as $\vec{\nabla} f(\vec{\boldsymbol{\theta}})$. The usual process to minimize the function is by updating $\vec{\theta}$ along the direction of steepest descent. This is represented mathematically as $\vec{\theta}_{n+1} = \vec{\theta}_n - \eta \vec{\nabla} f(\vec{\theta})$. η is the learning rate and is a hyperparameter which controls the step size for the next update. By continuing this, we hope to obtain convergence to a local optimum of $f(\vec{\theta})$.

Table 1 gives a summary of the gradient-based algorithms which are used for optimizing VQCs that are randomly initialized. Implementation details of some of the algorithms summarized in the table are available under the quantum machine learning course of https://qiskit.org/learn.

Table 1 Gradient-based optimization algorithms

Algorithm	Brief description
Finite difference derivative (or gradient) [8]	While this may seemingly be the simplest of the algorithms, for training a multi-parameterized VQC a formula which uses the central finite difference derivative is used. This is accurate to the second order in the finite increment ε
Analytic gradient [8]	In this algorithm, the approximate formula which is obtained using the central finite difference above is replaced by its original analytic expression, thus removing any bias
Stochastic gradient descent (SGD) [9]	For a high-dimensional dataset, it becomes prohibitively expensive to computationally determine the gradient for the entire dataset. In SGD, the gradient is determined for a randomly (or stochastically) chosen subset of the entire dataset. This reduces the number of iterations, but this is in exchange for a lower convergence rate
Natural gradient [9]	In the natural gradient algorithm, the way the next iteration $\vec{\boldsymbol{\theta}}_{n+1}$ is calculated from $\vec{\boldsymbol{\theta}}_n$ is changed by considering the sensitivity of the model. The sensitivity of the model depends on the loss landscape (which is the relationship between all the parameters in the parameter vector $\vec{\boldsymbol{\theta}}$). In the standard gradient descent, one uses the Euclidean length $\boldsymbol{d} = \left\| \left\langle \vec{\boldsymbol{\theta}}_{n+1} = \vec{\boldsymbol{\theta}}_n \right\rangle \right\|_2$. In natural gradients, the distance used depends on the sensitivity of the model and is given by $\boldsymbol{d} = \left\| \left\langle \Psi(\vec{\boldsymbol{\theta}}_{n+1}) \right\rangle \right\|^2$. In this expression, $\left\| \Psi(\vec{\boldsymbol{\theta}}) \right\rangle$ is the parameterized quantum state
Quantum version of the natural gradient [9]	In the quantum version of the natural gradient algorithm mentioned above, the optimization is considered as moving in the path of steepest descent while considering the real part of the Quantum Geometric Tensor. The Quantum Fisher Information Matrix (QFIM) is an important metric in this algorithm
Simultaneous perturbation stochastic approximation (SPSA) [10, 11]	SPSA is an optimization algorithm in which one randomly samples the gradient, for reducing the number of evaluations. This can be considered analogous to the SGD in terms of the process to reduce computational iterations
Variational quantum imaginary time evolution [11, 12]	In this framework, quantum imaginary time evolution is realized by mapping the evolution of a quantum state $\lvert\boldsymbol{\Psi}\rangle$ to the evolution of parameters $\vec{\boldsymbol{\theta}}$ in a VQC. This VQC is the ansatz (quantum circuit) for the evolution. Such algorithms can be used for combinatorial optimization problems including determining the ground states of Hamiltonians used in quantum chemistry or to prepare Gibbs states which are used in Quantum Boltzmann Machines (QBMs)

3.2 Natural Evolution Strategies

The term evolution strategies were introduced in the 1960s and 1970s [13]. These were designed to address optimization problems in high-dimensional continuous-valued domains. ***Natural evolution strategies (NES)*** are a set of black-box optimization algorithms in which parameters are updated iteratively by using a search gradient [14]. The procedure is the following: A batch of search points is identified through the parameterized search distribution. At each identified point the fitness function is evaluated. The parameters of the distribution help the distribution to capture the structure of the fitness function adaptively. As an example, in case of a Gaussian distribution, the mean and the covariance matrix constitute the structure. From these samples, a search gradient is estimated by the NES toward higher expected fitness. Next the NES carries out a gradient ascent step along the *natural gradient*. This is a second-order method, unlike the gradient descent and there is a renormalization of the update with regards to the uncertainty. Natural gradients help in mitigating the challenge of slow convergence when the plain gradient is used for optimization problem landscapes which involve ridges and plateaus.

Table 2 gives a summary of the various techniques of natural evolution strategies which can be used for optimizing VQCs. All the techniques listed are described in detail here [14].

The studies that are discussed below have applied NES to variational quantum circuits. In [15], the authors have used NES for optimization of VQCs which have been randomly initialized and are in the region of exponentially vanishing gradients. The study demonstrates that by using the NES gradient estimator, the exponential decrease in the variance is reduced. It is to be noted that exponential decrease in the variance is a sign of a barren plateau. This technique is applied to two problems. The first problem is estimating the ground state energy estimation using the variational quantum eigensolver (VQE). The second problem is regarding preparation of a quantum state containing varying number of length (which is the qubit number) and depth of the circuit (that corresponds to the number of layers).

In [16], the study involves applying NES to the task of binary classification as an alternative to the conventional gradient descent method for training hybrid quantum classical neural networks (HQCNN). In the study, they identify that the

Table 2 Evolutionary optimization algorithms

Technique	Brief description
Natural gradient	This technique applies primarily to the issue of scale-invariance
Fitness shaping	This technique applies to robustness
Adaptation sampling	This technique is primarily applicable to determine performance and sensitivity
Exponential parameterization	This applies to covariance constraints
Natural coordinate system	This applies to enhancing computational efficiency

performance of the HQCNN depends strongly on the number of hyperparameters. They also observe that as is the case with the gradient descent technique for training the network, using the NES still exhibits the problem of barren plateaus. An interesting observation that is mentioned in the study is that the number of evaluations of the cost function remains the same for increasing number of parameters. Secondly, since the evaluations of the cost function are independent, they can be carried out independently or in parallel. At present, this parallelization application is limited to the near-term noisy quantum computers (NISQ).

The next study [17] evaluates the optimization of two quantum heuristics, namely Quantum Alternating Operator Ansatz [18], a generalization of the well-known Quantum Approximate Optimization Algorithm (QAOA) [19] and the Variational Quantum Eigensolver (VQE) which are the most used VQCs in QML, with meta-learning, which is a class of optimization techniques. The meta-learning algorithm considered in the study is the Long Short-Term Memory (LSTM) and Recurrent Neural Network (RNN). This optimization is compared with three other optimizers including the NES, L-BFGS-B, and gradient-free Nelder–Mead approaches. These are carried out on three simulation environments, namely wavefunction, sampling, and noisy. It is shown in this study that in the noisy parameter setting, the meta-learning algorithm approaches the global optimum more frequently in comparison with other three optimizers. The study also shows that the meta-learner is more resistant to noise in comparison. The study also demonstrates that the meta-learner can be scaled to larger problems.

4 Causes Leading to Barren Plateaus in Variational Quantum Circuits

Before we discuss the various causes leading to barren plateaus in variational quantum circuits, we present a brief overview of what is a barren plateau. Barren plateaus are regions in the cost landscape of a VQC. In these regions, both the gradient and its variance approach zero. This results in the optimization algorithm getting stuck in a local minimum. It has also been discovered that as a metric for expressivity of a quantum circuit the closeness to the Haar measure is used [20]. By expressivity of a VQC or QNN once implies how close can the VQC or QNN approximate a target function. The target function may be a physical observable from the perspective of quantum mechanics [21, 22]. Expressivity of VQCs is connected to the existence of barren plateaus. A fundamental relation between expressibility and trainability of VQCs is presented here [20]. This is done by extending the form of VQCs which are usually in the exact 2-design, to arbitrary ansatzes. It is discovered that highly expressive ansatzes demonstrate flatter cost landscapes, which means that they are more prone to barren plateaus.

The first study identifying the problem of barren plateaus in variational quantum circuits was in 2018 [23]. The study considered random quantum circuits, exponential dimensions of Hilbert space, and hybrid quantum–classical algorithms which were being optimized using gradient-based methods as listed in sub-Sect. 3.1. It is to be noted that ***random circuit*** means selecting an arbitrary VQC as a starting point for QML problems. Usually, the random circuits are a combination of single-qubit parameterized rotation gates and two-qubit controlled NOT gates for achieving entanglement. The study demonstrated that for a significant number of randomly initialized circuits the average value of the gradient of the objective function exponentially went to zero. Further, the probability that any random circuit has a small nonzero average value ϵ is exponentially small as the qubit number increases. The study also shows that the zero gradient region is not limited to a local minimum of interest. A large plateau of states whose deviation of the objective value from zero is exponentially small. The circuit depth that is required to demonstrate this undesirable property is also not too large. The circuit depth required is $\mathcal{O}(n^{1/d})$ where d is the d-dimensional array. Such an issue of barren plateaus however does not happen in problems involving random circuits used for information security or demonstrating quantum supremacy.

Having explained the concept of barren plateaus, we now explore research studies that have since been carried out for identifying other reasons for barren plateaus to occur in VQCs and hybrid quantum–classical algorithms. In Table 3, we summarize potential causes for the occurrence of barren plateaus.

5 Mitigation Strategies to Avoid Barren Plateaus for Minimizing Cost Functions for QML Problems

In Sect. 4, we summarized the various ways in which barren plateaus can appear as part of theoretical and numerical experiments on variational quantum circuits. In this section, we look at the various options proposed by researchers to mitigate or remove barren plateaus in such quantum systems. These are summarized in Table 4.

6 Conclusions and Future Directions

Barren plateaus are a significant challenge in trainability and optimization of VQCs. This problem was discovered recently and had to be addressed urgently. The impact of barren plateaus is the challenge in building scalable quantum circuits which can help in solving larger real-life problems in quantum machine learning. In this study, we have posed three review questions relevant to barren plateaus, namely the various optimization techniques for VQCs, what are the various reasons for barren plateaus to manifest in VQCs and what are the mitigation strategies to address this challenge.

Table 3 Potential causes of occurrence of barren plateaus

Study	Description of the cause for barren plateaus
Cost function-dependent barren plateaus in shallow parametrized quantum circuits [24]	$V(\boldsymbol{\theta})$ is a parameterized quantum circuit. $\boldsymbol{\theta}$ is the set of parameters. C is the cost function, which is to be minimized using variational quantum algorithms (VQA). $V(\boldsymbol{\theta})$ is assumed to be an alternating layered quantum circuit (such as QAOA or VQE) which is created using blocks which form 2-designs. The study presents two results. The first result is that if we define C in terms of global observables it leads to vanishing gradients that are barren plateaus even for very low circuit depths. The second result is that if however, we define the cost function C in terms of local observables, we get in the worst case a polynomially vanishing gradient (i.e., appearance of barren plateau) in the case of the depth of $V(\boldsymbol{\theta})$ being $\mathcal{O}(\log \boldsymbol{n})$
Characterizing the loss landscape of variational quantum circuits [25]	In this work, the authors propose a tool to quantitatively evaluate the cost landscape and study the convergence of the VQCs. They propose computing the Hessian of the cost function of the VQCs and using it demonstrates how to characterize the cost landscape. The eigenvalues of the Hessian matrix provide information about the curvature of the local area. A comparison with classical neural networks is carried out to interpret the information. The authors also demonstrate using the Hessian to modify the learning rate during training of the VQCs for obtaining faster convergence
Noise-induced barren plateaus in variational quantum algorithms [26]	This study is regarding the impact of noise on any fundamental limitation on the performance of VQAs. Through a rigorous proof, they specify that noise results in barren plateaus. The type of noise considered is the Pauli noise. This case demonstrates that there is an exponentially vanishing gradient in the number of qubits n, as the circuit depth grows linearly with n. These results are obtained for two types of VQCs, namely Quantum Alternating Operator Ansatz (generalized QAOA) and Unitary Coupled Cluster Ansatz (UCCA). The noise-induced barren plateaus are shown to be fundamentally different from the noise-free barren plateaus (the latter case is essentially the case of random circuits usage as a VQC)
Entanglement-induced barren plateaus [27]	In this study the authors map the Quantum Neural Network (QNN) models to principles of quantum thermodynamics. Specifically, they suggest that the QNNs which satisfy a volume law based on entanglement entropy result in models that are not suitable for learning with high probability. Using the arguments from quantum thermodynamics the authors demonstrate that there exists barren plateaus in the landscape for optimization which is a result of entanglement

(continued)

Table 3 (continued)

Study	Description of the cause for barren plateaus
Effect of barren plateaus on gradient-free optimization [28]	In all the above studies, the optimization techniques have involved using gradient-based techniques. In this study, the authors focus on whether using gradient-free optimizers, the barren plateau problem vanishes. The study however concludes that using such optimizers does not address the challenge of barren plateaus. The study numerically proves this by using several gradient-free optimizers such as COBYLA, Powell, and Nelder–Mead. This demonstrates that the number of runs (also called shots) required for optimization grows exponentially as the number of qubits increases in the VQC
Dataset-induced barren plateaus [29]	In this study, the focus is on a general treatment of trainability and presence of barren plateaus in quantum machine learning models. The authors provide a rigorous connection between gradient scaling in variational quantum algorithm type of cost functions and the equivalent generic quantum machine learning cost functions. The reason for this is to examine if the various ways in which barren plateaus arise in VQAs (as given in this table) also extend to QML problems. Next the authors present analytical and numerical proofs that some features of datasets and embeddings may affect the training ability of QML models, since these result in so-called dataset-induced barren plateaus
Barren plateaus preclude learning scramblers [30]	Scrambling means increasing the amount of entanglement in VQCs to help in disseminating information of the initial conditions of a quantum system throughout all the degrees of freedom of the system. In this study, the authors propose using QML to study quantum scrambling. However, the study shows that barren plateaus are always obtained irrespective of any option of ansatz (VQC) or any parameter initialization

There are several open questions such as impact of data encoding schemes of classical data in QML and using nonlinear dynamics on mitigating barren plateaus and so on.

Table 4 Mitigation strategies to avoid barren plateaus

Study	Strategy for mitigating barren plateaus
Using PQCs to obtain large gradients through correlation [31]	Due to the existence of barren plateaus in VQCs, applying VQAs for problems of larger scale with gradient descent optimization algorithms becomes a challenge and it requires efficient optimization of the random VQCs. In this study, the authors propose the following alternative which is to reduce the parameter space dimensionality. This is done by using smaller circuit modules which contain gate layers which are correlated spatially or temporally. This can help in avoiding the barren plateau problem
Using an innovative initialization strategy for mitigating the barren plateau problem [32]	To mitigate the barren plateaus problem in VQCs initialized with random initial parameter values, this study proposes the following solution. The solution involves selecting some of the parameter values randomly. The other parameter values are chosen so that the circuit becomes a series of shallow groups with each group evaluating to unity. By doing this, the effective circuit depth is limited to the first parameter update, thus avoiding the barren plateau problem
Using quantum CNNs to mitigate the barren plateau problem [33]	In this study, to mitigate the barren plateau problem, the authors use quantum CNNs which consist of convolutional and pooling layers. Such a model reduces the qubits numbers while preserving the information about the features that the data provides. Examining such circuits, the study finds that the variance of the gradient only reduces polynomially. This implies that such a system does not exhibit barren plateaus. To prove their results, the study introduces graph-based methods for analyzing the quantum expectation values of Haar-distributed unitaries
Training VQAs optimally to mitigate barren plateaus [34]	In this study, the authors illustrate how to train VQAs optimally for getting to learn quantum states. VQCs can form Gaussian kernels and using this, adaptive learning rates are realized for gradient ascent which is a form of natural evolution strategies algorithm. Generalized quantum natural gradients are introduced, which exhibits stability and an optimized movement in the parameter space
Escaping barren plateaus in VQCs through Gaussian initializations [35]	In this study, which is similar to the previous study [34], the authors propose an initialization strategy which provides a theoretical guarantee of ensuring the VQC does not encounter vanishing gradients as the qubit numbers increase. By using parameters initialized through Gaussian probability functions, it is discovered that the gradient norm reduces only polynomially with increase in qubits and circuit depth. The results hold good for both the global and local observables
Mitigating barren plateaus in VQCs with initialization through Bayesian learning [36]	To address and reduce the barren plateau problem, this study takes recourse to fast and slow algorithms using Bayesian learning for the identification of an appropriate parameter space. Using this, a fast local optimizer is initialized for determining the global optimum in an efficient manner. The effectiveness of this study is demonstrated numerically on a Bars and Stripes (BAS) quantum generative model

References

1. Feynman RP (1982) Simulating physics with computers. Int J Theor Phys. https://doi.org/10.1007/BF02650179
2. Shankar R (1994) Principles of quantum mechanics. New York, NY: Springer US
3. Preskill J (2018) Quantum computing in the NISQ era and beyond. Quantum. https://doi.org/10.22331/q-2018-08-06-79
4. Perdomo-Ortiz A, Benedetti M, Realpe-Gómez J, Biswas R (2018) Opportunities and challenges for quantum-assisted machine learning in near-term quantum computers. Quantum Sci Technol 3(3):30502
5. Biamonte J, Wittek P, Pancotti N, Rebentrost P, Wiebe N, Lloyd S (2017) Quantum machine learning. Nature 549:195. [Online]. Available: https://doi.org/10.1038/nature23474
6. Garcia DP, Cruz-Benito J, Garcia-Peñalvo FJ (2022) Systematic literature review: quantum machine learning and its applications. arXiv Prepr. arXiv2201.04093
7. LaRose R, Coyle B (2020) Robust data encodings for quantum classifiers. Phys Rev A 102(3):32420. https://doi.org/10.1103/PhysRevA.102.032420
8. Guerreschi GG, Smelyanskiy M (2017) Practical optimization for hybrid quantum-classical algorithms
9. Stokes J, Izaac J, Killoran N, Carleo G (2020) Quantum natural gradient. Quantum 4:269. https://doi.org/10.22331/q-2020-05-25-269
10. Spall JC (1992) Multivariate stochastic approximation using a simultaneous perturbation gradient approximation. IEEE Trans Autom Control 37:332. https://doi.org/10.1109/9.119632
11. Gacon J, Zoufal C, Carleo G, Woerner S (2021) Simultaneous perturbation stochastic approximation of the quantum fisher information. Quantum 5:567
12. McArdle S, Jones T, Endo S, Li Y, Benjamin SC, Yuan X (2019) Variational ansatz- based quantum simulation of imaginary time evolution. NPJ Quantum Inf 5(1):75. https://doi.org/10.1038/s41534-019-0187-2
13. Rechenberg I (1973) Evolutionsstrategie. Optimierung Tech. Syst. nach Prinz. derbiologischen Evol
14. Wierstra D, Schaul T, Glasmachers T, Sun Y, Peters J, Schmidhuber J (2014) Natural evolution strategies. J Mach Learn Res 15:949
15. Anand A, Degroote M, Aspuru-Guzik A (2021) Natural evolutionary strategies for variational quantum computation. Mach Learn Sci Technol 2(4):45012. https://doi.org/10.1088/2632-2153/abf3ac
16. Friedrich L, Maziero J (2023) Evolution strategies: application in hybrid quantum-classical neural networks. Quantum Inf Process 22(3):132. https://doi.org/10.1007/s11128-023-03876-8
17. Wilson M, Stromswold R, Wudarski F, Hadfield S, Tubman NM, Rieffel EG (2021) Optimizing quantum heuristics with meta-learning. Quantum Mach Intell 3(1):13. https://doi.org/10.1007/s42484-020-00022-w
18. Hadfield S, Wang Z, O'Gorman B, Rieffel EG, Venturelli D, Biswas R (2019) From the quantum approximate optimization algorithm to a quantum alternating operator ansatz. Algorithms 12(2):1–51. https://doi.org/10.3390/a12020034
19. Farhi E, Goldstone J, and Gutmann S (2014) A quantum approximate optimization algorithm, pp 1–16. [Online]. Available: http://arxiv.org/abs/1411.4028
20. Holmes Z, Sharma K, Cerezo M, Coles PJ (2022) Connecting ansatz expressibility to gradient magnitudes and barren plateaus. PRX Quantum 3(1):10313. https://doi.org/10.1103/PRXQuantum.3.010313
21. Du Y, Hsieh M-H, Liu T, Tao D (2020) Expressive power of parametrized quantum circuits. Phys Rev Res 2(3):33125. https://doi.org/10.1103/PhysRevResearch.2.033125
22. Wu Y, Yao J, Zhang P, Zhai H (2021) Expressivity of quantum neural networks. Phys Rev Res 3(3):L032049. https://doi.org/10.1103/PhysRevResearch.3.L032049
23. McClean JR, Boixo S, Smelyanskiy VN, Babbush R, Neven H (2018) Barren plateaus in quantum neural network training landscapes. Nat Commun 9(1):4812. https://doi.org/10.1038/s41467-018-07090-4

24. Cerezo M, Sone A, Volkoff T, Cincio L, Coles PJ (2021) Cost function dependent barren plateaus in shallow parametrized quantum circuits. Nat Commun 12(1):1791. https://doi.org/10.1038/s41467-021-21728-w
25. Huembeli P, Dauphin A (2021) Characterizing the loss landscape of variational quantum circuits. Quantum Sci Technol 6(2):25011. https://doi.org/10.1088/2058-9565/abdbc9
26. Wang S et al (2021) Noise-induced barren plateaus in variational quantum algorithms. Nat Commun 12(1):6961. https://doi.org/10.1038/s41467-021-27045-6
27. Ortiz Marrero C, Kieferová M, Wiebe N (2021) Entanglement-induced barren plateaus. PRX Quantum 2(4):40316. https://doi.org/10.1103/PRXQuantum.2.040316
28. Arrasmith A, Cerezo M, Czarnik P, Cincio L, Coles PJ (2020) Effect of barren plateaus on gradient-free optimization
29. Thanasilp S, Wang S, Nghiem NA, Coles P, Cerezo M (2023) Subtleties in the trainability of quantum machine learning models. Quantum Mach Intell 5(1):21. https://doi.org/10.1007/s42484-023-00103-6
30. Holmes Z, Arrasmith A, Yan B, Coles PJ, Albrecht A, Sornborger AT (2021) Barren plateaus preclude learning scramblers. Phys Rev Lett 126(19):190501. https://doi.org/10.1103/PhysRevLett.126.190501
31. Volkoff T, Coles PJ (2021) Large gradients via correlation in random parameterized quantum circuits. Quantum Sci Technol 6(2):25008. https://doi.org/10.1088/2058-9565/abd891
32. Grant E, Wossnig L, Ostaszewski M, Benedetti M (2019) An initialization strategy for addressing barren plateaus in parametrized quantum circuits. Quantum 3:214. https://doi.org/10.22331/q-2019-12-09-214
33. Pesah A, Cerezo M, Wang S, Volkoff T, Sornborger AT, Coles PJ (2021) Absence of barren plateaus in quantum convolutional neural networks. Phys Rev X 11(4):41011. https://doi.org/10.1103/PhysRevX.11.041011
34. Haug T, Kim MS (2021) Optimal training of variational quantum algorithms without barren plateaus. arXiv Prepr. arXiv2104.14543
35. Zhang K, Liu L, Hsieh M-H, Tao D (2022) Escaping from the barren plateau via gaussian initializations in deep variational quantum circuits. In: Advances in neural information processing systems, 2022, vol 35, pp 18612–18627, [Online]. Available: https://proceedings.neurips.cc/paper_files/paper/2022/file/7611a3cb5d733e62808143l445cb01fd-Paper-Conference.pdf
36. Rad A, Seif A, Linke NM (2022) Surviving the barren plateau in variational quantum circuits with Bayesian learning initialization

Computational Modeling of Spiking in the Layer 5 Projection Neurons of the Mouse Motor Cortex

Arathi Rajendran, Navya Ajith, Aiswarya Chandrabhanu Nambiar, Giovanni Naldi, and Shyam Diwakar

Abstract Mathematical models have allowed the reconstruction of realistic behavior of large and complex neurons under various physiological functions. Understanding the electrophysiological behavior regarding neuronal gain, process, and connectivity of such complex pyramidal projection neurons in the motor cortex is essential for unraveling the neural basis of motor control and developing therapies for conditions related to motor dysfunctions. This modeling study analyzed the input/output properties of L5 cortico-striatal projection neurons and their sensitivity to abnormal firing behavior. The study reconstructs the excitability of layer 5 cortico-striatal neurons and suggests how they integrate the inputs temporally matching experimental observations. The study also showed variability in the output patterns in layer 5 cortico-striatal neurons with varying current stimuli. This model predicted that the firing rate is proportional to the delayed input currents, which may affect the neuron's feature integration property. The modeling methodology showed several single-neuron excitability characteristics relevant to current disease mechanism theories and recognized Parkinson's disease-associated symptoms. These results help to uncover the sensory integration at the single neuron level and pathophysiology of motor deficit along with other interconnected motor systems in the context of motor control and information processing.

A. Rajendran · S. Diwakar (✉)
Amrita Mind Brain Center, Amrita Vishwa Vidyapeetham, Amritapuri Campus, Clappana P.O., Kollam, Kerala 690525, India
e-mail: shyam@amrita.edu

N. Ajith · A. C. Nambiar
Amrita School of Biotechnology, Amrita Vishwa Vidyapeetham, Amritapuri Campus, Clappana P.O., Kollam, Kerala 690525, India

G. Naldi
Department of Environmental Science and Policy, Università Degli Studi Di Milano, Via C Saldini 50, 20133 Milano, Italy

S. Diwakar
Department of Electronics and Communication Engineering, School of Engineering, Amrita Vishwa Vidyapeetham, Amritapuri Campus, Clappana P.O, Kollam, Kerala 690525, India

S. M. Thampi et al. (eds.), *Fifth International Conference on Computing and Network Communications*, Lecture Notes in Electrical Engineering 1221,
https://doi.org/10.1007/978-981-97-4711-5_2

Keywords Spiking neuron model · Pyramidal projection neuron · Motor cortex · Gain · Computational neuroscience

1 Introduction

Cognitive computing aims to replicate the complex and adaptive functions observed in the brain, striving to emulate their capacity for learning, reasoning, and decision-making [1]. Using mathematical approximations such as the Hodgkin-Huxley model [2], the nonlinear neuronal properties may be modeled as a stochastic communication channel, which emphasizes the contributions made by individual cells and its dynamic characteristics. However, the spiking neuron models, such as the integrate-and-fire, and the Izhikevich models [3, 4] give a simpler justification for fundamental brain activities. A recent simulation study [5] on computational models of the mouse motor cortex showcases how to bridge the gap between the biological findings and computational models. Previous models [6] have considered modeling either single layers or two neurons or modeling without biophysically detailed morphological characteristics. A lack of detailed diversity analysis and the representation of disease conditions have been observed among the previous models of single projection neurons from the mouse motor cortex [7–9].

During the study of neuronal behavior, it is important to select the appropriate model, that can be used to understand the electro-responsive properties that include intrinsic firing and excitability of the biological neurons during current injections as stimuli.

Several regulatory processes such as inhibitory stimuli, neuromodulation, and feedback loops are involved in the maintenance of the electroresponsiveness of the neurons toward different stimuli, which is known as neuronal gain [10]. Gain is a fundamental characteristic of cortical neurons that supports a variety of brain information processing processes, from simple sensory perception to sophisticated cognitive processes [11]. The majority of neurons in the mammalian cerebral cortex are pyramidal cells, which are crucial for motor and cognitive functions [12]. Motor cortical layer 5 (M1 L5) pyramidal neurons integrate data from different sources and deliver to distant destinations such as spinal cord, basal ganglia, cerebellum, thalamus, and other brain areas involved in motor coordination [13–17]. Alterations in the excitation/inhibition on L5 pyramidal projection neuron have been known to cause different impairments associated with the motor deficits [18, 19]. Dopamine is the most important neuromodulator for the brain's regulation of movement [20, 21], which controls the excitability and I/O capabilities of neurons [22, 23]. The loss of dopamine in substantia nigra causes abnormal modulation in the cortico-basal ganglia circuit [24–26]. According to the classical theory related to the Parkinson's disease (PD), dopamine depletion in the basal ganglia increases their inhibitory output and decreases M1 activation, which may change the excitability [27–29]. Some of the factors which affect the gain during PD condition are the effects of dopamine depletion and altered excitability at the thalamocortical circuit, which affects the gain of cortical regions [30]. However, since individual point neuron models do not generate spatially distributed activities, which are required to compute the sensory-motor signal integration, estimating neuronal gain from spiking model neurons is a difficulty. But still, the knowledge about the large morphologically detailed pyramidal projection neurons should explain the quantitative estimation of neuronal gain based on delayed inputs.

In this work, we aimed to bridge the gap of knowledge between normal and disease behavior of M1 L5 single pyramidal projection neuron by using spiking neuron model to uncover the temporal integration of information in L5 projection neurons during normal and disease firing behaviors. To capture the variability of neuronal responses, varying current stimuli were provided in nanoampere (nA) scale with delays. The findings demonstrated the diverse neuronal behaviors of M1 L5 pyramidal projection neurons as well as how the delayed inputs alter input–output connections in L5. Additionally, this theoretical framework helps to anticipate the correlation between these interactions.

2 Methods

The study reconstructed the intrinsic firing activity of single pyramidal projection neurons of M1 L5 using adaptive exponential integrate and fire (AdEx) neuron model [4] and validated using experimental data [31]. The neuronal gain of cortico-striatal (Cst) neuron was calculated during normal and diseased conditions. The simulation was performed on the NEURON [32] simulation platform with varying durations.

2.1 *Modeling Single Projection Neurons of M1 Layer 5*

The single neuron dynamics were modeled using a simple spiking neuron model [4]. The model equations describe the membrane potential dynamics (Eq. 1) and adaptation. Both the variables will be reset if the action potential has triggered (Eq. 2). The membrane potential can be represented by

$$C\frac{dV}{dt} = -g_L(V - E_L) + g_L\Delta_T \exp\left(\frac{V - V_T}{\Delta_T}\right) - \omega + I \quad (1)$$

$$\tau\omega\frac{d\omega}{dt} = a(V - E_L) - \omega, \quad (2)$$

where V is the membrane potential, C is the capacitance, I is the input current, ω is the adaptation variable, g_L is leaking conductance, E_L is leaking reversal potential, VT is the threshold voltage, ΔT is the slope factor, a and b are adaptation coupling parameters, and $\tau\omega$ is the adaptation time constant (Fig. 1).

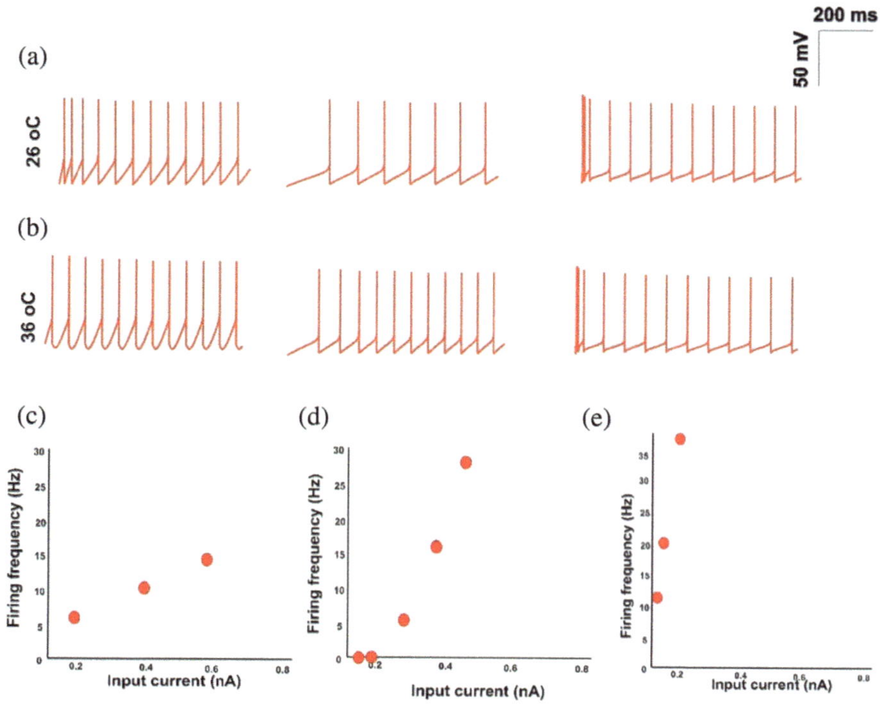

Fig. 1 Reconstructed firing patterns of layer5 projection neurons cortico-spinal (CS) (left), cortico-thalamic (CT) (middle), and cortico-cortical (CC) (right) for temperatures **a** 26 °C and **b** 36 °C. Validation of modeled neurons by analyzing the relationship between the firing frequency and amount of current. A linear relationship was observed; the firing frequency increased with increase in current for **c** cortico-cortical neuron, **d** cortico-spinal, and **e** cortico-thalamic neurons

The modeled neurons were validated by considering the firing activity of neurons based on the injected currents, which provided the current in nanoampere scale to the model to understand the electroresponsiveness of these neurons [32]. The diseased neuronal condition was reconstructed based on experimental evidence [18] (Fig. 2). The firing activity was modeled based on the fine-tuning of inter-spike interval (ISI) and inter-burst interval (IBI) by changing the AdEx parameters (Fig. 1) (Table 1).

Based on experimental evidence by Oswald et al. [31] which was performed during an in vitro-like condition, we have modeled four different projection neurons of the mouse motor cortex by considering its key features such as firing frequency, inter-spike interval, types of firing, and timing of spikes. The neuronal model was simulated by fine-tuning the parameters such as ISI to fit the data to the model which was best reproduced if the neuron has a regular/tonic firing. The CS neuron fires intrinsically with a firing rate of 12 Hz (initial burst), CT with 7 Hz (adaptation), and CC with 13 Hz (initial burst) at 36 °C. Similarly, the intrinsic firing frequency of CS neuron is 12 Hz with regular firing, CT is 11 Hz with regular firing, and CC is 14 Hz with an initial burst at 26 °C. To understand how the cortico-striatal projection

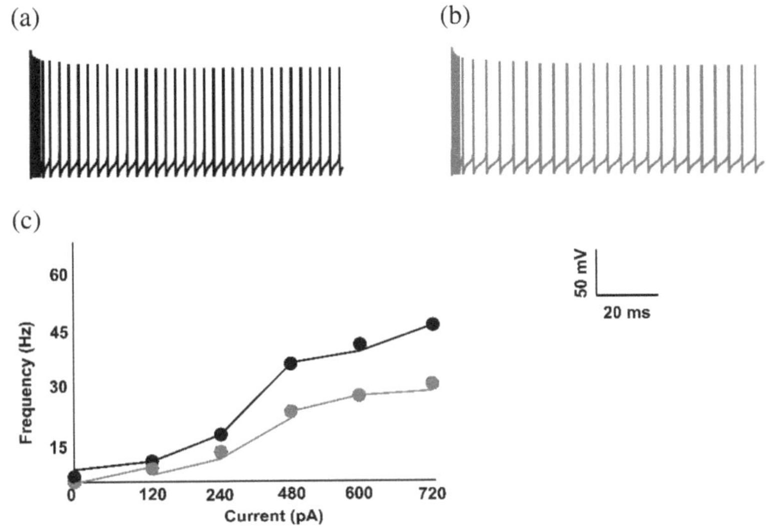

Fig. 2 Reconstructed firing activity of normal and disease conditions of the Cst pyramidal neuron based on the experimental evidence of [18]. **a** The cortico-striatal neuron's intrinsic excitability in the normal condition and **b** in the diseased condition. **c** Both the conditions were validated by examining the firing activity of the neuron during varying current injections (black color-control condition and grey color-disease condition)

Table 1 Tuned adex parameters for reconstructing layer 5 projection neurons at two temperatures

Parameters	26 °C				36 °C			
	CS	CT	Cst	CC	CS	CT	Cst	CC
C	170	2200	200	50	170	1400	83	50
gL	1.7	12	10	12	1.7	12	1.7	12
E_{IL}	−59	−59	−65	−70	−59	−70	−59	−70
Δ_T	2	2	5.5	2	2	2	5.5	2
V_T	−41.4	−50	−57	−50	−41.4	−50	−56	−50
τ	164	160	41	164	27.7	200	41	164
I	190	300	24	340	140	300	65	340
a	2	−8	2	2	0.01	−8	2	2
b	60	0	100	60	250	0	88	60

neuron responded to different input stimuli, the gain was calculated during changed current and input delays [11]. The neuronal gain can be calculated by dividing the firing frequency of the neuron (AP/s) with the injected input current (pA).

This study attempts to understand the sensitivity in the neuron based on the delayed input currents using a point neuron model that assumes electroresponsiveness without any spatial characteristics of biological neuron. From the study of [33], it was observed that the pyramidal projection neuron has an average a length of ~ 878 ± 684 μm, and the speed of action potential propagation is 11.48 ± 3.68 m/s [34]. It is evident that the time taken for the input to reach the pyramidal neuron's soma from the distal dendrite was ~ 1 ms [35] and from the apical dendrite was 0.017 ms based on the apical dendrite length of 200 μm from the soma. The neuronal sensitivity was also analyzed during diseased condition with the same delays (Fig. 2).

3 Results

Our study employed various delayed current stimuli to reconstruct the sensitivity of the layer 5 cortico-striatal projection neuron and to interpret how it affects its feature integration property. Four projection neurons were modeled based on the electrophysiological data from the mouse motor cortex in two different temperatures during *in vivo*-like conditions [36]. The reconstruction follows the changes in electrophysiological parameters of each neuron depending on temperature changes.

3.1 *Reconstructed Temperature-Dependent Intrinsic Firing Activity of Projection Neurons*

The simulation indicated that the firing activity of each projection neuron remained consistently steady with temperature changes. Each projection neuron had changing degrees of spike frequency adaptations [36]. Spike frequency adaptation in our model of M1 neurons could be discriminated against across cells because Cst and CC neurons showed rapid and gradual reduction of their firing rate, whereas CS and CT neurons did not (Fig. 2). The reconstruction proved that the CS model neurons are more regular when compared with other projection neurons. The CS showed doublet at the beginning at 26 °C and disappeared at 36 °C. The CT model neurons showed action potential firing frequency acceleration at both temperatures, but Cst and CC neurons' action potential firing rates often decreased over time. The firing frequency of CC neurons did not change during varying temperatures; hence, it did not depend on temperature changes. Based on the ability of action potential acceleration by CT neurons, we can distinguish it from CS neurons. It was observed that the amplitude (difference between action potential peak and action potential threshold) of Cst and CC is lower than other neurons. The observation proved that the largest CT and CS neurons' amplitude helps provide precise and accurate communication across large distances. Its morphologies are compatible with the same function: large soma size and fast-conducting axons [36–38]. Since their projection targets are closer, CC and

Cst neurons had slower, smaller amplitude action potentials, although the Cst group showed the widest action potentials. Possibly, this broadness lowers the threshold for calcium input to cause spike-timing-dependent synaptic plasticity.

The CS neuron showed a minimum of 7 spikes during 0.3 nA injection and a maximum of 2 spikes when injected with 0.5 nA. While comparing four neurons, CT neurons showed more firing frequency at 0.3 nA. A comparison of runtime and computation complexity revealed that the layer 5 projection spiking neuron model outperformed the biophysical model of a single pyramidal neuron [39]. The time taken to complete the simulation for the AdEx neuron was 0.09999905 ms and for the biophysical model was 0.57999992 ms.

3.2 Reduced Spontaneous Activity of Cortico-Striatal Neuron During Diseased Condition

The study reconstructed the intrinsic excitability of the single Cst pyramidal neuron during normal and disease conditions based on the analysis by [18]. The point neuron model recorded electrophysiological activity during 0.48 nA current injection. It was observed that the firing frequency is higher for the control condition (43 Hz) than for the diseased condition (30 Hz). The simulation showed that the Cst neuron showed minimal spike adaptation during normal conditions compared to disease conditions. A reduced intrinsic excitability of the Cst pyramidal neuron of M1 L5 (Fig. 2B) was observed.

3.3 Delayed Inputs Increased the Gain of Cortico-Striatal Pyramidal Projection Neuron During Normal and Diseased Conditions

The input currents were injected into the point neuron model of the M1 L5 Cst pyramidal neuron with three different delays from 0 ms, 0.017 ms, and 1 ms and determined the f/I relations during normal and disease conditions. This study attempts to understand the realistic cases of inputs distributed over the dendritic tree of pyramidal projection neurons Fig. 3

The peak current levels (in pA) were considered along with the increased delay to mimic the possibility of biological neurons amplifying the action potential propagation to reach the somatic region [36]. Current injection without any delays resulted in action potentials at a similar linear frequency that have already published for L5 pyramidal neurons [37]. It was observed that the neuron's gain increased with the simultaneous current injection of delayed inputs. The average gain of neurons without delay was 0.096 AP/s/pA and with a 1 ms delay was 0.1276 AP/s/pA. The average increase was 15% when the delayed inputs were combined with non-delayed

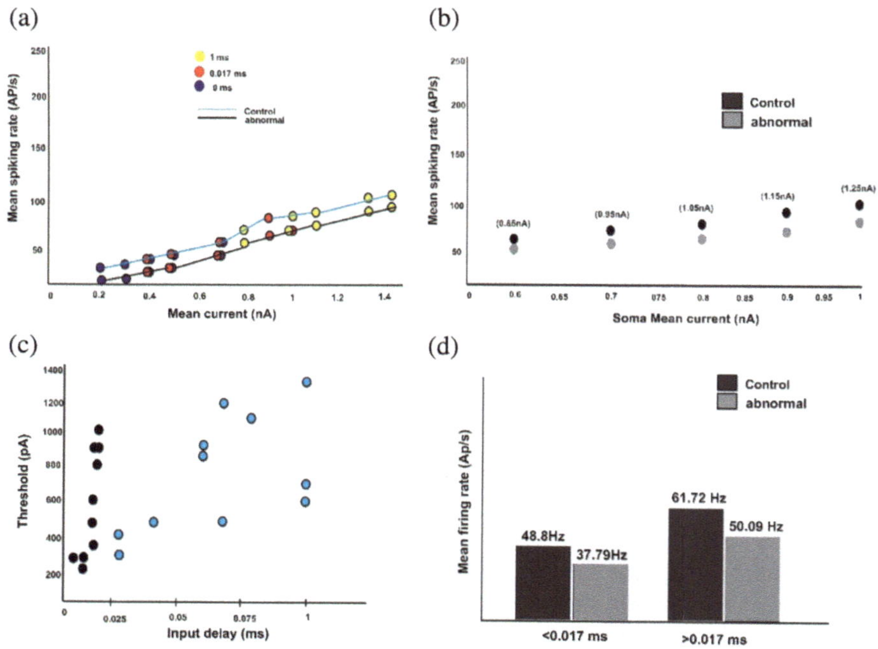

Fig. 3 Estimation of neuronal gain during normal and diseased conditions. **a** The firing rate of cortico-striatal pyramidal neuron while changing the input delay and injected current during normal and diseased conditions. **b** A combination of highest delayed input current (constant) and changing no delayed input currents were injected simultaneously during normal (black) and disease (grey) conditions. **c** A graphical representation of input currents and the delays. **d** The analysis of the average firing activities during normal and disease conditions

inputs. It's also evident that the delayed input current caused a much steeper rise in the initial part of the f/I relationship.

The study also compared and estimated Cst neuron intrinsic excitability gains during normal and disease conditions. It was observed that the average gain of the neuron during normal condition with an input delay of 0.017 ms was 0.087 AP/s/pA; during the disease condition, the gain was 0.061 AP/s/pA. With an input delay more significant than 0.025 ms, the average gain of neurons was 0.082 AP/s/pA and 0.057 AP/s/pA during disease conditions. Our simulations suggest that the Cst pyramidal projection neuron gain was reduced during the disease condition compared to the normal state, which could have affected the responses to the combination of input currents with delayed inputs.

4 Discussion

For each class of projection neurons, the simulation estimated the effectiveness of spike train variability and gain for Cst neuron during normal and disease conditions. Our findings suggest that the delayed input may be implicated to the increased neuronal gain and spike train variability. The pyramidal projection neurons could be sensitive to stimuli with delays, attributing to the changes in the temporal properties, and the milliseconds-scale changes may affect the encoding mechanism during sensory-motor processing [38, 39].

Our simulations show that the electroresponsiveness of each of these neurons varied depending on the environment, including temperature, current injections, and delayed inputs. This study adapted three types of firing including tonic, adaptation, and delayed accelerating firing to fine-tune the single spiking neurons. Simple spiking neuron models, which are computationally inexpensive and may be used to evaluate a wide variety of dynamic properties observed in real neurons, were employed to reconstruct the firing characteristics of a single projection neuron. Here, the intrinsic spiking dynamics were employed to reconstitute the intrinsic firing characteristics of pyramidal projection neurons. The characteristics of the point neuron may thus be inferred using this special interface. Through this model, a variety of activities, including subthreshold response and spiking activity, were captured by the neural responses [40].

The simulations showed that the firing frequency was increased for CT and Cst neurons, but the firing frequency was unaltered for CS and CC neurons [31] when temperature was 36 °C. In our simulations, it was observed that the reduced temperature could increase the firing frequencies, but some neurons increased frequencies at higher temperature as well [41]. The two types of firing patterns were observed in the projection neurons, the CT neuron is the burst firing type. In all the cells, except CC, burst firing is absent at 36 °C even though its present at 26 °C. To demonstrate the effectiveness of our model and determine the values, we analyzed run-time complexity between our spiking neuron model and HH model of pyramidal projection neurons [42].

The second part of the study specifically modeled and analyzed the Cst pyramidal projection neuron during normal and disease conditions. The simulations on gain analysis suggest that the Cst neuron had the capacity to integrate multiple information from different regions while responding to the delayed inputs. The gain increased with the increase in input delay as observed in previous experimental studies. Since the spiking neurons do not have any spatial properties, only the delay of current stimuli was considered and excluded other membrane as well as morphological properties like distance, length, diameter, and number of dendrites. This model-based observation may help augment our interpretations of neuronal signal processing.

The increase in the slope of the f/I curve as the input distance from the soma increases suggests that the neuron's responsiveness to input stimuli becomes more pronounced when located away from the soma. A steeper and more prominent increase was observed when the delayed stimuli combined with the non-delayed stimuli. During the disease condition, a similar rise in the firing rate was observed but the rate and gain were reduced when compared with the normal neuronal condition. This could be the result of neurotransmitter imbalance, ion channel deficit, and neuronal loss [43] during disease conditions.

This study reconstructed the electrophysiological properties of complex pyramidal projection neurons under different biological conditions to understand the intricate mechanisms underlying neuronal communication and integration and to know how layer 5 pyramidal neurons compute and transmit information during normal and diseased conditions.

5 Conclusion

Modeling layer 5 pyramidal neurons with simple approximation provides valuable insights into the foundational principles of neural computation, paving the way for a deeper understanding of how these building blocks contribute to the development of complex cognitive processes in large-scale neuronal network systems. The pyramidal projection neuron model simulated the firing behavior of neurons in in vivo-like condition, and its intrinsic property was fine-tuned. A linear correlation between firing frequency and injected current was observed in all neurons as in experimental data. The gain of cortico-striatal pyramidal projection neuron was calculated to understand how the delayed input contributes to the pyramidal output during normal and disease conditions. By focusing on the development of this specific model, the specific objective was to contribute to the field of large-scale network modeling, enabling researchers to study and understand the intricate dynamics and interactions within the mouse motor cortex. This model can be used in further analysis of complex network dynamics as well as the abstracted models to understand different motor learning schemes.

Acknowledgements This work derives direction and ideas from the Chancellor of Amrita Vishwa Vidya-peetham, Sri Mata Amritanandamayi Devi. This study was partially supported by Department of Science and Technology grant, Multi-Scale Brain Function India-Italy Network of Excellence INT/Italy/P-24/2022, Government of India, and Embracing the World Research-for-a-Cause initiative.

References

1. Kriegeskorte N, Douglas PK (2018) Cognitive computational neuroscience. Nat Neurosci 21(9):1148–1160. https://doi.org/10.1038/s41593-018-0210-5
2. Hodgkin AL, Huxley AF (1952) A quantitative description of membrane current and its application to conduction and excitation in nerve. J Physiol 117(4):500–544
3. Izhikevich E (2003) Simple model of Spiking Neurons. IEEE Trans. Neural Netw Publ IEEE Neural Netw. Counc 14:1569–72 https://doi.org/10.1109/TNN.2003.820440.

4. Naud R, Marcille N, Clopath C, Gerstner W (2008) Firing patterns in the adaptive exponential integrate-and-fire model. Biol Cybern 99:335–347. https://doi.org/10.1007/s00422-008-0264-7
5. Dura-Bernal S et al (2023) Multiscale model of primary motor cortex circuits predicts in vivo cell-type-specific, behavioral state-dependent dynamics. Cell Rep 42(6):112574. https://doi.org/10.1016/j.celrep.2023.112574
6. Hoshino O, Kameno R, Watanabe K (2019) Reducing variability in motor cortex activity at a resting state by extracellular GABA for reliable perceptual decision-making. J Comput Neurosci 47(2):191–204. https://doi.org/10.1007/s10827-019-00732-6
7. Bauer R, Clowry GJ, Kaiser M (2021) Creative destruction: a basic computational model of cortical layer formation. Cereb. Cortex N Y NY 31(7):3237–3253. https://doi.org/10.1093/cercor/bhab003
8. Mandge D. et al (eds) (2021) Computational modelling of a mouse layer 5 pyramidal neuron using genetic ion channels. J Comput Neurosci.
9. Moberg S, Takahashi N (2023) Neocortical layer 5 subclasses: from cellular properties to roles in behavior. Front Synaptic Neurosci 14. Accessed: 19 Nov 2022 [Online]. Available: https://www.frontiersin.org/articles/10.3389/fnsyn.2022.1006773
10. Slater C, Liu Y, Weiss E, Yu K, Wang Q (2022) The neuromodulatory role of the noradrenergic and cholinergic systems and their interplay in cognitive functions. A focused review Brain Sci 12(7):7 https://doi.org/10.3390/brainsci12070890
11. Ferguson KA, Cardin JA (2020) Mechanisms underlying gain modulation in the cortex. Nat Rev Neurosci 21(2):80–92. https://doi.org/10.1038/s41583-019-0253-y
12. Spruston N (2008) Pyramidal neurons: dendritic structure and synaptic integration. Nat Rev Neurosci 9(3):3https://doi.org/10.1038/nrn2286.
13. Borra E, Belmalih A, Gerbella M, Rozzi S, Luppino G (2010) Projections of the hand field of the macaque ventral premotor area F5 to the brainstem and spinal cord. J Comp Neurol 518(13):2570–2591. https://doi.org/10.1002/cne.22353
14. Morita K, Morishima M, Sakai K, Kawaguchi Y (2012) Reinforcement learning: computing the temporal difference of values via distinct corticostriatal pathways. Trends Neurosci 35(8):457–467. https://doi.org/10.1016/j.tins.2012.04.009
15. Palmer LM, Clark BA, Gründemann J, Roth A, Stuart GJ, Häusser M (2010) Initiation of simple and complex spikes in cerebellar Purkinje cells. J Physiol 588(Pt 10):1709–1717. https://doi.org/10.1113/jphysiol.2010.188300
16. Rouiller EM, Moret V, Tanné J, Boussaoud D (1996) Evidence for direct connections between the hand region of the supplementary motor area and cervical motoneurons in the macaque monkey. Eur J Neurosci 8(5):1055–1059. https://doi.org/10.1111/j.1460-9568.1996.tb01592.x
17. Theyel B.B, Llano D.A, Sherman S.M (2010) The corticothalamocortical circuit drives higher-order cortex in the mouse. Nat. Neurosci. 13(1):1 https://doi.org/10.1038/nn.2449.
18. Chen L, Daniels S, Kim Y, Chu H-Y (2021) Cell type-specific decrease of the intrinsic excitability of motor cortical pyramidal neurons in Parkinsonism. J Neurosci 41(25):5553–5565. https://doi.org/10.1523/JNEUROSCI.2694-20.2021
19. Underwood CF, Parr-Brownlie LC (2021) Primary motor cortex in Parkinson's disease: Functional changes and opportunities for neurostimulation. Neurobiol Dis 147:105159. https://doi.org/10.1016/j.nbd.2020.105159
20. Kida H et al (2016) Motor training promotes both synaptic and intrinsic plasticity of layer II/III pyramidal neurons in the primary motor cortex. Cereb Cortex 26(8):3494–3507. https://doi.org/10.1093/cercor/bhw134
21. Paz JT, Mahon S, Tiret P, Genet S, Delord B, Charpier S (2009) Multiple forms of activity-dependent intrinsic plasticity in layer V cortical neurones in vivo. J Physiol 587(13):3189–3205. https://doi.org/10.1113/jphysiol.2009.169334
22. Nadim F, Bucher D (2014) Neuromodulation of neurons and synapses Curr Opin Neurobiol vol 0, pp 48–56 https://doi.org/10.1016/j.conb.2014.05.003.
23. Bazzari AH, Parri HR (2019) Neuromodulators and long-term synaptic plasticity in learning and memory: a steered-glutamatergic perspective. Brain Sci 9(11):300. https://doi.org/10.3390/brainsci9110300

24. Hammond C, Bergman H, Brown P (2007) Pathological synchronization in Parkinson's disease: networks, models and treatments. Trends Neurosci 30(7):357–364. https://doi.org/10.1016/j.tins.2007.05.004
25. Bernheimer H, Birkmayer W, Hornykiewicz O, Jellinger K, Seitelberger F (1973) Brain dopamine and the syndromes of Parkinson and Huntington Clinical, morphological and neurochemical correlations. J Neurol Sci 20(4):415–455. https://doi.org/10.1016/0022-510X(73)90175-5
26. Dauer W, Przedborski S (2003) Parkinson's Disease: Mechanisms and Models. Neuron 39(6):889–909. https://doi.org/10.1016/S0896-6273(03)00568-3
27. Albin RL, Young AB, Penney JB (1989) The functional anatomy of basal ganglia disorders. Trends Neurosci 12(10):366–375. https://doi.org/10.1016/0166-2236(89)90074-X
28. McGregor MM, Nelson AB (2019) Circuit mechanisms of Parkinson's disease. Neuron 101(6):1042–1056. https://doi.org/10.1016/j.neuron.2019.03.004
29. Rajendran A.G, Thankamani A, Nirmala N, Nair B, Diwakar S (2017) Computational neuroscience of substantia nigra circuit and dopamine modulation during parkinson's disease. pp 1–5 https://doi.org/10.1109/ICACCI.2017.8125892
30. Galvan A, Devergnas A, Wichmann T (2015) Alterations in neuronal activity in basal ganglia-thalamocortical circuits in the parkinsonian state. Front Neuroanat 9:5. https://doi.org/10.3389/fnana.2015.00005
31. Oswald M, Tantirigama M, Sonntag I. Hughes S. Empson R (2013) Diversity of layer 5 projection neurons in the mouse motor cortex. Front Cell Neurosci vol 7 Accessed 21 Feb. 2023 [Online]. Available: https://www.frontiersin.org/articles/10.3389/fncel.2013.00174
32. Carnevale N.T, Hines M.L (2006) The NEURON book. Cambridge: Cambridge University Press, https://doi.org/10.1017/CBO9780511541612.
33. Morishima M, Kawaguchi Y (2006) Recurrent connection patterns of corticostriatal pyramidal cells in frontal cortex. J Neurosci 26(16):4394–4405. https://doi.org/10.1523/JNEUROSCI.0252-06.2006
34. Xi MC, Liu RH, Engelhardt JK, Morales FR, Chase MH (1999) Changes in the axonal conduction velocity of pyramidal tract neurons in the aged cat. Neuroscience 92(1):219–225. https://doi.org/10.1016/s0306-4522(98)00754-4
35. Zhu Y, Zhu JJ (2004) Rapid arrival and integration of ascending sensory information in layer 1 nonpyramidal neurons and tuft dendrites of layer 5 pyramidal neurons of the neocortex. J Neurosci off J Soc Neurosci. 24(6):1272–1279. https://doi.org/10.1523/JNEUROSCI.4805-03.2004
36. Crandall SR, Govindaiah G, Cox CL (2010) Low-Threshold Ca^2+ current amplifies distal dendritic signaling in thalamic reticular neurons. J Neurosci 30(46):15419–15429. https://doi.org/10.1523/JNEUROSCI.3636-10.2010
37. Oakley JC, Schwindt PC, Crill WE (2001) Dendritic calcium spikes in layer 5 pyramidal neurons amplify and limit transmission of ligand-gated dendritic current to soma. J Neurophysiol 86(1):514–527. https://doi.org/10.1152/jn.2001.86.1.514
38. Srivastava KH et al (2017) Motor control by precisely timed spike patterns. Proc Natl Acad Sci 114(5):1171–1176. https://doi.org/10.1073/pnas.1611734114
39. Putney J, Conn R, Sponberg S (2019) Precise timing is ubiquitous, consistent, and coordinated across a comprehensive, spike-resolved flight motor program. Proc Natl Acad Sci 116(52):26951–26960. https://doi.org/10.1073/pnas.1907513116
40. An L. et al. (2020) Intrinsic and synaptic properties shaping diverse behaviors of neural dynamics. Front Comput Neurosci. vol 14, Accessed: 23 Sep 2023. [Online]. Available: https://www.frontiersin.org/articles/10.3389/fncom.2020.00026
41. Buzatu S (2009) The temperature-induced changes in membrane potential. Riv Biol 102(2):199–217

42. Pospischil M et al (2008) Minimal Hodgkin-Huxley type models for different classes of cortical and thalamic neurons. Biol Cybern 99(4–5):427–441. https://doi.org/10.1007/s00422-008-0263-8
43. Juárez Olguín H, Calderón Guzmán D, Hernández García E, Barragán Mejía G (2016) The role of dopamine and its dysfunction as a consequence of oxidative stress. Oxid Med Cell Longev. vol 2016, p 9730467 https://doi.org/10.1155/2016/9730467.

Meta-learning Assisted Graph Representation Learning for Downstream Learning Tasks

Nischal R. Bhat, Prerana Sanjay Kulkarni, Ananya Menon, Smruthika S. Meda, and Bhaskarjyoti Das

Abstract The Graph Transformer Network (GTN) is a state-of-the-art solution for graph representation learning, which utilizes the transformer architecture to capture attention and long-range dependencies in a non-euclidean dataset such as a graph. However, even with GTN in a downstream learning task, there are usually a few data points that are harder to classify than others and are at risk of being wrongly classified. Metric-based meta-learning, which encompasses many meta-learning algorithms, proves promising in such scenarios. Through this work, we propose a novel 2-stage pipeline combining a Graph Transformer Network with metric-based meta-learning methodologies to make downstream classification tasks easier. The work described in this article investigates and evaluates this pipeline in two specific learning setups using two Metric Learning frameworks, i.e., a Siamese Network for binary classification of graphs and a Prototypical Network for multi-class classification of nodes, both appended to a representation learning block of Graph Transformer Network. The experiments described in this article conclusively prove the efficacy of such a pipeline both in full-shot and few-shot learning scenarios.

Keywords Graph transformer · Meta-learning · Siamese networks · Prototypical networks

---

N. R. Bhat · P. S. Kulkarni · A. Menon · S. S. Meda
Department of Computer Science and Engineering, PES University, Bengaluru, Karnataka, India
e-mail: nischal108bhat@gmail.com

P. S. Kulkarni
e-mail: prer.kulk@gmail.com

A. Menon
e-mail: ananya4am@gmail.com

S. S. Meda
e-mail: 2002smruthika@gmail.com

B. Das (✉)
Department of Computer Science and Engineering in AI & ML, PES University, Bengaluru, Karnataka, India
e-mail: bhaskarjyoti01@gmail.com

S. M. Thampi et al. (eds.), *Fifth International Conference on Computing and Network Communications*, Lecture Notes in Electrical Engineering 1221,
https://doi.org/10.1007/978-981-97-4711-5_3

1 Introduction

Graph Transformers provide context-aware embedding from graphs by capturing local and global dependencies. They are motivated by successful natural language processing (NLP) designs. Metric-based meta-learning, a family of meta-learning algorithms, is a potent strategy in downstream machine-learning tasks as it moves similar things closer and different things apart. This article investigates the fusion of Metric Learning and Graph Transformers. By combining these techniques, we hope to increase discriminative abilities and enhance similarity-based retrieval and anomaly detection in graph-structured data. We further the practical applications of graph representation learning through theoretical analysis and experimental validation on benchmark datasets.

The rest of the paper is organized as follows. Section 2 reviews the relevant existing work. Section 3 discusses the data sets used. The adopted methodology is explained in Sect. 4. The various experiments performed are explained in Sects. 5 and 6 discusses the results. Finally, Sect. 7 concludes with a summary and indication of future steps.

2 Existing Work

2.1 *Graph Transformers*

Graph representation learning seeks to preserve important structural and semantic aspects of a graph while converting complex graph-structured data into instructive and low-dimensional embeddings. Graph Transformers [14] are an extension of the Transformer architecture, which is widely used for Natural Language Processing (NLP) tasks. It is an add-on to the existing Graph Convolutional Network (GCN), due to its value addition in terms of capturing the attention of a node compared to its neighbors. The transformer architecture [5, 7] uses the common query, key, and vector concepts and extends them to the graph context. They are used to compute attention weights by capturing relationships between nodes in a graph.

The following equations represent the key components of the Graph Transformer:

1. Attention Mechanism:

$$\text{Attention}(Q, K, V) = \text{softmax}\left(\frac{QK^T}{\sqrt{d_\text{k}}}\right)V \tag{1}$$

where Q, K, and V are the query, key, and value matrices respectively, and d_k is the dimension of the key matrix.

2. Multihead Attention:

$$\text{MultiHead}(Q, K, V) = \text{Concat}(h_1, h_2, ..., h_n)W_O \tag{2}$$

where $h_i = \text{Attention}(QW_{Qi}, KW_{Ki}, VW_{Vi})$, n is the number of attention heads, and W_{Qi}, W_{Ki}, W_{Vi}, and W_O are learnable weight matrices.

3. Graph Transformer layer:

$$x' = W_1 x_i + \Sigma \alpha_{i_j} W_2 x_j \tag{3}$$

where alpha stands for the below:

$$\alpha_{i_j} = \text{softmax}\left(\frac{(W_3 x_i)^T (W_4 x_j)}{\sqrt{d}}\right) \tag{4}$$

Graph Transformer Network has been recently used in a few GNN-based architectures, i.e., document classification [18], hateful discussion detection [8], and fake news detection [13]. However, the Graph Transformer Network has not been used in conjunction with metric-based meta-learning, yet.

2.2 *Metric-Based Meta-learning*

Metric-based meta-learning represents a subset of meta-learning [9] strategies that strives to learn a suitable distance metric in the feature space, allowing for improved discrimination between data samples. In the context of graph data, Metric Learning offers a compelling advantage by embedding nodes into a lower-dimensional space, where similarity relationships among nodes are better preserved.

2.2.1 Siamese Network

Siamese Networks [11] is a popular metric-based meta-learning architecture that leverages its ability to learn a similarity metric that distinguishes between similar and dissimilar pairs. This approach makes it particularly valuable for learning from limited labeled data. By incorporating Siamese Networks into graph representation learning [10], we can leverage their effectiveness in learning similarity metrics and enhance the performance of graph-related tasks, while efficiently handling scenarios where obtaining ground truth labels is challenging or expensive.

2.2.2 Prototypical Learning

Prototypical Networks [16] are similar to Siamese networks for a multi-class scenario. By creating representative prototypes or embeddings for each node class, it can address challenges related to data scarcity and generalization on unseen graph structures. By incorporating Prototypical Networks into graph representation learning techniques [3], we can harness the strength of both metric learning and graph representation learning.

2.3 Meta-learning Using Graph Neural Networks

Since Graph Neural Networks and Meta-Learning are both relatively new areas of research, there exists a limited amount of work combining graph neural techniques with meta-learning for graph few-shot learning [6] in areas such as node classification [19], link prediction [1], and graph classification [12]. Disinformation such as fake news on social media is inherently multi-contextual [2] in nature and a graph-based learning approach is a natural fit for taking care of multiple contexts such as content, propagation, and user. Specifically for multi-modal fake news detection tasks, Wang et al. [17] have used meta-optimization as a meta-learning approach but this work does not use meta-learning with a graph neural network-based approach. As far as our knowledge extends, the approach introduced in this study has not been utilized yet for the task of fake news detection.

3 Datasets

In order to evaluate the performance of the introduced pipeline, we have made use of two datasets.

1. **UPFD Dataset**: The User Preference-aware Fake News Detection (UPFD) dataset [4] is a dataset of graphs and is used for binary classification of graphs, where each graph refers to news propagation network obtained from Twitter and validated by Politifact. Each node in the graph represents a user. The root node is the source of the news. An edge between users represents a retweet. There are a total of 314 graphs in this dataset, and each graph can represent either real news or fake news.
2. **Coauthor CS Dataset**: This dataset [15], used for node classification, consists of a single graph, where each node is represented by an author. The dataset has around 18,333 nodes and 1 graph with 163,788 edges. The labels in the range 0–14 represent the different scientific domains that the authors can belong to, i.e., Machine Intelligence, Human-Computer Interaction, etc. An edge exists between 2 nodes means that the two authors have co-authored a article. We will be using Coauthor CS dataset, from Pytorch Geometric, which is exclusively for Computer Science articles. Each node has can have 1 out of 15 labels, where each label represents a field of study.

4 Methodology

As mentioned earlier, the methodology employed in this work is a two-step process. The first step involves a Graph Transformer to obtain graph embeddings. In the second step, the learned embeddings are used as inputs to metric-learning-based meta-learning blocks to improve the accuracy of the downstream classification task.

1. **Graph Transformer—Siamese Network Pipeline for Binary Classification of Graphs**: Figures 1 and 2 show the pipeline used.
   - In the Graph Transformer Network, we use a pre-defined split for the training, testing and validation datasets (61-32-221 respectively). The transformer network is constructed with 6 layers in total. The sequence begins with a conventional graph convolutional layer, succeeded by a linear layer. Following this, a transformer layer featuring three attention heads is applied, succeeded by another linear layer. The subsequent step involves another transformer layer, also comprising three attention heads, culminating in a final linear layer. The graph's embeddings are obtained through a global max-pooling function applied to the node embeddings. It is a pooling operation that arrives at a single embedding for the entire graph. In our case, each graph has a 16-dimensional embedding.
   - These graph embeddings are used as input in the Siamese Network. For this, positive pairs are created from two instances of the same class (both real news or both fake news), and negative pairs are created from two instances of different classes (one real news and the other fake). The Siamese Network comprises two identical neural networks, which are trained using the pairs we create. The instances of a pair are passed through the twin neural networks in the training process. In order to compare the outputs of the neural networks, a similarity function is made use of. Specifically, a Contrastive Loss function built using Euclidian distance is used as the similarity metric. Pairs that are closer in the metric space exhibit higher similarity than pairs that are farther apart in the metric space. Using this similarity function and a certain threshold value to compare the calculated similarities, positive pairs are moved closer together in the metric space, and negative pairs are moved further apart. This training process when repeated multiple times with all the pairs created, tends to move the similar instances closer together, and dissimilar instances farther apart.
2. **Graph Transformer—Prototypical Network Pipeline for Multi-class Classification of nodes**: Figure 3 shows the pipeline used in detail.
   - First, the Graph Transformer Network is utilized to obtain the node embeddings in a graph. Here one Graph Convolutional Layer and two Graph Transformer Layers, with three convolutional layers, are used. These layers are followed by batch normalization, non-linear 'ReLU' activation, and dropout. The second Transformer Convolutional layer for multiheaded attention uses 2 heads and the third Transformer convolutional layer subsequently uses 4 heads. The output generated by the third layer is directed into the linear output layer to obtain the logits corresponding to each class. The model also returns embeddings, which holds the learned node embeddings obtained after the last convolutional layer. These embeddings are used for the few-shot multi-class node classification in the second step.
   - Construction of a Prototypical network involves splitting training and testing data into support and query sets. Computation or training is done on the sup-

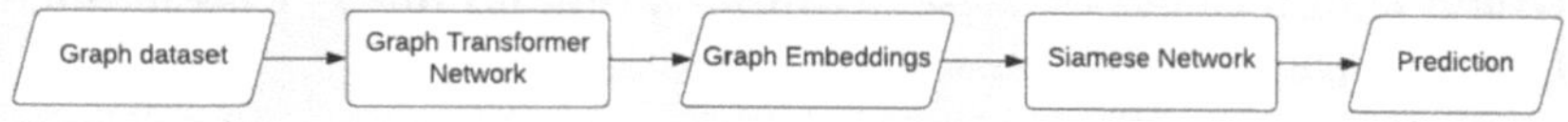

Fig. 1 Flowchart encapsulating the process

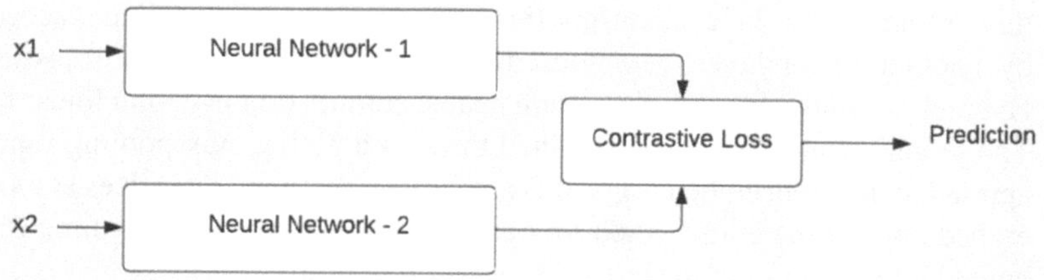

Fig. 2 Siamese network

port sets, and evaluation of the model is done on the query sets. The support set consists of a few samples from every label that the model is trained upon. This trained model is then evaluated on the query set that contains untrained/unseen examples. The objective is to generalize to this query set, through training with a limited number of examples drawn from the support set. During Prototypical Training, centroids or prototypes are computed from the support set of the train data, which are representations of every label that we are classifying to. The entire idea is to create a metric space, containing these centroids. Examples from the query set of the train data are then projected onto this metric space containing the centroids. The distance metric used here is "Euclidean distance". The Euclidean distance is measured between each of the query points in the query set, and all the prototypes present in the metric space. The shortest distance between the query point and the prototypes is the label that the query point is classified into. During Prototypical Testing, the model is trained for a certain number of episodes, and the trained model is used to do an evaluation of the query set of the testing data. The episodic training accuracy and the episodic testing accuracy are noted.

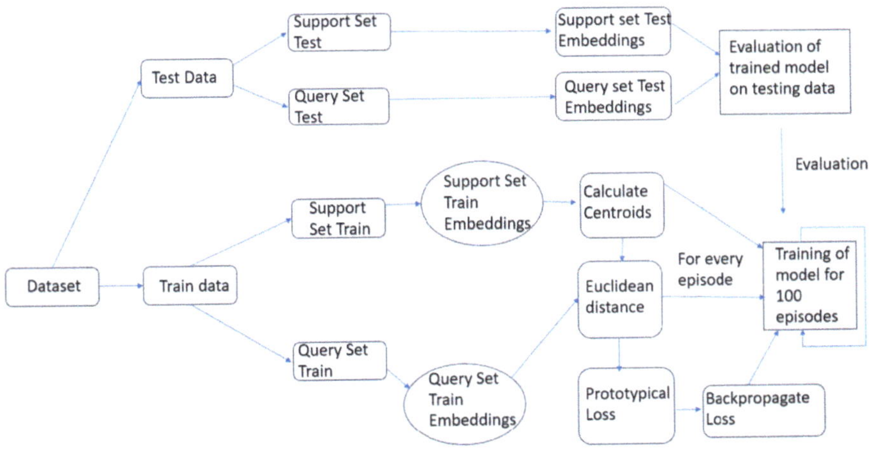

Fig. 3 Flow chart for prototypical learning

5 Experimentation

5.1 Graph Transformer Siamese Network Pipeline for Binary Classification of Graphs

1. The training data consists of 61 graphs, which is the default one provided by the dataset functionality. We use a combination of 1 GCN and 2 Transformer layers (each with 3 heads). The idea behind using a combination of different layers is based on observations from various configurations of layers. Any layers used in a model exceeding three are likely to result in over-smoothing, where all embeddings seem to converge to extremely similar values.
2. The model calculates node representations and then performs global max pooling to obtain the representations for a graph. We train the model, using Negative Log Likelihood Loss (NLLLoss), using and Adaptive Moment Estimation (ADAM). Passing all the graphs through the trained model provides us with embeddings for all graphs in the dataset. The dimension of embedding per graph in our case is 16 dim.
3. These embeddings are then split in multiple train-test ratios. We have performed this phase of the experiment for the test ratio from 0.1 to 0.9, in the increments of 0.1. We construct the required pairs for each anchor in the training data, i.e., one positive and one negative pair. Our objective in using Metric Learning is to exploit the similarities in instances of the same class, to differentiate between the two possible classes. The Siamese Network model is defined and we use the Contrastive Loss function and ADAM optimizer for this purpose. The similarity metric used here is Euclidean Distance, which needs to be greater to differentiate dissimilar classes.

4. We then test by creating pairs from the test data. Evaluation is done, in which accuracy is used to tell the quality of the pipeline. This is performed for all values of test splits.

5.2 Graph Transformer A Prototypical Network Pipeline for Multi-class Classification of Nodes

1. The train and test data have an 80–20 split. They are further split into support and query sets. In a prototypical learning scenario, the support set which is used for training, contains a few labeled nodes per class, and the query set which is used for testing comprises the remaining labeled nodes. The idea here revolves around allowing the model to adapt and learn from a constrained number of labeled nodes per class. Here, the few labeled nodes are experimented with for 1, 5, 10, 15, and 20. These nodes for the support set are chosen in a random fashion in order for the model to generalize better on query sets. The test data is split into support and query sets in a similar manner.
2. The criterion used here is the negative log-likelihood loss (NLLLoss), which is commonly used in classification tasks, and the optimizer used is Adam. The scheduler with step size 50 and gamma value 0.5 adjusts the learning rate every 50 episodes.
3. A training loop is performed for 100 episodes. In each episode,
   - It computes the centroids/prototypes for the support set using pre-computed support set embeddings.
   - Calculates the distances between query set embeddings and the class centroids.
   - Calculates prototypical loss using negative log-likelihood loss, from softmax probabilities.
   - Uses this loss for backpropagation and parameter updates.
   - Episodic train accuracy is calculated for the support set, while episodic test accuracy is computed for the query set.
   - The average episodic train accuracy is finally computed for over 100 episodes, as the sum of train accuracies divided by the total number of episodes.
4. The trained model is finally evaluated on the testing set. It samples a few labeled examples per class to create support and query sets, then computes class centroids and uses them to classify the query nodes. The average episodic test accuracy gives one a measure of how well a model can generalize to unseen, unlabeled, untrained samples (Tables 1 and 2).

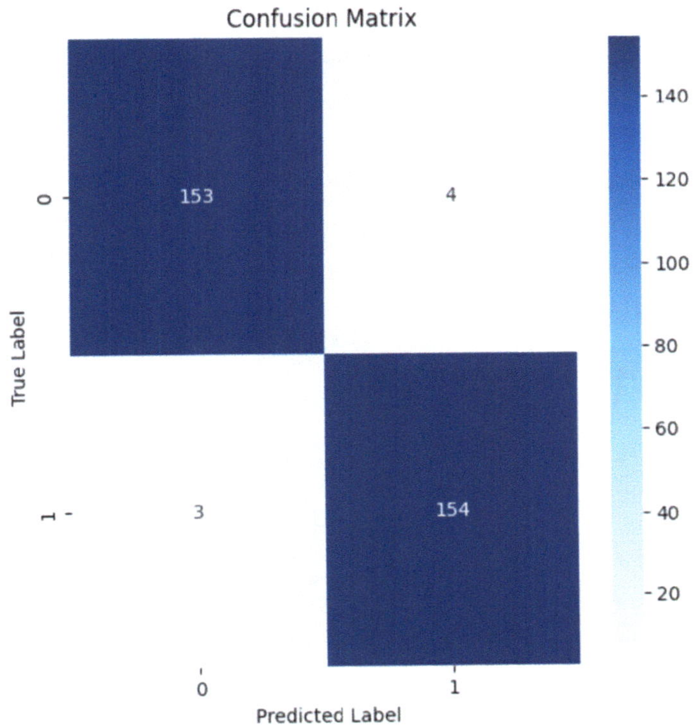

Fig. 4 Confusion matrix for 50–50 split

6 Results and Discussion

6.1 Graph Transformer Siamese Network Pipeline for Binary Classification of Graphs

The initial training on the transformer network model gives us a training accuracy of 1 and a testing accuracy of 0.855. This seems to be a good result, which can generate the required 16-dimensional embeddings.

The whole Siamese Network model training and testing was done with different train-test splits. As shown in the confusion matrix in Fig. 4, we observe that for a 50–50 split, the model gives a good accuracy of 0.978. The choice of 50–50 split has to do with that the quality of embeddings obtained was good to begin with after passing it through the transformer network architecture

As observed here in Table 1, we get an accuracy of 0.978. Here 0 means negative pair and 1 means positive pair in the context of the confusion matrix. The true positives and true negatives are very low compared to the correct predictions.

Table 1 Graph transformer network with 3 heads

Train accuracy	Validation accuracy	Test accuracy
1.0000	0.7742	0.8552

Table 2 Siamese network test performance

Accuracy	Precision	Recall	F1-score	Area under curve
0.9777	0.9747	0.9809	0.9777	0.9777

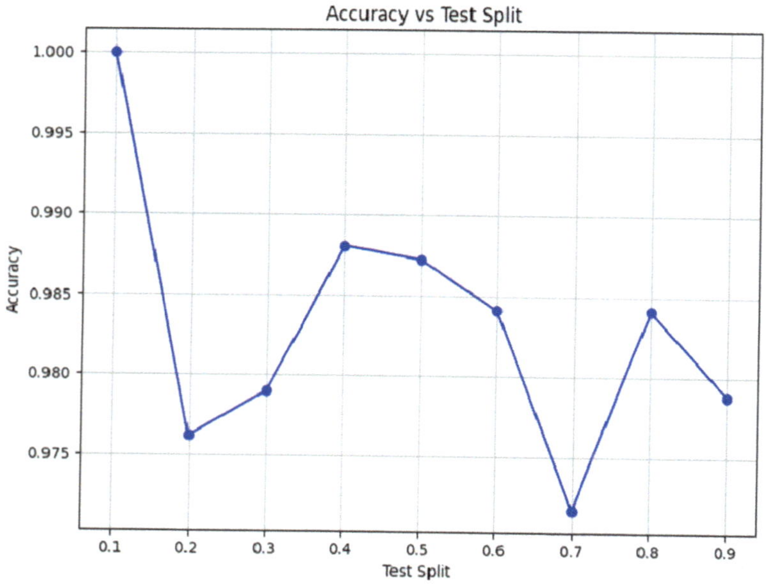

Fig. 5 Accurarcy values for various test splits

In Fig. 5, we notice that for test split value equal to 0.1 gives perfect accuracy. This however, is understandable because of 90% training data. The idea behind meta-learning is to learn with limited number of training examples. We chose 50–50 split because it is a midway point to test the benefits of meta-learning.

6.2 Graph Transformer A Prototypical Network Pipeline for Multi-class Classification of Nodes in a Graph

The trained model achieved an impressive average episodic test accuracy of 0.9062 during the testing/evaluation stage for 15 samples of every label of the support set,

Table 3 Table for different values of k

Support samples	AE train accuracy	AE test accuracy
1	0.9247	0.7615
5	0.9558	0.8939
10	0.9746	0.8993
15	0.9679	0.9062
20	0.9727	0.9038

over 100 episodes. This outcome demonstrates the model's effectiveness in generalizing to novel classes based on a limited number of support samples. The model accurately classified query samples from previously unseen classes, indicating its potential applicability in scenarios where data from novel classes is scarce. During the training process, the model achieved an average episodic train accuracy of 0.9679 for 15 samples of the support set during the training phase over 100 episodes. This high accuracy on the training set illustrates the efficient learning ability of the model during the training episodes. The model successfully leveraged the limited support samples to learn class prototypes, leading to better generalization to unseen classes. To gain deeper insights into the model's performance, we plotted the confusion matrix for the query test data. The confusion matrix visually represents the model's ability to distinguish between different research areas. The matrix showcases how well the model performs in predicting the true labels of the query samples. The figure below presents the confusion matrix.

The confusion matrix represents a side-by-side comparison between the actual and predicted labels by the model during the evaluation stage on query test data. The higher intensity in color signifies a high count and represents that the particular label/class has been accurately classified with the most number of nodes. In the figure, it is found that class or label '13' is the most accurately classified on the query test data with 148,400 query nodes. Class or label '3' is the second most accurately classified with 80,000 nodes.

In the confusion matrix shown in Fig. 6, the correctly classified elements are represented in the diagonal. The off-diagonal elements represent misclassifications. The high values on the diagonal indicate that our model performed well in classifying query samples from different research areas. Similarly, the figure below represents another confusion matrix on 15 samples of the support set, but on query train data during the training phase over 100 episodes.

Experimentation is done on values of k, i.e., 1, 5, 10, 15, 20, which represent the number of samples in support set during training. For 100 episodes, the value of k = 15 is found to have the highest episodic test accuracy. It is imperative to choose an optimal value of 'k' during training in order for the model not to overfit or underfit.

Table 3 represents the average episodic train and test accuracy for different values of 'k'.

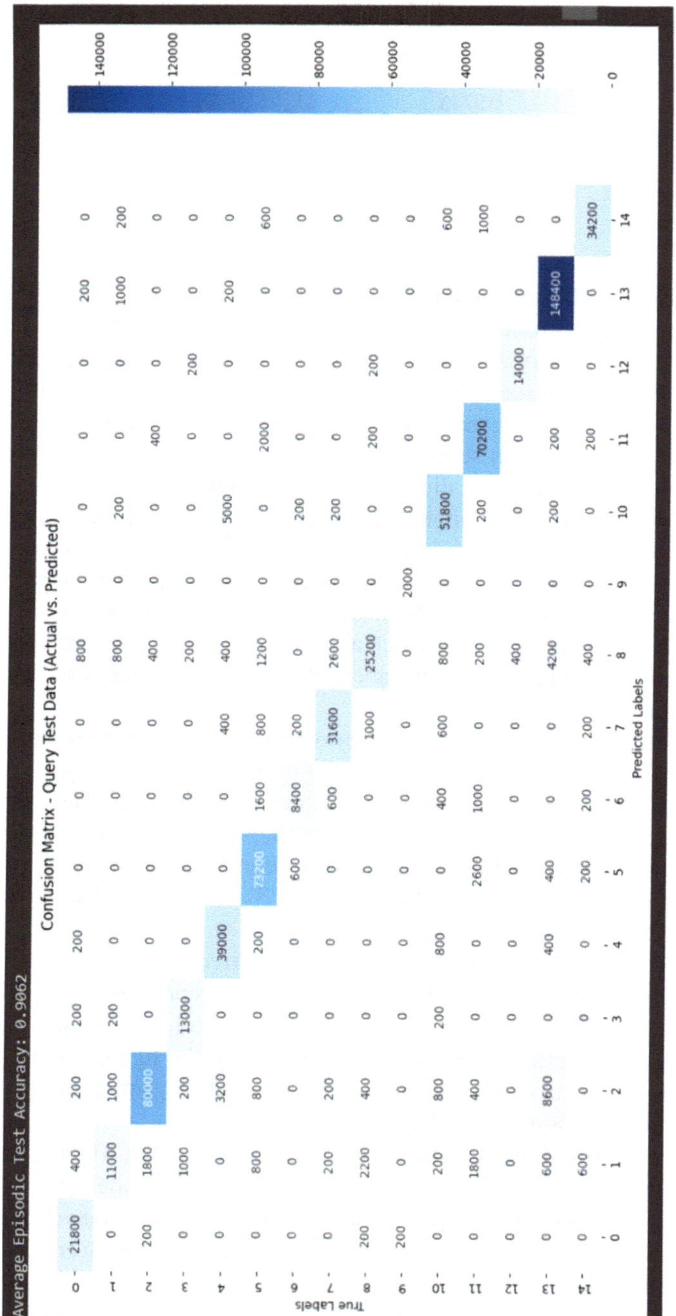

Fig. 6 Confusion matrix for query train data

- Support Set Size of 1: This represents an extreme type of learning scenario where the model must learn from just one labeled example per class. This setting is challenging and tests the model's ability to generalize from very limited information.
- Support Set Size of 5: This is a common choice for prototypical learning tasks. It provides a small but more diverse set of examples for each class, allowing the model to learn better representations and potentially improve its performance.
- Support Set Size of 10–20: As we increase the support set size, the model has more labeled examples to learn from, leading to better representations and potentially better results on a wide range of tasks.

Overall, the model shows consistent and strong performance both on the training and testing sets. The high average episodic test accuracy illustrates the model's proficiency in extrapolating to new classes with limited labeled data, which is a significant advantage in learning scenarios where the labeled data may be scarce or limited.

7 Conclusion and Next Steps

In this article, we proposed and demonstrated the efficacy of a two-step pipeline for Metric-based meta-learning using Siamese Networks and Prototypical Networks with Graph Transformers. Specifically on the Coauthor CS dataset, the pipeline is evaluated for a few-shot learning scenario from limited support samples during training. Thus the proposed pipeline not only showcases the high accuracy that can result from two factors, i.e., the quality of embeddings obtained from graph transformers and the metric learning-based meta-learning used by us. The proposed pipeline generalizes well to new classes by learning a metric space, in which similar samples are close to their respective class prototypes.

As a possible next step, the findings in this work encourage further exploration and application of these techniques in real-world scenarios with limited labeled data.

References

1. Bose AJ, Jain A, Molino P, Hamilton WL (2019) Meta-graph: few shot link prediction via meta learning. arXiv preprint arXiv:1912.09867
2. Das B et al (2023) Multi-contextual learning in disinformation research: a review of challenges, approaches, and opportunities. Online Soc Netw Media 34:100, 247
3. Ding K, Wang J, Li J, Shu K, Liu C, Liu H (2020) Graph prototypical networks for few-shot learning on attributed networks. In: Proceedings of the 29th ACM international conference on information & knowledge management, pp 295–304
4. Dou Y, Shu K, Xia C, Yu PS, Sun L (2021) User preference-aware fake news detection. In: Proceedings of the 44th international ACM SIGIR conference on research and development in information retrieval, pp 2051–2055
5. Dwivedi VP, Bresson X (2020) A generalization of transformer networks to graphs. arXiv preprint arXiv:2012.09699

6. Garcia V, Bruna J (2017) Few-shot learning with graph neural networks. arXiv preprint arXiv:1711.04043
7. Guo L, Zhang Q, Chen H (2022) Unleashing the power of transformer for graphs. arXiv preprint arXiv:2202.10581
8. Hebert L, Golab L, Cohen R (2022) Predicting hateful discussions on reddit using graph transformer networks and communal context. In: 2022 IEEE/WIC/ACM International joint conference on web intelligence and intelligent agent technology (WI-IAT). IEEE, pp 9–17
9. Huisman M, Van Rijn JN, Plaat A (2021) A survey of deep meta-learning. Artif Intell Rev 54(6):4483–4541
10. Jin M, Zheng Y, Li YF, Gong C, Zhou C, Pan S (2021) Multi-scale contrastive Siamese networks for self-supervised graph representation learning. arXiv preprint arXiv:2105.05682
11. Koch G, Zemel R, Salakhutdinov R et al (2015) Siamese neural networks for one-shot image recognition. In: ICML deep learning workshop, vol 2. Lille
12. Ma N, Bu J, Yang J, Zhang Z, Yao C, Yu Z, Zhou S, Yan X (2020) Adaptive-step graph meta-learner for few-shot graph classification. In: Proceedings of the 29th ACM international conference on information & knowledge management, pp 1055–1064
13. Matsumoto H, Yoshida S, Muneyasu M (2021) Propagation-based fake news detection using graph neural networks with transformer. In: 2021 IEEE 10th Global conference on consumer electronics (GCCE). IEEE, pp 19–20
14. Min E, Chen R, Bian Y, Xu T, Zhao K, Huang W, Zhao P, Huang J, Ananiadou S, Rong Y (2022) Transformer for graphs: an overview from architecture perspective. arXiv preprint arXiv:2202.08455
15. Shchur O, Mumme M, Bojchevski A, Günnemann S (2018) Pitfalls of graph neural network evaluation. arXiv preprint arXiv:1811.05868
16. Snell J, Swersky K, Zemel R (2017) Prototypical networks for few-shot learning. Adv Neural Inf Process Syst 30
17. Wang Y, Ma F, Wang H, Jha K, Gao J (2021) Multimodal emergent fake news detection via meta neural process networks. In: Proceedings of the 27th ACM SIGKDD conference on knowledge discovery & data mining, pp 3708–3716
18. Zhang H, Zhang J (2020) Text graph transformer for document classification. In: Conference on empirical methods in natural language processing (EMNLP)
19. Zhou F, Cao C, Zhang K, Trajcevski G, Zhong T, Geng J (2019) Meta-GNN: on few-shot node classification in graph meta-learning. In: Proceedings of the 28th ACM international conference on information and knowledge management, pp 2357–2360

Deep ResNet-Based Classification of Long-Tailed Data with Renyi Entropy as Loss Function

Salmaan Yousuf, Ranjeet Kumar, and Shahid Mehraj Shah

Abstract A long-tailed distribution in a dataset is a particular kind of imbalance where data samples from several classes are distributed according to a long-tail distribution. As a result, the predictive ability of deep learning algorithms is heavily biased toward those classes that have high data samples (head classes), which results in subpar performance when classifying minority class (tail classes) data samples. In literature, there are several methods to solve class imbalance problems like resampling inputs and reweighting model parameters. However, these methods suffer heavy overfitting and are hard-to-solve optimization problems during training. The existing technique uses Shannon cross entropy as a loss function. In our approach, we use different loss functions, in particular Renyi entropy, to compare the classification accuracy of ResLT-based deep learning models. The results depict that Renyi entropy loss function outperforms any other loss function when it comes to classification tasks on long-tailed data distributions. We test our proposed approach on the long-tailed version of two standard datasets, Cifar-10 and Cifar-100. We achieved an accuracy of 84.15% on the Cifar-10 and 71.78% on the Cifar-100 dataset, which outperforms any state-of-the-art method for class imbalance tasks. Experimental results demonstrate improvement in the accuracy of residual.

Keywords Renyi entropy · Cross entropy · ResNets · Long-tailed data distributions

S. Yousuf (✉) · S. M. Shah
Communication Control & Learning Lab (C2L2), Electronics & Communication Engineering Department, NIT, Srinagar, Jammu and Kashmir, India
e-mail: salmaanyousuf001@gmail.com

S. M. Shah
e-mail: shahidshah@nitsri.ac.in

R. Kumar
Computer Science Engineering Department, NIT, Srinagar, Jammu and Kashmir, India
e-mail: ranjeetkumarrout@nitsri.ac.in

S. M. Thampi et al. (eds.), *Fifth International Conference on Computing and Network Communications*, Lecture Notes in Electrical Engineering 1221,
https://doi.org/10.1007/978-981-97-4711-5_4

1 Introduction

Over the last few years, convolutional neural networks have had tremendous success in a number of deep learning domains like image classification [1–4], object detection [5, 6], and semantic segmentation [7–9]. The performance on classification tasks have been remarkably enhanced by the developement of deep convolutional models like VGGNET [1], GoogLeNet [10], ResNet [2], and DenseNet [11]. The accessibility to open source and accurately annotated datasets like ImageNet [12], Inaturalist [13], MS COCO [14], Places [15], and Cifar has made it possible to train deeper and more complex CNNs. As such, they have achieved cutting-edge performance in numerous computer vision tasks. These datasets are carefully designed in such a way that data samples belonging to different classes follow a balanced data distribution. However, this is not true when we work in raw real-life scenarios. In fact data could follow an unexpected long-tailed distribution where the majority of data samples belong to a few classes and the rest of the classes have relatively smaller sample counts. Such a dataset is referred to as long long-tailed dataset. If this long-tailed behavior of datasets is ignored then it is very probable that the head classes may dominate the training process, achieving high training accuracy but having subpar accuracy on the tail classes [16]. In this research approach, we improve the classification accuracy of deep residual networks on long-tailed datasets using different loss functions. We are aware that classification is one of the most common deep learning jobs, where each input must be categorized into a discrete category. Many real-life applications like autonomous driving [17], accident detection [18], speech recognition [19], medical image processing [20] etc., inherently involve classification tasks. In our proposed approach, we consider Renyi entropy loss function as the best choice for long-tailed data distributions. The Renyi entropy loss introduces the α parameter as a means of balancing the weights assigned to the more likely classes and the less likely ones in the loss calculation. An intuitive explanation of our proposed approach is to gradually improve classification accuracy on tail classes by changing the α parameter values. In machine learning, Cross entropy is a loss function, that corresponds to the limiting case of $\alpha = 1$ in the Renyi entropy loss function. The Renyi entropy loss function tends to place greater emphasis on the tail classes while giving the head class samples less weight when alpha is set to the range of (0, 1). On the other hand, the Renyi entropy loss function inherently assigns more weight to the head class samples when $\alpha > 1$. We can tailor the Renyi entropy loss function to better fit the specifics of the current classification task by modifying the alpha parameter. However, it should be emphasized that there is no ideal value of alpha that solves all classification problems, therefore choosing a suitable value for α involves considerable analysis and experimentation. Here are the main contributions we provide in our research approach:

- We did a comparative analysis of a long-tailed classification problem using different loss functions and labeled one loss function as the best choice for the task of long-tailed recognition by rigorous experimentation and the results as validation.

- We propose the use of Renyi smoothing factor α to regulate the balance between the importance of the head class data samples and tail class data samples and reducing the drawbacks of earlier techniques.
- We tested our suggested method on two commonly used datasets for comparison. Our outcomes considerably surpassed many state-of-the-art techniques.

2 Related Work

2.1 *Resampling Approach*

In deep learning, Resampling [21], involves manipulating an imbalance dataset to balance the class distribution, either by oversampling those classes that have relatively fewer data samples (tail class) or by undersampling those classes that have relatively more data samples (head class). The oversampling technique involves increasing the data sample count of tail classes by generating synthetic data points based on the existing tail class instances. Oversampling [22] is very useful on large datasets but on small datasets, it faces severe overfitting to the tail classes. The undersampling technique involves removing random data samples from the head class. Since undersampling [21], deletes a fair amount of informative data, as a result, the tendency of deep learning models to generalize on unseen data is limited. Also, it is worthy enough to mention that the undersampling method is inappropriate when the data is highly unbalanced. Reference [23] provides a detailed survey on deep learning tools dealing with data scarcity.

2.2 *Reweighting Approach*

Reweighting [24], is another mainline strategy to address class imbalance problems. It assigns different weights to data samples in different classes during the training process. The fundamental concept is to give lower weights to data samples in the head class and higher weights to the data samples in the tail classes to increase their importance in the loss calculations [25]. This makes the deep learning model's learning process focus more on correctly classifying the data samples of the tail classes. There are many approaches available for reweighting strategy and one of the most famous is inverse frequency weighting. In this method, each instance's weight is inversely proportional to how frequently the samples appear in the training set. Other approaches to reweighting methods include cost-sensitive learning, dynamic reweighting, and adaptive reweighting [26]. Cost-sensitive learning involves assigning different costs to mis-classification for each class, which is directly proportional to the class's relative occurrence in the training set. Cost-sensitive learning motivates the classifier to focus more on correctly categorizing the minority class by allocating higher mis-classification costs to the minority classes and lower costs to the majority classes. By adjusting the weights assigned to each instance during training, the cost-sensitive learning technique can help equalize the impact of each class on the loss function and reduce the bias toward the most prevalent class. Weights in dynamic reweighting methods are modified during training based on the model performance. High-confidence mis-classified samples are given larger weights, and low-confidence correctly classified samples are given lower weights. In this way, the model can improve its focus on the tail class samples, and can effectively handle complex and imbalanced datasets. Finally, in adaptive reweighting, for samples with a good classification accuracy, focal loss is given a reduced weight. The concept aligns with the understanding that instances with low gradients are generally classified correctly, whereas instances with high gradients are typically outliers beyond the distribution. Despite all of this, reweighting makes it challenging to optimize deep learning models during training when using large-sized datasets [27].

2.3 *Two Stage Approach for Handling Class Imbalance*

In [28], it is mentioned that reweighting and resampling strategies performed worse than the standard empirical risk minimization (ERM) algorithm before annealing the learning rate. In [28], the researchers suggested a delayed rebalancing optimization schedule comprising two stages. This schedular first trains using the vanilla ERM using the LDAM loss before annealing, and then deploys a reweighted LDAM loss with a significantly lower learning rate.

In [29], it is mentioned that class rebalancing techniques are important when simultaneously training a representation and classifier, researchers found that uniform sampling results in more generic representations. On the basis of this concept, in [29], they achieved significant results on imbalanced versions of ImageNet dataset [30]. Based on the concept of separating the representation and classifier learning, the deep learning model is trained initially with uniform sampling, and then the classifier is fine-tuned with balanced sampling while maintaining the representation learning parameters. Likewise, in [31], the researchers incorporated mixup into their novel cumulative learning approach, which allowed them to connect representation learning and classifier rebalancing. As a result, they were able to achieve the best performance on the long-tailed version of iNaturalist datasets [13]. The concept of a two-stage design contradicts the idea of end-to-end effectiveness that was previously held during the deep learning era. Despite this, in long-tailed classification, two-stage training significantly outperforms end-to-end training. Tang et al. evaluated the causal graph and discovered that negative momentum causal effects had a crucial role in explaining this in their paper [32]. In contrast, our approach addresses the long-tailed problem by focusing on the Renyi smoothing factor which is a tunable parameter of the Renyi loss function.

2.4 Meta Learning and Transfer Learning

Meta-learning is a concept in deep learning that refers to a deep learning model's tendency to draw insights from the past and apply them to quickly adapt to new tasks. In [33], the proposed MetaModelNet builds a classifier for low-frequency classes using a meta-regression network that is learned from high-frequency classes. In [34], Abdullah Jamal et al. examined unbalanced learning from the standpoint of domain adaptation and put forth a meta-learning paradigm for long-tailed recognition. Individual specialists were trained by Saurabh et al. in [35] who then imparted their knowledge to the student model. Although the training process is typically much more difficult, the effectiveness of meta-learning and transfer learning approaches still lags considerably below two-stage approaches.

2.5 ResLT

In [35], Jiequan Cui et al. proposed a novel method to solve the class imbalance problem in the context of parameter space. This method had two key insights, parameter specification and residual fusion mechanism. In parameter specification, the datasets were modified as in [35] and fed to a classifier via different branches and finally using the residual fusion method, outcomes of these branches were accumulated via additive shortcuts. This proved to be an effective approach to tackle the long-tailed distribution problem and provided significant results as well.

3 Methodology

Motivation The accuracy of a deep learning model is significantly diminished by undersampling the head classes as some important information is lost. This leads to heavy overfitting during training. Also, oversampling tail class data samples makes the model lose its generalization ability because, during training, the model learns from redundant data samples. With reweighting strategies, previous work depicts that it makes deep learning models hard to optimize. ResLT-based methods are by far the best to tackle the long-tailed distribution problem. It uses mean square error and cross entropy as the loss functions. However we in our research work observed that Renyi Entropy loss is a better choice of loss function when it comes to long-tailed data distributions and this statement is validated in the results.

3.1 Cross Entropy as a Loss Function

A common loss function utilized in machine learning, particularly for classification tasks, is cross entropy. It measures how different the target variable and predicted probability distributions are from one another. Suppose we have a classification problem where the dataset possesses k distinct classes, let $y_i \in [0, 1]^k$ be the one hot encoded label vector for the ith training example, and let $p_i \in [0, 1]^k$ be the predicted probability distribution for the same example. The cross entropy loss function L_{CE} is defined as:

$$L_{\text{CE}} = -\frac{1}{n}\sum_{i=1}^{n}\sum_{j=1}^{k} y_{i,j}\log(p_{i,j}) \tag{1}$$

where n is the total number of training examples. This function computes the average of the negative log-likelihoods of the correct class probabilities. It penalizes more heavily the predicted probabilities that deviate further from the true probabilities, and it tends to produce more confident and accurate predictions as the model learns. However, the cross entropy loss function does not give any specific emphasis to tail class data sampled in the classification tasks. Thus we need to be more specific about loss function when dealing with long-tailed data distributions.

3.2 Renyi Entropy as a Loss Function

Renyi entropy is used as a loss function in deep learning algorithms to incorporate classification tasks. It is a measurement of the disorder or uncertainty in a probability distribution. Given a probability distribution $p(x)$ over a discrete or continuous variable x, the Renyi entropy of order α is defined as:

$$H_\alpha(p) = \frac{1}{1-\alpha}\log\left(\sum_i p_i^\alpha\right) \tag{2}$$

In Renyi entropy, the parameter alpha α is a positive real number that determines the order of the Renyi entropy. It is a tunable parameter that controls the sensitivity of the entropy measure to the probabilities of different outcomes. When $\alpha = 1$, Renyi entropy reduces to the Shannon entropy, which is a widely used measure of the entropy of a probability distribution. For $\alpha > 1$, Renyi entropy gives more weight to the larger probabilities in the distribution and is more sensitive to the head of the distribution. For $\alpha < 1$, Renyi entropy gives more weight to the smaller probabilities in the distribution and is more sensitive to the tail events. The range of $(0, 1)$ for α is what we utilize in our approach to achieve good classification accuracy on tail data. The selection of alpha is influenced by the unique requirements of the given application and the characteristics of the probability distribution under investigation. In machine learning, α is often used as a hyperparameter that can be tuned during training to optimize the performance of the model. Given a set of training data with inputs X and corresponding labels Y, and a deep learning model f that predicts the probabilities of each class for a given input, the mathematical equation for Renyi entropy loss function of order α is defined as:

$$L_\alpha = \frac{1}{n(1-\alpha)} \log \left(\sum_{i=1}^{n} \sum_{j=1}^{k} \left(f(X_i)_j\right)^\alpha y_{ij}^{1-\alpha} \right) \tag{3}$$

Here, k represents the total number of categories in the dataset, while n indicates the total number of instances used for training, $f(X_i)_j$ is the predicted probability of class j for input X_i, and y_{ij} is the true label of input X_i for class j.

$$y_{ij} = \begin{cases} 1 & \text{if } X_i \text{ belongs to class } j \\ 0 & \text{otherwise} \end{cases}$$

There are obvious reasons for choosing Renyi entropy as a loss function for classification tasks with long-tailed data distributions, which are as under:

- It can manage class imbalance. Class-balanced datasets are not always available when working on real-life classification problems. So, by enabling a parameter to adjust how much each class contributes to the loss, the class imbalance problem can be effectively solved. The Renyi smoothing factor in the Renyi entropy loss function provides this flexibility.
- It forces the deep learning model to concentrate on the crucial elements which in our case are tail class data samples. Also, the regularization impact of the Renyi entropy loss function pushes the model to concentrate on the most crucial aspects of the data. This leads to a model that is more reliable and understandable.
- Renyi entropy may be computed effectively in a vectorized fashion, making it suited for use in deep learning models with large datasets.
- Renyi entropy loss function can promote the model to learn a more generalized decision boundary that distinguishes the classes, which can prevent overfitting and result in improved adaption to unseen data.

3.3 *Renyi Smoothing Factor as a Tunable Parameter to Solve Long-Tailed Recognition Problem*

ResLT [35] proposed by jiequan chi et al. is by far the most prominent method for tackling the class imbalance problem. Using this method as a base, we imple-

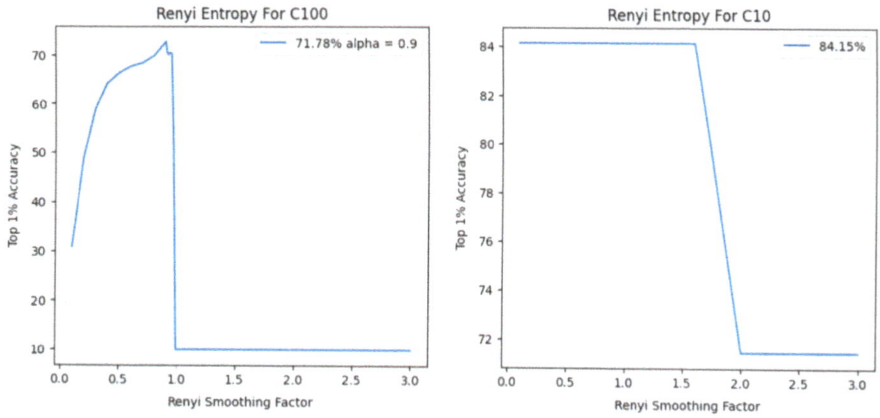

Fig. 1 Comparison of Top 1 Accuracy (%) with reny entropy loss on Cifar-10 and Cifar-100 datasets with varying values of α

mented various loss functions to achieve the best possible results. We made comparisons between training and test results generated by ResLT-based deep models with Renyi entropy and cross entropy as loss functions. After rigorous experimentation, it is found that Renyi entropy loss function performs better than the cross entropy loss function for long-tailed classification tasks. This is because the Renyi entropy loss function is less sensitive to the presence of rare classes, which are often more challenging to predict accurately in long-tailed datasets. In the Renyi entropy loss function, the order parameter α can be tuned to adjust the emphasis on tail classes. By setting α to a value less than 1, the Renyi entropy loss function places more emphasis on the rare classes, which can improve their classification accuracy. On the other hand, the cross entropy loss function treats all classes equally, which can lead to poor performance on the rare classes in long-tailed datasets. That being said, the decision about choosing the loss function ultimately boils down to the specific task and the dataset at hand. In some cases, the cross entropy loss function may perform just as well or even better than the Renyi entropy loss function, especially if the dataset is not heavily skewed toward certain classes. It is always a good idea to experiment with different loss functions and evaluate their performance on a specific task and dataset. Figures 1 and 2 shows the comparison between test accuracies of ResLT model using Renyi entropy loss function on Cifar 10 and Cifar 100 datasets. These outcomes perform much better than the standard cross entropy-based ResLT methods.

To update the weights of a deep learning model using the Renyi entropy loss function, we use gradient descent with the following updation rule:

$$w \leftarrow w - \eta \nabla L_{\alpha}(p, y) \tag{4}$$

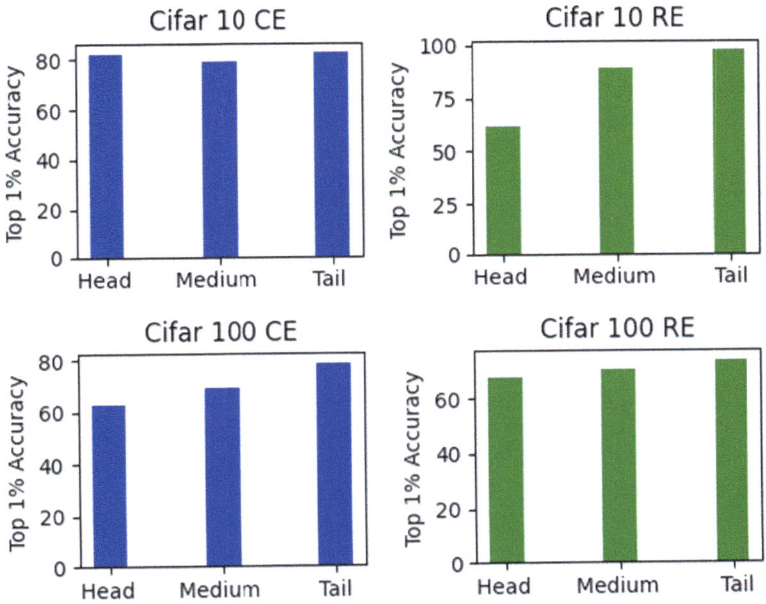

Fig. 2 Comparison of individual branch performance using Renyi entropy and cross entropy as loss function in ResLT-based deep models

where w is the weight vector, η is the learning rate, and $\nabla L_\alpha(p, y)$ is the gradient of the Renyi entropy loss function concerning the weight vector. The gradient can be computed using backpropagation, similar to other loss functions. To update the weights, the gradient of the Renyi entropy loss function is computed concerning the model parameters using backpropagation. The weights are then updated using an optimizer such as stochastic gradient descent (SGD) or Adam optimizer, based on the computed gradient and the learning rate. The goal of the optimization is to minimize the Renyi entropy loss, which would result in the model making better predictions on the imbalanced dataset.

The Renyi entropy loss function is known to be better than cross entropy for long-tailed data because it is less sensitive to outliers and can better handle the imbalanced class distribution. Consider the case of a two-class problem with a long-tailed distribution, where the majority class has many more samples than the minority class. For the majority class, let the number of samples be denoted by N_m, and for the minority class, let it be denoted by N_n, where $N_\mathrm{m} > N_\mathrm{n}$. Let p_m and p_n be the probabilities of the majority and minority classes, respectively. Then, the cross entropy loss function can be written as:

$$L_\mathrm{CE} = -[p_\mathrm{n}\log(p_\mathrm{n}) + (1 - p_\mathrm{n})\log(1 - p_\mathrm{n})] \quad (5)$$

when p_n is small, the $\log(1 - p_n)$ term dominates the loss, leading to a large value. This makes the model overemphasize the minority class, leading to poor performance of the majority class.

On the other hand, the Renyi entropy loss function can be written as:

$$L_\alpha = \frac{1}{N(1-\alpha)} \log \left(\sum_{i=1}^{N} \sum_{j=1}^{K} \left(f(X_i)_j\right)^\alpha y_{ij}^{1-\alpha} \right) \tag{6}$$

where α is a hyperparameter that determines the degree of emphasis on the tail of the distribution. When α is large, the Renyi entropy is more sensitive to the head of the distribution, allowing it to better capture the majority classes. However, when α is small, the Renyi entropy is more sensitive to outliers, allowing it to better capture the minority classes. Therefore, by tuning the hyperparameter α, the Renyi entropy loss function can be used to balance the emphasis on the minority and majority classes, leading to better performance on imbalanced datasets.

Thus we conclude that the Renyi entropy loss function is mathematically a better choice of loss function than cross entropy for long-tailed data distributions because it can better handle the imbalanced class distribution by allowing the degree of emphasis on the tail and head classes to be tuned via the hyperparameter α.

4 Experiments

4.1 Datsets and Experimental Setting

Cifar-10-LT is a standard benchmark dataset of 60,000 colored images with size of 32×32 and distributed in 10 distinct classes. There are 6000 images in each class. The training dataset possesses 50,000 sample images and the test dataset possesses 10,000 sample images [28]. The low-resolution photos in the dataset include typical objects from daily life. Machine learning and deep learning tasks involving image classification frequently employ the Cifar-10 as a benchmark dataset. In our approach, we used a long-tailed version of the Cifar-10 dataset.

Cifar-100 is a labeled subset of the 80 million tiny pictures dataset, possessing 60,000 colored images with the size of 32×32 and distributed in 100 distinct classes with 600 images per class. The 100 classes are grouped into 20 superclasses, each containing 5 subclasses [31]. The training dataset possesses 50,000 sample images and the test dataset possesses 10,000 sample images [28]. The low-resolution photos in the dataset include typical objects from daily life. Machine learning and deep learning tasks involving image classification frequently employ the Cifar-100 as a benchmark dataset. In our approach, we used a long-tailed version of the Cifar-100 dataset. Compared to Cifar-10, the images in Cifar-100 are more intricate and varied since they show fine-grained object categories like bird species, car models, etc.

Dataset Preparation Since both the benchmark datasets we use are balanced, we induce imbalance in them using code. We sort the datasets on the basis of class sample count. Inspired by [36], we divided the imbalanced dataset into three parts. Classes with more than 100 data samples form the head group, classes with 20 to 100 data samples form the medium group and classes with less than 20 data samples form the tail group.

Preprocessing and Training Details: Inspired from [28, 29], preprocessing steps applied on the input images are:

1. Randomly resized and cropped to a fixed size of 32 × 32 pixels.
2. Randomly flipped horizontally to increase the diversity of the training data.
3. Pixel values of images are randomly adjusted to change their brightness, contrast, saturation, and hue.
4. Pixel values of images are normalized to [0, 1].
5. To address the imbalance in the training data, we use random sampling to balance the number of samples in each class during each training epoch.
6. We adopt weights that vary by class to handle the disproportionate distribution of training data. The weights assigned are inversely proportional to the number of instances in each category, thereby giving greater weight to underrepresented categories during training.

By utilizing these preprocessing methods, the deep learning model's performance on data sets with long-tailed distributions can be enhanced.

Model Architecture and Parameters We have used Resnet-32 [35] as the backbone network with Renyi entropy as the loss function. The model is trained for 200 epochs, utilizing Stochastic gradient descent as the optimizer. It should be noted that for ablation studies, we have used different combinations of model parameters and settled for the best choices of parameters after empirical studies and deep experimentation. The model speciations adopted during experimentation are given in Tables 1 and 2.

Table 1 Individual branch performance

	Loss function	Head accuracy	Medium accuracy	Tail accuracy
Cifar 10 LT	Cross entropy	81.93	79.10	82.32
Cifar 10 LT	Renyi entropy	61.5	88.93	97.5
Cifar 100 LT	Cross entropy	62.8	69.37	78.48
Cifar 100 LT	Renyi entropy	67.72	70.31	73.79

Table 2 Model parameter specifications

	Resnet-32 Architecture		
Parameter	Value	Parameter	Value
Epochs	200	Weight decay	5e−4
Imbalance factor	0.02	Scale	1
Learning rate	0.1	Classes C10	10
β C10	1	Classes C100	100
β C100	0.8	Batch Size	128
Momentum	0.9	Random seed	42

Table 3 Top 1 accuracy(%) of ResNet-32 on long-tailed Cifar-10 and Cifar-100

Deep model	Cifar 10	Cifar 100	One class
Dataset	Cifar 100	Cifar 10	One Stage
LDAM	39.60	73.35	✓
BBN	42.56	79.82	✓
CB FOCAL	39.6	74.57	✓
ELF+DRW	43.10	78.10	×
LDAM+DRW	42.04	77.03	×
ResLT CE	69.90	81.3	✓
Result RE	71.78	84.15	✓

4.2 *Results and Discussion*

1. **Top 1 accuracy (%)**: Our strategy considerably outperformed ResLT and other existing methods on long-tailed recognition. We have conducted experiments on the long-tailed versions of Cifar 10 and Cifar 100. Table 3 provides information on the efficiency of our strategy. For Cifar 10 dataset, we achieved an impressive 84.15% testing accuracy and for the Cifar 100 dataset, we achieved 71.78% testing accuracy and both these results outperform the cross entropy-based ResLT methods and other two-stage methods. These results are verified in Table 3 and shown in Fig. 3.
2. **Individual branch performance**: Since we use three different branches to classify head, medium, and tail class data, the individual branch performance as shown in Table 1 depicts that we achieve significant results on tail class, which generally is a concern in long-tailed recognition. Later these results are merged in such a way to get a good overall accuracy. These results are shown Fig. 2.
3. **Impact of Renyi smoothing factor α on overall performance** We implemented our approach on imbalanced Cifar 10 and Cifar 100 datasets under different experimental settings. As far as the range of α is considered, we considered decimal values in the range [0+, 3] and obtained significant results. For Cifar 10 LT dataset,

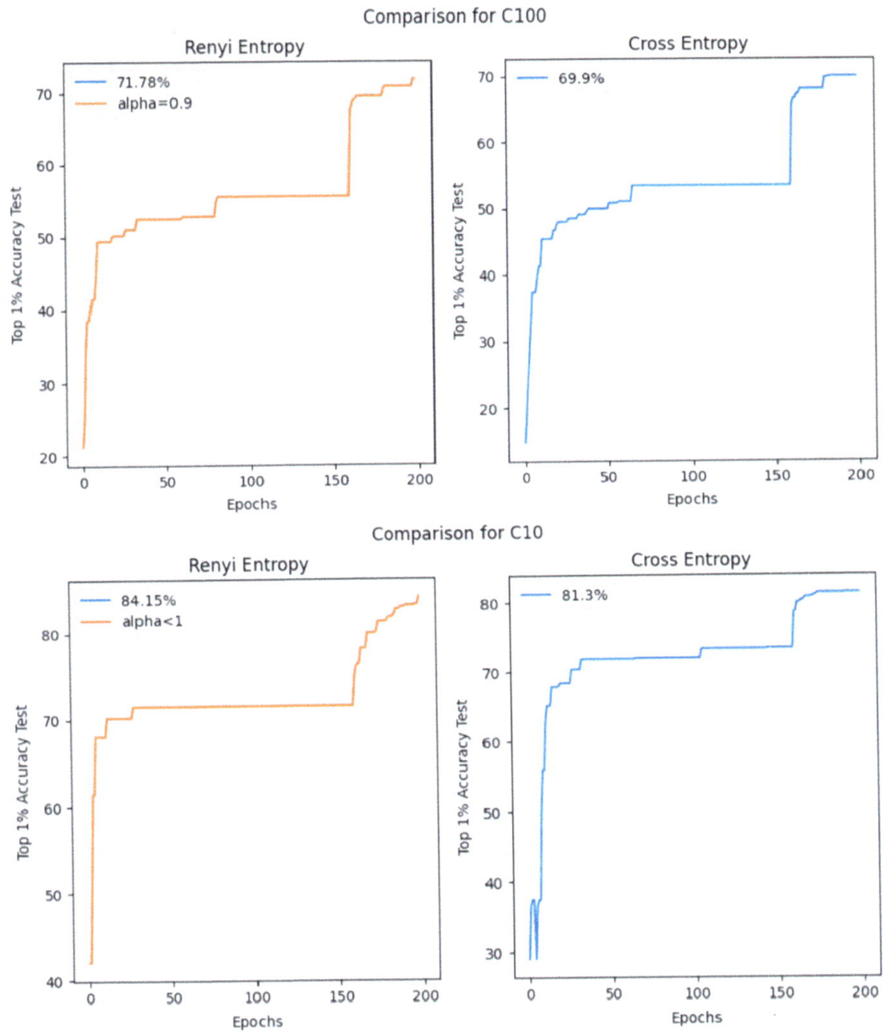

Fig. 3 Comparison of Top-1 accuracy (%) with Renyi Entropy and Cross Entropy as Loss function for Cifar 10 AND Cifar 100 datasets

for α in range (0, 1.6), we got a constant value for top 1% accuracy of 84.15% and for $\alpha > 1.6$, there is a sharp fall in accuracy and top 1% accuracy falls to 70%. This is in accordance with Sect. 3.3. For Cifar 100 LT dataset, we obtained the best accuracy of 71.78% at $\alpha = 0.9$ as shown in Fig. 1. However, for other values of α, the accuracy fall, s and for $\alpha > 1$, there is a significant dip in performance and top 1% accuracy falls to 10%. This result validates the fact that for classification tasks, the range of Renyi entropy smoothing factor is (0, 1).

5 Conclusion

In this paper, we made a comparative study of loss functions on ResLT-based deep learning models and provided concrete results that prove Renyi entropy with tunable smoothing factor α is a better choice of loss function over cross entropy loss, mean squared error loss, etc. for long-tailed data distribution as α can be tuned to fit complex imbalance distributions. Also, our results concretely depict that the range of values for Renyi smoothing factor α for classification tasks is (0, 1). Extensive experimental results on various challenging benchmark datasets like the long-tailed version of Cifar 10 and Cifar 100 under varying experimental settings manifest the effectiveness of our approach and our results significantly outperform any existing methods on long long-tailed recognition process. We leave more theoretical analysis of other loss functions as our future work.

References

1. Simonyan K, Zisserman A (2014) Very deep convolutional networks for large-scale image recognition
2. He K, Zhang X, Ren S, Sun J (2015) Deep residual learning for image recognition, 7
3. Ma N, Zhang X, Zheng H-T, Sun J (2018) ShuffleNet V2: practical guidelines for efficient CNN architecture design
4. Howard A et al (2017) MobileNets: efficient convolutional neural networks for mobile vision applications
5. Ren S, He K, Girshick R, Sun J (2015) Faster R-CNN: towards real-time object detection with region proposal networks. IEEE Trans Pattern Anal Mach Intell 39
6. Lin T-Y et al (2016) Feature pyramid networks for object detection
7. Cordts M et al (2016) The cityscapes dataset for semantic urban scene understanding
8. Chen L-C, Zhu Y, Papandreou G, Schroff F, Adam H (2018) Encoder-decoder with atrous separable convolution for semantic image segmentation
9. Zhao H, Shi J, Qi X, Wang X, Jia J (2016) Pyramid scene parsing network
10. Szegedy C et al (2014) Going deeper with convolutions
11. Huang G, Liu Z, van der Maaten L, Weinberger KQ (2017) Densely connected convolutional networks, pp 4700–4708
12. Zoph B, Vasudevan V, Shlens J, Le Q (2018) Learning transferable architectures for scalable image recognition, pp 8697–8710
13. Van Horn G et al (2018) The iNaturalist species classification and detection dataset, pp 8769–8777
14. Liu H, Simonyan K, Yang Y (2018) Differentiable architecture search, DARTS
15. Tan M, Chen B, Pang R, Vasudevan V, Le Q (2018) Platform-aware neural architecture search for mobile, MnasNet
16. Cui J, Chen P, Li R, Liu S, Shen X, Jia J (2019) Fast and practical neural architecture search
17. Bojarski M et al (2016) End to end learning for self-driving cars. arXiv preprint arXiv:1604.07316
18. Manogaran G, Varatharajan RK, Lopez D, Vijayakumar V (2017) Accident detection and reporting system using IoT. Wireless Pers Commun 95:423–439
19. Hinton G et al (2012) Deep neural networks for acoustic modeling in speech recognition: the shared views of four research groups. IEEE Signal Process Mag 29:82–97

20. Litjens G et al (2017) A survey on deep learning in medical image analysis. Med Image Anal 42:60–88
21. He H, Garcia E (2009) Learning from imbalanced data. IEEE Trans Knowl Data Eng 21:1263–1284
22. Shen L, Lin Z (2016) Relay backpropagation for effective learning of deep convolutional neural networks 9911:467–482
23. Alzubaidi L et al (2023) A survey on deep learning tools dealing with data scarcity: definitions, challenges, solutions, tips, and applications. J Big Data 10:46
24. Li F et al (2022) Decoupling representation learning for imbalanced Electroencephalography classification in rapid serial visual presentation task. J Neural Eng 19
25. Ren M, Zeng W, Yang B, Urtasun R (2018) Learning to reweight examples for robust deep learning
26. Lin T-Y, Goyal P, Girshick R, He K, Dollar P (2018) Focal loss for dense object detection. IEEE Trans Pattern Anal Mach Intell, p 1
27. Cui Y, Jia M, Lin T-Y, Song Y, Belongie S (2019) Class-balanced loss based on effective number of samples, pp 9260–9269
28. Cao K, Wei C, Gaidon A, Aréchiga N, Ma T (2019) Learning imbalanced datasets with label-distribution-aware margin loss
29. Kang B, Xie S, Rohrbach M, Yan Z, Gordo A, Feng J, Kalantidis Y (2019) Decoupling representation and classifier for longtailed recognition
30. Deng J et al (2009) ImageNet: a large-scale hierarchical image database. In: IEEE Conference on computer vision and pattern recognition (CVPR), pp 248–255
31. Zhou B, Cui Q, Wei X-S, Chen Z-M (2019) BBN: bilateral-branch network with cumulative learning for long-tailed visual recognition
32. Tang K, Huang J, Zhang H (2020) Long-tailed classification by keeping the good and removing the bad momentum causal effect
33. Yu H-T et al (2023) An in-depth study on adversarial learning-to-rank. Inf Retrieval J 26
34. Jamal M, Brown M, Yang M-H, Wang L, Gong B (2020) Rethinking class-balanced methods for long-tailed visual recognition from a domain adaptation perspective
35. Jiequan C, Liu S, Tian Z, Zhong Z, Jia J (2022) ResLT: residual learning for long-tailed recognition. IEEE Trans Pattern Anal Mach Intell, p 1
36. Liu Z, Miao Z, Zhan X, Wang J, Gong B, Yu SX (2019) Largescale long-tailed recognition in an open world

Leveraging ML Algorithms to Predict IT Job Opportunities and Inform Programme Approval

N. Sudharshana, J. Naren, and D. Sumathi

Abstract Unemployment is a serious concern around the world. Every year, millions of graduates are without a job. Students enrol in graduate schools without understanding what the job market expects. Upon graduation, the necessary skills to secure employment are often lacking. In such instances, it can be inferred that degree programmes fail to prepare students for job applications and interviews. A framework must be in place to educate graduates about available employment. Hence a job recommender system comes into the picture. Its main objective is to recommend jobs to students based on technical skillset, logical quotient rating, self-learning capability and a variety of other criteria. The system can also serve as feedback for programmes as it gauges the employability of the student passing out. It can also be used by regulatory councils (government organizations like AICTE which regulate and approve technical programmes) to decide whether the programme should be continued based on the number of job-ready graduates produced. The proposed work has employed classifier algorithms (machine learning), for categorizing student placement data based on features like 'Academic Percentage in Operating Systems', 'Percentage in Computer Networks', 'Hours working per day', etc. The performance of various classifiers is also analyzed.

Keywords Unemployment · Job recommender system · Classifier algorithms · AICTE

N. Sudharshana
Chennai, India

J. Naren (✉) · D. Sumathi
Department of Computer Science and Engineering, Alliance College of Engineering and Design, Alliance University, Bengaluru, India
e-mail: naren.jeeva3@gmail.com

S. M. Thampi et al. (eds.), *Fifth International Conference on Computing and Network Communications*, Lecture Notes in Electrical Engineering 1221,
https://doi.org/10.1007/978-981-97-4711-5_5

1 Introduction

The Centre for Monitoring Indian Economy (CMIE) reports an alarmingly high unemployment rate at 8.3% as of April 2023, comprising a significant population of graduates too. Between January and April 2022, it peaked at 17.8%. To make matters worse, disruption of economic activities during the pandemic spiked unemployment rate to 19.3% in 2021. Rajasthan, Bihar and Andhra Pradesh were not able to employ more than a third of the graduates.

A closer look tells that the main cause of unemployment among college (or university) graduates was the lack of information about employability in a particular programme. The All-India Council for Technical Education (AICTE) requires information on employability of various programmes in universities to decide upon introduction of new programmes, revision of curriculum and ensure adequate job market potential. Availability of information on employability would aid the AICTE in approving programmes and producing job-ready candidates. Eventually, students will succeed in recruitment after training in job-oriented programmes.

Therefore, a job prediction framework is needed to improve employability. The aim to develop an application to predict if a student will get a job based on the skills acquired at graduation, bridging gap between higher education institutions and the IT industry.

The dataset used for here (student placement data.csv) contains 38 columns, evaluating each aspect of a student. Some of the attributes are percentages in subjects like mathematics, algorithms, communication skills, computer networks; 'coding skills rating', 'participation in hackathons', 'suggested job role', and 'placements'.

2 Literature Survey

Elakia et al. [1] have harnessed the power of classification as a strategic approach to envision potential career paths suitable for students. Emphasis on understanding the nuances of student behaviour as a pivotal factor influencing subsequent professional phases underscored the significance of predictive models. One notable application highlighted was the forecasting of potential future violence among students, with the overarching goal of enabling organizations to provide targeted counselling interventions. Throughout the exploration, the authors meticulously evaluated various classifiers, scrutinizing student efficacy based on accuracy and performance metrics. The chosen tool for data mining endeavours was none other than the versatile RapidMiner.

Ganesan et al. [2] delved into the intricate relationship between academic performance and internet usage behaviour, employing machine learning algorithms as the investigative tool. The approach involved categorizing students based on academic performance and grades, and subsequently subjecting the data to the discerning eyes of decision trees, support vector machines and neural networks. The revelation of a substantial relationship between academic performance and internet usage behaviour hinted at potential avenues for refining student management skills.

Banu et al. [3] sought to unravel the perennial confusion faced by college students when navigating the labyrinth of elective course choices. Armed with an arsenal of various datasets, integrating data mining and natural language processing techniques, the outcome was a sophisticated collaborative recommendation system, meticulously designed to transform human-readable text into a format comprehensible to computers. The primary objective of this system was to serve as a guiding beacon, aiding students in making well-informed decisions about their elective courses.

Ashokkumar, K et al. [4] in the realm of workplace performance analysis for university students, proposed a comprehensive framework involving a psychological classification of students based on various traits such as emotional level, holistic approach, confidence, stress, and time management. The underlying belief was that enhancing these individual traits could catalyze improvements in students' workplace efficiency.

Naren et al. [5] have taken a proactive stance in predicting student performance, asserting that institutions bear the responsibility of identifying slow learners in advance and providing tailored training interventions. The research showcased a plethora of techniques, meticulously tailored to accommodate diverse learning needs and styles of individual learners.

Govindarajan et al. [6] rigorously explored the landscape of learning approaches and their applications in student behaviour analysis. The analytical study extended beyond academic performance, delving into behavioural nuances surrounding course preferences and online teaching modalities. The crux of the research lay in comparing the effectiveness of various machine learning models, shedding light on the evolving role of machine learning in deciphering student behaviour.

Srivats S.R. et al. [7] proposed a hybrid recommendation system in the expansive realm of big data, venturing into uncharted territory. The methodology developed here encompassed creating of an ontology for representing learners' knowledge, the construction of a framework involving big data and machine learning, and implementation of a module in Hadoop environment. The recommender system's output was a curated list of courses meticulously tailored to unique learning abilities of individual students.

Sivaneshwar et al. [8] proposed a hybrid recommendation system using big data and deep learning to address drawbacks in existing recommender systems. The approach includes creating ontology, constructing a framework and implementing a module in Hadoop for more accurate learning resource recommendations.

Madhumitha et al. [9] presented a comprehensive review on hybrid recommender systems, combining content-based and collaborative approaches. The study explores initial models, application of machine learning and deep learning and compares techniques for efficiency and accuracy, revealing that latter yields superior results.

Ragavendran et al. [10] in their study explored the significance of feedback in education, emphasizing technology integration, data mining and ontology-based applications for holistic student performance assessment and predictive analytics in organizational contexts.

Portugal et al. [11] have meticulously surveyed the landscape of recommender systems in the realm of machine learning, offering systematic review of literature, unravelling the performance of machine learning algorithms on various recommender systems. The insightful analysis not only identified trends in research but also pinpointed unanswered questions, guiding new researchers in navigating recommender system development.

Korayem et al. [12] have taken a novel approach by interpreting recommendation systems as a conduit for leveraging relationships between known features and content descriptions. The proposed methodology involving statistical relational learning algorithms aimed to construct a real-time, aggregated job recommender system. Content-based filtering and collaborative filtering were seamlessly integrated by this model to achieve the overarching goal of recommending jobs tailored to aspiring candidates.

Lahoud C et al. [13] scrutinized the efficacy of recommender systems in guiding university majors and career domains. A comparative analysis of different recommender systems conducted by them highlighted the effectiveness of hybrid approaches, particularly systems combining knowledge-based techniques with user collaborative filtering. The personalized nature of these recommendations, tailored to interests of high school students, garnered high satisfaction levels among students.

Mehzoudi et al. [14] provided a panoramic overview of data-driven and machine learning studies, positioning student employability as a crucial metric for educational institutions. Amidst job market transformed by globalization, automation and artificial intelligence, the research advocated for factors influencing employability. A holistic approach, it was argued, would empower students to strategically plan their careers, assist instructors in focusing on relevant skill sets and enable programme managers to fine-tune curricula to meet the evolving demands of the labour market.

Alibasic et al. [15] delved deep into the evolving job market within the oil and gas industry. The data-driven approach, employing tools such as LSI, LDA, factor analysis and NMF, shed light on the shifting dynamics of the job landscape. The study revealed a critical dissonance between skills taught by educational institutions and those demanded by job market, necessitating decision-makers to anticipate and prepare students for evolving job requirements, encompassing acquisition of new skills promoting collaboration between humans and machines.

Al-Qtaibi et al. [16] explored integrating content-based and collaborative filtering techniques in a hybrid model for job recommendation. By combining artificial immune system and collaborative filtering, the research is aimed to enhance recommendation quality, addressing design issues and presenting experimental evidence supporting its efficacy.

Liu, R. et al. [17] have observed that transitioning to a new job is challenging due to information overload. For students lacking work experience, job recommendation systems are crucial. The proposed framework utilized student profiling and past successful graduates' data, showing promising results in real recruitment data analysis.

The literature survey proves contemporary research places a premium on the deployment of machine learning techniques, ranging from data mining and classification algorithms to natural language processing, deep learning and convolutional neural networks. The multifaceted exploration delves into diverse facets of student behaviour, encompassing academic performance, course selection and career prospects. Proposed systems, ranging from hybrid recommender systems to predictive models integrating personality features, collectively support the notion of machine learning as a powerful tool for understanding and improving student behaviour and performance in university settings. Ultimately, efforts culminate to align with broader aspiration of successful job placement (Fig. 3.1).

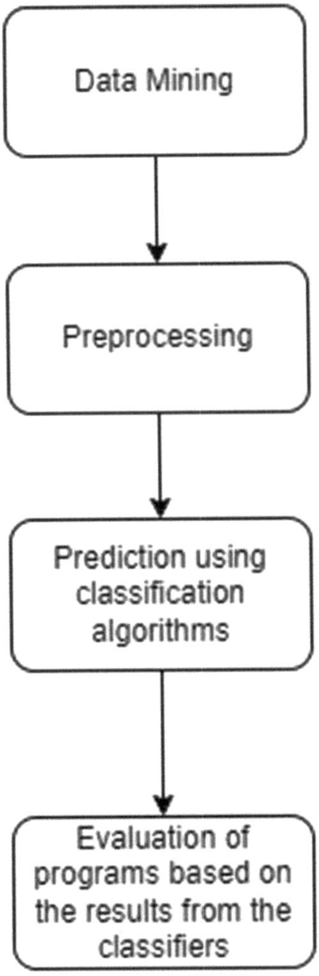

Fig. 3.1 Architectural framework of the proposed model

3 Architectural Framework

See (Fig. 3.1).

4 Description of the Framework

Here begins the description of the framework by first describing the dataset. It contains 500 rows and 41 columns. The columns are as follows: 'Academic percentage in Operating Systems', 'Percentage in Algorithms', 'Percentage in Programming Concepts', 'Percentage in Software Engineering, 'Percentage in Computer Networks', 'Percentage in Electronics Subjects', 'Percentage in Computer Architecture', 'Percentage in Mathematics', 'Percentage in Communication skills', 'Hours working per day', ' Logical quotient rating', 'hackathons', 'coding skills rating', ' public speaking points', 'can work long time before system?', 'self-learning capability?', 'Extra-courses did', 'certifications', 'workshops', 'talent tests taken?', 'Olympiads', 'reading and writing skills', 'memory capability score', 'Interested subjects', 'interested career area', 'Job/Higher Studies?', 'Type of company want to settle in?', 'Taken inputs from seniors or elders', 'interested in games', 'Interested Type of Books', 'Salary Range Expected', 'In a Relationship?', 'Gentle or Tuff behaviour?', 'Management or Technical', 'Salary/work', 'hard/smart worker', 'worked in teams ever?', 'Introvert', 'Suggested Job Role' and 'PlacedOrNot'.

The Column 'PlacedOrNot' indicates if a candidate will be hired after graduation, factoring in 40 other columns determining entry into the IT industry. Columns 'Percentage in.' represent scores of candidates in respective subjects and measure proficiency in a particular subject. The data was pre-processed thoroughly, ensuring that it was free of missing values, and outliers. Additionally, it was found that independent variables are not correlated.

A Python-based machine learning framework was developed to predict the academic performance of students. It was evaluated on a dataset of 500 students and was found to be effective in predicting the placement of a candidate in an IT firm.

5 Implemented Algorithms

The refined dataset undergoes analysis through a selection of machine learning classifier algorithms, specifically Naïve Bayes, K-nearest neighbours, decision tree, random forest and support vector machine.

5.1 Naïve Bayes

Naïve Bayes, grounded in Bayes' Theorem, functions as a classification algorithm by assuming independence among all features predicting the target. Calculating probability for each class, it selects the one with the highest probability. Applied extensively in diverse applications, Naïve Bayes exhibits notable efficacy, particularly in natural language processing (NLP) challenges.

5.2 K-Nearest Neighbours

K-nearest neighbours (KNN) stands as a straightforward and efficient algorithm proficient in addressing both classification and regression problems. In the training set, it classifies data points based on the majority class among its K-nearest neighbours, utilizing a distance metric such as Euclidean distance. As a nonparametric method, KNN refrains from making assumptions about the underlying data distribution, making it a preferred choice for computer vision and pattern recognition tasks.

5.3 Decision Tree

The decision tree, a tree-based decision-making model, partitions data into various branches according to the values of input features. Commencing with a single node, the root, it recursively divides data into subsets based on distinct conditions. Internal nodes and leaf nodes symbolize features and class labels, respectively. Final decision by considering feature values is reached at by traversing from root to leaf. Known for its simplicity and interpretability, this algorithm adeptly manages both categorical and numerical data.

5.4 Random Forest

Random forest, a powerful ensemble learning algorithm, operates by constructing a multitude of decision trees during training and outputs the mode of the classes for classification problems or the average prediction for regression tasks. This robust algorithm mitigates overfitting inherent in individual decision trees by aggregating diverse models. Each tree in the forest is trained on a random subset of data and features, enhancing its predictive accuracy and generalization. Renowned for its versatility and effectiveness, random forest finds applications across diverse domains, including finance, healthcare and image recognition, making it a cornerstone in machine learning.

5.5 Support Vector Machine

Support vector machine (SVM) stands out as a potent machine learning algorithm designed for classification and regression tasks. It constructs a hyperplane or a set of hyperplanes in a high-dimensional space to distinguish between different classes. The primary objective is to maximize the margin between the hyperplane and the nearest data points, identifying these points as support vectors. SVM's versatility extends to handling both linear and nonlinear problems through the utilization of kernel functions to transform data into higher-dimensional feature spaces positioning it as an effective tool across various applications, including text classification, image recognition, and bioinformatics.

6 Model Evaluation

The model undergoes evaluation using accuracy, a widely adopted metric for assessing performance. Accuracy is defined as the ratio of correctly predicted samples to the total number of samples.

$$\text{Accuracy} = \frac{\text{Number of Correct Predictions}}{\text{Total Number of Predictions}}$$

For instance, if a model makes 100 predictions and accurately predicts 80, its accuracy stands at 80%. While accuracy is straightforward, it may be deceptive. Models consistently predicting the most common class yield high accuracy, even if accuracy in predicting the correct class is lacking. Consequently, it is crucial to complement accuracy with other metrics like precision and recall for a comprehensive evaluation of the model's performance.

7 Conclusion

The application implements the classifier algorithms Naïve Bayes, K-nearest neighbours, decision tree, random forest, and support vector machine on the dataset and returns the accuracy of each algorithm in predicting the target variable 'PlacedOrNot'. The results are tabulated as follows (Table 1).

In accordance with the introduction, we have designed an application to predict if a recently graduated candidate with the relevant skillset is employable.

Table 1 Accuracy of each algorithm

Algorithm	Accuracy in %age
Naive Bayes	0.886
K-nearest neighbours	0.786
Decision tree	0.980
Support vector	0.840

8 Enhancing Future Projects

The current framework is based on data mining and machine learning techniques [1]. Analyzing the emotional behaviour of a student using fuzzy logic enhances the project [4]. A variety of educational settings could use the framework, if worked upon [3]. It is a valuable tool for educators and researchers who are interested in improving students' academic performance [2].

A model using semantic feedback and data mining can be developed [10] to identify students who are at risk of academic failure. It can be designed to provide personalized recommendations to help students succeed [8]. Enhancements to the framework can enable the identification of exceptional students and the provision of tailored advanced challenges [9]. By building upon the current model, a potent tool can be created to improve student success in multiple ways [12], becoming a valuable resource for educators and researchers who are exploring the applications of data mining and machine learning to improve academic outcomes [7]. Being a significant contribution to educational data mining and machine learning [1], the research can revolutionize teaching and learning, [6] thereby creating a skilled workforce, playing a vital role in the progress of the nation.

References

1. Elakia G, Aarthi N.J (2014). Application of data mining in educational databases for predicting behavioural patterns of the students. Elakia et al. (IJCSIT) International J Comput Sci Inf Technol 5(3):4649–4652
2. Naren J, Ganesan V, Gayathri P, Dhivyhalakshmi K.K, Ramalingam P (2021). A machine learning model for predicting academic performance of students through internet usage. In advances in electronics, communication and computing: Select proceedings of ETAEERE 2020 Springer Singapore, pp 315–324
3. Naren J, Banu M.Z, Lohavani S (2020) Recommendation system for students' course selection. In smart systems and IoT: Innovations in computing: proceeding of SSIC 2019 Springer Singapore, pp 825–834
4. Ashokkumar K, Naren J, Rakshanasri S.L, Vithya G (2020). Ontological framework for analysing student's emotional behaviour performance enhancement using fuzzy logic. Int J Psychosoc Rehabil. 24(2)
5. Naren J, Vithya D.G, Reddy B (2019). Predicting academic performance in students by an analytical study on big data machine learning techniques. International J Psychosoc Rehabil 23(1)

6. Ganesan V, Govindarajan V, Yellamma P (2021) Analysis, Prediction and Maintenance of Teaching learning process based on empathising Students' View of attending Online/Regular Class. EAI endorsed transactions on e-learning 7(20):e4
7. Ramanujam S.S, Naren J, Jayaseelan N, Vithya D.G (2019) Deep learning architecture based on behavioral and academic perspectives for classifying students in a university. International J Psychosoc Rehabil 23(01)
8. Ramanujam S.S, Sivaneshwar P, Naren J. Ganesan V (2019). A novel architecture for the development of an e-Learning hybrid recommender system with deep learning techniques in a big data environment. J Adv Res Dyn Control Syst. 11(Special Issue-08), 1124–1134
9. Ramanujam SS, Sivaneshwar P, Naren J, Madhumitha S, Vithya G (2019) A study on hybrid recommender systems with deep learning and deployment in big data. Test Eng Manag 81(11–12):1869–1875
10. Naren J, Ashokkumar K, Raghavendran V (2016). A semantic feedback on student's performance with data mining techniques: state of the art survey. In 2016 3rd international conference on computing for sustainable global development (INDIACom) pp 1605–1609 IEEE.
11. Portugal I, Alencar P, Cowan D (2018) The use of machine learning algorithms in recommender systems: A systematic review. Expert Syst Appl 97:205–227
12. Yang S, Korayem M, AlJadda K, Grainger T, Natarajan S (2017) Combining content-based and collaborative filtering for job recommendation system: a cost-sensitive statistical relational learning approach. Knowl-Based Syst 136:37–45
13. Lahoud C, Moussa S, Obeid C, Khoury H.E, Champin P.A (2022). A comparative analysis of different recommender systems for university major and career domain guidance. Education and information technologies, 1–27.
14. Mezhoudi N, Alghamdi R, Aljunaid R, Krichna G, Düştegör D (2023) Employability prediction: a survey of current approaches, research challenges and applications. J Ambient Intell Humaniz Comput 14(3):1489–1505
15. Alibasic A, Upadhyay H, Simsekler MCE, Kurfess T, Woon WL, Omar MA (2022) Evaluation of the trends in jobs and skill-sets using data analytics: a case study. J Big Data 9(1):32
16. Al-Otaibi S, Ykhlef M (2017) Hybrid immunizing solution for job recommender system. Front Comp Sci 11(5):511–527
17. Liu R, Rong W, Ouyang Y, Xiong Z (2017) A hierarchical similarity based job recommendation service framework for university students. Front Comp Sci 11:912–922

Mapping of City Development and Urban Flooding Using Deep Learning Model for Chennai City Study Area

Pavithraa Senthilkumar and Menaka Pushpa Arthur

Abstract People's lives and property are put in jeopardy by extreme weather occurrences such as floods, but key infrastructures, which must work even under subpar conditions, are also severely damaged. There have been a number of attempts to use deep learning algorithms to map the urban flood regions. In order to determine how vulnerable an area is to catastrophic flooding, it is critical to evaluate the flood-prone zones. Using Sentinel satellite photos, along with deep learning model, can help to map flood-prone areas and assess their vulnerability. Flood hazard mapping was evaluated using a Convolutional Neural Network (CNN) model. The response variable was a flood inventory of Chennai Corporation's flooded and non-flooded areas, while the predictor variables were flood-affecting factors. The flooded areas were then randomly divided into sections for the purpose of constructing flood models and testing them. According to the training and testing classification accuracy, the models were evaluated on their ability to predict outcomes. Flood protection and risk assessment might benefit from the use of the model. As a result, if a flood inventory is available, this method might be used to create a flood danger map for metropolitan areas.

Keywords Urban flooding · Deep learning · Flood prediction · Geographic Information System (GIS) · Computer vision · Urbanization · Distributed hydraulic modeling · Image processing · Synthetic aperture radar (SAR) images

1 Introduction

The majority of the world's population are vulnerable to numerous natural disasters. Floods are one of the most devastating natural catastrophes in the globe. The sudden, steady, and extensive precipitation causes waterlogging and flooding. In particular,

P. Senthilkumar · M. P. Arthur (✉)
School of Computer Science and Engineering, Vellore Institute of Technology, Chennai Campus, Chennai 600127, India
e-mail: menakapushpa.a@vit.ac.in

S. M. Thampi et al. (eds.), *Fifth International Conference on Computing and Network Communications*, Lecture Notes in Electrical Engineering 1221,
https://doi.org/10.1007/978-981-97-4711-5_6

in the region of south India, a strong El Nino and a depression over the southwest Bay of Bengal have caused 13 percent of Asia's population to live at a low sea level. The WHO estimates that 140 million people are displaced each year by floods, making these places particularly susceptible to severe weather events. In southern India, the city of Chennai is a major low-lying coastal metropolis [1]. Significant property, health, and infrastructure losses happened in Chennai six times between 1976 and 2010, with the most recent occurrences being in 2005. The collapse of large rivers and the lack of drainage infrastructure are two of the main causes of severe floods. Flood management and the proposal of preventive measures are vital in order to decrease the possibility of damage to agriculture, infrastructure, and other natural resources. Since an early warning system is vital for preventing future floods, a flood susceptibility assessment is critical.

Numerous studies employ the Analytical Hierarchy Process (AHP), Multi-criteria Decision Support Approach (MCDA), Weights of Evidence (WoE), adaptive neuro-fuzzy interface system, artificial neural networks (ANNs), and frequency ratio (FR) model to map flood susceptibility [1, 2]. Even though AHP is the most popular and commonly used approach, it is restricted owing to its reliance on researcher-provided information, which introduces uncertainty. Similarly, MCDA might be beneficial for flood mapping in regions with little available data, but only local planners utilize it effectively. It also tried to demonstrate the link between conditioning factors and the result using the ANN approach. This technique takes all inputs, which creates ambiguity when attempting to gather relevant data. Despite their widespread usage in mapping other natural hazards like landslides, flood susceptibility studies are the first time to applied deep learning algorithms. For the purpose of determining a flood-plain's vulnerability, both methods provide similar findings. But flood-vulnerable zones found by a deep learning model have proved useful in programs to reduce floods and associated impacts. There are several advantages to using a combination of remote sensing and a Geographic Information System (GIS) to identify potential hazard zones [3]. Chennai is considered a study area in [4] to detect flood zones using the frequency ratio model.

It is the major goal of this study to identify and map flood-prone locations inside the city of Chennai. The goal of this study is to collect as much data as possible to feed into the CNN model, which will help researchers, local governments, and planners develop flood mitigation strategies by identifying flood-vulnerable areas and mapping flood-prone zones using a deep learning model combined with Remote Sensing and Geographic Information System in Chennai Corporation. The paper is organized as follows. Section 3 explains the selected study area of our research. Section 4 discusses the methodology of the proposed system. Section 5 discusses the system implementation details and its generated results discussion. Section 6 concludes this work with highlighted future work.

2 Literature Review

It is well analyzed under different perceptions in the works [5–8] for unplanned urbanization and other flood-triggering factors along with improper urbanization. The models from numerical, analytical, hydrological [9, 10], machine learning, deep neural networks [11], and deep learning have been used for flood prediction. These literature survey works provide deep insight into the urban flood and its existing models. The study area [12] Haikou, China, has been considered for urban flood analysis with seven evaluation indices using the hydrological model. The hazard index in this work is calculated using the combination of the Analytic Hierarchy Process (AHP) method and the k-means cluster algorithm. Gangnam urban catchment located in Seoul, South Korea, is considered a study area in [9] to predict water levels and relevant inundation areas using sequential deep learning models along with 1D/2D urban hydrological models. This system is intended for urban flood prediction. In the work [13], water accumulating depth level in Tianjin Municipality, China, is rapidly predicted by the combined method of neural networks and numerical models. The urban flood happened in the two major cities of China predicted by the super-resolution mapping with the fusion of support vector machine and general regression neural network [14]. Synthetic Aperture Radar (SAR) [15] image is used to detect the urban flood using change detection (CD) and backscatter thresholding methods. The work in [16] used Bitemporal Multispectral Imagery Via Self-Supervised Learning to detect urban flooding.

3 Study Area

The Chennai Municipal Corporation is situated in the northeastern portion of Tamil Nadu on India's southeastern coast. It has an area of 426 km^2 (Fig. 1) and has a population density of 17,000 inhabitants per square kilometer. Approximately 88 micro watersheds of Adyar, Cooum, and Kosasthalaiyar rivers make up the Chennai Corporation. The average altitude of the region is 6, 7 m [1]. The average annual precipitation is around 140 cm, while the highest and lowest temperatures are approximately 35–40°C and 19–25°C, respectively. In the majority of the research region, the soil type is classified as clayey or gravelly clayey. About sixty percent of the land was occupied by newly constructed urban facilities. Lithologically, the coastline consists of marine sediments, whereas the interior region is composed of river sediments and Charnockite. Heavy rainfall connected with depressions and cyclones causes frequent flooding in the region under investigation. Figure 1 shows a map of the research region.

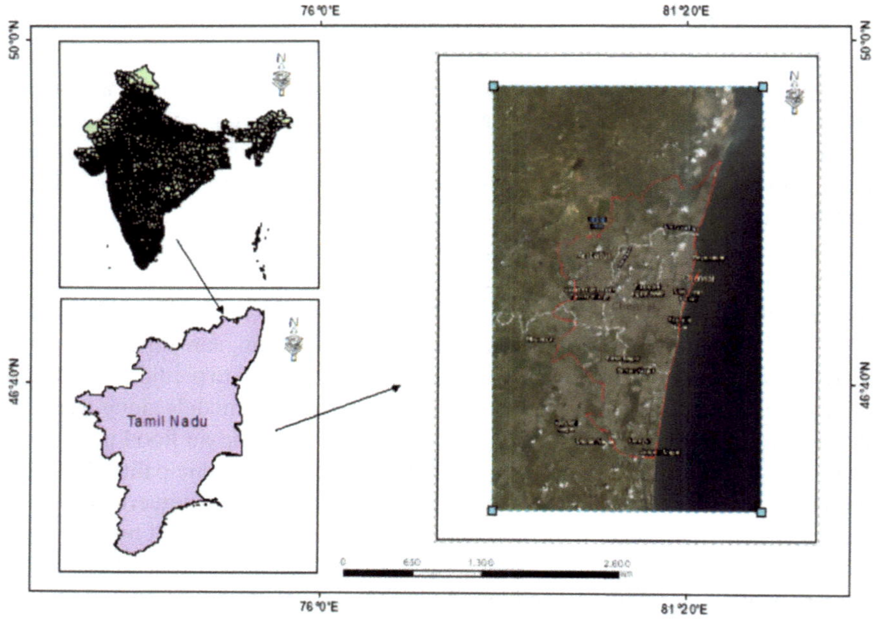

Fig. 1 Study area—Chennai metropolitan city [17, 18]

4 Methodology

The examination of flood vulnerability, using ten distinct criteria, including landforms, land use/land cover, slope, elevation, Topographic Wetness Index (TWI), Topographic Ruggedness Index (TRI), Topographic Position Index (TPI), Stream Power Index (SPI), aspect, and slope length factor, has been used in the flood susceptibility study. Data products utilized in this investigation are shown in Table 1. Arc GIS and SAGA-GIS software were used to creates the geodatabase and integrates the layers. Earth Explorer NASA MODIS-version 2 was used to create the land use/land cover (Lu/Lc) maps that were used in this interpretation. For mapping flood danger, height and slope are critical factors, because elevation variations have an impact on climatic variables. Surface runoff and vertical percolation may be affected by the slope of the terrain [2]. Elevation and slope maps were constructed using data from the Shuttle Radar Topographic Mission (SRTM). These data were made accessible to the general public on a worldwide scale [4].

Table 1 Flood factors and their software tools

Factors	Direct/indirect	Software used
Elevation [2, 11, 19, 20, 22]	Indirect	DEM
Slope [2, 4, 11, 16, 19, 20, 22]	Indirect	ArcGIS, SAGA-GIS
Stream power index [16, 20]	Indirect	SAGA-GIS
Topographic position index [20]	Indirect	SAGA-GIS
Terrain ruggedness index [20]	Indirect	SAGA-GIS
Topographic wetness index [4, 16, 19, 20, 22]	Indirect	SAGA-GIS
Land use/land cover [2, 4, 11 , 16, 19, 20, 22, 23]	Direct	ArcGIS
Aspect [4, 16]	Indirect	ArcGIS, SAGA-GIS
Slope length factor [20]	Indirect	SAGA-GIS

4.1 Urban Flood Factors

On the basis of field observations and published research, we chose ten predictor variables (also known as explanatory factors) to forecast flood likelihood in this setting. These factors were elevation, slope, slope length factor, topographic wetness index, curvature, stream power index, topographic position index, distance to the river, terrain roughness index, and land use.

- Slope—The direction that water flows is significantly affected by slope, which is a significant factor in flooding. Slope data were extracted from the DEM and analyzed using ArcGIS version 10.7 for this investigation.
- Elevation—The properties of rainfall and the course that runoff takes may be influenced by elevation. The topographical information was used to generate a digital elevation model, often known as a DEM.
- Aspect—The data were obtained from DEM by using ArcGIS 10.7. The aspect map displays the direction in which the terrain slopes, which in turn indicates the direction of the flow of floodwater.
- Land Use/Land Cover—Runoff conditions may be quite variable depending on the types of land use and land cover patterns that are present. More surface runoff is produced in regions that are built up as opposed to places that are vegetated. As a result, the way of land use significantly determines the flood risk it takes.
- Terrain Wetness Index—Estimating the hydrological responses to topographic factors or mapping the spatial patterns of soil moisture are also possible uses for the TWI. The total width index is a function that takes into account both the slope degree and the upstream contributing area. The flow of water across the ground is controlled by the Topographic Wetness Index (TWI). As a result, flood mapping benefits greatly from the TWI's work. For example, in Eq. 1, A represents the entire catchment area and B represents the slope in degrees [a = A/L, total area (A) divided by contour length (L)]. Water is gathered in catchments, such as creeks, rivers, and lakes, by the natural environment.

$$\text{TWI} = \text{Ln}\left(\frac{a}{\tan B}\right). \tag{1}$$

- Curvature—The term "curvature" refers to variations in the inclination of slopes and is used in the morphometry of local topography. Flat regions are shown by values of zero, whereas convex and concave regions are represented by positive and negative values, respectively. The water's tendency to either converge or diverge during downslope flow is affected by the slope's curvature. The danger of flooding will potentially increase when it comes to the concave curve as more water is retained. With the help of the ArcGIS program, a curvature map of the research region was created.
- Terrain Ruggedness Index—Inundation modeling during floods benefits from using TRI. A value of zero is assigned to flat locations, whereas mountainous regions with steep ridges tend to have TRI values that are significantly positive. The roughness of the surface, as measured by TPI, contributes to the hydrodynamic friction that is created by the things that are located on it (such as plants and structures). Through the use of SAGA-GIS, a TRI map was generated.

Before being included in the models, each of the input layers, also known as the predictor variables, was given a standard value.

4.2 Urban Flood Mapping

The system architecture of the urban flood mapping model proposed in this work is shown in Fig. 2. The system components are divided into data collection, model development, and flood map prediction and evaluation. In the first module, the selected region is extracted from the satellite image and preprocessed by a noise-filtering technique. Then nine urban flood factor maps are generated using GIS software tools. In the second module, the CNN architecture is trained and evaluated by the map with flooded and non-flooded zones. The susceptibility map is generated, and results are analyzed using the performance metrics of deep learning model.

The Convolutional Neural Network model has the potential for picture categorization and identification using deep learning. The core design of a Convolutional Neural Network model is composed of a feature-extraction layer and a feature-mapping layer. The pooling layer and the convolutional layer are both included in the feature-extraction layer (or fully connected layer) [19]. The convolutional layer is composed of convolution units with parameters that have been improved using a back-propagation technique. The convolution procedure computes the weights of predictive variables after taking into consideration the properties of the variables that are being predicted. Feature maps, which are essentially arrays of data, serve as both the input and output at each level of the model architecture. If the input is a two-dimensional picture, for instance, it is first decomposed into a sequential array of $x = x_1, x_2, \ldots, x_N$.

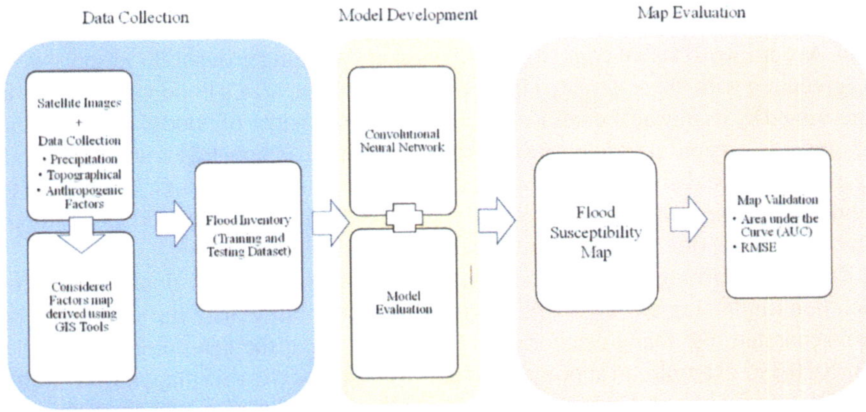

Fig. 2 System architecture diagram for urban flood mapping

The pooling layer reduces the output size of the convolutional layer's feature map by downsampling [20]. In the absence of changes to the image's structure, this procedure may finally solve the model's overfitting issue while dramatically decreasing its computational cost. The pooling layer maintains the most important information and minimizes the number of parameters when the quantity of input data is quite vast. There are a variety of pooling layer types (e.g., maximum, average, sum, etc.) [21]. By determining the maximum value for each region of the feature map, the frequent subsampling layer known as max pooling is able to reduce the amount of data that is output from the convolutional layer, which ultimately leads to an increase in the accuracy with which significant features can be identified. In point of fact, the most vital information from the layer that came before the max pooling layer is included in the output of that layer. The last layer in the architecture of a Convolutional Neural Network is referred to as the fully connected layer. This layer is composed of neurons that mix the characteristics that were acquired by the convolutional layer from a variety of feature maps. The output vector is converted to probabilities between 0 and 1 using an activation function.

5 Implementation Results Analysis

5.1 Urban Flood Susceptibility Mapping

The purpose of this study is to use deep learning methods to map the flood-prone areas in Chennai. Factor layers of nine flood-related factors were applied to 70% of the flooded areas in the training dataset that is elevation, slope, SLF, SPI, TPI, TRI, TWI, LS-factor, and land use. Locations were classified into five groups, respectively, on their flood potential on a flood risk map such as very high to very low classification.

Nearly 40% of the research region seems to be at risk of flooding, according to the map. According to visual comparisons, the test and training groups' flood spots were in agreement with the anticipated flood maps. A research area's flood vulnerability is determined by its physical qualities and a documented history of floods [1]. Assigning a degree of susceptibility to each site, from low to high, is how this is accomplished. As a result of the distribution of inputs, or "flood conditioning factors," in relation to previously documented flood episodes, the susceptibility is determined. Nine flood-affecting factors oriented maps are generated and shown in Fig. 3.

The deep learning model then calculates a score from 0 (not flooded) to 1 for each point in the region (flooded). To create a susceptibility map, the scores will be then separated into many categories in the way of using the natural (Jenks) breaks approach. For example, an approach that uses a GIS model for 465 mapping and trains its models to anticipate discharge levels. Regardless of the location of the documented flood occurrences, both scenarios show that the model works effectively when the predicted high-risk locations are hit [19]. All sorts of floods that can be well-measured may benefit from deep learning-related applications. Because susceptibility mapping provides a qualitative assessment of which sites are vulnerable to floods, this is

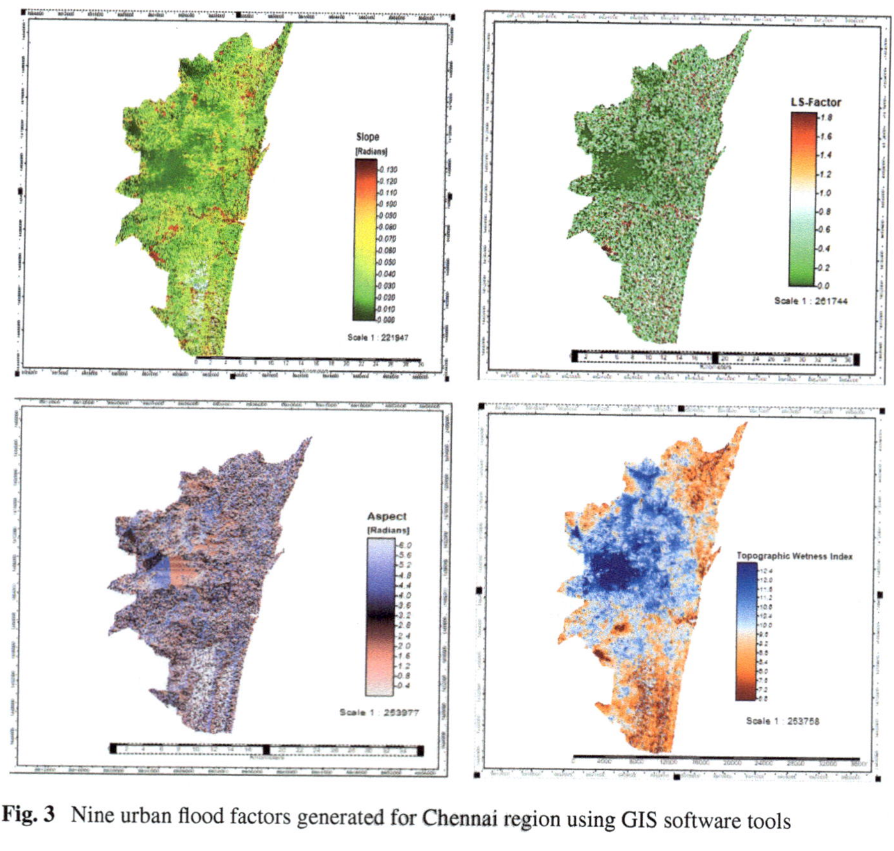

Fig. 3 Nine urban flood factors generated for Chennai region using GIS software tools

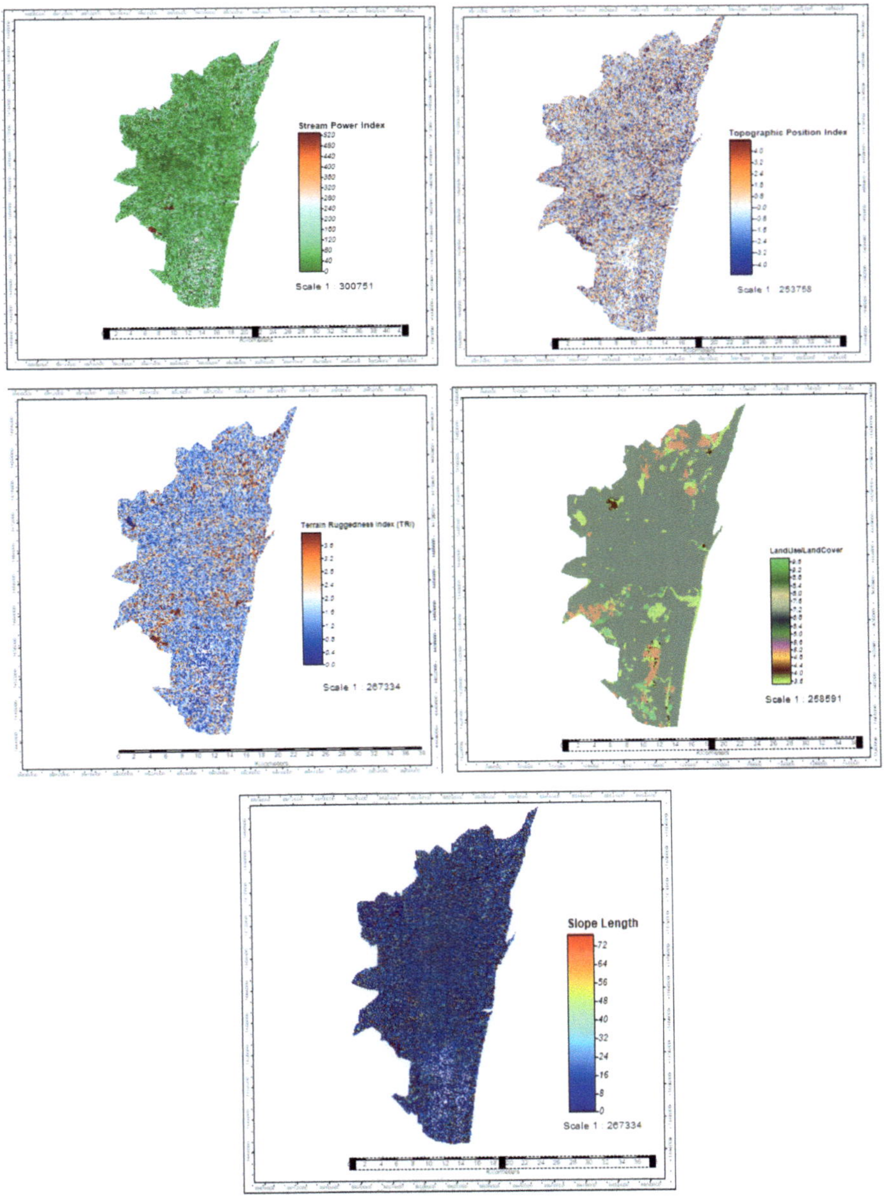

Fig. 3 (continued)

predicted to be the case. Thus, small-scale operations have limitations in both data availability and preventative strategy application. For a small region, the amount of data needed to provide an accurate assessment might be prohibitive [20].

5.2 Map Validation

It has become more important to pick an appropriate model because of the ever-increasing variety of algorithms with variable success rates. In the context of binary classification models, the ROC curve is a popular tool for analyzing the model's accuracy [22, 24]. Confusion matrix with four categories: True Negatives (TN), True Positives (TP, data properly viewed as positive), and False Negatives (FN, data wrongly seen as negative). An understanding of model performance indicators like as the true positive rate (TPR) and false positive rate (FPR) may be gained by looking at the confusion matrix. TPR and FPR represent the fraction of positively classified examples that were really mislabeled as negative. FPR is zero and TPR is one in the top left-hand corner of the ROC curve, making this point the best one on the ROC curve [20, 21]. The area under the receiver operating characteristic (AUC) curve is defined by the following equations:

$$\mathrm{TPR} = \frac{\mathrm{TP}}{\mathrm{TP} + \mathrm{FN}}, \tag{2}$$

$$\mathrm{FPR} = \frac{\mathrm{FP}}{\mathrm{FP} + \mathrm{TN}}. \tag{3}$$

5.3 Results Discussion

Flooding kills a large number of people and damages many properties across the world. Perhaps, it is the most widespread and destructive natural calamity in human history. Because so much rain ends up as runoff in cities and have heightened flood danger. Urban expansion has also repeatedly encroached on areas where there are rivers to divert surplus water for other purpose [25]. When it comes to living in metropolitan settings, there is a direct correlation between access to essential services as well as access to clean drinking water. By disrupting these services, the flood has a substantial secondary effect on urban people. Because floods cannot be prevented, effective flood control methods must be implemented [26]. Flood-prone regions may be found by analyzing previously flooded areas, and damages can be mitigated by taking mitigation measures. Using flood susceptibility mapping is a common method for preparing for future floods, and it is frequently employed. Maps of urban flood risk may also be used to design urban land uses. Engineers may also use these maps to locate places in need of drainage infrastructure. Extreme rainfall events have caused flooding in Chennai for most of the city's history. The political, economic, and other urban activities are concentrated in this megacity. Future floods are likely to cause significant financial damage because of the city's dense population, elaborate construction, and extensive subterranean and above-ground infrastructure. Some of the research regions have a high flood risk, as shown by the derived susceptibility maps. Scientists and government officials need to devote more time and resources to protecting the lives of people, property, and vast infrastructure against flooding. In the absence of this, even little floods might trigger enormous catastrophes [19].

The results obtained from the CNN-based flood susceptibility map model are given in Fig. 4a. The right side of the system-generated map is projected with Chennai city map [19] in Fig. 4b. From these two maps, we can analyze the flood areas identified by our systems such as Pallikaranai, Medavakkam, and Velacherry in south Chennai areas and areas nearer to Pozhal Lake, Kotivakkam, and Tiruvottiyur in north Chennai. These areas are usually submerged areas for every flood in Chennai city. The CNN model performance is highlighted in Table 2. The training accuracy for flooded and non-flooded areas is 96% and 98%, respectively. The classification accuracy is 82 and 89% for flooded and non-flooded regions. The system performs well as it is capable of well detecting both regions with less error.

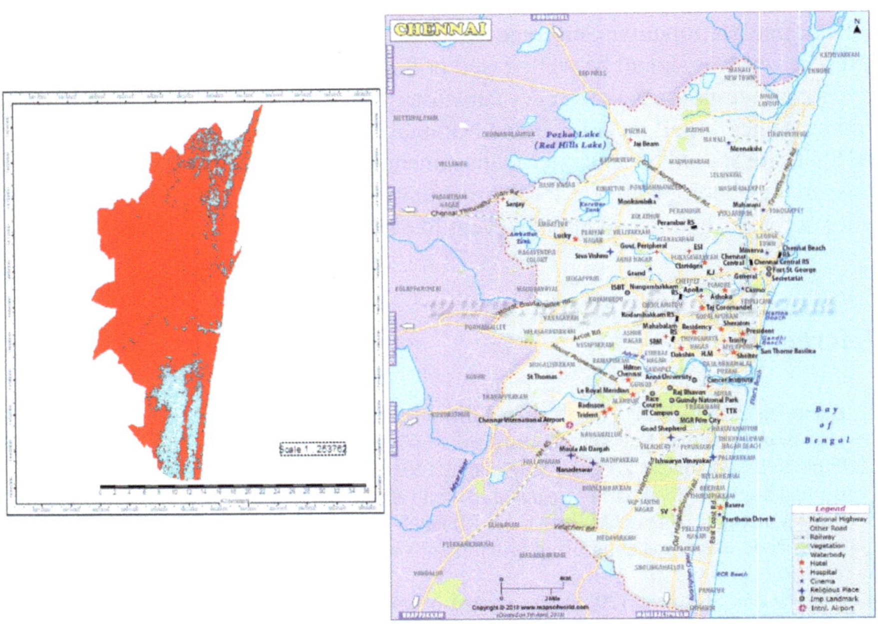

Fig. 4 Area identification of flood susceptibility map using Chennai city map. **a** CNN-generated flood susceptibility map, **b** Chennai city map [17, 18]

Table 2 CNN-based urban flood identification system

Model	Training samples	Validation samples	Training accuracy %	Prediction accuracy %
Flood areas	50	20	96	82
Non-flood areas	50	20	98	89

6 Conclusion

When it comes to preventing or lessening the potential dangers like floods, mitigation techniques play a critical role. Both structural and non-structural approaches might be used to accomplish this. The first comprises periodic de-silting of the three rivers that flow through the research area: Adyar, Cooum, and Kosasthalaiyar. Lakes like Chembarapakkam, Poondi, and Puzhal should have their water-holding capacity and the health of the concretionary structure to be periodically evaluated [1]. Buildings with raised platforms are allowed to be constructed in the medium and low flood plains, but no infrastructure should be authorized in the high flood plain. Non-structural methods, on the other hand, call for residents to move away from the high flood plain and into a region that is less at risk of flooding. At times of flooding, people should be able to use community infrastructures quickly and easily. People in moderate, high, and flood-prone locations should be made aware of the steps they should take before a flood occurs. Water Drainage Department-SWDD, Corporation of Chennai (Zonal Office), Chennai Metropolitan Development Authority CMDA, and Chennai Metropolitan Sewerage and Water Supply Board (CMWSSB) involved in rescue services should be well versed in flood map reading and prepared well in advance as a response to emergency situations. Immediate, medium-term, and short-term strategies are all included in the National Disaster Management Plan (NDMP, 2016). There should be a disaster management strategy that incorporates the use of geospatial technology to help local communities and government authorities make better decisions before and after an event like a flood [4].

References

1. Natarajan L et al. (2021) Flood susceptibility analysis in Chennai corporation using frequency ratio model. J Indian Soc Remote Sens. vol 49 p 1533–1543
2. Gigović L.D, Pamučar Z, Bajić S, Drobnjak (2017) Application of GIS-interval rough AHP methodology for flood hazard mapping in urban areas. Water vol 9
3. Peng B, Meng Z, Huang Q, Wang C (2019) Patch similarity convolutional neural network for urban flood extent mapping using bi-temporal satellite multispectral imagery. Remote Sens 11:2492
4. Pourghasemi H.R et al (2021) Assessment of urban infrastructures exposed to flood using susceptibility map and google earth engine. IEEE J Sel Top Appl Earth Obs Remote Sens. vol 14, pp 1923–1937

5. Mignot E, Li X, Dewals B (2019) Experimental modelling of urban flooding: a review. J Hydrol 568:334–342
6. Senthilkumar P, Arthur MP (2023) A review of intelligent models for mapping city development and urban flooding. Land Degrad Dev 34(13):3793–3809
7. Chan FKS, Yang LE, Scheffra J, Mitchell G, Adekol O, Griffiths J, Chen Y, Li G, Lu X, Qi Y, Li L, Zheng H, McDonald A (2021) Urban flood risks and emerging challenges in a Chinese delta: The case of the Pearl River Delta. Environ Sci Policy 122:101–115
8. Miller JD, Hutchins M (2017) The impacts of urbanisation and climate change on urban flooding and urban water quality: A review of the evidence concerning the United Kingdom. J Hydrol: Regional studies 12:345–362
9. Duc Hai N, Bae D H (2020) Correcting mean areal precipitation forecasts to improve urban flooding predictions by using long short-term memory network. J Hydrol vol 584 p 124710
10. Xing Y, Liang Q, Wang G et al. (2019) City-scale hydrodynamic modelling of urban flash floods: the issues of scale and resolution. Nat Hazards vol 96, pp 473–496
11. Lo SW, Wu JH, Chang J-Y, Tseng C-H, Lin M-W, Lin F-P (2021) Deep sensing of urban waterlogging. IEEE Access 9:127185–127203
12. Hongshi Xu, Chao Ma, Jijian Lian, Kui Xu and E. Chaima (2018) Urban flooding risk assessment based on an integrated k-means cluster algorithm and improved entropy weight method in the region of Haikou, Chin. J Hydrol vol 563, pp 975–986
13. Yan X, Xu K, Feng W et al (2021) A rapid prediction model of urban flood inundation in a high-risk area coupling machine learning and numerical simulation approaches. Int J Disaster Risk Sci. vol 12, pp 903–918
14. Li L. et al (2019) Enhanced super-resolution mapping of urban floods based on the fusion of support vector machine and general regression neural network. IEEE Geoscience and Remote Sensing Letters, vol 16, pp 1269–1273
15. Giustarini L, Hostache R, Matgen P, Schumann G.J, Bates P.D, Mason D.C (2013) A change detection approach to flood mapping in urban areas using TerraSAR-X. IEEE Transactions on Geoscience and Remote Sensing vol 51 pp 2417–2430
16. Peng B, et al (2021) Urban flood mapping with bitemporal multispectral imagery via a self-supervised learning framework. IEEE J Selected Top Appl Earth Obs Remote Sens. vol 14, pp 2001–2016
17. Flood susceptibility analysis in Chennai corporation using frequency ratio model
18. https://www.mapsofworld.com/india/tamilnadu/chennai-map.html
19. Zhao G, Pang B, Xu Z, Peng D, Zuo D (2020) Urban flood susceptibility assessment based on convolutional neural networks. J Hydrol 590:125235
20. Lei X, chen W, Panahi M, Falah F, Rahmati O, Uuemaa E, Kalantari Z, Ferreira C.S.S, Rezaie F, Tiefenbacher J.P, Lee S, Bian H (2021) Urban flood modeling using deep-learning approaches in Seoul, South Korea. J Hydrol vol 601, p 126684
21. Kalinicheva E, Sublime J, Trocan M (2019) Object-based change detection in satellite images combined with neural network autoencoder feature extraction 2019 Ninth international conference on image processing theory, tools and applications (IPTA) pp 1–6
22. Nsangou D, Kpoumié A, Mfonka Z, Ngouh A.N, Jourdan C, Foss D.H, Mbele H.Z, Mouncherou O.F, Vandervaere J.P, Ngoupayou J.R.N (2022) Urban flood susceptibility modelling using AHP and GIS approach: case of the Mfoundi watershed at Yaoundé in the South-Cameroon plateau. B.V. on behalf of African institute of mathematical sciences, Vol 15, p e01043
23. Nguyen H, Fox D, Dang D, Pham L, Viet Du Q, Nguyen T, Dang T, Tran V, Vu P, Nguyen Q.H, Nguyen T, Bui Q.T, Petrisor A.I (2021) Predicting future urban flood risk using land change and hydraulic modeling in a river watershed in the central province of Vietnam. Remote Sens
24. Guo Z, Leitão Joao P, Simoes Nuno E, Vahid M (2020) Data-driven flood emulation: speeding up urban flood predictions by deep convolutional neural networks. J Flood Risk Manag. vol 14 no 1
25. Kim D.j, Duc T.K. Dongkyun P (2020) Precipitation threshold for urban flood warning - an analysis using the satellite-based flooded area and radar-gauge composite rainfall data. J Hydro-environ Res vol 2020 pp 48–61

26. Dai W, Cai Z (2021) Predicting coastal urban floods using artificial neural network: the case study of Macau, China. Appl Water Sci 11:161
27. Heidari E, Mahmoudzadeh A, Mansouri Daneshvar M.R (2021) Urban flood susceptibility evaluation and prediction during 2010–2030 in the southern watersheds of Mashhad city, Iran. Environ Syst Res vol 10, p 41
28. Wang Y, Hong H, Chen W, Li S, Panahi M, Khosravi K, Shirzadi A, Shahabi H, Panahi S, Costache R (2019) Flood susceptibility mapping in Dingnan County (China) using adaptive neuro-fuzzy inference system with biogeography based optimization and imperialistic competitive algorithm. J Environ Manage 247:712–729

A Content-Based Online Recommender System Using AutoML Frameworks

Florin Stoica and **Laura Florentina Stoica**

Abstract In this paper we propose the development of an online movie recommender system, based on information extracted from user interaction with a Web application that allows creation of personal watchlists. Because we want the system to work in near-real-time mode, we will use prediction models generated and optimized by Automated machine learning (AutoML) frameworks, for which limited time budgets are set. AutoML provides methods and processes to build machine learning (ML) models with high scale, efficiency, and productivity all while sustaining model quality. The construction of ML pipelines will be made with two state-of-the-art AutoML tools: H_2O AutoML and Auto-sklearn. The comparative analysis aims at the accuracy of the prediction models but also at fitting into the allocated time budget. To evaluate and compare the results of the prediction models, we additionally used our own model, based on the XGBoost algorithm, optimized without a time limit. The obtained results confirm that each of the two AutoML frameworks can be selected as a solution for integration in the recommender system, due to the capabilities of generating optimized prediction models in a short time.

Keywords Content-based filtering · AutoML · Recommender system

1 Introduction

Recommender systems have changed the entire scenario of interacting with websites, helping the user to identify their favorite products more easily, movies, by recommending personalized content to that user. There are a multitude of platforms where

F. Stoica (✉) · L. F. Stoica
"Lucian Blaga" University, 550024 Sibiu, Romania
e-mail: florin.stoica@ulbsibiu.ro

L. F. Stoica
e-mail: laura.cacovean@ulbsibiu.ro

S. M. Thampi et al. (eds.), *Fifth International Conference on Computing and Network Communications*, Lecture Notes in Electrical Engineering 1221,
https://doi.org/10.1007/978-981-97-4711-5_7

these systems are currently in use and are constantly looking to improve the performance and accuracy of their recommender system to provide the user with the most enjoyable and tailored/personalized browsing experience.

Recommender systems play a very important role in many companies and applications worldwide. Companies such as Netflix, Amazon, Google, use these recommendation systems to help their customers to identify the movies or products that best fulfill their desires and preferences. These systems are essential for the business model followed by many companies to be a very successful one. Recommender systems deal with a very large volume of information. They filter the most meaningful information based on data taken from the user and other factors that consider the user's preferences and interests. The widespread implementation and utilization of these systems have brought benefits to both users and the services provided by companies.

Content-based filtering stands as one of the most widely employed approaches in recommender systems. This method relies on the premise that a user is likely to prefer items with features similar to their past preferences, thereby tailoring recommendations accordingly. Content-based methods treat all products and items as regular units and seek to construct a model based on user features and information about the items.

At the heart of content-based filtering is the idea of suggesting new items by assessing the similarity between the features of these items and a user's historical preferences. This is in contrast to collaborative filtering, which recommends new items based on the similarities between users who share similar tastes [1].

In this paper, we present the architecture of a content-based recommender system through which we aim to extend the functionality of a movies watchlist progressive web application (PWA). As a result, the users can include into their watchlists movie recommendations delivered to them via email or received through push notifications directly within the Web application. Also, users can request ad hoc recommendations, which they can view on a dedicated page.

The initial phase in developing a recommender system involves gathering and organizing data on users and items. After preparing the structured information, the next step is to construct a computational model capable of predicting user preferences for new items and generating recommendations [2].

The architectural framework of a content-based recommender system comprises three key components: (1) content analyzer, (2) profile learner, and (3) filtering component [3].

In [1, 4], the Rocchio algorithm, which was developed using the vector space model, was used to refine user profiles dynamically and efficiently. In our approach, profile learner collects user feedback and generalizes this data via machine learning techniques, in order to update the user profile for the most accurate recommendations. To streamline the construction of prediction models for efficiency and automation, in this paper we explore the feasibility of incorporating AutoML frameworks in our movie recommendation system.

One approach to handling recommendations is to utilize machine learning algorithms, wherein item descriptions serve as the data, and a user's ratings are treated as values for predictor variable. We followed the same approach as in [5–9] and propose a recommender engine that predicts top ratings for movies. To generate recommendations, we arrange the items in descending order based on the scores they have received.

The remainder of this paper is organized as follows. Section 2 presents related work on movie recommender systems or systems that have a content-based filtering component. Section 3 describes the architecture of the proposed recommendation system, with details for each of its modules. Section 4 describes the dataset used in the tests and the evaluation methods of the machine learning models provided by the AutoML tools or through own optimization. An analysis of the obtained results is also presented. Finally, Sect. 5 provides the conclusions and directions for future research.

2 Related Work

In [5], the rating of movie *m* is calculated as the weighted average of the similarity prediction scores between the current movie *m* and the two movies most similar to *m*, with the weights being the actual ratings of these similar movies. The similarity prediction score between two movies is calculated as weighted values of the feature vectors corresponding to the seven features considered: year, rating, genre, language, country, writer, and cast. The proposed mathematical model falls within the category of a linear programming model. Its objective is to determine optimal feature weights that minimize both positive and negative deviations from similarity values.

The paper [6] proposes a hybrid e-shopping recommender system which takes into account diverse user preferences and their interactions on the online store. The content-based filtering subsystem is based on the Term frequency-inverse document frequency (TF-IDF) algorithm. This algorithm offers the benefits of straightforward implementation and lower system resource requirements. However, a notable drawback is that the entire training process needs to be repeated when adding a new product, a challenge shared with the approach outlined in [5]. As in our approach, the system records whether the user has already rated the purchased products or not. Each purchased product is assigned a coefficient (ranging from 0.2 to 1) based on its rating, and this coefficient influences the TF-IDF algorithm calculations.

An adaptation of [6] for a movie recommendation system is presented in the work [7]. A collaborative filtering system is used to provide the list of 25 movies having the highest predicted evaluation. Movies that are among the users' unpopular genres are excluded from recommendations. Also, the system increases the predicted rating for movies with the user's preferred genres. In our methodology, we opted to decrease the rating of movies that do not align with the user's overall preferences before training the prediction model, and to exclude these movies as recommendation options. In [7], a content-based filtering system based on the TF-IDF algorithm is used to calculate for each movie in the list of 25, the maximum cosine similarity to the rated movies by the user, considering as inputs the descriptions of the movies. This value is used to rank the recommendations, but it is worth noting that it does not take into account the ratings given by the user to the evaluated films.

A big data approach for predicting movie ratings is presented in [8]. A recommendation engine that addresses the challenges of managing large volumes of data and scalability was built using Spark's machine learning (ML) libraries. The scalability of the system proposed by us can be ensured by the H_2O framework. H_2O can efficiently process and analyze large datasets distributed across multiple nodes in a cluster. Its scalability is achieved through a distributed computing framework, allowing it to handle big data tasks by leveraging the computing power of multiple machines in a cluster. This makes H_2O suitable for applications where the size of the dataset exceeds the capacity of a single machine.

The paper [9] proposes a movie recommendation system which examines preferences from a bias perspective, operating under the assumption that bias and preferences are intricately interconnected. The essence of a bias-based predictor (BBP) lies in identifying a pinpoint score s that serves as a fair moderator. It involves analyzing user and item biases with respect to this pinpoint score. The predicted score $P_{wt}(s, u, i)$ of item i of user u computed through s and a weight type wt is equal to the sum between s, $b_{wt}(s, u)$ (the bias score of user u computed through s and wt) and $b_{wt}(s, i)$ (the bias score of item i computed through s and wt). Three types of weights were employed: a weight based on rating frequency (RF), another on amplified RF (ARF), and a third on logarithmic RF (LRF). In order to enhance BBP, multiple bias analysis (MBA) was proposed, which includes two more terms in the formula for calculating the rating prediction: the generalized feature (GF) and the personalized feature (PF). GF means a global bias, computed from the ratings of other users, and PF means a local bias, calculated using only the user's own ratings. Since all predictions are based on the pinpoint score, its correct value is critical for the entire system. It is not specified how often the system must be retrained to maintain the accuracy of the predictions. The optimal value of the pinpoint score is determined by recursive search. A theoretical analysis of the complexity of the recursive learning algorithm is performed, but no experimental results of its efficiency are presented. The authors state that the feature weights used in GF and PF can be used to develop a recommendation system component that describes the reasons for the recommendations. In our approach, to ensure the interpretability of the generated predictions, can be employed the Shapley Additive explanations (SHAP) framework. This framework, introduced by Lundberg and Lee [10], serves as a unified approach designed to offer a comprehensive explanation of the outcomes produced by any machine learning model. Regarding the efficiency of the system, we propose the rapid construction of prediction models, within a limited time budget. For a dataset of approximately 15,000 movies, the H_2O AutoML framework provides five optimized models in less than 1 min.

3 Recommender System

The recommender system is integrated into a Web application, which enables users to search for movies and create personal watchlists. We use The Movie Database (TMDB) [11] as our data source, harnessing its extensive repository of information encompassing a wide array of movies and television shows. TMDB serves as a comprehensive source for movie and TV-related data, including details about films, TV series, actors, directors, producers, release dates, ratings, reviews, posters, and more. We use TMDB REST API to collect movie metadata and provide the list of most popular movies.

The creation of a user account is conditioned by the acceptance of the Terms of Use. In the Privacy Statement, the user is informed that his personal data is stored and processed according to the European General Data Protection Regulation (GDPR). Upon registration in the Web application, the user is presented with a statement of the recommender system's commitment to the Intermediary Guidelines and the Digital Media Code of Ethics [12]. Data collected from users and their feedback is used solely for the purpose of improving the predictions of the recommendation system to provide a personalized experience. The recommendation system does not incorporate demographic information, such as age or gender, in its decision-making process.

The architecture of the recommender system is shown in Fig. 1.

3.1 *User Profile Update*

Using the TMDB REST API, we automatically collect metadata about new movies (genre, country of production, movie release date, popularity, duration, number of votes, TMDB rating). The update of the user's profile is done based on the feedback obtained from him as a result of rating the movies viewed from his watchlist, but also based on the general preferences configured by the user in the recommendation system. In this way the recommender can handle the cold start problem, which

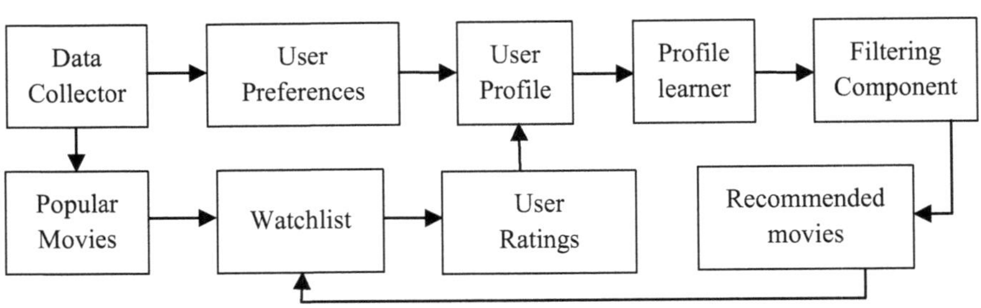

Fig. 1 Architecture of the proposed movie recommender system

denotes a state where the system cannot recommend any new movie to a new user due to lack of information.

The rating $R(i)$ assigned to a movie i by the recommendation system for training prediction models is done according to the formula:

$$R(i) = \begin{cases} R_{\mathrm{USER}}(i), & \text{if } w(i) = \text{true} \\ R_{\mathrm{TMDB}}(i), & \text{if } w(i) = \text{false and } p(i) = \text{true} \\ \alpha \cdot R_{\mathrm{TMDB}}(i), & \text{if } w(i) = \text{false and } p(i) = \text{false} \end{cases} \quad (1)$$

where $R_{\mathrm{USER}}(\mathrm{i})$ represents the rating given by the user to the movie i, $R_{\mathrm{TMDB}}(\mathrm{i})$ is the TMDB rating, $w(i)$ is *true* if the user watched the movie, and $p(i)$ is *true* if the movie fits the user's general preferences. In the current implementation, $\alpha = 0.7$, but the parameter can be adjusted in the future, after analyzing the results of several experiments. A movie that does not fit the user's preferences but has not been marked as undesirable either ($p(i)$ is *null*), will not be considered for the training of the ML model, but remains a possible candidate for recommendation.

3.2 Machine Leaning Models

To obtain robust and flexible predictive models, an approach based on machine learning algorithms and model ensembles is currently used. In the machine learning approach, the effort is shifted from a deep understanding of the application domain to the construction, optimization, and validation of models. Techniques like bootstrap aggregating, boosting, averaging, model stacking, or cascading combine many simpler models into one complex model to obtain better predictive performance.

Traditional machine learning model development is resource-intensive, requiring significant domain knowledge and time to produce and compare dozens of models. Automated machine learning, also referred to as automated ML or AutoML, is the process of automating the time-consuming, iterative tasks of machine learning model development. AutoML provides methods and processes to build ML models with high scale, efficiency, and productivity all while sustaining model quality.

With AutoML, the time needed to prepare ML models for production can be accelerated with great ease and efficiency. Popularized by Google in 2018 as a means of accelerating the task of determining the optimal neural network architecture to complete a given deep learning (DL) task, AutoML grown into a more widely applicable means of automating a wide range of ML tasks, including: data preparation, model selection, feature selection, and engineering, as well as hyperparameter tuning.

A generic AutoML pipeline is shown in Fig. 2.

Within the paper we describe and use two state-of-the-art AutoML tools: H_2O AutoML and Auto-sklearn. Since we want the recommender system to provide predictions in the shortest possible time, a time budget of 5 min was set for each AutoML framework. One may thus ask whether tuning the hyperparameters of a

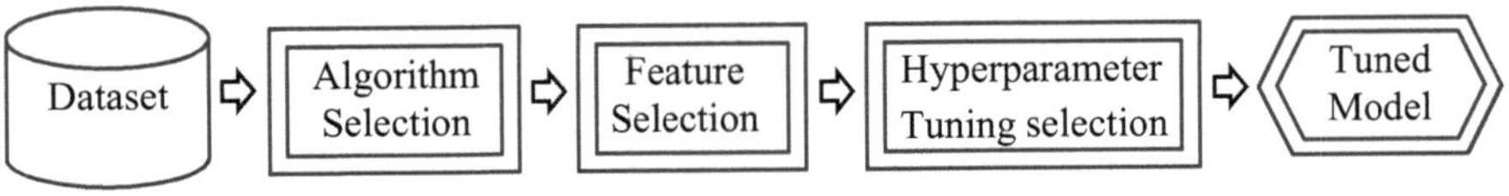

Fig. 2 Architecture of a generic AutoML pipeline

custom ML model without a strict time limit, a better prediction can be obtained. For this purpose, we built an XGBoost model [13] and used Optuna to optimize its hyperparameters.

Auto-Sklearn. The two major issues in AutoML are: the ranking of algorithms (considering that no machine learning method works best on all datasets) and hyperparameter optimization, a very important step for tuning some machine learning algorithms (e.g., K-Nearest Neighbors, Random Forest, Non-Linear Support Vector Machine). The two problems can be combined into one optimization problem. The formalization of the Combined Algorithm Selection and Hyperparameter optimization (CASH) problem is presented below.

Let $\mathcal{A} = \{A^{(1)}, \ldots, A^{(R)}\}$ be a set of algorithms and let the hyperparameters of each algorithm $A^{(j)}$ have domain $\Lambda^{(j)}$. Given a dataset D_{train}, it is split into K cross-validation folds $\{D_{\text{valid}}^{(1)}, \ldots, D_{\text{valid}}^{(K)}\}$ and $\{D_{\text{train}}^{(1)}, \ldots, D_{\text{train}}^{(K)}\}$ such that $D_{\text{train}}^{(i)} = D_{\text{train}} \backslash D_{\text{valid}}^{(i)}$ for $i = 1, \ldots, K$. Finally, let $\mathcal{L}(A_{\lambda}^{(j)}, D_{\text{train}}^{(i)}, D_{\text{valid}}^{(i)})$ denote the loss that algorithm $A^{(j)}$ achieves on $D_{\text{valid}}^{(i)}$ when trained on $D_{\text{train}}^{(i)}$ with hyperparameters λ. The CASH problem is to find the joint of the best algorithm A^* and hyperparameter setting λ^* that minimizes the average loss function [14]:

$$A^*, \lambda^* \in \operatorname*{arg\,min}_{\substack{A^{(j)} \in \mathcal{A} \\ \lambda \in \Lambda^{(j)}}} \frac{1}{K} \sum_{i=1}^{K} \mathcal{L}(A_{\lambda}^{(j)}, D_{\text{train}}^{(i)}, D_{\text{valid}}^{(i)}) \tag{2}$$

Auto-sklearn uses the random-forest-based Bayesian optimization method sequential model-based algorithm configuration (SMAC) [15] to solve the CASH problem.

The Auto-sklearn pipeline begins with a meta-learning step to warmstart the Bayesian optimization procedure [14]. In the meta-learning stage, for a large number of datasets, performance data and a set of meta-features are collected to allow Auto-sklearn to determine the appropriate algorithm for a new dataset. Meta-learning is performed offline, and for each ML dataset in the dataset repository (140 datasets from the OpenML [16] repository), an instance of the ML framework with the best results on that dataset is determined and stored. Meta-learning is used to warmstart Bayesian optimization of hyperparameters, by selecting k stored configurations and use their result to seed Bayesian optimization.

Auto-sklearn approach is to store all models trained during Bayesian hyperparameter optimization and to use an efficient post-processing method to build a weighted ensemble out of them. The adjustment of the respective weights is done using the predictions of all the individual models on a hold-out set. Following the experiments performed, it was chosen as the method of weight optimization *ensemble selection* [17].

H_2O AutoML. H_2O AutoML interface automates the machine learning workflow and includes as pipeline stages automatic training and tuning of a large selection of candidate models within a user-specified time limit. The result of the AutoML run is a "leaderboard", a ranked list of optimized models that can be stored for later use in a production environment.

A set $M_{(P)} = \{M_{(P,i)}|i = 1, .., k\}$ of pre-specified models are included to give reliable defaults for implemented algorithms $A_{(i)},\ i = 1, .., k$. For each algorithm $A_{(i)}$, the most important hyperparameters with their defined ranges are denoted by $\lambda^{\Delta}_{(P,i)}, i = 1, .., k$. Random search is utilized to generate models $M^*_{(P,i)}$ with hyperparameters $\lambda^*_{(P,i)} \in \lambda^{\Delta}_{(P,i)},\ i = 1, .., k$. The order of the algorithms, $P_i,\ i = 1, .., k$, is set to start with models that consistently provide good results (e.g., XGBoost, Generalized Linear Models (GLM) models). After this set of prescribed models is trained and added to the leaderboard, a random search and fine tuning is performed across those same algorithms. The proportion of time T_i spent on each algorithm $A_{(i)},\ i = 1, .., k$ is explicitly defined to give some algorithms (e.g., XGBoost Gradient Boosting Machines (GBM), H_2O GBM) more time than others (e.g., H_2O Deep Learning) according to the perceived or "estimated value" of each task. If a better model is found, it is added to the leaderboard.

After training the base models, the H_2O's Stacked Ensemble algorithm is used to train two Stacked Ensemble models [18]. Ensemble methods are machine learning techniques that can provide better accuracy by combining several base models instead of using a single model.

The *All Models* ensemble contains all the models, and the best of family ensemble contains the best performing model from each algorithm class/family. More precisely, that includes one H_2O GBM, XGBoost GBM, H_2O Extremely Randomized Tree Forest, H_2O Random Forest, H_2O GLM, and H_2O Deep Learning model [19].

XGBoost. eXtreme Gradient Boosting (XGBoost) is a popular and powerful machine learning algorithm used for regression and classification tasks. It belongs to the gradient boosting family of machine learning techniques. XGBoost is known for its efficiency, speed, and accuracy, and it has been widely used in various machine learning competitions and real-world applications.

XGBoost offers strong regularization techniques, such as L1 (Lasso) and L2 (Ridge), which help prevent overfitting and enhance the model's generalization performance.

Parameter tuning is a crucial step in getting the most out of machine learning algorithms like XGBoost. For this purpose, we used Optuna, an open-source Python library that can efficiently search for the best combination of hyperparameters to improve model performance [20]. Optuna uses various optimization algorithms to efficiently search the hyperparameter space and converge to the optimal set of hyperparameters based on the objective function. The optimization workflow with Optuna involves creating a study, specifying the search space for hyperparameters, defining an objective function that evaluates the model's performance with a given set of hyperparameters, and then running multiple trials to find the best hyperparameters.

Overcoming Model Drift. Machine learning models are trained using historical data, and when applied in real-world scenarios, they can become outdated and experience a decline in accuracy over time, a phenomenon which is known as drift. The concept of model drift refers to the degradation of a machine learning model's predictive performance over time due to changes in the environment, data distribution, or other external factors. To mitigate model drift, it's essential to monitor and retrain models regularly. In our approach, for retraining the ML models, a task is scheduled for execution at the designated time intervals or executed on demand.

3.3 Filtering Component

The filtering component uses the trained prediction models to recommend movies which are likely to be of interest to the users.

Figure 3 visually illustrates how the elements from a user's profile form the dataset used for training and testing the prediction model, as well as the items that are excluded from the recommendation process.

Figure 4 depicts the viewing interface for the recommended movies offered to the user. This interface allows for the application of various filters, the viewing of trailers, and the addition of approved films to the watchlist. After viewing a movie, the user can rate it on a scale from 1 to 10 and express whether he liked or disliked the movie.

The number of recommended movies can be set from the application, but to evaluate the performance of the prediction models, in Sect. 4 we considered sets of 5/10/20 recommendations.

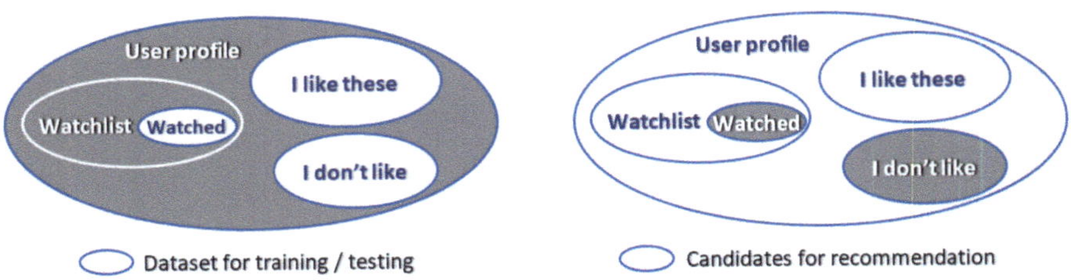

Fig. 3 Datasets used by the recommender system: dataset for training the ML model (left) and candidate items for recommendation (right)

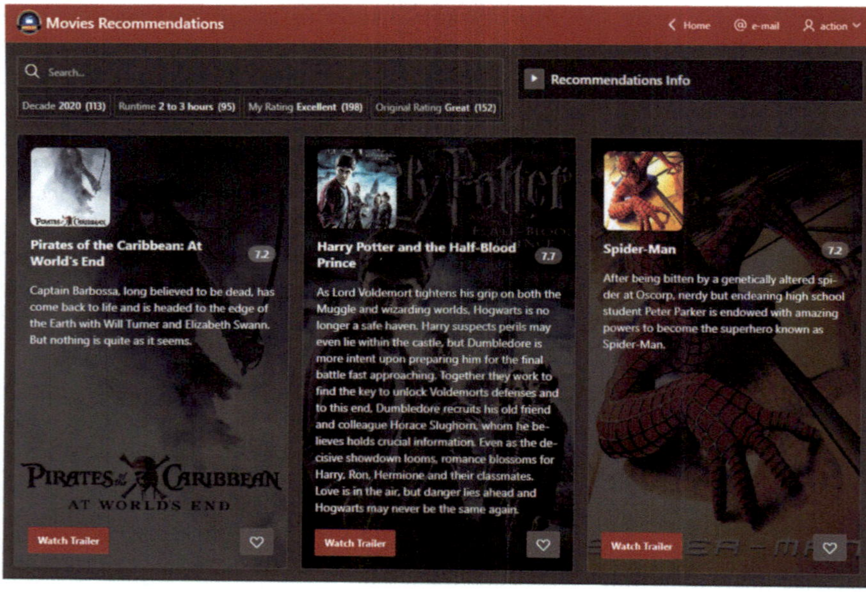

Fig. 4 The results of the recommendation process presented in the Web interface of the application

4 Experimental Results

This section presents the experimental results in building and evaluating prediction models with the tools presented in Sect. 3. For this study, all the recommender system algorithms were coded using the Python programming language and executed on a laptop computer equipped with an AMD Ryzen 7 5700U processor and 16 GB of RAM. Auto-sklearn heavily relies on the Python module "resource", which is part of Python's Unix Specific Services and is not available on a Windows machine. Consequently, running Auto-sklearn on a Windows platform is not feasible. To carry out the experiments in a unified environment, we opted for an Ubuntu distribution as operating system.

4.1 Dataset Description

The performance evaluation and analysis of the proposed prediction models was performed on a dataset taken from a user's profile, which contains 15,160 movies taken from TMDB via REST API. The dataset was split into two segments: 80% for training the machine learning models and 20% for testing them.

As feature variables we used the following metadata related to a movie: genre, country of production, movie release date, popularity, duration, number of votes, budget, the production company. The TMDB rating and user rating (if any) were

Table 1 Evaluation metrics for model validation and their formulas

Metric	Formula	Notations
RMSE	$\sqrt{\frac{1}{n}\sum_{i=1}^{n}(p_i - a_i)^2}$	• p_i—predicted data • a_i—observed (actual) data • n—the total number of observations in the dataset
MAE	$\frac{1}{n}\sum_{i=1}^{n}\lvert a_i - p_i\rvert$	
RMSLE	$\sqrt{\frac{1}{n}\sum_{i=1}^{n}(\ln(p_i + 1) - \ln(a_i + 1))^2}$	

used to calculate according to Eq. (1) the final value for movie rating, this being the target variable.

4.2 Evaluation Metrics

We used the following regression-type metrics to evaluate the performances of the prediction models: root mean squared error (RMSE), mean absolute error (MAE), and root mean squared logarithmic error (RMSLE).

RMSE quantifies the difference between predicted values and actual (observed) values in a dataset. A lower RMSE means that, on average, the model's predictions are closer to the actual values.

MAE uses the absolute value of the errors, making it less sensitive to outliers compared to RMSE. Higher MAE values indicate a larger average error between the model's predictions and the actual values.

RMSLE is a variant of the RMSE used in regression problems, particularly when the target values in the dataset cover a wide range and exhibit significant differences in scale. RMSLE penalizes underestimations and overestimations of the target values equally but in a way that scales with the magnitude of the values. This is the metric used for the ASHRAE Energy Prediction competition [21].

The formulas of the evaluation metrics are given in Table 1.

4.3 Prediction Models

To assess the prediction models, we utilized the top-performing model from H_2O AutoML, the best model provided by Auto-sklearn, and a custom XGBoost model that had been optimized with Optuna. The number of trials in the Optuna study was set to 300.

For a more in-depth analysis of the results, we also included in the comparison the H_2O XGBoost model and the H_2O Distributed Random Forest (DRF) model, taken from the H_2O AutoML leaderboard which contains the five best models. The DRF model averages multiple decision trees, each constructed using various random subsets of rows and columns. This approach enables it to capture non-linear relationships and provides an assessment of the importance of each predictor within the model. These attributes collectively render DRF one of the most resilient algorithms, especially well-suited for addressing noisy or intricate datasets.

The best models provided by each of the two frameworks are shown in Table 2.

The parameters of our XGBoost model, optimized with Optuna, have the following values: *eta*: 0.4, *learning_rate*: 0.018, *colsample_bytree*: 0.9, *colsample_bynode*: 0.7, *n_estimators*: 1342, *min_child_weight*: 9, *max_depth*: 10, *subsample*: 0.7.

Table 3 illustrates the detailed experimental results on the training dataset and the evaluation of model predictions on the testing dataset.

Table 4 shows the results obtained using Top 5, Top 10, and Top 20 recommendation tasks.

Table 5 contains information regarding the time needed to optimize the ML models, but also the way categorical data is handled by each analyzed framework. It can be noted that only Auto-sklearn requires programmatic preprocessing of the data.

It can also be noted the extraordinary efficiency of H_2O AutoML, which managed to generate and optimize five prediction models without consuming the allocated time budget (5 min).

Figures 5 and 6 show the values of the evaluation metrics for the test dataset and respectively for the first 5 /10 / 20 predictions. It can be seen that the best H_2O

Table 2 The best models built with H_2O AutoML and Auto-sklearn frameworks

AutoML framework	Best model	Combined models
H_2O AutoML	Stacked ensemble *all models*	Distributed random forest (DRF), Gradient boosting modeling (GBM), eXtreme gradient boosting (XGBoost), Generalized linear model (GLM)
Auto-sklearn	Weighted ensemble	Adaptive boosting (AdaBoost), Liblinear support vector regression (SVR), K-nearest neighbors (KNN), Random forest (RF)

Table 3 Experimental results on the training/testing datasets

Model	Training dataset			Testing dataset		
	MAE	RMSE	RMSLE	MAE	RMSE	RMSLE
H_2O-Ensemble	0.279	0.413	0.064	0.423	0.645	0.096
H_2O DRF	0.221	0.332	0.052	0.429	0.655	0.101
H_2O XGB	0.271	0.404	0.067	0.461	0.698	0.109
Custom-XGB	0.202	0.309	0.048	0.523	0.763	0.111
Auto-sklearn	0.410	0.590	0.096	0.478	0.691	0.119

Table 4 Experimental results for Top 5, Top 10 and Top 20 recommendation tasks

Model	Top 5 recommendations			Top 10 recommendations			Top 20 recommendations		
	MAE	RMSE	RMSLE	MAE	RMSE	RMSLE	MAE	RMSE	RMSLE
H_2O-Ensemble	1.061	1.336	0.142	1.458	1.731	0.197	1.440	1.848	0.233
H_2O DRF	1.153	1.434	0.151	1.529	1.824	0.206	1.650	2.010	0.245
H_2O XGB	1.070	1.282	0.123	1.347	1.597	0.165	1.396	1.736	0.199
Custom-XGB	1.256	1.415	0.142	1.127	1.294	0.131	1.309	1.648	0.210
Auto-sklearn	1.737	2.274	0.305	1.614	1.937	0.245	1.498	1.814	0.212

Table 5 The time required and the handling of categorical variables

Framework	Handling categorical data	Time used
H_2O AutoML	Automatically	< 60 s
Auto-sklearn	Requires data preprocessing for conversion into numerical values	≈5 min
Optuna (XGBoost)	Using DMatrix from XGBoost library	≈16 min

model (H_2O-Ensemble) on the training dataset is outperformed by other H_2O models when we evaluate only the most important predictions. The best H_2O model for these particular cases is the H_2O XGB (XGBoost).

Analyzing the predictions in Fig. 7 corresponding to the H_2O XGB and Custom XGBoost models respectively, one can observe a better behavior of the custom model. Thus, it de-points the films with high ratings, but they still receive the best evaluations. The H_2O model assigns high scores to movies that in reality have a rating of 7.

Comparing the two AutoML frameworks, it can be seen that H_2O provides better predictions, with one exception. The Auto-sklearn model demonstrates competitiveness only in the context of Top 20 recommendations. For this case, the higher RMSE value for the H_2O-Ensemble and H_2O DRF models may be due to some outlier predictions penalized by this metric. It should be investigated whether Auto-sklearn performance can increase by expanding the allocated time budget.

The ability to set a time limit for the generation of prediction models by AutoML frameworks is crucial, especially in the context of generating ad hoc recommendations. Nevertheless, a challenge arises in determining the minimum value of the time budget to ensure that the provided models are sufficiently optimized to generate high-quality predictions. The issue can be addressed by dynamically adjusting the time budget in correlation with the metrics used to assess the performance of prediction models, for which threshold values are set as references. These threshold values can be obtained from a pre-trained model offline, without any time constraints.

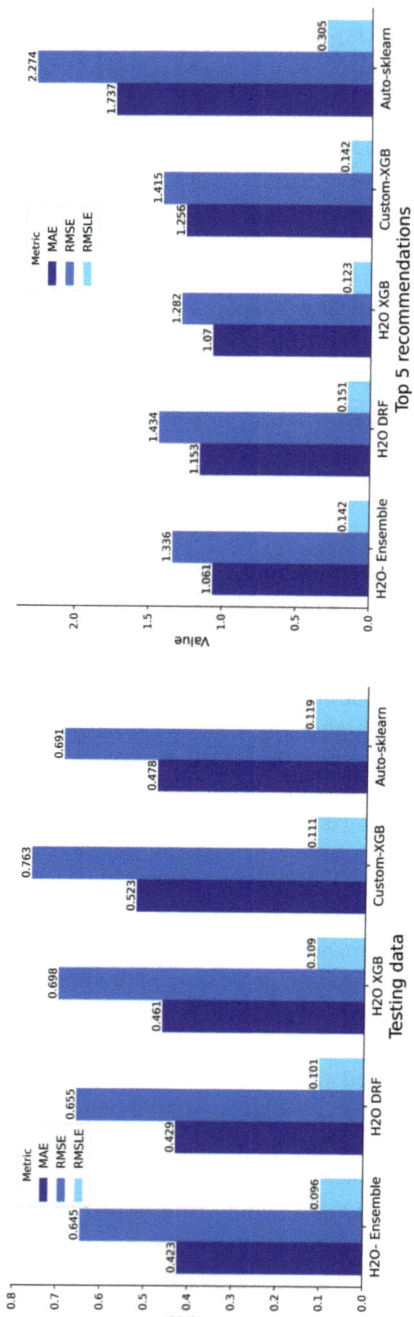

Fig. 5 The evaluation metrics plot for testing data and for Top 5 recommendations

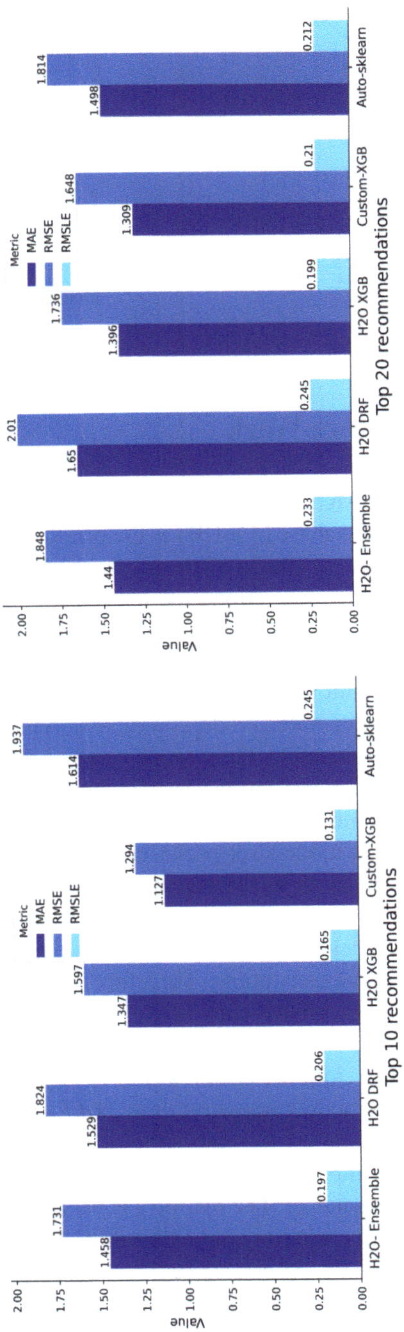

Fig. 6 The evaluation metrics plot for both Top 10 and Top 20 recommendations

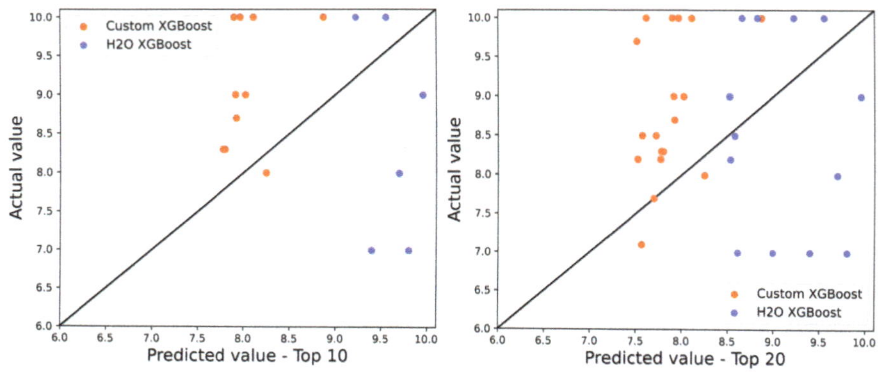

Fig. 7 Scatter plots for Top 10 and Top 20 recommendations respectively (XGBoost)

5 Conclusions and Future Work

We propose a Web-based movie recommender system designed to predict top ratings for movies, using The Movie Database (TMDB) as the primary data source. To build the prediction models, we used two state-of-the-art AutoML frameworks, namely H_2O and Auto-sklearn. For the generation and optimization of the models, the maximum execution time was limited to five minutes. An XGBoost model, optimized with Optuna in 300 trials (without imposing additional time constraints) was used for comparisons with the AutoML models. Based on the experiments conducted, it can be concluded that H_2O generates prediction models remarkably quickly, and the accuracy of predictions produced by these models makes this framework highly suitable for ad hoc recommendations that necessitate frequent model updates. Our custom XGBoost model is recommended to be used in offline updating of ML models.

Our approach, based on ML models for providing recommendations, addresses the model drift issue much better than alternative solutions [5–7, 9], discussed in Sect. 2. However, to prevent model drift, it is crucial to consistently monitor and retrain models. An AutoML pipeline contains three main tasks: algorithm selection, feature selection, and hyperparameter tuning. Therefore, in our approach, updating the prediction model does not consist only in its retraining, but the prediction algorithm can even be changed.

In this paper, we used two AutoML frameworks recognized for their performance in providing optimized models in a short time. H_2O is a distributed machine learning platform specifically crafted to scale effectively with extensive datasets. Leveraging in-memory data compression, H_2O can manage billions of rows of data in-memory, even with a modest cluster. The platform is versatile, capable of running on a local desktop with a small cluster or scaling seamlessly across multiple nodes using technologies like Hadoop, an Amazon Elastic Compute Cloud (EC2) cluster, or Spark. Auto-sklearn is a robust AutoML framework built upon the scikit-learn machine learning framework [22]. Modified versions of Auto-sklearn 1.0 secured the top positions in the first and second ChaLearn AutoML challenges. These challenges systematically evaluated AutoML systems under rigid time and memory constraints, requiring predictions to be delivered in less than 20 min. In our experiments, H_2O demonstrated superior accuracy in generating predictions within the shortest timeframe.

As future directions of development, we intend to add an online evaluation component of the predictions. With the increase in the number of users, we will start the transition of the recommendation system to a hybrid version, which also contains a collaborative component [23]. Additionally, we will explore the extraction of semantic features from plot summaries and digital images (movie posters), to enhance the quality of recommendations.

References

1. Van Meteren R, Van Someren M (2000) Using content-based filtering for recommendation. In: Proceedings of the machine learning in the new information age: MLnet/ECML2000. Workshop
2. Gupta A, Shivhare H, Sharma S (2015) Recommender system using fuzzy c-means clustering and genetic algorithm based weighted similarity measure. In: 2015 International Conference on computer, communication and control, 1–8, pp. 10–12, Indore, India
3. Lops P, Gemmis M, Semeraro G. (2011) Content-based recommender systems: State of the art and trends. In: Ricci F, Rokach L, Shapira B, Kantor PB (eds) Recommender systems Handbook, pp. 10–73. Springer
4. Kuo RJ, Cheng H.R (2022) A content-based recommender system with consideration of repeat purchase behavior. Appl Soft Comput 127:109361. ISSN 1568-4946
5. Inan E, Tekbacak F, Ozturk C (2018) Moreopt: a goal programming based movie recommender system. J Comput Sci 28: 43–50. ISSN 1877-7503
6. Walek B, Fajmon P (2023) A hybrid recommender system for an online store using a fuzzy expert system. In: expert systems with applications 212:118565. ISSN 0957-4174
7. Walek B, Fojtik V (2020) A hybrid recommender system for recommending relevant movies using an expert system. Expert systems with applications.158:113452. ISSN 0957-4174
8. Awan MJ, Khan RA, Nobanee H, Yasin A, Anwar SM, Naseem U, Singh VP (2021) A recommendation engine for predicting movie ratings using a big data approach. Electronics 10(10):1215
9. Hwang TG, Kim SK (2021) Movie recommendation through multiple bias analysis. Appl Sci 11(6):2817 https://doi.org/10.3390/app11062817
10. Lundberg S.M, Lee Su-In (2017) A unified approach to interpreting model predictions. In: Proceedings of the 31st international conference on neural information processing systems (NIPS'17), pp 4768–4777. Curran Associates Inc., Red Hook, NY, USA.
11. The Movie Database (MDB). https://www.themoviedb.org/. Last accessed 30 Oct 2023
12. Information Technology (Intermediary Guidelines and Digital Media Ethics Code) Rules, updated 2023, https://www.meity.gov.in/content/information-technology-intermediary-guidelines-and-digital-media-ethics-code-rules-2021. Last accessed 10 Nov 2023
13. Chen T, Guestrin C (2016) XGBoost: A scalable tree boosting system. In: 22nd SIGKDD conference on knowledge discovery and data mining
14. Feurer M, Klein A, Eggensperger K, Springenberg JT, Blum M, Hutter F (2015) Efficient and robust automated machine learning. In: Proceedings of the 28th international conference on neural information processing systems. NIPS, 2:2755-2763. MIT Press, Cambridge, MA, USA
15. Hutter F, Hoos HH, Leyton-Brown K. (2011) Sequential model-based optimization for general algorithm configuration. In: Proceedings of the 5th international conference on learning and intelligent optimization. pp 507–523

16. Vanschoren J, van Rijn J, Bischl B, Torgo L (2013) OpenML: networked science in machine learning. SIGKDD Explor 15(2):49–60
17. Caruana R, Niculescu-Mizil A, Crew G, Ksikes A (2004) Ensemble selection from libraries of models. In: Proceedings of the twenty-first international conference on machine learning (ICML '04). Association for computing machinery, New York, NY, USA, 18.
18. H_2O stacked ensembles. https://docs.h2o.ai/h2o/latest-stable/h2o-docs/data-science/stacked-ensembles.html. Last accessed 30 Oct 2023
19. LeDell E, Poirier S (2020) H_2O AutoML: scalable automatic machine learning. In: 7th ICML workshop on automated machine learning
20. Akiba T, Sano S, Yanase T, Ohta T, Koyama M (2019) Optuna: a next-generation hyper-parameter optimization framework. In: KDD '19: Proceedings of the 25th ACM SIGKDD international conference on knowledge discovery data mining, pp 2623–2631
21. ASHRAE Energy Prediction competition. https://www.kaggle.com/c/ashrae-energy-prediction, Last accessed 30 Oct 2023
22. Pedregosa F, Varoquaux G, Gramfort A, Michel V, Thirion B, Grisel O, Blondel M, Prettenhofer P, Weiss R, Dubourg V, Vanderplas J, Passos A, Cournapeau D, Brucher M, Perrot M, Duchesnay E (2011) Scikit-learn: Machine learning in python. J Mach Learn Res (JMLR) 12(85):2825–2830
23. Talha MM, Khan HU, Iqbal S, Alghobiri M, Iqbal T, Fayyaz M (2023) Deep learning in news recommender systems: a comprehensive survey, challenges and future trend. Neurocomputing 562:126881

Integrated AI-Based Pipeline for Law Enforcement Application: FIR Assistance

A. Amruth, R. Ramanan, Rhea Paul, R. Pranav, Deepa Gupta, Susmitha Vekkot, and Priyanka C. Nair

Abstract This paper presents an integrated AI-based pipeline for law enforcement applications, specifically targeting the creation of first information report (FIR) in the Kannada language. The pipeline comprises a speech processing block utilizing automatic speech recognition (ASR) models to convert spoken Kannada into text, and a natural language processing block employing advanced AI techniques for information extraction and generating concise FIR summaries. The research focuses on optimizing and streamlining the FIR creation process by leveraging AI technologies and overcoming data scarcity in the Kannada language through data augmentation techniques. OpenAI's Whisper is used for ASR, resulting in two-word error rate in this use case. For information extraction, a Bi-LSTM neural network is employed for named entity recognition (NER) with a performance of 94%. The proposed NER model utilizes three tags, which can be expanded to accommodate additional tags as per FIR requirements.

Keywords Speech processing · Whisper · Natural language processing · NER · Information extraction · Fine-tune · Kannada · Low-resource language · Law enforcement · FIR

1 Introduction

The adoption of AI technologies in law enforcement agencies has been constrained by factors such as resource limitations, ethical considerations, and concerns regarding bias and privacy. However, there is a growing recognition of the benefits that AI

A. Amruth · R. Ramanan · R. Paul · R. Pranav · D. Gupta · P. C. Nair
Department of Computer Science and Engineering, Amrita School of Computing, Amrita Vishwa Vidyapeetham, Bengaluru, India

S. Vekkot (✉)
Department of Electronics and Communication Engineering, Amrita School of Engineering, Amrita Vishwa Vidyapeetham, Bengaluru, India
e-mail: v_susmitha@blr.amrita.edu

S. M. Thampi et al. (eds.), *Fifth International Conference on Computing and Network Communications*, Lecture Notes in Electrical Engineering 1221,
https://doi.org/10.1007/978-981-97-4711-5_8

can bring to law enforcement such as investigations and decision-making processes. AI can assist law enforcement agencies in identifying patterns, detecting anomalies, and generating actionable insights from available data, thereby enhancing operational efficiency and effectiveness. Additionally, AI algorithms can analyze large volumes of textual data, such as social media posts and online communications, to identify potential criminal activities or threats. AI-powered data analytics systems can also assist in detecting patterns and trends in crime, aiding law enforcement agencies in resource allocation and implementing preventive measures.

This research paper aims to address the underutilization of AI in law enforcement by presenting an integrated AI-based pipeline focused on the creation of first information report (FIR) Summary in the Kannada language. The pipeline consists of two key components: a speech processing block, which uses the OpenAI's Whisper model [1] for accurate speech-to-text conversion in Kannada [2], and a natural language processing block, which employs named entity recognition (NER) via a Bi-LSTM neural network for effective information extraction from the transcribed FIRs. This pipeline aims to automate the identification and categorization of entities such as names, locations, and dates within the FIR text by integrating the NER model [3]. The proposed NER model utilizes three tags, which can be expanded further to accommodate additional tags as per FIR requirements.

A significant challenge is the scarcity of Kannada language data for law enforcement. To address this, data augmentation techniques are applied in both the speech processing and natural language processing blocks [4, 5]. In the speech module, data is augmented in the time and frequency domains, using techniques like Gaussian noise addition and generative adversarial networks (GANs) [3]. For the text-based NLP component, data augmentation occurs at multiple levels, including word, character, phrase, and document levels.

By leveraging state-of-the-art AI techniques, this research aims to improve the accuracy and efficiency of FIR summary creation in Kannada, thus enhancing decision-making and investigative processes in law enforcement [6–8]. The study serves as a step toward exploiting AI's full potential in law enforcement applications.

Our paper presents pioneering contributions in the challenging domain of natural language processing (NLP) and automatic speech recognition (ASR) in Kannada, a low-resource language with limited prior research. These contributions are essential for enhancing language processing capabilities in Kannada for law enforcement applications.

- Tailored AI-Based Pipeline for Kannada: Our paper introduces a specialized AI pipeline for creating FIR summaries in Kannada, addressing the unique challenges posed by this low-resource language.
- Dual-Component System in Kannada: Within this pipeline, we employ OpenAI's Whisper for Kannada automatic speech recognition (ASR) and a Bi-LSTM-based named entity recognition (NER) for information extraction from Kannada text. This demonstrates our commitment to advancing NLP and ASR capabilities in Kannada, where limited work has been done.

- Adaptive NER Model for Kannada: Our NER model is designed to adapt and expand its tagging capabilities as per FIR-specific entity recognition requirements in Kannada, showcasing its versatility and contribution to Kannada NLP.
- Data Augmentation Techniques Addressing Kannada Data Scarcity: Recognizing the scarcity of Kannada language data, we employ advanced data augmentation techniques for both the speech and text components, featuring a custom GAN model. These contributions are pivotal in enhancing the availability of data resources for Kannada NLP and ASR, facilitating robust model training and performance.

While this research is centered on the Kannada language, the methodologies and techniques used in the AI pipeline have potential applicability in other languages, particularly those that are underrepresented in technology. This adaptability of our approach could be beneficial for a broader audience, including linguists, AI researchers, and law enforcement agencies in different regions.

Understanding the importance of this work, it's crucial to acknowledge that its impact extends beyond the technical community. The law enforcement agencies, which constantly seek efficient and reliable tools for data processing and analysis, stand to gain significantly from our findings. Moreover, policy-makers and regulatory bodies involved in AI governance can also find value in our research, as it addresses critical issues like bias reduction and ethical AI use.

To underscore the novelty of our work, it's important to highlight our unique contributions to the field of AI in law enforcement, particularly in a low-resource language like Kannada. Our integrated AI-based pipeline for FIR summary creation and the use of advanced data augmentation techniques represent pioneering steps in this domain. These contributions not only address the immediate challenges in Kannada language processing but also lay the groundwork for future advancements in AI applications for other underrepresented languages.

While developing the AI-based pipeline for creating FIR summaries in Kannada, our focus has primarily been on the integration of OpenAI's Whisper for Kannada automatic speech recognition (ASR) and a Bi-LSTM-based named entity recognition (NER) model for effective information extraction from Kannada text. However, certain alternative options were not explicitly pursued. For instance, the exploration of pre-trained transformer-based language models, such as BERT, for NER in Kannada could have been considered. Transformer-based models have demonstrated success in various NLP tasks, but their adoption was constrained by computational requirements and the availability of large-scale pre-trained models for Kannada. Additionally, the feasibility of transfer learning from languages with more abundant data to Kannada is explicitly addressed. The decision to opt for a Bi-LSTM-based NER model has been influenced by factors such as interpretability, data availability, and alignment with the unique characteristics of the Kannada language. These considerations have been highlighted as potential areas for further exploration and comparative analysis in future research.

This paper is structured into several sections to provide a comprehensive overview of the research conducted. We begin by discussing the related work in the field, which serves as a foundation for our study. Following that, we delve into a detailed description of the dataset used in our research, providing insights into its composition and relevance. The methodology section outlines the techniques and approaches employed in our study, shedding light on the research process. Subsequently, we present the results, analyzing the findings and their implications. The paper concludes with a summary of the key takeaways and potential directions for future research. Lastly, we provide a list of references.

2 Related Work

In the dynamic realm of natural language processing (NLP) and speech recognition, particularly for resource-constrained languages, there's a growing body of research addressing the nuances of text and speech processing. This analysis synthesizes insights from scholarly papers in this field, highlighting their implications for law enforcement, especially in processing multilingual legal texts and speech.

Advancements in Low-Resource Language Models:

The progress in Chinese language processing, notably in automatic speech recognition (ASR) and named entity recognition (NER), is substantial. Yang's work with the Whisper system [2], evident in the AISHELL datasets, significantly improved Chinese ASR and NER. Likewise, Wang's integration of Char2Vec with Bi-LSTMs [1] in telecommunications NER shows the effectiveness of character embeddings in processing Chinese text. These advancements are vital for understanding Chinese legal texts and assisting law enforcement in multilingual settings. This evolution in Chinese language models is more than a technological leap; it's an important step toward equalizing the treatment of non-English legal texts and speech, matching the precision given to English equivalents.

Innovations in Biomedical NLP:

In the biomedical domain, Ullah Miah and colleagues have made significant strides, especially regarding Covid-centric literature [3]. Their deep learning model for Medical NER, with its remarkable 98% F1-score, exemplifies the potential of NLP in the biomedical field. This innovation is crucial for extracting relevant information from vast biomedical texts, which can be instrumental in legal cases involving medical documentation. The accuracy and efficiency of this model underscore the potential of NLP in transforming how legal entities approach complex biomedical data, making it more accessible and interpretable.

Developments in Kannada Language Processing:

Kannada, a language widely spoken in India, has made significant strides in natural language processing (NLP) and speech processing. Dhananjaya's text-to-speech conversion algorithm [4] showcases the societal benefits of NLP, aiding those with reading difficulties. P. N. and team's work on a Kannada speech dataset [9] has been instrumental in advancing ASR systems. Gladston's study of TTS techniques [10], along with Prakash's unified HMM-based speech synthesis model for Indian languages [11], are crucial for enhancing Kannada's digital processing, particularly important for law enforcement in regions where Kannada is prevalent. These advancements ensure that language does not hinder justice and law enforcement in areas where regional languages like Kannada dominate.

Data Augmentation in NLP:

Data augmentation in NLP has seen considerable exploration, with Lucas Francisco Amaral Orosco Pellicer analyzing various methods [5]. Ding et al. language model-based augmentation [12] and Issifu et al. adaptation of easy data augmentation for named entities [13] mark significant progress in boosting model performance, especially in low-resource tagging and medical NER tasks. These techniques are crucial for increasing the accuracy of NLP models in legal contexts, where precision is critical. Effective data augmentation enables the development of robust models capable of managing the complexity and nuances of legal language, ensuring precise data interpretation in challenging scenarios [14].

Kannada Language-Specific Techniques:

In Kannada language processing, the emphasis on POS tagging by Anami and [15] and the Kannada POS Tagger developed by Todi Mishra et al. [16] using machine learning and neural networks are noteworthy. Deepamala and Ramakanth Kumar's research on Kannada stemming [17] lays essential groundwork in NLP, vital for interpreting legal texts in Kannada. These methods not only enhance the processing of Kannada texts but also set a precedent for innovations in other lesser-known languages, fostering linguistic inclusivity in legal contexts.

Contextualizing Our Work:

Our project, within the realm of NLP progress, focuses on processing multilingual legal texts for law enforcement. Leveraging the groundwork in languages like Kannada and Chinese, we've developed an AI-based pipeline for generating FIR summaries in Kannada [18]. This initiative not only tackles Kannada language processing challenges but also serves as a model for similar progress in other lesser-represented languages. Our work marks a significant advancement in applying NLP techniques to practical law enforcement and legal scenarios, ensuring fair and accurate representation of non-English speakers and their legal matters.

3 Dataset Description

For the audio data, we are using a custom dataset which consists of two folders, audio and transcript. The audio folder consists of .wav files of people reporting an incident in the Kannada language obtained from local police stations and manually combined. The audio dataset comprises approximately two hours of recorded content in total.

The transcript folder has the corresponding transcripts. The data will be made available on request. The textual data is a multilingual named entity recognition (NER) dataset called WikiANN [19] available in the TensorFlow Datasets catalog. It is designed to support research in natural language processing (NLP) and information extraction tasks. The 'wikiann' dataset for Kannada has seven named entity annotations designed exclusively for the Kannada language. These annotations contain Kannada sentences with labels for named entities like 'PER' (person), 'ORG' (organization), and 'LOC' (place). ['O', 'B-PER', 'I-PER', 'B-ORG', 'I-ORG', 'B-LOC', 'I-LOC'] are the tags in the dataset.

4 Proposed Methodology

The proposed methodology is a two-module pipeline designed to process audio input data in Kannada for FIR summary generation (as seen in Fig. 1). The Kannada ASR module employs data augmentation, fine-tuning Whisper for transcription, and error correction to convert spoken Kannada FIRs into text. The entity recognition (NER) module utilizes data augmentation, data preprocessing, and a Bi-LSTM model for NER to identify entities in the transcribed text. This methodology aims to automate and enhance FIR summary creation in Kannada, bolstering the efficiency and accuracy of law enforcement procedures.

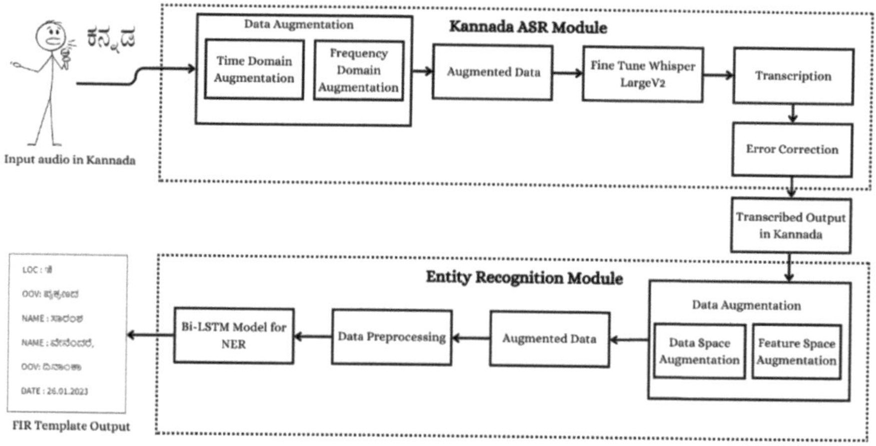

Fig. 1 Integrated AI-based pipeline for law enforcement application: FIR assistance pipeline

4.1 Kannada ASR Module

The Kannada automatic speech recognition module referenced as a part of Fig. 1 accurately converts audio signals in Kannada language to Kannada text. There are plenty of resources for languages such as English, Spanish, but Kannada is classified as a low-resource language [10]. To improve model performance, various data augmentation tasks are conducted. Several sub-modules are involved to achieve these tasks. Figure 2 mentioned here is a detailed flow diagram of the Kannada ASR module as seen in Fig. 1.

The input audio is taken as .wav files and data augmentation is performed to artificially expand the size and diversity of a training dataset [20].

In the context of speech, this research focuses on enhancing Kannada ASR systems by employing time and frequency domain data augmentation techniques due to limited data availability. Time domain methods, including time stretching, pitch shifting, and others, introduce variations in speech data, while frequency domain techniques manipulate spectral characteristics for dataset augmentation. In the context of audio data augmentation, GANs can be used to generate new audio samples with augmented spectral characteristics [21]. GANs can learn to generate audio samples with variations in spectral shapes, such as different vocal tract configurations, noise characteristics, or frequency modulation effects. These synthesized

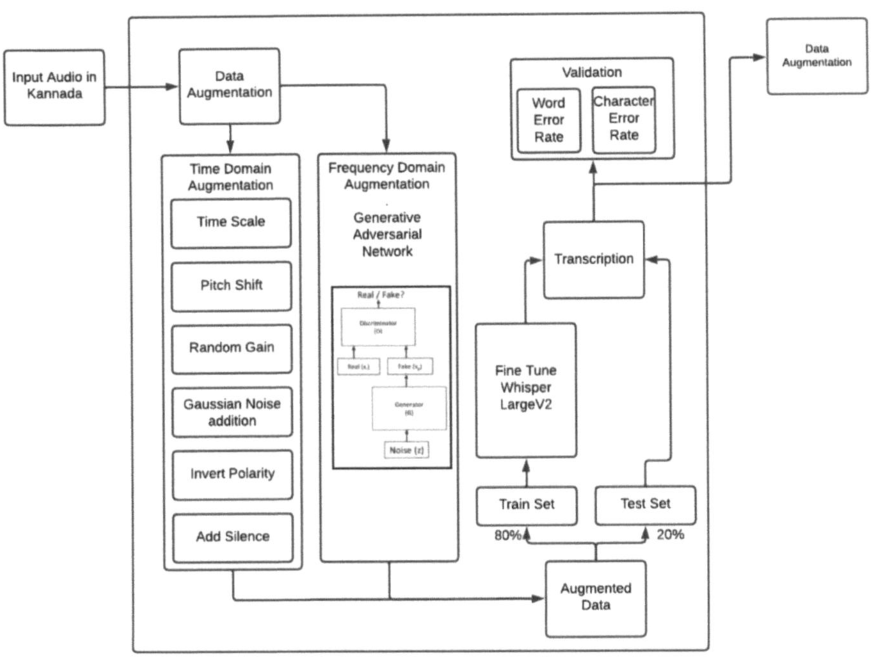

Fig. 2 Kannada automatic speech recognition module

samples can then be combined with the original data to create an augmented dataset that captures a broader range of spectral characteristics and variations.

Whisper large-v2 model is used to fine-tune the model for its application in law enforcement domain [22]. The data used to fine-tune the model is mentioned in data description section. After fine-tuning, we validate the transcription. To do so, we use the word error rate (WER) metric that measures the percentage of word-level errors in the ASR output compared to a reference transcription. It provides an important measure of accuracy, considering substitutions, deletions, and insertions of words. By analyzing the WER, we can evaluate the performance of our ASR system and assess its effectiveness in real-world applications. The lower the WER, the higher the accuracy of the system's transcriptions. The model is then used to transcribe audio files from the test data. These transcriptions act as input to the natural language processing module. The transcripted data is then passed on to the entity recognition module.

4.2 *Entity Recognition Module*

The primary intention of the entity recognition module referenced as part of Fig. 1 is to extract certain information from a passage of texts and arrange it in the format of an FIR summary. In this research study, we focus on using named entity recognition (NER) by using neural networks. There are several sub-modules used to achieve this task. Figure 3 is a detailed flow diagram of the entity recognition module as seen in Fig. 1.

Data augmentation in texts involves creating new training samples by modifying existing textual data, enhancing the diversity of the training set and boosting model performance [5]. The primary methods are data space augmentation and feature space augmentation. In the data space, augmentation introduces variations at the character, phrase, and word levels [12, 13]. Within the feature space, augmentation techniques center around the vector representation of words or sentences. Techniques include the use of generative adversarial networks (GANs), specifically conditional GANs for creating Doc2Vec representations [6–8].

The data is then preprocessed by extracting and flattening tokens and tags from the original dataset. The input audio is taken as .wav files and data augmentation is performed to artificially expand the size and diversity of a training dataset. In the context of speech, data augmentation plays a crucial role in enhancing the performance of automatic speech recognition (ASR) systems especially given the limited availability of Kannada language data [23–26]. In this research study, augmentation is performed in time domain and frequency domain. Time domain augmentation techniques such as time stretching, pitch shifting, inverting polarity, random gain, adding silence, and Gaussian noise addition is done to create variations in Kannada speech data. Frequency domain data augmentation involves manipulating the spectral characteristics of an audio signal to augment the dataset. Post-flattening, unique tokens and tags are further processed to form the token and tag vocabulary, which

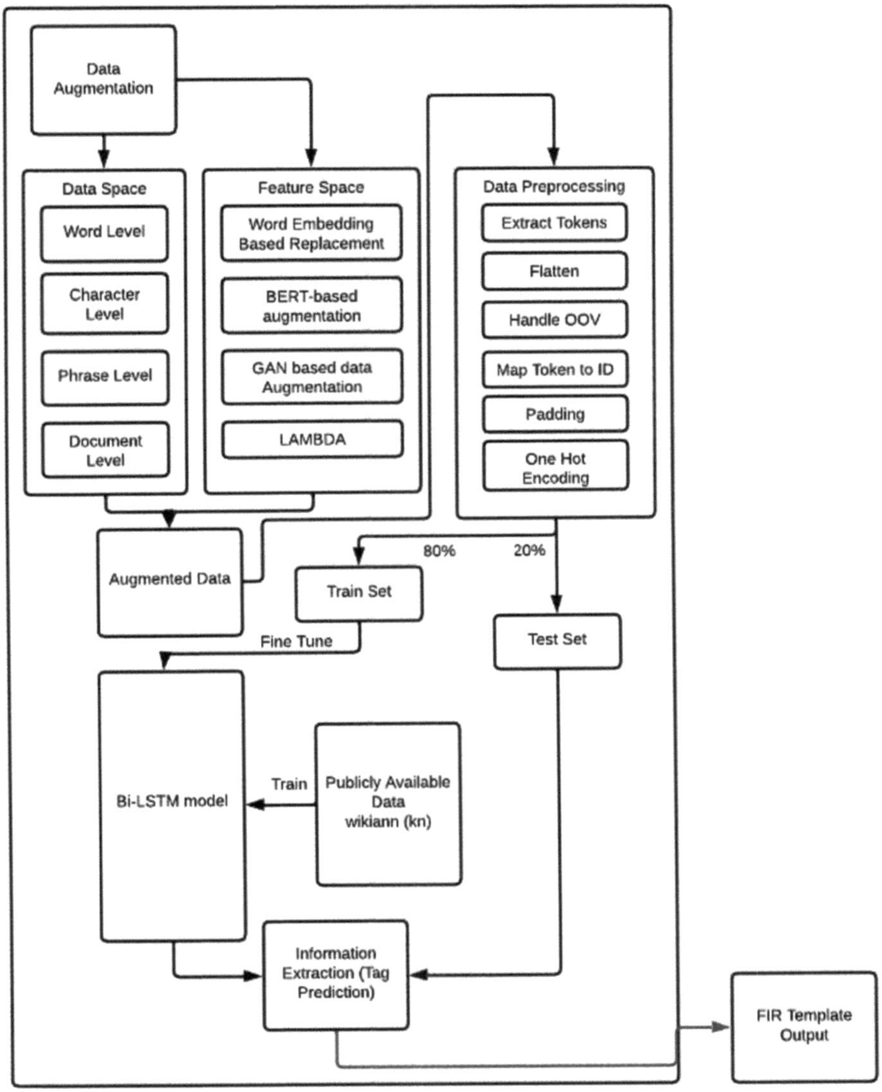

Fig. 3 Entity recognition module

captures all distinct tokens and tags from the data [15]. A unique identifier, such as tag 7, is designated for infrequently encountered words. To ensure uniformity, the training, validation, and test datasets undergo padding. This step ensures each sequence is of identical length, optimizing it for training. Concurrently, tokens and tags undergo numerical encoding [16]. For tags, one-hot encoding is utilized, where each tag gets a binary vector representation with a size equal to the tag vocabulary

[17]. This format is vital for modeling token-tag relationships during the training phase.

For training and fine-tuning the model, we have used a Bi-LSTM model. It consists of three LSTM layers and two dense layers. This model trains using tokens from the dataset as input and their matching tags as the output. The input dimension correlates with the number of distinct tokens, while the output dimension relates to the unique tag count, predicting the correct tag for each token. Within the model architecture, the first dense layer uses the rectified linear unit (ReLU) activation function for learning intricate patterns. The subsequent dense layer utilizes the SoftMax activation function to achieve a probability distribution over tag classes. The Adam optimizer, known for its adaptive learning rate based on the loss function gradient, is employed for training. Notably, the model undergoes training on the Wikiann kn dataset.

To create the FIR summary with the entities [1], our focus is on three specific tags: date, name, and location. The model's capabilities can be extended to accommodate additional tags as required by the first information report (FIR) requirements. Essential tags for FIR documentation include name, age, sex, religion, caste, profession, nationality, contact number, date, day, taluk, district, address, accused name, accused address, accused sex, accused caste, accused profession, and damage intensity. We annotate a substantial amount of data by labeling the text with the mentioned tags. This annotated data trains the model, ensuring it learns the relationships and patterns associated with each tag [3]. A large dataset ensures robust model performance. Once trained, the model can be used for automatic information extraction from FIR documents. This pipeline offers a solution for real-time deployment and usage.

5 Results and Discussion

The integrated AI-based pipeline for law enforcement applications, specifically targeting the creation of FIR summary in the Kannada language, yielded promising results. Experiments were run on a computer with AMD Ryzen 5 4600 H processor, integrated with Radeon Graphics, and fortified with 16.0 GB of RAM. The system utilized a 64-bit operating system with a $\times$ 64-based processor, providing a stable and efficient computational environment under a Windows platform with an execution time of approximately 8.46 s. For the proposed approach, the Whisper ASR model demonstrated excellent performance with a word error rate (WER) of 2% in the specific use case of Kannada speech recognition. Comparing this to WER of different ASR models trained with low-resource languages as seen in Table 1, we find that our proposed model performs astonishingly well. This indicates high accuracy in converting spoken Kannada into text, a crucial step in the FIR creation process. A few sample transcribed outputs have been compared with the original transcriptions to find out how accurate the transcription is as seen in Table 2. The model utilized three tags initially, namely date, name, and location, but can be expanded to accommodate additional tags as per the requirements of FIR documentation. The entity recognition accuracy of the fine-tuned Bi-LSTM model is measured by taking multiple sample

Table 1 Difference in WER for open-source models and proposed model

Year	Model	Language	WER (%)
2020	*Kaldi*	Punjabi	62
2022	*Wav2Vec*	Bhojpuri	29.7
2021	*Kaldi*	Malayalam	32.21
2022	*Wav2Vec*	Dogri	36.5
2023	*Whisper large V2*	Bengali	26.7
2023	*Proposed model*	Kannada	2

transcriptions and the average accuracy is calculated as shown in Table 3. The results as seen in Fig. 4 highlight the capabilities of the Whisper large-v2 model, which was fine-tuned for its application in the law enforcement domain. Whisper's array of models, including Tiny, Base, Small, Medium, Large, and Large-v2, offer a range of functionalities such as speech-to-text transcription, speaker diarization, keyword spotting, and noise classification. The specific use of the Whisper large-v2 model in this research study demonstrated its suitability for law enforcement applications.

The custom Bi-LSTM model is trained, and the training loss and accuracy is shown (see Fig. 5).

Table 2 Transcribed output accuracy

Sample	Original transcription	Obtained transcription	Accuracy (%)
1	ಒಬ್ಬ ವ್ಯಕ್ತಿ ಬಾವಿಯ ಕಡೆಗೆ ಹೋಗುತ್ತಿರುವುದನ್ನು ನಾನು ನೋಡಿದೆ	ಒಬ್ಬ ವ್ಯಕ್ತಿ ಬಾವಿಯ ಕಡೆಗೆ ಹೋಗುತ್ತಿರುವುದು ನಾನು ನೋಡಿದೆ	94.44
2	ಹುಡುಗರು ಬೆಳೆಗಳನ್ನು ಕದಿಯುತ್ತಿದ್ದರು	ಹುಡುಗರು ಬೆಳೆಗಳನ್ನು ಕದಿಯುತ್ತಾರೆ	84.85
3	ಇದು ಸಂಭವಿಸಿದಾಗ ಸುಮಾರು ರಾತ್ರಿಯಾಗಿತ್ತು	ಇದು ಸಂಭವಿಸಿದಾಗ ಸುಮಾರು ರಾತ್ರಿಯಾಗಿದೆ	88.89
4	ನನ್ನ ಎದುರಿಗೆ ಅಪಘಾತ ಸಂಭವಿಸುವುದನ್ನು ನಾನು ನೋಡಿದೆ	ನನ್ನ ಎದುರಾಗ ಅಪಘಾತ ಸಂಭವಿಸುವುದನ್ನು ನಾನು ನೋಡಿದೆ	95.56

Table 3 Entity recognition accuracy for an average sample

Entity	No. of original entities	No. of obtained entities	Accuracy (%)
Date	*2*	2	100
Name	*3*	2	66.6
Location	*3*	5	60

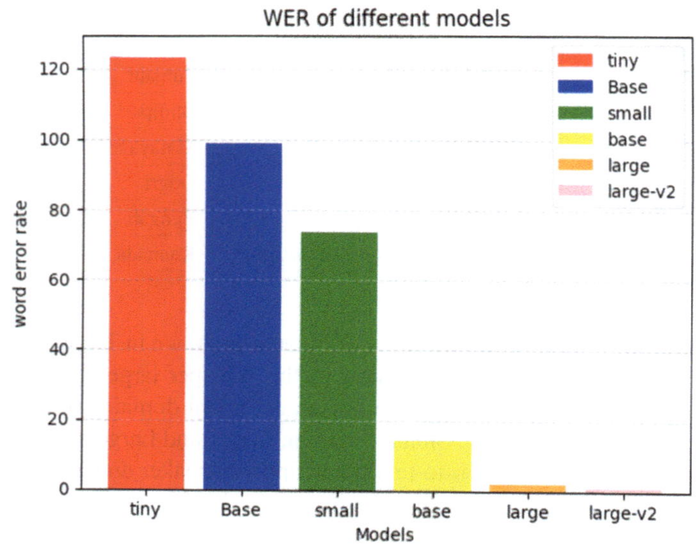

Fig. 4 WER for different Whisper models

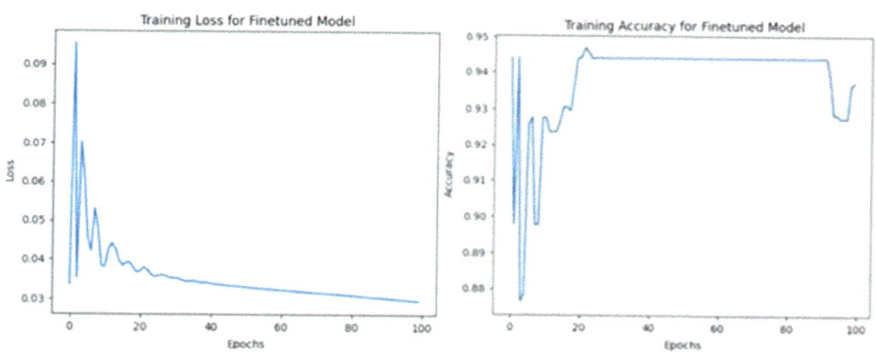

Fig. 5 Training loss and accuracy for Bi-LSTM fine-tuned model

6 Conclusion and Future Work

The integrated AI pipeline utilizes OpenAI's Whisper ASR models for Kannada speech-to-text conversion and advanced AI techniques for precise FIR summary generation, achieving a remarkable 2%-word error rate. In information extraction, a Bi-LSTM NER model demonstrates its effectiveness with an impressive 94% accuracy in identifying named entities in FIR text. The fine-tuned Whisper large-v2 model proves to be an asset for law enforcement applications.

Future work includes expanding NER tags, further fine-tuning models for the law enforcement domain, integrating multi-modal data, enabling real-time processing, and conducting rigorous evaluations with law enforcement personnel to gather essential feedback for continuous improvements. Additionally, privacy-preserving techniques and responsible AI guidelines will be implemented to enhance the pipeline's capabilities and efficiency in Kannada FIR summary creation for law enforcement.

References

1. Wang Y, Xia B, Liu Z, Li Y, Li T (2017) Named entity recognition for Chinese telecommunications field based on Char2Vec and Bi-LSTMs, 2017 12th International conference on intelligent systems and knowledge engineering (ISKE), Nanjing, China
2. Yang H, Zhang M, Tao S, Ma M, Qin Y (2023) Chinese ASR and NER improvement based on whisper fine-tuning, 2023 25th International conference on advanced communication technology (ICACT), Pyeongchang, Korea, Republic of
3. Ullah Miah MS, Sulaiman J, Sarwar TB, Islam SS, Rahman M, Haque MS (2023) Medical named entity recognition (MedNER): a deep learning model for recognizing medical entities (drug, disease) from scientific texts, IEEE EUROCON 2023—20th International conference on smart technologies, Torino, Italy
4. Dhananjaya MS, Krupa BN, Sushma R (2016) Kannada text to speech conversion: a novel approach, 2016 International conference on electrical, electronics, communication, computer and optimization techniques (ICEECCOT), Mysuru, India
5. Orosco Pellicer LFA, Ferreira TM, Reali Costa AH (2023) Data augmentation techniques in natural language processing. Appl Soft Comput 132:109803
6. Pellicer LF, Ferreira TM, Costa AH (2023) Data augmentation techniques in natural language processing. Appl Soft Comput 132:109803
7. Ding B et al (2020) DAGA: data augmentation with a generation approach for low-resource tagging tasks. arXiv preprint arXiv:2011.01549
8. https://huggingface.co/datasets/wikiann
9. PN and Kini S (2022) Phoneme based Kannada speech corpus for automatic speech recognition system, 2022 IEEE International conference on distributed computing and electrical circuits and electronics (ICDCECE), Ballari, India
10. Gladston R, Pradeep KV (2023) Exploring solutions for text-to-speech synthesis of low-resource languages, 2023 4th International conference on signal processing and communication (ICSPC), Coimbatore, India
11. Prakash A, Reddy MR, Nagarajan T, Murthy HA (2014) An approach to building language-independent text-to-speech synthesis for Indian languages, 2014 Twentieth national conference on communications (NCC), Kanpur, India, 1–5, https://doi.org/10.1109/NCC.2014.6811356.
12. Ding B, Liu L, Bing L, Kruengkrai C, Nguyen TH, Joty S, Si L, Miao C (2020) DAGA: data augmentation with a generation approach for low-resource tagging tasks. Proceedings of the 2020 conference on empirical methods in natural language processing (EMNLP), 6045–6057
13. Issifu M, Ganiz MC (2021) A simple data augmentation method to improve the performance of named entity recognition models in medical domain, 2021 6th International conference on computer science and engineering (UBMK), Ankara, Turkey
14. Prabhakar P, Gupta D, Pati PB (2022) Abstractive Summarization of Indian Legal Judgments, 2022 OITS International conference on information technology (OCIT), Bhubaneswar, India
15. Todi KK, Mishra P, Sharma DM (2018) Building a Kannada POS tagger using machine learning and neural network models. ArXiv. /abs/1808.03175
16. Deepamala N, Ramakanth Kumar P (2015) Kannada stemmer and its effect on Kannada documents classification. In: Jain L, Behera H, Mandal J, Mohapatra D (eds) computational intelligence in data mining—Vol 3. Smart Innovation, systems and technologies, vol 33. Springer, New Delhi

17. Trishala G, Mamatha HR (2021) Implementation of stemmer and lemmatizer for a low-resource language—Kannada. In: Pandian AP, Palanisamy R, Ntalianis K (eds) Proceedings of international conference on intelligent computing, Information and control systems. Advances in intelligent systems and computing, vol 1272. Springer, Singapore
18. Jayashree R, Anami BS, Poornima BK (2021) Text document summarization using POS tagging for Kannada text documents, 2021 11th International conference on cloud computing, Data science & engineering (Confluence), Noida, India
19. Jeeva Priya K, Sree SS, Navya T, Gupta D (2018) Implementation of phonetic level speech recognition in Kannada using HTK, 2018 International conference on communication and signal processing (ICCSP), Chennai, India
20. Qian Y, Hu H, Tan T (2019) Data augmentation using generative adversarial networks for robust speech recognition. Speech Commun 114:1–9
21. Rizos G, Baird A, Elliott M, Schuller B (2020) Stargan for emotional speech conversion: validated by data augmentation of end-to-end emotion recognition, ICASSP 2020–2020 IEEE International conference on acoustics, Speech and Signal Processing (ICASSP), Barcelona, Spain
22. Whisper https://doi.org/10.48550/arXiv.2212.04356
23. Praveen Kumar PS, Thimmaraja Yadava G, Jayanna HS (2020) Continuous Kannada speech recognition system under degraded condition. Circuits Syst Signal Process 39:391–419
24. Vancha P, Nagarajan H, Inakollu VS, Gupta D, Vekkot S (2022) Word-level speech dataset creation for Sourashtra and recognition system using Kaldi," 2022 IEEE 19th India Council International Conference (INDICON). Kochi, India
25. Sheshadri SK, Sai Bharath B, Hari Naga Sree Chandana Sarvani A, Reddy Vijaya Bharathi Reddy P, Gupta D (2022) Unsupervised neural machine translation for english to kannada using pre-trained language model, 2022 13th International conference on computing communication and networking technologies (ICCCNT), Kharagpur, India
26. Sharma RS, Paladugu SH, Priya KJ, Gupta D (2019) Speech recognition in Kannada using HTK and Julius: a comparative study, 2019 International conference on communication and signal processing (ICCSP), Chennai, India, 0068–0072, https://doi.org/10.1109/ICCSP.2019.8698039

Review on Emotion-Centric Music Recommender Systems

L. Kamatchi Priya, Alekhya Sundari R. Nanduri, D. Mrudula, Harshitha Golla, and Reshmi Pradeep

Abstract In the rapidly evolving landscape of music recommender systems, this literature paper embarks on a comprehensive exploration of recommender systems tailored for music, specifically emphasizing content-based recommendations with an emotional orientation. The paradigm of music suggestion has witnessed remarkable growth, placing a heightened emphasis on delivering music that resonates with the emotional state of listeners. This review encapsulates the progress and utilization of content-based music recommendation systems centered on emotions, drawing insights from a wide spectrum of academic contributions. The study meticulously scrutinizes a diverse array of approaches and algorithms employed in analogous systems, encompassing techniques such as Extreme Gradient Boosting (XGBoost), Support Vector Machines (SVM), Decision Trees, and Random Forests. In addition to providing a synthesis of the current research landscape, this study also identifies nascent patterns and envisages potential trajectories for future investigations in this dynamic domain.

Keywords Recommendation systems · XGBoost · SVM · Decision Trees · Random forests

1 Introduction

The need for music recommendation systems is growing as more online businesses that handle huge music libraries restrict user access to such libraries. Therefore, users are recommended music based on different queries that they provide. Many users find it challenging to specify their musical interests as queries [1]. Content-based filtering recommender systems recommend items based on their characteristics or content, building user profiles from their interactions [2]. These algorithms, which

L. K. Priya (✉) · A. S. R. Nanduri · D. Mrudula · H. Golla · R. Pradeep
PES University, Bangaluru, India
e-mail: priyal@pes.edu

S. M. Thampi et al. (eds.), *Fifth International Conference on Computing and Network Communications*, Lecture Notes in Electrical Engineering 1221,
https://doi.org/10.1007/978-981-97-4711-5_9

do not rely on other users' ratings or preferences, recommend goods whose characteristics match a user's profile as well as items that are comparable to those the user has expressed interest in. A music recommendation system built on emotions revolutionizes the music listening experience by automating the painstaking task of curating playlists and categorizing songs [3]. This approach allows one to effortlessly generate personalized playlists tailored to one's moods. It not only transforms the music listening experience but also caters to diverse user needs. Mood-based listeners can effortlessly discover music aligned with their current emotional states, whether they seek relaxation or energy. Moreover, music therapists can leverage the system to access therapeutic music that aids clients in relaxation, focus, or emotional management during therapy sessions. Additionally, the system can be integrated with existing music platforms such as Spotify or Apple Music, significantly enhancing the user experience by providing highly personalized playlists that take into account individual preferences, moods, and other relevant factors for a more comprehensive and tailored listening experience.

Despite their success in analyzing musical elements, content-based algorithms frequently have trouble capturing the nuanced emotional content found in songs. Since emotions are complex and individualized, it has been difficult to create algorithms that can reliably identify them in musical compositions [4]. Greater research is necessary into how content-based systems may include user-provided emotional feedback since emotions are temporary and vary between short periods [5]. Research gaps exist in understanding how to smoothly combine user-provided emotional cues with content-based music recommendations. This includes determining the weight and relevance of emotional input, as well as refining the algorithms that interpret and act upon this user feedback [6].

Different approaches and techniques have arisen in recent years to fill in these research gaps. These include integrating affective computing and machine learning models to identify emotional content in music and developing novel methods [7, 8], some combine user-provided emotional input with content-based recommendation algorithms or use collaborative filtering-based recommendation algorithms alongside to improve personalization [9]. Other methods include utilizing natural language processing techniques to mine emotional cues from user-generated text and extracting low-level features to further find the levels of emotion that a song might trigger [10]. These innovative techniques may overcome the emotional gap in content-based music suggestions, but they come with their own set of difficulties and need to be thoroughly tested and improved.

In the sections that follow, we'll conduct a thorough investigation of content-based algorithms that could be used to recommend music based on emotions. First, we begin with a thorough literature study to illuminate the main ideas, strategies, and conclusions of the field's current state. After that, we move on to an experimental comparison of approaches where we evaluate the performance of different recommendation algorithms, to see how well they can detect and react to user emotions. Following the methodological assessment, we present the results of our study and research, which will shed light on the advantages and disadvantages of various content-based recommender systems in this specific field.

2 Related Work

To provide an overview of the comparative analysis of various machine learning models, focusing on their performance when encountering missing values, flexibility, computational power, and scalability, various research papers were studied and summarized. Each model exhibits unique characteristics and trade-offs in different scenarios. The effectiveness of XGBoost in handling missing values through its adaptive learning approach is highlighted, while Random Forest and Decision Trees are noted for their robustness in the presence of missing data. For large datasets, the scalability, efficiency, and predictive performance of XGBoost and Random Forest are emphasized. Decision Trees, while suitable for smaller datasets, can be prone to overfitting, which can be mitigated through pruning techniques. SVM is also acknowledged to have scalability issues, making it less suitable for handling large datasets. These comparative insights provide the framework for more investigation and analysis of the strengths and weaknesses of different machine learning models in various scenarios.

2.1 Decision Trees

Decision trees are used to build recommender systems as they offer several advantages like flexibility, efficiency, and interpretability of different input data types, but can be prone to overfitting [11]. Based on the attributes of the input, the decision tree builds a predictive model that links the input to a forecasted value. Every attribute is represented by an internal node in the tree, and each arc from a parent to a child node denotes a potential value or a set of values for that particular attribute. In the context of emotion-based music recommendation systems, decision trees are well-suited for capturing the intricate and subjective nature of emotional content in music because they offer an intuitive understanding of the impact of various musical features on emotional outcomes. Sharma et al. [12] use different classifiers which include decision trees to predict emotions of music which can then be used for recommending music, but the accuracy was found to be suboptimal. Decision trees can also be used together with a classifiability measure that appropriately reflects the complexity of the subproblem at each node and can catch potential sources of misclassification with a reasonable degree of ease. This approach is faster and achieves better classification accuracy [13, 14].

2.2 Extreme Gradient Boosting (XGBoost)

XGBoost algorithm can be implemented for predicting emotions based on various music features, exhibiting high performance and suitability for music recommendation. In emotion-based music recommendation systems, XGBoost excels in discerning intricate emotional nuances, due to its ensemble learning nature and the incorporation of regularization terms to prevent overfitting. But despite its strengths, the computational complexity of XGBoost may pose challenges, especially in real-time applications or with extensive datasets. Also, the intricate nature of XGBoost could make it more susceptible to overfitting in scenarios involving smaller datasets in emotion-based music recommendation systems. When discussing how XGBoost functions, it is mentioned that it employs a boosting technique that makes use of multiple shallow-depth decision tree classifiers [15]. Following the creation of several decision tree classifiers that process the learning method to improve classification accuracy while selecting the best musical features, learning is processed in the sequence of each classifier. The processing of huge datasets is aided by distributed and parallel computing, which is made possible by the use of XGBoost. Additionally, by repeatedly reinforcing the training data in a sequential manner, errors are steadily decreased contributing to its robust performance in predicting emotions in music. XGBoost classifier can be used after selecting optimum parameter ranges for it and an accuracy of 77–85% was achieved [16]. Another method of implementation is to use an ensemble of XGBoost, KNN, and Random Forests which observed an accuracy of 93.69 [17]. In low-duty cycle wireless networks, XGBoost was utilized to suggest neighbors. It achieved accurate and precise neighbor recommendations along with a more favorable trade-off between discovery latency and energy use. The authors also state that it has been frequently utilized in machine learning scenarios with regard to both performance and speed because it is scalable, computationally efficient, and sparsity-aware [18, 19].

2.3 Random Forest

Random Forest is a supervised ensemble approach that uses several decision trees called estimators to collectively provide predictions for classification and regression tasks. This proves to be a robust solution for emotion-based music recommendation systems. Employing bagging and random feature selection, Random Forests outperform expectations in terms of accuracy and efficiency, making them preferable for extensive databases with gaps [11]. It addresses the overfitting problem associated with individual decision trees, enhancing its suitability for emotion prediction in music [12]. The method offers advantages such as large dataset readability, high accuracy in predicting missing data gaps, and robust data processing, but also has disadvantages like high complexity and overfitting [20]. Emotion classification of information received through psychological sensors was performed using random forest and was shown to give the best results out of the other models that were used [21]. An intelligent recommender system based on product reviews can likewise be created using Random Forest [22]. Sentiment analysis was performed to find the polarity of reviews using machine learning algorithms including random forest. Recommendations to the user were made based on polarity and product features. Random forest was observed to have high accuracy and it was also observed that the accuracy is directly proportional to the number of decision trees in the random forest. By eliciting and comparing self-reported emotions versus physiologically observed emotions, gathered via skin conductance response and heart rate data, random forest was utilized to detect emotions in trading decisions [23]. Therefore, it can predict positive emotion categories based on physiological signals and behavioral responses [24].

Random forests can also be used to classify different text that users give to different sentiment categories [25] and they can also classify emotions based on detecting faces through a web camera [26]. The development of a fully automated, sophisticated system for facial expression detection and classification is highlighted in [27]. It achieved high performance using a proprietary face detector and a cutting-edge classifier composed of a number of Random Forests linked with an SVM labeler. Random forest can also be used to build a sad music preference prediction model with audio attributes and individual attributes as inputs [28]. An enriched random forest method that employs weighted random sampling for subset selection, enhancing performance in high-dimensional datasets for both regression and classification is used in [29]. It uses balanced leave-one-out cross-validation to lighten the computational burden and boost precision, especially when working with a small number of pertinent characteristics [30, 31].

2.4 Support Vector Machine (SVM)

SVM is a machine learning algorithm that is supervised, it excels in binary classification and outlier detection tasks for both regression and classification. Its effectiveness in capturing complex relationships aligns well with the diverse set of musical features associated with emotions, which can further be used in emotion-based music recommendation. However, its sensitivity to noise may present a challenge in the context of subjective emotional content in music. SVM comes in a variety of forms, including linear, RBF, poly, and sigmoid [12]. The SVM algorithm is a potent tool for classification in machine learning, data mining, and predictive analytics. It is employed in multi-class scenarios to assign scores to mixed emotions, ultimately considering the highest score [32]. It is found that SVM sigmoid has the highest accuracy of 67.167% for feature extraction but is usually not preferred as it reduces accuracy making it unfit for further research [12]. The valence classification, which is used to categorize emotions as positive or negative, is found to have the highest accuracy with Gaussian RBF SVM, with a maximum accuracy of 61.979% [12]. It performs optimally with smaller, complex datasets and therefore not preferred for large datasets with increased noise.

SVM was utilized for granularity-adapted emotion categorization using audio data. The audio data was processed to extract features, which were then utilized to train the SVM classifier [33]. SVM was also used to recommend activities to people for mental health and well-being [34]. It was used to categorize the severity of users' depression, and depending on the stage of depression, various pre-specified actions were advised, such as music therapy or ideas for making plans for social events. SVM demonstrated the best performance when measured against the other models. SVM can also be used to identify a person's personality traits by examining the brain waves that are produced while that person views emotionally charged content using EEG signals [35]. The SVM classifier is then used to predict five dimensions of personality traits using features that have been derived from these signals together with subjective assessments. Using machine learning methods, [36] identified emotion in speech. After the features were retrieved, SVM was given the data, and it was found to perform well [14, 19, 30, 37].

3 Results and Analysis

In the following sections, we provide a comprehensive account of our experimental setup and the dataset employed, which forms the foundation of our exploration into emotion-based music recommendation systems. This setup encompasses the methodologies, models, and data sources that have fueled our research, culminating in the presentation of final results, findings, and comparative performance analysis.

3.1 Dataset

The Moodify Dataset is a comprehensive and meticulously curated collection of music data that was specifically created to include emotional categorization as a key component in song classification. This dataset consists of about 278,000 songs that were selected from Spotify's enormous music collection and then rigorously labeled with four primary emotional characteristics: happiness, sadness, energetic, and calmness. The dataset's classification algorithm successfully categorizes these songs into four main emotional categories: happy, melancholy, energetic, and quiet, drawing inspiration from [38]. The primary objective of the Moodify project is to classify songs not only based on their lyrical and musical features but also by incorporating emotions as a key factor in the categorization process.

3.2 *Experimental Setup*

To choose the most suitable machine learning model for the dataset, a range of different models were thoroughly compared. The data was initially divided into 8:2 ratios for training and testing, respectively. We began our model analysis using the SVM and a linear kernel function as a hyperparameter. However, the first findings lacked the requisite level of accuracy. The decision tree classifier, which performed better than the SVM in the experiments that followed, was used. Nevertheless, throughout the training period, this method was somewhat slower. We looked more closely at the Random Forest model, a potent ensemble method that also uses decision trees. By tuning hyperparameters to values of 20 for max depth, 13 for random state, and 150 for the number of estimators, we achieved a notable increase in accuracy.

Confusion matrices help in assessing how well a model performs in terms of correctly classifying data points, which can be essential for model improvement and fine-tuning. They were used for each model as shown in Fig. 1. To provide insights into their accuracy and misclassifications. The diagonal elements of each confusion matrix correspond to the correct predictions for each class made by the respective model, while the off-diagonal elements represent instances of misclassification performed by each model.

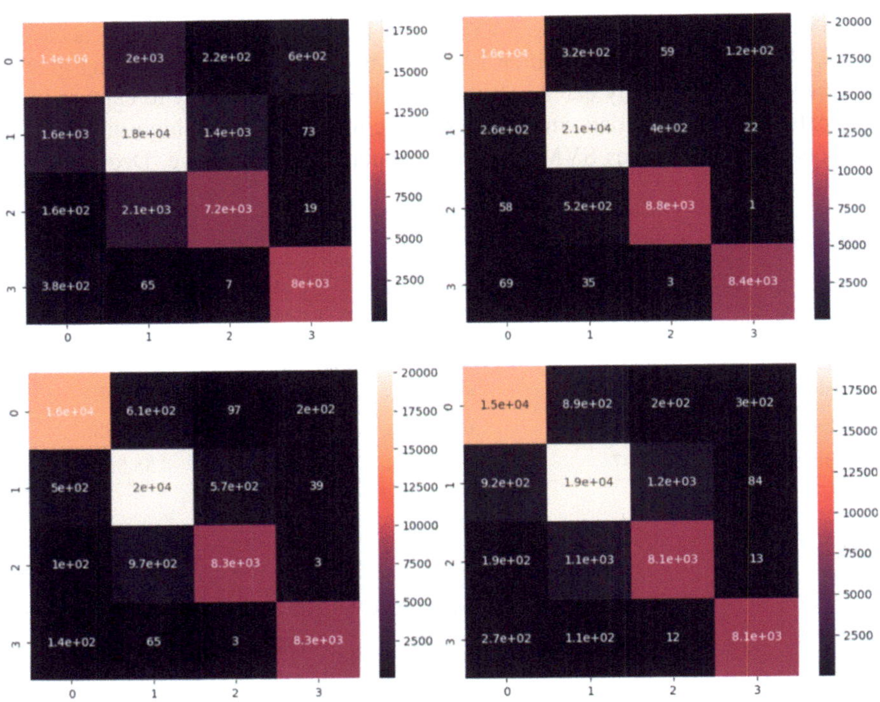

Fig. 1 Confusion Matrices {'sad': 0, 'happy': 1, 'energetic': 2, 'calm': 3}

As demonstrated in earlier studies, our experiments showed that Random Forest and SVM are both suitable for classification applications. We determined the aforementioned parameters to be the best ones by fine-tuning the hyperparameter values.

3.3 Results

As demonstrated in earlier studies, our experiments showed that Random Forest and SVM are both suitable for classification applications. We determined the aforementioned parameters to be the best ones by fine-tuning the hyperparameter values.

From Table 1. It is observed that XGBoost and Random Forest have the highest precision, recall, and F1 scores indicating that they perform well in terms of classification accuracy on the given dataset. They demonstrate a consistent and balanced performance across all evaluated metrics which makes them reliable choices for accurate and well-rounded classification tasks on the specified dataset.

Despite the remarkable performance of Random Forest, fine-tuning the XGBoost classifier not only increased accuracy by 2% but also outperformed Random Forest in terms of training and prediction speed, as can be observed in Table 2.

Table 2. Also shows that the SVM with a linear kernel achieves an accuracy of 84.48%, while the SVM with an RBF kernel outperforms it, exhibiting greater accuracy. This superiority can be attributed to the latter's capacity to capture intricate patterns within the data, making it suitable for complex and high-dimensional datasets. This also makes SVM models computationally intensive. Decision trees

Table 1 Comparing metrics for the models tested

Models	Evaluation metrics			
	Precision	Recall	F1 score	Support
SVM	0.84	0.84	0.84	55,588
Decision tree	0.91	0.91	0.91	55,588
Random forest	0.94	0.94	0.94	55,588
XG Boost	0.97	0.97	0.97	55,588

Table 2 Model accuracies

Models		Accuracy
Support vector machine	Linear kernel	84.48
	RBF kernel	89.01
Decision tree classifier		90.57
Random forest		94.04
Extreme gradient boosting (XG boost)		96.63

are known for their computational efficiency but carry the risk of overfitting. Ensemble methods like Random Forest manage to strike a balance by combining multiple decision trees, offering flexibility and competitive accuracy. The extreme gradient boosting method, XGBoost, consistently outperforms Random Forest, predominantly due to its distributed computing capabilities and advanced feature engineering.

Using the XGBoost algorithm and taking into account the confidence scores, it finds and delivers five songs that are specifically suited to the given emotion. The confidence score can be chosen to fit a variety of needs and domains.

From Fig. 2a. We can observe that different models exhibit variations in their precision for classifying the "sad" emotion. This implies that the choice of model should depend on the specific emotional category of interest. Figure 2b. Helps select the model with the highest weighted average which can ensure accurate classification across various emotional categories. Figure 2c. Consistently highlights XG Boost as a superior performer. This is attributed to its proficiency in handling complex data, feature ranking, and its ability to train and predict swiftly. From Fig. 2d. We can infer that we find that XG Boost excels in explaining and predicting variations in the dependent variable, indicating its prowess in modeling relationships between variables.

4 Conclusion

In conclusion, this research review study has examined the topic of emotion-based music recommendation systems, shedding light on the evolution of this innovative field. Our thorough examination of the literature, methods, and datasets has revealed the significance of incorporating user emotions into music suggestions. The many machine learning algorithms that we looked at, especially XGBoost, have demonstrated potential for improving accuracy and effectiveness. These results highlight how the way people interact with music platforms could change as a result of emotion-based song suggestions.

Looking ahead, there is an urgent need for more study to enhance and broaden these systems, maybe investigating additional emotional dimensions and improving algorithms. Additionally, it is crucial to take into account actual user experiences and these systems' ability to adapt to different domains.

This research opens the door to future investigations that could unlock even more sophisticated and personalized music recommendation mechanisms. In this dynamic landscape, the fusion of technology, emotion, and music promises to continue to captivate researchers and music enthusiasts alike.

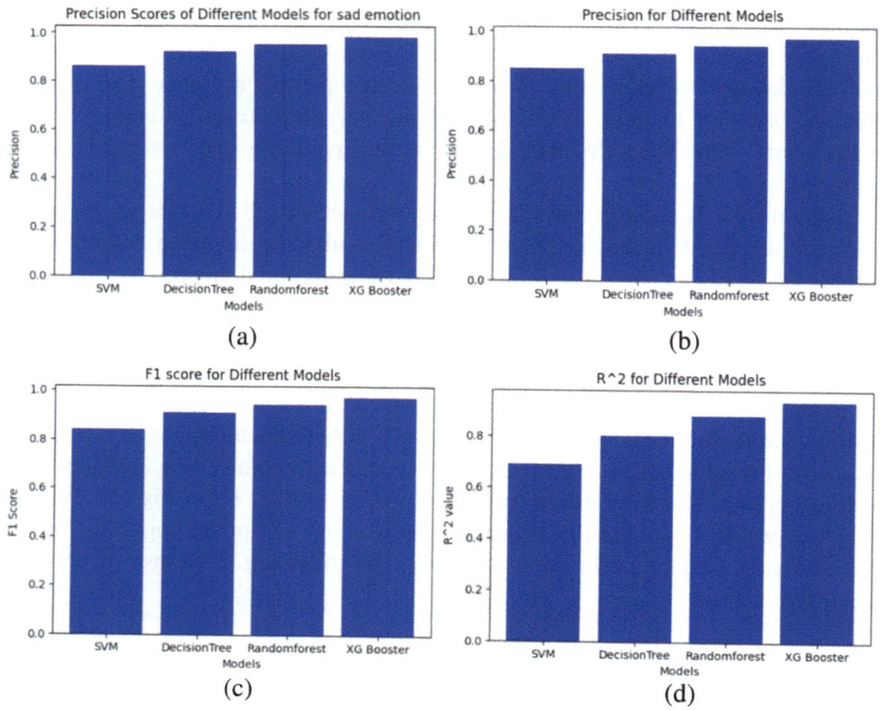

Fig. 2 **a** The precision score of sad emotion for different models. **b** The weighted average precision score for different models. **c** F1-score for different machine learning models. **d** R^2 for different models

5 Future Research

As we conclude our investigation of emotion-driven content-based music recommendation systems, we acknowledge the potential for further development in this area. Utilizing user social media data to obtain deeper insights into their emotional states and musical tastes represents one encouraging direction for ongoing and future study. We will be able to make music recommendations that are extraordinarily in line with users' emotional context thanks to the integration of social media signals and emotional data, which will enable a more sophisticated knowledge of users' emotions in real time. We will be able to offer highly customized music suggestions thanks to the use of XGBoost and collaborative filtering, which will improve the user experience overall. These ongoing and future research endeavors are geared towards elevating the emotional resonance of music discovery and ensuring that our recommendation systems remain on the cutting edge of user-centricity.

References

1. Yoshii K, Goto M, Komatani K, Ogata T, Okuno HG (2008) An efficient hybrid music recommender system using an incrementally trainable probabilistic generative model. IEEE Trans Audio Speech Lang Process 16(2):435–447. https://doi.org/10.1109/TASL.2007.911503
2. Jain S, Grover A, Thakur PS, Choudhary SK (2015) Trends, problems and solutions of recommender system, International Conference on Computing, Communication & Automation, Greater Noida, India. 955–958, https://doi.org/10.1109/CCAA.2015.7148534
3. Sharma VP, Gaded AS, Chaudhary D, Kumar S, Sharma S (2021) Emotion-based music recommendation system, 2021 9th International conference on reliability, Infocom technologies and optimization (Trends and Future Directions) (ICRITO), Noida, India, 1–5, https://doi.org/10.1109/ICRITO51393.2021.9596276
4. Zhu X, Shi Y-Y, Kim H-G, Eom K-W (2006) An integrated music recommendation system. IEEE Trans Consum Electron 52(3):917–925. https://doi.org/10.1109/TCE.2006.1706489
5. Lin Q, Niu Y, Zhu Y, Lu H, Mushonga KZ, Niu Z (2018) Heterogeneous knowledge-based attentive neural networks for short-term music recommendations. IEEE Access 6:58990–59000. https://doi.org/10.1109/ACCESS.2018.2874959
6. Rosa R, Rodriguez DZ, Bressan G (2015) Music recommendation system based on user's sentiments extracted from social networks. IEEE Trans Consum Electron 61:359–367. https://doi.org/10.1109/TCE.2015.7298296
7. Nanopoulos A, Rafailidis D, Symeonidis P, Manolopoulos Y (2010) MusicBox: personalized music recommendation based on cubic analysis of social tags. IEEE Trans Audio Speech Lang Process 18(2):407–412. https://doi.org/10.1109/TASL.2009.2033973
8. Lee K (2014) Using dynamically promoted experts for music recommendation. Multimed IEEE Trans 16:1201–1210. https://doi.org/10.1109/TMM.2014.2311012
9. Shao B, Wang D, Li T, Ogihara M (2009) Music recommendation based on acoustic features and user access patterns. IEEE Trans Audio Speech Lang Process 17(8):1602–1611. https://doi.org/10.1109/TASL.2009.2020893
10. Yoon K, Lee J, Kim M-U (2012) Music recommendation system using emotion triggering low-level features. IEEE Trans Consum Electron 58(2):612–618. https://doi.org/10.1109/TCE.2012.6227467
11. Gundabatini DS, Bindu Sneha R, Vamsi M, Yagna Yaswanth N, Krishna Vamsi K (2023) Music recommendation system using machine learning. Int J Innov Eng Manag Res, 12, 04
12. Sharma H et al (2020) A new model for emotion prediction in music. 2020 6th International conference on signal processing and communication (ICSC). IEEE
13. Li Y, Dong M, Kothari R (2005) Classifiability-based omnivariate decision trees. IEEE Trans Neural Netw 16(6):1547–1560
14. Jaishankar B, Anitha R, Shadrach FD, Sivarathinabala M, Balamurugan V (2023) Music genre classification using african buffalo optimization. Comput Syst Sci Eng, 44(2)
15. Lee S, Jeong H, Ko H (2021) Classical music specific mood automatic recognition model proposal. Electronics 10(20):2489
16. Ulleri P et al (2021) Music recommendation system based on emotion. 2021 12th International conference on computing communication and networking technologies (ICCCNT). IEEE
17. Joshi D et al (2023) Song recommendation based on user's activity using ensemble learning and clustering. ITM Web of conferences, EDP Sciences. Vol 56
18. Wang Y et al (2020) XgBoosted neighbor referring in low-duty-cycle wireless sensor networks. IEEE Int Things J 8(5):3446–3461
19. Wang T, Bian Y, Zhang Y, Hou X (2023) Classification of earthquakes, explosions and mining-induced earthquakes based on XGBoost algorithm. Comput Geosci 170:105242
20. Bhareti K et al (2020) A literature review of recommendation systems, 2020 IEEE International conference for innovation in technology (INOCON). IEEE
21. Ayata D, Yaslan Y, Kamasak ME (2018) Emotion based music recommendation system using wearable physiological sensors. IEEE Trans Consum Electron 64(2):196–203

22. Khanvilkar G, Vora D (2019) Smart recommendation system based on product reviews using random forest. 2019 International conference on nascent technologies in engineering (ICNTE). IEEE
23. Hariharan A, Adam MT (2015) Blended emotion detection for decision support. IEEE Trans Human-Mach Syst 45(4):510–517
24. Zhang Y et al (2021) Cped: a chinese positive emotion database for emotion elicitation and analysis. IEEE Trans Affect Comput 14(2):1417–1430
25. Huang Y et al (2021) Sentiment classification of crowdsourcing participants' reviews text based on LDA topic model. IEEE Access 9:108131–108143
26. Subramanian B et al (2022) Digital twin model: a real-time emotion recognition system for personalized healthcare. IEEE Access 10:81155–81165
27. Abd El Meguid MK, Levine MD (2014) Fully automated recognition of spontaneous facial expressions in videos using random forest classifiers. IEEE Trans Affect Comput 5(2):141–154
28. Xu L et al (2021) Predicting the preference for sad music: the role of gender, personality, and audio features. IEEE Access 9:92952–92963
29. Ghosh D, Cabrera J (2021) Enriched random forest for high dimensional genomic data. IEEE/ACM Trans Comput Biol Bioinf 19(5):2817–2828
30. Zheng Y (2023) Network traffic classification in the visual communication of music based on support vector machine. Int J Syst Assur Eng Manag, 1–11
31. Velankar M, Kulkarni P (2023) Music feature extraction for machine learning. In: Computer assisted music and dramatics: possibilities and challenges. Singapore: Springer Nature Singapore. 59–70
32. Revathy VR, Pillai AS (2022) Multi-class classification of song emotions using Machine learning, 2022 2nd International conference on advance computing and innovative technologies in engineering (ICACITE). IEEE
33. Shepstone SE, Tan ZH, Jensen SH (2016) Audio-based granularity-adapted emotion classification. IEEE Trans Affect Comput 92:176–190
34. Rohani DA et al (2020) Recommending activities for mental health and well-being: insights from two user studies. IEEE Trans Emerg Top Comput 93:1183–1193
35. Zhao G et al (2017) Emotion analysis for personality inference from EEG signals. IEEE Trans Affect Comput 93:362–371
36. Aljuhani RH, Alshutayri A, Alahdal S (2021) Arabic speech emotion recognition from Saudi dialect corpus. IEEE Access 7(9):127081–127085
37. Prabhakar SK, Lee SW (2023) Holistic approaches to music genre classification using efficient transfer and deep learning techniques. Expert Syst Appl 211:118636
38. Michael N, Electrical and computer engineering design handbook—an introduction to electrical and computer engineering and product design by TUFTS ECE Students. Music mood classification https://sites.tufts.edu/eeseniordesignhandbook/author/mnuzzo02/

Computational Models and Neural Insights in Music Neuroscience

Satchithananthi Aruljothi and G. Malu

Abstract Understanding the intricate dynamics between the human brain and music is a formidable yet captivating pursuit in contemporary neuroscience. This comprehensive review delves into the art and science of music, exploring cutting-edge models and technologies used to decode the intricacies of music perception and production in the human brain. It highlights Rule-Based Interactions, Agent-Based Models, Generative Models, Reinforcement Learning Models, Neural Oscillation Models, Graph Theory, and Statistical Models for Connectivity Analysis. These models offer insights into the neural mechanisms that underlie music cognition, from rule-based interactions to neural oscillation patterns. Music neuroscience integrates these models to unravel the complex relationship between music and the brain, bringing us closer to understanding the mysteries of this universal art form.

Keywords Music neuroscience · Computational modeling · Brain–computer interfaces · Neural oscillations · Generative models · Reinforcement learning · Cognitive processes

1 Introduction

Music, often regarded as a product of cultural acquisition and a profound art form, engages in a distinctive interplay with the human brain that transcends cultural boundaries. While music creation typically demands extensive formal training, music perception represents a nearly universal and innate human phenomenon that emerges early in life, contingent on a conducive environment [1]. This perceptual process is fundamentally rooted in biological foundations, rendering it a captivating subject of inquiry. Computational models play a pivotal role in unraveling the intricacies of this interrelationship between music and the brain. These models facilitate our comprehension of how neural systems handle the perception, processing, and generation of music, disentangling key attributes such as pitch, note arrangement, and neural reactions to consonance and dissonance [2].

S. Aruljothi (✉) · G. Malu
Digital University Kerala, Trivandrum, Kerala 695317, India
e-mail: satchithananthi.a@duk.ac.in

S. M. Thampi et al. (eds.), *Fifth International Conference on Computing and Network Communications*, Lecture Notes in Electrical Engineering 1221,
https://doi.org/10.1007/978-981-97-4711-5_10

Exploring the multifaceted landscape of music perception and cognition reveals gaps in understanding the intricate processes within the human brain across tonal, rhythmic, emotional, and cognitive dimensions. While music's universal appeal is evident, the specific neural mechanisms involved in its processing remain elusive. Computational models offer a promising avenue to untangle these complexities, providing insights into the underlying neural foundations of musical perception and creation. While existing research has delved into facets like emotional responses, neural correlates of musical experiences, and predictive coding paradigms, a comprehensive synthesis of diverse computational models from rule-based interactions to statistical frameworks remains fragmented within the literature. This highlights the need for a unified exploration to bridge these gaps and offer a more holistic understanding of music's neural underpinnings [3].

Moreover, while individual studies have explored specific computational models in music neuroscience, a comprehensive synthesis that amalgamates diverse computational tools and their applications in this domain is limited. This review seeks to bridge these gaps by providing a holistic overview of different computational approaches. It aims to fill the void by offering a unified narrative that elucidates how these models collectively unravel the intricate interplay between music and the brain. The investigation dives into unexplored territories by interlinking models like Neural Oscillation Models, and Graph Theory, fostering a deeper understanding of how these computational frameworks collectively elucidate the nuanced processing of music in the brain. Ultimately, this study endeavors to create a comprehensive roadmap that not only consolidates existing knowledge but also paves the way for future research avenues in music neuroscience.

The models also hold potential for clinical applications, particularly in personalized music therapy for neurological and psychiatric conditions, leveraging the therapeutic capabilities of music [4]. Moreover, within the domain of brain–computer interfaces (BCIs), computational models facilitate real-time neurofeedback during music consumption to regulate emotional responses.

The exploration commences with an examination of Rule-Based Interactions, wherein we scrutinize the intricate dynamics of cognitive processes interacting and mutually modulating during music perception, drawing insights from various studies within the existing literature. Transitioning to Generative Models, we delve into the functioning of Variational Autoencoders (VAEs) and Generative Adversarial Networks (GANs) in the realm of music composition. Reinforcement Learning Models furnish a perspective on how musical rewards influence learning and decision-making processes within the brain, as documented in pertinent literature. Neural Oscillation Models elucidate rhythmic brain activity and its synchronization with musical rhythms.

Furthermore, the analysis extends to Graph Theory and Network Models, facilitating the investigation of functional brain networks during musical experiences, unveiling patterns of connectivity and information flow. Finally, through a review of pertinent research in the field, we explore Statistical Models for Connectivity Analysis, which includes Dynamic Causal Modeling (DCM) and Granger Causality. These models assist in inferring directional interactions between brain regions during music processing. Our objective is to shed light on the contributions of these computational tools to our understanding of the intricate interplay between the human brain and music.

2 Modular Framework for Music Processing

A fundamental cognitive model of music processing that highlights various elements of musical processing was shown Peretz and Coltheart [1]. A modular framework for music processing is depicted. Each box symbolizes a processing element, while arrows indicate the channels of information exchange between these elements. Neurological abnormalities can harm a processing element (box) or disrupt information flow between two boxes. The components explicitly tailored to music are marked in green, while those potentially shared with other domains are shown in blue. Additionally, three components-rhythm analysis, meter analysis, and emotion expression analysis are uniquely identified (in italics) but remain uncertain in their exclusive association with music. This model of music processing illustrates two parallel, semi-independent subsystems for pitch and temporal content. The pitch-focused pathway encodes melodic contours and tonal functions of pitch intervals, while the temporal pathway represents the metric and rhythmic structure. It involves 'rhythm analysis' for durational segmentation and 'meter analysis' for detecting temporal regularity, defining strong and weak beats. The outputs from these pathways influence the 'musical lexicon,' containing representations of familiar musical phrases. Recognition depends on task requirements; for singing, the melody pairs with lyrics in the phonological lexicon, while other tasks access 'associative memories.' Simultaneously, perceptual modules feed into an 'emotion expression analysis' component, allowing emotional responses to music, with features like mode and tempo driving this emotional pathway. This is shown in Fig. 1. The model raises questions about the component's specificity to music in emotional processing and suggests further exploration of the neural mechanisms involved in music processing.

In summary, this modular architecture for music processing incorporates several component modules and pathways of information flow. It derives from the detailed study of brain-damaged patients exhibiting selective impairments or preservations in musical abilities. The inclusion of new output modules stems from neurological patient studies, providing a foundation for investigating the neural underpinnings of music processing.

3 Computational Approaches in Modeling Brain–Music Interactions

Computational models have emerged as invaluable tools to shed light on how our neural systems respond to and interact with musical stimuli. This section delves into

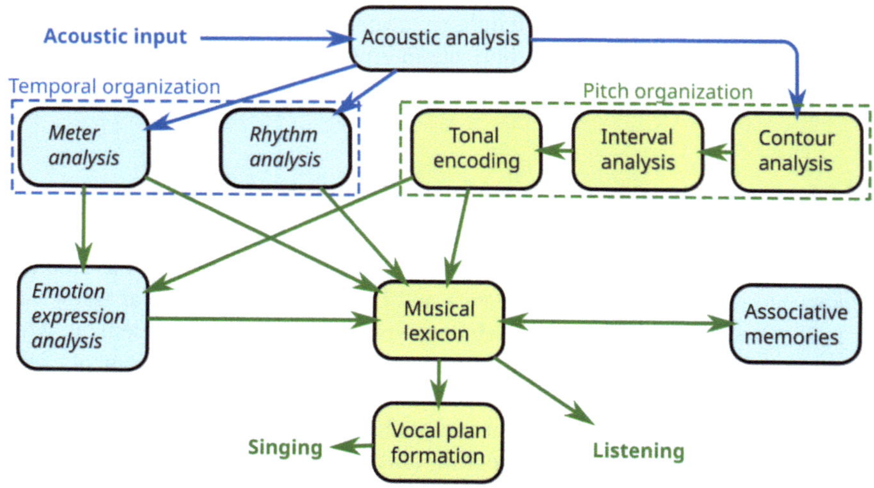

Fig. 1 Modified Musical cognition model by Peretz and Coltheart [1]. The model depicts a modular framework for music processing. Boxes represent processing elements, arrows denote information exchange. Music-specific components are in green, potentially shared elements in blue

various computational approaches employed to model brain–music interactions, elucidating their significance in music neuroscience research. A network of specialized areas in the human brain coordinates the processing and enjoyment of music. A visual representation illustrating the sequence of events, commencing with the auditory perception of music, followed by the brain's processing of the musical signal, and culminating in its transformation into a computational model is shown in Fig. 2. The different brain regions involved in musical experience are represented is also represented in the same figure. The superior temporal gyrus houses the primary auditory cortex, which is the entrance point for auditory information and is in charge of the first decoding of sound waves into basic components like pitch and timbre [5]. As musical intricacies deepen, auditory association areas, spanning the temporal and frontal lobes, come into play. These regions process higher-level features such as harmony, melody, and rhythm, allowing for the recognition of complex musical patterns [6]. The limbic system, including the amygdala and hippocampus, contributes significantly to the emotional resonance of music. It underlies the profound ability of music to evoke powerful emotions and memories [7]. Simultaneously, the prefrontal cortex, known for its role in executive functions, contributes to the cognitive aspects of musical engagement, including attention, anticipation, and decision-making [8]. Lastly, the cerebellum, often associated with motor coordination, also plays a crucial role in rhythmic processing, facilitating synchronization with music's temporal patterns [9]. These areas work together to weave the complex web of musical perception and experience.

The Venn diagram in Fig. 2 visually represents the dynamic interplay of computational, cognitive, and analytical models within the realm of music neuroscience. In

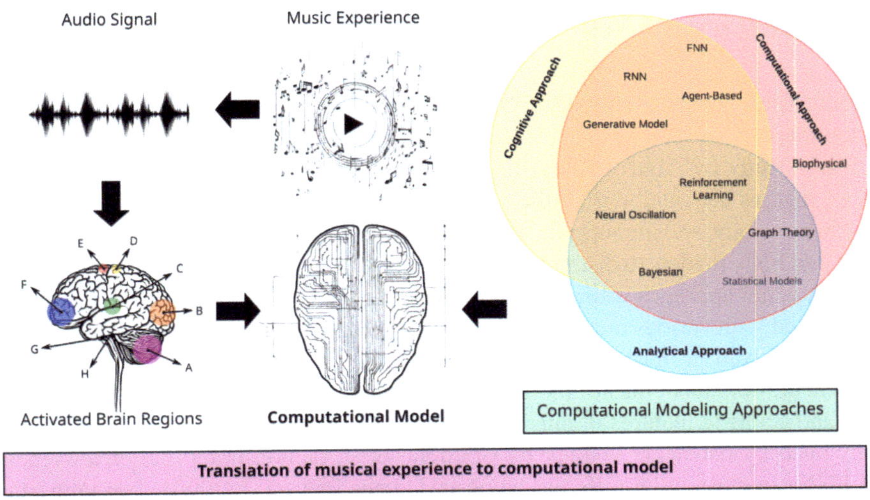

Fig. 2 Schematic representation of the flow of process modeling brain–music interactions: This diagram illustrates the sequence of steps taken by musical perception as it traverses the human brain, beginning with an audio signal that embodies various musical elements. Important brain areas linked to various facets of music processing are indicated: The brain regions responsible for perceiving sound are the (A) Auditory cortex for sound perception; (B) Cerebellum for rhythm and coordination; (C) Somatosensory cortex for the tactile response to music; (D) Motor cortex for body movement coordination; (E) Prefrontal cortex for higher-order cognitive processing; (F) Amygdala for emotional response; (G) Hippocampus for memory integration. This representation is followed by the integration of a computational approach (right) aimed at deciphering the intricate neural mechanisms underlying music cognition

the diagram, the three primary circles labeled "Computational approach," "Cognitive approach," and "Analytical approach" represent these modeling approaches. The intersections between these circles emphasize the collaborative and interdisciplinary nature of music neuroscience. These shared areas signify that the models may not be confined to a single category; they often transcend the boundaries, enriching research by combining elements from different approaches.

For instance, models like Recurrent Neural Networks (RNNs) may not directly fall under the "Analytical approach" category since they primarily focus on modeling and processing data rather than providing specific analytical methods. Analytical frameworks typically refer to tools and methods used for data analysis, whereas RNNs are primarily employed for modeling and generating data. This overlap between categories demonstrates the versatility and adaptability of these models in addressing complex questions related to music and the brain.

This figure serves as a visual representation of the flexible and collaborative nature of music neuroscience research, where computational, cognitive, and analytical models come together to provide a comprehensive framework for exploring the intricate relationship between the human brain and music.

3.1 Feedforward Neural Network (FNN)

Feedforward Neural Networks (FNNs) have found valuable applications in music brain mapping, aiding in the analysis of complex neural responses to musical stimuli. In a study by Alluri et al. [10], FNNs were employed to classify and map the brain regions activated during the processing of different musical features, such as rhythm, melody, and timbre [10]. The authors used magnetoencephalography (MEG) data and FNN models to distinguish between brain responses to different music conditions. This approach allowed for the identification of distinct neural signatures associated with various musical elements, shedding light on the intricate neural processes involved in music perception and cognition. Furthermore, FNNs have been used in the domain of music emotion recognition. Through the analysis of physiological and acoustic features in music, FNN models can predict the emotional content perceived by listeners. A study by Coutinho et al. demonstrated the effectiveness of FNNs in classifying different emotional states induced by music, revealing the potential of this technology to map the neural correlates of music-induced emotions [11].

In addition to mapping music-induced emotions, FNNs have also been employed to study individual differences in music perception and cognition. In a study by Omigie et al. [12], FNN models were used to predict listeners' perceptual judgments of musical intervals based on their neural responses. By considering the variations in neural responses across participants, this approach provides insights into how individual differences in music processing are reflected in the brain. FNNs have thus proven to be a versatile tool in music brain mapping, enabling researchers to uncover the neural underpinnings of various aspects of music perception and cognition. These findings contribute to a deeper understanding of the complex relationship between music and the human brain.

3.2 Recurrent Neural Network (RNN)

Recurrent Neural Networks (RNNs) have gained prominence in the realm of music-related tasks, particularly in the field of music generation. These neural network architectures have demonstrated their efficacy in creating novel compositions. In music generation tasks, RNNs are highly regarded for their ability to capture temporal dependencies and generate sequences of musical notes that exhibit coherence. A seminal work by Eck and Schmidhuber [13] introduced the concept of using RNNs for music generation. The authors demonstrated the capacity of RNNs to produce melodious sequences by learning patterns from existing music datasets. This research paved the way for subsequent studies exploring RNNs in various music composition genres, from classical to electronic music. The potential of RNNs to capture musical nuances and generate compositions with both short-term and long-term structure has made them invaluable tools for composers and musicians alike.

Additionally, there is a growing interest in employing RNN-based models in music recommendation systems and personalized music playlists. For instance, an investigation by [14] showcased how RNNs, particularly the Long Short-Term Memory (LSTM) variant, can be utilized for music recommendation. RNN-based models analyze listening behaviors and contextual cues, refining music suggestions for a personalized experience. Leveraging sequential data handling, they unveil intricate musical connections within user preferences, significantly influencing music recommendation systems, elevating accuracy, and enhancing creativity in music generation.

Utilizing an Emotion Recognition Neural Network (ERNN), this study by Khosrowabadi et al. [15] developed a six-layer feedforward neural network replicating biological emotion perception mechanisms in the brain. Leveraging spectral filtering, short-term memory, coherence estimation, feature selection, and radial basis function learning, the ERNN discerned emotions from EEG signals. EEG data collected from 57 healthy subjects exposed to varied valence and arousal stimuli were assessed via the Self-Assessment Manikin (SAM) questionnaire. Recorded from eight channels using the 10/20 system, the EEG signals underwent comparison with diverse feature extraction methods and feedforward learning algorithms. Demonstrating robust performance, ERNN achieved 86.29% accuracy in valence level recognition and 83.19% in arousal level recognition through fivefold cross-validation. Additionally, ERNN discerned the most effective EEG channels and frequencies for precise emotion recognition

3.3 Bayesian Modeling

Bayesian models have provided valuable insights into the neural processing of music, particularly in understanding how the human brain constructs and anticipates rhythmic structures. The study by Temperley et al. explores how Bayesian models, based on probabilistic reasoning, can be employed to represent and analyze various aspects of music, including harmony, melody, rhythm, and more [16]. It focuses on how listeners make predictions about the structure of music based on the information they have encountered, emphasizing the role of expectations and surprise in musical experiences.

The study by Asadzadeh et al. [3] focuses on enhancing emotion recognition from EEG signals. They introduce a novel approach that leverages a Bayesian model to map emotional responses from scalp sensors to brain sources, employing the sLORETA method for initial source signal initialization. This Bayesian model-based approach results in significantly improved emotion recognition accuracy, demonstrating a 99.25% classification accuracy for distinguishing positive and negative emotions. Notably, the Bayesian model plays a central role in accurately localizing the sources of emotional responses within the brain, contributing to the success of the proposed method in emotion classification.

3.4 Agent-Based Models

Agent-Based Models (ABMs) stand as a robust computational framework for delving into the intricacies of music processing by simulating the dynamic interplay of various cognitive processes. This comprehensive review navigates the terrain of ABMs, offering an in-depth understanding of their mechanisms and how they unveil the interactive nature of musical experiences, all while grounding the discussion in relevant scientific literature. Agent-based models (ABMs) have proven instrumental in music neuroscience for delving into the intricacies of cognitive processes during music perception. ABMs employ digital agents as surrogates, representing cognitive facets such as perception, memory, emotion, and decision-making, mirroring the complex web of interactions that occur during musical experiences [17]. Governed by rule-based interactions, these models simulate how cognitive processes interact within the music perception environment, providing a dynamic lens into the interplay of mental faculties [18]. A remarkable feature of ABMs is their capacity to produce emergent behavior, where complex patterns arise as agents interact over time, offering insights into the dynamic and evolving nature of music cognition [19]. Researchers analyze the outcomes of these simulations to unravel the intricate dynamics of musical experiences.

A linear regression model was used in an experimental study by Kirke et al. [20] aimed to gauge emotional states in musical performances, leveraging key features: loudness, tempo, articulation, and tempo variability. Trained on 8192 MIDI files reflecting Director Musices' rules for happy, sad, and angry emotions, the model underwent testing across five randomized sets of 8192 MIDI files, achieving an average accuracy of 68% in identifying intended emotional states. However, this accuracy trailed human recognition rates (approximately 77%) and notably favored happiness, particularly influenced by tempo variations, indicating room for refinement in emotional recognition models.

3.5 Generative Models for Music Composition

Generative Models in the context of music have shown promise in unraveling the intricacies of musical composition and creativity. One notable study by Welikala et al. (2016) introduced a novel approach that employed Variational Autoencoders (VAEs) for modeling musical data, enabling the generation of highly structured and creative compositions [21]. Similarly, the work of Yang et al. [22] demonstrated the use of GANs to generate music that mimics the style of renowned composers, illustrating the potential of these models to capture and replicate the artistic nuances of musical creation. These generative approaches not only contribute to music composition but also have implications for understanding how the human brain processes and appreciates complex musical structures.

Table 1 Comparison of surprisal models

Model	R^2	Log-likelihood	AIC	BIC
IDyOM	0.240	– 1869.0	3744	3761
Audio-diffusion	0.062	– 1860.5	3727	3744

Furthermore, Generative Models have been employed to explore the emotional aspects of music perception. A study by van den Oord et al. [14] showcased the effectiveness of WaveNet, a generative model based on deep neural networks, in capturing the emotional content of music through audio synthesis. This approach allows for the generation of music that can elicit specific emotional responses, shedding light on the emotional dimensions of music processing in the brain. These studies collectively emphasize the role of Generative Models in music research, from composition and style replication to the emotional impact of music on listeners, offering valuable insights into the neural underpinnings of musical experiences.

The study by Masclef et al. [23] introduces a novel music expectancy model employing deep generative models, specifically diffusion models, to gauge musical sequence likelihood and assess surprise or expectancy. Testing the Wundt effect, the authors contrast surprisal values derived from the diffusion model with human preference ratings for musical pieces. An anticipated outcome is an inverted U-shaped curve, indicating a preference for intermediate surprisal levels. Comparative analysis with the Information Dynamics of Music (IDyOM) model-a melodic structure probabilistic model using hand-crafted features-reveals that the diffusion model achieves superior likelihood and AIC/BIC values, suggesting its enhanced data fitting capability over IDyOM1 (Table 1).

3.6 Reinforcement Learning Models

The integration of Reinforcement Learning (RL) Models into music neuroscience offers a powerful lens for understanding the profound impact of musical rewards on the brain's learning and decision-making processes [24]. In this context, the agent (representing the brain) dynamically navigates the musical landscape, taking actions aimed at maximizing a cumulative reward signal that reflects the pleasure derived from music [25]. These actions are guided by learning policies, which represent the brain's strategies for selecting actions in response to different musical stimuli, all with the ultimate goal of maximizing expected musical rewards. Gold et al. [26] investigated how music influences the process of reinforcement learning. Their findings reveal that the pleasurable quality of music can significantly affect reinforcement learning in listeners. This suggests that the emotional and cognitive responses induced by music can modulate one's ability to learn from reinforcement, shedding light on the complex interplay between music, emotions, and cognitive processes. The study underscores the intricate relationship between music and cognitive functions, highlighting that music can have a meaningful impact on learning processes. These findings provide valuable insights into the potential therapeutic and educational applications of music, emphasizing its role in enhancing or modulating reinforcement learning based on the individual's emotional responses to the music they are exposed to.

The study on physiological responses to music segments involved a listening test with 50 participants across three classical compositions, dissected into happy and tense segments. Analyzing changes in blood pressure, heart rate, and oxygen saturation, the study revealed increased oxygen saturation and diastolic blood pressure in happy segments compared to tense ones, while heart rate showed no variance. Composer variance within segments remained statistically insignificant.

In a study by Phatnani et al. [27] using reinforcement learning model of musical expectancy, researchers constructed a dopamine-based model employing ϵ-greedy and Upper Confidence Bound algorithms to predict pitch and note sequences across 15 compositions. Results indicated 22.48% accuracy in pitch prediction, 64.17% in note prediction, and 73.84% in simultaneous prediction. Notably, accuracy demonstrated an inverse relationship with the variance of edge distributions in the compositions' directed graphs.

3.7 Graph Theory and Network Models

Graph Theory and Network Models have significantly contributed to the understanding of brain mechanisms in music processing. Studies applying these models have unveiled intricate patterns of functional and structural connectivity related to music perception and cognition. Functional connectivity networks have shown that musical stimuli engage a distributed set of brain regions with specialized roles. This includes nodes dedicated to pitch processing, rhythm, emotional response, and memory [10]. The use of graph theory metrics, such as clustering coefficient and network hubs, has helped characterize the efficiency and integration of these networks, providing insights into the orchestration of cognitive functions during music processing [28, 29].

Moreover, the application of Network Models to structural brain data has revealed associations between individual differences in musical abilities and variations in network topology. Such models enable us to investigate the relationship between music and general cognitive abilities [4]. These findings represent a substantial step toward uncovering the neural underpinnings of music perception and the structural and functional networks that enable our profound engagement with music. Research in this domain continues to expand our understanding of how the human brain interacts with the world of music, shedding light on the intricate interplay between neural systems and music perception, ultimately enhancing our appreciation of this universal human phenomenon [30, 31].

A study by Aydin et al. [32] explores a study scrutinizing brain connectivity during music-induced emotions. Utilizing EEG data from 30 individuals exposed to 40 music pieces evoking fear, anger, happiness, and sadness, the investigation deployed four connectivity estimators and six network measures across various frequency bands and time segments. Partial Directed Coherence (PDC) in the gamma band emerged as pivotal, effectively discerning fear and anger-traditionally akin in arousal-valence models-showing high modularity for fear and low transitivity for anger. Similarly, PDC in gamma bands aptly differentiated happiness (associated with low clustering coefficients (CC), local efficiency (LE), and assortativity) and sadness. Long-Short Term Memory Networks (LSTM) networks validated these findings, achieving 96% accuracy for happiness versus sadness and 91.79% for fear versus anger classification.

3.8 Statistical Models for Connectivity Analysis

Statistical models like Dynamic Causal Modeling (DCM) and Granger Causality decode intricate brain interactions in music processing. DCM, employing candidate models, dissects brain region interactions during tasks like music perception, enabling inference of causal influences between regions based on brain activity data [33]. On the other hand, Granger Causality evaluates causal links in brain activity during music perception, uncovering influential brain regions and directional influences, enriching the understanding of music's temporal dynamics [34].

These statistical models for connectivity analysis not only provide insights into how different brain regions interact but also contribute significantly to our comprehension of the neural mechanisms underlying music cognition. DCM has been employed in research to explore how the auditory cortex, hippocampus, and prefrontal cortex collaboratively process rhythm in music, highlighting their distinct roles in temporal encoding, rhythmic memory, and cognitive integration [33]. Meanwhile, studies employing Granger Causality have illuminated the cross-modal integration involved in music perception, showing how the auditory and visual cortices interact to enhance the processing of both auditory and visual information during music experiences [34]. Furthermore, DCM has been utilized to investigate the motor coordination of brain regions during music production, revealing how the motor cortex, premotor cortex, and supplementary motor area work together to control muscle movement and execute intricate sequences required for creating music [35]. These studies underscore the utility of statistical models in exploring the multifaceted landscape of brain functions entwined with music, promising continued revelations about the neural underpinnings of musical perception and production.

The study by Farahani et al. [36] employed regression dynamic causal modeling (rDCM) to assess connectivity among 18 brain areas during varied emotional states triggered by an audio-visual stimulus. rDCM, operating in the frequency domain via a linear model, offers computational efficiency and enhanced precision compared to conventional DCM. Their model encompassed auditory, emotional processing, and

Table 2 Brain regions and cognitive functions associated with each oscillation frequency

Frequency band (Hz)	Brain regions	Associated cognitive functions
Delta (0.5–4)	Thalamus, hippocampus, cortex	Deep sleep, unconsciousness
Theta (4–8)	Hippocampus, prefrontal cortex, cingulate cortex	Memory, attention, learning
Alpha (8–12)	Parietal cortex, occipital cortex	Relaxation, cognitive processing
Beta (12–30)	Frontal cortex, parietal cortex, sensorimotor cortex	Attention, processing, movement
Gamma (30–80)	Sensory cortices, association cortices	Sensory processing, cognitive binding

visual regions, integrating input across auditory areas with assumed full interconnections. Notably, distinct connectivity patterns emerged, linking emotion to auditory, insular, and hippocampal regions. A non-parametric permutation test revealed significant differences in connectivity, notably between emotions like happiness and fear, happiness and love, and happiness and anger, providing nuanced insights into emotion-specific brain network dynamics.

3.9 *Neural Oscillation Models in Human Brain Mapping*

Neural Oscillation Models focus on the rhythmic oscillatory activity in the brain and its relationship to musical rhythms and timing. They represent a fundamental theoretical framework for understanding the rhythmic and synchronized electrical activity of neurons in the brain. These cyclical patterns of neural firing play a pivotal role in various cognitive processes and have substantial applications in the fields of human brain mapping, music, and neuroscience.

The investigation of neural oscillations is a cornerstone of human brain mapping, offering valuable insights into the brain's functional connectivity and network dynamics. Techniques such as electroencephalography (EEG) and magnetoencephalography (MEG) enable the precise measurement of oscillatory activity. Neural oscillations in the brain are associated with a diverse range of cognitive functions. For example, alpha oscillations (8–12 Hz) have been linked to states of relaxation and are the focus of studies exploring meditation and mindfulness [37] (Table 2).

Neural oscillations are rhythmic fluctuations in the electrical activity of the brain. They are thought to play an essential role in a variety of cognitive functions, including memory, attention, and perception.

In a study by Chen et al. [38] investigated the neural representation of auditory duration in working memory (WM) using electroencephalogram (EEG) recordings.

EEG analyses revealed that the maintenance of auditory duration in WM is associated with alpha band amplitudes. Specifically, the alpha band amplitudes during 3-s and 4-s auditory duration conditions were lower than during the 1-s and 2-s conditions, and the alpha amplitude positively correlated with accuracy in the 2-s condition. This suggests that the neural representation of auditory duration is segmented, with a critical threshold point of 2 s. The study's results emphasized the involvement of the alpha band in auditory duration maintenance in WM and indicated distinct internal representations of auditory durations below and above the critical threshold point. The findings also suggested differences in the neural representations of auditory and visual durations and highlighted the role of the alpha band in successful maintenance of auditory duration information in WM. The study effectively addresses the neural representation of auditory duration in WM and highlights the significance of modality-specific representations of temporal information.

4 Brain Signal Recording Techniques for Music Studies

Various brain recording methods complement computational models in understanding music cognition. EEG aids Reinforcement Learning Models and Rule-Based Interactions, fMRI suits Neural Oscillation Models, Graph Theory, and Statistical Models. MEG benefits Agent-Based Models, PET aligns with Generative Models, and Connectivity Analysis Techniques support related computational models. Biometric Measures enhance Rule-Based Interactions, and Eye-Tracking contributes to visual-musical interaction exploration. These techniques along with the computational model used in are tabulated in Table 3. Invasive techniques such as intracranial electroencephalography (iEEG), microelectrode recordings, and functional electrical stimulation (FES) have played a crucial role in unraveling the neural intricacies of music perception and production. iEEG studies have offered insights into the neural processing of musical elements like tonal harmony and rhythm, shedding light on how the brain deciphers complex harmonic structures and tracks musical beats [39, 40]. Microelectrode recordings have delved into the neural representation of music-evoked emotions, revealing the involvement of specific neurons in emotional reactions to music [41]. These recordings have also elucidated how the auditory cortex encodes fundamental musical features like pitch and melody. Meanwhile, FES investigations have explored the impact of electrical stimulation in motor brain regions on musical motor skills and coordination, providing valuable knowledge for enhancing musical performance and motor abilities, particularly in instrumental music [42]. While these invasive techniques offer profound insights, their application is limited due to ethical concerns, invasiveness, and the need for specialized equipment and expertise.

Table 3 Recording techniques and corresponding computational models

Recording technique	Computational models
EEG	Reinforcement learning models, rule-based interactions
fMRI	Neural oscillation models, graph theory, statistical models for connectivity analysis
MEG	Agent-based models
PET	Generative models
Connectivity analysis techniques	*Computational models for connectivity analysis*
Biometric measures	Rule-based interactions
Eye-tracking	Reinforcement learning models, rule-based interactions

5 Limitations

As with any scientific endeavor, the application of computational models in music neuroscience has limitations. One notable challenge lies in the quality and representativeness of the data used to train and test these models. Biases or insufficiencies in the data can lead to model inaccuracies and hinder generalizability [43]. The inherent complexity of the human brain and music perception presents another formidable obstacle. Models, by their nature, simplify these intricate processes, potentially overlooking crucial nuances. Additionally, the computational resources required for training and running these models can be substantial, posing barriers to entry for researchers with limited access to high-performance computing facilities [44]. Collaborative efforts among experts from diverse disciplines, spanning neuroscience, psychology, computer science, and musicology, are essential for successful interdisciplinary research. Moreover, ethical considerations, particularly concerning neural data acquisition and privacy, must be diligently addressed to ensure responsible research practices [35].

6 Conclusion

In this review we have explored a wide range of models and technologies that provide distinctive insights into the neurological mechanisms underlying music cognition, the models have been tabulated below (see Table 4). Each method advances our knowledge of the significant effects of music on the brain, from rule-based interactions that clarify the dynamics of cognitive processes to generative models that reveal the meeting point of computational creativity and brain activity.

Bayesian modeling emerges as a key tool in music neuroscience, decoding music perception's neural underpinnings-unveiling emotional processing, feature decod-

Table 4 Summary and applications of the computational models in music neuroscience

Computational model	Description and key aspects	Applications in music neuroscience
Feedforward neural network (FNN)	Sequential processing of input data, weighted connections, and activation functions	Modeling brain perception and processing of musical features. Recognizing patterns like melody, rhythm, and harmony in music [12]
Recurrent neural network (RNN)	Feedback connections, modeling sequences, dynamic temporal dynamics in music perception and generation	Simulating brain processing of music over time, predicting next notes, capturing temporal aspects of music perception [15]
Deep learning architectures	Deep generative models like VAEs and GANs for music generation, learning patterns and structures in music data	Music composition, emotion recognition, genre classification, capturing underlying musical patterns in data [23]
Dynamic causal modeling (DCM)	Quantifying interactions between brain regions, data collection from fMRI/EEG, parameter estimation, model comparison	Investigating dynamic brain interactions during music perception, melody, rhythm, and emotional responses [45]
Bayesian modeling	Probabilistic models connecting brain states and musical features, parameter estimation, inference, uncertainty quantification	Inferring brain states and their relation to musical features like emotions, decoding brain activity related to music [46]
Agent-based models	Simulation of cognitive processes, rule-based interactions, emergent behavior, application in music perception	Investigating cognitive interactions in music perception (e.g., attention, emotion, memory) and their dynamics [47]
Generative models for music composition	Latent space representation, training on data, sampling and creativity, brain signal integration	Understanding the relationship between neural activity patterns and creative music composition [21]
Reinforcement learning models	Agent-environment interaction, learning policies, musical rewards, neuroscientific insights	Investigating how musical rewards influence learning, decision-making, and music preferences in the brain [48]

ing, and expertise signatures. Its probabilistic approach systematically uncovers neural mechanisms governing music perception. ABMs dynamically explore cognitive aspects in music, dissecting attention, emotion, and memory's role. This computational lens unravels music cognition's complexity, revealing the neural tapestry behind musical experiences. Generative models, coupled with brain signals, elucidate how neural dynamics drive musical creativity, bridging neuroscience and musical composition. RL models delve into neural bases of musical preference, decoding reward-driven musical behaviors, offering insights into musical choices' neural roots. Graph theory outlines core brain networks crucial for music integration, showcasing expertise-induced dynamics. Neural oscillations synchronize across brain regions, notably in rhythm processing, evident in functional MRI studies, shedding light on music perception's neural orchestration and broader neuroscientific contexts [49–51].

Computational approaches have played a pivotal role in unraveling the mysteries of how the human brain perceives, processes, and interacts with music. From neural network models to graph theory and statistical connectivity analysis, these models offer diverse perspectives on music neuroscience. They have provided profound insights into the neural mechanisms underlying music perception, creative composition, decision-making, and reward-driven learning. While facing limitations, these computational models continue to drive innovation in the study of brain–music interactions. The future of music neuroscience holds the promise of further advancements in computational modeling, leading to a deeper understanding of the cognitive, emotional, and neural dynamics of music in the human experience.

References

1. Peretz I, Coltheart M (2003) Modularity of music processing. Nat Neurosci 6(7):688–691
2. Kaliakatsos-Papakostas M, Floros A, Vrahatis MN (2020) Artificial intelligence methods for music generation: a review and future perspectives. Nat-Inspired Comput Swarm Intell, pp 217–245
3. Asadzadeh S, Yousefi Rezaii T, Beheshti S, Meshgini S (2022) Accurate emotion recognition using Bayesian model based EEG sources as dynamic graph convolutional neural network nodes. Sci Rep 12(1):10282
4. Loui P, Li HC, Hohmann A, Schlaug G (2011) Enhanced cortical connectivity in absolute pitch musicians: a model for local hyperconnectivity. J cogn Neurosci 23(4):1015–1026
5. Olszewska AM, Gaca M, Herman AM, Jednoróg K, Marchewka A (2021) How musical training shapes the adult brain: predispositions and neuroplasticity. Front Neurosci 15:630829
6. Zatorre RJ, Peretz I (2005) Music, the food of the brain. Trends Cogn Sci 9(1):31–35
7. Pant U, Frishkopf M, Park T, Norris CM, Papathanassoglou E (2022) A neurobiological framework for the therapeutic potential of music and sound interventions for post-traumatic stress symptoms in critical illness survivors. Int J Environ Res Public Health 19(5):3113
8. Cahart M-S, Amad A, Draper SB, Lowry RG, Marino L, Carey C, Ginestet CE, Smith MS, Williams SC (2022) The effect of learning to drum on behavior and brain function in autistic adolescents. Proc Nat Acad Sci 119(23):e2106244119
9. Kasdan AV, Burgess AN, Pizzagalli F, Scartozzi A, Chern A, Kotz SA, Wilson SM, Gordon RL (2022) Identifying a brain network for musical rhythm: a functional neuroimaging meta-analysis and systematic review. Neurosci Biobehav Rev 136:104588
10. Alluri V, Toiviainen P, Jääskeläinen IP, Glerean E, Sams M, Brattico E (2012) Large-scale brain networks emerge from dynamic processing of musical timbre, key and rhythm. Neuroimage 59(4):3677–3689
11. Coutinho E, Cangelosi A (2011) Musical emotions: predicting second-by-second subjective feelings of emotion from low-level psychoacoustic features and physiological measurements. Emotion 11(4):921

12. Omigie D, Pearce MT, Williamson VJ, Stewart L (2013) Electrophysiological correlates of melodic processing in congenital amusia. Neuropsychologia 51(9):1749–1762
13. Eck D, Schmidhuber J (2002) Learning the long-term structure of the blues. In: International conference on artificial neural networks, Springer, pp 284–289
14. van den Oord A, Dieleman S, Zen H, Simonyan K, Vinyals O, Graves A, Kalchbrenner N, Senior A, Kavukcuoglu K (2016) Wavenet: a generative model for raw audio, arXiv preprint arXiv:1609.03499
15. Khosrowabadi R, Quek C, Ang KK, Wahab A (2013) ERNN: a biologically inspired feedforward neural network to discriminate emotion from EEG signal. IEEE Trans Neural Netw Learn Syst 25(3):609–620
16. Temperley D (2004) Bayesian models of musical structure and cognition. Musicae Sci 8(2):175–205
17. Cohen M (2017) Agent-based modeling: understanding music cognition and music behavior. Front Hum Neurosci 11:235
18. Addyman C, French RM (2012) Computational modeling in cognitive science: a manifesto for change. Top Cogn Sci 4(3):332–341
19. Jordan MI, Rumelhart DE (1992) Forward models: supervised learning with a distal teacher. Cogn Sci 16(3):307–354
20. Kirke AJ, Miranda ER (2009) Artificial social composition: a multi-agent system for composing music performances by emotional communication
21. Welikala MD, Fernando T (2021) Komposer v2: a hybrid approach to intelligent musical composition based on generative adversarial networks with a variational autoencoder. In: Proceedings of the future technologies conference (FTC) 2020, vol 1, Springer, pp 413–425
22. Yang L-C, Chou S-Y, Yang Y-H (2017) Midinet: a convolutional generative adversarial network for symbolic-domain music generation. arXiv preprint arXiv:1703.10847
23. Masclef NL, Keller TA (2023) Deep generative models of music expectation. arXiv preprint arXiv:2310.03500
24. Sutton RS, Barto AG (2018) Reinforcement learning: an introduction. MIT Press
25. Kaelbling LP, Littman ML, Cassandra AR (1996) Planning and acting in partially observable stochastic domains. Artif Intell 101(1–2):99–134
26. Gold BP, Frank MJ, Bogert B, Brattico E (2013) Pleasurable music affects reinforcement learning according to the listener. Front Psychol 4:541
27. Phatnani KS, Patil HA (2023) Modeling musical expectancy via reinforcement learning and directed graphs. Multim Tools Appl, pp 1–25
28. Bassett DS, Bullmore E (2006) Small-world brain networks. Neuroscientist 12(6):512–523
29. Rubinov M, Sporns O (2010) Complex network measures of brain connectivity: uses and interpretations. Neuroimage 52(3):1059–1069
30. Groussard M, La Joie R, Rauchs G, Landeau B, Chetelat G, Viader F, Desgranges B, Eustache F, Platel H (2010) When music and long-term memory interact: effects of musical expertise on functional and structural plasticity in the hippocampus. PloS One 5(10):e13225
31. Janata P (2009) The neural architecture of music-evoked autobiographical memories. Cereb Cortex 19(11):2579–2594
32. Aydın S, Onbaşı L (2023) Graph theoretical brain connectivity measures to investigate neural correlates of music rhythms associated with fear and anger. Cogn Neurodyn, pp 1–18
33. Friston KJ, Harrison L, Penny W (2003) Dynamic causal modelling. NeuroImage 19(4):1273–1302
34. Chomsky N, Janata P, Halleck MD, Warren ER (2003) Neural entrainment and integration of auditory and visual rhythms in music perception. Nat Neurosci 6(12):1147–1151
35. Cohen L, Penhune P, Peretz I (2007) Neural correlates of musical performance in pianists. J Neurosci 27(33):8979–8990
36. Farahani N, Fatemizadeh E, Nasrabadi AM (2019) Using RDCM method in the mixed model in order to inference effective connectivity in emotions. Front Biomed Technol
37. Klimesch W (1999) EEG alpha and theta oscillations reflect cognitive and memory performance: a review and analysis. Brain Res Rev 29(2–3):169–195

38. Yu X, Chen Y, Qiu J, Li X, Huang X (2017) Neural oscillations associated with auditory duration maintenance in working memory. Sci Rep 7(1):5695
39. Molholm S, Martinez A, Ritter W, Javitt DC, Foxe JJ (2015) The neural circuitry of pre-attentive auditory change-detection: an FMRI study of pitch and duration mismatch negativity generators. Cereb Cortex 15(5):545–551
40. Nozaradan S, Peretz I, Missal M, Mouraux A (2011) Tagging the neuronal entrainment to beat and meter. J Neurosci 31(28):10234–10240
41. Zatorre R, Salimpoor V (2013) From perception to pleasure: music and its neural substrates. Proc Nat Acad Sci 110(2):E76–E84
42. Altenmüller E, Marco-Pallarés J, Münte TF, Schneider S (2009) Neural reorganization underlies improvement in stroke-induced motor dysfunction by music-supported therapy. Ann N Y Acad Sci 1169(1):395–405
43. Massimini M, Huber R, Ferrarelli F, Hill S, Tononi G, Laureys S (2005) The sleep slow oscillation as a measure of consciousness in humans. Brain Res Bull 66(5):416–429
44. Zuo Y-F (2012) Individual variability in the brain's functional connectivity network for music processing. J Neurosci 32(39):13644–13656. e. a
45. Bastos AM et al (2015) Visual areas exert feedforward and feedback influences through distinct frequency channels. Neuron 85(2):390–401
46. Bishop CM (2006) Pattern recognition and machine learning. Springer 42(6):1265–1282
47. Maes P et al (2014) Modeling the impact of cognitive and affective processes in sound quality evaluation. IEEE Trans Affect Comput 5(2):198–210
48. Menon V, Levitin DJ (2005) The rewards of music listening: response and physiological connectivity of the mesolimbic system. NeuroImage 28(1):175–184
49. Miranda ER (2021) Handbook of artificial intelligence for music. Springer
50. Zatorre RJ, Levitin PJ (2010) Cortical connections for music processing. Ann N Y Acad Sci 1252:211–229
51. Grahn JA, Rowe JB (2009) Feeling the beat: premotor and striatal interactions in musicians and nonmusicians during beat perception. J Neurosci 29(23):7540–7548

Enhanced Road Safety Through AI-Enabled Voice Assistance

Mahit Venkat Gautam Kumpatla, Gandivalasa Keerthi Tej, Gottimukkala Sahil, and Sruthy Anand

Abstract Minor and major road accidents are a critical problem in urban and rural transportation. Enhancing road safety is one of the primary necessities in cities to ensure citizen's quality of life. This research proposes an AI-enabled Voice assistant to provide warnings to drivers to improve road safety in urban areas. The system combines deep learning models, specifically the YOLO (You Only Look Once) object detection algorithm, and an AI voice assistant using natural language processing techniques. The AI voice assistant provides timely warnings and instructions to the driver based on the detected objects and the information gathered by the Vehicular Networks. The experiment shows that the deep learning model accurately predicts objects on the road from the given video input. The work also presented the various scenarios in which voice assistance helps drivers ensure better security.

Keywords YOLO · Road safety · Voice assistance · Deep learning · Vehicular networks

1 Introduction

Road safety is essential in urban cities as pedestrians and vehicles face many risks while utilizing the roads. According to the World Health Organisation, road accidents kill around 1.3 million people each year [1]. The transportation sector takes many

---

M. V. G. Kumpatla (✉) · G. Keerthi Tej · G. Sahil
School of Computing, Amrita Vishwa Vidyapeetham, Amritapuri, India
e-mail: amenu4aie20043@am.students.amrita.edu

G. Keerthi Tej
e-mail: amenu4aie20027@am.students.amrita.edu

G. Sahil
e-mail: amenu4aie20128@am.students.amrita.edu

S. Anand
Center for Wireless Networks & Applications (WNA), Amrita Vishwa Vidyapeetham, Amritapuri, India
e-mail: sruthyanand@am.amrita.edu

S. M. Thampi et al. (eds.), *Fifth International Conference on Computing and Network Communications*, Lecture Notes in Electrical Engineering 1221,
https://doi.org/10.1007/978-981-97-4711-5_11

measures to improve road safety, such as indicators, speed limits, traffic rules, etc. [1]. However, these couldn't prevent the accidents caused by careless driving and road crossing by pedestrians. The transportation section must prioritize road safety to avoid the risk posed to road users. Many factors add to the road risk, such as inadequate sidewalks and crosswalks. This results in pedestrians being compelled to stroll on the road, putting them at risk for accidents. Other significant problems that contribute to road accidents are reckless driving, excessive speed, wrong-way driving on divided highways [2], and distracted driving, such as using mobile phones while driving. Additionally, inadequate traffic signals and poor lighting conditions at night further exacerbate the issue. These problems emphasize the necessity for new approaches to increasing road safety and have to look into IoT and AI solutions to assist drivers and pedestrians.

This work focuses on building AI solutions to assist drivers by making them aware of possible dangers. The primary goal of this work is to propose an AI voice assistant system that seamlessly integrates with a vehicular network to increase road safety. The AI voice assistant system needs to utilize technologies such as deep learning, vehicular networks, and NLP techniques to achieve the envisioned safety warnings. For this, it needs to collect real-time data from numerous sources, including roadways, traffic sensors, and vehicular networks. By this, the system will be able to analyze the driving environment continuously, recognize possible dangers and hazards, and deliver crucial real-time information for drivers to anticipate and change their actions, enabling more informed and responsible judgments. Implementing this system in the real world will enhance safety and driving experience. For instance, if a pedestrian is crossing the road, then the voice assistant should inform the driver to "drive slowly; a pedestrian is crossing the road". Thus, its primary purpose is to facilitate safer trips, warn drivers about pedestrian presence, traffic conditions, and possible dangers, and minimize the chance of accidents while promoting responsible driving behavior. Integrating an AI voice assistant with vehicle networks offers a critical connection between multiple vehicles, boosting road safety. By 2050, it is estimated that around 70% of the total population will be living in the urban regions [3] and this type of safety system is necessary to reduce accidents on roads.

The major component of the AI voice assistant system is the object detection component, which is used to identify objects effectively. The YOLO (You Only Look Once) object detection model is used for experimentation. The model can process the camera inputs and produce detected objects as the output. The output of the object detection model is being given as input to the AI voice assistant. The detected object and related information can be transferred to the neighboring vehicles using the VANET, which enables vehicle-to-vehicle communication. This real-time communication enhances vehicle data exchange, enhances situational awareness, and promotes safer driving practices. Integration of more IoT devices and sensors, along with object detection models, is required to improve road safety. Collecting data from heterogeneous nodes capable of communicating with each other [4] is necessary to realize the proposed system.

The rest of the paper is organized as follows. Section 2 discusses the literature review. Section 3 shows the need for intelligent AI driver assistance and presents different scenarios. Section 4 discusses the architecture of the proposed voice assistant. Section 5 discusses experimentation and results and Sect. 6 draws the conclusion and future work.

2 Literature Review

Many existing works on road safety focus on vehicle trajectory prediction for road safety and vehicular network applications for road safety. One of the works [5] presents a mechanism for predicting a vehicle's trajectory utilizing Long Short-Term Memory (LSTM) models in addition to spatial-temporal attention mechanisms. The suggested method surpasses existing approaches for trajectory prediction and can be used to improve autonomous vehicle safety and efficiency. Another work [6] discusses a pedestrian safety index (PSI) for urban areas and presents pedestrian guidelines. The PSI is used to evaluate the facilities alongside roads. This method is used to identify problems and propose solutions for improving safety. The recent evaluation methods, such as regression analysis, only process up to a specific range of street conditions and do not apply to all situations. The process of PSI is to rate the streets to examine the safety facilities in urban areas.

Distracted walking using smartphones in a street or crossroad can cause road accidents. Work presents a safety service solution to warn about possible dangers [7] to pedestrians. They propose a Safety-Aware Navigation Application (SANA) for smartphone users. This SANA calculates collision possibility and warns vehicles and pedestrians to avoid possible collision.

Vehicular Ad Hoc Networks (VANET) is widely used in Intelligent Transportation System(ITS) to enhance road safety. The work [8] discusses a thorough examination of the systems, protocols, and applications utilized in automotive networking for safe and efficient driving in smart transportation. The work covers a wide range of issues, such as communication protocols, network architectures, and intelligent transportation systems, and provides insights into the current state of the art and future directions of vehicular networking for enhancing road safety. Another work utilizes VANET to efficiently exchange location data between users on the road, such as vehicles, cyclists, and pedestrians [9]. They proposed a specific message format for encapsulating position data of pedestrians and two-wheelers to ensure safety. The applicability of VANET to provide services such as safety, driving assistance, collision avoidance and traffic control has been discussed in [10]. The paper also details IoT, virtual network functions, etc., to support new services and applications for V2X communication.

The work [11] discusses a comprehensive analysis of the application of machine learning techniques in intelligent vehicular networks. The authors discuss the challenges faced by traditional vehicular networks and propose a specialized machine-learning framework to address these challenges. The framework encompasses various machine-learning algorithms like Support Vector Machines (SVM), K-Nearest Neighbors (KNN), Neural Networks (Deep Learning), and even techniques for traffic management, resource allocation, security, and communication protocols. These machine-learning models can predict congestion patterns, optimize traffic flow, allocate network resources dynamically, detect anomalies and attacks, and optimize communication protocols.

Another trend in research to improve road safety focuses more on road structures and indicator-based warnings for drivers. The work discusses the importance of crosswalks for urban transportation, and it needs to be designed carefully to ensure safety. A comprehensive framework for crosswalks evaluation has been presented in [12]. This work shows that the provision of a zebra crossing is the leading indicator for road crossings, and they also pointed out that drainage near crosswalks is the least important. Another work [13] explains how indicators are reflective requirements and necessities for roads in daily liveable traffic situations. Another work [14] presented a unique idea of joining geographical data of the locations where road crash problems occur or may exist in future. The Geographical data consists of the analysis of crash types, severity, road age, and other factors. The data also updates the road crashes spontaneously from time to time. Collisions have been a serious threat to roads and motor vehicles. To overcome this, the database has the identification of interactions and road segments with a huge set of total crashes.

3 AI Driver Assistant for Road Safety

The requirement of AI driver assistance is discussed in this section with the help of a few common scenarios where the assistant has the potential to significantly enhance road safety, assist the driver, and help to prevent road accidents that may be encountered in various situations.

3.1 Junction Scenario

A common scenario observed on the Indian road in a semi-urban area is shown in Fig. 1. The two vehicles approach an intersection without any knowledge of each other's presence or intentions. With poor visibility, the absence of traffic signals, and limited awareness of surrounding vehicles, both drivers continue toward the junction, assuming it is clear to proceed. However, as they reach the intersection simultaneously in attempting to cross the junction, without any information about the oncoming vehicles in such cases the unawareness may lead to a collision or near-miss accident. This scenario underscores the necessity of intelligent vehicular networks and advanced technologies that facilitate information sharing and enable vehicles to make informed decisions.

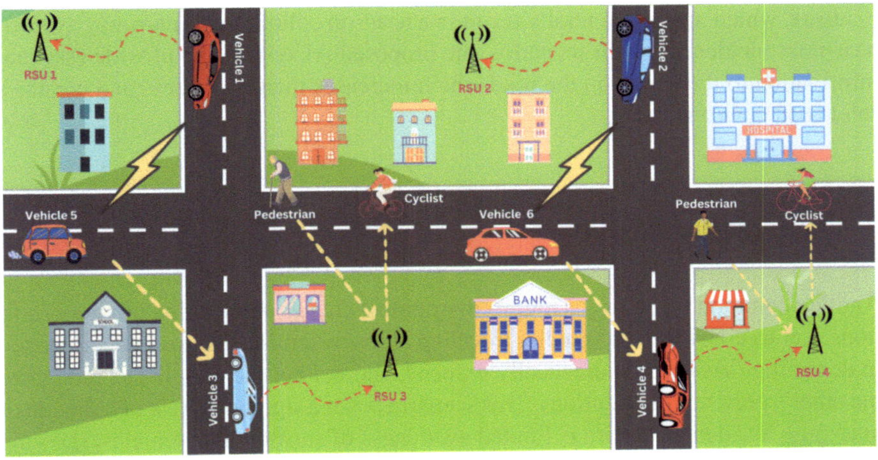

Fig. 1 Junction road scenario

Fig. 2 Hilly road scenario

3.2 Hilly Road Scenario

In this scenario Fig. 2, consider two vehicles navigating a winding hill road-one descending (Car A) and the other ascending (Car B)-with limited visibility due to blind corners and obstructed vision. Lacking prior awareness of each other's presence or approach, Car A proceeds downhill, assuming an unobstructed path, while Car B ascends, unaware of the potential oncoming vehicle. The absence of communication or signaling heightens the risk, with both drivers assuming a clear roadway, which might potentially result in a head-on collision or a narrowly avoided near-miss incident. In such scenarios, the proposed AI system can provide essential information to the driver on the road which increases their awareness and creates a safer road environment.

3.3 Foggy Road Scenario

Another common scenario is a foggy environment creates challenging driving conditions. Let us consider two vehicles traveling across a semi-urban road encompassed in dense fog. Visibility is drastically reduced, and neither the drivers can recognize the other's presence or intentions. This scenario poses a significant risk of collision or near-miss accidents as there is limited awareness of surrounding vehicles, showing the importance of systems for foggy conditions. The proposed AI system can make driving safer by giving real-time info about nearby vehicles and helping navigate safely in fog, reducing risks.

3.4 Other Scenarios

Apart from this, AI voice assistance is suitable for alerting users on various scenarios in rural and Urban areas. For instance, it can be used to alert a driver who is exceeding the speed limit in the school zone and can give the driver the need for a speed limit as a warning. Similarly, suppose the vehicle passes through a landslide-prone area. In that case, the system can alert the driver on that, making the driver more aware of the location and drive carefully.

4 Architecture of Proposed AI Voice Assistant

The proposed architecture comprises the YOLO-based object detection model, AI Voice Assistant, and data transmission module. The object detection model recognizes roads, automobiles, and traffic signs. The model used for object detection to make quick decisions must be fast so that the driver gets real-time object detection according to the dynamic driving situations. We chose the YOLO algorithm for this application because it is renowned for its capacity to carry out real-time object identification due to its speed [15].

The AI Driver Assistant provides real-time voice notifications to drivers, alerting them about potential threats and preventing drivers from negligence. For the prompt interchange of information about traffic congestion, accidents, and road conditions, this real-time capacity of VANET data transmission is essential. Vehicle-to-vehicle and vehicle-to-roadside infrastructure communication is required for real-time data

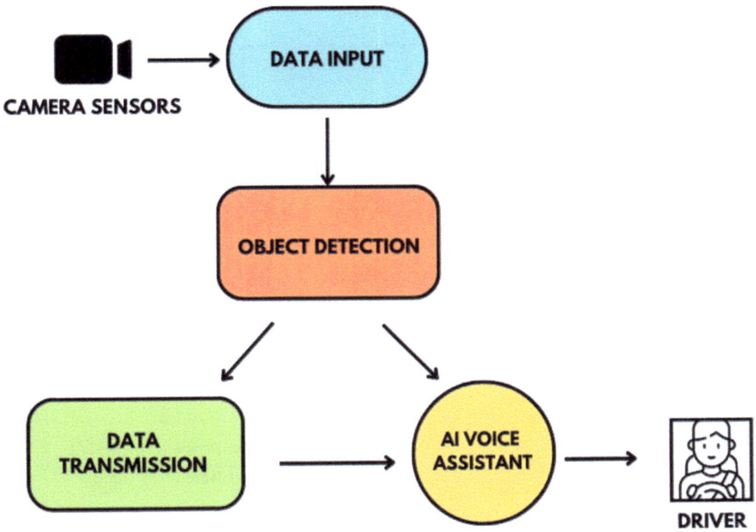

Fig. 3 Flow process of voice assistant

transmission. In scenarios like the hilly road shown in Fig. 2, where the object detection module is not relevant, in such cases, the location of the vehicles needs to be transmitted to the roadside units through vehicle-to-infrastructure connection and the neighbor nodes within the RSU will get the information and that will act as input to the voice assistant.

An overview of the step-by-step process of the proposed AI model for enhancing road safety is shown in Fig. 3. The process begins with data collection, where crucial information like on-road activities, Other Vehicle information, and traffic flow is gathered through camera sensors on the vehicle as raw video input. This collected data is then subjected to the object detection algorithm, enabling the system to identify various objects on the roads. The output of the object detection algorithm consists of the information about detected objects in a video frame.

The next step is disseminating the detected data from the object detection algorithm to the AI voice Assistant and all other nearby vehicles through VANETS. A suitable protocol is also required to transmit data between the vehicles [16]. The voice assistant utilizes this transferred information from the object detection module and the received data from other vehicles, which serves as a supportive co-pilot, assisting the driver in making decisions while driving on the road. Integrating vehicular networks to transfer data between vehicles helps to communicate road information to drivers and promotes safe driving practices.

For instance, let us assume three vehicles in a daily-life scenario, "A", "B", and "C" which utilize this solution. All three vehicles are traveling on a road, one behind the other. Assume a situation where there is a pothole in the road surface. Vehicle 'A' travels ahead of the other two vehicles and detects this pothole presence through

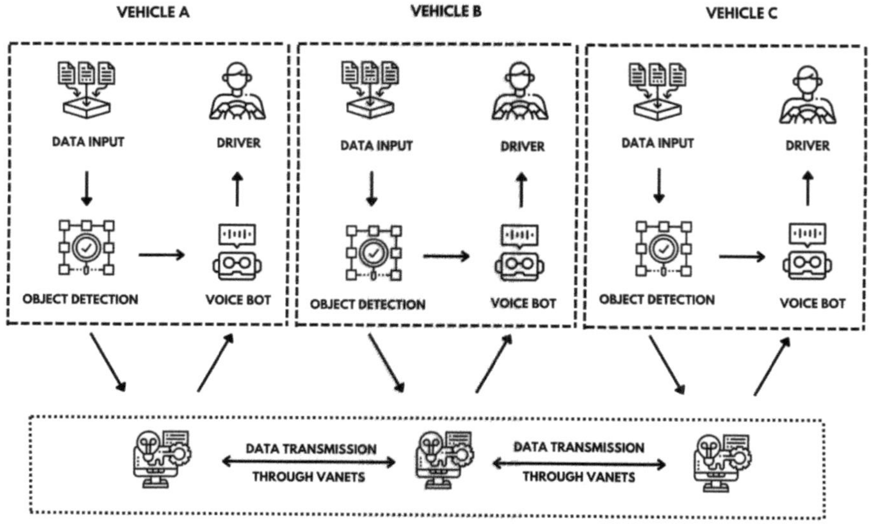

Fig. 4 AI voice assistance architecture

the object detection algorithm. Now, Vehicle 'A' gives this detected information parallelly to the AI voice Assistant of the same vehicle and other Vehicles 'B' and 'C' through VANETS.

5 Experimentation

As part of this work, we have developed two key components of the architecture: (i) the Object Detection Model and (ii) the Voice Bot Model, which play crucial roles in the functionality of the proposed system.

5.1 YOLO-Based Object Detection Model

Many existing works use CNN for object detection [17, 18]. In this work, the deep learning model YOLO is used for recognizing different objects on roads, such as cars, traffic signs, and humans. YOLO is capable of detecting any number of object classes [19]. The YOLO has object detection characteristics and a straightforward method. It outputs the object location and its classification through a neural network.

The YOLO architecture consists of three main components (Fig. 4).

- Neck: The Neck is a collection of neural network layers that merge and combine the features to pass the next stage of prediction.

- Head: The Head consumes features from the neck and creates prediction outputs.
- Backbone: The Backbone is a type of CNN that creates image features.

For this experimentation, we used YOLOv7 with a modified architecture to improve its performance on road object detection, which has increased accuracy without affecting the inference speed. YOLOv7 algorithm has architectural enhancement that makes it possible to learn much more efficiently with its extended efficient layer aggregation network (E-ELAN), allowing the model to control the learning path effectively, which can positively impact convergence and accuracy. YOLO v7 also achieves a higher mean average precision (mAP) of 94.2% compared to YOLO v6's mAP of 92.1%. It shows superior performance and accuracy for YOLO v7 [20]. The architecture changes of YOLOv7 compared to the YOLO architecture described above are:

- Extended Efficient Layer Aggregation Network (E-ELAN): E-ELAN is a method that allows a deep model to learn to control the path with the smallest longest gradient to converge more effectively. The proposed work for YOLO v7 incorporates a significantly larger number of stacked computational blocks into the models.
- Model scaling for concatenation-based models: By changing specific model properties, scaling creates models of various sizes. The YOLOv7 architecture is a concatenation-based design in which depth scaling and other common scaling methods adjust the ratio between a transition layer's input channel and output channel [21].

The YOLO v7 improves speed and accuracy with a few architectural reforms, such as E-ELAN and Computational modeling. The E-ELAN is the computational block of our YOLO backbone. The E-ELAN can expand, shuffle, and merge the data to enhance the network's learning ability continuously. In general terms, E-ELAN architecture enables the framework to learn better. We require parameters like resolution, width, depth, and stage for scaling the model. Our approach uses Network Architecture Search (NAS) to scale the model, as it is a widely used method. It helps to find the best scaling factors [22].

The YOLO Object detection model implements object detection using a single neural network that is applied to the entire image. The network divides the image into a grid of cells and predicts bounding boxes and probabilities for each object class for each cell. The predicted probabilities are used to weight the bounding boxes, and the bounding boxes with the highest weights are selected as the final detections.

The network is implemented using a CNN backbone, a neck, and a head. The backbone extracts features from the image, the neck combines and mixes these features, and the head predicts the bounding boxes and classes of objects in the image. The network is trained using deep supervision, which helps to ensure that the model is learning to extract useful features at all levels of the network. Only a trained network can be used to detect objects in real time by simply applying the network to the input image and selecting the bounding boxes with the highest weights as the final detections [23].

In scenarios involving foggy roads, we have incorporated a de-fogging algorithm into the YOLO system to enhance its performance when identifying and tracking objects in foggy conditions. This enhancement involves the integration of a de-fogging module at the conclusion of the YOLO architecture. The de-fogging module is implemented using a convolutional neural network (CNN), a suitable choice due to its ability to learn and extract features from images that are resilient to fog. Our algorithm is trained using a dataset comprising both foggy and non-foggy images to ensure its effectiveness. For this we used another dataset taken from the Dense-Haze [24]. This trained CNN is seamlessly integrated into the YOLO pipeline. The de-fogging module is applied to the image before being processed by YOLO, thereby significantly improving YOLO's object detection accuracy in foggy conditions.

5.2 AI Voice Bot

The final section of the proposed model is the AI voice bot. This model enables the driver to get real-time voice information about the detected objects and other pertinent information. By this, the voice bot improves the driver's situational awareness by issuing necessary instructions or alerts.

The core functionality of our voice assistant relies on two important steps. The primary step interprets the output from YOLO object detection. It processes each detected object, extracting its label, confidence score, and positional information. The system is specifically designed to identify vehicles labeled as cars, trucks, and bikes. Suppose a car is detected with a confidence score surpassing a predefined threshold (e.g., 0.6). In that case, the method calculates the distance between the vehicle and the camera based on its position in the image. Subsequently, it generates a text command, such as the presence of a car and its estimated distance. These text commands are then appended to a list for further processing. The second step is text-to-speech conversion, and we used Google's Text-to-Speech (gTTS) library. It generates the voice and saves it as an MP3 audio file by using the Pydub library. The same library is also used to play the generated audio, ensuring the driver receives verbal instructions from the bot.

5.3 Dataset

Our dataset has been sourced from the Common Objects in Context (COCO) dataset, a comprehensive collection used for object detection, segmentation, and captioning tasks in deep learning. It consists of an extensive repository of resources, comprising over 330,000 images, encompassing 1.5 million object instances across 80 distinct object categories and 91 categories for various materials. Five captions accompany each image, and there are annotations for 250,000 individuals with key points. This dataset categorizes objects into multiple classes, such as people, animals, vehicles, and household items, making it highly suitable for training and evaluating object detection models [25].

Additionally, we've used another dataset focusing on foggy road scenarios extracted from Dense-Haze. Dense-Haze represents a specific type of haze characterized by severely reduced visibility, often attributed to pollution or smoke [26]. This collection from Dense-Haze comprises 22 indoor and 33 outdoor images, some of which exhibit dense haze while others do not. This dataset serves as a valuable resource for investigating the impact of haze on scene visibility, enabling the evaluation of object radiance within the camera's proximity, extending up to a maximum distance of 20 m [27].

5.4 *Results and Analysis*

Object Detection Model The model is designed to identify objects from raw video input captured by a camera. In this particular experiment, we utilized a video recorded on an Indian road as the input for the algorithm. Each detected object is accompanied by annotations that include its name, confidence score, and bounding box coordinates, as demonstrated in Fig. 5. The object classification process involves the analysis of the input data to generate anchor boxes for multiple objects. The confidence score, which falls within the range of 0–1, is determined by multiplying the probability of an object's presence within the bounding box (objectness) by the probability that the bounding box accurately contains the object (localization). In foggy road conditions, where the input video may be unclear and dim, we initially apply the de-fogging procedure before the object detection process. The output of our detection algorithm in a foggy road scenario is illustrated in Fig. 6.

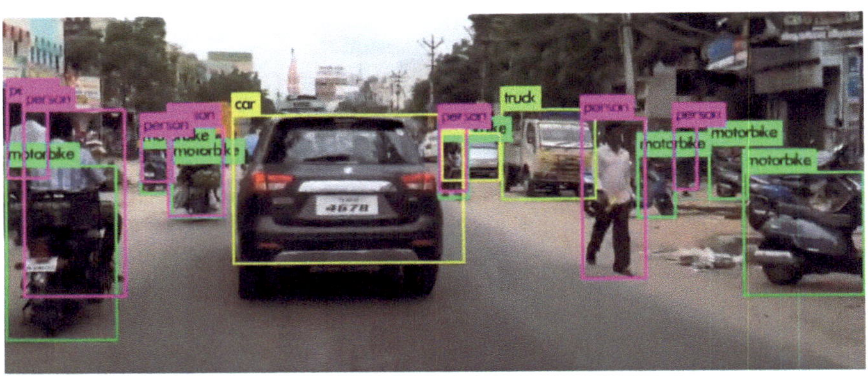

Fig. 5 Output of object detection

Fig. 6 Input and output of object detection for foggy scenario

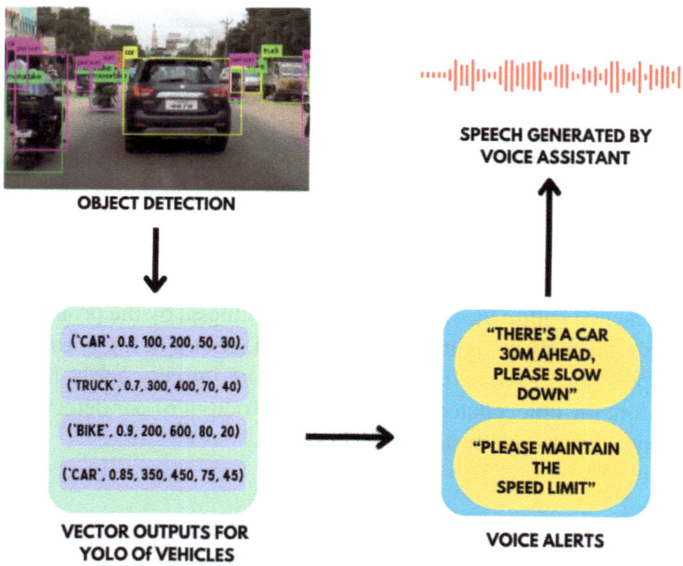

Fig. 7 Output for AI voice assistant

Voice Assistant Model The YOLO model's detection results are utilized as input data for the AI voice bot, as shown in Fig. 7. Based on the bounding boxes of the YOLO model, it classifies the scenario and produces output in the form of vectors. The voice alert generation module then interprets this data, applying predefined conditions to determine the type of object detected and its significance. The messages generated by the voice alert generation module are context-specific and tailored to enhance the driver's awareness. The generated text is used to produce the mp3 file using Google's text-to-speech library.

6 Conclusion

In this research, we presented an AI-enabled voice assistant system to enhance road safety in urban areas. The work also explained how the AI voice assistant can serve as a support system to the drivers by providing situational awareness. The experiments show that the object can be detected using the deep learning model, which can be used as an input to the voice assistant module. Overall, this research showcases the promising impact of AI-enabled systems in promoting safer and more efficient transportation in urban settings. As a future work, we plan to experiment with vehicular networks to increase the data transmission speed between the various vehicles that need to share the detected object details with each other to make this AI voice assistant for road safety.

Acknowledgements We would like to express our immense gratitude to our beloved Chancellor Dr. Sri. Mata Amritanandamayi Devi (AMMA) for being the guiding light and inspiration to do this work.

References

1. World Health Organization (2022) WHO accident report
2. Mohan D, Tsimhoni O, Sivak M, Flannagan MJ (2009) Road safety in India: challenges and opportunities
3. Anand S, Ramesh MV (2021) Multi-layer architecture and routing for internet of everything (IOE) in smart cities. In: 2021 sixth international conference on wireless communications, signal processing and networking (WiSPNET). IEEE, pp 411–416
4. Anand S, Ramesh MV, et al (2018) Performance analysis of delay tolerant network routing protocols in a heterogeneous vehicular network. In: 2018 IEEE International conference on computational intelligence and computing research (ICCIC). IEEE, pp 1–6
5. Lin L, Li W, Bi H, Qin L (2021) Vehicle trajectory prediction using LSTMS with spatial-temporal attention mechanisms. IEEE Intell Transp Syst Mag 14(2):197–208
6. Asadi-Shekari Z, Moeinaddini M, Shah MZ (2015) Pedestrian safety index for evaluating street facilities in urban areas. Saf Sci 74:1–14
7. Hwang T, Jeong J (2015) SANA: safety-aware navigation application for pedestrian protection in vehicular networks. In: Internet of vehicles-safe and intelligent mobility: second international conference, IOV 2015, Chengdu, China, December 19–21, 2015, Proceedings 2. Springer, pp 127–138
8. Shahwani H, Shah SA, Ashraf M, Akram M, Jeong JP, Shin J (2022) A comprehensive survey on data dissemination in vehicular ad hoc networks. Veh Commun 34:100420
9. Abid MA, Chakroun O, Cherkaoui S (2013) Pedestrian collision avoidance in vehicular networks. In: 2013 IEEE International conference on communications (ICC). IEEE, pp 2928–2932
10. Hussein NH, Yaw CT, Koh SP, Tiong SK, Chong KH (2022) A comprehensive survey on vehicular networking: communications, applications, challenges, and upcoming research directions. IEEE Access 10:86127–86180
11. Liang L, Ye H, Li GY (2018) Toward intelligent vehicular networks: a machine learning framework. IEEE Internet of Things J 6(1):124–135
12. Ahmed T, Moeinaddini M, Almoshaogeh M, Jamal A, Nawaz I, Alharbi F (2021) A new pedestrian crossing level of service (PCLOS) method for promoting safe pedestrian crossing in urban areas. Int J Environ Res Public Health 18(16):8813

13. Amoroso S, Castelluccio F, Maritano L, et al (2012) Indicators for sustainable pedestrian mobility. In: Urban transport XVIII: urban transport and the environment in the 21st century, 18:173
14. Schneider RJ, Khattak AJ, Zegeer CV (2001) Method of improving pedestrian safety proactively with geographic information systems: example from a college campus. Transp Res Record 1773(1):97–107
15. Redmon J, Farhadi A (2015) You only look once: unified, real-time object detection. arXiv preprint arXiv:1506.02640
16. Anand S, Ramesh MV et al (2019) Impact of network density on the performance of delay tolerant protocols in heterogeneous vehicular network. In: 2019 International conference on wireless communications signal processing and networking (WiSPNET). IEEE, pp 293–298
17. Toby T, Gopalakrishnan U, Rao SN (2022) A survey of deep learning techniques based on drone images for the search and rescue of victims from collapsed structures. In: 2022 IEEE 19th India Council international conference (INDICON). IEEE, pp 1–6
18. Aloysius N, Geetha M (2017) A review on deep convolutional neural networks. In: 2017 International conference on communication and signal processing (ICCSP). IEEE, pp 0588–0592
19. Vallimeena P, Gopalakrishnan U, Nair BB, Rao SN (2019) CNN algorithms for detection of human face attributes—a survey. In: 2019 International conference on intelligent computing and control systems (ICCS). IEEE, pp 576–581
20. Boesch G (2022) Yolov7: The most powerful object detection algorithm (2023 guide). https://viso.ai/deep-learning/yolov7-guide/#:~:text=The%20YOLO%20v7%20algorithm%20achieves,higher%20using%20a%20GPU%20V100.&text=Compared%20to%20the%20best%20performing,speed, 20(509):25
21. Terven J, Cordova-Esparza D (2023) A comprehensive review of YOLO: from yolov1 and beyond. arxiv 2023. *arXiv preprint* arXiv:2304.00501
22. Wang C-Y, Wang H-Y, Chen Y (2022) Yolov7: trainable bag-of-freebies sets new state-of-the-art for real-time object detectors. arXiv preprint arXiv:2207.02696
23. Jiang P, Ergu D, Liu F, Cai Y, Ma B (2022) A review of YOLO algorithm developments. Procedia Comput Sci 199:1066–1073
24. "NTIRE" (2019) Dataset
25. Lin T-Y, Maire M, Belongie S, Hays J, Perona P (2014) Microsoft COCO: common objects in context. arXiv preprint arXiv:1405.0312
26. Ancuti CO, Ancuti C, Sbert M, Timofte R (2019) Dense haze: a benchmark for image dehazing with dense-haze and haze-free images. In: IEEE International conference on image processing (ICIP), IEEE ICIP 2019
27. Ancuti CO, Ancuti C, Timofte R, Van Gool L, Zhang L, Yang M-H (2019) Ntire 2019 image dehazing challenge report. In: Proceedings of the IEEE/CVF conference on computer vision and pattern recognition workshops

RFM Analysis Using K-means Algorithm for Customer Segmentation

S. Lalitha, M. Uday Reddy, G. Nasir Hussain, and N. Vishnu Lokhesh Reddy

Abstract Customer Segmentation is one of the most crucial areas in the field of targeted marketing. This proposed work aims to develop an environment between customers and retailers to set up a beneficial scheme that can help achieve maximum benefit for both customer groups and retailers, for the purpose of creating a win–win situation between customers and retailers. This work focuses on applying RFM Analysis in the field of e-commerce with an overview to segment the customers into different groups. Here, a simple clustering technique using K-means in combination with the Elbow method is applied for the proposed task. Initially, to calculate the ideal number of clusters, the Elbow approach is used, followed by K-means to cluster the group of customers. The customers are divided into groups viz Champions, Can't Lose, Loyal, promising, require activation, and followed by customers divided into optimal clusters for categorizing the customers into different groups. The experimental results depicted the formation of 3 optimal clusters, in which cluster 1 is more reliable and efficient in terms of loyal customers which contains high monetary amounts and frequency, but low recency values when compared to the other two clusters.

Keywords RFM analysis · K-means clustering · Data pre-processing · Unsupervised algorithm · Customer segmentation · Heatmap representation of customer groups

1 Introduction

RFM analysis is a method used for analyzing customer value. In precise it is a technique used quantitatively to rank and cluster customers based on the recency [R], frequency [F], and monetary values[M] of their transactions and purchases detail over

---

S. Lalitha (✉) · M. U. Reddy · G. N. Hussain · N. V. L. Reddy
Department of Electronics and Communication Engineering, Amrita School of Engineering, Amrita Vishwa Vidyapeetham, Bengaluru, India
e-mail: s_lalitha@blr.amrita.edu

S. M. Thampi et al. (eds.), *Fifth International Conference on Computing and Network Communications*, Lecture Notes in Electrical Engineering 1221,
https://doi.org/10.1007/978-981-97-4711-5_12

a certain time duration (maybe for more than one year). This method is frequently applied in direct marketing and database marketing. RFM talks about Recency—How recently did the consumer make a purchase, Frequency—How frequently do they make purchases, Monetary Value—How much do they spend in terms of money? [1].

RFM segmentation is effective, but it has limitations. It is prone to human error when done manually. Additionally, due to the fact that RFM analysis is based on a limited set of behavioral traits, it cannot compare to modern sophisticated predictive analytics in terms of power. This analysis may be used by some businesses as an excuse to send messages to high-ranking customers, lowering response rates on campaigns that could otherwise be very effective. Then again, it can make advertisers disregard clients with low rankings even though many of them might be worth developing. For instance, the RFM model might not consider how seasonality or previous promotions affect the analysis. In a similar vein, a customer might spend only one month of exposure to your brand but be ready to buy the next month due to a birthday or anniversary [2].

This work proposes the use of simple K-means to categorize the customer base into required clusters, for experiencing better marketing and sales revenue by the retailers. The work aims at the accurate distribution of customer groups into the correct cluster.

2 Literature Survey

Essayem et al. enlisted the procedure used for customer segmentation. RFM-based features and refined features are the extracted features. The classification model used is the K-means clustering decision boundary. Five client clusters were created as a consequence of the outcomes of the business analysis and silhouette analysis [1]. The work by Mariani et al. specified RFM variables as extracted features and the decision boundary applied by K-means algorithm. This resulted in the formation of 4 clusters to classify customers and expected that the company shall take the right decision in marketing strategy [2]. Shihab et al. used the RFM variables and R-score, F-score, and M-score features for RFM table creation. Classification and analysis are done using K-means, advanced K-means, and agglomerative approaches. Compared to traditional K-means and agglomerative techniques, advanced versions of K-means can successfully cut the overall running time by between 27.8 and 97.8% [3]. Ranggadara et al. included the usage of rate loyalty status brand funnel consumer attitudes RFM values and scores. Naïve Bayes and K-means classification models are used. The results obtained are 62% feasible and 38% not feasible by the naïve Bayes method and then assisted by the RFM method to increase the feasible results of customer loyalty [4]. Yoseph et al. Enlisted the use of RFM summary, Avg. age, Avg. gender, and Avg. nationality as the feature taken to cluster the group of customers. The models used for classification are the RFM model, Hybrid Regression, and K-means for most decision boundary-making in the model. The comparison of methods employing cluster quality evaluation revealed that EM clustering with a size of the cluster quality of (2.45967329) is more accurate than the K-means algorithm with a size of the cluster quality of (0.001661999) [5]. Riana et al. included the features of RFM parameters/variables, with normalization of R, F, and M. The classification model used is the standard K-means method and the decision tree boundary method is performed. The results obtained are as follows, i.e., to create 4 optimal clusters, and data were retrieved using K-means clustering methods. Afterward, a decision tree technique was used to classify the four groupings [6]. Sayan et al. discussed about the features that are extracted, Avg. number of transactions by the user, avg. frequency, avg. redundany, avg. monetary value, and last activity [5, 7]. The model mentioned is K-means Affinity Propagation, PAM, and 7 more clustering techniques. The results obtained are from different clustering algorithms used for customer segmentation and compared their performance [7]. Pradhan et al. accumulated a total of 29 different attributes to analyze customer behavior. The elbow technique, PCA, K-means clustering, and the classification model were all utilized to construct the decision border. Four clusters are produced as a consequence, employing RFM values and other consumer behavioral variables to create more accurate and compact segments [8]. Ait Daoud had included RFM variables, i.e., R-score, F-score, and M-score as the features extracted for customer segmentation. The classification model used is the K-means clustering method, an unsupervised decision-making algorithm. The study produced 9 groups of consumers, and the findings show that group 7 is the most dependable cluster since its average R, F, and M values are higher than the overall average value [9]. Hu et al. included the details of Avg. of the single-order transaction amount and the number of order transactions done by customers by contributing their time are the features taken for classification. The models used for comparing the performance of each other are mean-shift clustering, the DBSCAN algorithm, agglomerative clustering, and K-means clustering techniques. The results taken into consideration are in the form of a snake plot and heat map of customer segments [10]. Faran et al. enlisted the features for the segmentation and used the agglomerative ward's method for the segmentation of donors. The result obtained by the author concludes that his research replaces the conventional process [11]. Hariram et al. Tried to compare the segmentation results from different algorithms such as analysis and clustering algorithms. The author obtained the results and described that RFM clustering gave the best visualizations to segment customers in a better and faster way [12].

3 Methodology

The methodology of the proposed work is shown in Fig. 1, which first deals with data collection, data cleansing, integration, and visualization are all parts of the process of preparing data, and is followed by obtaining individual customers' RFM variables, then the RFM variables are passed on to the creation of an RFM table, RFM segment values, and RFM scores. These RFM tabular columns are used by elbow method to get the optimal clusters, these clusters are used to cluster the customers into desired

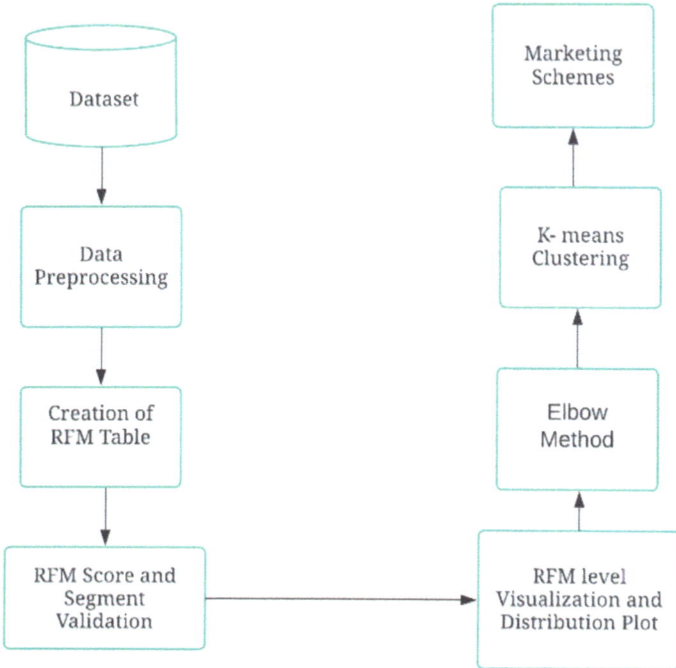

Fig. 1 Schematic diagram for proposed RFM analysis

categories according to the taken threshold values. The desired categories in which the customers are segmented are as follows—can't lose them, Champions, Potential, Promising, Needs Attention is Loyal, and Require Activation. The threshold values that are dependent factors to segment customers are taken by the author in the proposed work for over five hundred thousand samples, here the author considered the threshold values over a scale of 0–9 and tried to categorize the customer based on threshold values.

For threshold value $> = 9$ the customer comes under Can't lose them category. For threshold values $> = 8$ and < 9 the customer comes under the Champions category. For threshold values $> = 7$ and < 8 the customer comes under the Loyal category. For threshold values $> = 6$ and < 7 the customer comes under the Potential category. For threshold values $> = 5$ and < 6 the customer comes under the Promising category. For threshold values $> = 4$ and < 5 the customer comes under the Needs attention category. If certain threshold values don't come under these categories, then it goes to Requires Attention.

Let T be the threshold value, and the categories are defined as follows:

1. Can't lose them: $T > = 9$
2. Champions: $8 = < T < 9$
3. Loyal: $7 = < T < 8$
4. Potential: $6 = < T < 7$

5. Promising: $5 = < T < 6$
6. Needs attention: $4 = < T < 5$
7. Requires attention: $T < 4$

These inequalities provide a mathematical representation of the customer categories based on the given threshold values.

Dataset:

The first block describes the customer details, whether it may be their transactions or behavior details to get the data pre-processed. The data set, which was obtained from the Kaggle website, includes 500 thousand samples of customers Those samples consist of 8 variables such as invoice no, quantity, Unit price, country, and customer ID [13].

Data Pre-processing:

The pre-processing methods used to improve the quality of the data are described in the block. Data cleansing, data integration, data reduction, and data transformation or visualization are the pre-processing techniques employed.

Creation of RFM Table:

The block discusses about generating the RFM table for the group of customers taken in the dataset. The individual customer RFM table is obtained by equating the recency value to the customer invoice date, frequency is equated to invoice no, and monetary value is equated to unit price value this procedure is followed to build a RFM table for the whole dataset.

RFM Score and Segment Validation:

The RFM segment describes the conversion of categorical data to string data type and is represented as "RFM" (e.g., a customer has R = 1, F = 1, M = 4, and the segment is "114"). RFM score is obtained by summing the values of the segment (e.g., RFM segment = 114 and the score is 1 + 1 + 4 = 6).

RFM Level Visualization and Distribution Plot:

The block discusses the visualization of the RFM level of the customer groups according to their densities around the values of statistical data, whereas the distribution graph plots the customer RFM against their density of country and year-wise. These visualizations help to analyze the difference in the size of each cluster and describe the distributions of customers into various groups.

Elbow Method:

To calculate the number of ideal clusters, the elbow approach is described in the block. Due to K-means' unsupervised nature and requirement for pre-defined clusters, this technique was used. There were three ideal clusters produced by the elbow approach.

K-Means Clustering: The block discusses the division of customer groups by K-means algorithm depending upon the similarities between the customers. Clustering usually goes around to the given data to separate the data with more similarities and cluster them together, so there is no need for the involvement of classes for classification.

Marketing Schemes: The cluster data of customers is used by the retailers to provide exclusive services to the top-placed customer and to attract the other group of customers. These marketing strategies are the key schemes that are used by the retailers to keep an engaging environment for their customer base. So, it results in a win–win situation for both customers and retailers.

A. *Overview of customer segmentation:*

The steps in this work involved for customer segmentation in RFM analysis is as follows: Data Collection-Gather the required information about the client transactions, such as the date of their most recent purchases, the amount they spent overall, and the frequency of their purchase. Calculating RFM Scores: Determine the consumers' RFM Scores based on their transactional records. Customers should be ranked for each metric in a range of 1–9 or 10, with 9 or 10 representing the highest score. Segment Customers—Use the RFM scores to divide your customer base into various categories based on how they behave, such as high-value clients, devoted clients, or clients who risk leaving. Analyze Customer Behavior—Utilize the categories to examine consumer behavior and spot trends and patterns. This analysis can profit out places where marketing and sales efforts need to be strengthened. Develop Marketing Strategies—For creating targeted marketing strategies for each client segment, use the insights from RFM analysis. For instance, unique offers or rewards may be given to high-value customers to encourage their continued loyalty and to hold the customer group that is least interested in a particular retailer.

4 Experimental Results and Analysis

In this proposed work, the K-means clustering technique has been used in RFM analysis for customer segmentation. A heat map of the RFM quantile and a heat map of K-means are shown for the RFM analysis and K-means clustering. Therefore, groups that are most lucrative as well as groups that require attention may be discovered by analyzing these heat maps and quantile plots.

Prior to using the clustering algorithms, various important RFM factors are observed, including the methods utilized for data pre-processing. First, the statistical descriptions of data are presented to analyze the mean, std median, min, and max values of the chosen data set. Figure 2 describes the statistical variables of the data set. The most common procedure to be followed while performing data pre-processing is data cleaning, i.e., to remove any outliers, or to handle any missing data present in the data set. RFM analysis mostly consists of customers' transactional, purchasing, and behavioral details that are present in the numerical data, so the noise component in the numerical data is missing data or Null values. It is essential to remove these null values or handle the data that is missing. So, the null values

Fig. 2 Statistical descriptions of data

	InvoiceNo	Quantity	UnitPrice	CustomerID
count	29.000000	25.000000	29.000000	29.000000
mean	536366.965517	7.720000	4.331379	14489.586207
std	1.523219	7.585732	2.107007	2298.404532
min	536365.000000	2.000000	1.650000	12583.000000
25%	536366.000000	3.000000	2.750000	13047.000000
50%	536367.000000	6.000000	3.750000	13047.000000
75%	536368.000000	6.000000	4.950000	17850.000000
max	536370.000000	32.000000	9.950000	17850.000000

are found, and the empty slot is represented with a dummy variable known as True. After finding the null values the model tries to replace that with the name true and if it doesn't find a null value in the slot it is then represented as false. These changes in the terminologies are clearly displayed in Fig. 3.

Finding the missing data or null values involves the procedure to remove the values. The author performed 3 techniques to handle missing data, although there are various techniques to handle missing data like statistical methods, prediction algorithms, and ML algorithms. Missing Completely at Random (MCAR), Missing at Random (MAR), and Not Missing at Random (NMAR) are the three categories for the missing data. So, the 3 techniques used are (1) Preceding Filling Strategy, (2) Back Filling Strategy, (3) User-Defined Strategy, these techniques require substituting the missing value with a certain value before using it, which is required to have domain knowledge of the data set, these are used to populate the MAR and MCAR values.

	InvoiceNo	StockCode	Description	Quantity	InvoiceDate	UnitPrice	CustomerID	Country
0	False	False	False	False	False	False	False	False
1	False	False	False	False	False	False	False	False
2	False	False	False	False	False	False	False	False
3	False	False	False	False	False	False	False	False
4	False	False	False	True	False	False	False	False
5	False	False	False	False	False	False	False	False
6	False	False	False	False	False	False	False	False
7	False	False	False	True	False	False	False	False
8	False	False	False	False	False	False	False	False
9	False	False	False	False	False	False	False	False
10	False	False	False	True	False	False	False	False

Fig. 3 Checking for missing values

The preceding Filling Strategy works in the view of taking the preceding mentioned values are the null values, which is a vice-versa situation in the Back Filling Strategy where the function takes the next values that are mentioned after the null values and user-defined techniques take the value given by the user (in the model the user-defined value is 100).

Figure 4 shows that serial number 4, 6, and 10 have null values and their places are replaced by their preceding values.

The preceding Filling Strategy considers the preceding values that exist in front of missing values in the given data samples. The insights observed form Fig. 4 are previously the customer with I'd 536,367 had a null value which is replaced by the preceding customer value, i.e., (6.0).

The preceding Filling Strategy considers the preceding values that exist in front of missing values in the given data samples. The insights observed from Fig. 4 are previously the customer with I'd 536,367 had a null value which is replaced by the preceding customer value, i.e., (6.0).

Figure 5 shows the Back Filling Strategy in which customer IDs 4, 6, and 10 have their null values replaced by successor values, and Fig. 6 shows the user-defined technique in which the null values are replaced by the user-defined value, i.e., 100.

16	536367	22622	BOX OF VINTAGE ALPHABET BLOCKS	6.0	2010-12-01 08:34:00
10	536367	22745	POPPY'S PLAYHOUSE BEDROOM	6.0	2010-12-01 08:34:00
6	536365	21730	GLASS STAR FROSTED T-LIGHT HOLDER	6.0	2010-12-01 08:26:00
12	536367	22749	FELTCRAFT PRINCESS CHARLOTTE DOLL	8.0	2010-12-01 08:34:00
2	536365	84406B	CREAM CUPID HEARTS COAT HANGER	8.0	2010-12-01 08:26:00
4	536365	84029E	RED WOOLLY HOTTIE WHITE HEART.	8.0	2010-12-01 08:26:00
22	536368	22913	RED COAT RACK PARIS FASHION	3.0	2010-12-01 08:34:00
13	536367	22310	IVORY KNITTED MUG COSY	6.0	2010-12-01 08:34:00
7	536366	22633	HAND WARMER UNION JACK	6.0	2010-12-01 08:28:00

Fig. 4 Preceding filling strategy

10	536367	22745	POPPY'S PLAYHOUSE BEDROOM	6.0	2010-12-01 08:34:00
6	536365	21730	GLASS STAR FROSTED T-LIGHT HOLDER	6.0	2010-12-01 08:26:00
12	536367	22749	FELTCRAFT PRINCESS CHARLOTTE DOLL	8.0	2010-12-01 08:34:00
2	536365	84406B	CREAM CUPID HEARTS COAT HANGER	8.0	2010-12-01 08:26:00
4	536365	84029E	RED WOOLLY HOTTIE WHITE HEART.	8.0	2010-12-01 08:26:00
22	536368	22913	RED COAT RACK PARIS FASHION	3.0	2010-12-01 08:34:00
13	536367	22310	IVORY KNITTED MUG COSY	6.0	2010-12-01 08:34:00
7	536366	22633	HAND WARMER UNION JACK	6.0	2010-12-01 08:28:00

Fig. 5 Back filling strategy

10	536367	22745	POPPY'S PLAYHOUSE BEDROOM	100.0	2010-12-01 08:34:00
6	536365	21730	GLASS STAR FROSTED T-LIGHT HOLDER	6.0	2010-12-01 08:26:00
12	536367	22749	FELTCRAFT PRINCESS CHARLOTTE DOLL	8.0	2010-12-01 08:34:00
2	536365	84406B	CREAM CUPID HEARTS COAT HANGER	8.0	2010-12-01 08:26:00
4	536365	84029E	RED WOOLLY HOTTIE WHITE HEART.	100.0	2010-12-01 08:26:00
22	536368	22913	RED COAT RACK PARIS FASHION	3.0	2010-12-01 08:34:00
13	536367	22310	IVORY KNITTED MUG COSY	6.0	2010-12-01 08:34:00
7	536366	22633	HAND WARMER UNION JACK	100.0	2010-12-01 08:28:00

Fig. 6 User-defined strategy

The Back Filling Strategy includes the fundamental rule of taking the after value that exists after missing values. Figure 5 describes these fundamental rules. The inference observed from the Fig. 5 is that the customer with I'd 536,365 had an null value and it is replaced by the next customer value.

Figure 6 depicts the rule of user-defined, as the null value is replaced by the author's desired value, i.e., (100). The insights observed from the Fig. 6 are that the customers with Id's 536,367, 536,365, and 536,366 had null/ missing values that are replaced by the author defined number (100). These techniques are used to remove the null values present in the data set, so the purpose of performing these methods results in accurate clustering of the customer base. The author used the 3 techniques in order to identify the most reliable method, so the insights resulted that user-defined method is more reliable in terms of complex database because the other two methods resulted in discrepancies in the data cleaning process over the complex database. After the process of data cleaning is completed the model ends up dealing with data integration—integrating data from several sources into a cohesive database, so the problem faced here is data redundancy which often occurs when.

integrating multiple databases, Correlation Analysis, which measures how strongly one feature indicates the other based on the data given, may be able to uncover these redundant attributes, for the taken numerical data correlation co-efficient and covariance can be used, both of which explore how the values of one characteristic differ from those of another.

The equation for evaluation of Correlation Analysis is defined as:

$$\chi^2 = \sum\left((\textbf{observed} - \textbf{Expected})^2 \div \textbf{Expected}\right) \quad (1)$$

The correlation between the attributes of Customer ID, Unit Price, Quantity, and Invoice Number is represented in the heat map distribution over a range of – 0.2–1. The heat map is briefed in Fig. 7. Correlation Analysis provides insights into the relationship between two or more variables/attributes. It gives the details regarding the strengths and direction of the association between variables, so the variables taken in Fig. 7 are Invoice No, Quantity, Unit Price, and Customer I'd. The insights observed from Fig. 7 are the grouping of values of invoice no, quantity, unit price,

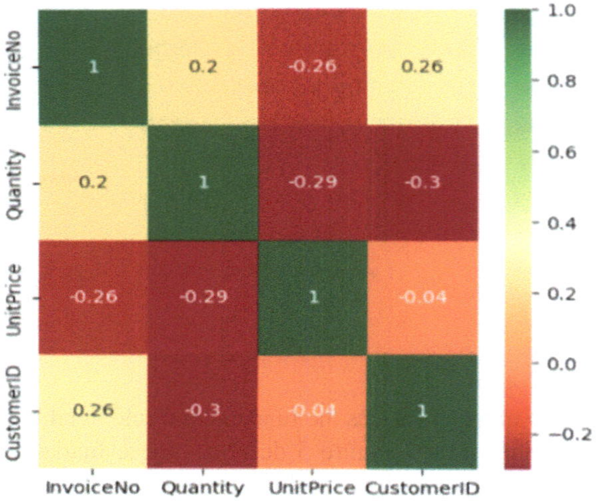

Fig. 7 Heat map representing correlation

and customer id resulted in the formation of a diagonal cluster with more customers having similar attribute values. The relation between invoice no and quantity is high (0.26) and a weak relation exists between quantity and customer id (− 0.3).

Data Visualization:

Integrated data is passed on to visualize the entire data in different visualization methods, the most known method is Histogram, attributes like Invoice No and Unit Price are represented in the histogram for better visualization the visualization of these two attributes is depicted in Fig. 8.

Figure 8 gives the graphical representation of the attribute invoice no, it is displaying the frequencies into different intervals or bins. The histogram has distributions with many several peaks, which represent the data varies from each customer. The uneven peaks of the histogram represent the variation in the values of invoice no per customer I'd country-wise-wise. The invoice no from 00 to 78 is considered and the uneven length of invoice indicates the number of customers over the range of invoice no around different countries.

Figure 9 gives the graphical representation of the attribute invoice no, and it is displaying the frequencies into different intervals or bins. The histogram has distributions with many several peaks, which represent the data varies from each customer. The unit price graph varies according to the increasing scale of 2.5 from all customer's unit price and the range of 11.5–15.5 has the lowest bin length indicating less number of customers with unit price around this range.

Scattered plots are yet another technique that is widely employed to present data. The data for two variables are displayed using points that are parallel to the horizontal and vertical axes. This type of data visualization may be used to identify patterns or correlations in the data as well as effectively illustrate the relationships between

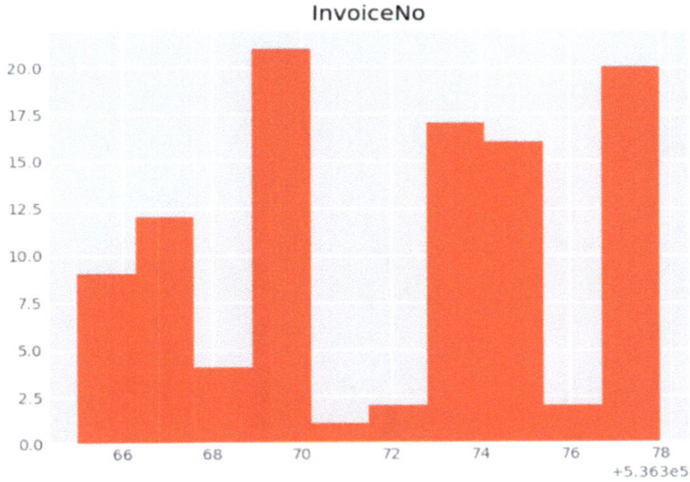

Fig. 8 Histogram representation of Invoice No

Fig. 9 Histogram representation of Unit Price

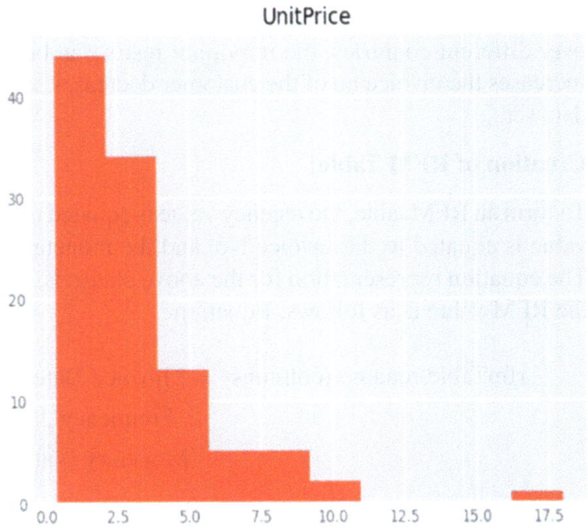

variables. Because it is generally easier to see trends when there are more data points available, scatter plots perform well with rather big data sets. The stronger the correlation or trend, the closer the data points are clustered together. The scatter plot between Invoice No and Unit Price is represented in Fig. 10.

Figure 10 describes the graphical representation of two continuous variables. It plots a two-dimensional graph, with invoice no variable on the horizontal axis and unit price on the vertical axis. The scatter play plot provides insights into the strength, direction, linearity, outliers, and possible clusters within the data of two variables.

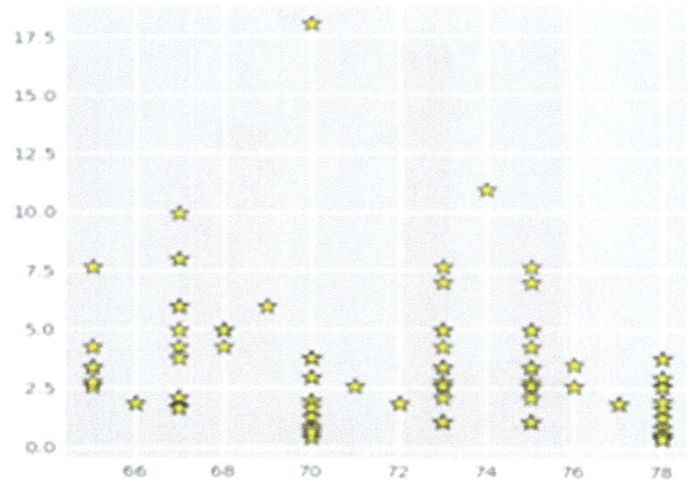

Fig. 10 Scatter plot

The insights observed from the scatter plot are the nature of invoice no and unit price over different countries, the important result can be observed that as the unit price increases the invoice no of the customer decreases according to the data given by the data set.

Creation of RFM Table:

To form an RFM table, the recency value is equated to the Invoice Date, the frequency value is equated to the Invoice No, and the monetary value is equated to the Price. The equation representation for the above clause is, a snippet of the code to evaluate the RFM value is as follows: Equation-

$$\begin{aligned} \text{rfmTable.rename (columns } &= \{'\text{Invoice Date}': '\text{Recency}', \ '\text{Invoice No}' \\ &: '\text{Frequency}', \ '\text{Price}': \\ &'\text{Monetary\_Value}'\}, \text{ inplace } = \text{ True}) \end{aligned} \tag{2}$$

The RFM table formed by the given descriptions is depicted in the Table 1.

Table 1 Performance analysis of the resulting RFM values using the proposed method

Customer id	Recency	Frequency	Monetary value
12,346	325	1	77,183.60
12,347	2	182	4310.00
12,348	75	31	1797.24
12,349	18	73	1757.55
12,350	310	17	334.40

RFM Segment: Convert Categorical to String Data Type

RFM segment is described as the string data type that is obtained by combining the individual R, F, and M values. So, the RFM segment is represented in the Table 2.

RFM Score:

The RFM score is obtained by the sum of individual R, F, and M values to get a single score to represent customer value according to the RFM score. This RFM score is used to cluster the customer groups by taking the RFM score as the threshold value to segment customers.

RFM score is represented by the Equation—

$$\text{rfmTable}['\text{RFM\_Score}'] = \text{rfmTable}['\text{R}','\text{F}','\text{M}'].\text{sum}(\text{axis} = 1) \quad (3)$$

The RFM score is tabulated below Table 3 to display the individual customer RFM score.

Table 2 Performance analysis of the resulting RFM segment

Customer id	Recency	Frequency	Monetary value	R	F	N	RFM Segement
12,346	325	1	77,183.60	1	1	4	114
12,347	2	182	4310.00	4	4	4	444
12,348	75	31	1797.24	2	2	4	224
12,349	18	73	1757.55	3	3	4	334
12,350	310	17	334.40	1	1	2	112

Table 3 Performance analysis of the resulting RFM scores using the proposed method

Customer id	Recency	Frequency	Monetary value	R	F	M	RFM Segment	RFM Score
12,346	325	1	77,183.60	1	1	4	114	6
12,347	2	182	4310.00	4	4	4	444	12
12,348	75	31	1797.24	2	2	4	224	8
12,349	18	73	1757.55	3	3	4	334	10
12,350	310	17	334.40	1	1	2	112	4
…	…	…	…	…	…	…	…	…
18,280	277	10	180.60	1	1	1	111	3
18,281	180	7	80.82	1	1	1	111	3
18,282	7	12	178.05	4	1	1	411	4
18,283	3	756	2094.88	4	4	4	444	12
18,287	42	70	1837.28	3	3	4	334	10

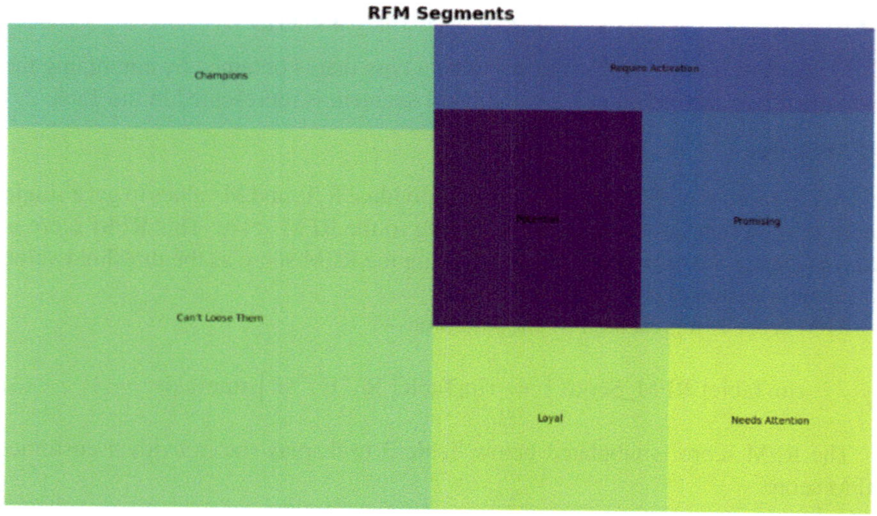

Fig. 11 RFM level visualization

RFM Level Visualization:

This visualization allows us to relate the different levels of customer groups which depend on the mean, and mean count values, the RFM level visualization is represented in Fig. 11. These visualizations help to analyze different clusters of customer groups, the difference in the size of each cluster describes the distributions of customers into various groups.

The insights observed from Fig. 11 are the variation in cluster size represents that more customer comes under the cluster of 'can't lose them' and the same goes for the other clusters. The size of the cluster represents the number of customers goes to the cluster from the given data set.

RFM Distribution Plot:

The individual customer RFM values are plotted against their density in the country and year-wise so that we get a clear idea of how the drop or gain is happening in the RFM values. Figure 12 represents the RFM distribution against their density.

The insights observed from the Fig. 12. are the plot built up against the density vs RFM values, the variation in the peaks of each graph according to its RFM values, the recency graph has a lot of ups and downs as its recency increases, there is a decreasing graph between the frequency and density, and where are monetary values goes on decreases.

Elbow Method:

The ideal value of K must be known before feeding the characteristics to the K-means algorithm. The Elbow method, which is the most common, was used to analyze the stages listed below.

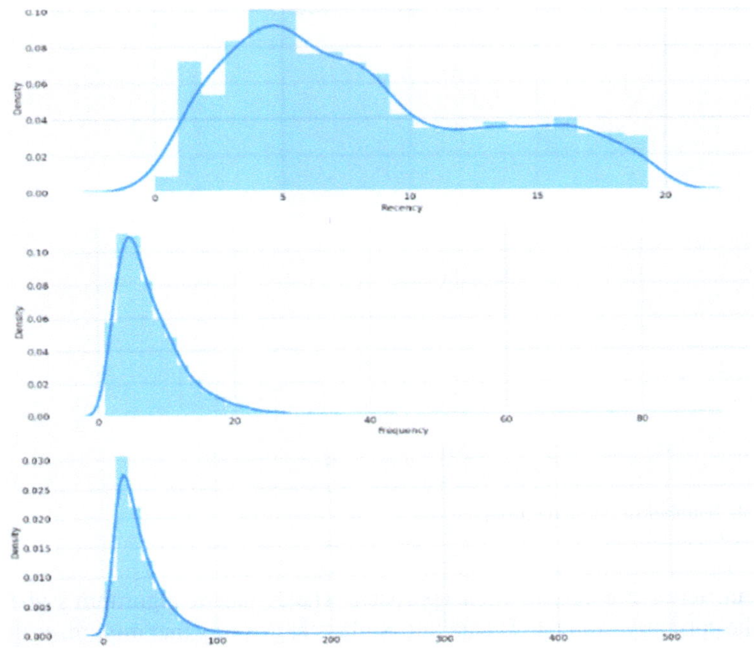

Fig. 12 RFM distribution

Step 1: Change the value of K between 1 and 10.
Step 2: Compute the within-cluster sum of square (WCSS) using the following formula.

$$\text{WCSS} = \sum (x - C) \wedge 2 \tag{4}$$

and the sum (*xi*—*Ci*) 2 is the result of squaring the distance between the centroid and each data point in the cluster.

Step 3: The graph of WCSS versus *K* value should be plotted in this step to create the elbow-shaped figure that gives the Elbow method its name.
Step 4: Find the 'turning-point', or the point where the behavior plot almost exactly parallels the *x*-axis. K should be set to this value.

Figure 13 shows the elbow curve, it is observed that at point 3 the elbow curve starts to move parallel to the *x*-axis and that point is knowns as a turning point. So the turning point here is three (3), so the required optimal clusters are 3 for the proper segmentation of customer groups for the given data set.

K-Means Clustering:

This algorithm's fundamental step is to record *n* observations for dataset D, which is represented by the notation D = (D1, D2, D3,..., Dn). For Lloyd's NP-hard problem,

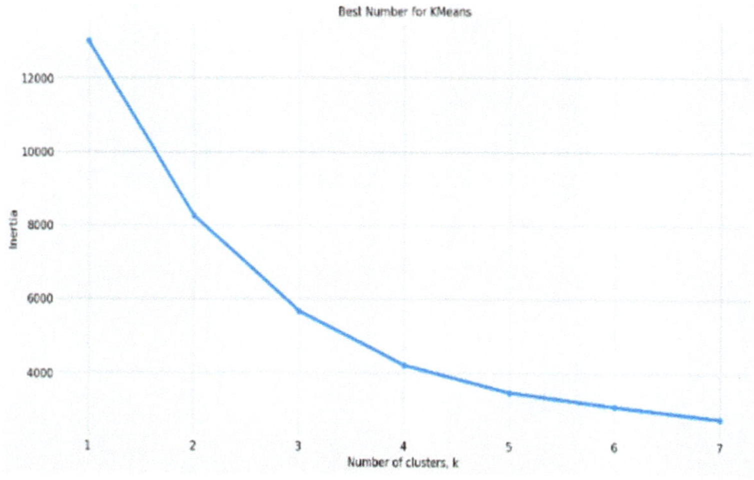

Fig. 13 K-Number of optimal clusters

this is an iterative approximation approach. The K-means algorithm's goal is to group the currently available D data into K clusters in a way that minimizes the total Euclidean distance between each cluster's centroid and its associated data points [11]. The customer groups are divided into 3 optimal cluster groups and are represented in the form of heat maps. The clusters are displayed corresponding to their respective RFM scores and segments, which can be observed from the Table 4.

Table 4 shows different RFM variables like individual RFM scores, RFM values for each customer, and RFM level. RFM segment values are obtained by combining the RFM score values (For e.g., For customer—12,346 individual RFM scores are 1 1 4, from which the RFM segment value is 114).

The heat map of quantile representation gives a detailed report about the customer segmentation in terms of different levels like champions, loyal, etc. The details are enlisted in Fig. 14.

Table 4 Performance analysis of the resulting cluster values using the proposed method

Customer	Recency	Frequency	Monetary value	R	F	M	RFM Segment	RFM Score	RFM level	K cluster
12,346	325	1	77183.60	1	1	4	114	6	Potential	1
12,347	2	182	4310.00	4	4	4	444	12	Can't lose them	1
12,348	75	31	1797.24	2	2	4	224	8	Champions	0
12,349	18	73	1757.55	3	3	4	334	10	Can't lose them	0
12,350	310	17	334.40	1	1	2	112	4	Needs Attention	2

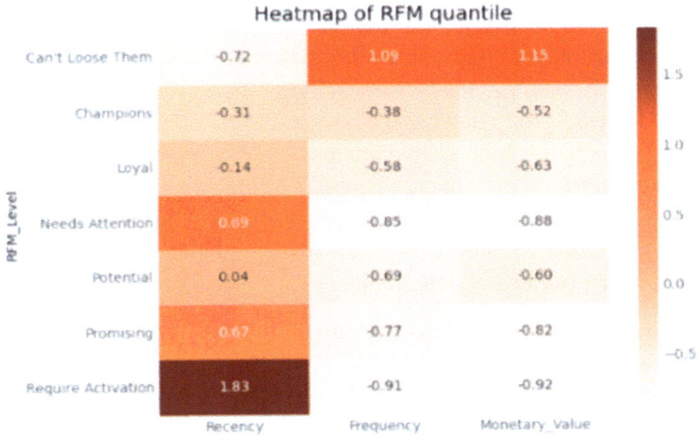

Fig. 14 Heat map of RFM quantile

K-means clustering resulted in the formation of 3 optimal clusters, which are used to segment the customer groups into different levels. So, the final cluster of customers is represented in the heat map for K-means. This is represented in Fig. 15.

According to the ideal number of clusters found using the elbow approach, the K-means clustering result is divided into 3 groups based on the heat map.

The analysis that can be observed from the heat map in Fig. 15 is:

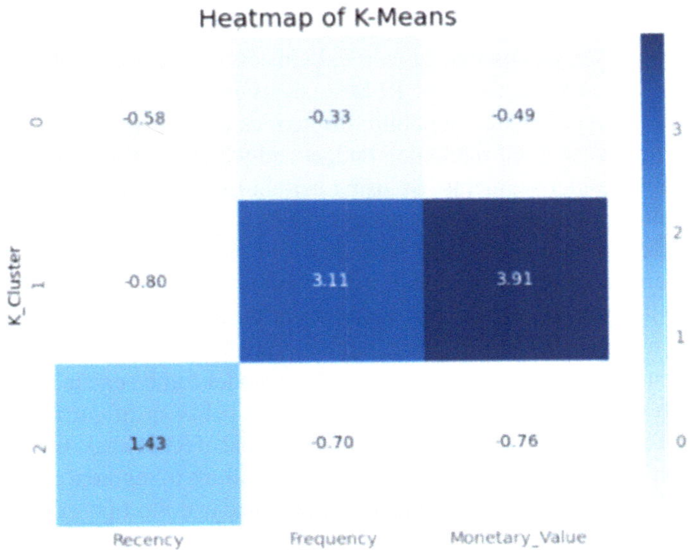

Fig. 15 Heat map of K-means

Cluster 0—This is a promising group of customers, the reason for this description of promising customers comes from the reasoning of having moderate recency [R], frequency [F], and monetary values [M].
Cluster 1—This is a group that has the best and most loyal customers because this group of customers has high frequency and monetary values, but they have low/ poor recency values.
Cluster 2—which is a group of new customers that is in precise they are customers that require attention because this group of customers has high recency values, but they have low frequency and monetary values.

The customer base may be classified into the necessary clusters based on their recentness, frequency, and monetary values using the observations and descriptions provided by both heat maps (Figs. 14 and 15) and various cluster tables (Tables 1, 2, 3, and 4). These results can describe the behavior of the customer base based on which cluster they belong and according to the cluster, the retailer can supervise their customer base.

The cluster that is more reliable and efficient in terms of loyal customers—is obtained as cluster 1 which, in contrast to the other two clusters, has low recency values but high frequency and monetary values.

5 Conclusion and Future Work

This work aimed at applying RFM analysis and clustering techniques in the field of E-commerce with an overview to segment the customers into different groups. To divide consumer groups into three ideal clusters, the k-means clustering technique is applied. Accordingly, the RFM analysis is used to assess and compare these 3 ideal clusters for their suitability to various client groups. Businesses may decide how to allocate their marketing budget and which clients to target by identifying several customer segments. By using certain methods (K-means, altering the threshold values) on specific client groups (RFM, Champion segment, etc.), their values can grow proportionally and have an effect like staying pleased or retaining sliding consumers. Companies might use this method (case study) to increase sales of their current items by releasing many iterations of the same product, other clustering methods such as mean-shift clustering, agglomerative clustering, and the DBSCAN algorithm can be utilized in future work, and the model can be enhanced by include classification algorithms to forecast the purchasing behavior of current consumers and potential customers. The existing work on the topic of RFM analysis for customer segmentation uses either clustering or analysis algorithms to segment customers into the desired cluster of groups. The case study mentioned in this proposed methodology describes about the clustering method that utilizes heat map to divide customer group into desired cluster of segments from the comparison of respective RFM values.

References

1. Essayem W, Bachtiar FA, Priharsari D (2022) IEEE International conference on cybernetics and computational intelligence (Cybernetics Com), Malang, Indonesia, pp 23–27
2. Mariani I, Riana D, Astuti RD, Ishaq A, Sutrisno, Pratama EA (2018) Customer segmentation based on RFM model and clustering techniques with K-means algorithm, 2018 Third international conference on informatics and computing (ICIC), Palembang, Indonesia, pp 1–6
3. Shihab SH, Afroge S, Mishu SZ (2019) RFM Based market segmentation approach using advanced K-means and agglomerative clustering: a comparative study, 2019 International conference on electrical, computer and communication engineering (ECCE), Cox'sBazar, Bangladesh, pp 1–4
4. Ranggadara I, Wang G, Kaburuan ER (2019) Applying customer loyalty classification with RFM and Naïve Bayes for better decision making, 2019 International seminar on application for technology of information and communication (I Semantic), Semarang, Indonesia, pp 564-568
5. Yoseph F, Heikkila M (2018) Segmenting retail customers with an enhanced RFM and a hybrid regression/clustering method, 2018 International conference on machine learning and data engineering (include), Sydney, NSW, Australia, pp 108–116
6. Maryani I, Riana D (2017) Clustering and profiling of customers using RFM for customer relationship management recommendations, 2017 5th International conference on cyber and IT service management (CITSM), Denpasar, Indonesia, pp 1–6
7. Sayan IU, Demirdag M, Yuceturk G, Yalcinkaya SM (2022) A review of customer segmentation methods: the case of investment sector, 2022 IEEE 5th International conference on big data and artificial intelligence (BDAI), Fuzhou, China, Intelligence (BDAI), pp 200–204
8. Saraf E, Pradhan S, Joshi S, Sountharrajan S (2022) Behavioral segmentation with product estimation using K-means clustering and seasonal ARIMA, 2022 6th International conference on trends in electronics and informatics (ICOEI), Tirunelveli, India, pp 1641–1648
9. Ait Daoud R, Amine A, Bouikhalene B, Lbibb R (2015) Combining RFM model and clustering techniques for customer value analysis of a company selling online, 2015 IEEE/ACS 12th International conference of computer systems and applications (AICCSA), Marrakech, Morocco, pp 1-6
10. Hu YH, Wu F, Yen TW (2010) Considering RFM-values of frequent patterns in transactional databases, The 2nd International conference on software engineering and data mining, Chengdu, China, pp 422–427
11. Faran J, Triayudi A, Aldisa RT (2023) Combination of RFM's (Recency Frequency Monetary) method and agglomerative ward's method for donors segmentation, 2023 International conference on computer science, information technology and engineering (ICCoSITE), Jakarta, Indonesia, pp 962–967
12. Hariram S, Raja Subramanian R, Kamalakannan P, Reddy MT, Dhanasekaran S, Ishvarya M (2023)Customer segmentation analysis leveraging machine learning algorithms, 2023 IEEE International conference on contemporary computing and communications (InC4), Bangalore, India, pp 1–8
13. https://www.kaggle.com/code/sarahm/customer-segmentation-using-rfm-analysis/input?select=data.csv

University Timetable Scheduling: A Heuristic and Machine Learning Approach

L. Kamatchi Priya, Tushar Bhat, Tenzin Tsephel, Voolla Sai Rethwik, and Tanya Bansal

Abstract The University Class Scheduling Problem stands as a pivotal and intricate challenge within academia. A proficient university timetable holds the key to seamless academic operations, necessitating the delicate balance of multiple constraints like teacher availability and subject-hour allocations. This NP-Hard problem's complexity arises from its diverse limitations. Manual timetable creation presents a formidable task for educators and evaluators alike, underscoring the need for a system that fosters a balanced and equitable learning environment. This study introduces a novel university timetable scheduling algorithm adept at accommodating dynamic constraints, thus addressing the demand for adaptable and responsive scheduling solutions. By harnessing the potential of heuristic and machine learning algorithms, our Timetable Scheduling System aims to efficiently generate schedules while adhering to specified constraints that should not be compromised, ensuring the production of high-quality timetables. Ultimately, this endeavor strives to streamline the timetable creation process, liberating educators to focus on delivering superior education and optimizing student outcomes.

Keywords Dynamic constraint · NP-Hard · Heuristics · Machine learning

1 Introduction

Back then at various universities, the timetable scheduling was done manually by the faculty timetable coordinator along with some group of teachers [1–3] involved but there were some problems and issues in this system. Over the past few years, educational institutions have struggled to construct a timetable manually while also dealing with dynamic constraints which makes it an NP-Hard problem [2, 4, 5].

L. K. Priya (✉) · T. Bhat · T. Tsephel · V. S. Rethwik · T. Bansal
PES University, Bangaluru, India
e-mail: priyal@pes.edu

S. M. Thampi et al. (eds.), *Fifth International Conference on Computing and Network Communications*, Lecture Notes in Electrical Engineering 1221,
https://doi.org/10.1007/978-981-97-4711-5_13

Collecting data, drafting, and cross-checking for any clashes is indeed very time-consuming. Furthermore, in the event of any modifications, it is imperative to communicate these changes to all relevant stakeholders. [1, 3, 6, 7]. To keep in mind the professors' needs and requirements and ensure the students benefited in every efficient way possible, a Genetic Algorithm for Timetable Scheduling which allocates the resources fairly [2, 8] among the slots is created. The genetic algorithm is a heuristic approach within the realm of machine learning, enabling it to efficiently solve complex optimization problems.

The timetable for the IT engineering department of the university is generated using Genetic Algorithm and Adaptive Elitist Method for Optimization with constraints that can be adjusted to match the requirements of various institutions [2, 8].

Previous scheduling algorithms have presented several disadvantages when dealing with dynamic constraints [9]. Traditional methods often struggle to accommodate last-minute changes, leading to inefficient schedules and disruptions in the learning process. Additionally, these algorithms do not adequately consider student and faculty preferences, resulting in suboptimal timetables that do not align with the needs and expectations of the university community [4]. These limitations underscore the urgency of developing an algorithm capable of handling dynamic constraints effectively, thus prompting the need for our research.

The existing body of research on university timetable scheduling has shown notable gaps in addressing dynamic constraints [4]. Many prior studies have focused primarily on static scheduling scenarios, neglecting the evolving nature of academic requirements. Furthermore, while some research has explored the incorporation of dynamic constraints, there remains a lack of comprehensive solutions that consider various factors simultaneously [2, 9]. Our research aims to bridge these gaps by proposing an algorithm that not only handles dynamic constraints but also optimizes schedules by considering multiple variables, including student preferences and faculty availability.

The coded program produced a high-performance scheduling solution by optimizing the number of courses in each department and scheduling them based on their credit priority. The suggested approach outperformed manual design and existing systems during the test phase and displayed excellent outcomes.

In the following sections of this research paper, we will conduct a thorough literature review, examining established scheduling algorithms, their merits, drawbacks, and their relevance to our research. Through this analysis, we aim to lay the groundwork for our proposed solution and highlight the innovations and improvements our approach brings to the field of university timetable scheduling. Our research ultimately seeks to contribute to the development of a more adaptable and efficient scheduling system for academic institutions, improving the overall educational experience for students and faculty alike.

2 Literature Review

2.1 *Timetable Scheduling Using Evolutionary Algorithm*

Abduljabbar, Israa Abdulameer, and Sura Mahmood Abdullah, in their research paper aimed to optimize the algorithm employed in generating timetable systems. Its objective is to craft timetables that are free from scheduling conflicts, thus enhancing operational efficiency. Additionally, the research endeavors to broaden the scope of course scheduling systems, aiming for universality. This approach promotes consistency in timetable creation across diverse campuses and facilitates the tailoring of timetables to meet the specific requirements of any educational institution.

The proposed algorithm has several advantages. First, it offers a high degree of flexibility, as Genetic Algorithm (GA) generates multiple solutions, allowing users to request changes as needed. Second, it operates with impressive computational efficiency, significantly reducing the time required for crafting optimal timetables. Third, it provides a streamlined data entry and revision process, enhancing user-friendliness. The resulting high productivity, along with the elimination of paperwork, simplifies administrative tasks associated with timetable management.

However, the study identifies notable gaps in previous research. Prior research treated data as constants, restricting the generated timetable to a fixed number of courses. In contrast, this paper presents data as dynamic factors, allowing for the creation of multiple timetables with the flexibility to add or remove courses as necessary. The complexity of scheduling timetables for educational purposes remains a challenge, requiring further research to address its intricacies and refine scheduling algorithms [6].

2.2 *Institute Timetable Scheduler Using Genetic Algorithm*

Gore et al. in their research paper aimed to minimize human effort while maximizing efficiency, offering a generic solution adaptable to various educational institutions. It caters to user-defined constraints and prioritizes ease of use for system users, allowing for automatic timetable creation. Notably, the focus remains on resource optimization, encompassing teachers, labs, and rooms, among others. The system's inclusivity is further demonstrated by providing a user-friendly interface for timetable access and generating multiple useful views from the timetable.

The advantages of employing Genetic Algorithm (GA) in this context are manifold. GA, as an adaptive heuristic search algorithm, can produce high-quality solutions for optimization and search problems, making it an intelligent tool for exploiting historical data to direct searches toward better-performing solutions. Crucially, GA's capacity to handle both hard and soft constraints [10] is vital in the context of academic timetable scheduling, allowing for an efficient allocation of resources and adherence to college rules [11]. Moreover, GA can generate feasible timetables based on diverse inputs such as student details, subjects, and classroom availability.

However, the authors acknowledge the persisting challenge of manual lecture timetable scheduling across universities due to its inherent complexities. The scheduling problem is rooted in constraint satisfaction, where computational time increases exponentially with the growth of variables while consistently trying to deliver high-quality solutions [7].

2.3 Timetable Scheduling Using Dynamic Chromosomes

Alnowaini and Aljomai, in their research paper, highlight the introduction of a Genetic Algorithm (GA) as an effective solution for the University Course Timetabling Problem. This algorithm stands out for its notable flexibility, enabling it to adapt to varying course numbers across different departments, a feature absent in traditional manual scheduling systems. The GA demonstrates resource efficiency, demanding minimal processing and computing power, thereby rendering it accessible to institutions with diverse computational resources. Moreover, it significantly reduces the time required for generating near-optimal schedules, ultimately enhancing overall productivity. The user-friendly interface simplifies data input and review, streamlining the timetabling process.

Additionally, the GA achieves an impressive optimality rate of 93%, surpassing the performance of manual scheduling and other available systems while simultaneously reducing paperwork. However, the study acknowledges certain constraints, such as the system's limitations in handling complex constraints, which may necessitate manual intervention and the need for manual review of generated timetables. Notably, not all facets of the genetic algorithm have been fully implemented, suggesting room for further optimization and refinement in future iterations of the system.

In summary, the Genetic Algorithm-based solution presents a promising approach to dynamic university course timetabling, with opportunities for ongoing improvement in algorithm completeness and automation. [2]

2.4 Timetable Scheduling by Dividing Constraints into Soft and Hard

Kutty Mammi and Ying Ying in their research paper aimed to create a Timetable Generator software using Genetic Algorithm (GA). The approach innovatively divided constraints into two categories, namely Soft and Hard constraints, which successfully addressed the complex challenges associated with timetable scheduling. This software has a notable advantage of remarkable efficiency in timetable generation. In contrast to the time-consuming manual methods that can take up to 5–6 days to complete, this project achieved an impressive time frame, averaging just 5 min and 34 s to generate a timetable. This substantial reduction in time not only streamlines the scheduling process but also significantly enhances productivity and resource utilization. However, the system has its limitations. One of the key challenges lies in the division of constraints into hard and soft categories. Hard constraints, which are mandatorily fulfilled, are easier to manage. However, soft constraints, which are optional, present a greater challenge [11, 12]. They are often neglected, leading to clashes in schedules and a lower number of users being satisfied. This limitation underscores the need for careful consideration of soft constraints to improve the overall quality of generated timetables.

In summary, the research project presented in the cited paper introduces an innovative approach to timetable generation using Genetic Algorithm, effectively dividing constraints into Soft and Hard categories. While it significantly reduces timetable generation time, it also highlights the complexity of managing soft constraints and the importance of addressing them to optimize scheduling outcomes. This software offers a valuable tool for institutions seeking to expedite and enhance their timetable scheduling processes [8, 10].

2.5 *Multi-stage Class Timetable System*

Wong et al. [10], in their research paper aimed to develop an efficient college course timetable through a multi-stage approach that incorporated heuristics and grouping strategies. The system achieved remarkable efficiency in generating college course timetables, taking only 1–10 s using standard commodity hardware and a relatively simple algorithm. This achievement underscores the system's feasibility for widespread adoption and its potential to reduce the time and resource investments traditionally associated with timetable generation. However, the algorithm's dynamic constraints are relatively straightforward, initialized with a non-empty timetable already filled with ongoing semester classes. It may not account for more complex and evolving scheduling scenarios that institutions may face. In conclusion, the research offers an efficient and practical multi-stage approach to college course timetable generation, incorporating heuristics and grouping strategies. The system's ability to generate timetables quickly holds promise for institutions seeking to streamline their scheduling processes. However, it is crucial to consider the system's limitations in handling more intricate dynamic constraints and evolving scheduling requirements [13].

2.6 *Particle Swarm Optimization*

Ahmad, Azmi, and Faizah Shaari aimed to solve the challenging task of creating optimal exam timetables for universities using Particle Swarm Optimization (PSO).

Particle Swarm Optimization (PSO) is a population-based evolutionary computation technique that simulates the movement and interaction of particles in search of an optimal solution, inspired by the social behavior of birds and fish. Particle Swarm Optimization (PSO) efficiently explores the solution space to create timetables that meet all criteria in the context of exam timetabling. The use of Particle Swarm Optimization (PSO) in this task has many advantages. It is fast to converge, allowing it to quickly find high-quality solutions to exam timetabling problems. Its minimal parameter settings simplify implementation. Particle Swarm Optimization (PSO) adeptly handles constraints, making it suitable for scenarios where the same student cannot take two different exams simultaneously. Its adaptability to various problem domains and the incorporation of neighborhood search techniques for refining solutions enhance its flexibility and solution quality. However, it is also important to note the potential disadvantages when using Particle Swarm Optimization (PSO). Particle Swarm Optimization (PSO) may occasionally get trapped in local optima [14], potentially missing the global optimal solution. Sensitivity to parameter choices, especially in complex problem spaces, can impact its performance. Computational complexity may become a concern for large-scale timetabling problems, leading to longer processing times [14]. Additionally, Particle Swarm Optimization (PSO)'s tendency to converge to solutions lacking diversity may limit its ability to explore the entire solution space. Finally, Particle Swarm Optimization (PSO) may face challenges in scalability for exceedingly large timetabling problems, where the required number of particles can become impractical.

In conclusion, Ahmad et al. paper demonstrates that Particle Swarm Optimization (PSO) offers an efficient and effective optimization technique for generating optimal exam timetables while accommodating various constraints. However, users should be aware of its potential limitations, including local optima, parameter sensitivity, computational complexity, diversity issues, and scalability concerns, when applying Particle Swarm Optimization (PSO) in large-scale and complex scheduling scenarios [1].

2.7 *Priority-Based Scheduling*

Yadav et al. the central focus of their research revolves around achieving conflict-free resource allocation and balanced workload management [12] in the context of cloud computing. The primary objective of this paper is to address the challenges associated with resource allocation and workload optimization in cloud computing environments. To achieve this, the authors employ a priority scheduling algorithm, which categorizes tasks into high, medium, and low priorities. This categorization enables the utilization of various scheduling techniques, such as round-robin or Weighted Fair scheduling algorithms, to efficiently manage tasks within the cloud infrastructure.

Among the notable advantages of the proposed approach is its support for backfilling, facilitated by algorithms like CMBF (Conservative Migration-supported

Backfilling) and AMBF (Aggressive Migration-supported Backfilling). These algorithms enable the identification and tracking of backfilling jobs, enhancing overall scheduling efficiency. Additionally, the system accommodates parallel workloads and considers user task deadlines, ensuring timely and responsive task execution.

However, certain disadvantages are associated with this approach. The division of tasks into different priorities may introduce subjectivity, as the prioritization can vary based on individual needs and requirements. This variability can potentially lead to buffer overflow issues, where tasks may need to be dropped from the queue due to resource constraints.

In summary, their research addresses the critical aspects of resource allocation and workload management in cloud computing through a priority-based task scheduling approach. The incorporation of backfilling algorithms, support for parallel workloads, and consideration of task deadlines are among its strengths. Nevertheless, the subjective nature of task prioritization and the potential for buffer overflow challenges should be carefully considered when implementing this approach in cloud computing environments [15].

2.8 Ant Colony Optimization

Gore and et al. in their research paper, the primary objective of applying the Ant Colony Optimization (ACO) algorithm is to develop a heuristic solution for solving the university timetabling problem. This heuristic approach extends to addressing the highly constrained real-world instance of the college course timetabling problem. The Ant Colony Optimization (ACO) algorithm draws inspiration from the Traveling Salesman Problem (TSP), a classic combinatorial challenge, and aims to find efficient schedules for academic courses.

The advantages of employing Ant Colony Optimization (ACO) for timetable generation are noteworthy. First, Ant Colony Optimization (ACO) is recognized as a metaheuristic algorithm well-suited for solving combinatorial optimization problems, particularly the complex college course timetabling problem. Second, its population-based nature enables the exploration of multiple solutions simultaneously, enhancing the potential for higher-quality timetable solutions. Third, Ant Colony Optimization's adaptability to various problem domains and constraints makes it a flexible choice. Additionally, its robustness in handling noisy and incomplete data adds to its appeal. Last, Ant Colony Optimization's scalability allows it to effectively manage large-scale scheduling instances, a crucial feature for real-world educational institutions [16].

However, it is essential to acknowledge that ACO, like other metaheuristic algorithms, does not guarantee optimal solutions [17] but strives to find good solutions within a reasonable timeframe. The algorithm's performance can be influenced by parameter choices, such as the number of ants, pheromone update rates, and heuristic functions [18]. Therefore, effective tuning of these parameters is imperative to achieve optimal performance. In summary, the utilization of ACO for timetable generation offers the potential for high-quality solutions that satisfy both stringent and flexible constraints, all while maintaining efficiency and scalability in addressing complex academic scheduling challenges.

ACO is not inherently designed to handle dynamic scheduling problems or last-minute changes. Universities often face unexpected events or scheduling adjustments, and ACO may struggle to adapt quickly.

While the algorithms and techniques mentioned above are accurate and efficient in their own ways, they are chastised by limitations and effectiveness.

Hence, in our system, we have modified the underlying objective function implemented using Genetic Algorithm and combined it with Adaptive Elitist Method to suit the specific constraints to improve the accuracy while also avoiding getting trapped in local optima [3].

3 Proposed Methodology

In this section, the overall methodology of each step is explained accordingly. The application takes input parameters related to class scheduling including working days of the week, class duration, number of sections, timeslots, and teacher details from the graphical user interface. We have implemented a fitness function to evaluate the fitness of a schedule using a penalty-based method. The fitness score is calculated by assigning a penalty score to the constraints and then tallying up the final penalty score based on the number of constraints violated. The quality of the schedule is determined by the fitness score. Once the algorithm receives the input parameters, it generates an initial population of schedules (individuals).

The fitness of this population is calculated and then genetic operators like mutation and crossover are used to create new solutions. These operators ensure the exploration and diversity of the search space. To ensure the satisfaction of the constraints, a repair function is called after every genetic operation to repair the solutions. The algorithm dynamically changes the genetic operators (Adaptive) based on the mean and variance of the population. This helps balance the exploitation and exploration of the search space. The algorithm runs for a fixed set of generations and produces the timetable with the lowest fitness score as the best solution for the user (Fig. 1).

3.1 Adaptive Crossover Rate

This feature dynamically adjusts the crossover rate based on population fitness statistics. It balances exploration and exploitation by increasing the crossover rate when fitness mean is higher and variance is lower.

Algorithm Our Algorithm for Timetable Scheduling

```
population ← initialize_population()
mutation_rate ← 0.01
crossover_rate ← 0.9
for generation = 1 to max_generations do
    Evaluate fitness of population
    offspring ← select(population)
    offspring ← clone(offspring)
    Retain elite individuals
    for each pair (child1, child2) in offspring do
        if random() < crossover_rate then
            crossover(child1, child2)
            repair(child1), repair(child2)
        end if
    end for
    for each mutant in offspring do
        if random() < mutation_rate then
            mutate(mutant)
            repair(mutant)
        end if
    end for
    Evaluate fitness of offspring
    population ← merge(population, offspring)
    Adjust mutation_rate and crossover_rate based on fitness variance and
    fitness mean
end for
best_individual ← select_best(population)
Update timetable with best_individual
```

Fig. 1 Algorithm for our approach

$$\begin{aligned}\text{Crossover Rate } &= \text{Max}(1,\ \text{Max}(0.01,\ \text{Crossover Rate})) \\ &\quad + (\text{Fitness Mean}/(\text{Fitness Variance} * *2))\end{aligned} \tag{1}$$

3.2 Adaptive Mutation Rate

The mutation rate dynamically changes based on the variance and mean of the population's fitness values. It encourages a variety of solutions at the start and gradually converges towards better solutions.

$$\begin{aligned}\text{Mutation Rate } &= \text{Min}(1,\ \text{Max}(0.01,\ \text{Mutation Rate})) \\ &\quad + (\text{Fitness Variance}/(\text{Fitness Mean} * *2))\end{aligned} \tag{2}$$

3.3 Specific Constraints Addressed

The fitness function evaluates various specific constraints relevant to the scheduling problem, including teacher over-teaching, teacher consecutive hours, timetable completeness, etc.

3.4 Weighted Fitness Function

The fitness function assigns higher penalties (weight 100) to hard constraints, reflecting their critical importance in the scheduling problem.

3.5 Clamping of Adaptive Rates

Both the adaptive crossover and mutation rates are clamped to remain within certain bounds, preventing extreme changes in the rates.

3.6 Use of Repair Function

A repair function is employed after mutation and crossover to ensure constraint satisfaction. It fills empty slots with classes that adhere to constraints, significantly improving solution feasibility.

These unique features and corresponding formulas showcase the specialized and innovative aspects of the research in the context of genetic algorithm-based scheduling (Fig. 2).

4 Results and Discussions

For experimental purposes, we have chosen our university data as input to our algorithm. The inputs provided include the identification of section names, specification of working days, determination of class durations, and the available time slots. Furthermore, the input data comprise crucial information regarding the teacher's name, including their respective subjects, the total number of hours a subject needs to be taught per week, as well as the maximum number of class hours any given teacher can deliver per week. Additionally, the specification of the maximum allowable consecutive classes is also given.

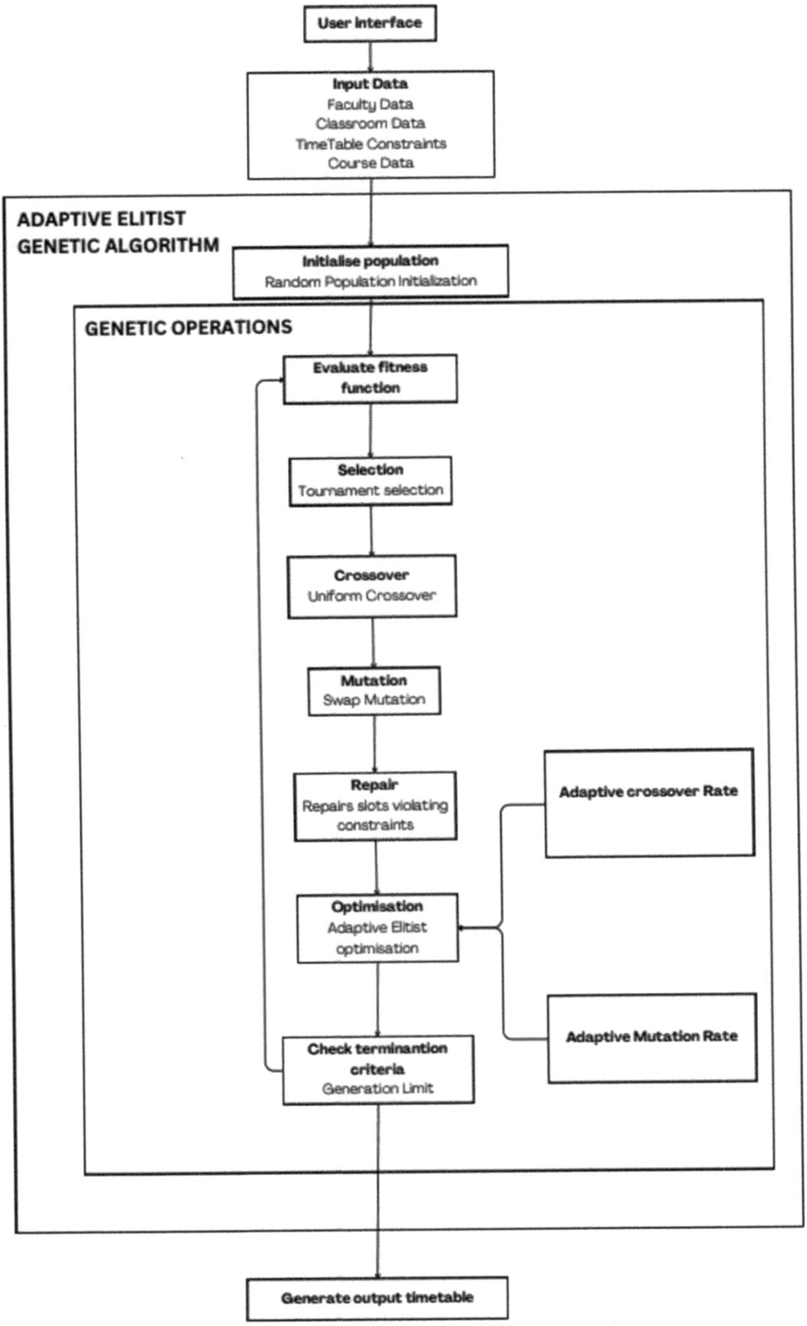

Fig. 2 Architecture of the timetable generation using GA

The experimental setup used consisted of 8 GB of RAM (Random Access Memory) and a 6-core CPU (Central Processing Unit) with Windows 10 Operating System. The program was executed using the Python programming language, version 3.10.

The experiment was conducted for a total of three distinct algorithms across four varied input configurations for three different datasets.

The result includes instances where hard constraints are not met, serving as the benchmark to evaluate the algorithms' performance. The hard constraints considered are teachers teaching multiple sections at the same time (C1) and Empty slots encountered in the scheduled timetable (C2) (Table 1).

The first dataset included a standard workweek of five days, with each day comprising 60 min of class duration and six time slots. For the first set of inputs, designated for four sections, the parameters encompassed four section names, five subjects, and a total of 20 teachers distributed evenly across these sections. The second set, intended for five sections, featured five section names, five subjects, and a total of 25 teachers. Moving to the third set, tailored for six sections, the input specifications included six section names, five subjects, and a total of 30 teachers. Last, for the fourth set catering to seven sections, the parameters entailed seven section names, five subjects, and a total of 35 teachers (Table 2).

We then conducted the experiment for a second dataset, that differed from the first dataset only in the class duration and the number of classes in a day. Our new dataset had a class duration of 45 min and consisted of nine classes in a day (Table 3).

Table 1 Comparison of number of constraint violations in Dataset 1

	Ant colony optimization		Particle swarm optimization		AEGA	
Clash	C1	C2	C1	C2	C1	C2
4 sections	12	0	11	0	0	0
5 sections	22	0	15	0	0	0
6 sections	11	0	12	0	1	0
7 sections	9	0	8	0	2	1

Table 2 Comparison of number of constraint violations in Dataset 2

	Ant colony optimization		Particle swarm optimization		AEGA	
Clash	C1	C2	C1	C2	C1	C2
4 sections	9	0	12	0	0	0
5 sections	8	0	11	0	0	0
6 sections	18	0	12	0	0	0
7 sections	21	0	25	0	1	0

Table 3 Comparison of number of constraint violations in Dataset 3

	Ant colony optimization		Particle swarm optimization		AEGA	
Clash	C1	C2	C1	C2	C1	C2
4 sections	7	0	13	0	0	0
5 sections	14	0	19	0	1	0
6 sections	6	0	21	0	1	0
7 sections	11	0	8	0	2	0

The third dataset includes a workweek of six days, with each day comprising 60 min of class duration and five time slots. Aside from these input parameters, the dataset is similar to the first dataset.

As seen from the above observations, our proposed algorithm has demonstrated remarkable efficacy when compared to techniques such as Ant Colony Optimization (ACO) and Particle Swarm Optimization (PSO). Specifically, our algorithm exhibits superior performance in minimizing clashes related to two critical hard constraints.

In our empirical study, we systematically evaluated the performance of our algorithm alongside ACO and PSO across different scenarios characterized by varying numbers of sections. The results showed the effectiveness of our approach in mitigating clashes. For instance, in the case of scheduling for four sections, our algorithm yielded substantially fewer clashes (0 for C1 and 0 for C2) for all the datasets in comparison to both ACO and PSO. Similarly, across scenarios involving 5, 6, and 7 sections and more, our algorithm demonstrated superior clash reduction capabilities, highlighting its robustness and practical applicability in real-world scheduling challenges.

Our algorithm across various scenarios with 4, 5, 6, and 7 sections, consistently outperformed ACO and PSO, yielding significantly fewer clashes. This can be credited to the fact that ACO and PSO are more likely to get stuck in a local optima and not give the most optimal solution [19]. Our algorithm's oversight of the other two, ACO and PSO, marks a significant advancement in the field of timetable scheduling optimization. These findings not only emphasize the practical applicability of our approach but also reinforce its potential to revolutionize scheduling paradigms, offering an innovative solution to real-world scheduling challenges.

Based on the above results, it is evident that our algorithm displays many advantages that show its superiority over other algorithms. Our algorithm provides good results not just at a lower scale, but also at a higher number of parameters, thereby demonstrating its potential and scalability. Being able to generate timetables without any empty slots, showcases its ability to efficiently utilize all the resources given to it.

5 Conclusion

The research presented in this study has contributed significantly to the field of timetable scheduling, particularly in the context of dynamic constraints. It is evident that our approach surpasses alternative optimization techniques such as Particle Swarm Optimization (PSO) and Ant Colony Optimization (ACO). This assertion is reinforced by several key findings. First, as the number of sections increases, the number of clashes in the timetables generated using our algorithm is significantly lower compared to other algorithms. This observation proves the robustness and scalability of our approach, making it particularly suitable for complex scheduling scenarios.

Second, the minimality of empty slots in the timetables generated through our algorithm highlights its ability to efficiently utilize available resources. This ensures optimal utilization of time and facilities, ultimately leading to cost savings and improved operational efficiency.

Furthermore, the high level of consistency in teacher assignments achieved through our algorithm is a crucial advantage. Maintaining teacher consistency is essential for providing a stable and conducive learning environment [1, 3, 6], and our algorithm has demonstrated its capability to achieve this objective. Finally, the accuracy and efficiency of the timetables generated by our algorithm cannot be overstated. The algorithm consistently outperformed other methodologies, producing timetables that not only adhere to dynamic constraints but also optimize various relevant criteria, including faculty allocation, course distribution, and student preferences.

In essence, this research has set a new benchmark in the domain of timetable scheduling using multiple constraints. The application of Genetic Algorithms has not only addressed the limitations of previous methods but has also proven to be the superior choice in terms of effectiveness, accuracy, and practicality [2, 8]. These achievements underscore the significance of this research in enhancing scheduling processes across various educational institutions and operational settings. As such, the findings presented here contribute to the ongoing advancement of scheduling algorithms and have the potential to significantly impact real-world applications in educational institutions and beyond [20].

References

1. Ahmad A, Shaari F (2016) Solving university/polytechnics exam timetable problem using particle swarm optimization, Proceedings of the 10th international conference on ubiquitous information management and communication
2. Alnowaini G, Aljomai AA (2021) Genetic algorithm for solving university course timetabling problem using dynamic chromosomes, 2021 International conference of technology, science and administration (ICTSA), Taiz, Yemen, 1-6, https://doi.org/10.1109/ICTSA52017.2021.9406539
3. Gore P, Sonawane P, Potdar S (2017) Timetable generation using ant colony optimization algorithm. Int J Innov Res Comput Commun Eng 5(3):6033–6039

4. Fiarni C et al (2015) Automated scheduling system for thesis and project presentation using forward chaining method with dynamic allocation resources. Proced Comput Sci 72:209–216
5. Limkar S et al (2014) Genetic algorithm: paradigm shift over a traditional approach of timetable scheduling. Proceedings of the 3rd international conference on frontiers of intelligent computing: theory and applications (FICTA), Cham: Springer International Publishing, Vol 1
6. Abduljabbar IA, Abdullah SM (2022) An evolutionary algorithm for solving academic courses timetable scheduling problem. Baghdad Sci J 19(2):0399
7. Gore B, Shirdhankar D, Belanekar G (2020) Institute timetable scheduler
8. Kutty Mammi H, Ying Ying L (2021) Timetable scheduling system using genetic algorithm for school of computing (tsuGA). Int J Innov Comput 11(2):67–72
9. Cooper TB, Kingston JH (1996) The complexity of timetable construction problems. Practice and theory of automated timetabling: first international conference Edinburgh, UK, August 29–September 1, 1995 Selected Papers 1. Springer Berlin Heidelberg
10. Wong CH, Goh SL, Likoh J (2022) A genetic algorithm for the real-world university course timetabling problem, 2022 IEEE 18th International colloquium on signal processing & applications (CSPA), Selangor, Malaysia, 46-50, https://doi.org/10.1109/CSPA55076.2022.9781907
11. Adrianto D (2014) Comparison using particle swarm optimization and genetic algorithm for timetable scheduling. J Comput Sci 10(2):341
12. Chunbao Z, Nu T (2012) An intelligent interactive efficient exam scheduling system (IIEESS v1.0). Proceeding of the Practice and Theory of Automated Timetabling (PATAT) Norway. 2012:437-50
13. Emily Siew SK, Sze SN, Goh KL, Wong JJ (2022) Grouping and heuristics for a multi-stage class timetabling system, 2022 International conference on digital transformation and intelligence (ICDI), Kuching, Sarawak, Malaysia, 111–116, https://doi.org/10.1109/ICDI57181.2022.10007351
14. Schmitt M, Wanka R (2013) Particle swarm optimization almost surely finds local optima, Proceedings of the 15th annual conference on genetic and evolutionary computation
15. Yadav A, Rathod SB (2016) Priority based task scheduling by mapping conflict-free resources and optimized workload utilization in cloud computing. 2016 International conference on computing communication control and automation (ICCUBEA). IEEE
16. Dowsland KA, Thompson JM (2005) Ant colony optimization for the examination scheduling problem. J Oper Res Soc 56:426–438
17. Wen X, Huang M, Shi J (2012) Study on resources scheduling based on ACO allgorithm and PSO algorithm in cloud computing, 2012 11th International symposium on distributed computing and applications to business, Engineering & Science, Guilin, China, 219-222, https://doi.org/10.1109/DCABES.2012.63
18. Kakkar MK et al (1950) (2021) Class schedule generation using evolutionary algorithms. J Phys Conf Series. 1:012067
19. Guéret C et al (1996) Building University timetables using constraint logic programming. Practice and theory of automated timetabling: first international conference Edinburgh, UK, August 29–September 1, 1995 Selected Papers 1. Springer Berlin Heidelberg
20. Czibula O et al (2017) A multi-stage IP-based heuristic for class timetabling and trainer rostering. Ann Oper Res 252:305–333

Speech Emotion Recognition Using Machine Learning

Asritha Veeramaneni, V. Samitha, Talluri Charitha, Sagi Shriya, and Tripty Singh

Abstract Speech emotion recognition (SER) is a technique for accurately determining a person's emotion from their speech. Enhancing communication between people and machines. Even though emotions are arbitrary, it can be hard to annotate audio and it's still hard to gauge someone's emotional state. It now plays a crucial part in technological research as new technologies for human machine or medical applications are developed. It is obvious that surprise and rage feelings are more accurately predicted by our model than other emotions, It is understandable given how significantly the pitch, speed, and other qualities of the audio files for these emotions differ from those of other audio files. The system can capture and analyze various acoustic elements contained in voice signals using the SVM-based architecture and enabling the identification and categorization of various moods. Mel frequency cepstral coefficients (MFCCs), pitch, energy, and spectral features are some of these features, and they all work together to reveal important details about the emotional content that is present in speech. The deployed models attain a remarkable accuracy of 80% in identifying various emotional states from speech data after extensive training and optimization. This level of accuracy shows how well the suggested method works in properly classifying emotions based on auditory characteristics.

A. Veeramaneni (✉) · V. Samitha · T. Charitha · S. Shriya · T. Singh
Department of Computer Science and Engineering, Amrita School of Computing, Amrita Vishwa Vidyapeetham, Bengaluru, India
e-mail: bl.en.u4cse20196@bl.students.amrita.edu

V. Samitha
e-mail: bl.en.u4cse20152@bl.students.amrita.edu

T. Charitha
e-mail: bl.en.u4cse20173@bl.students.amrita.edu

S. Shriya
e-mail: bl.en.u4cse20148@bl.students.amrita.edu

T. Singh
e-mail: tripty_singh@blr.amrita.edu

S. M. Thampi et al. (eds.), *Fifth International Conference on Computing and Network Communications*, Lecture Notes in Electrical Engineering 1221,
https://doi.org/10.1007/978-981-97-4711-5_14

Keywords MFCC (Mel Frequency Cepstral Coefficients) · Zero-crossing rate · Mel spectrogram

1 Introduction

Speaking is one of the simplest and most natural means that people may express themselves. We count on it so much that we understand how important it is even when we utilize different kinds of communication channels, such as emails and text messages, where emotions are frequently used in expressing our emotions. Emotion detection and analysis are essential because emotions are a key component of communication in the current digital era of distant communication. Emotions are random, which makes it difficult to distinguish between them. It is debatable as to the appropriate way to quantify or categorize them. The term "SER system" refers to a group of techniques used to classify and analyze speech data in order to detect any embedded emotions. The analysis of caller-agent interactions and interactive speech-based assistants are just two examples of uses for such a system. We seek to by examining the acoustic properties of the audio data of recordings in the speech that was recorded for this investigation, identify the underlying emotions.

It entails evaluating speech in order to pinpoint the affective and emotional traits of individuals. This makes use of the way that the tone and pitch of the voice frequently express underlying emotion. A computer is used in voice emotion detection technology to glean emotional information from spoken signals. The distinctive qualities and emotional change that emerge from this are then compared, examined, and contrasted. Speech emotional states were evaluated as soon as the legislation governing speech and emotion was finished. Current hot subjects include pattern recognition, signal processing, and voice emotion detection, a brand-new discipline that merges artificial intelligence and artificial psychology. The research was widely used in various fields, including security, interactive learning, entertainment, and human–computer interaction. Speech signal acquisition, emotion recognition and feature extraction are the three key components of speech-emotion processing and recognition systems. The diagram that follows shows the system framework. Nearest Neighbor (KNN), Hidden Markov Model (HMM), Gaussian Mixture Model (GMM), Artificial Neural Network (ANN), Support Vector Machine (SVM) and other classifiers are frequently used for emotion recognition (Fig. 1).

There have been several well-known research on nonverbal communication that convey emotions in addition to basic ones. Studies of laughter and the identification of spoken phrases, which can signal extremely angry or aggressive discourse, are two examples. In a separate study, nonverbal speech sounds have been thoroughly investigated and analyzed. Paralinguistic noises, emotional speech, and expressive voices were used to categorize them. These studies suggest that nonverbal communication plays a significant role in conveying emotions. By understanding and analyzing paralinguistic sounds, we can gain a deeper understanding of the emotions being conveyed in speech.

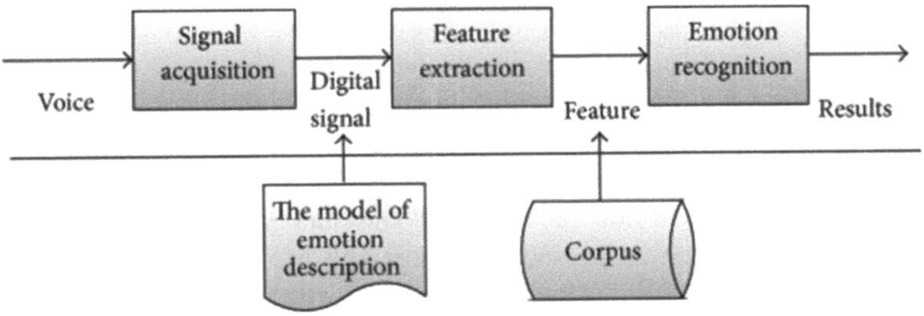

Fig. 1 System block diagram

Signal energy, zero-crossing rate (ZCR), instantaneous fundamental frequency (FO), Formant frequencies, and dominant frequencies are all used in this study. To determine instantaneous FO and ZCR, signal processing, short-time energy, and zero-frequency filtering techniques were utilized. ZCR illustrates the traditional method of identifying emotions. In this study, the speech signals' features are subtracted before the signals are assessed in light of the subtracted features. Formant and Dominant frequencies were carefully evaluated using box and whisker graphs. The sections that follow are arranged as follows: Concentrating on the use of machine learning and voice emotion recognition, Sect. 2 provides a thorough analysis of pertinent work. The proposed methodology, including data pretreatment methods and the design of SVM-based implementation, are presented in Sect. 3. The experimental evaluation discussed in Sect. 4 examines how well the technique performs on test datasets. The paper ends in Sect. 5 with a discussion of the contributions and a preview of possible future paths in the field of machine learning-based emotion recognition.

2 Literature Survey

Analyses the number of supervised learning techniques used in speech emotion recognition, including decision trees, k-nearest neighbour (KNN) and support vector machines (SVM). The authors describe approaches for dimensionality reduction and clustering as well as unsupervised learning strategies for emotion recognition [1].

It examines the various deep learning architectures used in speech emotion recognition, such as CNNs, RNNs, and hybrid models like CNN-RNNs. The authors talk about how well these models work at extracting important elements from speech data and understanding temporal connections [2].

It delves into deep learning training techniques including stochastic gradient descent (SGD) and backpropagation. The authors also go over unsupervised pretraining, regularisation methods, and the use of activation functions in deep networks [3].

It concentrates on several machine learning modalities. The study looks at numerous learning paradigms, including reinforcement learning, transfer learning, unsupervised learning, semi-supervised learning, and supervised learning, and discusses their properties and uses in diverse fields. The paper is a useful tool for comprehending the various machine learning methodologies, as well as their advantages and disadvantages [4].

The focus is on the application of support vector machines (SVM), and it discusses the subject of speech emotion recognition. The survey analyses several feature extraction strategies, such as prosodic features and Mel frequency cepstral coefficients (MFCCs), and highlights the importance of extracting emotions from speech data. The authors review various methods for classifying emotions using SVM and highlight how well they work for precise emotion recognition [5].

It analyses many approaches and strategies utilised in this area, including feature extraction techniques using prosodic, spectral and speech content features. It also examines the importance of emotion recognition. The authors examine various emotion models and classification algorithms, pointing out their benefits and drawbacks in reliably identifying emotions from speech [6].

It explains how to recognise emotions in speech. The survey highlights the importance of accurately deciphering speech signals for emotions by analysing various approaches and methodologies used in this field. The support vector machines (SVM), k-nearest neighbours (KNN), and decision trees are some of the categorization techniques used. Prosodic features, spectral features and speech content features are just a few of the feature extraction approaches they cover [7].

Explores the many strategies and techniques used in this field and demonstrates the importance of accurately detecting emotional indicators in voice communications. The authors discuss various feature extraction methods, including prosodic features, spectral features, and speech content features, and review the use of machine learning algorithms for emotion classification, such as hidden Markov models (HMM), support vector machines (SVM), and artificial neural networks (ANN) [8].

The importance of accurately identifying emotions from voice data is examined, and several neural network topologies and techniques used in this context are discussed. The authors highlight the potential of neural networks for speech emotion recognition by reviewing numerous elements, such as feature extraction techniques, training algorithms, and performance evaluation metrics [9].

Identifying emotions in student speech during tutoring conversations. The survey explores several methods and approaches that have been employed in this environment and emphasises the significance of appropriately identifying emotions in educational contexts. The authors highlight the difficulties and possible uses of emotion recognition in tutoring discussions as they evaluate various features, classifiers, and evaluation approaches [10].

A review of the literature on speech emotion recognition examines several methods and techniques for identifying emotions in speech signals. It includes techniques including machine learning methods and acoustic feature extraction. The survey emphasizes the advancements made in this area and suggests areas for additional investigation and development.

3 Methodology

The architecture is specifically designed to interpret audio inputs, extract features to measure human emotions, and anticipate emotional output based just on speech characteristics. This enables speech emotion detection (Fig. 2).

For voice emotion recognition, the system design often has numerous steps. The audio input is first pre-processed, which could involve actions like segmentation, noise removal, and resampling. Voice activity detectors (VAD) assist in this by classifying the speech signal into unvoiced or voiced segments, speech segments, and non-speech segments. These features are fed into an artificial intelligence system that learns to categorize emotions using the features it has retrieved. The model can be reviewed and fine-tuned to increase performance after being trained on labelled data. To accurately recognize speech emotions, the final system architecture comprises the pre-processing, feature extraction, and classification components.

3.1 Data Description

The dataset consists of 1440 speech audio-only files in 16-bit, 48 kHz. Wav format is taken from the Ryerson Audio-Visual Database of Emotional Speech and Song (RAVDESS). In a neutral North American accent, 24 professional actors 12 women and 12 men vocalize statements that span a range of emotions, including anger, happiness, and sadness. Every emotion has two intensity levels normal and strong as

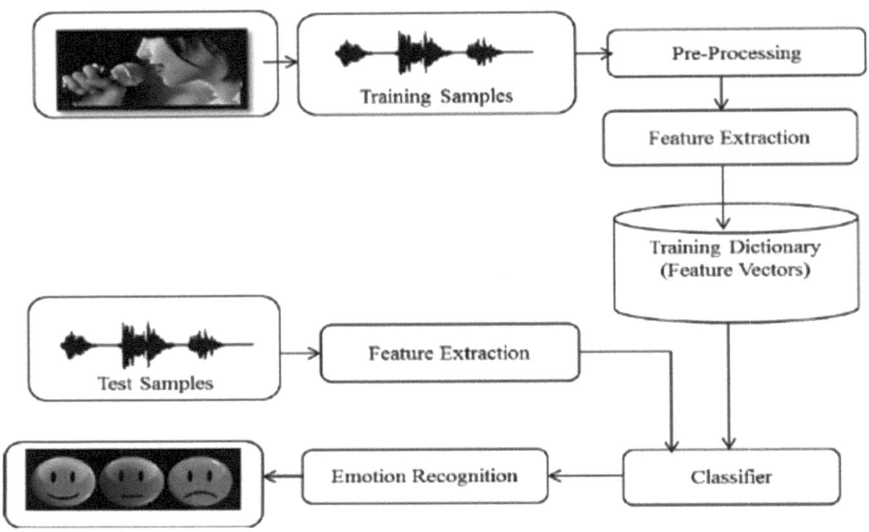

Fig. 2 System Architecture

well as a neutral expression. There is a corresponding Kaggle emotion dataset, and the complete dataset which includes audio and video is accessible on Zenodo.

3.2 Experimental Setup

The system can be broadly separated into the following processes: dataset creation, feature extraction, classification and pre-processing. Programming for the entire system is done in MATLAB. Three emotions, namely anger, happiness, and sadness, were tested using the Ryerson Audio-Visual Database of Emotional Voice and Song (RAVDESS), which included male and female voice audio samples. First, training and testing sets are created from the audio database. To prepare each signal from both sets for data collection and analysis, pre-processing is applied. The pre-processed signal's features are successively extracted. The multiclass support vector machine (SVM) classifier uses the feature vectors as its input to create a model for each emotion.

3.3 Data Pre-processing

For the RAVDESS dataset, pre-processing steps for speech emotion recognition may include Pre-emphasis, Framing, Voice Activity Detection, Normalization, and Noise reduction.

Pre-emphasis: Raw vocal signals can soon become lifeless, which could make them difficult to use in practical situations. After performing a mathematical transform, different parts of the raw signal would have varying degrees of precision. Pre-emphasis is a typical pre-processing method that emphasizes higher frequencies to adjust for the average spectral shape. To get the desired outcome, the signal is routed via a finite impulse response (FIR) filter, often a first-order FIR filter.

Framing: Framing is the division of the voice signal into a number of frames, each carrying a few samples. Since the voice signal is non-stationary and its temporal features change quickly, it is separated into brief frames. As a result of choosing.

With a small frame size, we assume that the speech signal will be stationary and that its characteristics won't change greatly within the frame. We may view voice signals as periodic thanks to frames, which makes Fourier analysis easier.

Voice Activity Detection: The primary issue for a speech recognizer is identifying the endpoints of a signal in a speech. Inaccurate endpoint detection will impair the voice recognizer's performance. Though it might seem simple, speech recognition systems have found it extremely challenging to determine the endpoints of a voice utterance. In a dynamic context, it is difficult to precisely describe quiet and noise; removing voice and noise frames will make it easier to simulate speech. The huge majority of quiet and noisy frames that can be found in speech adds to the computational complexity. Accuracy is raised while complexity is reduced by deleting these

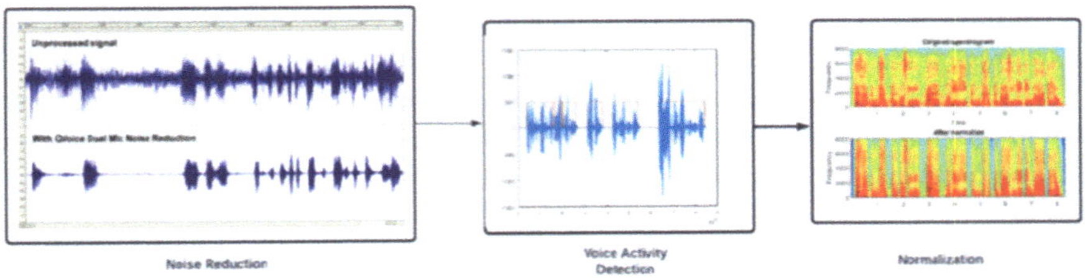

Fig. 3 Audio-signal Pre-processing steps

frames. By categorizing the speech signal into unvoiced or voiced segments, speech segments, and non-speech segments, voice activity detectors (VAD) aid in this.

Normalization: Feature normalization is a crucial step in order to reduce speaker and recording variability without compromising feature discriminative strength. The ability of characteristics to be generalized is improved by feature normalization. The effectiveness of ASR systems can be increased by applying various levels of normalization.

Noise reduction: In the beginning stages of an audio analysis system process, the ability to extract significant voice signal segments from a data stream may be crucial. Any signal that is not the detected signal is referred to as "ambient noise". It is a form of noise pollution or interference. Actually, the amount of background noise has a big impact on how loud ASR systems are. The performance of voice recognition systems is noticeably reduced when training and testing data with various noise levels are employed. To reduce the impact of noise in the speech signals within the identification system, it is essential to control loud speech. Using a close-talk microphone is the most popular method for minimizing the impact of background noise on speech recognition (Fig. 3).

3.4 *Feature Extraction*

In this work, we have utilized the MFCC (Mel Frequency Cepstral Coefficients), Chroma, and Mel Spectrogram, three key components from the audio data.

Mel Frequency Cepstral Coefficients: Any periodic component, such as echoes, is discovered as sharp peaks in the accompanying frequency spectrum, or Fourier spectrum, in the traditional analysis of temporal signals. The temporal signal is subjected to a Fourier Transform to produce this result. Fourier Transform and a spectrogram can be used to find any cepstrum feature. To quantify MFCC, the Mel scale is utilized, which differs from other frequency measurement techniques by linking a tone's perceived frequency to its actual measured frequency. The frequency range chosen is more closely aligned with human hearing. MFCC simulates the vocal tract by carefully reproducing the temporal power spectrum envelope of the voice stream.

Chroma: A chroma vector is a 12-element feature vector that reflects the energy of each pitch class in the signal in a conventional chromatic scale.

Mel Spectrogram: By applying a Fast Fourier Transform on overlapping windowed portions of the data, the spectrogram is produced. In this spectrogram, the only amplitude that has been Mel-mapped is displayed.

3.5 Feature Selection

Reducing the amount of features used to characterize a dataset is the goal of feature selection in machine learning (ML), which aims to improve the performance of a learning algorithm on a particular job. Less features will be required to develop the final classification model because the goal will be to maximize classification accuracy in a specific task using a certain learning approach. Feature selection (FS) chooses a subset of the original features that are relevant based on a predetermined relevance evaluation criterion, which frequently results in increased recognition accuracy. It can significantly cut down on how long learning algorithms take.

The prediction strength of each feature is then established after the estimator has been trained on the original set of features. The least significant features are then eliminated, reducing the size of the current set of features. Once the collection has been pruned, the procedure is then repeated until the ideal amount of features is determined. In this study, basic linear regression (LR-RFE) was used to build for eliminating features recursively, to rank the features. In addition to RFE, another linear model called SVM-RFE, which was based on an SVM, is used in research. Using SVM-RFE, Guyon et al. selected relevant feature sets. It can reduce computing time while accelerating calculations for classification accuracy.

3.6 Classifiers

The linear classification algorithm supported vector machine is used for classification. A Non-probabilistic linear classifier is what SVM does. It is an algorithm that divides categories into two. As a result, we create models that contrast each feeling with the others. The feature vectors are displayed in space, and a hyperplane that is as wide as possible divides the two classes. The collection of points x that satisfy (1), which is how the hyperplane is represented.

$$\vec{w} \cdot \vec{x} - b = 0_1 \tag{1}$$

In this case, w is the hyperplane vector normal. The set y_j, which consists of the categories of the training signals, is another component of the training dataset. From the support vectors, one can deduce the hyperplane's margin or width. They create the x_i boundaries of the hyperplane that satisfy the condition in Eq. (1, 2)

$$y_i(< w_i x_i > +b) = 1 \quad (2)$$

There are three different groups of emotions: happiness, sadness, and anger. Each of the training signals is given a label. The training and test feature sets, along with the label set, are provided to the classifier. For every class, the classifier creates a model, and the signal is compared to each model. The signal's tested emotion is the classifier's output.

4 Results

The explanation provided gives details on how the American English speech corpus was used to calculate the accuracy of emotion classification. A 20% testing set and 80% training set were created from the dataset. Accuracy was also determined for input in real-time (Table 1).

The provided percentages represent the accuracy of a Speech Emotion Recognition (SER) system in identifying specific emotions in speech data. It achieved 92.307% accuracy for anger, 76.923% for happiness, 69.230% for sadness, and an overall accuracy of 79.487% across all emotions. Anger contains higher percentages among all it indicates better precision in recognizing corresponding emotions. The system shows a respectable performance in identifying emotions across the supplied dataset, with an overall accuracy of 79.487%. But there is still opportunity for growth, especially in identifying. The overall accuracy of the system might be improved by more feature set analysis and refining, as well as by investigating cutting-edge categorization algorithms (Fig. 4).

4.1 Qualitative Analysis

These techniques improve the quality of the signal Pre-emphasis boosts the high-frequency components of the signal, boosting some characteristics and enhancing the signal's overall quality. The spectral content of the pre-emphasized signal is altered, higher frequencies are amplified, and any increases in speech clarity and

Table 1 Accuracy comparison

Emotion	Feature activity (%)
Anger	92.307
Happiness	76.923
Sadness	69.230
Overall	79.487

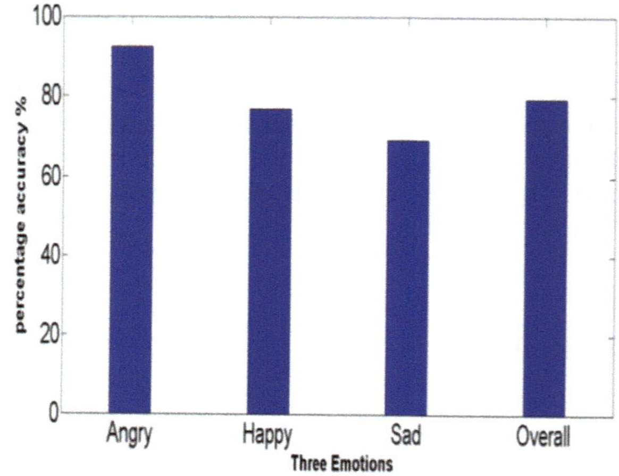

Fig. 4 Accuracies for emotions

understandability are all observed during the qualitative examination of the signal (Fig. 5).

The window function applied to the frame and its impact on the signal are both examined as part of the qualitative analysis of the windowed frame. It entails keeping an eye on the signal's attenuation as it approaches the frame's edges, maintaining the spectral qualities therein, and minimizing transition artefacts (Fig. 6).

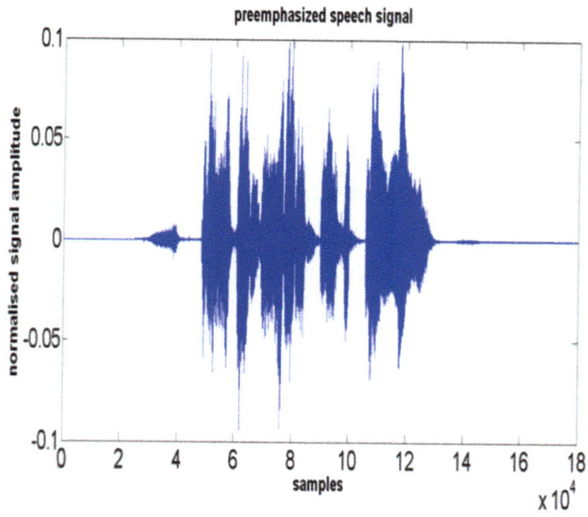

Fig. 5 Speech signal for pre-emphasized

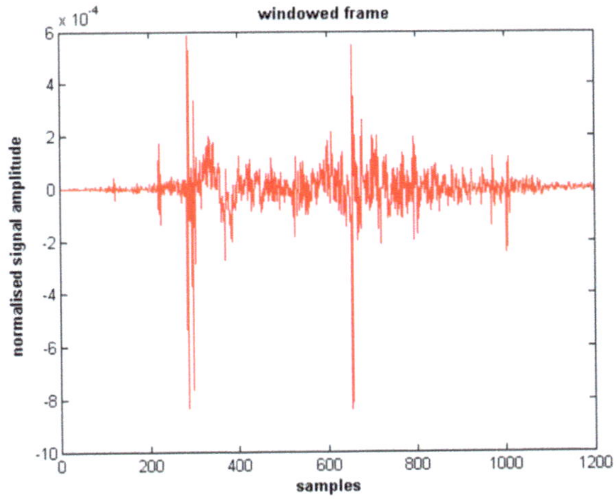

Fig. 6 Windowed frame

5 Conclusion and Future Work

The investigation and analysis of spoken emotion recognition (SER) methods was the main goal of this work. By reaching an accuracy of 80% utilizing Support Vector Machine (SVM) and conventional machine learning techniques, we reached a significant milestone. We used a variety of speech traits and an expertly curated collection of emotional speech to achieve this. We discovered during our research that speech emotion identification holds a lot of promise and potential for practical applications. It does, however, also present a number of problems and opportunities for development. Despite the accuracy that has been attained, SER system performance and capabilities can yet be improved. While traditional machine learning techniques have produced encouraging results, investigating deep learning models for SER may yield notable advancements. Convolutional and recurrent neural network architectures, particularly Long Short-Term Memory (LSTM) networks, have shown exceptional performance in a variety of natural language processing tasks and may also be successful in SER.

References

1. Anusha R, Subhashini P, Jyothi D, Harshitha P, Sushma J, Mukesh N (2021) Speech emotion recognition using machine learning, 2021 5th International conference on trends in electronics and informatics (ICOEI), Tirunelveli, India, 1608–1612, https://doi.org/10.1109/ICOEI51242.2021.9453028
2. Aouani H, Ayed YB (2020) Speech emotion recognition with deep learning. Proced Comput Sci 176:251–260. https://doi.org/10.1016/j.procs.2020.08.027
3. LeCun Y, Bengio Y, Hinton G (2015) Deep learning. Nature 521:436–444. https://doi.org/10.1038/nature14539

4. Sah S (2020) Machine learning: a review of learning types. https://doi.org/10.20944/preprints202007.0230.v1
5. Chavhan Y, Dhore M, Pallavi Y (2010) Speech emotion recognition using support vector machines. Int J Comput Appl. https://doi.org/10.1007/978-3-642-21402-8_35
6. Koolagudi SG, Rao KS (2012) Emotion recognition from speech: a review. Int J Speech Technol 15:99–117
7. Pathak S, Kulkarni A (2011) Recognizing emotions from speech, 2011 3rd International conference on electronics computer technology, Kanyakumari, India, 107–109, https://doi.org/10.1109/ICECTECH.2011.5941867
8. Basu S, Chakraborty J, Bag A, Aftabuddin M (2017) A review on emotion recognition using speech, 2017 International conference on inventive communication and computational technologies (ICICCT), Coimbatore, India, 109–114, https://doi.org/10.1109/ICICCT.2017.7975169
9. Bhatti MW, Wang Y, Guan L (2004) A neural network approach for human emotion recognition in speech, 2004 IEEE International symposium on circuits and systems (ISCAS), Vancouver, BC, II-181, https://doi.org/10.1109/ISCAS.2004.1329238
10. Basu S, Bag A, Mahadevappa M, Mukherjee J, Guha R (2015) Affect detection in normal groups with the help of biological markers, 2015 Annual IEEE India conference (INDICON), New Delhi, India, 1-6, https://doi.org/10.1109/INDICON.2015.7443733
11. Basu S, Bag A, Aftabuddin M, Mahadevappa M, Mukherjee J, Guha R (2016) Effects of emotion on physiological signals, 2016 IEEE Annual India conference (INDICON), Bangalore, India, 1-6, https://doi.org/10.1109/INDICON.2016.7839091
12. Basu S et al (2015) Emotion recognition based on physiological signals using valence-arousal model, 2015 Third international conference on image information processing (ICIIP), Waknaghat, India, 50–55, https://doi.org/10.1109/ICIIP.2015.7414739
13. Litman D, Forbes K (2003) Recognizing emotions from student speech in tutoring dialogues, 2003 IEEE Workshop on automatic speech recognition and understanding (IEEE Cat. No.03EX721), St Thomas, VI, USA, 25-30, https://doi.org/10.1109/ASRU.2003.1318398
14. Bajpai A, Yegnanarayana B (2004) Exploring features for audio clip classification using LP residual and AANN models, International conference on intelligent sensing and information processing, 2004. Proceedings of, Chennai, India, 305–310, https://doi.org/10.1109/ICISIP.2004.1287672
15. Chauhan A, Koolagudi SG, Kafley S, Rao KS (2010) Emotion recognition using LP residual, 2010 IEEE Students technology symposium (TechSym), Kharagpur, India, 255-261, https://doi.org/10.1109/TECHSYM.2010.5469162
16. Krishna KV, Sainath N, Posonia AM (2022) Speech emotion recognition using machine learning, 2022 6th International conference on computing methodologies and communication (ICCMC), Erode, India, 1014–1018, https://doi.org/10.1109/ICCMC53470.2022.9753976
17. Arya R, Pandey D, Kalia A, Zachariah BJ, Sandhu I, Abrol D (2021) Speech based Emotion recognition using machine learning, 2021 IEEE Mysore sub section international conference (MysuruCon), Hassan, India, 613-617, https://doi.org/10.1109/MysuruCon52639.2021.9641642
18. Cao J (2021) Application of machine learning algorithms in speech emotion recognition, 2021 International conference on signal processing and machine learning (CONF-SPML), Stanford, CA, USA, 113–116, https://doi.org/10.1109/CONF-SPML54095.2021.00031
19. Prasanna YL, Tarakaram Y, Mounika Y, Palaniswamy S, Vekkot S (2022) Comparative deep network analysis of speech emotion recognition models using data augmentation, 2022 International conference on disruptive technologies for multi-disciplinary research and applications (CENTCON), Bengaluru, India, 185–190, https://doi.org/10.1109/CENTCON56610.2022.10051557
20. Lalitha S, Madhavan A, Bhushan B, Saketh S (2014) Speech emotion recognition, 2014 International conference on advances in electronics computers and communications, Bangalore, India, 1–4, https://doi.org/10.1109/ICAECC.2014.7002390

21. Poorna SS, Jeevitha CY, Nair SJ, Santhosh S, Nair GJ (2015) Emotion recognition using multi-parameter speech feature classification, 2015 International conference on computers, communications, and systems (ICCCS), Kanyakumari, India, 217–222, https://doi.org/10.1109/CCOMS.2015.7562904
22. Saranya G, Tamilvizhi T, Tharun SV, Surendran R (2023) Speech emotion recognition with high accuracy and large datasets using convolutional neural networks, 2023 8th International conference on communication and electronics systems (ICCES), Coimbatore, India, 859–864, https://doi.org/10.1109/ICCES57224.2023.10192891
23. Avula H, Ranjith R, Pillai AS (2022) CNN based recognition of emotion and speech from gestures and facial expressions, 2022 6th International conference on electronics, Communication and Aerospace Technology, Coimbatore, India, 1360–1365, https://doi.org/10.1109/ICECA55336.2022.10009316

LAYAMRITAM: A Scalable and Secure Multimedia Delivery Platform for Devotional Songs

Ramesh Guntha

Abstract The creation of a platform for delivering multimedia content poses distinct challenges, especially when focusing on devotional music. In contrast to text-based platforms, multimedia delivery necessitates meticulous management of latency, augmented storage space, and bandwidth needs. Although the market is saturated with mainstream multimedia streaming services like Spotify, Apple Music, and Amazon Music, there is a noticeable absence of a platform dedicated to devotional music, despite its deep cultural and spiritual importance. This paper introduces Layamritam, a platform thoughtfully crafted to meet the specific needs of devotional music enthusiasts and artists. Layamritam distinguishes itself through its dedication to providing an immersive musical experience, complemented by sing-along lyrics in 26 languages. We delve into the distributed server architecture of Layamritam, which utilizes adaptive and resilient algorithms to offer a scalable music listening experience while safeguarding user privacy and digital content security. Investigating the user experience of Layamritam, this paper emphasizes its intuitive interface, which organizes songs by deity, mood, speed, tune, and beat style, ensuring a frictionless experience for users across all age groups and digital proficiency levels. Not only does this paper reveal Layamritam, but it also scrutinizes its potential to transform the preservation and distribution of devotional music in the digital age.

Keywords Multimedia streaming · Distributed computing · Client-side caching · Offline mode

R. Guntha (✉)
Center for Wireless Networks and Applications (WNA), Amrita Vishwa Vidyapeetham, Amritapuri, Vallikavu, India
e-mail: rameshg@am.amrita.edu

S. M. Thampi et al. (eds.), *Fifth International Conference on Computing and Network Communications*, Lecture Notes in Electrical Engineering 1221,
https://doi.org/10.1007/978-981-97-4711-5_15

1 Introduction

Music universally captivates beings of all kinds. The evolution of music consumption has transitioned through various technological phases: from cassettes to CDs, mp3 players, online music sites, and currently, streaming platforms. Online streaming services like Spotify, Amazon Music, and Apple Music have notably transformed the music industry [1]. Listeners can effortlessly access a vast array of music from any artist, spanning the oldest to the latest songs, without the hassle of finding, storing, maintaining, and organizing individual music libraries [2]. Artists, in turn, are relieved from the struggle of finding producers or marketing their music. Furthermore, through advanced algorithms, streaming platforms curate daily music recommendations that align with listeners' preferences and explorative tendencies. However, despite these advancements, a conspicuous gap persists, necessitating a dedicated focus on devotional music.

Devotional music holds a unique and special place in the hearts of many. It transcends mere melody, touching our souls, filling us with contentment, and elevating us to realms untouched by contemporary music. Whether through the meditative chants of bhajans, melodious spiritual hymns, or compositions dedicated to deities, devotional music uniquely bridges the earthly with the divine. Its power not only resides in its harmonious melodies but also in its ability to inspire reverence, evoke emotions, and deepen spiritual connections. While folk music has its place, a significant portion of ancient music centered around devotion, only to be overshadowed by pop-culture music in modern times. Although streaming platforms host devotional music alongside mainstream tracks, an exclusive platform for devotional music is desirable for numerous reasons. A dedicated platform allows listeners to focus solely on devotional music, organized into specific categories such as deity, raaga, taala, mood or bhava, speed, language, festivals, and chants, without distraction from secular tracks. A dedicated platform allows for more search options to allow search by deities, festivals, etc. Also the search algorithm can be tailored to take care of phonetic variations in the search query.

A multimedia streaming platform exclusively for devotional songs must be designed with considerations for scalability, bandwidth, latency, and security, among other factors [3]. Devotional music appeals to a global audience of all age groups and digital proficiency levels; therefore, the platform must support a multilingual interface with straightforward screen flows, adhering closely to standard music interfaces. In devotional music, lyrics hold paramount importance, enabling listeners to understand, sing along, and contemplate deeper spiritual meanings. As songs may span multiple languages, lyrics require transliteration and meaning translation. Streaming systems must deliver music with low latency, necessitating servers with sufficient bandwidth and computational power to meet demand. One strategy is to split the music servers from the transactional servers, allowing many music download servers to be distributed globally while keeping one or two transactional servers to handle the registration, subscription, and authentication [4, 5]. Servers must be globally distributed to ensure users in specific geographies can stream songs from the nearest server, with distributed architecture allowing for load balancing during peak times and fault tolerance to provide uninterrupted service even in the event of partial system failure. Given that an exclusive devotional music platform like Layamritam will host significantly fewer songs compared to platforms like Spotify, it can enhance user experience by storing most-listened-to songs on user devices, saving server bandwidth, and enabling low latency and offline listening. However, users should have the option to set a limit on the number of stored songs, considering potential device space constraints. Unlike Spotify or Apple Music, which support operations through advertisement revenue due to their large user bases, exclusive devotional streaming platforms like Layamritam must charge a subscription fee to sustain operations. To safeguard against misuse of downloaded and streamed songs, they must be encrypted and decrypted only in memory while playing.

In subsequent sections of this paper, we explore related work in the fields of music streaming platforms, music subscriptions, music encryption, and distributed content delivery. We delve into the design and user interface of the Layamritam devotional music streaming platform, discussing how it satisfies the considerations outlined above in Sects. 3, 4, 5 and 6. We will also discuss user adoption and feedback, concluding with final thoughts and acknowledgments.

2 Related Works

Streaming music platforms afford listeners a vast array of choices across various genres, artists, periods, languages, and countries. However, such an immense selection can sometimes perplex users, making decisions about what to listen to challenging. Streaming platforms leverage AI algorithms, considering users' interests and listening history to curate personalized playlists [6]. A key objective of these algorithms is to recommend less popular songs, artists, and tracks from genres that listeners might not typically explore, thereby avoiding a filter bubble and popularity bias. This ensures all songs and artists have an equitable opportunity to be heard and rewarded [7]. In the case of Layamritam, implementing AI algorithms for song recommendations is not yet feasible due to the limited number of users and songs. However, we facilitate song discovery by offering the latest and recently popular songs, which are aggregated from all listener community plays. We also showcase an aggregated list of favorites, summarizing selections from the entire listener community, providing a means to tune into the general mood during festival days and special events. Streaming platforms have recognized that beyond musical content and AI-based algorithmic curation, immersive music listening experiences are crucial for attracting and retaining users [8]. Layamritam engineers an immersive devotional experience through various functionalities, such as displaying sing-along lyrics in multiple languages, showing relevant deity images with the playing song, regularly updating albums, enabling auto-logins, restoring the app to its previous state upon login (including navigation view, song queue, and songs) [9], and offering numerous personalization options like theme, color scheme, sort order, and song list view. Layamritam also allows offline mode through caching the song metadata on the mobile device and automatically synchronizing with the server to keep it up to date with latest changes [10]. Despite the availability of AI-generated curated playlists, research indicates that users often prefer to create personal playlists based on theme, mood, and intended use [11]. Layamritam allows users to quickly view songs through various filters like deity, raaga, taala, mood or bhava, speed, year, and language. It also enables searching within these categories and easily converting these search results into personal playlists, thereby simplifying the creation of customized playlists for listeners.

Streaming music platforms have traditionally monetized their content through advertisements and subscription models. Paying customers typically enjoy unlimited access to content without advertisements, while non-paying users encounter a mix of restricted content, features, and advertisements. Over time, content providers have recognized that advertisement revenue is not only undependable but also does not proportionally grow with an expanding user base. In contrast, subscription revenue grows in tandem with the number of paying customers. Consequently, platforms are exploring attractive and flexible subscription plans to encourage users to adopt subscription models [12]. Layamritam, offering exclusive devotional music, operates with a limited user base and song selection. Relying solely on minimal advertising revenue would make it exceedingly difficult to support operations and compensate creators. Therefore, we provide purely subscription-based access. However, we do offer flexible recurring subscription options, such as monthly, 3-month, half-yearly, and yearly plans, as well as one-time payment options for a month or a year. Research indicates that, compared to free users, paying users tend to listen to more songs, create more playlists, and are more active in social forums, sharing their playlists and answering questions [13]. Among Layamritam customers, we have observed that subscribing customers often exhibit a sense of ownership toward the platform. They also demonstrate greater loyalty and actively participate by reporting bugs and requesting new features.

To safeguard artists' interests, streaming platforms must shield music from unauthorized piracy and distribution. Digital Rights Management (DRM) [14] technology is the most used technique by multimedia streaming providers to protect music and video content from piracy. Nevertheless, DRM protection can be bypassed in various ways, such as by stealing encrypted keys or accessing decrypted content in memory [15]. To enhance content protection mechanisms, numerous implementations amalgamate different content protection strategies like DRM, encryption, and watermarking with Blockchain technologies. Reference [16] provides a survey of these implementations, aligning them with a taxonomy organized into seven subcategories, including types of blockchains, transactions, consensus mechanisms, and protection methods. It's crucial to note that many DRM systems, which encrypt the content once and serve the same content to all users, can be easily compromised; if one user's key is compromised, it can be utilized by others to decrypt the content. DRM systems that encrypt content separately for individual users, while secure, are inefficient and are typically limited to protecting lightweight content, such as eBooks. Implementing Blockchain can further impair performance, rendering the content unsuitable for mass consumption. Layamritam introduces an innovative rapid multi-layer encryption mechanism, uniquely encrypting each song with three keys: the system key, song key, and user key, making it impossible to use the same keys across all songs or users. The innovative decryption allows for rapid one-shot decryption, ensuring no compromise on performance.

Content Delivery Networks (CDN) are widely preferred and utilized by multimedia streaming platforms and content-rich websites to deliver bandwidth-intensive content with minimal latency to a global audience [17]. A CDN comprises numerous interconnected servers distributed worldwide, organizing content from providers and delivering it to requesting users, with options to shield the content from unauthorized access. When new content is added or modified, providers can propagate the changes to all CDN servers. Although Layamritam employs the pop server concept from the content delivery network architecture in its download servers [18], we opted to implement our own CDN instead of utilizing established providers like Akamai. The primary reason revolves around the security of the songs; while established CDN providers can deliver commonly encrypted content, they cannot provide individually encrypted content, which is a pivotal requirement for Layamritam, as previously mentioned. Upon receiving a download request, one of the nearest download servers contacts the transaction server to authenticate the user and verify the subscription's validity. Subsequently, it encrypts the song based on a user-specific key before delivering it to the user. This method ensures that even if a user's key is compromised, other users will not be able to access the song.

Before the advent of streaming music providers in the digital music landscape, earlier content providers enabled listeners to purchase and download songs onto their devices. These downloaded songs could be freely shared and used by others and allowed users to listen offline without an internet connection. The download option was less demanding on the service providers in terms of server performance and bandwidth. However, when streaming providers entered the scene, they initially did not offer a download option and only permitted songs to be streamed. This approach allowed content providers to safeguard songs from being shared. Nonetheless, streaming necessitates that service providers maintain low latency to ensure a smooth listening experience, requiring a complex distributed server architecture and imposing substantial bandwidth costs on both listeners and servers [19]. Spotify now allows its paid customers the option to download songs onto their devices for offline listening [20]. In Layamritam, songs are automatically downloaded to the local device for playback, ensuring they are readily available for offline listening. This approach is particularly effective given that Layamritam hosts a collection of 2500 devotional songs, and users frequently select from a consistent set of tracks. Even if they download the entire collection, it will not require more than 10 GB of storage. Nevertheless, we provide an option for users to set an upper limit for stored songs. Once this limit is reached, the app will automatically delete the song that was listened to the earliest.

3 Introduction to Layamritam

Devotional music serves as a spiritual and cultural wellspring, enriching the lives of myriad individuals worldwide. Beyond the meditative chants of bhajans, melodious spiritual hymns, and soul-stirring compositions dedicated to deities, devotional music uniquely bridges the earthly with the divine. It transcends captivating melodies, embedding rich lyrics that invite contemplation on the nuanced facets of devotion and spirituality. The power of devotional music not only resides in its harmonious melodies but also in its ability to inspire reverence, evoke emotions, and deepen one's spiritual connection. While popular streaming platforms like Spotify, Apple Music, and YouTube Music offer a vast selection of devotional songs, they often juxtapose them with materialistic tracks from popular culture, potentially diverting the listener's attention. The visual allure of popular culture and movie album covers might tempt the listener. To preserve the devotional mood and sanctity, continuous and uninterrupted listening of devotional music for extended periods is preferred. A dedicated platform allows listeners to focus exclusively on devotional music, organized into specific categories such as deity, raaga, taala, mood or bhava, speed, language, festivals, and chants, which might be challenging if interspersed with secular songs. A specialized platform also facilitates more search options, allowing users to search by deities, festivals, etc., and the search algorithm can be tailored to accommodate phonetic variations in the search query.

Layamritam enables songs to be organized by various devotional and regional categories, including deity, language, raaga, taala, bhaava, year, speed, and difficulty. It arranges content into diverse albums based on artists, events, festivals, and other special categories like instrumentals, folk songs, and chants. Layamritam allows users to add a list of songs to the current playing queue, offering the ability to reshuffle the queue and remove or reorder songs, thereby enabling a customizable playing experience. The entire queue can be saved as a playlist for future access. Layamritam also features various deity images corresponding to the currently playing song, assisting listeners in entering deeper meditative states while enjoying the music. Layamritam showcases two types of playing screens: a compact player displayed as a thin strip at the bottom and a larger player that occupies the entire screen. The expansive player provides a view of a larger deity image and allows for volume adjustment, displays the current playing time, and offers options for looped or sequential playing (Fig. 1).

Layamritam enables users to mark songs as their favorites and, beyond this, allows the creation of an unlimited number of personal playlists for future listening. Favorites and playlists are stored on the server, ensuring access across various devices and synchronization to the current device upon login. In addition to these features, Layamritam allows listeners to engage with the broader user community by tuning into their current listening trends. The platform aggregates everyone's favorites, presenting a list of songs sorted from most to least favored. Layamritam maintains a history of recently played songs and albums, extending back to 1000 entries. Moreover, it aggregates songs played by the entire user community within the last 1 and 7-day intervals, presenting a sorted list with the most-played songs at the top (Fig. 2).

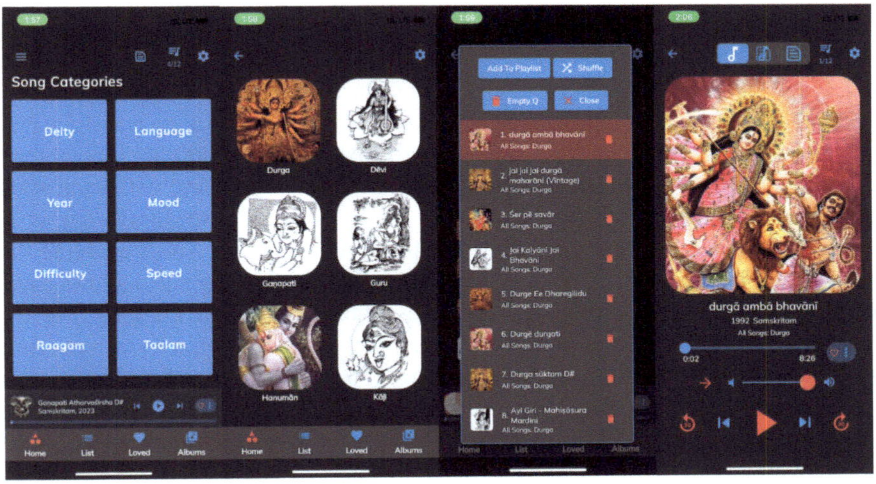

Fig. 1 Layamritam screens, showing categories, deities, playlist, and the player screen

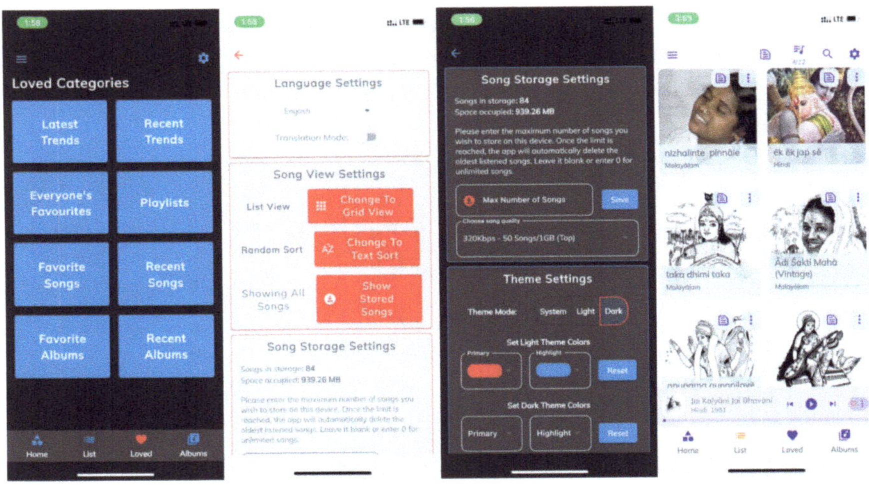

Fig. 2 Playlists, trends, language settings, personalization, and grid view

Layamritam offers a variety of personalization options. Users can modify the theme to System, Light, or Dark, and select primary and highlight colors for both light and dark themes, adding a fun and customized touch to Layamritam. The default sorting of songs across all categories (excluding playlists) can be set to either alphabetical or shuffled sort. Songs can be displayed in a grid or list format, with the list view accommodating more songs, while the grid view offers a visually appealing experience by showcasing larger deity images. Furthermore, Layamritam provides an interface viewable in 26 languages, catering to users across Bharath and the world (Fig. 2).

Layamritam enhances user engagement by featuring sing-along lyrics for all songs in 26 languages, each accompanied by their meanings. The lyrics are transliterated, and the meanings are translated, offering a deeper understanding of the content. Users can opt to view the lyrics in a document mode or as slides, with each slide presenting one stanza of lyrics along with its corresponding meaning. The lyric stanzas are manually tagged with timestamps correlating to the song, enabling the lyrics to automatically progress in sync with the song. This feature also allows users to click on a lyric slide and jump to the corresponding position in the song (Fig. 3).

Given its specificity to devotional songs, which encompass tracks in numerous phonetic languages native to Bharath, searching for a song title on Layamritam can be challenging. A single word might be spelled in multiple ways yet carry the same meaning. For instance, 'Shiva' and 'Siva', 'Kripa' and 'Krupa', or 'Ganesha' and 'Ganesa', as well as 'Jai', 'Jay', and 'Jaya' are considered equivalent. Users searching for 'Siva' might also expect results for 'Shiva'. Layamritam adeptly manages such spelling variations arising from the transliteration of Bharath languages into English, providing comprehensive search results that align with the user's intent.

In addition to its music-related features, Layamritam endeavors to ensure a delightful app usage experience by automatically logging users in upon app opening and restoring the app view to the state it was in prior to the previous exit, whether that

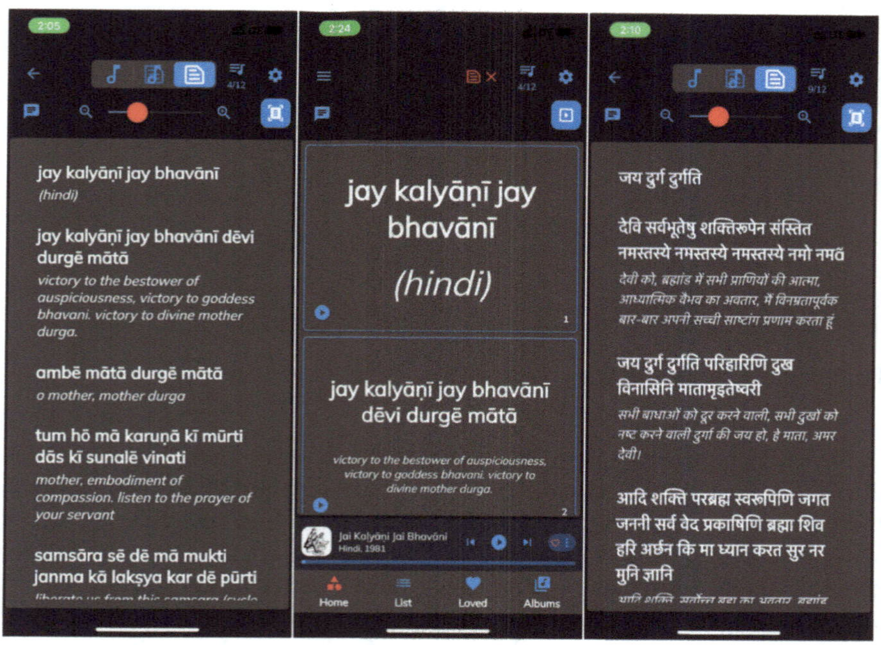

Fig. 3 Lyrics screens. Document view, slide view, and multilingual lyrics

be a specific playlist, album, or deity category. Alongside the view, settings such as theme, color, sort order, and view style are all preserved and restored. Furthermore, the song queue and the song that was being played are also reinstated. All these features afford listeners a seamless experience across app logins.

4 Distributed Multimedia System

Layamritam, at its essence, transcends being merely a multimedia delivery platform; it embodies a transformative journey dedicated to revitalizing the devotional music landscape. Recognizing the unique challenges intrinsic to this genre, such as sensitivity to latency, storage demands, and bandwidth consumption, Layamritam has been meticulously crafted as a comprehensive solution to meet the nuanced needs of devotional music enthusiasts. Fundamental to Layamritam's capabilities is its distributed server architecture, which encompasses two core server types—transactional servers and download servers.

4.1 Transactional Server

The transactional servers act as the pivotal backbone of Layamritam, overseeing crucial functions such as user registration, authentication, subscription management, song details, lyrics, and category organization. This architecture is underpinned by a strategic approach to scalability, incorporating load-balancing configurations to manage varying loads. In a single-server configuration, Layamritam employs an internal caching mechanism to enhance performance. This cache, which stores frequently accessed data, includes vital elements like user profiles, deity information, mood classifications, tunes, song metadata, and albums (Fig. 4). The cache adeptly mitigates latency concerns by reducing the need for repeated database queries. However, as the platform expands and deploys multiple transactional servers, cache synchronization introduces new complexities. For instance, if a song metadata update request is received at server1, it updates the database and its server cache, but server2's cache remains outdated with the previous song data. This inconsistency also affects user data, depending on which server they download the song metadata from. This challenge can be addressed through three alternative solutions: eliminating server-side caching, implementing inter-server notifications for updates, or scheduling periodic data synchronization from the database. The first solution involves avoiding server-side metadata caching, which, while simple, places stress on the database and increases the latency of user-read requests. The second solution entails having servers notify each other upon updates, enabling them to refresh the corresponding data either from the database or through notifications from the other server. While slightly more complex, this solution is highly efficient in terms of performance. The third solution, which is currently adopted by Layamritam, involves scheduling

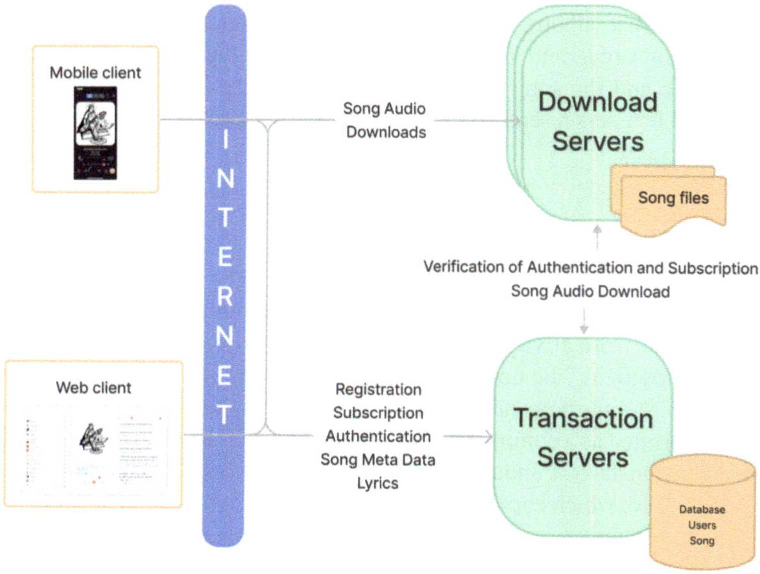

Fig. 4 Layamritam architecture showing download and transactional servers

periodic data syncs from the database, with customizable refresh frequencies based on the nature of the change. Since most cached data do not change frequently, a 5-min refresh rate would typically be more than sufficient, offering a straightforward implementation without sacrificing performance.

4.2 Download Server

Complementing the transactional servers, the download servers are entrusted with the intricate task of delivering devotional songs to users' devices. Before streaming, songs automatically undergo a transcoding process, resulting in multiple quality levels—from the highest quality at 320 Kbps to lower bit rates of 192, 128, and 96 Kbps. This nuanced approach ensures users with varied bandwidth capabilities and device specifications can enjoy a seamless listening experience. There may be more than one download server, distributed across various physical locations. When a request to download a song is received, it is directed to the download server with a song ID, download quality, and auth-token. Subsequently, the download server sends a request to validate the request by transmitting the auth-token, song ID, and desired quality in Kbps to the transactional server. The transactional server verifies the validity of the login and subscription through the auth-token, then checks the timestamp of the requested song file denoting the quality and sends back information related to validity and timestamp. The download server then checks whether it has the song on the local server; if not, it will download the song from the transactional server and then serve it to the requester. If the song is already present at the required quality on the download server, it ensures it is the most recent version by comparing timestamps. If the song is not the latest, it downloads from the transactional server and sends it to the user (Fig. 4).

4.3 Server Caching and Client Caching

Device or browser cache emerges as a pivotal component in effortlessly achieving the requisite scalability of a streaming platform. Local storage not only enables app usability in the absence of a network but also ensures that downloaded songs play instantly, thereby saving bandwidth. Metadata related to songs and other entities, such as deities and albums, is stored on the client device. Each time the app is opened, it automatically updates the local storage with the latest changes, identifying alterations by comparing the last updated timestamp of the storage to the most recent timestamp on the server metadata (Fig. 5). Songs, once listened to, are also downloaded and stored on the mobile device. Users can set a maximum number of songs to be stored on the mobile device or opt for unlimited storage. When a new song is downloaded, the app checks against this limit; if the limit is exceeded, it deletes the least recently listened to songs until the storage is within the limit. Users can also select the default song quality for download and play, affecting latency, bandwidth, and local storage space.

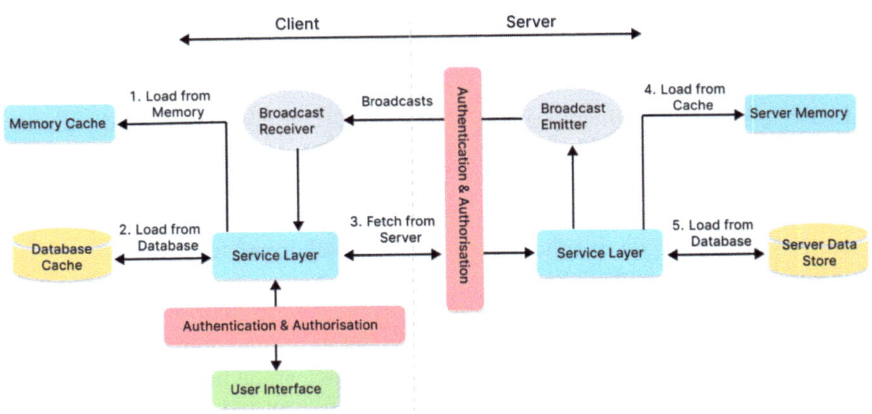

Fig. 5 The cache mechanism on client and server

4.4 Adaptive and Resilient Algorithms

Layamritam places paramount importance on adaptivity and resilience. The adaptive aspect meticulously aligns app features with user needs and conditions, ensuring a user-centric approach rather than imposing a one-size-fits-all solution. Resilience, on the other hand, involves adept recovery from network and bandwidth constraints and failures. As delineated in previous sections, various elements of Layamritam's architecture are crafted with these principles in mind. The offline mode of the client-side mobile application, facilitated by downloaded songs to the mobile client, provides a buffer against unstable network conditions. Multi-bit rate audio empowers users to select song quality that aligns with their bandwidth and space constraints and to accordingly set storage limits on their device. Auto-logins and state restorations guarantee a seamless app experience across listening sessions, while the availability of multiple categorizations and views cater to a spectrum of needs and use cases. Phonetic search functionality further simplifies song location tasks. By embracing these principles, Layamritam not only delivers a robust and user-friendly platform but also ensures a spiritually enriching and uninterrupted devotional music experience for all its users, seamlessly bridging the divine and the everyday through technological finesse.

5 Privacy and Security

Layamritam steadfastly anchors its architecture on the pivotal pillars of security and privacy. Security is fortified by mandating each user to register using a unique email address or, exclusively for customers in Bharath, a mobile number, coupled with a password. Upon registration, the app navigates the user through an email address and mobile number verification process via OTP (One-Time Password), thereby validating the user's contact details. To utilize any feature of the app, user login is requisite, which can be executed either manually or automatically. In instances where the user forgets their password, it can be reset through an OTP process, which dispatches a 6-digit code to either the registered email or phone number.

In a bid to safeguard the interests of artists, the Layamritam app levies a nominal subscription fee, offering options for both one-time and recurring subscriptions. Recurring subscriptions are available in various durations: monthly, quarterly, semi-annually, and annually, while one-time subscriptions can be procured for a month or a year. Subscriptions can be purchased via iOS or Android apps, or through the Amrita Books website.

To ensure that streamed and downloaded song audio and lyrics are not misused or shared with unauthorized users, the content undergoes a proprietary multilayer obfuscation process. Just before a song is played, it is loaded into memory following a swift clarification process that takes merely a sub-second, ensuring both security and user-friendly experience in the spiritual journey that Layamritam offers.

Table 1 Summary of architectural objectives and solutions

Architectural objective	Solution
Usability	Song categorization into deity, beat/taalam, tune/raagam, mood/bhavam, etc. Sing-along lyrics. Playlists and favorites. Community features like 'everyone's favorites', recent and latest trends
Scalability	Distributed architecture comprising of transaction and download servers. Supports multiple download servers and transactional servers behind a load balancer. Download servers implement a store-and-forward architecture for propagating the updated content from the transactional servers
Resiliency	Offline mode on the mobile app to support against unstable networks. Automatic downloading and storing of songs on the mobile device. Supports multiple download servers and transactional servers behind a load-balancer
Adaptability	Manage the storage consumed by the downloaded songs. Song quality settings. Theme and song layout settings
Privacy	Song audio and metadata encryption
Security	Registration, authentication based auth-token, role based and data-based access controls, OTP based email verification, and password recovery

6 Summary

Table 1 presents a comprehensive summary of the Layamritam architecture, meticulously aligned with the overarching objectives of the software application. This alignment illustrates how each architectural component contributes to fulfilling the general goals and functionalities expected from the application. The table serves as a crucial reference point for understanding the synergy between Layamritam's design elements and the application's intended outcomes.

7 Comparative Analysis

Table 2 offers a high-level comparison of Layamritam with other well-known music streaming services, focusing on key differentiators such as content variety, categorization methods, and subscription models. This comparative analysis provides a clear perspective on how Layamritam stands out in the competitive landscape of music streaming platforms.

Table 2 Comparative analysis among Layamritam, YouTube, Spotify, and Apple Music

Topic	Layamritam	YouTube	Spotify	Apple Music
Content	2500 songs	Millions of songs		
Categorization	Devotional: Deity, Mood, Beats, Pitch, Difficulty, Year	Devotional and regular Genres: Mood, Beats, Pitch, Difficulty, Year		
Subscription	Paid	Paid or Advertisements		Paid
Download	For all	Paid users only		

8 Adoption

Layamritam made its initial debut to the public in October 2021 through Android and Web clients, followed by the release of the iOS version in February 2022. Since its inception, Layamritam has experienced steady growth, primarily through word-of-mouth and promotions within WhatsApp groups.

Table 3 showcases a collection of aggregate statistics and pertinent metrics, illustrating Layamritam's adoption over the previous two years. Notably, a substantial segment of users who have subscribed to Layamritam have sustained their loyalty to the platform. The retention percentage, calculated as 71% by determining the ratio of the number of active subscriptions to the number of users who have ever subscribed, clearly indicates a broad appreciation for both the platform and its content among most users. Additionally, Fig. 6 elucidates the correlation between song downloads and plays, as well as user registration and subscriptions. A notable correlation between songs downloaded and played is evident (Fig. 6a), indicating that while users are enjoying previously downloaded songs, they are also exploring new tracks. This suggests the success of initiatives like 'everyone's favorites' and 'latest trends', as previously discussed. Figure 6b demonstrates the efficacy of the word-of-mouth marketing strategy, revealing a consistent correlation between the addition of new users and new orders.

Table 3 Aggregate statistics of Layamritam worldwide adoption and usage

Statistic	Measure	Statistic	Measure
Number of registered users	9001	Number of users with verified email or phone	7053
Number of active subscriptions	2458	Number of users who ever had a subscription	3457
Number of orders	8824	Total number of songs	2541
Number of songs played in september, 2023	205,804	Number of song downloaded in september, 2023	68,375
Number of playlists	4469	Number of countries paying users belonged to	45

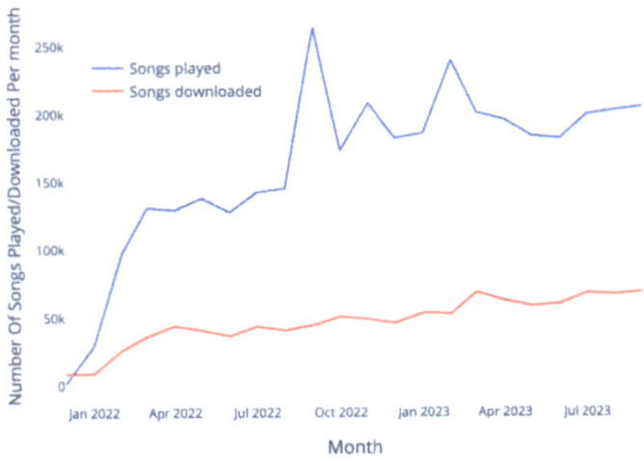

(a)

(b)

Fig. 6 **a** Number of songs played versus downloaded for each month (Left). **b** Monthly number of cumulative registered users versus orders

The songs featured in Layarmritam, despite being freely accessible on platforms like Spotify and YouTube, have garnered significant appreciation and adoption following the release of the Layarmritam app, which offers exclusive devotional tracks. Table 4 provides insights into this positive reception, showcasing a selection of user feedback that highlights the unique appeal and impact of the Layarmritam app in the realm of devotional music.

Table 4 User feedback

Platform	Feedback
Android	A treasure trove. All of Amma's songs—in multiple languages—organized and accessible in many different ways. Vintage songs from the early days to the immense flowering of bhajans in 2020 and 2021. Truly a dream come true. It would be great to have a playlist creation facility too. Thank you
Android	Marvelous App, for thirsty devotees like us. No words to describe the joy felt within my heart when I got this app, with lyrics
Android	Layamritam app is truly a treat for all Amma's Children and the new feature which is added to acces the lyrics of the bhajans is truly commendable. The whole Layamritam team is also very eager to solve the discrepancies faced by the users. Lastly, I congratulate and Thank the whole Layamritam team for their consistent efforts
iOS	Superb collection of Amma's bhajans! Very insightful interface. The latest update has even lyrics and meanings, which is an awesome much-awaited feature!
iOS	Amma's music is just delightful to listen to and extremely well organized and easy to find your favorites!
iOS	This is an essential app for anyone interested in learning, or simply enjoying, Amma's divine bhajans. Amma has recorded literally thousands of songs and this app helps to organize them and guide us so we can enjoy them in our daily life

9 Conclusion

In this paper, we have advocated for the essentiality of establishing a dedicated music streaming platform specifically for devotional songs. We illuminated various technical considerations imperative for constructing such a platform, including scalability, security, usability, and the provision of engaging experiences. A detailed exploration of Layamritam's rich feature set was undertaken, elucidating its architecture and implementation details, and demonstrating how it adeptly meets the aforementioned considerations. Additionally, we summarized user adoption through several salient statistics.

Looking ahead, future work will focus on further enhancing the interface by incorporating advanced AI features, such as voice-based searches, commands, and alerts, to facilitate hands-off operations. We also aim to expand Layamritam to other desktop platforms and intend to integrate it with additional devices and AI agents, such as Alexa, Siri, and Google Agent, to broaden its accessibility and user convenience.

Acknowledgements We extend our heartfelt gratitude to Sri Sadguru Mata Amritanandamayi Devi, affectionately known as Amma, who is not only the Chancellor of Amrita Vishwa Vidyapeetham but also a revered spiritual master and humanitarian leader. Our thanks extend to Amma and her disciples, who have penned and composed most of the songs featured on Layamritam. Additionally, we express our appreciation to the devotee community for warmly embracing Layamritam and remaining loyal customers. Our sincere thanks also go to Pujya Swami Amritaswarupanandamrita Puri Ji for providing us with the opportunity and guidance to participate in developing the Layamritam solution.

References

1. Webster J (2020) Taste in the platform age: music streaming services and new forms of class distinction. Inf Commun Soc 23(13):1909–1924
2. Morris, Jeremy Wade, Devon Powers. (2015) Control, curation and musical experience in streaming music services. Creat Ind J 8(2):106–122
3. Malecki, Pawel, Dorota Czopek, Katarzyna Sochaczewska (2018) Music streaming platforms—quality and technical comparison. In: Audio engineering society convention 145 Audio Engineering Society
4. Bijlani, Kamal, Jayahari KR, Ancy Mathew. (2011) A-view: real-time collaborative multimedia e-learning. In: Proceedings of the third international ACM workshop on multimedia technologies for distance learning. pp 13–18
5. Hariharan, Balaji, Neethu Joy, Sreedevi A.G. (2013) Multimedia: Video sharing for e-classroom. In: 2013 22nd wireless and optical communication conference. IEEE, pp 372–377
6. Prey R (2018) Nothing personal: algorithmic individuation on music streaming platforms. Media Cult Soc 40(7):1086–1100
7. Hansen, Christian, Rishabh Mehrotra, Casper Hansen, Brian Brost, Lucas Maystre, Mounia Lalmas (2021) Shifting consumption towards diverse content on music streaming platforms. In: Proceedings of the 14th ACM international conference on web search and data mining. pp 238–246
8. Hracs BJ, Webster J (2021) From selling songs to engineering experiences: exploring the competitive strategies of music streaming platforms. J Cult Econ 14(2):240–257

9. Guntha, Ramesh, Balaji Hariharan, Venkat Rangan P (2016) Multi-agent framework for automatic deployment and state restoration in iterative software development process. In: 2016 International conference on advances in computing, communications and informatics (ICACCI) IEEE, pp. 1833–1838.
10. Guntha, Ramesh, Aryadevi Remanidevi Devidas, Maneesha Vinodini Ramesh. (2016) High-performance fault-tolerant data caching and synchronization architecture for smart-home mobile application. In: Mobile and wireless technologies 2016 Springer Singapore, pp 135–144
11. Pichl Martin, Eva Zangerle, Günther Specht. (2016) Understanding playlist creation on music streaming platforms. In: 2016 IEEE International symposium on multimedia (ISM) IEEE pp 475–480
12. Rußell R, Berger B, Stich L, Hess T, Spann M (2020) Monetizing online content: digital paywall design and configuration. Bus Inf Syst Eng 62:253–260
13. Bapna, Ravi, Jui Ramaprasad, Akhmed Umyarov. (2016) Monetizing freemium communities: does paying for premium increase social engagement? Available at SSRN 2885681
14. Liu Qiong, Reihaneh Safavi-Naini, and Nicholas Paul Sheppard (2003) Digital rights management for content distribution. In: conferences in research and practice in information technology series vol 34, pp 49–58
15. Wang, Ruoyu, Yan Shoshitaishvili, Christopher Kruegel, and Giovanni Vigna. (2013) Steal this movie: automatically bypassing DRM protection in streaming media services. In: USENIX security symposium pp 687–702
16. Qureshi, Amna, David Megías Jiménez (2020) Blockchain-based multimedia content protection: review and open challenges. Appl Sci 11(1):1
17. Pathan Mukaddim, Rajkumar Buyya, Athena Vakali (2008) Content delivery networks: State of the art, insights, and imperatives. Content delivery networks 3–32
18. Wang, Meisong, Prem Prakash Jayaraman, Rajiv Ranjan, Karan Mitra, Miranda Zhang, Eddie Li, Samee Khan, Mukkaddim Pathan, and Dimitrios Georgeakopoulos. (2015) An overview of cloud based content delivery networks: Research dimensions and state-of-the-art. Transactions on large-scale data-and knowledge-centered systems XX: special issue on advanced techniques for big data management. 131–158
19. Nguyen, Godefroy Dang, Sylvain Dejean, François Moreau (2014) On the complementarity between online and offline music consumption: the case of free streaming. J Cult Econ 38:315–330.
20. Gomes, Inês, Inês Pereira, Inês Soares, Mariana Antunes, Manuel Au-Yong-Oliveira (2021) Keeping the beat on: a case study of spotify. In: Trends and applications in information systems and technologies: Vol 2 9 Springer International Publishing , pp 337–352

Detection of Aphasia Using Speech Characteristics: A Comparative Analysis of Learning Models

Hima Varshini Surisetty, Sarayu Varma Gottimukkala, Susmitha Vekkot, Deepa Gupta, and Yousef A. M. Alotaibi

Abstract Aphasia is one of the brain disorders that affects the part of the brain that is responsible for communication. This work deals with the detection of aphasia from speech features. The features extracted from the speech samples are MFCC, mel spectrogram, zero crossing rate, chroma, energy, spectral centroid, spectral roll-off, fundamental frequency and utterance duration. These different speech features were then ranked based on the SHapley Additive exPlanations (SHAPs), and the most important features were retained in an optimized feature set that was given to various machine learning models like support vector machine and decision trees and deep learning models like LSTM, LSTM with GRU, Bi-LSTM and Bi-LSTM with GRU. These models were then evaluated based on their train and test set accuracy, $F1$ score, recall, precision, plots of training loss and validation loss and confusion matrix. Based on the analysis of the performance metrics, Bi-LSTM with GRU was found to be the best model among all the models used. The model's performance was evaluated using metrics like accuracy, $F1$ score, recall and precision when trained and tested with the optimal feature set which were 87.5%, 79.7%, 90% and 75%, respectively.

Keywords Aphasia · Brain disorder · Communication · Aphasia TalkBank · Shapley Additive exPlanations · Accuracy · Loss · Confusion matrix

---

H. V. Surisetty (✉) · S. V. Gottimukkala · D. Gupta
Department of Computer Science and Engineering, Amrita School of Computing, Bengaluru
Amrita Vishwa Vidyapeetham, Bengaluru, India
e-mail: surisettyhima@gmail.com

S. Vekkot
Department of Electronics and Communication Engineering, Amrita School of Engineering, Amrita Vishwa Vidyapeetham, Bengaluru, India
e-mail: v_susmitha@blr.amrita.edu

Y. A. M. Alotaibi
Department of Computer Engineering, College of Computer and Information Sciences, King Saud University, Riyadh 11451, Kingdom of Saudi Arabia

S. M. Thampi et al. (eds.), *Fifth International Conference on Computing and Network Communications*, Lecture Notes in Electrical Engineering 1221,
https://doi.org/10.1007/978-981-97-4711-5_16

1 Introduction

Aphasia may be a complex dialect clutter that impacts an individual's capacity to comprehend and deliver talked and composed dialect. It is regularly caused by stroke, brain harm, or neurological conditions influencing the dialect regions of the brain. Individuals with aphasia confront noteworthy challenges in their day-to-day lives, battling to communicate successfully and take part in social intuitive, which can lead to dissatisfaction, segregation and decreased quality of life. Detecting and evaluating aphasia are pivotal for successful determination, treatment arranging and recovery. Aphasia discovery includes the recognizable proof and assessment of dialect shortfalls through different strategies. One promising approach is the utilization of discourse records, which contain important etymological and acoustic data that can be analysed to identify and characterize aphasic symptoms.

Speech plays a principal role in the identification and understanding of aphasia, as it serves as a primary means for communication. People with aphasia may encounter difficulties in various aspects of speech, including articulation, comprehension, word retrieval, sentence formation, reading and writing [1]. These challenges significantly hinder their ability to express thoughts, participate in conversations and engage in everyday activities. Current work in aphasia detection involves developing machine learning models that analyse speech data to automatically classify individuals with aphasia. These models extract acoustic parameters, prosody and linguistic patterns from speech recordings to identify anomalies and distinguish between people with and without aphasia. Researchers are also exploring multimodal approaches and cross-lingual adaptation techniques to improve accuracy and applicability in different languages and populations [2]. One of the gaps identified in the work done so far is the absence of a mechanism to select the most important features for the detection of aphasia that has been identified and acted up in the proposed work using SHAP.

Despite promising advancements, there are still significant gaps in the existing work. One key gap lies in the limited generalization of current models across diverse populations and languages, making it challenging to ensure their effectiveness in real-world scenarios. Another crucial aspect is the need for bigger and varied datasets to improve the efficiency and applicability of the detection models. Moreover, the ethical considerations surrounding data privacy and informed consent in speech-based research also require careful attention. By addressing these gaps and leveraging speech data and advanced machine learning approaches, researchers can work towards creating more precise and accessible tools for detecting and monitoring aphasia, ultimately enhancing the lives of individuals with aphasia and aiding their journey towards regaining communication abilities. The contributions of this work to the existing work done in aphasia detection are as follows:

- Design and development of an aphasia detection model by using speech features.
- Design of an optimal feature set by selecting the top three features that contribute the most to the classification using an explainable AI tool, namely SHAP.
- Different deep learning models are trained and tested using all the different extracted features and the optimized feature set.

- These different models are evaluated using benchmark performance metrics like accuracy, precision, recall and $F1$ score.

2 Related Work

Brain injury or stroke is the main causes of aphasia which is a language disorder, and this has been extensively studied in research. Various tools have been developed to detect aphasia in stroke patients [3]. Another type of speech disorder is Parkinson's disease that affects the vocal folds causing the patient to have communication problems [4]. One way of detecting Parkinson's disease is by using VFSS parameters and SDQ evaluation metrics [5]. Impairments in recognizing verbal and auditory feedback errors and performing motor corrections have been found in post-stroke aphasia [6]. Automated detection of aphasia, a common symptom of aphasia, has been studied [7]. Additionally, temporal visual facial features have been studied to identify language deficits in aphasic patients [8]. Sensorimotor aspects of speech-acoustic feedback processing have also been studied [9]. People with aphasia tend to have stress and depression which can be recognized through their speech [10]. Endo language in aphasia is an interesting topic whose evidence, clinical implications and future directions are being explored [11]. Self-monitoring skills in fluent aphasia have been studied [12]. Associated language abnormalities have been studied in non-semantic variants of aphasia [13]. Multi-lingual training is important especially one that is speaker-independent to ensure that the model can handle any type of data [14].

A new benchmark for aphasic speech detection and recognition using advanced techniques has been proposed wherein features like MFCC, chroma, energy, etc., are being used to process different speech samples [15]. An important application of speech features such as prosodic, voice quality and cepstrum features is dementia detection [16]. Dementia detection is done using deep learning models like Bi-LSTM, GRU, LSTM, etc., and evaluated using metrics like accuracy and $F1$ score [17]. The translingual aspects of aphasia detection have been investigated with a focus on methods such as optimal transport domain matching [18]. Some pre-trained language-based models were used to study aphasia in Cantonese and Mandarin-speaking patients which were evaluated using metrics such as accuracy and precision [19]. Zero-shot detection of cross-language aphasia using automatic speech recognition has also been studied [20]. In addition, researchers have studied language segmentation in aphasia to understand how people with aphasia organize linguistic units [21]. Successful syllable recognition in aphasia, despite processing deficits, was described using event-related potentials [22]. The clinical utility of the Spanish version of the Mississippian aphasia screening test has been validated in stroke patients [23].

Existing literature primarily focuses on a single modality, such as speech analysis or neurological imaging. Several papers propose aphasia detection models for only specific languages. Though existing work analyses different types of features from speech, there lies a necessity to develop an optimal feature set. For the development of an optimal dataset that can work across multiple languages, the focus should be on feature engineering rather than the monolingual perspective. In this work, the aim is to address these issues by developing an optimized feature set using tools like SHAP and XGBoost algorithms that help in selecting the best features from a set of numerous speech features. This optimal feature set should be such that it can work on all aphasia affected individuals regardless of the language they speak.

3 Methodology

In order to build machine learning models and deep learning models that can help in the detection of aphasia from speech characteristics, the data were collected from Aphasia TalkBank [24] for both healthy individuals and individuals with aphasia, as a first step. Figure 1 depicts the diagrammatic representation of the workflow of the model proposed in this study.

3.1 Dataset Details

Table 1 represents the information about the data used for this work. The data for implementing the models using learning was obtained from the Aphasia TalkBank

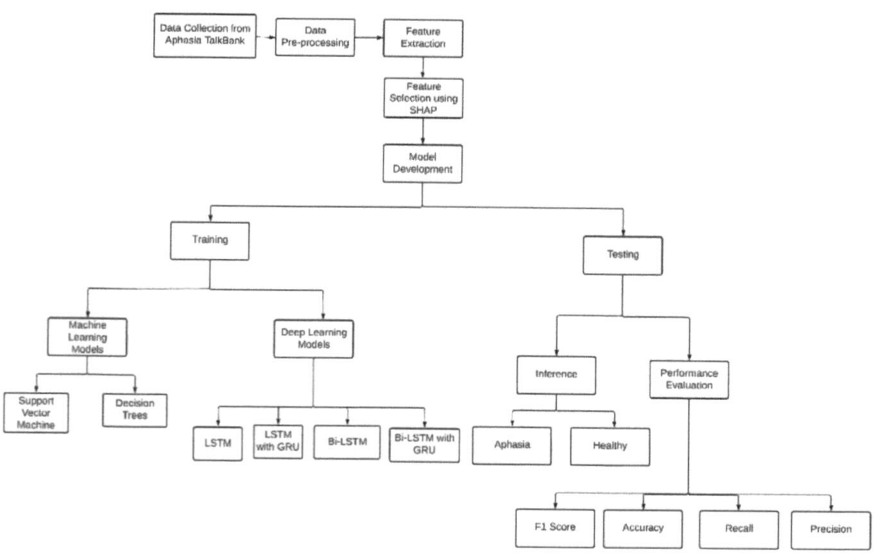

Fig. 1 Flowchart depicting the proposed workflow

Table 1 Description of the dataset used

Gender	Total number of records	Length of each record (in sec)	Number of healthy samples	Number of aphasia samples
Female	65	20	25	40
Male	35	20	25	10

[24] which is a database for the study of communication in aphasia. The Aphasia TalkBank [24] has a set of interviews of healthy individuals and individuals with aphasia with a speech therapist. All the data that were collected were from English-speaking individuals. The interviews were all videos which were all gathered and converted to audio files. The sub-types of aphasia were not analysed for any of the individuals

3.2 Data Preprocessing and Feature Extraction

Once the data were collected from the Aphasia TalkBank [24] and preprocessed, the feature extraction was performed using the Librosa library. To get a little more insight about each feature obtained during the feature extraction, some information [25] regarding them is listed in Table 2.

Table 2 Description of the speech features

Feature	Description
MFCC	Mel frequency analysis in the cepstral domain
Mel Spectrogram	A visual representation of an audio signal that shows the strength of frequencies converted to the mel scale over time
Chroma	The chroma function represents, within the musical octave, the 12 different pitch classes
Energy	The energy of an audio signal indicates the strength or volume of sound
ZCR	Measures the speed at which an audio signal fluctuates from negative to positive or vice versa. Useful for differentiating voiced and unvoiced speech
Fundamental frequency	It represents the perceived pitch of a tone and is important for speech analysis and music processing to identify the paralinguistic aspects
Spectral centroid	It indicates the centroid of the spectral distribution and gives an impression of the “brightness” of the sound in audio analysis
Spectral roll-off	It describes the frequencies below which a certain percentage of the total spectral energy lies
Utterance duration	It refers to the length of spoken words, phrases and sentences

3.3 Feature Selection

After the feature extraction was done, there were nine different features that were extracted, namely, MFCC's, Mel Spectrogram, Chroma, Energy, Zero-Crossing Rate (ZCR), Fundamental Frequency, Spectral Centroid, Spectral Roll-Off and Utterance Duration which will be referred to as the complete feature set from here on. However, all these features might not all be contributing factors for determining whether an individual is affected by aphasia or whether an individual is healthy. Therefore, feature selection is performed by using the Shapley Additive exPlanations (SHAP) values of feature importance with an underlying XGBoost algorithm. The XGBoost algorithm gave an accuracy of 95%, $F1$ score of 94.9%, recall of 90% and precision of 100% when the complete feature set was given as input.

Figure 2 represents a part of the plot that was obtained for the top 1000 most important features among all the features. The majority of components of the features that are of top priority were found to be MFCC's, mel spectrogram and ZCR which will be referred to as the optimal feature set from here on. Therefore, only these three features were utilized in order to build all the different models. Different combinations of these top three features was tested, and it was observed that the model's performance is the best when all the three features are used together. Also, when the models were tested with more than three best features, the model's performance degraded.

3.4 Model Development

The complete feature set and the optimal feature set were utilized for training all the different models. The implementation details of all these different models are illustrated below under their respective sub-sections. The machine learning models that were implemented were the support vector machine and decision trees. These machine learning models were imported from the sklearn package in python. Since the models were pre-trained, no hyperparameter tuning was done for these models. The deep learning models that were implemented were LSTM, LSTM with GRU, Bi-LSTM and Bi-LSTM with GRU. The architectures for all models are shown below. More information regarding the deep learning models can be obtained from [26]. Table 3 represents the architectures for the deep learning models. The output shape and activation function for each layer are mentioned in the table for every deep learning model.

Apart from building these models using the optimal feature set, the complete feature set was also used to build the same models in order to analyse the effectiveness of the SHAP feature selection on the efficiency and performance of the models. In order to implement the models with the complete and optimal feature set as the input, the data was split into an 80% training set and a 20% testing set, and the training set was further divided into a 70% training set and a 30% validation set.

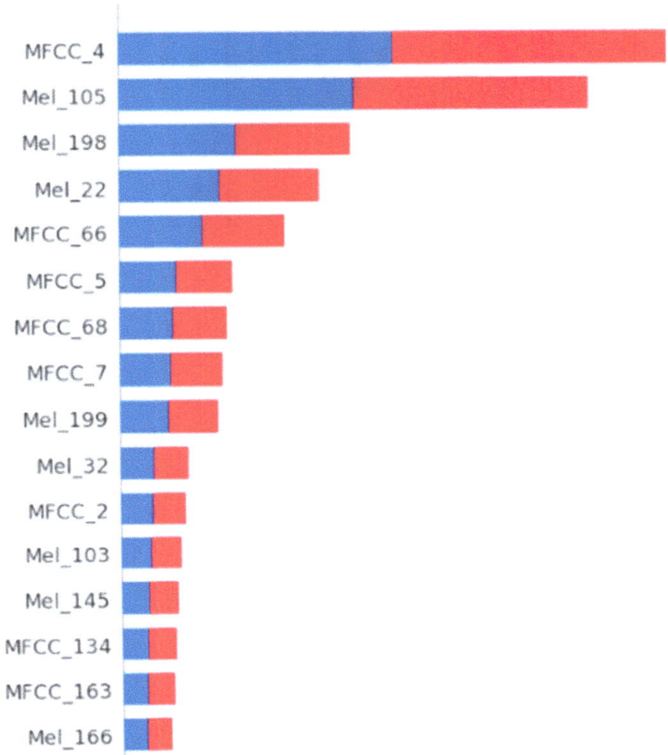

Fig. 2 Illustration of the importance of the features

4 Results and Discussion

The machine learning-based models, namely, support vector machine and decision trees, were implemented by giving the complete feature set and the optimal feature set as the input to these models. The feature sets were passed through different layers of preprocessing such as flattening, padding, standardizing and reshaping before being converted into a data frame. This data frame was given to the input layer in each model's architecture. The accuracy, recall, $F1$ score and precision scores were used to determine better-performing models. The support vector machine gave an accuracy of 75%, recall of 73.33% and precision of 66.66% when the optimal feature set was given as an input which indicates that the model does not work as well on unseen data while the decision trees model gave a 100% accuracy and precision. This means that this model was overfitted when the complete and optimal feature set was given as input while that was not the case for the support vector machine. In the case of the support vector machine model, the accuracy of the model decreased when the optimal feature set was given as input. The deep learning models, namely, LSTM,

Table 3 Architecture of deep learning models

Model	Bi-LSTM	LSTM	GRU	Dense	Flatten	Dense	Dense	Dropout	Dense	Dense
LSTM	–	(None, 3, 512, None)	–	–	(None, 4608, None)	(None, 128, Relu)	–	(None, 128, None)	–	(None, 1, Sigmoid)
LSTM with GRU	–	(None, 3, 512, None)	(None, 256, Relu)	–	(None, 256, None)	(None, 128, Relu)	–	(None, 128, None)	(None, 64, Relu)	(None, 1, Sigmoid)
Bi-LSTM	(None, 3, 256, None)	–	–	(None, 3, 256, Relu)	(None, 768, None)	(None, 128, Relu)	–	(None, 128, None)	(None, 64, Relu)	(None, 1, Sigmoid)
Bi-LSTM with GRU	(None, 3, 256, None)	(None, 256, Relu)	–	–	(None, 256, None)	(None, 128, Relu)	(None, 64, Relu)	(None, 64, None)	(None, 64, Relu)	(None, 1, Sigmoid)

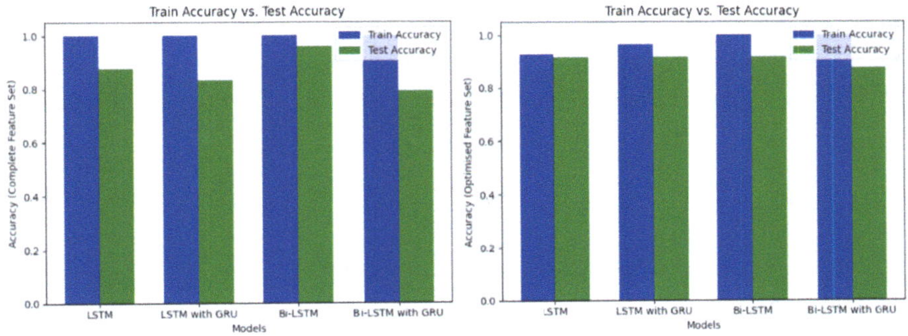

Fig. 3 Train versus test set accuracy

LSTM with GRU, Bi-LSTM and Bi-LSTM with GRU, were implemented by giving the complete feature set and the optimal feature set as the input.

From Fig. 3a, b, it can be inferred that all the deep learning models were overfitted when the models were built using the complete feature set, whereas, when the optimal feature set was utilized in order to build these models, the test accuracy of all the models increased and also the overfitting in the LSTM model and LSTM with GRU model decreased. Though the Bi-LSTM model and Bi-LSTM with GRU models still overfit, the test set accuracy of the Bi-LSTM with GRU model has increased.

Table 4 represents the $F1$ score, recall and precision scores that were obtained for these models when the complete and optimal feature set was used as input. From Table 4, it can be inferred that the models achieved competitive $F1$ scores when the models were built using the optimal feature set, with Bi-LSTM having the highest overall performance. LSTM models demonstrated high recall rates, while Bi-LSTM models achieved perfect precision when the complete feature set was used. LSTM models showed higher precision when the optimal feature set was used. Evaluating the trade-offs amidst precision score, recall score and $F1$ score is crucial when assessing the performance of these models when the models were built using the optimal feature set. The loss plots and confusion matrices have been plotted for all the deep learning models as represented in Figs.4, 5, 6 and 7.

Table 4represents the confusion matrix and loss plots of the LSTM with GRU model when the complete feature set and the optimal feature set were used in order to build the model, respectively. From Fig. 4a, b, it can be inferred that the number of false negatives decreased when the optimal feature set was used to build the LSTM with the GRU model. But the observations from Table 4 must also be taken into consideration along with this observation from the confusion matrix. From Fig. 4c, d, it can be inferred that when the complete feature set is being used, the LSTM with GRU model has a validation loss that is greater than the training loss. However, when the optimal feature set was used as the input, the validation loss was continuously fluctuating. But in this case, the gap amidst the training loss and the validation loss

Table 4 Performance metrics values for the deep learning models

Deep learning model	Complete feature set			Optimized feature set		
	$F1$ score (%)	Recall (%)	Precision (%)	$F1$ score (%)	Recall (%)	Precision (%)
LSTM	94.9	100	90.9	73.3	100	66.6
LSTM with GRU	62.6	90	60	84.6	100	76.9
Bi-LSTM	89.8	80	100	84.9	90	81.8
Bi-LSTM with GRU	60.1	100	58.8	79.7	90	75

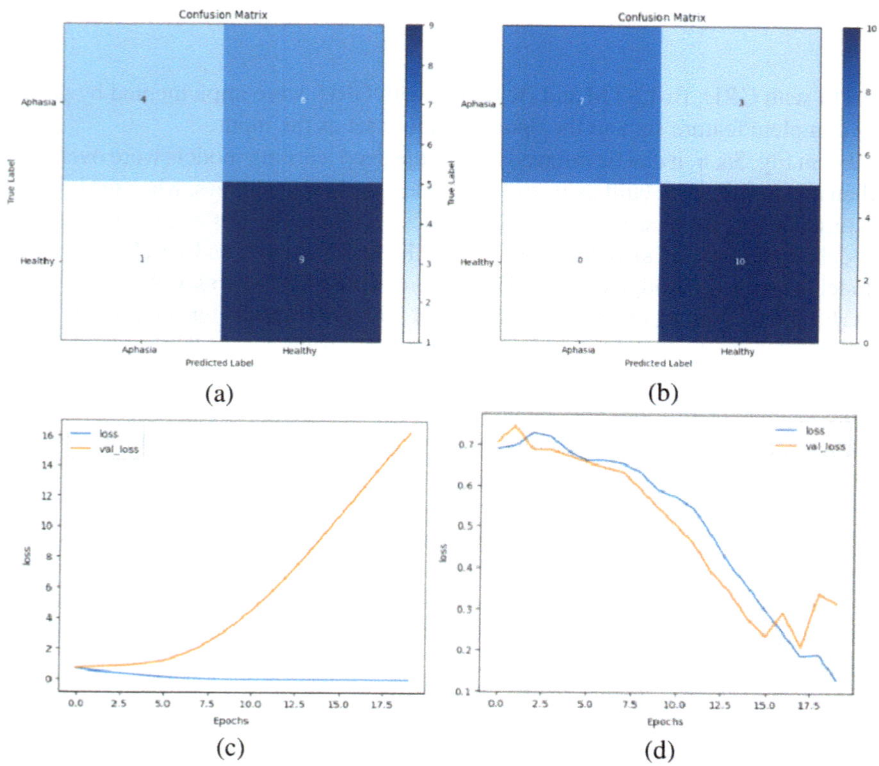

Fig. 4 Illustration of the confusion matrix and loss plots of the LSTM with GRU model

was low which is why there is not much overfitting, whereas, in the case with a complete feature set, the gap amidst both the curves kept increasing.

Figure 5 represents the confusion matrix and loss plots of the Bi-LSTM model when the complete feature set was used in order to build the model and when the optimal feature set was used in order to build the model, respectively. From Fig. 5a,

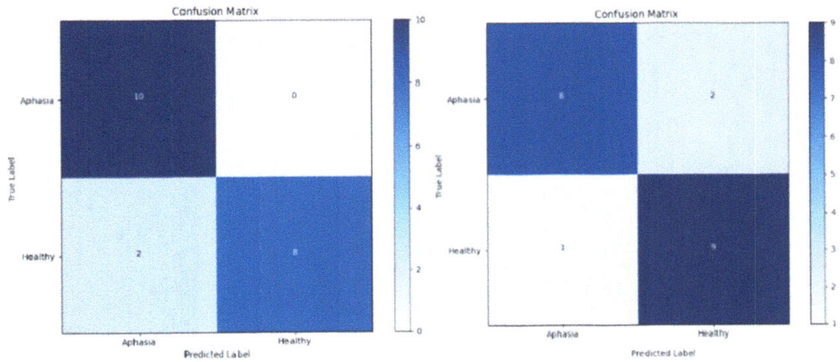

(a) (b)

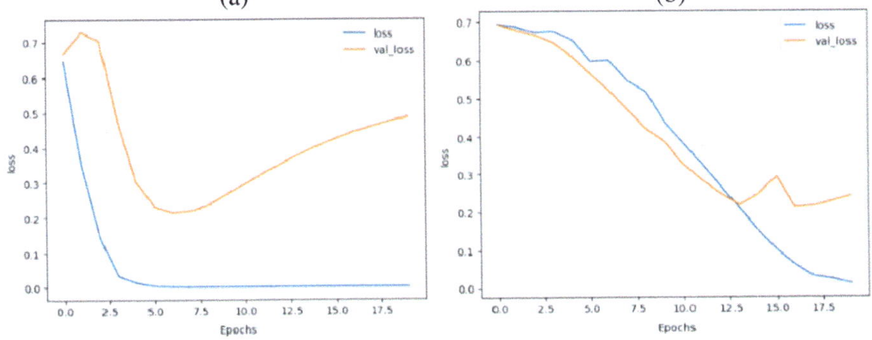

(c) (d)

Fig. 5 Illustration of the confusion matrix and loss plots of the Bi-LSTM model

b, it can be inferred that the number of false negatives increased when the optimal feature set were used to build the Bi-LSTM model. But, the observations from Table 4 must also be taken into consideration along with this observation from the confusion matrix. From Fig. 5c, d, it can be inferred that in the case where the complete feature set is being used, the Bi-LSTM model has a validation loss that was initially close to the training loss and later on was decreasing along with the training loss, but the gap between the losses is larger as compared to when optimal feature set was used. But, when the optimal feature set was used as the input to the models, the validation loss was observed to be lower than the training loss and later on was observed to be greater than the training loss. But in this case, the gap amidst the training loss and the validation loss was low which is why there is not much overfit.

Figure 6 represents the confusion matrix and loss plots of the LSTM model when the complete feature set and the optimal feature set were used in order to build the model, respectively. From Fig. 6a, b, it can be inferred that the number of false negatives increased when only the selected features were used to build the LSTM model. But, the observations from Table. 4 must also be taken into consideration along with this observation from the confusion matrix. From Fig. 6c, d, it can be inferred that in the case where all the features are being used, the LSTM model has

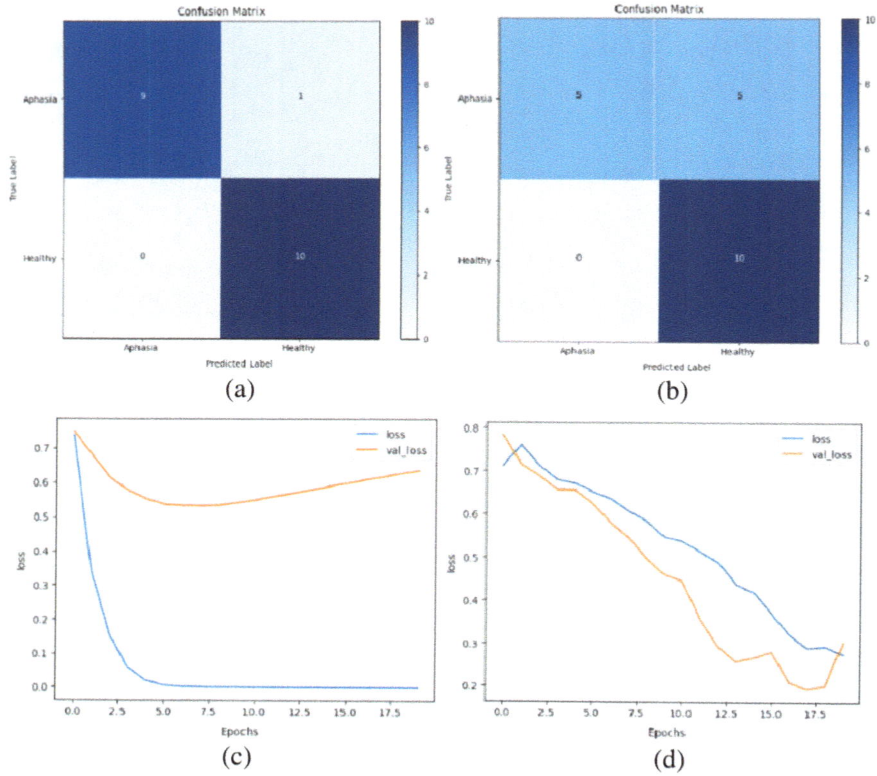

Fig. 6 Illustration of the confusion matrix and loss plots of the LSTM model

a validation loss that is lower than the training loss which should not be the case. With selected features used as the inputs, the validation loss was higher than the training loss and they were separated by an acceptable amount of gap amidst each other which is the reason why this model is not overfit as reflected in Table 4.

Figure 7 represents the confusion matrix and loss plots of the Bi-LSTM with GRU model when the complete feature set and the optimal feature set were used in order to build the model, respectively. From Fig. 7a, b, it can be inferred that the number of false positives increased when only the selected features were used to build the Bi-LSTM with the GRU model. From Fig. 7c, d, it can be inferred that in the case where all the features are being used, the Bi-LSTM with GRU model has a validation loss that was greater than the training loss and the gap amid both the curves kept on increasing over the epochs. But, when only the selected features were used as the inputs, the validation loss that was greater than the training loss and the gap amid both the curves kept on increasing over the epochs, but this increase in gap was not as high as in the previous case.

Among the models, the best-performing model in terms of precision was Bi-LSTM with a perfect precision score of 100% when the complete feature set was used. On

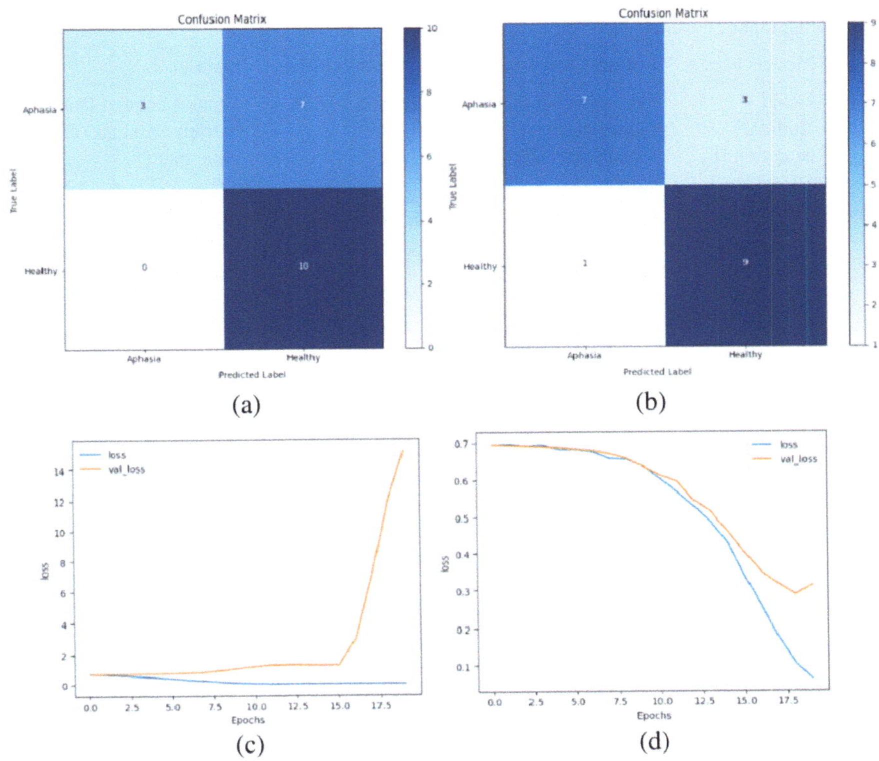

Fig. 7 Illustration of the confusion matrix and loss plots of the Bi-LSTM with GRU model

the other hand, the worst-performing model in terms of precision was LSTM with GRU, which achieved the lowest precision score of 60% when the optimal feature set was used. The best-performing model in terms of F1 score was LSTM when the complete feature set was used, achieving the highest score of 94.9%. When all scores were taken into consideration, LSTM with GRU and Bi-LSTM were found to be the best models. In the case of the LSTM model, the number of false negatives increased when the optimal feature set was used to build the model. But, when the complete feature set is being used, the LSTM model has a validation loss that is lower than the training loss which should not be the case. But, when the optimal feature set was used as the input, the validation loss was higher than the training loss and they were separated by an acceptable amount of gap amidst each other which is the reason why this model is not overfit. Therefore, all the deep learning-based models except for the Bi-LSTM with model saw a rise in the test set accuracy and a reduction in the overfit when the models used the optimal feature set. However, considering the loss plots and other evaluation metrics, LSTM with GRU and Bi-LSTM were found to be the best models.

Table 5 represents the comparison of the proposed model with the state-of-the-art. From Table 5, it can be seen that the proposed model outperforms the state-of-the-art

Table 5 Comparative analysis of the proposed work with state-of-the-art

Year	Models used	Features involved	Evaluation metrics	Dataset
2019	Facial landmark detection [8]	Pause frequency, mouth pattern repetitions and pattern variety	Accuracy—80%	Dataset created from therapy sessions of patients
2020	SVM, Random Forest, Vector PLDA [2]	Prasody, Jitter, pitch, intensity and statistical functions	Accuracy—90.3%	Speech language pathology department dataset and aphasia bank dataset
2023	Deep neural networks [15]	Filler words, word pauses and word phones	Word Error Rate—22.1	Aphasia bank and dementia bank
2023	**Proposed work**	**MFCC, mel spectrogram and ZCR**	**Precision—81.8%** **Recall—90%** ***F*1 score—84.9%** **Accuracy—100%**	**Aphasia bank dataset**

in all indices. This can be attributed to the feature selection and careful design of the model based on explainable AI tools. However, it can be inferred that there is scope for improvement, but as compared to existing work, more features and different deep learning models have been tried out in the proposed work. This proposed work has also used several evaluation metrics to evaluate the deep learning models. In Table 5, the precision, recall, $F1$ score and accuracy of the best model is shown which the Bi-LSTM model when the model was trained with the optimal feature set which include the features mentioned in Table 5.

5 Conclusion and Future Scope

Based on the literature survey, it can be inferred that the other works in this area do not analyse the importance of different features from speech. Hence, the main aim of this work is to extract different features from speech and select the best ones based on their importance ranking that was generated by SHAP and XGBoost. Also, the current works do not analyse the efficiency of different deep learning models for aphasia detection. Hence, this work covers that gap by analysing the performance of four different deep learning models and selecting the best model.

This work can be extended further by doing a deeper analysis on the data from the aphasia TalkBank [24] and distinguishing aphasia into more sub-levels and extending this problem of aphasia detection into a multi-class problem. Also, more deep learning models and architectures can be explored in this problem domain. Also, the more data can be gathered on which type of speech therapy is to be employed based on the type and the level of aphasia that is identified.

References

1. Kohlschein C, Schmitt, M, Schüller B, et al. (2017) A machine learning based system for the automatic evaluation of aphasia speech. 2017 IEEE 19th international conference on e-health networking. Applications and Services (Healthcom)
2. Jothi K.R, Sivaraju S.S, Yawalkar P.J (2020) AI based speech language therapy using speech quality parameters for aphasia person: A comprehensive review. 4th international conference on electronics, communication and aerospace technology (ICECA)
3. Choi Y.H, Park H.K, et al. (2015) A telescreening tool to detect aphasia in patients with stroke. Telemedicine and e-Health, Vol 21
4. Jahnavi B, Supraja B, Lalitha S (2020) A vital neurodegenerative disorder detection using speech cues. J Intell Fuzzy Syst vol 38
5. Aarthi R.S, Radhakrishnan S. Kumar et al. (2021) Dysphagia in Parkinson's disease: analysis of screening questionnaire and videofluoroscopy findings. Annals of movement disorders
6. Sangtian S, Wang Y, Fridriksson J, et al. (2021) Impairment of speech auditory feedback error detection and motor correction in post-stroke aphasia. J Commun Disord vol 94
7. Le D, Licata K, Provost E.M (2017) Automatic Paraphasia detection from aphasic speech: a preliminary study. Interspeech
8. Einfalt M, Lienhart R, Lee M, Kennedy L (2019) Detecting speech impairments from temporal visual facial features of aphasia patients. IEEE conference on multimedia information processing and retrieval (MIPR)
9. Behroozmand R, Phillip L, Johari K, et al. (2018) Sensorimotor impairment of speech auditory feedback processing in aphasia. NeuroImage vol 165
10. Gillespie S, Moore E, Laures-Gore J, Farina M, Russell S, Logan Y.Y (2017) Detecting stress and depression in adults with aphasia through speech analysis. IEEE international conference on acoustics. Speech and signal processing (ICASSP)
11. Fama M.E, Turkeltaub P.E (2020) Inner speech in aphasia: Current evidence, clinical implications, and future directions. American J Speech-Lang Pathol
12. Sampson M, Shah Y.F (2011) Investigation of self-monitoring in fluent aphasia with jargon. Aphasiology, Vol 25
13. Sajjadi S.A, Patterson K, et al. (2012) Abnormalities of connected speech in the non-semantic variants of primary progressive aphasi, anda. Aphasiology. Volume 21
14. Vekkot S, Gupta D (2022) Fusion of spectral and prosody modeling for multilingual speech emotion conversion. Knowledge-based systems
15. Tang J, Chen W, et al. (2023) A new benchmark of aphasia speech recognition and detection based on e-branchformer and multi-task learning. Audio and speech processing
16. Kumar M.R, Vekkot S, Lalitha S, Gupta D., et al. (2022) Dementia detection from speech using machine learning and deep learning architectures. Sensors 22(23)
17. Nambiar A.S, Gupta D, Vekkot S, et al. (2022) Comparative study of deep classifiers for early dementia detection using speech transcripts. 19th India council international conference (INDICON)
18. Balagopalan A, Novikova J, Matthew et al. (2020) Cross-Language Aphasia Detection using Optimal Transport Domain Adaptation. Machine Learning for Health neurips Workshop
19. Qin Y, Lee T, Kong A.P.H, Lin F (2022) Aphasia detection for cantonese-speaking and mandarin-speaking patients using pre-trained language models. 2022 13th international symposium on chinese spoken language processing (ISCSLP)
20. Chatzoudis G, Plitsis M, Stamouli S, et al. (2022) Zero-shot cross-lingual aphasia detection using automatic speech recognition. Machine learning
21. Peñaloza C, Benetello A, Tuomiranta L, et al. (2015) Speech segmentation in aphasia. Aphasiology
22. Becker F, Reinvang I (2007) Successful syllable detection in aphasia despite processing impairments as revealed by event-related potentials. Behavioral and brain functions 3

23. Romero M, Sánchez A, Marín C, et al. (2012) Clinical usefulness of the Spanish version of the mississippi aphasia screening test (MASTsp): validation in stroke patients. Neurology, volume 27
24. AphasiaBank https://aphasia.talkbank.org/
25. Rao K.S, Koolagudi S.G (2013): Emotion recognition using speech features. Springer-Briefs in Speech Technology
26. Goodfellow I, Bengio Y, Courville A (2016) Deep learning. MIT Press

VTSON: Various Techniques of Survivability in Optical Networks—A Review

Ira Nath, Purba Chatterjee, Rajib Roy Choudhury, Dharmpal Singh, and Sumit Das

Abstract The survivability of optical networks is a very important issue to address. Due to these networks' high speeds and capacity for handling big data volumes, it is essential. In optical networks, path or link failures are highly frequent occurrences. A significant loss of data, time, and money could result from these errors. Restoring the downed network as soon as possible is necessary to reduce this loss. If the link is repaired so that the impacted traffic is swiftly and precisely rerouted using the available network resources, the impact of the failure may be reduced. Various survivability mechanisms are employed based on the specific requirements and network architecture. Managing failures of single links and multiple links is the primary focus. The path failure may be caused by several broken links. The current network survivability schemes have been thoroughly examined. Network protection schemes with regenerators make up the majority of these schemes (Zhou and Subramaniam in IEEE Network 14:16–23, 2000).

Keywords Survivability · Protection · Restoration

1 Introduction

It is well known that optical networks are essential for cost-effectively supporting the steadily increasing global network traffic intensity. For the purpose of guaranteeing optimal network operation, issues pertaining to optical network security therefore assume great significance. Numerous physical-layer attack techniques and optical-layer security flaws, especially in transparent optical networks, have been noted in the literature. Attacks have the potential to cause network-wide service disruptions or financial losses for clients, which could result in massive data losses and revenue

I. Nath (✉) · P. Chatterjee · R. R. Choudhury · S. Das
JIS College of Engineering, Kalyani, India
e-mail: ira.nath@gmail.com

D. Singh
JIS University, Agarpara, India

S. M. Thampi et al. (eds.), *Fifth International Conference on Computing and Network Communications*, Lecture Notes in Electrical Engineering 1221,
https://doi.org/10.1007/978-981-97-4711-5_17

losses for operators. Attacks may result in network-wide service disruption or monetary losses for clients, which could cost operators a significant amount of money in lost data. A common tactic used in attacks aimed at degrading service quality, like power jamming, is the introduction of harmful signals into the network, which may spread through configured connections and result in extensive harm. Operators' growing concerns about optical-layer security are fueling the development of intrusion detection systems as well as other safety results meant to either identify anomalies or make it harder for users. Those results could be additionally integrated with jamming attack-related. These well-established paradigms allow for attack detection and identification of source, but the question of survivability presence of attacks has not been raised. Conventional optical network protection strategies focus on cuts of cables as well as defects at nodes as the primary causes of disruption of continuous facility, typically establishing alternative paths or links. While these methods work well in the event of failures, they might not offer sufficient defense against attacks that have the ability to promulgate as well as disrupt more than one link disjoint paths at once.

Extremely fast network communication is possible with optical networks. A network system frequently experiences failure. There are several reasons why it could occur. Any failure might mean losing money, time, and data. Millions of apps might be disrupted by a single little failure. Both network operators and end users may suffer significant data and money losses as a result. Thus, it is crucial to protect data and communication systems. Failures of links happen frequently. Ensuring dependable and continuous communication for all users is the fundamental goal of network and data protection. A restoration procedure is necessary in case of failure in order to recover the data. There are two types of restoration schemes: proactive and reactive. Additionally, these schemes may be path-based or link-based. These strategies must guard against both physical-layer attacks and component failures.

Regarding Survivability Plans, when it comes to optical networks, link failure is a very common failure. To deal with link failures, there are primarily two approaches. These strategies are restoration and protection. Backup routes are preplanned in the protection method and are utilized in case of a link failure. These routes already have the necessary resources assigned to them and are designed. The backup routes in the restoration method are not preplanned. By identifying free wavelength channels, these pathways are computed as and when some failure is actually happening. Approaches to protection and restoration include benefits and drawbacks depending on a number of factors, including recovery time, resource usage, and spare capacity utilization.

Backup resources are pre-scheduled in the protection approach to avoid failures. Preplanned backup routes are utilized in case of an outage. In optical networks, this method is frequently applied to handle link failures. Protection can provide significantly quicker recovery from failure if it is applied in a capacity-efficient manner. Even so, this method wastes some resources. Both path and link failures can be handled with a protection strategy.

Every link's backup path between every pair of nodes is pre-calculated in the link protection approach. Whenever a link failure is identified, data is always retransmitted using this backup path. Consequently, in the event that a link fails, each link has a designated backup path that should be used.

The whole backup path route is preplanned in the path protection approach. Using this pre-calculated backup path, all of the failed path's traffic is retransmitted. Although this flexibility in path protection may result in a reduced capacity for protection, it necessitates that each failed path influence recovery separately. The backup path completely safeguards every path.

Network survivability can also be achieved through restoration. This approach does not allocate spare resources in advance. Following the failure, the backup resources are set up online. Finding a new backup plan after a failure is a constant challenge for restoration-based recovery. Therefore, nothing is lost by saving resources for a backup route. Restoration can take the form of reestablishing links, paths, or sub-paths.

The failed link is located in Link Restoration. In order to dynamically identify a restoration path, the failed link's end nodes do take part in a distributed algorithm. The traffic is transmitted via the newly found route that circumvents the failed link. The amount of time spent choosing a backup plan is vital to the recovery.

Under the path restoration technique, every source–destination pair for every connection independently finds a backup path from beginning to end. It is possible for these backup routes to use a separate wavelength channel. Data transmission between the source and destination pairs occurs via this backup path.

When a link fails in the Sub-Path Restoration technique, the upstream node of the failed link recognizes the failure and finds a backup route from itself to the corresponding destination node for each interrupted connection [2, 3].

In this paper, in Sect. 2, a literature survey is depicted. In Sect. 3, overview of various survivability schemes is tabled. Conclusion is drawn in Sect. 4.

2 Literature Survey

In 2000, Zhou et al. [1] discussed about the methods utilized to take a look at the techniques used to attain survivability in optical networks. It had been discussed about the methods that were developed to create high level of survivability in optical networks.

In 2002 Maier et al. [5] surveyed some survivability schemes for optical networks. A new algorithm had also been invented regarding survivability of mesh optical networks. In order to minimize the total used capacity, an ILP formulation for the dedicated path protection problem was put forth.

In 2004, Pin Han Ho et al. [6] established an ILP for shared path protection scheme.

In 2008, Pan et al. [7] studied protection plans in relation to failures that are geographically correlated and can result from physical infrastructure attacks (like weapons of mass destruction or electromagnetic pulse attacks) or natural disasters. Nevertheless, there hasn't been much research done on protection against subtler signal degradation attacks like power jamming, which may impact several connections that don't always have the same physical components. A preliminary method for resilient routing and regenerator positioning in the face of power jamming attacks was put forth for translucent optical network.

In 2010, Drid et al. [8] suggested a novel approach to multi-domain optical networks using p-cycles. Additionally, it was suggested a topology aggregation model tailored to p-cycle computations for reasons of scalability and security. With the help of this aggregation model, this suggested approach could balance the need for both short running times and effective backup resource usage. The results of the simulation indicated that, in comparison to other options, the suggested method offered a decent trade-off between running time and resource usage.

In 2013, Mohan et al. [9] developed several methods to deal with multiple failures. Several network resources, such as fibers sharing a fiber conduit, were assumed to fail collectively. This concept of groups of shared risk had been taken into consideration in duplicate path-finding methods by making sure that the backup and working paths were SRG-disjoint.

In 2015, Furdek et al. [10] invented that a significant number of resources would be needed in transparent optical networks to guarantee complete defense against service degradation attacks. To limit the number of connections susceptible to a single jamming attack, attack-awareness has been integrated into the assignment of wavelength and routing procedures in optical networks. In order to offer defense, against such attacks, the topic of survivability was covered in this paper. A heuristic method for survivable WA was created.

In 2018, Ashraf et al. [11] discussed that network operators desperately required solutions that not only solve the survivability problem but also got rid of problems like network congestion and cascade failures. The remedies for both large-scale (disaster-based) and small-scale (network component) failures were critically examined in this research. A framework with capacity optimization that was based on several techniques might offer enhanced catastrophe protection and survival. This would help the network become more resilient to disasters through the use of fault tolerance management.

In 2020, Mohan et al. [4] suggested to use p-cycles in a dual-link failure recovery method. The p-cycle circle was replicated using this technique. The initial failure independent path protection p-cycle technique had been improved upon. The nodes had been protected using the same p-cycle that was available thanks to the replica properties of p-cycle. The cost of the network increased with the creation of a new p-cycle, but it decreased dramatically when an existing p-cycle was used as a replica. Three stages of network simulator were used to accomplish the suggested technique.

3 Overview of Various Survivability Schemes

3.1 Survivability

It measures a system's, subsystem's, piece of machinery, process, or procedure's capacity to carry on operating both during and after a natural or man-made disruption, such as an optical fiber breaking or a system failing.

In order to ensure survivability for a particular application, it is necessary to define the minimum acceptable level of post-disturbance functionality, the maximum acceptable downtime, and the range of conditions over which the entity will survive.

In Fig. 1. A Generalized loopback protection has been shown. Two light paths are represented before (a) and after (b) the failure of link AB.

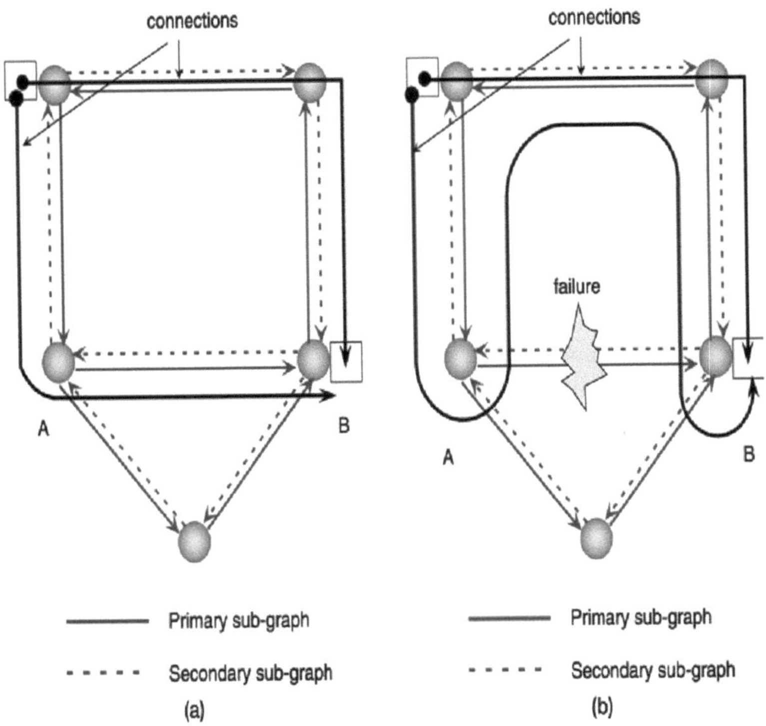

Fig.1 Generalized loopback protection. Two light paths are represented before **a** and after **b** the failure of link AB

3.1.1 Advantages and Limitations of Network Survivability

The optical layer can handle some faults more efficiently than the client layers. Optical-layer protection can provide an additional degree of resilience in the network to protect against multiple failures.

Advantages

- Utilizing the optical layer allows users to share protection resources regardless of format or source.
- Electronic networks without protection and restoration features are provided (PDH, ESCON, etc.).
- It makes use of less complex methods, which suggest fewer network entities.
- Higher layers' electronic coordination is not required for them.

Limitations

- Detecting high BER, for instance, is one problem that optical protection cannot resolve in higher layer equipment.

3.2 Concept About Path Protection

Restoration must be completed as quickly as possible because more data is lost the longer a system is unavailable. In order to prevent the loss of frame synchronization, SONET/SDH has set a maximum time of 60 ms for traffic restoration.

- This time includes time of fault detection and location, time of signal propagation, delay, and switching times.

Working paths: Optical paths designed for carrying traffic during regular operation (SDH, ATM, IP, etc.).

Protection paths: When a failure occurs, optical paths are set up to offer a backup route and reroute the work traffic.

Then protection can be performed by:

- Path switching

  Traffic is rerouted end to end along a different path between the source and the destination.

- Line switching

  It happens in between the two nodes that indicate the location of the failure.
  It can also be carried out via a protection line or span-basis protection, which is an alternate path connecting the two nodes.
  We can make use of protection resources based on their location.

In order to address this issue, the notion of "path protection with regenerators present" is a solution under research. Depending on the unique spread features of the bout(s) under consideration, this identifies connection paths that may have an impact on one another should they carry malicious signals or cause link breakage. If a connection's primary and backup paths share a source and destination, it is presumed to be protected against attacks and errors.

3.2.1 Dedicated Protection Schemes

Every working connection has a fixed bandwidth allotted to it that can be used in the event of a failure (working fiber).

Typically, the second fiber—the protection fiber—is used to send a working copy of the traffic.

- Keep as a backup plan in case it becomes necessary to resume traffic. Furthermore.
- Low-priority traffic is also transmitted using this method.

3.2.2 Shared Protection Schemes

It is considered impossible for all links or nodes to fail at the same time.

- This makes it possible to carefully identify and allocate several paths so that the same protection bandwidth can be shared.
- Transporting low-priority traffic can use protection bandwidth as long as there are no faults (Figs. 2 and 3).

3.3 Link-Based Protection

Each link has a protection path set aside for it, and traffic is redirected (looped back) around the failed link. This is the basic idea behind link-based protection. As for illustration, the impacted traffic in Fig. 4b is redirected via the backup path 5–2–6 following a link failure between nodes 5 and 6. Recoveries in this case are handled by the failed link's end nodes, or nodes 5 and 6.

3.4 Path-Based Protection

A protection path and wavelength are reserved for each working wavelength path and link failure in path-based protection. It might not be simple to accomplish this as it calls for the cooperation of multiple network nodes. Normally, during connection setup, protection wavelength paths are reserved for each failed link, and they ought to be disconnected from the failed link. The wavelength paths set aside for this

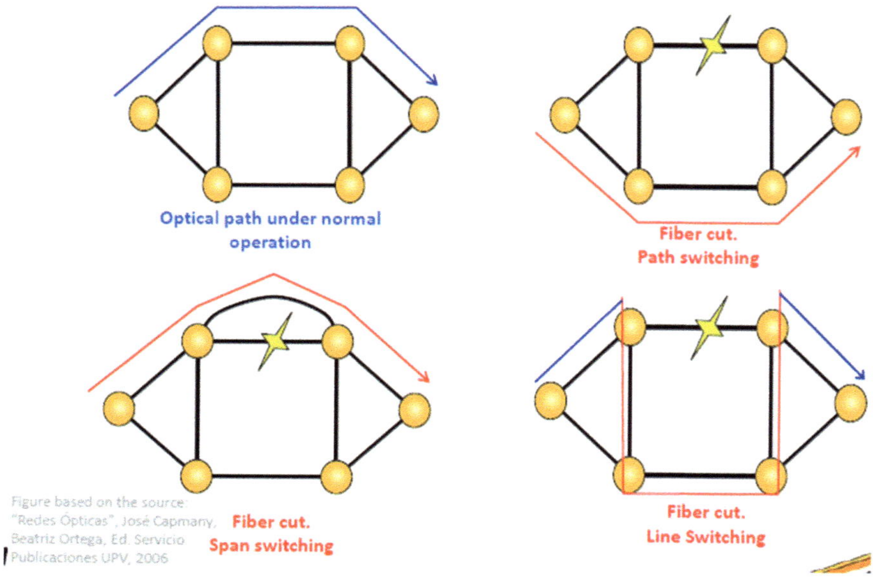

Fig. 2 Protection schemes when path failures

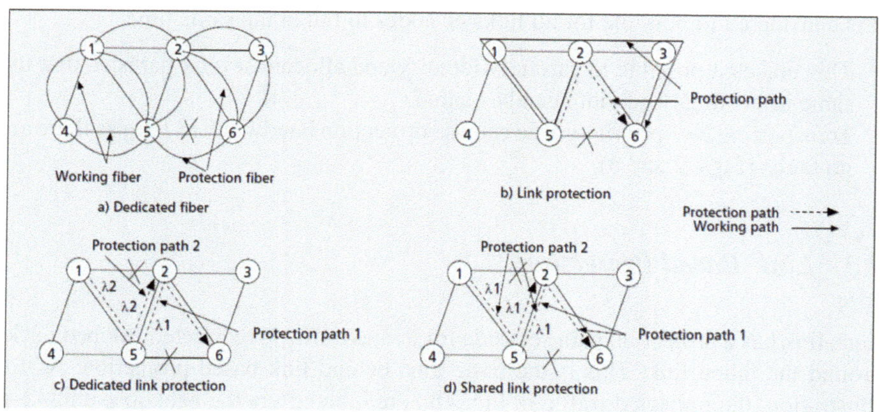

Fig. 3 Link-based protection

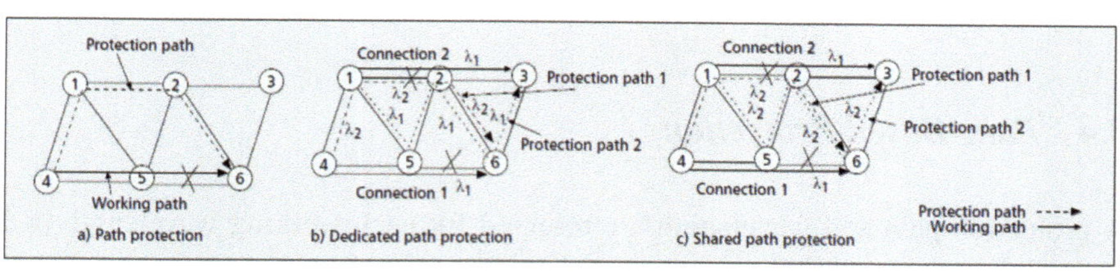

Fig. 4 Path-based protection

failure scenario are activated upon link failure. In the unlikely event that a single-link failure occurs along the working path, a connection can be restored using the same wavelength path if the protection wavelength path is disjoint with every link of the working path. Take note that in this instance, recovery can proceed without the need to identify the broken link. The working wavelength and the wavelength used on the protection path may or may not match. Additionally, different connections using the same working path may use different protection paths. Path-based protection is dedicated or shared, just like link-based protection.

4 Conclusion

With the development of technology and the internet's rapid growth, communication at larger data rates has become essential for both current and future generations. Failure for any reason can result in significant financial or revenue losses, therefore survivability is crucial. Many studies have been conducted on survivability in the past, especially on single-link failures, and many more are being conducted now. There are essentially two approaches to protecting and restoring survivability, and each has the advantages and disadvantages that have been mentioned. The survival solution that minimizes both the backup path length and the network restoration time after a failure is the most suitable.

References

1. Zhou D, Subramaniam S (2000) Survivability in optical networks. IEEE Network 14(6):16–23
2. Wang D (2010) Path protection in translucent WDM optical networks
3. Sivakumar M, Shenai R.K, Sivalingam K.M (2005) A survey of survivability techniques for optical WDM networks. Emerging optical network technologies: architectures, protocols and performance. 297–331
4. Mohan N, Gupta S, Bhandari A (2020) Link failure recovery using p-cycles in wavelength division multiplex (WDM) mesh networks. J Opt Commun (0) 000010151520200099
5. Maier G, Pattavina A, De Patre S, Martinelli M (2002) Optical network survivability: protection techniques in the WDM layer. Photon Netw Commun 4:251–269
6. Pin-Han Ho, Janos Tapolcai1, Hussein T, Mouftah Chi-Hsiang Yeh (2004) Linear formulation for path shared protection. IEEE Communications Society
7. Pan Z, Chatelain B, Plant D.V, Gagnon F, Tremblay C, Bernier E (2008) Tabu search optimization in translucent network regenerator allocation. In: 2008 5th international conference on broadband communications, networks and systems IEEE pp 627–631
8. Drid H, Lahoud S, Cousin B, Molnár M (2010) Survivability in multi-domain optical networks using p-cycles. Photon Netw Commun 19:81–89
9. Mallika, Neeraj Mohan (2013) Multiple link failure in optical network. International J Res Eng Technol (IJRET) Vol 2 No 2 ISSN 2277–4378
10. Furdek M, Skorin-Kapov N, Wosinska L (2015) Attack-aware dedicated path protection in optical networks. J Lightwave Technol 34(4):1050–1061
11. Ashraf M.W, Idrus S.M, Iqbal F, Butt R.A, Faheem M (2018). Disaster-resilient optical network survivability: a comprehensive survey. In Photonics 5 (4):35 MDPI

Harnessing the Power of Human-AI Fusion in Urban Disaster Management Chatbot

Sruthy Anand and A. M. Abhishek Sai

Abstract Disasters are inevitable in human life, and the impact is more significant when a disaster hits a crowded place like cities. Artificial Intelligence significantly impacted urban society in various ways to support their quality of life. It provides opportunities to learn about the environment and delivers systems to help individuals, organizations, and communities improve their well-being. Similarly, urbanization has led to increased economic opportunities and new opportunities to co-locate to provide services that impact each other's lives. Digital transformation of society is necessary to bring an inclusive opportunity to meet the needs of evolving circumstances and events. In this work, we proposed an AI chatbot for disaster-related information dissemination in urban settings. We present the various example scenarios in which the chatbot can serve users to disseminate legitimate information. The work also shows the human-chatbot relationship factors and chatbot framework. We used the Bidirectional-GRU deep learning model for the chatbot experimentation.

Keywords Chatbots · Smart city · Disaster management · BGRU · Human-chatbot relationship

1 Introduction

Artificial Intelligence (AI) and its applicability is growing in every nook and corner of our lives in every way. The recent advancements and relevance of deep learning have enhanced how AI can support human lives. Leveraging the increased penetration of intelligent IoT devices, 5G and 6G communication advancement, now the importance of AI and its capabilities to impact societies are more compared to how it was a

S. Anand (✉)
Center for Wireless Networks & Applications (WNA), Amrita Vishwa Vidyapeetham, Amritapuri, India
e-mail: sruthyanand@am.amrita.edu

A. M. Abhishek Sai
School of Computing, Amrita Vishwa Vidyapeetham, Amritapuri, India
e-mail: amenu4aie20101@am.students.amrita.edu

S. M. Thampi et al. (eds.), *Fifth International Conference on Computing and Network Communications*, Lecture Notes in Electrical Engineering 1221,
https://doi.org/10.1007/978-981-97-4711-5_18

decade ago. This work sheds light on how the recent advancement in neural networks and chatbots can impact the lives of city residents and visitors to provide timely information on disaster-related information to make prompt and informed decisions. Chatbots are software agents that engage with users through text or voice, thus giving access to services and information through interaction [1]. Real-time information about the ground truth is necessary from any information services dissemination system.

Urban cities experience various situations that require information management and information dissemination as one of the core requirements to meet the smart city objectives [2]. Some risks, including accidents, attacks, robbery, etc., always threaten crowded places like cities. The city also poses various dangers, such as fire, gas vehicle leakage, cascading accidents, and other man-made attacks. Also, other threats, such as robbery and attacks on individual homes, are higher in urban areas. This work focuses on a chatbot design to support urban residents in receiving information during emergency scenarios. A chatbot is a computer program capable of conversing through informal chat with a human in the natural language [3]. When it comes to the design of a chatbot for disaster management or for receiving information during an emergency, we need to focus on two significant factors: (a) the dialogue management of the chatbot needs special care, and (b) ensuring the chatbot response is real-time with the help of authorities and all other stakeholders participation. The work also focuses on the human-chatbot relationship for better adoption by users.

Both functionally and from a temporal perspective, disaster management has four phases: (a) preparedness (pre-disaster preparedness, i.e., the individual measures or activities taken before the disaster events), (b) response (i.e., the actions or activities taken during the onset of an event to save lives and prevent further damage), (c) recovery, and (d) mitigation[4]. Information flow and reliable information availability play a significant role in disaster management in all these four phases. Disaster-related information spreads through mass media through television channels such as news, social media updates, and instant messaging services such as WhatsApp. But the reliability of such messages is always a question. There is a high chance of fake news spreading as well. During a disaster, social media and instant messaging services help a lot with information spreading. At the same time, they also spread many fake messages, which can negatively impact the rescue and relief services [5]. This work is an extension of the chatbot proposed for the disaster management chatbot[6] and tourism which was designed to provide information to community actors about disaster information to the coastal communities and for tourist-related information [7] dissemination for tourist visiting cities. However, the chatbot envisioned to provide disaster-related information has to focus on various aspects, such as the user's emotions and the words the chatbot produces in response to the user's query. Once established and involved by different human agents, this chatbot can act as a holistic approach.

Transparent information for all users based on their current location and the probable places they are supposed to visit is necessary to enhance disaster risk reduction in cities. If any problem is identified within the city, it must alert users with relevant notifications. If there is any issue, irrespective of the impact, they will be notified to the users to make informed decisions by the chatbot users. For example, flood usually affects urban areas for several reasons, such as blocked drainage and excessive rain. If the chatbot is linked to any of the AI prediction models, it can inform the user about the chances of clogged roads and potential risks the user might pose if the rain continues. The chatbot can also be used to alert the users to take the necessary steps to move from their current location and other information required to make decisions with the help of stakeholder participation. For this, the users must prompt the chatbot with their location. Similarly, a chatbot like this helps tourists make informed travel decisions. Usually, tourists are less aware of the city, and a chatbot can guide them with the information required to stay safe in any disaster and inform them about the impending risk.

This work explores the opportunities generative AI can bring to urban disaster management in the context of various urban societal risks affecting city residents and tourists. The work also discusses the essential factors that need to be considered to establish the human-chatbot relationship for a better chatbot experience. The work also looks into the state of the art of chatbot developments based on generative AI. The recent popularity of the chatGPT shows that the conversational nature is one thing that people adore due to its ease of use. Work needs to consider heterogeneous users and diverse digital literacy and language. This work also emphasizes the need for Persuasive Systems Design (PSD) [8] and Technology Adoption Model (TAM) [9] to adopt this technology widely in the urban community.

The rest of the paper is organized as follows. Section 2 discusses the literature review. Section 3 discusses the urban disaster management (UDM)-chatbot framework catering to the needs of the heterogeneous users. Section 4 discusses the chatbot architecture and implementation challenges. Section 5 discusses the experimentation and result. Finally Sect. 6 concludes the paper.

2 Literature Review

Frequent occurrences of natural disasters have disturbed life and property of people. Emerging technologies can be used to minimize their impact. A deep learning-based computer vision and crowd-sourcing methodology for detecting and estimating flood depths are proposed by [10]. Similarly, another work reported how crowd-sourcing played an important role in delivering resources during an earthquake by increasing the speed for obtaining resources and providing them quickly [11]. The extent of damage and effect of natural disasters depends on the hazard's intensity, duration, scale, and the human community's capability to resist, respond to, adapt to, and adjust to such disaster [12]. Generally, an emergency event gives various risks and damages to the community and needs immediate response [13]. Enhancing urban community resilience requires a timely response, rescue exposure assessment, and quick information dissemination, which can save lives [14].

The design of an urban disaster management chatbot needs to consider many factors, and when it comes to the Indian scenario, multi-lingual, regional characteristics also need to be assessed to devise a solution. Multi-lingual properties in the Indian context have been discussed in the [15] and show the importance of incorporating the local language for the chatbot. Chatbots use NLP to respond to the users by mapping to the most suitable response available to the system [16]. The recent widespread use of the ChatGPT shows the potential of the Generative Pre-training Transformer(GPT) architecture. Because of its human style of interactive capability, it has widespread adoption. Since the generative AI models are trained using huge amounts of data, their output is identical to human-generated content. ChatGPT uses both unsupervised pre-training and supervised fine-tuning [16].

The chatbot's ability to support deep-felt human needs and provide variety in interactions drive human-chatbot relationships (HCR) forward [17]. Similarly, self-disclosure in HCR depends on trust formation, and this trust in HCR has a practical and effective basis [1]. Generative means it can create new content by analyzing the situation. Hence, the generative power should be utilized to produce the data required for the situation. Compared to the existing ML, which is suitable for analyzing the patterns, ChatGPT is one step advanced; it recognizes patterns and uses them to create new data [16]. In recent years, many huge companies have shown interest in developing chatbots, such as "Apple's Siri, Google Assistant, and so on, and their objective is to obtain relevant information from several Knowledge Bases (KBs) with the help of Natural Language Understanding (NLU) and semantic [3]. The work also discusses speed, scalability, interoperability, and Turing Test to increase chatbot efficiency.

Research on the social characteristics of chatbot [18] reported several factors that determine the social capabilities of a chatbot because chatbots are designed to converse with the user, emulating a person-to-person conversation style. Making a chatbot acceptable to the intended users is not confined to a technical or social problem. Much of the existing research focuses on two-party chat [19]. Because current dialogue research focuses on two-user conversations, it doesn't address multi-party chats. However, in a study on multi-party discussion [19], each participant has an assigned character to role-play. They evaluated the ability of language models to act as one or more characters in such conversations, and the two requirements for such a model as per that research are (1) the ability to decide when to talk, (2) creating readable utterances grounded on the multiple characters [19].

3 Urban Disaster Management Chatbot Framework

This section details the components and features required for the Urban Disaster Management (UDM) chatbot to enable disaster management support during an emergency. Disaster management requires integrating various stakeholders, experts, and citizens [20]. Integrating them and fostering a collaborative working environment requires communication mediums to disseminate timely information and alerts by considering the ground realities. When a disaster hits a place, the first responders are always the local people, known as spontaneous volunteers. These volunteers will actively participate in the rescue and relief operations and provide local ground-level information to the disaster managers. From the literature, it is clear that crowd-sourcing will play a significant role in supporting a disaster response. Common challenges in developing an information dissemination system for urban disaster management (UDM) must consider demographic diversity, city resources, crowd-sourcing, information processing, context awareness, and information validation. For this, context awareness is a must. Some factors that need to be considered for context awareness are the context dimension, situation, space, event, agent, location, software agents, and activity [21]. For instance, take the context of a flood; the context dimension includes the user's location, and the situation can be modeled as safe, unsafe, or critical. Critical means they are in danger, and rescue has to happen immediately. Here, the event is a flood; agents are the human agents who can be victims, volunteers, and disaster managers. Location includes the spatio-temporal area of the various agents. Here, activities describe the actions the agents participate in, and software agents are the software components that help to provide information during the event.

Prior research shows that victims can provide valuable information to the disaster managers (DM) from the impacted area, such as real-time information on food, communication channels, the status of roads, water, availability, and other essential details [11]. The work also reported that the victims are motivated to provide information about the situation to emergency managers to regain control of their lives. When the chatbot is used as a medium, the user can input the other disaster managers (DM) in the chat, and the information must be processed and analyzed to redirect the information to the DM. Verifying information received on disaster requests for assistance is critical before a resource is dispatched in response to the request; otherwise, the valuable and scarce resources can be wasted, diverted, or go unused [11].

Chatbots can assist disaster managers in four emergency management phases, such as preparedness by giving real-time warning information dissemination during the response phase by providing valuable information from the different ground-level users who can be victims or spontaneous volunteers. Similarly, during recovery and mitigation phases, it can be used to disseminate standard disaster-related information. The chatbot can also provide information from the DM on how to prepare for any impending disasters. The city also exhibits a low social cohesion compared to the rural community. Similarly, the urban community's socio-economic characteristics are diverse compared to rural areas, where many rural populations have similar socio-economic conditions.

Chatbot for information dissemination in disaster management activities is possible only if emergency managers, AI, and citizens are properly amalgamating in analyzing and disseminating accurate situational awareness and informed action decisions. The system will have inputs from various government departments and software services. Figure 1 shows the high-level framework and components required for designing, developing, and implementing an urban disaster management chatbot. The different types of stakeholders and systems that need to be integrated into the chatbot include:

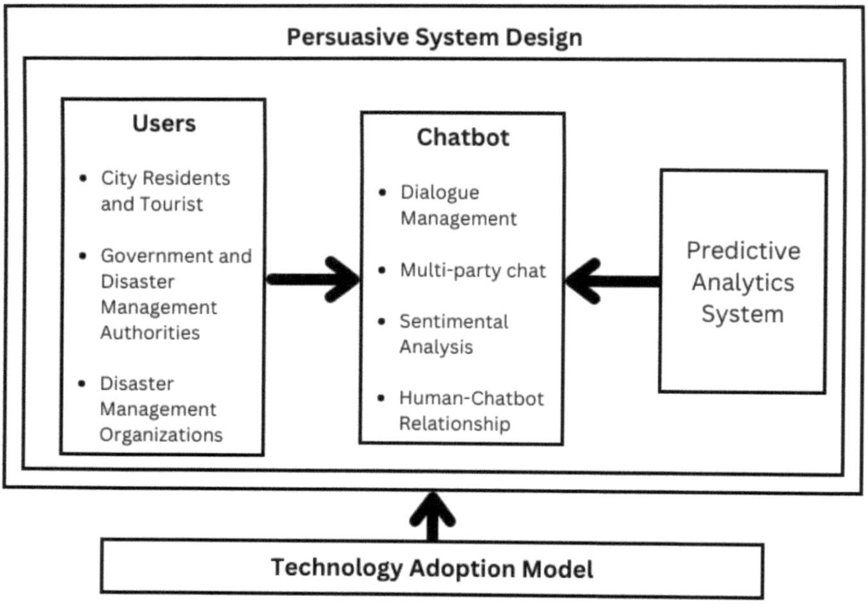

Fig. 1 Urban disaster management chatbot framework

- Users: The users category includes both city residents and visitors(tourists). The information received by both classes of users remains the same. However, multilingual capability is required to ensure the users receive the information in their preferred language. During disasters, local volunteers work in the group in response to a disaster, and often, these local members are the first responders. Generally, these groups form during the event and disperse after the response phase. Many volunteers help without skill or expertise during emergencies and are known as spontaneous volunteers. These untrained volunteers often provide valuable assistance during response and recovery periods, especially at the local level [22]. However, in this work, we are yet to consider the role of spontaneous volunteers, and we consider them as users, as often the local city residents who voluntarily take up some response activities to support the rescue and relief operations when a disaster occurs. Many disaster response solutions use crowd-sourcing information to help with rescue and relief operations [23].
- Disaster Management and Government Authorities: Integration of state-level and district-level disaster management authorities is inevitable in developing and implementing disaster management-related services in cities. State authorities and representatives of various government departments and sections must be connected to the system to provide emergency-related information and help during a disaster response and recovery phase. Similarly, the involvement of local city administration is the core to the sustainable future of any city-related information service systems.

- Predictive Analytic System: It applies to flood monitoring services, which can monitor real-time flood situations and predict future floods. This information can be disseminated as a warning message to residents to take precautions. Similarly, other systems that help to alert users about the pollution level or any other type of monitoring system integration would be helpful.
- Persuasive System Design(PSD): A participatory approach is required along with policy-making to make the system persuasive by enabling collaborative information sharing and working between the various stakeholders. Similarly, the system has to consider the usefulness of the system.
- Technology Adoption Model: A chatbot adoption model must be designed to effectively disseminate the chatbot to the community and for their adoption. Factors such as trust and ease of use [9] need to be considered along with the system's usefulness.

4 Human–Chatbot Relationship

A human–chatbot relationship is a significant factor determining the successful adoption of the disaster management chatbot by its different human agents. The major factors determining the adoption of a disaster management chatbot are shown in Fig. 2. Even though all the elements must be considered in designing a disaster management chatbot, this paper focuses only on a few factors, such as multi-party chat, ground-level information, real-time information, and sentiment analysis. This chatbot is intended to improve the communication between ordinary people, disaster

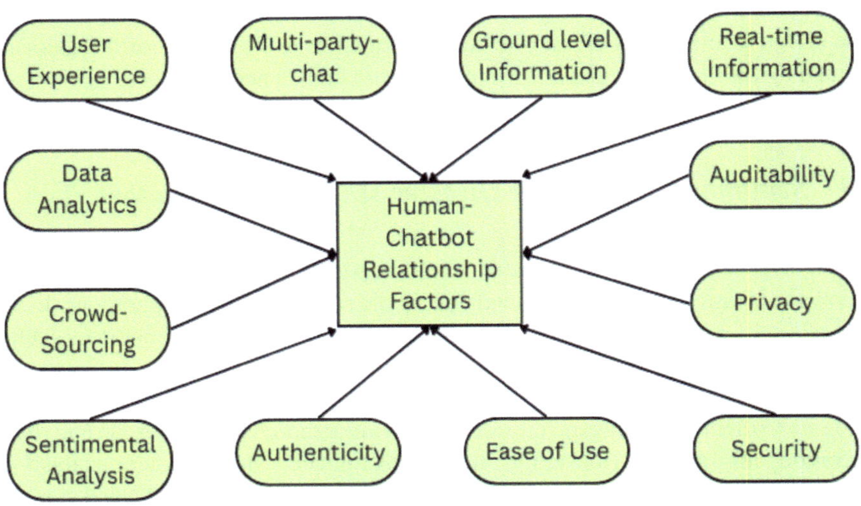

Fig. 2 Factors affecting human–chatbot relationship in UDM chatbot

managers, spontaneous volunteers, and other decision-makers during a natural or man-made disaster.

During an emergency, many existing volunteer groups and organizations with altruistic motives may come forward to lend their help wherever possible to support rescue and relief operations [5]. However, they often rely on phone calls and social media updates to collect information, which may be time-consuming. In this scenario, a chatbot can handle more victims and actors to receive ground-level vital information. The design and development of the chatbot have to consider various challenges, including user query understanding, intent classification, and multi-lingual aspects.

Depending on the possible user's query, the chatbot has to develop. The training of chatbots creates a colossal data requirement challenge. Understanding text and context is the primary responsibility that needs to be taken care of in designing the chatbot. Dialogue management interactive user interface design also needs to be addressed. Similarly, understanding the context narrative nature, bringing in a human-like conversation style, and getting more details are possible if we effectively integrate the GPT into the bot. However, how to analyze the data received through the chatbot is not considered in this work, which remains a future scope for improving the work.

When it comes to the urban environment, compared to the rural and sub-urban, the bonding between the community is generally low. Hence, with the involvement of various users, hybrid chatbots are required to produce various information services. A human-AI fusion is necessary to deliver the desired information dissemination. For disaster managers, manually satisfying and answering the queries of different individuals requires considerable resources in terms of both human and technical resources. Similarly, handling multiple disaster management queries requires greater mental strength, especially handling distress queries. Here, AI can be used to deliver information in a human style. In the case of a chatbot for disaster information dissemination, the chatbot has to generate and present the content to suit the situation. The effectiveness of any disaster-related information is the quality of the message created by the chatbot, which motivates the users to take necessary actions.

4.1 Sentiment Analysis for Disaster Management Chatbot

In an emergency scenario, it is natural to feel anxiety; hence, the victims might need to follow the disaster managers frequently about the rescue and relief operations. In the case of a call center, the users have to call them often and may only sometimes get the desired behaviors from them as they might be handling several such calls. Compared to a standard procedure, introducing a chatbot will give the victims a sense of someone listening to them. It can engage the user by providing more valuable rescue and relief operations information. Hence, the chatbot has to handle the sentiment analysis of the users' chat. Ideally, the chatbot should identify the user's mood and have to reciprocate to the user by selecting suitable words. The focus is to identify sentiments and user emotions from the user's text, and the chatbot has to produce a conversation that can reciprocate the user's emotions. It will give them a sense of comfort and make them feel good. More research is needed on how the victims feel when the chatbot attends to them during an emergency. Generally, sentiment analysis is intended to capture the emotions and moods of the user by analyzing their text, and it is a valuable functionality during a hazard because understanding user emotions and moods are direct indicators of their resilience and overall well-being [24]. The work also mentioned that positive-thinking people are happier and more resilient.

4.2 Challenges and Research Opportunities

For implementing this solution in the community, many real-world challenges need to be addressed, including social parameters, policy-making, multi-party chat, sentiment analysis, and integration of predictive analytics systems. First and foremost, the collaboration of multiple stakeholders needs to be established with proper policy-making and collaboration strategies. Multi-party chatbots with the fusion of different human agents, including disaster managers, users, and other stakeholders, need to be defined. It also needs to include the flow of information between the multiple agents and how to build the conversation between the agents, which are critical in a multi-party chatbot. Apart from this, many other factors must be taken care of while designing a chatbot for disaster management and emergency response, including building awareness and risk perception by the community. Other factors, including content regulation, data protection, content moderation, etc., must be considered before introducing it in any urban setting. Similarly, most of this work needs to be in tandem with other authorities in the urban areas, such as the local governing body and disaster management authorities. The future scope includes all the above-listed social features for effectively implementing an urban disaster management chatbot in the community. From the technical perspective, integrating the predictive analytics part, multi-party chat, and sentiment analysis remain the most critical challenges to be addressed. Similarly, how to bring in human-like reasoning and reply construction considering the situation and sentiment analysis is one of the significant future works.

5 Experimentation

This section details the deep learning model designed for the UDM chatbot. For the experimentation, this work used a Bidirectional Gated Recurrent Unit (BGRU), which is a variant of a Gated Recurrent Unit (GRU). GRU is a specialized type of recurrent neural network (RNN) architecture designed to address the limitations of traditional RNNs, which are weak in handling long-range dependencies. GRU also helps to mitigate the vanishing gradient problem. Unlike standard RNNs, GRUs employ gating mechanisms that regulate the flow of information within the network. A typical GRU structure consists of an update gate, a reset gate, a candidate hidden

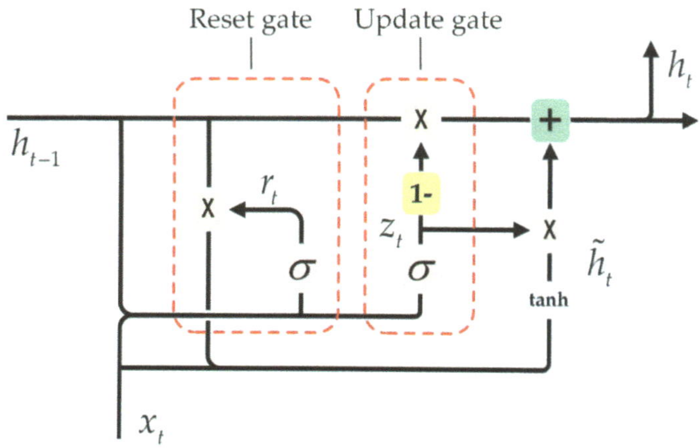

Fig. 3 Bidirectional gated recurrent unit [25]

state, and the final hidden state, as shown in Fig. 3. The update gate decides how much of the prior concealed state should be preserved and how much new candidate information should be incorporated. Based on the current input and the prior hidden state, the update gate determines whether the cell ought to remember or forget particular information. On the other hand, the reset gate decides how much of the previous hidden state should be discarded and how much of the new candidate information should be considered. This gate allows the cell to decide how much of the previous state is still relevant, given the new input. This architecture enables GRUs to capture and propagate information over extended sequences while effectively mitigating the vanishing gradient problem.

The advantage of GRUs lies in their computational efficiency compared to other RNN variants like Long Short-Term Memory (LSTM) networks. This efficiency makes GRUs suitable for processing large datasets or resource-limited environments. GRUs excel at learning both short-term and long-term dependencies in sequences, making them valuable for tasks such as language modeling, sentiment analysis, machine translation, and time-series analysis. The gating mechanisms in GRUs also provide a built-in form of regularization, reducing the risk of overfitting. Additionally, GRUs are often easier to train and require less hyperparameter tuning than more complex architectures like LSTMs.

5.1 Dataset

The dataset consists of a series of intents, patterns, and responses designed for UDM chatbot. These intents cover a wide range of user queries and topics, aiming to provide relevant and informative responses. Intents are the high-level categories or topics that the chatbot can understand and respond to. Each intent is defined by a "tag" which serves as its identifier. Patterns are user inputs or queries that correspond to a specific intent. These are the phrases or questions that users might type or say to the chatbot when they want information or assistance related to a particular topic. Responses are the messages or answers generated by the chatbot when it recognizes a specific intent based on the user's input. These responses provide information, guidance, or assistance related to the user's query.

5.2 *Process-Flow*

When a user submits a query or question, these neural networks employ natural language processing techniques to analyze the text, identifying the user's intent by comparing it to predefined patterns and intents in the dataset. Once the intent is recognized, the neural network selects an appropriate response from a predefined pool associated with that intent. This response is then provided to the user, addressing their query or providing relevant information.

5.3 *BGRU Model*

In this work, we have utilized a Bidirectional Gated Recurrent Unit (BGRU) architecture. BGRUs enhance the capabilities of traditional GRUs by introducing bidirectionality, allowing the model to capture context from both past and future sequences simultaneously. This architectural enhancement makes BGRUs particularly effective for tasks involving sequential data, where understanding dependencies in both directions is crucial. In a standard GRU, information flows only in one direction through the recurrent connections. However, in a BGRU, the data is processed by two separate GRUs, one in the forward direction and the other in the backward direction. This setup results in a comprehensive understanding of each time step, considering both its historical context and its potential future impact. The outputs from both directions are often combined or concatenated, providing a richer representation of the sequence's context. The model begins with an Embedding layer, which transforms the input integer sequences into dense vectors, helping to capture semantic relationships between words. Following the Embedding layer, the architecture employs multiple Bidirectional Gated Recurrent Units (GRUs) layers. These layers introduce bidirectional processing, allowing the model to consider the context of words preceding and following the current word. The first GRU layer, consisting of 128 units, serves as a powerful feature extractor, uncovering relevant information within the sequence. To prevent overfitting and enhance the model's generalization ability, dropout, and recurrent dropout techniques with a rate of '0.2' are integrated, ensuring that the network learns robust data representations. The subsequent two Bidirectional GRU layers, each consisting of 64 units, refine the extracted features further. These layers

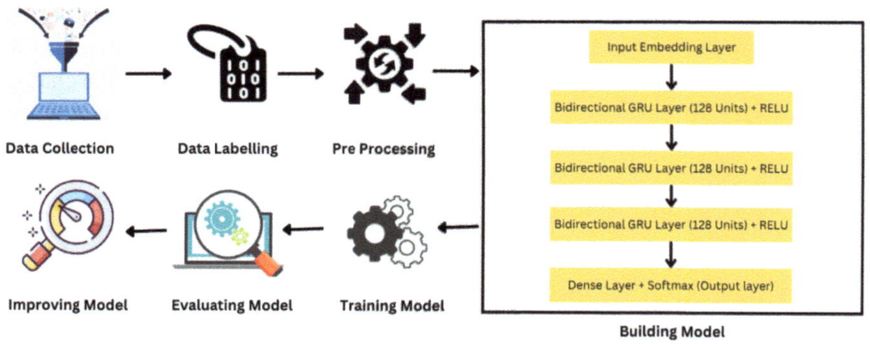

Fig. 4 Model architecture

aim to capture complex patterns within the text. To prevent overfitting, dropout, and recurrent dropout are again applied, with rates of 0.2 and 0.3, respectively.

The Rectified Linear Unit (ReLU) activation function plays a vital role in introducing non-linearity to the model. This activation function, applied after each Gated Recurrent Unit (GRU) layer, transforms the output of each unit to be zero for negative values and linear for positive values. The complete Architecture of UDM chatbot is depicted in Fig. 4. The model concludes with a Dense layer and uses the softmax activation function to generate class probabilities. It transforms the network's raw outputs into probability scores, categorizing the input text into its most likely class. During training, categorical cross-entropy is the loss function, guiding the model toward better predictions. The Adam optimizer refines the model's parameters based on the computed loss. The model's accuracy on the training and validation data is tracked throughout the process.

Suppose the stakeholders are updating the information on time. Users can rely on the UDM chatbot to obtain real-time information on diverse aspects, including weather updates, road conditions, evacuation orders, emergency shelter locations, and the availability of essential services. Figure 5 depicts a sample conversation between the chatbot and a sample user. The user can ask queries related to the crisis. The user gets the proper updates from the chatbot. The user receives real-time updates only if the stakeholders are updating the details on time and the intent is getting updated. For example, suppose a sudden change in weather patterns or new evacuation orders are issued. Stakeholders can promptly update the chatbot to reflect these changes and provide users with accurate and timely information. It ensures that users receive the most recent and pertinent information, enhancing the chatbot's role as a reliable source during emergencies.

The different evaluation metrics used to evaluate the model are accuracy, precision, and recall. For the experiment, the dataset we created for this chatbot received an accuracy of 97.23, and precision of 0.943, and a recall of 0.928. One of the most prominent strengths of BGRU lies in its ability to perform bidirectional processing. While LSTM processes data sequentially from the past to the future, BGRU

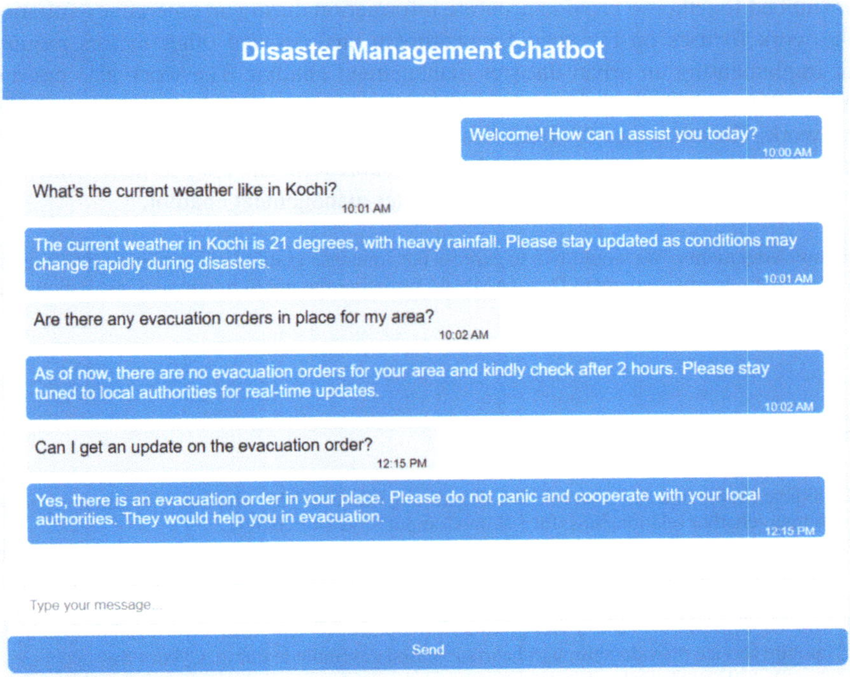

Fig. 5 Sample conversation between chatbot and user

considers information from past and future time steps. This bidirectional approach allows BGRU to capture contextual information more comprehensively, making it particularly effective for tasks where understanding the context from both directions is essential, such as natural language understanding in conversation. Additionally, BGRU's architecture is simpler and has fewer gating mechanisms than LSTM, resulting in computational efficiency and potentially faster training times.

6 Conclusion

Even though cities are populated compared to other areas due to the comfort and economic opportunities available, they also possess many risks such as accidents, natural disasters like floods due to excess rain, and man-made attacks. In this work, we addressed the need for collaborative information management during an urban emergency using a chatbot for information dissemination in an urban scenario. The work presented why the cities need a chatbot and how this urban area differs from the rural areas, where they have better social cohesion than urban areas. The work also presented the various components, including the agents' crowd-sourcing, that must be clubbed together to provide accurate information during an emergency situation. The work focuses on the need for sentiment analysis and other factors required for implementing an urban disaster management chatbot. The work also presents the BGRU model, which is used for the urban chatbot model experimentation for this work. The advantage of GRUs lies in their computational efficiency compared to other RNN variants. It also excels at learning short-term and long-term dependencies in sequences, making it valuable for a disaster management chatbot.

Acknowledgements We would like to express our immense gratitude to our beloved Chancellor Dr. Sri. Mata Amritanandamayi Devi (AMMA) for being the guiding light and inspiration to do this work.

References

1. Skjuve M, Følstad A, Fostervold KI, Brandtzaeg PB (2021) My chatbot companion-a study of human-chatbot relationships. Int J Hum-Comput Stud 149:102601
2. Anand S, Ramesh MV (2021) Multi-layer architecture and routing for internet of everything (IOE) in smart cities. In: 2021 Sixth international conference on wireless communications, signal processing and networking (WiSPNET). IEEE, pp 411–416
3. Magoo C, Singh M (2022) Design and development of new interactive chatbot system for mobile service providers through heuristic-based ensemble learning. Cybern Syst, pp 1–33
4. Ng SL (2023) The role of risk perception, prior experience, and sociodemographics in disaster preparedness and emergency response toward typhoons in Hong Kong. Nat Hazards 116(1):905–936
5. Amritanand S, Anand S, Amritesh AR (2020) Dynamic and time critical emergency management for level three disaster: a case study analysis of Kerala floods 2018. In: Proceedings of the 21st international conference on distributed computing and networking, pp 1–6
6. Anand S, Raj D, Abhishek Sai AM, Rao SN, Ramesh MV (2023) Techno-social synergy for disaster resilience in coastal communities: a sustainable approach. In: Proceedings of the 24th International conference on distributed computing and networking, pp 366–371
7. Anand S, Abhishek Sai AM, Karthikeya M (2023) Chatbot enabled smart tourism service for indian cities: an AI approach. In: 2023 11th International conference on internet of everything, microwave engineering, communication and networks (IEMECON). IEEE, pp 1–7
8. Oinas-Kukkonen H, Harjumaa M (2018) Key issues, process model and system features 1. Routledge handbook of policy design
9. Davis FD (1989) Perceived usefulness, perceived ease of use, and user acceptance of information technology. MIS Q, pp 319–340
10. Nair BB, Krishnamoorthy S, Rao SN et al (2021) Machine vision based flood monitoring system using deep learning techniques and fuzzy logic on crowdsourced image data. Intell Decis Technol 15(3):357–370
11. Riccardi MT (2016) The power of crowdsourcing in disaster response operations. Int J Disaster Risk Reduction 20:123–128
12. Deng Y, Su G, Gao N, Sun L (2019) Perceptions of earthquake emergency response and rescue in China: a comparison between experts and local practitioners. Nat Hazards 97:643–664
13. Abbasi A (2014) Link formation pattern during emergency response network dynamics. Nat Hazards 71:1957–1969
14. Tzavella K, Fekete A, Fiedrich F (2018) Opportunities provided by geographic information systems and volunteered geographic information for a timely emergency response during flood events in Cologne, Germany. Nat Hazards 91:29–57

15. Kolar S, Kumar R (2023) Multilingual tourist assistance using ChatGPT: comparing capabilities in Hindi, Telugu, and Kannada. arXiv preprint arXiv:2307.15376
16. Dwivedi YK, Kshetri N, Hughes L, Slade EL, Jeyaraj A, Kar AK, Baabdullah AM, Koohang A, Raghavan V, Ahuja M et al (2023) "So what if ChatGPT wrote it?" Multidisciplinary perspectives on opportunities, challenges and implications of generative conversational AI for research, practice and policy. Int J Inf Manag 71:102642
17. Skjuve M, Følstad A, Fostervold KI, Brandtzaeg PB (2022) A longitudinal study of human–chatbot relationships. Int J Hum-Comput Stud 168:102903
18. Chaves AP, Gerosa MA (2021) How should my chatbot interact? A survey on social characteristics in human-chatbot interaction design. Int J Hum-Comput Interact 37(8):729–758
19. Wei J, Shuster K, Szlam A, Weston J, Urbanek J, Komeili M (2023) Multi-party chat: conversational agents in group settings with humans and models. arXiv preprint arXiv:2304.13835
20. Shah AA, Wu W, Gong Z, Pal I, Khan J (2021) Multidimensional six-stage model for flood emergency response in schools: a case study of Pakistan. Nat Hazards 105:1977–2005
21. Pradeep P, Krishnamoorthy S, Pathinarupothi RK, Vasilakos AV (2021) Leveraging context-awareness for Internet of Things ecosystem: representation, organization, and management of context. Comput Commun 177:33–50
22. Daddoust L, Asgary A, McBey KJ, Elliott S, Normand A (2021) Spontaneous volunteer coordination during disasters and emergencies: opportunities, challenges, and risks. Int J Disaster Risk Reduction 65:102546
23. Guntha R, Rao SN, Ramesh MV (2020) Deployment experiences with Amrita Kripa: a user-friendly feature rich crowdsourced humanitarian application. Procedia Comput Sci 171:302–311
24. Zander KK, Nguyen D, Mirbabaie M, Garnett ST (2023) Aware but not prepared: understanding situational awareness during the century flood in Germany in 2021. Int J Disaster Risk Reduction, pp 103936
25. Li P, Luo A, Liu J, Wang Y, Zhu J, Deng Y, Zhang J (2020) Bidirectional gated recurrent unit neural network for Chinese address element segmentation. ISPRS Int J Geo-Inf 9(11):635

Healthcare and Medical Applications

Deep Learning Classification of Subtypes of Luminal Breast Cancer from Immunohistochemistry Slides

Vishnu Shaji, Mridul Sharma, Merin Mathew, K. Malavika, Dhanya Mary Louis, Dehannathparambil Kottarathil Vijaykumar, Lakshmi Malavika Nair, Archana George, and Georg Gutjahr

Abstract This paper proposes deep-learning models to assist in classifying subtypes of luminal breast cancer from immunohistochemistry slides. The classification has traditionally been done by pathologists who evaluate Ki67 indexes by manually counting positive and negative cells that are dyed in different colors in the immunohistochemistry slides. This manual process is error-prone and time-consuming. Therefore, this paper explores the possibility of automatically classifying immunohistochemistry slides. The performances of 4 pre-trained, deep convolutional networks are compared. Based on the good performance of the models, we conclude that the task of Ki67 classification for determining subtypes of luminal breast cancer is well-suited for deep learning models and that for this reason, manual counting of Ki67-positive cells are neither needed nor desirable.

Keywords Luminal breast cancer · Ki67 index · Image classification · Convolutional neural networks

V. Shaji (✉) · M. Sharma
Department of Computer Science and Engineering, Amrita School of Computing, Amrita Vishwa Vidyapeetham, Amritapuri, India
e-mail: amenu4aie21167@am.students.amrita.edu

M. Mathew · K. Malavika
Department of Mathematics, Amrita Vishwa Vidyapeetham, Amritapuri, Kollam, Kerala, India

D. M. Louis
Department of Pharmacy Practice, Amrita School of Pharmacy, Amrita Institute of Medical Sciences And Research Centre, Kochi, Kerala, India

D. M. Louis · D. K. Vijaykumar · L. M. Nair
Department of Breast Oncology, Amrita Institute of Medical Sciences And Research Center, Amrita Vishwa Vidyapeetham, Kochi, Kerala, India

A. George
Department of Pathology, Amrita Institute of Medical Sciences And Research Center, Amrita Vishwa Vidyapeetham, Kochi, Kerala, India

G. Gutjahr
Center For Research In Analytics And Technologies For Education (CREATE), Amrita Vishwa Vidyapeetham, Amritapuri, Kollam, Kerala, India

S. M. Thampi et al. (eds.), *Fifth International Conference on Computing and Network Communications*, Lecture Notes in Electrical Engineering 1221,
https://doi.org/10.1007/978-981-97-4711-5_19

1 Introduction

Ki67 is widely used as a proliferation marker in breast cancers [11]. A higher Ki67 index indicates a higher tumor cell proliferation rate, which in turn indicates a more aggressive tumor type [5]. The proliferation rate is defined the percentage of positively stained Ki67 cells among the total number of malignant cells. Usually, pathologists evaluate the Ki67 index from an immunohistochemistry (IHC) slide. An IHC slide is a type of histopathological slide where immunostaining is applied. Figure 1 shows two images of IHC slides with high and low Ki67 indexes.

To prepare an IHC slide for studying breast cancer, the first step is to take a tissue sample from the tumor specimen, typically obtained through biopsies or surgical resections. This sample is sent to a pathology lab, where the sample is embedded in paraffin after formalin fixation. The resulting paraffin block is then cut into fine slices that are mounted in glass. These glass-mounted slides are called histopathological slides. To bring out the properties of the cells, the tissue needs to be dyed with stains that highlight the structures of interest. The two most common staining protocols are hematoxylin and eosin (H&E) and IHC straining. In H&E straining, hematoxylin colors the nuclei blue and purple by binding to the DNA, and the eosin colors cytoplasm and some other structures pink by binding to certain proteins. On the other hand, in IHC straining, specific proteins can be detected by using dyes that bind to specific primary antibodies. Veta et al. [18] give an overview of the issues in tissue preparation and staining for breast cancer histopathology slides.

The evaluation of Ki67 starts with an IHC slide where negative cells are dyed in blue and positive cells are dyed in brown shades. Several methods have been proposed to determine the Ki67 value. Among these, the three most common approaches are following [6].

1. In the first approach, one estimates the percentage of positive Ki67 cells among a fixed number of tumor cells within a hotspot region. Such a hotspot region is an

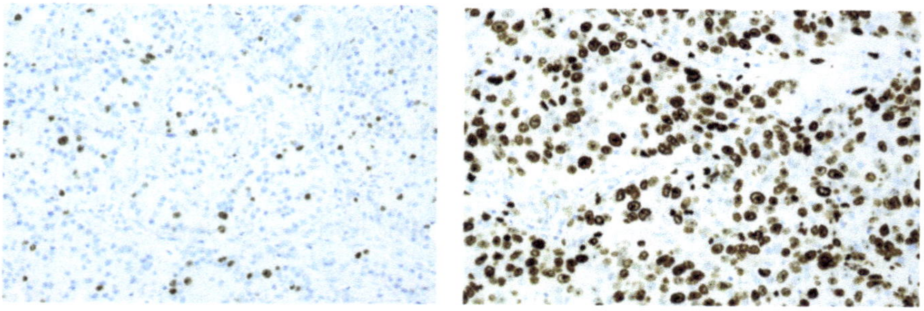

(a) Ki67 - Low (b) Ki67 - High

Fig. 1 Zoomed views on two IHC stained slides showing different intensities of Ki6, where brown color indicates positively stained Ki67 cells and blue color indicates negatively stained cells. **a** Low Ki67 value (less than 10%). **b** High Ki67 value (almost 80%)

area in the slide with a high density of Ki67-positive tumor cells relative to adjacent areas. This is done by counting the number of positive cells among a fixed number of cells (for example, a thousand cells). The selection of the hotspot area and the predefined number of cells may considerably impact the estimated Ki67 value.
2. In the second approach, multiple hotspots are selected, and each hotspot's average percentage of positive cells is estimated. The overall estimate of the Ki67 index is then obtained by averaging the percentages from each of the selected hotspot areas. Unlike in the first approach, the average percentage of positive cells is estimated within all tumor cells within the hotspot. However, the selection of hotspot regions is still subjective, and this selection might still lead to different estimates on the same slide by different pathologists.
3. In the third approach, the faction of positive cells within the whole slide is estimated. This includes the whole slide, unlike in the previous two approaches. While this removes the subjective selection of hotspot areas, it introduces a new problem: manually counting all cells is time-consuming and error-prone.

Lashen et al. [6] recommend using the first method as the most feasible approach. However, in practice, none of these approaches might be strictly followed, and the Ki67 index is often estimated using various techniques such as "eye-balling estimation" and "vision counting". In summary, all the methods described above are tedious and involve human judgment, which makes them error-prone and subjective with limited reproducibility [6].

As mentioned before, Ki67 is of great clinical relevance. Among other uses [9, 10], Ki67 assessment is one of the criteria to distinguish between Luminal A and Luminal B cancers [17], where Luminal B is a high-risk type of cancer, and those patients may need additional treatment like chemotherapy [12]. Several cutoff points for distinguishing between the luminal subtypes are in common use; in the following, we will use 20 percent [7]. The clinical guidelines for early-stage breast cancer published by the European Society for Medical Oncology (ESMO) in 2019 also proposes a cut-off of 20%. Since the manual assessment of Ki67 from IHC slides has the issue of reproducibility between different observers and is a time-consuming process, this paper aims to develop a method to automatically process and analyze IHC slides for the classification of the genetic breast cancer subtypes Luminal A and Luminal B. One approach to solve this problem would be to start by automatically counting the Ki67-positive cells, either by thresholding (implemented in software like ImageJ [1] and Aperio Image Scop [14]) or using machine learning methods [21], and in a second step compare the count with the threshold value. Reliable, automated counting of Ki67-positive cells is an ongoing research area, and none of the solutions are completely satisfactory [3]. In this paper, we take a different approach, where instead of segmenting and counting cells first and classifying later, we use deep learning to classify images into a high Ki67 (positive) and low Ki67 (negative) group directly.

The outline of the paper is as follows. Section 2 describes the background of the data set, the data preprocessing, the validation strategy, and the deep learning models that are used for the classification. Section 3 presents the results, and Sect. 4 ends with a discussion and gives conclusions.

2 Methodology

2.1 Data

A total of 129 histopathologic slides from breast cancer patients were prepared for evaluation, with 80 positive cases and 49 negative controls. The resolution of the images is 300 pixels per inch. Data were collected at the Amrita Institute of Medical Sciences (AIMS), a super-specialty tertiary-care hospital in South India. The study was reviewed and approved by the Institutional Review Board of AIMS.

Two expert pathologists assessed the percentage of invasive cancer cells expressing Ki67 in each slide. In more detail, the pathologists independently scored 500 tumor cells in three to five hot spots at 400x magnification. Nuclear staining of any intensity was considered positive. If the two pathologists came to different assessments on whether Ki67 exceeded 20 percent in a slide, the judgment was made by a third pathologist who independently made an additional assessment. The average Ki67 value of the two pathologists was recorded, and the classification of whether or not Ki67 exceeded 20 percent was taken for the analysis.

2.2 Preprocessing

During the preprocessing phase, various transformation techniques were used to prepare the data for deep learning using the PyTorch library. From the available dataset, new images were generated using data augmentation techniques to increase the size of the training set (see description of the validation strategy in the next section). In total, 400 images were generated. For each image in the training, the below augmentation methods were used to create new images which had different orientations, contrast levels, to create a challenging dataset which can be used to train the models efficiently.

Resize() This transformation is commonly used in image preprocessing and computer vision tasks. It is used to resize input images to a specified target size. It ensures that all the input images are of the same size so that the deep learning models can work on a uniform input size.

RandomRotation() The transformation rotates input images. The degree of rotation is chosen randomly from a specified range. This introduces diversity into the training data, ensuring that the model does not depend on the specific orientation of the slides.

ColorJitter() This transformation applies random variations to an image's brightness, contrast, saturation, and hue within specified limits. By doing so, the model can adapt to different lighting issues that may be caused when a raw input is given.

RandomHorizontalFlip() This transformation randomly flips the images horizontally. It horizontally mirrors the image, thereby shifting the orientation of the images.

Including augmented images that are randomly flipped into the training data ensures that the model is robust with respect to unfamiliar orientations.

RandomVerticalFlip() Similar to RandomHorizontalFlip, this transformation randomly flips the input images vertically. It vertically mirrors the image, thereby again changing the orientation of the images.

ToTensor() The deep learning models expect their input as tensor objects. ToTensor transforms an image to such a PyTorch tensor. The tensor can then be fed into the model for training.

2.3 *Models*

Deep learning is a sub-field of artificial intelligence and more specifically, machine learning. Artificial neural networks are the most prominent class of deep learning models. Such networks are computational models inspired by a human brain's functions and structure. Deep learning models have been successfully applied across various domains, from computer vision and natural language processing [16] to intrusion detection [20], and healthcare [19]. This section will briefly summarize the concepts of deep learning, neural networks, and convolutional neural networks for image processing.

A neural network consists of different interconnected nodes, which are also known as artificial neurons. These neurons are organized in layers, typically consisting of an input layer, one or more hidden layers, and one output layer. The strength of each connection between two neurons is determined by the weight associated with the connection. Neural networks are highly versatile because of their capability to learn complex patterns and representations from data. The process of training a neural network involves adjusting the weights of its connections to minimize the difference between the predicted output and the observed outcomes. Different iterative algorithms are used for this optimization, such as backpropagation and gradient descent. In the context of fitting parameters of neural networks, each iteration is called an epoch.

Convolutional Neural Networks (CNNs) are specially designed deep neural networks which are used to process grid-like data, such as images and video frames. CNNs have revolutionized computer vision tasks by incorporating convolutional layers that apply learnable filters to the input data. These filters capture the local patterns and features, and then the filters build a hierarchical representation of the input. The ability of CNNs to preserve the spatial relationships between pixels makes them well-suited for image-related tasks. Such networks consist of convolutional layers, pooling layers, and the fully connected layers for making predictions. Due to this architecture enables CNNs to automatically detect the edges, textures, and higher-level features like object parts and shapes, which is essential for image classification and segmentation. CNNs have been used in a wide range of applications, including medical image analysis, remote sensing, autonomous driving, and many others.

Instead of training CNNs from scratch on each new dataset, it has now become common to use pre-trained models that determine the network architecture and contain pre-trained neuron connections. These pre-trained models have been created on large datasets, and transferring them to the specific application domain generally improves the performance for tasks such as image classification and segmentation.

We have used four suitable pre-trained models for the classification of breast cancer IHC slides. The pre-trained models are ResNet-50, DenseNet-121, MobileNet V2, and Inception V3. We will now briefly discuss the capability and performance of these four models.

ResNet-50 [2] The Deep Residual Network 50 (ResNet-50) is one of the variants of the ResNet family. ResNet-50 has a 50-layer convolutional neural network in which 48 are convolutional layers, then one MaxPool layer, and one average pool layer. The architecture features residual blocks with identity shortcuts. The convolutional layers of ResNet-50 have already learned a wide range of low to high-level features from the massive datasets available, such as ImageNet. ResNet-50 main advantage is its capability to serve as a powerful feature extractor for a broad range of variety of computer vision tasks. The residual blocks are the blocks that allow for shortcuts, i.e., skip connections, which make it easier for the gradients to flow and allow networks to learn complex transformations and stabilize the learning process, as it also enhances the ability of the blocks to learn identity mapping. The deep architecture of ResNet-50 makes it well-suited for feature extraction in complicated images, such as breast cancer IHC slides.

DenseNet-121 [4] The Densely Connected Convolutional Network: The architecture of DenseNet-121 is known for its densely connected blocks. It contains a total of 121 layers in which 117 convolutions, 3 transitions, and 1 classification layers are present. In the traditional convolutional neural networks, the layers are connected sequentially. Whereas in the DenseNet model, each layer is connected to every other layer. As the name suggests, this dense connectivity of the layers promotes feature reuse and facilitates gradient flow during training, resulting in a compact and highly efficient network. Transition blocks help reducing the the spatial dimension, which refers to the height and width of the activation maps. It reduces this spatial dimension using max pooling or average pooling and convolutions. Reduction in spatial dimension reduces the computational cost and makes sure that the model does not overfit. While pre-training the DenseNet-121 models, they capture a very rich hierarchy of features. Their performance in tasks requiring precise feature discrimination, such as medical image analysis and other fields, is impressive. Dense connectivity of the layer allows for better gradient propagation through the network; this helps in the mitigation of the vanishing gradient problem, making DenseNet-121 a valuable pre-trained model in deep learning. The densely connected blocks of the model enable the efficient extraction of fine-grained features. In Ki-67 breast cancer cell recognition also, this model works with good accuracy at capturing variations in cell morphology and staining patterns.

MobileNet V2 [13] MobileNet V2 is a popular deep-learning model used in the field of computer vision; as the name suggests, it is primarily used for mobile devices and edge devices. MobileNet-v2 is a convolutional neural network that is 53 layers deep. It has depth-wise separable convolutions, which reduce the computational cost at the same time maintaining high accuracy. Depth-wise convolution is used in lightweight models, which splits the convolution process into 2 steps. First is depth-wise convolution, in which each input channel is convolved with its own set of filters and point-wise convolution which combines the output channel using linear combinations. This can reduce the computational cost and act as regularization, which makes it harder for the model to overfit. This makes the model lightweight and easy to train and adjust to new data as compared to other models available for the same set of tasks.

Inception V3 [15] It is a CNN model created by Google for computer vision tasks. It is 48 layers deep. It uses inception modules that incorporate multiple filter sizes that run in parallel to capture the different features. Compared to GoogleNet, another CNN model developed at Google, Inception incorporates label smoothing and factorized convolutional. Label smoothing is a regularization technique that generalizes the one-hot labels by redistributing some probability mass. This ensures that the model is not overconfident in its prediction and reduces the chance of the model overfitting. Inception V3 is widely used in the medical field, where its segmentation capability captures fine-grained features.

2.4 *Validation Scheme*

Initially, the data were divided into training and test data, with 70% for training and the remaining 30% for testing. After preprocessing and augmentation, 20% of the augmented training set was set aside as a validation set. During each epoch of the iterative training of the models, the validation set was used to validate the model's current performance. Changes in the validation loss from one epoch to the next were used as stopping criteria. Based on the validation loss, we also adjusted the learning rate to optimize the model's training speed. This iterative validation and adjustment cycle ensured that the model constantly improved its performance.

After training the models,the following metrics were used to evaluate the models' performances on the test data:

Accuracy This measures the overall accuracy of a model's predictions. Put more precisely, the measure gives the percentage of correct classifications. It is calculated as follows:

$$\text{Accuracy} = \#\{\text{correct predictions}\}/\#\{\text{all predictions}\}$$

Precision This is another important indicator of a machine-learning model's performance. It measures the percentage of correct positive model predictions. It is calculated as follows:

$$\text{Precision} = \#\{\text{true positive predictions}\}/\#\{\text{true positive predictions}\} + \#\{\text{false positive predictions}\}$$

Recall Recall is also called sensitivity. It measures the proportion of the true predictions among positive cases and thereby provides a measure of the model's sensitivity in detecting positive cases:

$$\text{Recall} = \#\{\text{true positive predictions}\}/\#\{\text{positive cases}\}$$

F1-Score The F1-Score is the harmonic mean of precision and recall. It balances precision and recall, making it a valuable metric for assessing model performance in imbalanced data sets. It helps to analyze false positives and false negatives. It is calculated as follows:

$$\text{F1-Score} = (\text{Precision} \times \text{Recall})/(\text{Precision} + \text{Recall})$$

2.5 *Implementation*

We implemented the models in Python due to its widespread popularity and the extensive ecosystem of libraries and frameworks, simplifying the complexities associated with deep learning tasks. More specifically, we used the PyTorch and Torchvision for working with the deep learning models. PyTorch is one of the most efficient deep learning frameworks for Python, and it seamlessly integrates with GPUs. Torchvision is a popular package in the PyTorch framework, which has a wide range of tools and utilities for working with computer vision tasks. It contains pre-trained models we have used in this project, transformation techniques used to create the augmented images, and a data loader, which makes it easy to batch-load the data for training, validation and testing the model. Hyperparameter tuning is the process of finding the optimal hyperparameter. We adjusted hyperparameters during the training phase by monitoring the validation loss on the validation set.

3 Results

Figure 2 shows the validation loss of the different models during the model training. We see that each model converged very quickly and that all models' validation loss goes to zero. This means that, at least on the training data, the models can make perfect predictions. The stopping criteria for the models were if the validation loss was not improving with each epoch when compared to the past epoch, the model training would stop because overtraining the model would cause it to overfit and make the accuracy way lower than expected. However, to assess the models' true

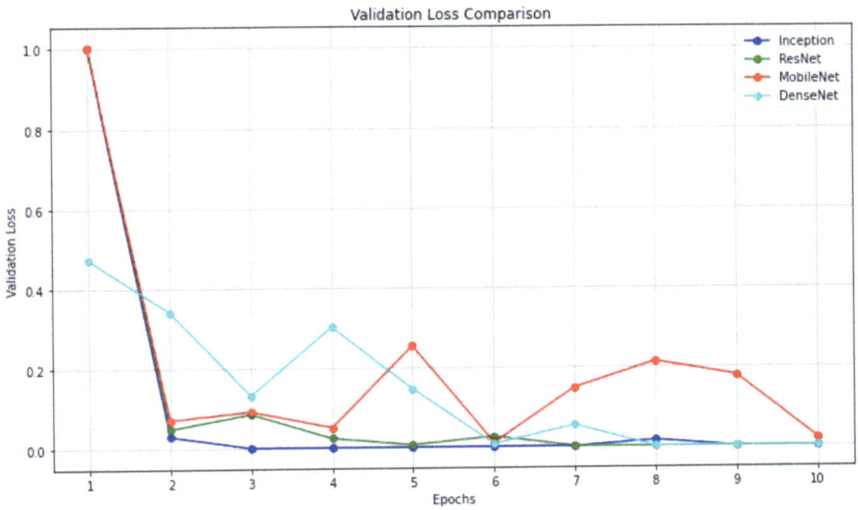

Fig. 2 Validation loss comparison for different models

Table 1 Percentages of performance comparison of deep learning models on various metrics for the test data

	Accuracy	Precison	Recall	F1-score
ResNet	90.9	90.9	90.9	90.9
Inception	89.1	89.1	89.1	89.1
MobileNet	100.0	100.0	100.0	100.0
DenseNet	92.7	92.7	92.7	92.7

performance on new data, we compared the models on the test data; the results are shown in Table 1, which lists the model's results on the test set and displays all the key metrics, including accuracy, precision, recall, and F1-Score, for each of the luminal cancer subtype classification models. We can observe here that all 4 models have excellent performance, and we conclude that the task of classifying luminal cancer subtypes can be done efficiently and reliably by deep learning models.

4 Discussion

The traditional method used to evaluate the percentage value of Ki67 is time-consuming and subjective. To manually find the number of positively stained cells from the stained slide requires a lot of concentration and takes a long time. The problem of automatic classification of IHC slides is of considerable relevance. Multiple approaches have been proposed to determine the Ki67 value from IHC slides [18].

Traditional image-segmentation techniques cannot adequately count overlapping cells in slides [21]. Nielson et al. details the recommendations from International Ki67 in Breast Cancer Working Group(IKWG) [11].

Instead of direct assessment of Ki67, some papers consider the classification of images based on features that would traditionally have been derived from a manual assessment of Ki67, for example, the diagnosis of breast cancer in the early stages [22].

This paper considers another such problem that currently requires the manual assessment of Ki67. Instead of segmenting cells and estimating Ki67, the problem is formulated as a binary classification problem. The results show that deep learning models can solve this problem with high accuracy.

In more detail, we find that the specific network architecture is not of paramount importance because all the pre-trained networks can achieve very high accuracy within a relatively short training period. Regarding the feasibility of this approach, we note that the trained models can be applied with little computational demand and a relatively small memory footprint.

However, even with robust feature extraction and classification mechanisms, the challenges of integrating such an automated image classification approach in laboratory practice are multifaceted. Automated histopathologic analysis likely involves laboratory-specific preprocessing of images, and integrating deep learning methods might, therefore, require adaptations to the workflows in pathology laboratories. Furthermore, the lack of standardized Ki67 assessment protocols poses a challenge to the training and validating of the methods in a new laboratory.

Future work might elicit domain-specific and laboratory-specific demands and extend the deep learning models accordingly. Furthermore, future work should explore to what extent the method can be generalized to other laboratories that might follow different guidelines for analyzing IHC slides.

In conclusion, the relevance of the proposed approach lies in its potential to automate routine tasks in breast cancer pathology. Accurate classification of IHC slides is required to diagnose and treat breast cancer patients. Standardizing the preanalytical and analytical variabilities in IHC Ki67 evaluation would minimize the inconsistencies in Ki67 estimation [8]. In this way, this paper provides another small step to bridge the gap between traditional pathology and cutting-edge machine learning, paving the way for efficient, reproducible, and objective diagnoses.

References

1. Collins TJ (2007) ImageJ for microscopy. Biotechniques 43(S1):S25–S30
2. Deng J, Dong W, Socher R, Li LJ, Li K, Fei-Fei L (2009) Imagenet: A large-scale hierarchical image database. In: 2009 IEEE conference on computer vision and pattern recognition. IEEE (2009)
3. Fulawka L, Halon A (2016) Proliferation index evaluation in breast cancer using ImageJ and ImmunoRatio applications. Anticancer Res 36(8):3965–3972

4. Huang G, Liu Z, Van Der Maaten L, Weinberger KQ (2017) Densely connected convolutional networks. In: Proceedings of the IEEE conference on computer vision and pattern recognition, pp 4700–4708
5. Kreipe H, Harbeck N, Christgen M (2022) Clinical validity and clinical utility of Ki67 in early breast cancer. Therapeutic Adv Med Oncol 14:17588359221122724
6. Lashen A, Toss MS, Green AR, Mongan NP, Rakha E (2022) Ki67 assessment in invasive luminal breast cancer: a comparative study between different scoring methods. Histopathology 81(6):786–798
7. Lombardi A, Lazzeroni R, Bersigotti L, Vitale V, Amanti C (2021) The proper Ki67 cut-off in hormone responsive breast cancer: a monoinstitutional analysis with long-term follow-up. Breast Cancer: Targets Therapy 213–217
8. Louis DM, Nair LM, Vallonthaiel AG, Narmadha M, Vijaykumar D (2023) Ki67: a promising prognostic marker in early breast cancer-a review article. Ind J Surgical Oncol 14(1):122–127
9. Louis DM, Nair LM, Vijaykumar D, Vallonthaiel AG, Yesodharan J, Bhaskaran R, Narmadha M (2023) Unchartered waters: significance of fall in Ki67 index after short-term preoperative endocrine therapy in early breast cancers. Breast Disease 42(1):27–36
10. Louis DM, Vijaykumar DK, Nair LM, Narmadha M, Vallonthaiel AG, Yesodharan J, Bhaskaran R (2023) Fall in Ki67 index after short-term preoperative letrozole: a gateway to assess the response in hormone-positive early breast cancers. Ind J Surgical Oncol 14(1):208–214
11. Nielsen TO, Leung SCY, Rimm DL, Dodson A, Acs B, Badve S, Denkert C, Ellis MJ, Fineberg S, Flowers M et al (2021) Assessment of Ki67 in breast cancer: updated recommendations from the international Ki67 in breast cancer working group. J Natl Cancer Inst 113(7):808–819
12. Orrantia-Borunda E, Anchondo-Nuñez P, Acuña-Aguilar LE, Gómez-Valles FO, Ramírez-Valdespino CA (2022) Subtypes of breast cancer. Breast Cancer [Internet]
13. Sandler M, Howard A, Zhu M, Zhmoginov A, Chen LC (2018) Mobilenetv2: Inverted residuals and linear bottlenecks. In: Proceedings of the IEEE conference on computer vision and pattern recognition, pp 4510–4520
14. Staniszewski W (2009) Virtual microscopy, data management and image analysis in Aperio ScanScope system. Folia Histochemica et Cytobiologica 47(4):699–701
15. Szegedy C, Liu W, Jia Y, Sermanet P, Reed S, Anguelov D, Erhan D, Vanhoucke V, Rabinovich A (2015) Going deeper with convolutions. In: Proceedings of the IEEE conference on computer vision and pattern recognition, pp 1–9
16. Tamuly S, Jyotsna C, Amudha J (2019) Deep learning model for image classification. In: Computational vision and bio-inspired computing: ICCVBIC 2019. Springer, pp 312–320
17. Tang P, Tse GM (2016) Immunohistochemical surrogates for molecular classification of breast carcinoma: a 2015 update. Arch Pathol Laboratory Med 140(8):806–814
18. Veta M, Pluim JP, Van Diest PJ, Viergever MA (2014) Breast cancer histopathology image analysis: a review. IEEE Trans Biomed Eng 61(5):1400–1411
19. Vikranth CS, Jagadeesh B, Rakesh K, Mohammad D, Krishna S, AS RA (2022) Computer assisted diagnosis of breast cancer using histopathology images and convolutional neural networks. In: 2nd International conference on artificial intelligence and signal processing (AISP), pp 1–6. https://doi.org/10.1109/AISP53593.2022.9760669
20. Vinayakumar R, Alazab M, Soman K, Poornachandran P, Al-Nemrat A, Venkatraman S (2019) Deep learning approach for intelligent intrusion detection system. IEEE Access 7:41525–41550
21. Zawisza A, Tabakov M, Karanowski K, Galus K (2021) Morphological analysis of histopathological images using deep learning. In: Advances in computational collective intelligence: 13th international conference, ICCCI 2021, Kallithea, Rhodes, Greece, September 29–October 1, 2021, Proceedings 13. Springer, pp 134–145
22. Zerouaoui H, Idri A (2022) Deep hybrid architectures for binary classification of medical breast cancer images. Biomed Signal Process Control 71:103226

Applied Health Informatics for Diabetes Risk Prediction from Food Consumption Behavior of Rural Communities in Northern Thailand with Data Analytics

Pratya Nuankaew, Apatcharaporn Kadkasame, Kunthida Sawasit, Patchara Nasa-Ngium, Thapanapong Sararat, and Wongpanya S. Nuankaew

Abstract This research has three critical research objectives. The first objective is to study the context and risk of diabetes from the food consumption behavior of Thai people in the northern region. The second objective is to develop a model for predicting diabetes risk from the food consumption behavior of Thai people in the northern region. The last objective is to evaluate the efficiency of the model for predicting diabetes risk from the dietary behavior of Thai people in the northern region. The population and sample in this research were 370 people in Chiang Muan District, Phayao Province, aged between 20 and 99 years. The research tools include an IOC-approved questionnaire and data mining analysis using machine learning and artificial intelligence tools known as CRISP-DM. Model performance was evaluated using data partitioning and testing known as split-validation methods and confusion matrix performance. The research results found that the context of rural people in northern Thailand toward food consumption behavior was normal. Overall, the risk of people developing diabetes was at an acceptable level. The developed model has the highest level of accuracy with an accuracy value of 81.80%. It is given in Tables 11 and 12 that people have good food consumption habits. Moreover, the model for predicting the risk of diabetes in rural communities was found to have a high level of accuracy by comparing various techniques, as given in Table 4. This research, therefore, concludes that all objectives were achieved. The results and expectations

P. Nuankaew (✉) · A. Kadkasame · K. Sawasit · T. Sararat · W. S. Nuankaew
School of Information and Communication Technology, University of Phayao, Phayao 56000, Thailand
e-mail: pratya.nu@up.ac.th

W. S. Nuankaew
e-mail: wongpanya.nu@up.ac.th

P. Nasa-Ngium
Faculty of Science and Technology, Rajabhat Maha Sarakham University, Maha Sarakham 44000, Thailand

S. M. Thampi et al. (eds.), *Fifth International Conference on Computing and Network Communications*, Lecture Notes in Electrical Engineering 1221,
https://doi.org/10.1007/978-981-97-4711-5_20

from the research are to know the food consumption context of rural people in the northern region of Thailand, which can be used to design and plan strategies to prevent the increase in people at risk of diabetes in Thailand.

Keywords Applied informatics · Diabetes risk prediction · Food consumption behavior · Health forecasting model · Health informatics

1 Introduction

Diabetes can occur in all genders and all ages, with the majority being found in the elderly. Diabetes is a disease in which blood sugar levels are higher than usual because the hormone produced by the pancreas, insulin, is responsible for transporting glucose from the blood into the cells of various organs, not functioning correctly. In addition, it also affects changes in the arteries throughout the body, thus causing many complications such as diabetic retinopathy, diabetic kidney disease, or diabetic neuropathy. Diabetes is a chronic non-communicable disease and a significant public health problem in all countries worldwide, including Thailand. In 2019, the International Diabetes Federation (IDF) reported and predicted trends in people with diabetes between the adults' ages of 20–79 years. It is estimated that the number of people with diabetes will reach 578 million by 2030 and 700 billion by 2045, accounting for 10.9 percent of the world's population [1]. The number of people with diabetes increases with age, with 19.9% (111.2 million) of people aged 65–79 years suffering from diabetes worldwide. For Thailand, the Department of Disease Control of the Ministry of Public Health reported the rate of patients with type 2 diabetes in the age group 60 years and over from 2018 to 2020 at 19.07%, 19.67%, and 20.14%, respectively. From the above statistics, it can be seen that the elderly population with type 2 diabetes in Thailand tends to increase. Moreover, the Department of Disease Control released a report on Thai public health statistics showing that the incidence of diabetes continually increases. The registry has 300,000 new patients per year and 3.3 million patients. In 2020, there were 16,388 deaths (death rate 25.1 per 100,000 people). Public health costs for treatment average as high as 47,596 million baht per year.

The causes of diabetes in Thailand include many factors [2–5], including food consumption behavior, stress, emotions, family environment, and etc. Food consumption is a daily activity that provides the body with enough energy to meet its needs. But sometimes, people with diabetes consume various foods according to their preferences, which arise from appetite and taste. Factors such as stress and multiple emotions expressed affect appetite. Diabetic patients, therefore, need to control their diet appropriately, especially foods such as starches, sugar, desserts, and fat. People with diabetes must maintain their diet in an appropriate amount continuously to prevent risk factors that cause diabetes to worsen. The Department of Disease Control reports that deaths from diabetes in Thailand increased by 2.59 percent in

5 years (2017–2021). The number of deaths in 2017 was 14,322, accounting for 21.96 percent. In 2021, there were 16,008 deaths, accounting for 24.55 percent.

From this problem, it is necessary to study the food consumption behavior of people to prevent death from diabetes in Thailand in the future. It has three essential objectives: (1) to study the context and risk of diabetes from the food consumption behavior of Thai people in the northern region, (2) to develop a model for predicting diabetes risk from the food consumption behavior of Thai people in the northern region, and (3) to evaluate the efficiency of the model for predicting diabetes risk from dietary behavior of Thai people in the northern region. The population and sample in this research were 370 people in Chiang Muan District, Phayao Province, aged between 20 and 99 years. The research tools include an IOC-approved questionnaire and data mining analysis using machine learning and artificial intelligence tools known as CRISP-DM. Model performance was evaluated using data partitioning and testing known as split-validation methods and confusion matrix performance.

The results and expectations from the research are to know the food consumption context of rural people in the northern region of Thailand, which can be used to design and plan strategies to prevent the increase in people at risk of diabetes in Thailand. There are an additional three hundred thousand people every year. Moreover, concrete research results can be used to predict the risk of people with food consumption behavior in each area through mathematical models and artificial problems, which are standard and have high acceptance.

2 Materials & Methods

2.1 Population and Sample

The research area is the rural communities in northern Thailand in the Sa Subdistrict, Chiang Muan District, Phayao Province. Most of the people work in agriculture and general labor. There are 6035 people in the area living in thirteen villages, as detailed in Table 1.

With the research objectives, the researchers aim to study diabetic patients and people at risk for diabetes to develop a model for predicting the risk of diabetes from the food consumption behavior of rural communities in northern Thailand. The researchers have determined to collect data from the population and sample according to Krejcie and Morgan's calculation method, which is equal to 361 samples. Sampling was chosen using stratified random sampling in equal proportions. The calculation method divides the population by the desired sample size, as calculated in Eq. (1). The data collection and sample details are as follows in Table 1.

$$\frac{\text{Population}}{\text{Sample Size}} = \text{proportions} \tag{1}$$

Table 1 Data collection

No.	Name	Population	Sample by proportions	Sample by targets		
				Diabetics	General	Total
V.01	Ban Tha Fah Nuea	385	23	12	12	24
V.02	Ban Tha Fah Tai	676	40	20	20	40
V.03	Ban Sa	472	28	14	14	28
V.04	Ban Sa Tai	746	45	23	23	46
V.05	Ban Thung Nong	612	37	19	19	38
V.06	Ban Tha Fah Lai	213	13	7	7	14
V.07	Ban Fah Mai	405	24	12	12	24
V.08	Ban Rat Phatthana	488	29	15	15	30
V.09	Ban Na Bua	483	29	15	15	30
V.10	Ban Lao Phatthana	619	37	19	19	38
V.11	Ban Fah Se Thong	384	23	12	12	24
V.12	Ban Sa Klang	288	17	9	9	18
V.13	Ban Huai Kang Pla	264	16	8	8	16
Total		6035	361	185	185	370

$$\frac{6035}{361} = 16.72$$

2.2 *Research Design and Data Collection*

The research design focuses on three types of data analytics: descriptive, diagnostic, and predictive. Descriptive analytics aims to study the context of the research sample, where the researchers want to know the overall behavior of the representatives of the study population. Diagnostic analytics focuses on finding root causes and problems from initially collected data to develop more advanced analytics. Finally, predictive analytics is taking the causes and problems found and analyzing them using the historical data collected to propose a model to predict the behavior and risks of the sample group according to the research objectives.

Effective data collection is an important aspect that facilitates the research to achieve its research design. The researchers created a data collection tool, which was a questionnaire. It is used to collect data through field research from April 2023 to August 2023 and was conducted in Thai. These face-to-face meetings were performed, and the questions were based on the study objectives. The results of the research data collection are given in Table 1.

The information in Table 1 consists of a list of thirteen villages, classifying the population according to the village and showing the sample data that must be collected by calculating the specified proportions. Nine samples, totaling 370 samples, increased the actual data collection in the research because the data must be used for analysis and development of behavior prediction models. The researchers classified the data collection into two goals: diabetic patients and the general people.

2.3 Research Tools and Model Development

Research tools are divided into questionnaires and model development tools, which use machine learning techniques. In the questionnaire development process, the researchers collected relevant research in Thailand through the Thai-Journal Citation Index Center (TCI) database at https://tci-thailand.org.

The researchers collected data using a consent form and questionnaire divided into three parts. Part 1 is the general information of the respondents, including gender, date of birth, weight, height, status, education level, occupation, duration of diabetes, type of diabetes, complications, or other diseases.

Part 2 is the food consumption behavior interview form. There are 20 questions. The answer options are on a Likert scale with five levels: five means consumed regularly (every day), four means consumed very often (5–6 times/week), three means consumed often (3–4 times/week), two means consumed little (1–2 times/week), and one means not consuming. In addition, the researcher divided the questions into ten positive questions, including items 1, 3, 5, 6, 7, 9, 11, 14, 15, and 18, and ten negative questions, including items 2, 4, 8, 10, 12, 13, 16, 17, 19, and 20. Both types of questions were mixed in the questionnaire, as given in Table 2. The criteria for calculating scores indicating good dietary behavior are given in Table 3.

Part 3 is the recommendations. All clarifications and questions in the questionnaire were prepared in Thai. Before the researchers used the questionnaire to collect data, the researchers put the questionnaire through the ICO: Index of item objective congruence process by seven experts, consisting of four academics and three nutritionists.

The machine learning technique used to develop the diabetes risk prediction model uses a data mining analysis principle called CRISP-DM [6, 7], which consists of six steps. The first step is business understanding. Researchers found that the trend of diabetes in Thailand is increasing, with more than 300,000 new diabetes patients increasing each year. The leading cause of diabetes comes from food consumption habits. The second step is data understanding. Understanding the data allows researchers to achieve all their research goals. Moreover, understanding the data must be relevant to the objective, so the researchers conducted an ICO involving up to seven nutrition experts and academics. The third step is data preparation. Excellent and efficient data preparation allows for accurate data analysis, precision, and adequate confirmation of research results. The primary tasks in research data preparation are as follows: data discretization, data cleaning, data integration, data transformation, and data reduction. The data preparation phase is the most time-consuming.

Table 2 Questions

No	Questions	Types of questions	
Item 1	How frequently do you consume all five food groups?	/	
Item 2	How frequently do you primarily consume starchy foods?		/
Item 3	How frequently do you consume a variety of vegetable foods?	/	
Item 4	How frequently do you regularly consume fatty meats?		/
Item 5	How often do you consume at least half a plate with vegetables?	/	
Item 6	How frequently do you consume low-fat meat foods?	/	
Item 7	How frequently do you consume food at regular times?	/	
Item 8	How frequently do you consume fried food?		/
Item 9	How much water do you drink, eight glasses of water per day?	/	
Item 10	How frequently do you use vegetable oils in cooking?		/
Item 11	How frequently do you consume adequate meals?	/	
Item 12	How frequently do you consume dessert?		/
Item 13	How frequently do you consume beverages and alcohol?		/
Item 14	How frequently do you consume artificial sweeteners?	/	
Item 15	How frequently do you consume milk or low-fat products?	/	
Item 16	How frequently do you consume grilled, charred food?		/
Item 17	How frequently do you regularly consume sugary beverages?		/
Item 18	How frequently do you consume sweet fruit?	/	
Item 19	How frequently do you consume fermented and dried foods?		/
Item 20	How frequently do you regularly consume foods high in sugar?		/

Table 3 Question scoring criteria

Consumption frequency	Positive questions	Negative questions
(5) Consumed regularly (every day)	5 points	1 point
(4) Consumed very often (5–6 times/week)	4 points	2 points
(3) Consumed often (3–4 times/week)	3 points	3 points
(2) Consumed little (1–2 times/week)	2 points	4 points
(1) Not consuming	1 point	5 points

Table 4 Five levels of interpretation criteria for positive questions

Level/Average range	Interpretation
4.21–5.00	Food consumption behaviors stand at the daily level. (Excellent)
3.41–4.20	Food consumption behaviors stand at a continuous level. (Good)
2.61–3.40	Food consumption behaviors stand at a frequent level. (Normal)
1.81–2.60	Food consumption behaviors stand at a low level. (Critical)
1.00–1.80	Food consumption behaviors stand at a non-consumption level. (Serious crisis)

After the data has been prepared, it is the model development step known as "Modeling". The development of this model uses a combination of unsupervised and supervised learning techniques to provide the highest efficiency and accuracy of the model. The methods and steps used to develop the model include three main parts: appropriate clustering, selection of consistent factors, and model development.

For proper clustering, the researchers used k-means clustering as a simple and popular technique. In addition, the tool used to determine the appropriate *k*-values is the elbow technique, where the shape of the graph is changed from the vertical axis to the horizontal axis. The researchers then proceeded to consider the 20 factors used in this study from the second part of the questionnaire, classified into ten positive questions and ten negative questions. The collected data was used to research factors using the Pearson correlation technique to select significant characteristics for model development.

The final process of this phase is model development. The researchers drew on two-aspect model development techniques to compare the top-performing models. The methods used include single model and ensemble model development. A single machine learning technique to develop a single model includes *K*-nearest neighbor (KNN), support vector machine (SVM), Naïve Bayes (NB), artificial neural network (ANN), decision tree (DT), and generalized linear model (GLM) [8, 9]. The techniques used to develop the ensemble model include three techniques: majority vote, random forest, and Bayesian booting. The performance analysis results from all model developments are presented in Table 15.

Step five is testing the constructed model, known as "Evaluation". It was two essential tools. The first part is to divide the data for testing, called "Split-validation".

The second part measures model performance, called "the Confusion Matrix", which consists of four indicators: accuracy, precision, recall, and $F1$-score [10, 11].

The final step is deployment. This step involved researchers in three parts: finalizing the follow-up and care plan for the assistant and at-risk group for diabetes, creating a plan for the follow-up and care of the person with diabetes, and creating the final report. These three parts are the next phase of work after completing a prototype for predicting the risk of diabetes from food consumption behavior.

2.4 Research Analysis and Interpretation

Analysis and interpretation of research results, the researchers divided it into two phases consistent with the research design. The first part involves analyzing and interpreting research results using descriptive statistics tools, including frequency, percentage, mean, and standard deviation. In addition to interpreting the results, the researchers established a five-level scale that classified interpretations according to the nature of positive and negative questions, as given in Tables 4 and 5

The second part of the research analysis and interpretation is the analysis of model performance using the confusion matrix technique, which consists of four indicators, as detailed in Table 6 and below:

The accuracy value is an evaluation of the model's performance. It is calculated from the total amount of data that can correctly be predicted divided by the total amount of data, as shown in Eq. 2.

Table 5 Five levels of interpretation criteria for negative questions

Level/Average range	Interpretation
4.21–5.00	Food consumption behaviors stand at the daily level (serious crisis)
3.41–4.20	Food consumption behaviors stand at a continuous level (critical)
2.61–3.40	Food consumption behaviors stand at a frequent level (normal)
1.81–2.60	Food consumption behaviors stand at a low level (good)
1.00–1.80	Food consumption behaviors stand at a non-consumption level (excellent)

Table 6 Confusion matrix

		Actual Class		Precision
		True class 1	True class 2	
Predicted	Pred. class 1	True class 1 (TC1)	False class 2 (FC2)	$\frac{TC1}{TC1+FC2}$ (2)
	Pred. class 2	False class 1 (FC1)	True class 2 (TC2)	$\frac{TC2}{TC2+FC1}$ (3)
Recall		$\frac{TC1}{TC1+FC1}$ (4)	$\frac{TC2}{TC2+FC2}$ (5)	

$$\text{Accuracy} = \frac{(\text{TC1} + \text{TC2})}{\text{TC1} + \text{FC1} + \text{TC2} + \text{FC2}} \tag{2}$$

The precision value is the predicted ability of the model to make accurate predictions for each class, as shown in Eqs. (2, 3). The recall value measures how accurately the model predicts by class, as shown in Eqs. (4, 5). The $F1$-score is a harmonic mean between precision and recall values. The purpose of creating an $F1$-score is to be a single metric that measures a model's ability, as shown in Eq. (3).

$$F1 - \text{Score} = 2 * \frac{(\text{Precision} * \text{Recall})}{(\text{Precision} + \text{Recall})} \tag{3}$$

3 Research Results

The researchers divided the content of the findings to align with the research objectives. The content in this results summary consists of three main points: (1) summarizing the context of the population and research sample, (2) summarizing the results of model development from machine learning techniques, and (3) model selection and model performance testing.

3.1 *Context of Food Consumption Behavior in Rural Communities*

Results of the study: Researchers collected data from diabetic patients and the general people with 370 respondents, as presented in Table 1. The study results are divided into two parts: Part 1: General information of the respondents, as concluded in Tables 7, 8, 9, 10 and 11. Part 2: Food consumption behavior interview form, as completed in Tables 11 and 12.

Table 7 gives the results of data collection classified by gender. It was found that the majority of the subjects were female, with 217 samples (58.65%). However, both genders had similar proportions of diabetic and non-diabetic patients.

Table 7 Number of respondents with diabetes and non-diabetes was classified by gender

Gender	Respondents		Total
	Diabetic patients (%)	General people (%)	
Male	77 (20.81)	76 (20.54)	153 (41.35)
Female	108 (29.19)	109 (29.46)	217 (58.65)
Total	185 (50.00)	185 (50.00)	370 (100)

Table 8 Number of respondents with diabetes and non-diabetes was classified by age

Age	Respondents	General people (%)	Total (%)
	Diabetic patients (%)		
20–29 years old	0 (0.00)	10 (2.70)	10 (2.70)
30–39 years old	7 (1.89)	21 (5.68)	28 (7.57)
40 – 49 years old	44 (11.89)	12 (3.24)	56 (15.14)
50 – 59 years old	52 (14.05)	47 (12.70)	99 (26.76)
60 – 69 years old	61 (16.49)	70 (18.92)	131 (35.41)
70 – 79 years old	20 (5.41)	21 (5.68)	41 (11.08)
80 – 89 years old	0 (0.00)	2 (0.54)	2 (0.54)
90 – 99 years old	1 (0.27)	2 (0.54)	3 (0.81)
Total	185 (50.00)	185 (50.00)	370 (100)

Table 9 Number of respondents with diabetes and non-diabetes was classified by education

Education level	Respondents	General people (%)	Total (%)
	Diabetic patients (%)		
Elementary school	123 (33.24)	93 (25.14)	216 (58.38)
Junior high school	18 (4.86)	43 (11.62)	61 (16.49)
Senior high school	43 (11.62)	39 (10.54)	82 (22.16)
Diploma/High vocational certificate	0 (0.00)	9 (2.43)	9 (2.43)
Bachelor's degrees	1 (0.27)	1 (0.27)	2 (0.54)
Master's degrees	0 (0.00)	0 (0.00)	0 (0.00)
Doctor's degrees	0 (0.00)	0 (0.00)	0 (0.00)
Total	185 (50.00)	185 (50.00)	370 (100)

Table 10 Number of respondents with diabetes and non-diabetes was classified by occupation

Occupation	Respondents	General people (%)	Total (%)
	Diabetic patients (%)		
Business owner	0 (0.00)	10 (2.70)	10 (2.70)
Private employee	0 (0.00)	2 (0.54)	2 (0.54)
General employee	51 (13.78)	32 (8.65)	83 (22.43)
Merchant	12 (3.24)	24 (6.49)	36 (9.73)
Farmer	97 (26.22)	78 (21.08)	175 (47.30)
Housekeeper	0 (0.00)	2 (0.54)	2 (0.54)
Do not work	25 (6.76)	37 (10.00)	62 (16.76)
Total	185 (50.00)	185 (50.00)	370 (100)

Table 11 Summary of consumption behavior from positive questions

Items	Consumption frequency					Means	S.D	Interpretation
	(5)	(4)	(3)	(2)	(1)			
Item 1	160	127	74	9	0	4.18	0.81	Good
Item 3	55	161	129	24	1	3.66	0.80	Good
Item 5	38	126	163	42	1	3.43	0.84	Good
Item 6	52	214	95	7	2	3.83	0.70	Good
Item 7	157	112	87	13	1	4.11	0.91	Good
Item 9	276	33	57	4	0	4.57	0.79	Excellent
Item 11	263	77	23	7	0	4.61	0.70	Excellent
Item 14	0	0	0	0	370	1.00	0.00	Serious crisis
Item 15	42	31	139	144	14	2.85	1.03	Normal
Item 18	8	30	163	162	7	2.65	0.75	Normal

Table 12 Summary of consumption behavior from negative questions

Items	Consumption frequency					Means	S.D.	Interpretation
	(5)	(4)	(3)	(2)	(1)			
Item 2	270	72	25	3	0	4.65	0.67	Serious crisis
Item 4	3	33	124	194	16	2.49	0.76	Good
Item 8	6	18	109	208	29	2.36	0.77	Normal
Item 10	121	101	108	38	2	3.81	1.02	Critical
Item 12	1	11	54	251	53	2.07	0.65	Good
Item 13	6	25	74	72	193	1.86	1.05	Good
Item 16	0	61	128	139	42	2.56	1.14	Good
Item 17	9	84	81	112	84	2.52	0.75	Good
Item 19	1	18	97	202	52	2.23	0.75	Good
Item 20	8	27	99	215	21	2.42	0.80	Good

Table 8 gives the results of data collection classified by age. It was found that most of the sample were between 60 and 69 years old. Moreover, researchers found that most diabetes patients were between 40 and 69 years old, with 44 examples (11.89%), 52 instances (14.05%), and 61 subjects (16.49%), respectively. When all three groups were joined, it was found that there were as many as 157 people (42.43%) with diabetes in this group. The researchers, therefore, concluded that the large group at risk for diabetes was between the ages of 40–69 years old.

Table 9 gives the results of data collection classified by education. It was found that the majority of the subjects were at the elementary school level, with 216 samples (58.38%). What is worrying is that most people with diabetes have primary school education. Therefore, providing knowledge and having a higher level of education is necessary to prevent the increase in diabetes cases. Moreover, researchers found that the majority of patients with diabetes were caused by dietary habits rather than genetics, with a proportion of 161 samples (87.03%) to 24 samples (12.97%).

Table 10 gives the results of data collection classified by occupation. It was found that the majority of the subjects were farmers, with 175 samples (47.30%). Another cause for concern is the high proportion of people with diabetes, who account for about one-third of the data collection, with 97 instances (26.22%).

From Tables 7, 8, 9 and 10 we conclude the descriptive data. The summary of essential points includes the overall data collection: the majority of the sample was female, with an average age between 60 and 69 years, and the majority status was married. In addition, the sample had a primary education level, and many were employed. These findings require researchers to answer their research questions to develop risk prediction models for people with diabetes in the next step. However, the researchers summarized the data collection results in Part 2, as given in Tables 11 and 12.

Table 11 gives the collection of food consumption behavior data from positive questions. The researchers found that most of the informants had good consumption habits. The only issue where the informant had any strange behavior was the consumption of artificial sweeteners. The possibility of not consuming artificial sweeteners may be because the study area is rural, and the majority of the population is not aware of it.

Table 12 gives the collection of food consumption behavior data from negative questions. The overall picture of the informants was positive, with only the issue of consumption of starchy foods being consumed by the majority of the people. The cause is probably due to the food consumption behavior of Thai people, most of whom consume starch as the primary source. In addition, most of the sample group are farmers, so they must consume flour to create energy during their careers.

3.2 *Model for Predicting the Risk of Diabetes*

The model development process consists of three phases. The first phase is the optimal instance clustering. The second phase is the factors analysis and attributes selection, and the last phase is the model-building operations.

Optimal instance clustering aims to create clusters to predict risk groups of diabetic patients. The clustering procedures used were K-means and elbow techniques to determine the number of k-values. Factors used in the cluster analysis included BMI and diabetes status. The clustering results are presented in Fig. 1 and Table 13.

Figure 1 shows the analysis of clustering using the elbow technique. It was found that the optimum k-value was $k = 4$, which changed the graph from vertical to horizontal and had an optimal computational speed.

Table 13 gives the clusters, cluster members, and percentage of cluster members. It was found that cluster_0 had the most significant number of members, with 205 members (55.41%), classified as 111 diabetic patients (30.00%) and 94 general persons (25.41%). In addition, it was found that cluster_3 had the least number

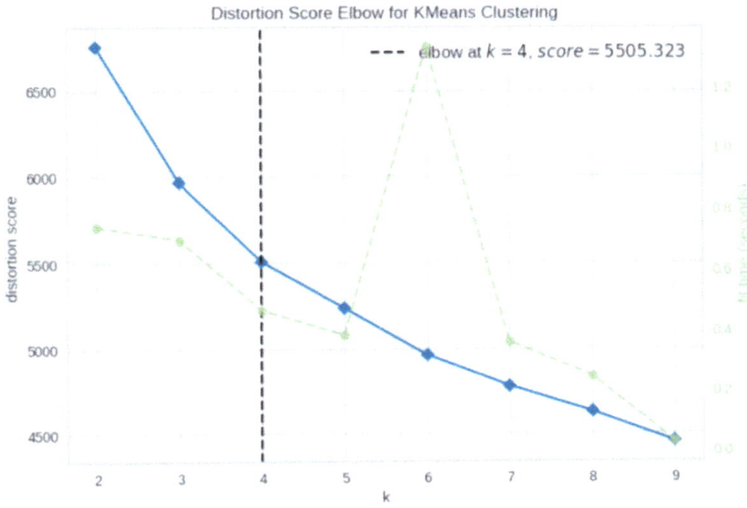

Fig. 1 Consideration of k-values using the elbow technique

Table 13 Cluster, cluster members, percentage of cluster members

Cluster	Cluster members	General people (%)	Total (%)
	Diabetic patients (%)		
Cluster_0	111 (30.00)	94 (25.41)	205 (55.41)
Cluster_1	63 (17.03)	74 (20.00)	137 (37.03)
Cluster_2	6 (1.62)	11 (2.97)	17 (4.59)
Cluster_3	5 (1.35)	6 (1.62)	11 (2.97)
Total	185 (50.00)	185 (50.00)	370 (100)

of members, with 11 members (2.97%), classified as 5 diabetic patients (1.35%) and 6 general persons (1.62%).

After finding the appropriate clusters, the second step is to select variables to analyze the consistency of the variables that have been collected. The technique used is Pearson correlation method, which results in a summary of fourteen variables used in the analysis, consisting of seven positive questions, including items 1, 3, 5, 6, 7, 11, 15, and seven negative questions, including items 4, 8, 10, 13, 16, 17, 19. The results of the factors analysis and attributes selection are shown in Fig. 2.

The last phase is the model-building operations. The researchers classified the model development results into single and ensemble model development, as given in Table 14.

Table 14 reports the results of single and ensemble model development. It concludes the testing of model efficiency by split-validation testing and estimating efficiency values from the confusion matrix with an accuracy indicator. It was found that the model with the highest accuracy was constructed using the ensemble model

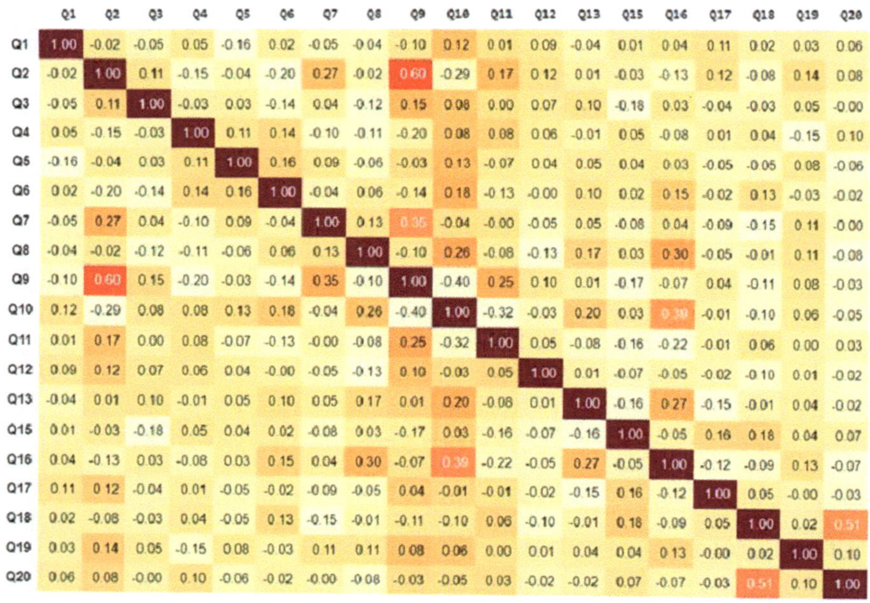

Fig. 2 Factors analysis and attributes selection

Table 14 Single and ensemble model development

Techniques	Split-validation/accuracy			
	60:40 (%)	70:30 (%)	80:20 (%)	90:10(%)
Single Models				
K-nearest neighbor (KNN)	67.57	69.09	73.97	72.97
Support vector machine (SVM)	70.95	67.72	68.49	64.86
Naïve Bayes (NB)	72.97	70.00	78.08	70.27
Artificial neural network (ANN)	68.92	67.27	71.23	72.97
Decision tree (DT)	70.95	60.00	68.49	67.57
Generalized linear model (GLM)	67.57	68.18	69.86	72.96
Ensemble Models				
Majority vote*	73.33	76.85	81.08*	78.38
Random forest	71.62	73.64	76.71	70.27
Bayesian booting	72.30	70.91	71.23	67.57

with majority vote technique and split-validation by 80:20, with an accuracy of 81.08%. The details of the model performance testing results with the four confusion matrix indicators are presented in Table 15.

Table 15 Model performance for top-performing model

Acc. = 81.80%	True cluster_0	True cluster_1	True cluster_2	True cluster_3	Precision (%)
Pred. cluster_0	36	12	0	0	75.00
Pred. cluster_1	2	19	0	0	90.48
Pred. cluster_2	0	0	3	0	100.00
Pred. cluster_3	0	0	0	2	100.00
Recall (%)	94.74	61.29	100.00	100.00	
$F1$-score (%)	83.72	73.08	100.00	100.00	

3.3 *Performance of the Most Efficient Model*

Table 15 provides an overview of the model performance analysis with the confusion matrix with four indicators. It has been shown that the selected models are robust and reliable. The following section will discuss the findings from this study.

4 Research Discussions

The discussion summarizing the findings is divided into two points. The first point is the general context of the respondents. From the data collection, it was found that the majority of the samples were female, totaling 217 samples (58.65%), with an average age between 60 and 69 years, totaling 131 samples (35.41%), and had marital status, calculating 271 instances (73.24%). In addition, most of the sample were farmers, with 175 samples (47.30%), and had primary school education, totaling 216 respondents (58.38%). The critical point discovered is that the majority of the sample group with diabetes had the cause of diabetes from food consumption, accounting for 87.00%. It suggests that people in the area need more education to promote correct dietary behavior solutions to prevent new cases of diabetes in Thailand. Moreover, Tables 12 and 13 indicate that rural people have inappropriate consumption behaviors. Most of the sample group mainly consumed starchy foods and did not use sweeteners or sugar to account for sugar consumption, which is the leading cause of diabetes.

The second point is consistent with the second and third objectives of the research. Researchers have developed a diabetes risk prediction model by setting the classification model into three main steps. The first step is appropriate clustering. The researchers obtained the proper number of clusters using the k-means technique and the elbow technique, in which the fair number of clusters has a k-value of 4, as shown in Fig. 1. After obtaining the number of clusters, the researchers selected factors using the Pearson correlation method, which resulted in 14 variables that were significant for model development, consisting of seven positive questions, including items 1, 3, 5, 6, 7, 11, 15, and seven negative questions, including items 4, 8, 10, 13, 16, 17, 19, as detailed in Fig. 2.

The final step is to develop the model using two techniques consisting of a single model including *K*-nearest neighbor (KNN), support vector machine (SVM), Naïve Bayes (NB), artificial neural network (ANN), decision tree (DT), and generalized linear model (GLM), and ensemble model technique include three techniques: majority vote, random forest, and Bayesian booting. The model development results revealed that the most effective model was developed with the ensemble model with majority vote technique and split-validation by 80:20, with an accuracy equal to 85%, as shown in the details of every model development in Table 15. Moreover, the details of the model performance test with the confusion matrix consisting of four indicators are presented in Table 16

from the success of this research, which was consistent with and achieved the research objectives. The researchers are confident that the research results will be a tool for improving the quality of life and supporting the further development of Thailand.

5 Conclusion

This research has three critical research objectives. The first objective is to study the context and risk of diabetes from the food consumption behavior of Thai people in the northern region. The second objective is to develop a model for predicting diabetes risk from the food consumption behavior of Thai people in the northern region. The last objective is to evaluate the efficiency of the model for predicting diabetes risk from dietary behavior of Thai people in the northern region. The population and sample in this research were 370 people in Chiang Muan District, Phayao Province, aged between 20 and 99 years. The research tools include an IOC-approved questionnaire and data mining analysis using machine learning and artificial intelligence tools known as CRISP-DM. Model performance was evaluated using data partitioning and testing known as split-validation methods and confusion matrix performance.

The research results found that the context of rural people in northern Thailand toward food consumption behavior was normal. Overall, the risk of people developing diabetes was at an acceptable level. It is given in Tables 12 and 13 that people have good food consumption habits. Moreover, the model for predicting the risk of diabetes in rural communities was found to have a high level of accuracy by comparing various techniques, as given in Table 15. This research, therefore, concludes that all objectives were achieved. The results and expectations from the research are to know the food consumption context of rural people in the northern region of Thailand, which can be used to design and plan strategies to prevent the increase in people at risk of diabetes in Thailand.

6 Limitations

The prominent limitations of this research are: this research is a project of undergraduate students, which is a project that encourages students to learn and have experience in doing research. Therefore, in some situations, including data collection, some steps are inaccurate. However, all researchers did their best to achieve all the objectives of this research.

Acknowledgements This research project was supported by the Thailand Science Research and Innovation Fund and the University of Phayao. In addition, this research was supported by many advisors, academics, researchers, students, and staff. The authors would like to thank all of them for their support and collaboration in making this research possible.

Conflict of Interest The authors declare no conflict of interest.

References

1. Saeedi P, Petersohn I, Salpea P, Malanda B, Karuranga S, Unwin N, Colagiuri S, Guariguata L, Motala A.A, Ogurtsova K, Shaw J.E, Bright D, Williams R. (2019) Global and regional diabetes prevalence estimates for 2019 and projections for 2030 and 2045: Results from the international diabetes federation diabetes Atlas, 9th edition. Diabetes Res Clin Pract 157:107843. https://doi.org/10.1016/j.diabres.2019.107843
2. Thepwongsa I, Muthukumar R, Sripa P, Piterman L (2023) Uptake and effectiveness of online diabetes continuing education: The perspectives of Thai general practitioner trainees. Heliyon. 9:e13355. https://doi.org/10.1016/j.heliyon.2023.e13355
3. Aung T.N.N, Thaikla K, Wiwatkunupakarn N, Aramrat C, Pinyopornpanish K, Jiraporncharoen W, Quansri O, Papachristou Nadal I, Kinra S, Angkurawaranon C (2023) Development of a tool to estimate sugar and caloric contents in alcoholic beverages for a diabetes self-management program in Thailand. Heliyon. 9: e21162. https://doi.org/10.1016/j.heliyon.2023.e21162
4. Tangjittipokin W, Thanatummatis B, Wardati F, Narkdontri T, Teerawattanapong N, Boriboonhirunsarn D (2023) The genetic polymorphisms and levels of adipokines and adipocytokines that influence the risk of developing gestational diabetes mellitus in Thai pregnant women. Gene 860:147228. https://doi.org/10.1016/j.gene.2023.147228
5. Nuankaew P, Chaising S, Temdee P (2021) Average weighted objective distance-based method for type 2 diabetes prediction. IEEE Access. 9:137015–137028. https://doi.org/10.1109/ACCESS.2021.3117269
6. Cazacu M, Titan E (2021) Adapting CRISP-DM for social sciences. BRAIN Broad Res Artif Intell Neurosci 11:99–106. https://doi.org/10.18662/brain/11.2Sup1/97.
7. Venter J, de Waal A, Willers C (2007) Specializing CRISP-DM for evidence mining. In: Craiger P, Shenoi S (eds.) Advances in digital forensics III. pp 303–315. Springer, New York, NY https://doi.org/10.1007/978-0-387-73742-3_21
8. Nuankaew W.S, Thipmontha R, Jeefoo P, Nasa-ngium P, Nuankaew P (2023) Using text mining and tokenization analysis to identify job performance for human resource management at the university of Phayao. In: Nguyen N.T, Boonsang S, Fujita H, Hnatkowska B, Hong T.P, Pasupa K, Selamat A (eds.) Recent challenges in intelligent information and database systems. pp 572–584 Springer Nature Switzerland, Cham. https://doi.org/10.1007/978-3-031-42430-4_47
9. Nuankaew P, Jeefoo P, Nasa-Ngium P, Nuankaew S W (2023) Hybrid learning and blended learning in the perspective of educational data mining and learning analytics. In: a systematic

literature reviews. Int J Eng Trends Technol IJETT. 71:115–132. https://doi.org/10.14445/22315381/IJETT-V71I10P211
10. Deng X, Liu Q, Deng Y, Mahadevan S (2016) An improved method to construct basic probability assignment based on the confusion matrix for classification problem. Inf Sci 340–341:250–261. https://doi.org/10.1016/j.ins.2016.01.033
11. Luque A, Carrasco A, Martín A, de las Heras A (2019) The impact of class imbalance in classification performance metrics based on the binary confusion matrix. Pattern Recognit. 91:216–231. https://doi.org/10.1016/j.patcog.2019.02.023

Enhancing Breast Cancer Lesion Segmentation with an Ensemble of U-Net and SeqNet

B. H. Shekar and Mohammad Zaher Taljeh

Abstract Accurate segmentation of breast cancer lesions is vital for early detection, precise diagnosis, and effective treatment planning. In this research work, we propose an ensemble-based segmentation approach that combines the power of the U-Net model and the SeqNet model for breast cancer segmentation. By leveraging the complementary strengths of these components, which capture both spatial and sequential information, we aim to improve the accuracy and reliability of breast cancer lesion segmentation, thereby facilitating better clinical decision-making and patient care. Experiments are conducted on BUSI dataset which is one of the standard datasets used by many researchers. A comparative analysis is also provided with the other approaches to justify the suitability of the proposed ensemble model for breast cancer segmentation task.

Keywords Ultrasound images · U-net · SegNet · Ensemble · Breast cancer segmentation

1 Introduction

Breast cancer represents a considerable health issue that impacts women globally, underscoring the crucial importance of early detection for effective treatment and better patient outcomes [1]. Breast ultrasound imaging has become a valuable diagnostic instrument for the detection and characterization of breast abnormalities [2]. Precise lesion segmentation is pivotal for diagnosing, planning treatment, and monitoring disease progression in breast ultrasound images; nevertheless, the intricate nature of these images, featuring factors like speckle noise, low contrast, and varying lesion appearances, poses significant challenges to achieving accuracy [3]. In recent times, deep learning models such as U-Net and SegNet have exhibited promising

These authors contributed equally to this work.

B. H. Shekar (✉) · M. Z. Taljeh
Department Of Computer Science, Mangalore University, Mangalagangotri,
Mangalore 574199, Karnataka, India
e-mail: bhshekar@gmail.com

S. M. Thampi et al. (eds.), *Fifth International Conference on Computing and Network Communications*, Lecture Notes in Electrical Engineering 1221,
https://doi.org/10.1007/978-981-97-4711-5_21

outcomes in the field of medical image segmentation. In this research work, we propose an ensemble-based segmentation approach that combines the strengths of U-Net and SegNet to enhance the accuracy and reliability of breast lesion segmentation in ultrasound images considering the Breast Ultrasound Image (BUSI) dataset.

Traditional segmentation methods have shown limitations in accurately delineating breast lesions, which is a fundamental step in the early detection of cancer and subsequent treatment planning. To bridge this knowledge gap, our work focuses on by developing an ensemble approach that leverages cutting-edge deep learning technologies. This work is driven by the urgency of early cancer detection and the need for more reliable, accurate, and robust segmentation methods tailored to the unique challenges presented by breast ultrasound images such as low contrast, varying lesion shapes and sizes, and the presence of acoustic shadows, making them inherently complex for automated analysis. By harnessing the capabilities of deep learning models like U-net and SeqNet, our research aims to significantly improve the accuracy and reliability of segmenting breast lesions, ultimately contributing to the early detection of cancer. The BUSI dataset, specifically designed for breast ultrasound image analysis, provides a diverse range of cases, enabling thorough evaluation and validation of our proposed ensemble segmentation approach.

U-Net and SegNet are widely recognized deep learning architectures that have demonstrated exceptional performance in tasks related to segmentation of medical images [4]. U-Net utilizes an encoder-decoder structure featuring skip connections, enabling the integration of multiscale information and precise localization of lesions [5]. SegNet, on the other hand, leverages an encoder-decoder framework with pooling indices for efficient upsampling, effectively recovering detailed spatial information [6]. By combining these two architectures, we aim to leverage their unique strengths and improve the accuracy and robustness of breast lesion segmentation.

Our proposed ensemble segmentation framework involves training individual U-Net and SegNet models using the BUSI dataset. Each model is trained to learn distinctive features and capture specific aspects of breast lesions. During the inference phase, the predictions of both models are combined using ensemble techniques, such as weighted averaging or voting, to generate the final segmentation map. The objective of employing this ensemble approach is to capitalize on the complementary capabilities of U-Net and SegNet, thereby elevating the overall accuracy and dependability of the segmentation outcomes.

To assess the efficacy of our ensemble segmentation approach, we conducted extensive experiments on the BUSI dataset. We conducted a performance comparison of our ensemble method against individual U-Net and SegNet models, additionally with other state-of-the-art segmentation algorithms. Assessment measures, such as the Dice coefficient, Jaccard index, sensitivity, and specificity, were utilized for the assessment of the reliability and accuracy of the segmentation results.

We anticipate that our proposed ensemble segmentation approach using U-Net and SegNet will yield superior segmentation accuracy and robustness for breast ultrasound images from the BUSI dataset. By leveraging the strengths of these deep learning architectures, we aim to enhance the clinical utility of breast ultrasound imaging, supporting accurate diagnosis, treatment planning, and monitoring of breast lesions. Our research contributes to the growing field of medical image analysis, providing valuable insights and techniques for improved breast cancer management.

2 Related Work

Medical image segmentation plays a pivotal role in enhancing the visualization of human tissue and pathological structures, making them more vibrant and intuitive. Moreover, the segmentation results facilitate the modeling of relevant tissue, enabling further manipulation and analysis. Initially, image segmentation methods were primarily categorized into threshold-based segmentation techniques [7], region-based segmentation methods [8], and segmentation methods based on specific theories [9]. Each of these approaches have contributed to advancing the realm of medical image segmentation Notable segmentation methods are SegNet [10] and U-net [5] which directly extract features from the target and have demonstrated exceptional segmentation performance.

Meyer et al. [11] introduced an automated pipeline that utilizes deep convolutional neural networks to segment areas of COVID-19 infection in CT scans. They conducted an evaluation to assess how ensemble learning techniques such as bagging and augmenting impact the performance of the segmentation process.

Li et al. [12] propose a method called Bagging Ensemble Deep Segmentation (BEDs) for training multiple U-Nets using fractional training data for segmenting dense nuclei in medical images.

Khaled et al. [13] proposed integrating three U-Net models into an ensemble technique, with each model using a distinct set of inputs for segmentation of the breast lesion from DCE-MRI.

El Adoui et al. [14] utilized two distinct and complete CNN encoder-decoder architectures SegNet and U-Net for segmenting breast lesion from DCE-MRI data and observed that U-Net outperformed SegNet.

Cao et al. [15] constructed a densely linked segmentation network by integrating a set of hybrid dilated convolutions into the D2U-Net's dense blocks and this segmentation algorithm has been evaluated on ABUS dataset.

Yan et al. [16] introduced an attention improved U-Net model incorporating hybrid dilated convolution for the segmentation of breast ultrasound images.

Liu et al. [17] introduced a Multiscale Nonlocal Feature Extraction Network (MNFE-Net) comprising three essential components: a parallel encoder (PE), a multiscale feature module (MFM), and a global feature guidance module (GFGM). This network architecture was designed to attain accurate segmentation of breast abnormalities.

Corsi et al. [18] introduced a segmentation method that combines region- and edge-based information within a robust graph-based (RGB). This method efficiently segments the filtered image by employing a generative adversarial network (GAN) framework.

Xue et al. [19] proposed an innovative deep (CNN) architecture featuring specialized modules for detecting breast lesion boundaries along with a global guidance block. This network leverages integrated feature maps across multiple layers as informative guidance, allowing for the effective capture of long-range nonlocal dependencies in both spatial and channel domains. The primary goal is to significantly improve the segmentation of breast ultrasound lesions.

Lyu et al. [20] proposed a new method called AMS-PAN (Improved Pyramid Attention Network) for segment breast ultrasound image. AMS-PAN is designed to enhance segmentation accuracy by combining multiscale features and Attention mechanism.

Thus, we have seen many approaches for segmentation tasks and found that none of the approaches explained the power of U-Net and SeqNet which is the prime motivation for our work, i.e., to integrate these two models and come out with better segmentation approach. The particulars of the proposed approach are outlined in the subsequent sections.

3 Methodology

Numerous studies have demonstrated the significant impact of integrating both preprocessing and post-processing techniques within a deep learning pipeline, leading to substantial improvements in the model's performance in comparison to utilizing the network independently [21].

3.1 Preprocessing

The BUSI dataset underwent several preprocessing steps to ensure the data was prepared appropriately for analysis. The following preprocessing techniques were applied to the dataset:

Resizing: The images in the BUSI dataset were resized to a standardized dimension. Resizing helps to ensure uniformity in the dataset, making it easier to process and compare the images. A common size for the images, 192x256 pixels, was chosen to maintain consistency throughout the dataset.

Normalization: Normalization is a crucial step in preprocessing to standardize the intensity values of the images. By applying normalization, the range of pixel values is adjusted to improve the comparability and convergence of different images. Studies have demonstrated that batch normalization [22] can significantly improve neural networks' ability to capture data bias in their internal latent space [23].

3.2 Data Augmentation

Data augmentation plays a crucial role in enhancing the robustness and generalization capabilities of deep learning models [24]. In the context of this research work, we employ two data augmentation techniques: random rotation and horizontal flip. These techniques introduce variations in the training dataset through the application of geometric transformations to the images. The following sections describe each augmentation technique:

Random rotation: We take pictures and apply a random spin to them. The rotation angle is generated randomly in the range of -40 to 40 degrees.

Horizontal flip or horizontal mirror or left-right flip: We take image and flipping an image horizontally along the vertical axis, resulting in a mirror image of the orignal. Before augmentation, our dataset consisted of 647 images, and post-augmentation, the dataset expanded to 1455 images.

3.3 The U-Net

We build UNet model [5] with some customization options for the double convolution layer and input shape as follows:

Encoder:
It starts with an input layer (input) that takes the input images with shape `(192, 256, INPUT_CHANNELS)`. The encoder part consists of several convolutional Layers texttt(double_conv_layer) followed by max-pooling (MaxPooling2D) operations. The quantity of filters in the convolutional layers progressively increases, starting from filters and doubling at each stage ($2\times$, $4\times$, $8\times$, $16\times$, $32\times$) compared to the previous layer.

Decoder:
The decoder part involves upsampling (UpSampling2D) layers. Subsequently, followed by the merging of these feature maps with their corresponding encoder counterparts. Each upsampling step is accompanied by a convolutional layer `(double_conv_layer)`, which reduces the number of filters ($16\times$, $8\times$, $4\times$, $2\times$, $1\times$) compared to the concatenated input.

Final Layers:
After the decoder, the feature maps are further processed by a series of convolutional layers and activations. The last convolutional layer (Conv2D) with a 1×1 kernel produces the final output mask with the number of channels specified by `OUTPUT_MASK_CHANNELS`. To obtain the predicted masks, the output undergoes an activation function (sigmoid) application. The predicted masks are reshaped to the original image size using Reshape.

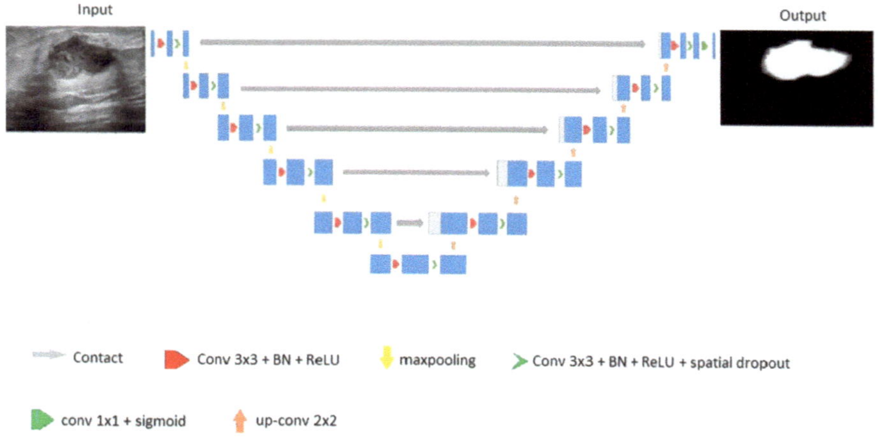

Fig. 1 Segnet architecture

Model Compilation and Training:
The model is configured using the Adam optimizer, utilizing a learning rate of 0.003. The loss function employed is `jaccard_distance`. Multiple evaluation metrics are specified such as the Jaccard index, Dice coefficient, precision, recall, and accuracy. After training, the model is saved and returned. The U-Net architecture is frequently used for image segmentation tasks and has been customized for diverse applications, including medical image analysis (Fig. 1).

3.4 The SegNet

The Seqnet model, constructed as per [25], is comprised of 64 layers, excluding the final activation layer. The architecture and training configuration are detailed below.

The encoding layer begins with an input dimension of (192, 256, 3) and consists of consecutive convolutional layers, each followed by batch normalization and ReLU activation. These layers extract features at various abstraction levels, with max-pooling layers reducing spatial dimensions. The encoding layer further includes multiple sets of convolutional layers, progressively increasing filters from (64, 64) to (512, 512).

Following the encoding layers, two dense layers with 1024 units and ReLU activation capture global information. The decoding layer processes the output from dense layers, passing it through transpose convolutional layers, batch normalization, and ReLU activation. These transpose convolutional layers upsample feature maps to the original image size, employing multiple sets with decreasing filters from (512, 512) to (64, 1). The final layer uses a sigmoid activation function, yielding a segmentation mask reshaped to dimensions (192, 256).

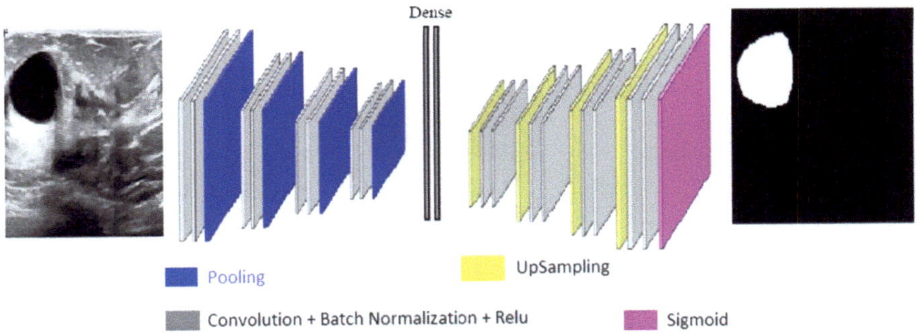

Fig. 2 Segnet architecture

For model compilation and training, stochastic gradient descent (SGD) is utilized with a learning rate of 0.001, momentum of 0.9, and a decay rate of 0.0005, excluding Nesterov acceleration. The loss function is binary cross-entropy, and multiple evaluation metrics, including the Jaccard index, Dice coefficient, precision, recall, and accuracy, are specified. Training involves a set number of epochs and a batch size of 18, with validation data used to monitor performance.

After training, the model is saved with the provided filename. This architecture and training setup are tailored for image segmentation tasks, with the goal of producing accurate segmentation masks for input images of dimensions (192, 256, 3) (Fig. 2).

3.5 *Ensemble Bagging*

In this work, we leverage ensemble bagging as a core strategy to elevate the performance of our cancer segmentation models. Ensemble bagging is a technique that capitalizes on the synergy of multiple base models to enhance predictive accuracy and robustness. Our ensemble approach involves training independent instances of two renowned convolutional neural network architectures, the U-Net and SegNet models, with the overarching objective of refining the segmentation of cancerous regions within medical images. During the segmentation process, we employ a straightforward averaging strategy, whereby the predictions from these base models are element-wise averaged to generate a consolidated segmentation map. The advantages stemming from our ensemble bagging approach are highly pertinent to our specific cancer segmentation task. By embracing diversity among base models, our method significantly bolsters the robustness of cancer segmentation, rendering it less susceptible to overfitting and more dependable when analyzing medical images (Fig. 3).

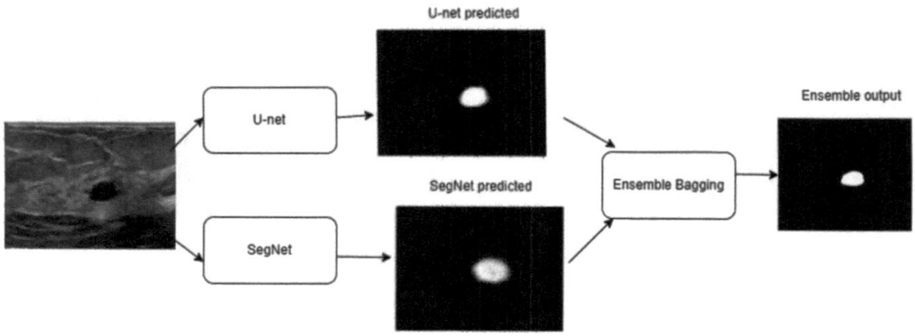

Fig. 3 .

3.6 *Post Processing*

Thresholding: After obtaining the ensemble predictions, a post-processing technique called thresholding is applied. Thresholding is a simple technique involving the establishment of a threshold value and the subsequent categorization of pixels based on whether their predicted values are above or below the threshold. In this case, a threshold value of 0.7 is chosen arbitrarily. For each pixel, the ensemble prediction value is compared to the threshold value of 0.7. If the prediction value is greater than 0.7, the pixel is assigned the value 1, indicating a positive classification. On the other hand, if the prediction value is less than or equal to 0.7, the pixel is assigned the value 0, indicating a negative classification. By applying this thresholding process to the ensemble predictions, the final binary output is obtained, which helps to identify and distinguish the relevant objects or regions of interest in the image.

4 Dataset and Experimental Settings

4.1 *Dataset*

The dataset used for the research, referred to as Breast Ultrasound Images (BUSI), was constructed by Al-Dhabyani et al. It consists of a total of 780 images that were obtained at the Baheya Hospital. The images were acquired using two types of ultrasound equipment, namely the LOGIQ E9 ultrasound and the LOGIQ E9 Agile ultrasound system [26].

The samples in the BUSI dataset are categorized into three classes based on their characteristics. The distribution of samples across these classes is as follows:

Benign: There are 437 images classified as benign. Benign breast lesions refer to non-cancerous or non-malignant abnormalities in breast tissue.

Malignant: There are 210 images classified as malignant. Malignant breast lesions represent cancerous or cancer-related abnormalities in breast tissue.

Normal: There are 133 images classified as normal. Normal breast images represent healthy breast tissue without any identified abnormalities.

4.1.1 Evaluation Metrics

In this section, we introduce the evaluation metrics employed for assessing the performance of our image segmentation model. We employed the following metrics to evaluate the segmentation results:

True Positive (TP): This metric represents the pixels that were correctly classified as belonging to the target object category. These are the pixels in the predicted segmentation mask that align with the ground truth mask.

False Positive (FP): The false positive metric denotes the pixels that were wrongly classified as belonging to the target object category. These are pixels in the predicted segmentation mask that do not align with the ground truth mask.

True Negative (TN): The true negative metric signifies the pixels that were correctly classified as not belonging to the target object category. These are pixels in the background or non-object regions that are correctly identified as such.

False Negative (FN): The false negative metric indicates the pixels that were incorrectly classified as belonging to the background or non-object category. These are pixels in the ground truth mask that are not captured in the predicted segmentation mask.

- **Intersection over Union (IoU)**:

$$\mathrm{IoU} = \frac{G \cap P}{G \cup P} \tag{1}$$

- **Dice Coefficient(Dice)**:

$$\mathrm{Dice} = \frac{2 * \mathrm{TP}}{2 * \mathrm{TP} + \mathrm{FP} + \mathrm{FN}} \tag{2}$$

- **Precision**:

$$\mathrm{Precision} = \frac{\mathrm{TP}}{\mathrm{TP} + \mathrm{FP}} \tag{3}$$

- **Recall**:

$$\mathrm{Recall} = \frac{\mathrm{TP}}{\mathrm{TP} + \mathrm{FN}} \tag{4}$$

- **Accuracy**:

$$\mathrm{Accuracy} = \frac{\mathrm{TP} + \mathrm{TN}}{\mathrm{TP} + \mathrm{FP} + \mathrm{FN} + \mathrm{TN}} \tag{5}$$

Table 1 Modules training parameters

Parameter	Learning-rate (LR)	Activation	Loss Function	Batch-size	Epoch
U-net	0.003	Sigmoid	Jaccard_distance	18	100
Segnet	0.001	Sigmoid	Cross-ent ropy	18	100

Table 2 Comparison of the proposed method with recent studies

Method reference	IOU	Dice	Precision	Recall	Accuracy
Chen et al. [27]	68.82	77.51	79.61	81.10	–
Tang et al. [28]	73.27	–	84.81	84.26	97.33
Lyu et al. [20]	68.53	80.71	83.50	79.30	97.13
Liu et al. [18]	77.72	85.49	–	88.18	98.18
Ours	**95.87**	**72.57**	**84.96**	**70.46**	**96.05**

4.2 Results

In this work, we present a ensemble bagging method that is based on the U-Net model and the SegNet model for segment breast ultrasound image, where the training was done after preprocessing the images (resize, normalize, and histogram equalization) and image augmentation. During the training phase, we adopted a rigorous train-validation split, allocating a portion of our augmented dataset exclusively for model validation. We conducted several experiments to evaluate our work, varying the number of training iterations for both the U-Net and SegNet models over 100 epochs to identify the optimal parameters, as presented in Table 1. Subsequently, we applied a simple post-processing technique using thresholding. The training time was 253,342.56750035285 s.

We utilized Table 2 for comparing the proposed approach with both the best and most recent methods available. This comparison enables us to assess the effectiveness and performance of our approach in relation to existing techniques. We evaluated the following metrics for comparison: Intersection over Union, Dice Coefficient, Precision, Recall, and Accuracy.

5 Conclusions

This research work presents a method for cancer segmentation using Ensemble Bagging, which combines the U-Net and SegNet architectures. Our study is centered on the BUSI dataset, with the objective of enhancing the accuracy of cancer segmentation.

The evaluation of our method against previous works demonstrates its superior performance in terms of the Intersection over Union (IoU) metric, which is employed for assessing the accuracy of segmentation algorithms. Our method outperformed both the best and the last work in this specific domain, indicating its efficacy in accurately identifying cancer regions within breast ultrasound images.

The success of our ensemble methodology can be attributed to the synergistic combination of U-Net and SeqNet, which excel in capturing different aspects of the underlying patterns in breast ultrasound images. U-Net, known for its excellent segmentation capabilities, complements SeqNet's proficiency in handling sequential data. The ensemble effectively mitigates the limitations of individual models, resulting in improved overall performance

However, it's crucial to emphasize that our method exhibited slightly lower performance in a specific metric, indicating the potential need for further refinement and optimization. Nevertheless, the high IoU scores attained by our approach underscore its potential for precise cancer segmentation, underscoring its value in aiding medical professionals for the identification and diagnosis of breast cancer.

Future research can focus on addressing the limitations identified in the alternative metric and exploring ways to enhance the overall performance of the proposed method. Additionally, efforts can be made to expand the dataset to include a more diverse range of breast ultrasound images, facilitating a more comprehensive assessment of the method's performance across various scenarios.

In summary, our research demonstrates the effectiveness of the Ensemble Bagging approach using U-Net and SegNet for cancer segmentation. Despite the slight shortcomings in one metric, our method exhibits superior performance in terms of IoU scores, promising improved accuracy in cancer identification. This work contributes to the field of medical image analysis and provides a foundation for further advancements in the early identification and therapy of breast cancer.

References

1. Wilkinson L, Gathani T (2022) Understanding breast cancer as a global health concern. Br J Radiol 95(1130):20211033
2. Gay H, Allen SD (2013) 19 role of ultrasound in the diagnosis of breast cancer. In: Early breast cancer: from screening to multidisciplinary management, p 169
3. Liu X, Huo Z, Zhang J (2006) Automated segmentation of breast lesions in ultrasound images. In: 2005 IEEE engineering in medicine and biology 27th annual conference. IEEE, pp 7433–7435
4. Du G, Cao X, Liang J, Chen X, Zhan Y (2020) Medical image segmentation based on u-net: a review. J Imaging Sci Technol **64**(2) (2020)
5. Ronneberger O, Fischer P, Brox T (2015) U-net: convolutional networks for biomedical image segmentation. In: Medical image computing and computer-assisted intervention–MICCAI 2015: 18th international conference, Munich, Germany, 5–9 Oct 2015, Proceedings, Part III 18. Springer, pp 234–241
6. Liu X, Song L, Liu S, Zhang Y (2021) A review of deep-learning-based medical image segmentation methods. Sustainability 13(3):1224

7. Tang lB, Kong J, Wu S, (2017) Review of surface defect detection based on machine vision. J Image Graph 22(12):1640–1663
8. Deng L, Gong Y, Lin Y, Shuai J, Tu X, Zhang Y, Ma Z, Xie M (2019) Detecting multi-oriented text with corner-based region proposals. Neurocomputing 334:134–142
9. Liu X, Wang Z, Li Y-L, Wang S (2022) Self-supervised learning via maximum entropy coding. Adv Neural Inf Process Syst 35:34091–34105
10. Chen L-C, Papandreou G, Kokkinos I, Murphy K, Yuille AL (2017) DeepLab: semantic image segmentation with deep convolutional nets, atrous convolution, and fully connected CRFs. IEEE Trans Pattern Anal Mach Intell 40(4):834–848
11. Meyer P, Müller D, Soto-Rey I, Kramer F (2021) Covid-19 image segmentation based on deep learning and ensemble learning. In: Public health and informatics, pp 518–519. IOS Press ???
12. Li X, Yang H, He J, Jha A, Fogo AB, Wheless LE, Zhao S, Huo Y (2021) Beds: bagging ensemble deep segmentation for nucleus segmentation with testing stage stain augmentation. In: 2021 IEEE 18th International symposium on biomedical imaging (ISBI). IEEE, pp 659–662
13. Vidal J, Vilanova JC, Martí R et al (2022) A u-net ensemble for breast lesion segmentation in DCE MRI. Comput Biol Med 140:105093
14. El Adoui M, Mahmoudi SA, Larhmam MA, Benjelloun M (2019) MRI breast tumor segmentation using different encoder and decoder CNN architectures. Computers 8(3):52
15. Cao X, Chen H, Li Y, Peng Y, Wang S, Cheng L (2021) Dilated densely connected u-net with uncertainty focus loss for 3d ABUS mass segmentation. Comput Methods Programs Biomed 209:106313
16. Yan Y, Liu Y, Wu Y, Zhang H, Zhang Y, Meng L (2022) Accurate segmentation of breast tumors using ae u-net with HDC model in ultrasound images. Biomed Signal Process Control 72:103299
17. Luo Y, Liu L, Huang Q, Li X, et al (2017) A novel segmentation approach combining region-and edge-based information for ultrasound images. BioMed Res Int
18. Liu G, Wang J, Liu D, Chang B (2023) A multiscale nonlocal feature extraction network for breast lesion segmentation in ultrasound images. IEEE Trans Instrum Measure
19. Xue C, Zhu L, Fu H, Hu X, Li X, Zhang H, Heng P-A (2021) Global guidance network for breast lesion segmentation in ultrasound images. Med image Anal 70:101989
20. Lyu Y, Xu Y, Jiang X, Liu J, Zhao X, Zhu X (2023) AMS-PAN: Breast ultrasound image segmentation model combining attention mechanism and multi-scale features. Biomed Signal Process Control 81:104425
21. Salvi M, Acharya UR, Molinari F, Meiburger KM (2021) The impact of pre-and post-image processing techniques on deep learning frameworks: a comprehensive review for digital pathology image analysis. Comput Biol Med 128:104129
22. Ioffe S, Szegedy C (2015) Batch normalization: accelerating deep network training by reducing internal covariate shift. In: International conference on machine learning, pp 448–456 (PMLR)
23. Li Y, Wang N, Shi J, Liu J, Hou X (2016) Revisiting batch normalization for practical domain adaptation. arXiv:1603.04779
24. Nalepa J, Marcinkiewicz M, Kawulok M (2019) Data augmentation for brain-tumor segmentation: a review. Front Comput Neurosci 13:83
25. Badrinarayanan V, Kendall A, Cipolla R (2017) SegNet: a deep convolutional encoder-decoder architecture for image segmentation. IEEE Trans Pattern Anal Mach Intell 39(12):2481–2495
26. Al-Dhabyani W, Gomaa M, Khaled H, Fahmy A (2020) Dataset of breast ultrasound images. Data Brief 28:104863
27. Chen G, Li L, Dai Y, Zhang J, Yap MH (2022) AAU-net: an adaptive attention U-net for breast lesions segmentation in ultrasound images. IEEE Trans Med Imaging
28. Tang F, Wang L, Ning C, Xian M, Ding J (2022) CMU-Net: A strong convmixer-based medical ultrasound image segmentation network. arXiv:2210.13012

Multiclass Classification of Breast Lesion Using Histopathology Images

Swetha Kulkarni, Shrinivas Desai, Vishwanath P. Baligar, and S. R. Nirmala

Abstract Breast cancer ranks as the second most common cancer affecting women. Early prediction of breast lesions significantly enhances patient survival rates. In India, however, obtaining an early medical diagnosis poses considerable challenges. Often, by the time the initial screening takes place, the lesion has progressed to a grade 2 or higher level. The classification of different stages of breast lesion histopathology images is a complex task, as it may lead to confusion in interpreting certain lesion images. Most of the existing literature has extensively explored various neural network models for multiclass classification of different stages of breast lesion histopathology images. In such cases, pre-trained deep neural network models have been commonly utilized, but they have yielded comparatively lower accuracy. Therefore, in this paper, we propose a novel approach involving the fine-tuning of the standard Inception V3 model and optimizing hyperparameters for categorizing histopathology breast lesion images stained with the BACH dataset. To enhance the deep neural networks (DNN) performance, several data augmentation techniques are employed. Our proposed network architecture, utilizing the fine-tuned Inception V3 model, achieves an impressive 97% test accuracy, a 76% $F1$-score, 87% precision, and 80% recall. When compared with recently reported literature, our model exhibits slightly superior performance.

Keywords Histopathology images · Breast cancer · BACH dataset · Fine-tuning · Inception V3 · Multiclass · Binary class

---

S. Kulkarni · S. Desai (✉) · V. P. Baligar · S. R. Nirmala
School of Computer Science and Engineering, KLE Technological University, Hubballi, India
e-mail: sd_desai@kletech.ac.in

S. Kulkarni
e-mail: swetha.kulkarni@kletech.ac.in

V. P. Baligar
e-mail: vpbaligar@kletech.ac.in

S. R. Nirmala
e-mail: nirmala.s@kletech.ac.in

S. M. Thampi et al. (eds.), *Fifth International Conference on Computing and Network Communications*, Lecture Notes in Electrical Engineering 1221,
https://doi.org/10.1007/978-981-97-4711-5_22

1 Introduction

According to the World Health Organization (WHO), breast cancer is the most frequent type of cancer occurring in women between the ages of 20–59 globally [1]. In 2022, there were 2, 87,250 million women diagnosed with breast cancer and 43,250 deaths worldwide, as reported by the World Health Organization. The number of cancer cases is predicted to rise from 13.9 lakh in 2020 to 15.7 lakh in 2025, representing an almost 20% increase according to the NCRP [2]. Tumors are classified into four categories: benign, invasive, in situ, and normal. These classifications are based on tumor size and location. Detecting breast lesions in their early stages is crucial for improving the survival rate of patients.

In the present era, numerous advanced medical imaging modalities are available, including digital X-rays, CT scans, MRI, and digital microscopy. However, when it comes to breast lesion images, we primarily rely on digital microscopy and histopathology images. These modalities allow us to study clinical signs through microscope examination of biopsy or surgical samples that have been processed and placed on glass slides. Consequently, the multi-classification of various stages of breast lesion histopathology images is highly relevant in the current context. The performance of standard pre-trained models is notably poorer when compared with the results presented in Table 1. Many researchers have proposed corrective methods to enhance model performance. Machine learning and deep neural network models play a significant role in multiclass classification and the prediction of various diseases in the medical domain. It is important to note that DNN or ML models should complement the diagnostic decisions made by medical experts and should never be considered as replacements for experts. A DNN model can classify both binary and multiclass classification by fine-tuning hyperparameters for the DNN or ML model includes optimizing the loss function and activation function.

2 Related Study

In our related studies, we discuss literature survey, motivation and contribution, and research gap.

2.1 Literature Survey

In recent years, medical imaging has seen significant advancements due to technology. Here, we provide a review of the literature in Table 1, which reports various methods for improving the classification of breast cancer histopathology images. From the literature, it is evident that preprocessing histopathology images before

Table 1 Literature survey

Author	Proposed model	Dataset	Results	Remarks
Alruwaili et al. 2022 [3]	ResNet50	Mammographic image analysis society (MIAS)	89%	Fine-tuning on multiple pre-trained models
Heenave Mamode khan et al. 2021 [4]	Enhanced CNN model	BreaKHis dataset	88%	Segmentation and classification of various types of breast abnormalities, such as calcifications, masses, asymmetry, and carcinomas
Dabas et al. 2021 [5]	ResNet-18	Mini-MIAS dataset	98.91	Hesitancy-based Hanman transform classifier was used on small datasets and found better results compared to deep learning models
Kassani et al. 2019 [6]	Xception	ICIAR 2018	92.50%	Deep convolutional neural network (DCNN) models and pooling operation
Nawaz et al. 2018 [7]	DenseNet	BreaKHis dataset	96%	Classification and prediction of a subclass of the tumors like Fi-broad enema, lobular carcinoma, etc.

feeding them into the model for training is essential. Consequently, in this research, we have incorporated breast lesion histopathology images into the models.

2.2 *Motivation and Contribution*

Deep neural network models have demonstrated promising results in various medical imaging applications. Furthermore, it has been shown that preprocessing techniques, such as data augmentation and stain normalization, are beneficial for achieving a more robust and accurate performance. These findings motivated our investigation into the performance of different pre-trained models and the effectiveness of preprocessing and data augmentation techniques in assessing breast cancer from histopathology images. In the following section, we outline the main contributions of this study: since the dataset provided with this study is limited, there is a need to increase the dataset size and enhance the performance of our model. To achieve this, we employed several data augmentation techniques, including rescaling, shearing, rotation, width and height shifting, horizontal flipping, and zooming.

1. Fine-tuning the standard Inception_V3 model involves adding an additional fully connected dense layer with 512 features and a ReLU activation function, creating an output layer with 4 dense layers and a softmax activation function, and selecting which layers need to be updated during training.
2. Fine-tuning hyperparameters for the Inception_V3 model includes optimizing the loss function and activation function.

2.3 Research Questions

From the literature's review we arrived at following research questions.

1. Can a DNN model classify both binary class and multiclass breast cancer histopathology images?
2. Which hyperparameters influence the classification of breast cancer histopathology images?

3 Methodology

In this section, we explain the proposed method based on a DNN architecture for training and predicting breast cancer. The process begins with data augmentation techniques to address the issue of a limited dataset size and to optimize the DNN model's performance. In the second step, high-level features are extracted from pre-processed images using a network architecture derived from well-established DNN models. These collected features are then input into a typical multilayer perceptron classifier. Finally, we evaluate and report the performance of the proposed model on test images. The methodology includes details about the dataset, the trained models, the proposed models, loss functions, and class activation functions.

3.1 Network Architecture

The Inception V3 model has undergone fine-tuning, as illustrated in Fig. 4. During the fine-tuning process, we added a fully connected dense layer. In a neural network, fully connected dense layers connect all inputs from one layer to every activation unit in the following layer. Additionally, we replaced the flattening layer with a global average pooling layer.

The flattening layer converts any tensor into a one-dimensional tensor while preserving all values within the same tensor. In contrast, the global average pooling layer applies average pooling to one spatial dimension while leaving the other dimensions unchanged. The difference between the flatten layer and the global average pooling layer is shown (see Fig. 1). After fine-tuning the Inception V3 model, we achieved a 92% test accuracy.

The fine-tuned Inception_V3 model (see Fig. 2.) is a fully connected dense layer with 512 features and the ReLU activation function is added, followed by a sigmoid activation function. Immediately after that, a four-density output layer with softmax activation is added, followed by a global average pooling layer. The extracted features are then fed into a multilayer perceptron classifier for prediction. Illustration of the proposed architecture (see Fig. 3) BACH dataset (a) benign, (b) in situ, (c) invasive, and (d) normal for breast cancer classification. This fine-tuning of the Inception_

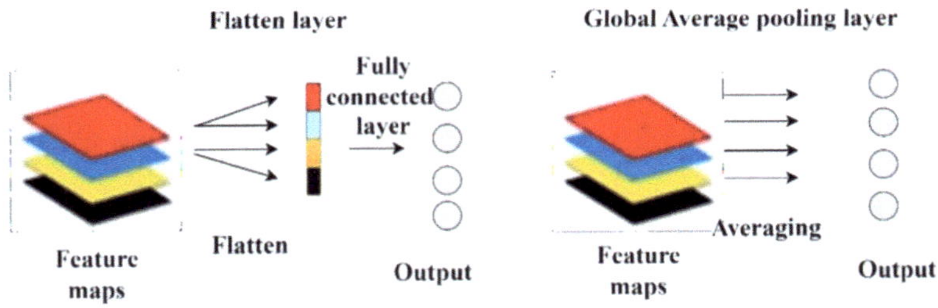

Fig. 1 Flatten versus pooling layers

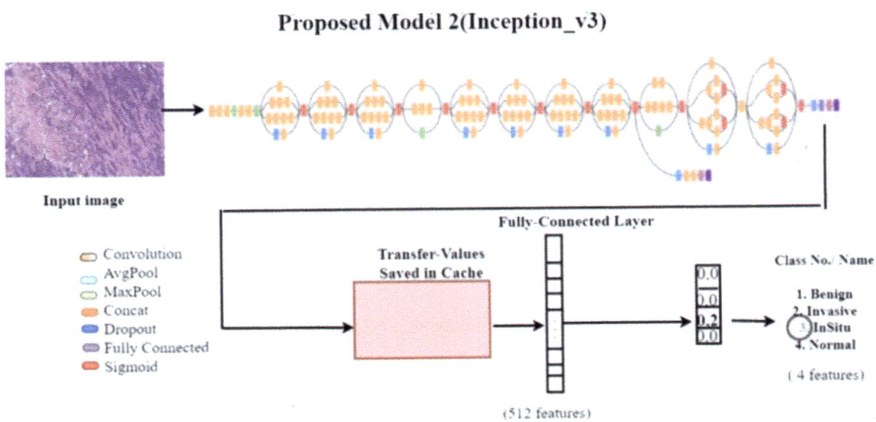

Fig. 2 Fine-tuned Inception_V3 model

V3 model yielded better results when compared with the standard or pre-trained Inception_V3 model and the proposed model 1 in this paper. Proposed model 2 achieved a test accuracy rate of 97%.

3.2 *Data Augmentation*

The limited size of the training dataset, the performance of CNN predictive models may decrease. Therefore, various data augmentation techniques are employed to augment the dataset and enhance classification performance. These techniques include rescaling, shearing, rotational adjustments, width and height shifting, horizontal flipping, and zooming.

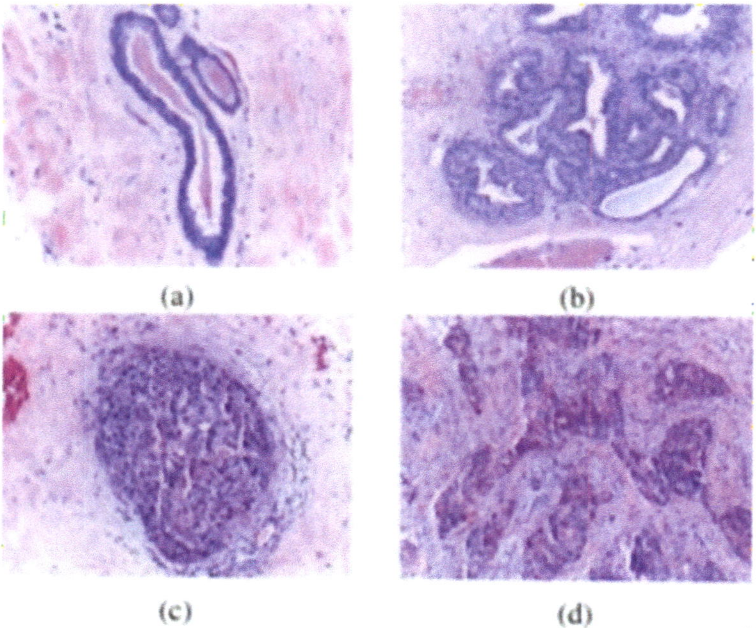

Fig. 3 BACH dataset, **a** benign, **b** in situ, **c** invasive, and **d** normal

3.3 Hyperparameters Fine-Tuning

The fine-tuning of hyperparameters includes the adjustment of both the loss function and the activation function. The available loss functions are Kullback-Liebler divergence and categorical cross-entropy, while the available activation functions are softmax and sigmoid.

Loss Functions

Categorical Cross-entropy

The categorical cross-entropy loss function is employed for multiclass classification involving four classes: benign, in situ, invasive, and normal. It measures the difference between probability distribution functions using the softmax function.

$$\text{Loss} = -\sum_{i=1}^{\text{output size}} y_i \cdot \log \hat{y}_i \tag{1}$$

$\hat{y}_i$ = i-th scalar value in the model output
y_i = Corresponding target value
Output size = the number of scalar values in the model output.

Kullback-Leibler Divergence

The Kullback-Liebler divergence (relative entropy) loss function is employed for multiclass classification, specifically for four classes: benign, in situ, invasive, and normal. It quantifies the relative difference between two probability distributions for a given random variable or set of occurrences and is measured using KL divergence.

$$D_{\mathrm{KL}}(p||q) = \int_x p(x) \log \frac{p(x)}{q(x)} \mathrm{d}x \tag{2}$$

$P(i)$ and $Q(i)$ = Two probability distributions.
X = is the set of all possible variables for x.

The log function is sharp close to zero, so this may allow sensitive detection of small probability distribution changes.

Activation Function

An activation function in a neural network describes how a node or nodes in a layer of the network transform the weighted sum of the input into an output. Two activation functions, sigmoid and softmax, are used.

Sigmoid

The sigmoid activation function is applied to the weighted sum of inputs, and the result is used as an input to the following layer. In a neural network, when a neuron's activation function is sigmoid, the unit's output will always fall between 0 and 1.

$$\sigma(X) = \frac{1}{(1 + e^{-z})} \tag{3}$$

$\sigma(X)$ = sigmoid function
$\hat{e}$ = Euler's number.

Softmax

In neural network models that predict a multinomial probability distribution, the softmax function is used as an activation function in the output layer [8]. Softmax is employed as the activation function for multiclass classification problems involving more than two class labels.

Table 2 Original images and augmented images

Classes	Original training dataset	Augmented training dataset	Original dataset	Augmented test data
Normal	270	3400	430	570
Benign	770	8400	930	070
In situ	270	3400	430	570
Invasive	770	8400	930	070
Total	2280	32,400	4120	5280

4 Experiment and Results

4.1 Dataset Description

The datasets used are publicly available datasets from Kaggle which contains HE-stained breast histology microscopy images of (a) normal, (b) benign, (c) in situ, and (d) invasive cases from ICIAR 2018 grand challenge on breast cancer histology (BACH) dataset. The dataset is randomly divided into three parts training, testing, and validation method. The dataset contains 400 HE-stained histological breast tissue photos uniformly divided into four categories: benign, normal, in situ, and invasive cancer (each with 100 images). The dataset images are shown in Fig. 3. All images are in TIFF format, with a magnification factor of 200 and pixel sizes of 0.42 * 0.42 m. All of the images have the same dimensions of 2048 1536 pixels. We split the dataset into two parts at random, using 300 images for training and 100 images for testing. We used several data augmentation techniques to expand the training dataset. The dataset's class distributions before and after data augmentation are given in Table 2.

4.2 Results and Discussion

ResNet-50, Inception_V3, VGG-16, and VGG-19 are four deep convolutional neural network (DCNN) architectures used to train the proposed system. The total test accuracy is used to assess the performance of all architectures. Table 3 summarizes the outcomes of these architectures. The Inception_V3 architecture, as well as both fine-tuned Inception V3 designs, achieved superior average classification accuracy, scoring 90%, 92%, and 97%, respectively, as demonstrated in Table 3.

Table 3 Results

Model	Test accuracy	Validation accuracy	Precision	Recall	$F1$ score	Support
VGG-16	0.87	0.87	0.89	0.88	0.88	40
VGG-19	0.72	0.72	0.85	0.72	0.76	40
Resnet-50	0.34	0.30	0.57	0.30	0.20	40
Inception-V3	0.89	0.89	0.91	0.90	0.90	40
Proposed Model 1	**0.92**	**0.86**	**0.87**	**0.80**	**0.76**	**40**
Proposed Model 2	**0.97**	**0.97**	**0.83**	**0.80**	**0.73**	**40**
Model2 (IDC data)	**0.81**	**0.81**	**0.84**	**0.81**	**0.81**	**400**

Table 4 Comparative analysis

Sl. No.	Authors	Model	Test Accuracy	Validation Accuracy
1	Guanglili 2021 [12]	–	93.57	95.18
2	Pratik Bhowal 2022 [13]	Xception	91	91
3	Peter Bokor 2021 [14]	Inception ResNetV2	89	–
4	**Proposed model 2022**	**Fine-tuned Inception_ V3**	**97**	**97**

4.3 Comparative Analysis

Table 4 compares the findings of some previously published work from the same dataset with those of the suggested approach to assess its efficacy. Table 3 gives that the approaches in [9–11] have accuracy of 93.57%, 91%, and 89%, respectively, but the findings produced utilizing the network design presented here have a precision of 97%. These findings show that the accuracy of the proposed model is higher than the other similar methods presented in literatures.

5 Conclusions and Future Scope

The performance of ResNet-50, VGG-16, VGG-19, and Inception_V3 is explored for the classification of breast lesion histopathology images, both for binary classification and multiclass classification, in the study. The major contribution lies in incorporating fine-tuning of the Inception_V3 model and optimizing hyperparameters. This technique has proven to be effective in improving accuracy. The approach also involves training the model iteratively to gradually transition from general ImageNet dataset images to the classification of breast lesion histopathology images from the BACH dataset, and ultimately to multi-classification of breast lesion images.

The proposed network architecture achieves an average classification accuracy of 97% when utilizing a pre-trained Inception_V3 model. Furthermore, the experiment reveals that the DNN model, which performs well in multiclass classification, may not perform as well in binary class classification.

Through comparative studies, it is evident that the proposed method's accuracy is reasonable and acceptable within the medical research community. In future, alongside publicly available datasets from web sources, efforts should be made to collect datasets from local hospitals in the Indian context. This will contribute to the prediction of multi-stage cancer in histopathology images using segmentation methods.

References

1. https://my.clevelandclinic.org/health/diseases/3986-breast-cancer
2. https://www.who.int/news-room/fact-sheets/detail/breast-cancer
3. Alruwaili M, Gouda W (2022) Automated breast cancer detection models based on transfer learning. Sensors 22(3):876. https://doi.org/10.3390/s22030876
4. Heenaye-Mamode Khan M, Boodoo-Jahangeer N, Dullull W, Nathire S, Gao X et al (2021) Multi- class classification of breast cancer abnormalities using deep convolutional neural network (CNN). PLoS ONE 16(8):e0256500
5. Dabass J et al. (2021) Multi-class classification of mammograms with hesitancy based Hanman transform classifier on pervasive information set texture features
6. Kassani SH, Kassani PH, Wesolowski M, Schneider K, Deters R (2019) Breast cancer diagnosis with transfer learning and global pooling. https://doi.org/10.1109/ICTC46691.2019.8939878
7. Nawaz M et al (2018) Multi-class breast cancer classification using deep learning convolutional neural network. Int J Adv Comput Sci Appl 9
8. Nawaz M, Sewissy AA, Soliman THA (2018) Multi-class breast cancer classification using deep learning convolutional neural network. Int J Adv Comput Sci Appl (IJACSA) 9(6). https://doi.org/10.14569/IJACSA.2018.090645
9. McDonald BR et al (2019) Personalized circulating tumor DNA analysis to detect residual disease after neoadjuvant therapy in breast cancer. Sci Transl Med 11(504):eaax7392. https://doi.org/10.1126/scitranslmed.aax7392
10. Li G, Li C, Wu G, Ji D, Zhang H (2021) Multi-view attention-guided multiple instance detection network for interpretable breast cancer histopathological image diagnosis. IEEE Access 9:79671–79684. https://doi.org/10.1109/ACCESS.2021.3084360
11. Bhowal, P et al (2022) Fuzzy ensemble of deep learning models using choquet fuzzy integral, coalition game, and information theory for breast cancer histology classification. Expert Syst Appl
12. Li G et al (2020) Effective breast cancer recognition based on fine-grained feature selection. IEEE Access 1–1. https://doi.org/10.1109/ACCESS.2020.3046309
13. Bokor P et al (2021) Weighted multi-level deep learning analysis and framework for processing breast cancer WSIs. arXiv:abs/2106.14708
14. Bokor P et al (2021) Weighted multi-level deep learning analysis and framework for processing breast cancer WSIs. arXiv 2021. https://doi.org/10.48550/arxiv.2106.14708

15. Shahidi F, Mohd Daud S, Abas H, Ahmad NA, Maarop N (2020) Breast cancer classification using deep learning approaches and histopathology image: a comparison study. IEEE Access. 8:187531–187552. https://doi.org/10.1109/ACCESS.2020.3029881.@inproceedingsinproceedings

Medical Diagnosis Through Conversational Intelligence Using Large Language Models

S. Akilesh, R. Suganya, Swarup Ravi, R. Tulasi Raman, Vignesh Srinivasakumar, and Girish H. Subramanian

Contents

Abstract This paper investigates the utility of GPTs in interactive medical diagnosis through natural language processing (NLP). The model dynamically engages

---

S. Akilesh (✉) · R. Suganya · R. Tulasi Raman
Vellore Institute of Technology, Chennai, Tamil Nadu 600127, India
e-mail: akilesh.s2021@vitstudent.ac.in

R. Suganya
e-mail: suganya.ramamoorthy@vit.ac.in

R. Tulasi Raman
e-mail: tualsiraman.r2021@vitstudent.ac.in

S. Ravi
PES University, Bengaluru, Karnataka 560100, India
e-mail: pesug20ec096@pesu.pes.edu

V. Srinivasakumar
University of Illinois, Urbana-Champaign, IL 61801, USA
e-mail: vs59@illinois.edu

G. H. Subramanian
Penn State University, Harrisburg, PA 17057, USA
e-mail: ghs2@psu.edu

S. M. Thampi et al. (eds.), *Fifth International Conference on Computing and Network Communications*, Lecture Notes in Electrical Engineering 1221,
https://doi.org/10.1007/978-981-97-4711-5_23

patients in diagnostic discourse, generating contextually relevant follow-up questions grounded in symptomatology, test results, and attempted remedies. Preliminary experimentation showcases the model's capacity to craft pertinent inquiries, offering nuanced insights into health concerns. The study emphasizes the necessity for ongoing development, addressing ethical dimensions associated with AI-guided diagnoses. The integration of GPTs into medical diagnostics not only accentuates improved patient interactions but also signifies a potential paradigm shift in diagnostic support within the healthcare domain. This research lays the groundwork for future innovations in diagnostic methodologies, positioning NLP-driven interactions as a promising avenue for enhancing the effectiveness of medical diagnosis.

Keywords Medical diagnosis · OpenAI · GPT · NLP · Healthcare · Large Language Models · BioBERT

1 Introduction

In the healthcare domain, the cornerstone of effective patient care rests on the accuracy and timeliness of medical diagnoses. As medical knowledge expands and patient cases grow more intricate, there is a burgeoning demand for precise diagnostic tools. Addressing this need, Artificial Intelligence (AI) emerges as a promising solution, leveraging its computational prowess to enhance diagnostic processes. GPT-3.5, a cutting-edge natural language processing model, stands out among AI advancements. Its unique ability to comprehend and generate human-like text makes it particularly suited for engaging in intricate conversations, sparking interest in its application for medical diagnosis.

The potential of GPT-3.5 is rooted in its capacity to facilitate interactive patient engagement through dynamic conversations. This approach not only enhances patient involvement but also enables the extraction of nuanced information that might be overlooked in conventional diagnostic systems. This paper undertakes a preliminary exploration of GPT-3.5 for interactive medical diagnosis. Through structured conversations, the model generates follow-up questions based on initial symptoms, inquiries about conducted tests, and details about attempted home remedies. The primary objective is to investigate the model's prowess in refining diagnostic understanding through iterative questioning, ultimately providing more accurate and contextually informed diagnoses.

This study illuminates the potential advantages of integrating AI, specifically GPT-3.5, into the medical diagnostic process. The adoption of an interactive approach empowers healthcare providers to unveil hidden symptoms, access comprehensive patient histories, and deliver tailored recommendations. However, it is vital to recognize the inherent limitations, including concerns about privacy, data security, and the essential role of medical professionals in validating AI-generated insights.

In consideration of these factors, this paper contributes to the ongoing discourse surrounding AI's pivotal role in healthcare. It presents a preliminary investigation into the application of GPT-3.5 for interactive medical diagnosis. Subsequent sections delve into the methodology, experimental results, and ethical implications, providing insights into potential benefits and challenges. As AI continues to reshape healthcare practices, this research serves as a catalyst for further innovation in diagnostic techniques and patient engagement strategies.

2 Literature Survey

The use of Artificial Intelligence (AI) and expert systems in the field of medical diagnosis has been a topic of extensive research and application over the years. Early pioneering efforts, such as the MYCIN system, laid the foundation for the integration of AI into healthcare. Here, we provide an overview of the evolution of AI and expert systems in medical diagnosis, emphasizing their impact and potential.

2.1 Early AI Systems: The MYCIN Era

The MYCIN system, developed in the 1970s by Shortliffe et al. [1], was one of the first expert systems designed for medical diagnosis. MYCIN was specifically crafted to diagnose bacterial infections and recommend appropriate antibiotic treatments. It utilized a rule-based approach, wherein a knowledge base containing expert-elicited rules was employed to make inferences about patient conditions. The system's success in diagnosing infectious diseases demonstrated the potential of AI to assist medical professionals.

2.2 Advancements in Expert Systems

Subsequent developments in expert systems witnessed a proliferation of rule-based systems in various medical domains. Dendral [2], another pioneering system, focused on chemical mass spectrometry and laid the groundwork for rule-based expert systems' application in diverse fields.

Internist-1, developed by Shortliffe [3], was among the earliest comprehensive expert systems capable of diagnosing a broad range of medical conditions. The expert knowledge encoded in such systems enabled them to mimic human diagnostic reasoning, offering valuable assistance to healthcare practitioners.

2.3 Contemporary AI Applications in Medical Diagnosis

In contemporary healthcare, AI has evolved beyond rule-based systems. Machine learning techniques, including deep learning and natural language processing (NLP), have enabled AI to analyze vast datasets and detect patterns that may elude human clinicians. AI-driven medical imaging, for instance, has achieved remarkable success in diagnosing conditions like cancer from radiological scans [4].

Moreover, the integration of AI with electronic health records (EHRs) has facilitated real-time patient data analysis and predictive modeling for disease progression and risk assessment [5]. These advances highlight the continued evolution of AI in medical diagnosis, from rule-based expert systems to data-driven machine learning models.

2.4 Recent Advancements: Natural Language Processing and Text-Based Medical AI

Recent years have witnessed a surge in the development of natural language processing (NLP) and text-based AI models in the realm of medical diagnosis. These models leverage large-scale healthcare text data, electronic health records (EHRs), and medical literature to enhance diagnostic accuracy and patient care.

BERT and Biomedical NLP. Bidirectional Encoder Representations from Transformers (BERT), a breakthrough in NLP, has been adapted for biomedical applications. Models like BioBERT [6] are fine-tuned on biomedical text, enabling them to extract relevant information from medical records and scholarly articles. These models contribute to more precise diagnostic insights by understanding medical narratives and terminology.

Clinical Chatbots. Building on the concept of interactive AI-driven diagnosis, clinical chatbots are becoming increasingly sophisticated. Chatbots like Ada Health [7] engage users in detailed symptom-based conversations, providing preliminary diagnoses and recommendations. They employ NLP techniques to understand and respond to user inputs, contributing to the accessibility of healthcare information.

Diagnosis from Unstructured Text. AI models, such as Deep Patient [8], are designed to extract diagnostic information from unstructured medical narratives. By parsing doctor's notes and patient records, these systems assist healthcare providers in diagnosing diseases, suggesting treatment options, and tracking patient progress.

Medical Literature Analysis. AI tools like IBM Watson for drug discovery [8] are designed to analyze vast volumes of medical literature to identify potential drug candidates and treatment options. These applications rely on advanced NLP and machine learning techniques to extract knowledge from scientific texts, facilitating drug discovery and personalized treatment strategies.

Generative AI. Large Language Models (LLMs) have emerged as transformative tools in healthcare, offering a range of promising applications across education, research, and practice [9]. Their ability to process and analyze vast amounts of medical data, coupled with their natural language processing capabilities, has opened up new avenues for improving patient care and advancing clinical knowledge [10, 11]. In education, LLMs can serve as personalized tutors, providing tailored instruction and feedback to medical students [9]. In research, they can accelerate drug discovery and disease modeling by efficiently synthesizing and analyzing scientific literature [12]. And in practice, LLMs can assist healthcare providers in clinical decision-making, risk stratification, and patient communication [9]. While LLMs hold immense potential, their integration into healthcare must be accompanied by careful consideration of ethical, regulatory, and societal implications [10].

The literature survey illuminates the evolving landscape of AI in medical diagnosis. Prior efforts, encompassing rule-based systems, clinical chatbots, and biomedical NLP, have significantly contributed to healthcare technology. The advent of GPT-based diagnosis signals a transformative shift with unparalleled capabilities in dynamic patient interactions and nuanced information extraction. Its contextual awareness and extensive medical knowledge position GPT as a versatile tool.

3 Methodology and Implementation

The proposed method centers on utilizing an advanced Large Language Model (LLM), renowned for its exceptional proficiency in generating human-like text prompted by inputs. This LLM employs an extensive pre-trained neural network architecture designed for encoding linguistic patterns and semantics. Functioning interactively, the system engages in iterative conversations with patients, harnessing the capabilities embedded in its neural network to adapt dynamically to diverse linguistic contexts.

The OpenAI Playground facilitates the integration of this NLP framework into medical diagnosis. The system accepts symptom descriptions and other medical information as input prompts, triggering the neural network to generate contextually coherent follow-up questions. These generated questions are based on the learned relationships within the neural network, allowing the system to probe deeper into the patient's condition. Patient responses to these questions guide subsequent inquiries, creating an informative dialogue. This dynamic interaction goes beyond conventional diagnostic tools by capturing intricate patient information that might otherwise be missed. The NLP model's ability to understand context, semantic nuances, and syntactic patterns enhances its question formulation, leading to more precise diagnostic insights. The system's adaptability and linguistic expertise contribute to a more comprehensive assessment of the patient's condition, potentially leading to more accurate diagnostic outcomes.

The data flow for GPTs begins with user input, undergoes message preprocessing, and engages natural language processing (NLP) to generate an action plan. Subsequently, the system retrieves pertinent information from databases or external sources, utilizing knowledge bases. The implementation of efficient data structures and algorithms facilitates information retrieval. The system may extend its search to the internet for additional insights. Finally, a response is generated, and dialogue management ensures a coherent interaction. This streamlined process (Fig. 1) enables GPTs to dynamically respond to user inputs and provide contextually relevant information, showcasing their versatility in information retrieval and dialogue handling (Fig. 2).

OpenAI's APIs facilitate a streamlined process involving retrieval, dialogue, action plan, and response generation. Users interact by providing a message, initiating the retrieval of relevant information. The dialogue system incorporates this input, creates an action plan, and generates a response using the OpenAI model, specifically "gpt-3.5-turbo." The code snippet, **ask_gpt()**, takes a user message, incorporates a system prompt, and uses OpenAI's ChatCompletion API to obtain a generated message in response. This message is then processed and returned as a JSON object, showcasing the practical application of OpenAI's APIs in dynamic conversation generation.

4 Experimental Results and Discussion

The initial phase of the experiment involved the deployment of GPT-2 and BioBERT, utilizing training and validation datasets (Fig. 3) scraped and organized from *X* (formerly Twitter) medical forums. Despite the promising capabilities of GPT-2 and BioBERT, the results were underwhelming in comparison to GPT-3. Both GPT-2 and BioBERT struggled to maintain coherence in their generated responses, often failing to conclude sentences adequately. The limitations of GPT-2 and BioBERT became pronounced as they exhibited a tendency to hallucinate massively, generating content that lacked factual accuracy and contextual relevance. The experiment shed light on the challenges faced by these models in grasping the intricacies of medical conversations and producing meaningful outputs. Consequently, the inadequacies observed in GPT-2 and BioBERT underscored the transformative leap achieved by GPT-3 in natural language processing tasks, particularly within the specialized domain of medical discourse on social media platforms. This comparative analysis highlights the superior performance of GPT-3, emphasizing its advanced language understanding and contextual coherence, thereby positioning it as a potent tool for extracting valuable insights from medical discussions in online forums. The entirety of these experiments with BioBERT, GPT-2 and GPT-3 were primarily performed in the English language although other languages like Hindi, Tamil, and Telugu were tested, they didn't yield satisfactory results acquired using English text prompts.

The next experimental phase was conducted using GPT-3 with an 8 K context. A diverse array of simulated medical cases, spanning a spectrum of symptoms and

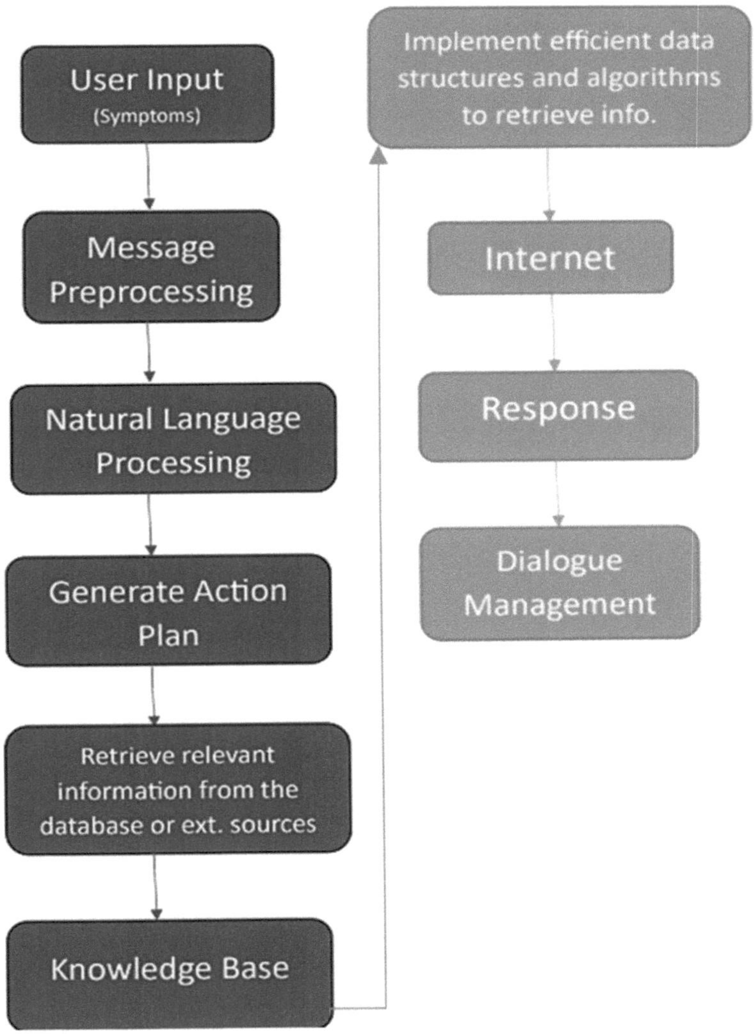

Fig. 1 Data and control flow of the API used

complexities, was employed for evaluation. A total of 200 cases were included, varying in severity and specificity. The outcomes of the experimentation unveiled insightful findings into the diagnostic capabilities of the system. On average, the interactive system successfully generated 12 follow-up questions per case, probing patients for detailed information. Notably, in instances where patients described atypical or rare symptoms, the system demonstrated adaptability by tailoring its questions to the unique context.

```
def ask_gpt():
    message = request.form.get('Message')

    completion = openai.ChatCompletion.create(
        model="gpt-3.5-turbo",
        messages=[
            {"role": "system", "content": system_prompt},
            {"role": "user", "content": message}
        ]
    )

    generated_message = completion.choices[0].message['content']
    generated_message= generated_message.encode().decode('unicode_escape')

    return jsonify({'message': generated_message})
```

Fig. 2 How the API is used in a flask server to generate responses

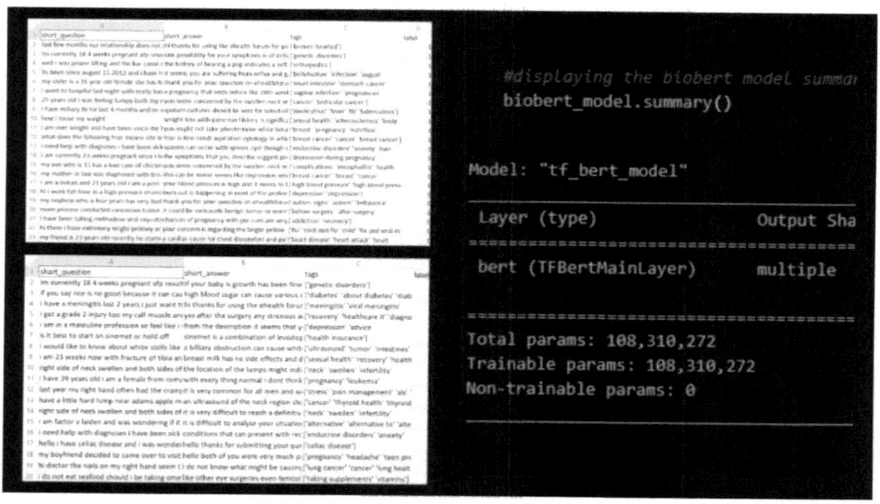

Fig. 3 Training and validation datasets with the Biobert model

In complex cases, where traditional diagnostic tools might struggle to pinpoint the underlying condition, for example, the interactive AI system excelled by consistently posing inquiries like, "Can you describe the frequency and intensity of your headaches?" The patient responded with detailed information, allowing the system to explore potential factors, contributing to a more accurate diagnosis. Moreover, the system exhibited a remarkable ability to identify connections between seemingly unrelated symptoms, potentially uncovering previously undetected conditions.

In scenarios where patients provided incomplete or unclear responses, the system demonstrated resilience by asking clarifying questions and guiding patients to provide more informative details. This iterative process, akin to a clinician's probing, contributed significantly to the overall diagnostic accuracy.

Table 1 Experimental metrics versus performance

Experimental metric	Performance
Follow-up questions per case	Average: 12 Questions
Diagnostic accuracy	High accuracy in complex cases, outperforming traditional tools
Adaptability to rare symptoms	Effective adaptation, tailoring questions appropriately
Identifying underlying factors	Consistent and effective in exploring potential underlying factors
Connection between symptoms	Successfully uncovered connections in seemingly unrelated symptoms
Handling incomplete responses	Resilience through clarifying questions, improving information completeness
Interpretation of symptom descriptions	Occasional misinterpretations, indicating potential for refinement

However, the experimentation also unveiled limitations. The system occasionally misinterpreted certain symptom descriptions, highlighting the importance of refining its understanding of medical jargon and context-specific language. Additionally, privacy concerns emerged, underscoring the need for robust data security measures.

The experimental evaluation of the AI-driven medical diagnosis system using GPT-3 yielded promising results (Table 1). The system demonstrated an average of 12 follow-up questions per case, showcasing adaptability and effectiveness in identifying underlying factors. It exhibited high diagnostic accuracy, particularly in complex cases, outperforming traditional tools. While occasionally misinterpreting symptom descriptions, the system displayed resilience through clarifying questions, contributing to improved information completeness and overall diagnostic capabilities.

Emphasizing the integral role of the OpenAI Playground and its fine-tuning tools, the strengths of the AI-driven interactive medical diagnosis system stem from the seamless integration of rule-based prompt engineering and machine learning-based adjustments. This intricate fusion yields a diagnostic tool with unparalleled adaptability, context-awareness, and diagnostic acumen. These refined features, hold the promise of reshaping medical diagnosis through advanced natural language processing-driven interactions.

5 Future Advancements

Several avenues for improvement exist to augment the effectiveness of the interactive medical diagnosis tool. While the current tool excels in natural language processing, it has limitations in interpreting visual information, hindering its ability to analyze symptoms with visible manifestations. Integrating GPT-4's enhanced vision capabilities could address these shortcomings, allowing the tool to assimilate and interpret visual cues in addition to textual information. This improvement would broaden the tool's diagnostic scope, enabling a more comprehensive understanding of patients' conditions. However, careful consideration of ethical concerns, data privacy, and potential biases in visual analysis should accompany these advancements to ensure responsible and equitable healthcare applications.

Further advancements in the interactive medical diagnosis tool could involve integrating real-time patient data from wearable devices. This addition would enable the tool to dynamically incorporate physiological metrics, enhancing the accuracy of diagnostic assessments. Moreover, considering a port of the application for devices like Apple Vision Pro, Meta Quest Pro or similar extended reality (XR) and spatial computing headsets could revolutionize vision-assisted diagnosis [13, 13]. Such immersive technologies could offer a three-dimensional perspective, aiding in the analysis of medical conditions with visual components. However, these improvements should be implemented cautiously, addressing privacy concerns, ensuring data security, and prioritizing user accessibility to guarantee responsible and inclusive healthcare solutions.

6 Conclusion

In the evolving landscape of healthcare, the fusion of Large Language Models (LLMs) and Generative Artificial Intelligence (AI) to create an interactive medical diagnosis assistant represents a promising leap toward more precise, accessible, and efficient healthcare services. This research has demonstrated the remarkable potential of such an assistant to revolutionize the way medical diagnoses are conducted. However, as we envisage the practical deployment of this technology, it is imperative to consider both its strengths and ethical considerations.

The AI-powered medical diagnosis assistant has showcased remarkable diagnostic capabilities. Its adaptability empowers it to navigate intricate clinical scenarios, generate context-aware questions, and offer refined diagnostic insights. This not only augments the diagnostic accuracy but also holds the potential to address the global shortage of healthcare professionals, especially in underserved regions. Additionally, by streamlining the diagnostic process, this technology has the potential to significantly reduce healthcare costs, optimize resource allocation, and enhance patient care.

While the advantages of AI-driven medical diagnosis are profound, there are noteworthy challenges. The misinterpretation of medical jargon, and occasional privacy breaches underscore the necessity for continued refinement in language understanding and stringent data security measures. Moreover, ethical questions arise concerning the dependency on GenAI for diagnoses, emphasizing the importance of striking a balance between GenAI-assisted diagnosis and human expertise. Ensuring that the technology augments rather than replaces the role of healthcare professionals remains a crucial ethical consideration.

As we navigate this exciting frontier of AI-driven healthcare, we recognize that the synergy between technology and human expertise will be paramount. The deployment of AI as a diagnostic assistant, rather than a standalone decision-maker, ensures that the compassionate, empathetic, and intuitive aspects of medical practice, inherent to human clinicians, are retained.

In conclusion, this research lays the foundation for a brighter future in healthcare, where AI and LLM-driven medical diagnosis assistants can work hand-in-hand with healthcare professionals. The potential to enhance diagnostic accuracy, improve healthcare accessibility, and optimize resource utilization is profound. By addressing the ethical concerns and embracing a collaborative approach, we can look forward to a healthcare landscape that seamlessly integrates cutting-edge technology and human expertise, ultimately benefiting patients worldwide. This journey, marked by its technical innovation and ethical considerations, holds the promise of transforming healthcare as we know it.

References

1. Shortliffe EH, Buchanan BG (1975) A model of inexact reasoning in medicine. Math Biosci 23(3–4):351–379
2. Buchanan BG, Feigenbaum EA (1978) Dendral and Meta-Dendral: Their applications dimension. Artif Intell 10(3):198–208
3. Shortliffe EH, Scott AC (1978) Computer-based medical consultations: MYCIN Elsevier
4. Litjens G, Kooi T, Bejnordi BE, Setio AAA, Ciompi F, Ghafoorian M, Sánchez CI (2017) A survey on deep learning in medical image analysis. Medical image analysis. 42: 60–88.Rajkomar A, Oren E, Chen K, Dai AM, Hajaj N, Hardt M, Dean J (2018) Scalable and accurate deep learning with electronic health records. NPJ Digital Medicine. 1(1): 1–10
5. AI chatbots not yet ready for clinical use Joshua Au Yeung, Zeljko Kraljevic, Akish Luintel, Alfred Balston, Esther Idowu, Richard J. Dobson, and James T. Teo https://doi.org/10.3389/fdgth.2023.1161098
6. Health TLD (2019) Ada Health: AI symptom assessment apps—an introduction and comparison. The Lancet Digital Health 1(1):e20–e21
7. Mullenbach J, Wiegreffe S, Duke J, Sun J, Eisenstein J, Recht B (2018). Explainable prediction of medical codes from clinical text. arXiv preprint arXiv:1802.05695.
8. Rizvi NF, Palade V, Rios P, Harris S, Ursu M (2021) Applying artificial intelligence to drug discovery: a comparative analysis of AI-driven drug development strategies. Pharmaceutics 13(3):313
9. Sallam M et al (2023) ChatGPT Utility in healthcare education, research, and practice: systematic review on the promising perspectives and valid concerns. Healthcare 11:887
10. Meskó B, Topol EJ (2023) The imperative for regulatory oversight of large language models (or generative AI) in healthcare. NPJ Digit. Med. 6:120
11. Harrer Stefan (2023) Attention is not all you need: the complicated case of ethically using large language models in healthcare and medicine. EBioMedicine 90
12. Tamkin A, Brundage M, Clark J, Ganguli D (2021) Understanding the capabilities, limitations, and societal impact of large language models. arXiv preprint arXiv:2102.02503.

13. Vyas, Sonali (2023) Extended reality and edge AI for healthcare 4.0: systematic study. Extended reality for healthcare systems. Academic Press 229–240
14. Gasteiger Norina et al. (2022) Exploring how, for whom and in which contexts extended reality training works in upskilling healthcare workers: a realist review. IEEE

Automated Medicinal Plant Recognition for Ayurvedic Healthcare Solutions

B. S. Vandana, S. Moksith, M. Mahammad Afnan, and Akash C. Bhat

Abstract In today's fast-paced and modernized world, the connection between humanity and nature has significantly dwindled. As a result, the vast reservoir of knowledge present in nature, especially pertaining to medicinal plants, is gradually being overlooked. Our ancestors relied on these natural remedies, collectively known as "Ayurveda," to address various health issues effectively. However, the current generation tends to prioritize chemical-based medicines, commonly referred to as "English Medicine," due to their rapid recovery rates. Unfortunately, these chemical remedies often come with side effects and steep costs, limiting their long-term viability. To bridge the gap between traditional wisdom and modern healthcare, we propose an automated system for identifying medicinal leaves using computer vision technology. The system requires an extensive dataset comprising diverse images of different medicinal plant species to accurately learn essential features for recognition. By automating this task, we aim to empower individuals to quickly identify medicinal plants and understand their inherent therapeutic values. The proposed work will greatly benefit current and future generation as it will help them to know what kind of a plant it is. In this way it might reduce the gap between the human and the nature.

Keywords Ayurveda · Computer vision · Medicinal plants · Automated recognition

1 Introduction

B. S. Vandana (✉)
Department of Information Science and Engineering, NMAM Institute of Technology, Nitte (Deemed to be University), Karnataka, India
e-mail: vandanamonappa07@gmail.com

S. Moksith · M. Mahammad Afnan · A. C. Bhat
Department of MCA, Vivekananda College of Engineering and Technology, Puttur, India

S. M. Thampi et al. (eds.), *Fifth International Conference on Computing and Network Communications*, Lecture Notes in Electrical Engineering 1221,
https://doi.org/10.1007/978-981-97-4711-5_24

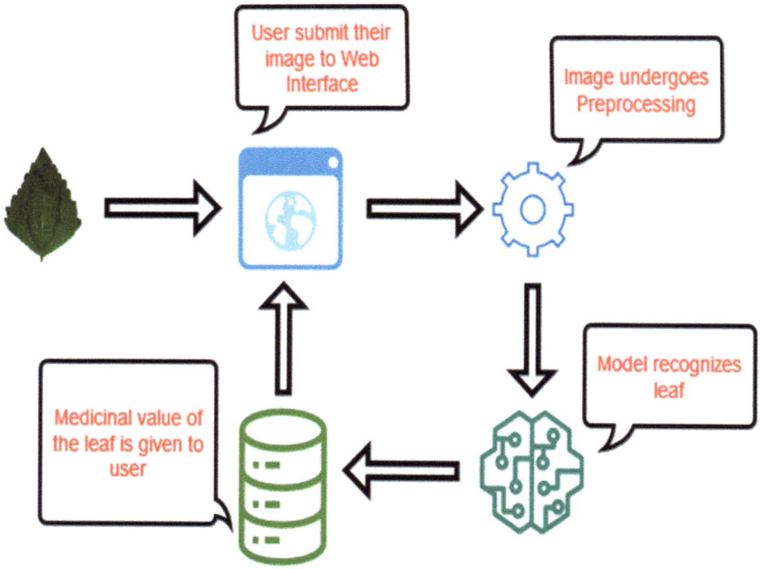

Fig. 1 Automated medicinal plant recognition

Plants play a vital role in maintaining ecological balance, serving as both a food source for animals and providing potential health benefits. However, human actions have disrupted this delicate equilibrium, leading to the endangerment of several plant species. Consequently, people are gradually losing touch with the significance of plants and their various uses. To preserve this valuable knowledge, creating a comprehensive database of plants and their properties is essential. Computers, renowned for their learning and memory capabilities, offer a solution to store and manage such information. Automating tasks has become indispensable in many domains, and one such application lies in automatic leaf recognition. This research focuses on utilizing digital images of leaves to develop a recognition system. Digital images encompass a wealth of information that can serve as features for the model to learn from. The model assimilates leaf shapes to comprehend their structures, considers color intensity and types for further insights, and performs mathematical operations, such as calculating area and perimeter, as additional distinguishing features. Accurate feature extraction is vital for the model to effectively differentiate between different types of leaves. The initial step involves collecting a dataset required for training our MobileNetV2 model. The dataset may contain noise, which can impact the model's performance. To address this, a process is employed to identify the region of interest, namely the leaf, resulting in a binary image that eliminates the noise. Once all images in the dataset undergo these preprocessing steps, they are utilized to train the MobileNetV2 model. This model extracts pertinent features from the images, such as color, texture, edges, and lines, enabling it to effectively recognize various plant species based on their distinctive characteristics (Fig. 1).

1.1 Objective

The proposed model capable of recognizing given leaf images and providing the corresponding leaf name based on the features it has learned from the dataset. Additionally, the system will display the medicinal value associated with the recognized leaf.

1.2 Scope

In our everyday surroundings, we encounter a multitude of diverse plants, making it challenging to remember each one's name and properties. When we observe a plant, our immediate associations often revolve around its leaves, roots, flowers, and stems.Some of the medicinal plants are illustrated in Fig. 2 The proposed work aims to address this issue by employing leaf images for plant recognition. Through this approach, individuals can easily identify plants and access comprehensive details about their properties, leading to a greater understanding of their significance. Additionally, this work plays a crucial role in preserving and passing on plant-related knowledge to future generations. Specifically, the focus of this project is on recognizing medicinal plants. The system has been designed to excel in identifying leaves with known medicinal value. However, it is important to note that the system may not accurately recognize leaves without medicinal properties or those with undiscovered medicinal value. As a result, the system is tailored to best serve the recognition of plants with established medicinal importance, ensuring its efficacy in providing valuable information about such plants and their uses.

2 Related Work

This comprehensive review focuses on various works related to automated plant leaf recognition using Artificial Neural Networks (ANN) and other machine-learning

Fig. 2 Sample medicinal leaves (Eupartorium odoratum, Piper nigrum, Garcinia indica)

techniques. The primary objective across these studies is to accurately identify different plant species based on their distinctive leaf features. The articles explore different models, preprocessing techniques, and feature extraction methods to improve the recognition capabilities of the systems.

In one study, Huixian [1] used the KNN-neighborhood classification algorithm to classify 50 leaf images based on parameters like leaf shape, color, and texture. Another article by Li et al. [2] addressed disease detection in plants using neural networks, utilizing various models and preprocessing steps to achieve accurate results. Cut flower classification was also explored by Fei et al. [3], where the ShuffleNet V2 model was used with transfer learning to determine flower freshness based on color.

Morphological features like the minor axis, major axis, and perimeter were considered in another work by Kumar et al. [4] to detect plant types, achieving high accuracy with the KNN classifier and multilayer perceptron tree with Adaboost technique. Additionally, Bakhshipour [5] used RFCN for disease detection in plants, dividing the object into regions for more accurate detection.

Disease detection in mango leaves was addressed using CNN and ANN models by Pham et al. [6], with feature selection methods to improve accuracy. Another study by Sun et al. [7] focused on selective weeding by identifying the right leaf to remove, employing ensemble learning with Adaboost and LogitBoost algorithms. Singh et al. [8] conducted a systematic literature review of machine learning techniques deployed in agriculture, with a case study of banana crop.

Other works include Vishnoi et al. [9] detecting apple plant diseases using leaf images through a Convolutional Neural Network (CNN), Singh et al. [10] using a multilayer CNN for the classification of mango leaves infected by anthracnose disease, and Zhang et al. [11] proposing a high-quality rice leaf disease image data augmentation method based on a Dual GAN.

Yuan et al. [12] developed SPEDCCNN, a spatial pyramid-oriented encoder-decoder cascade convolution neural network for crop disease leaf segmentation. Barburiceanu et al. [13] used convolutional neural networks for texture feature extraction in leaf disease classification in precision agriculture. Sunday Samuel [10, 14] worked on early cocoa blackpod pathogen prediction with a machine learning ensemble algorithm based on climatic parameters.

The reviewed works collectively contribute to the advancement of automated plant recognition systems, showcasing the potential of ANN and other machine learning approaches in addressing various challenges in the domain. The incorporation of diverse models, preprocessing techniques, and feature extraction methods has shown promising results in accurately identifying plant species and detecting diseases. The review sheds light on the importance of preserving traditional plant knowledge and promoting sustainable healthcare practices through automated plant recognition systems.

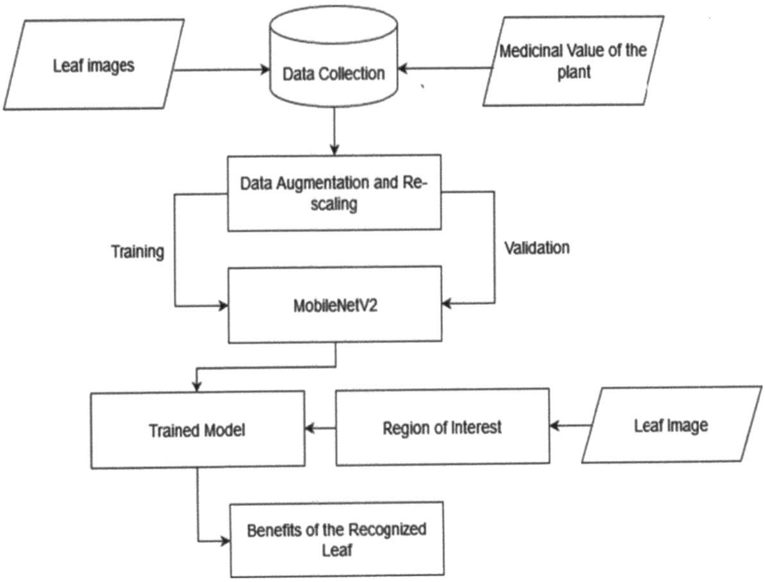

Fig. 3 Proposed model of medicinal plant recognition system

3 Methodology

The implementation process for the automated medicinal plant recognition illustrated in Fig. 3 using the MobileNetV2 model involves several key steps:

3.1 Dataset Preparation

The first step is to prepare the dataset of leaf images for training the MobileNetV2 model. The dataset should be diverse, containing images of various medicinal plant species with labeled annotations. Each image should represent a different plant leaf, capturing a range of shapes, sizes, and colors.

3.2 Preprocessing

The dataset undergoes preprocessing to enhance the quality of the images and remove any noise or irrelevant elements. One important step is to generate binary images for all the leaf images, effectively segmenting the leaf from the background. This binary image aids in extracting the region of interest (ROI), which contains the essential leaf features required for recognition.

3.3 Model Configuration

During the implementation, the MobileNetV2 model is configured and set up for training. The model architecture and hyperparameters are carefully chosen to achieve optimal performance. The process may involve trial and error to find the best combination of parameters, ensuring the desired accuracy and efficiency of the model.

3.4 Training Process

The model is trained using the prepared dataset and the extracted features from the ROI of each leaf image. The training process involves passing the images through the MobileNetV2 model and iteratively adjusting the model's weights and biases to minimize the error between the predicted output and the actual labels.

3.5 Feature Extraction

As the model is trained, it learns to extract relevant features from the leaf images, such as leaf shape, texture, and color intensity. These extracted features are crucial in differentiating one plant species from another and enabling accurate recognition.

3.6 Web Interface

To provide wider accessibility to the model, a user-friendly web interface is developed. Through this interface, users can interact with the model, inputting their leaf images for recognition. The interface will pass the images to the trained MobileNetV2 model, which will then process and analyze the images to determine the plant species.

3.7 Evaluation

Once the model is implemented and the web interface is ready, the system is evaluated to ensure its accuracy and effectiveness in recognizing medicinal plant species based on their leaf images. A validation dataset may be used to assess the model's performance, and any necessary refinements or adjustments are made to improve its accuracy. By following this implementation process, the automated medicinal plant recognition system using the MobileNetV2 model becomes a valuable tool in identifying and understanding different plant species and their medicinal properties. The user-friendly web interface enhances accessibility and encourages the sharing of plant knowledge for the benefit of current and future generations.

4 Result and Discussion

The implemented system, Plant Reco, provides a user-friendly platform for automated medicinal plant recognition using the MobileNetV2 model. Users can conveniently capture an image of a leaf and upload it to Plant Reco for analysis. The MobileNetV2 model, which has been trained on a diverse dataset of medicinal plant leaves, processes the uploaded image and accurately identifies the plant species. Upon successful recognition, the system displays valuable information about the identified plant, including its common name, scientific name, a brief description, and its medicinal usage. This feature allows users to gain insights into the potential health benefits and traditional uses of the recognized plant.To evaluate the performance of the Plant Reco system, the average accuracy for recognizing specific plant species, such as bitter bush, black pepper, kokum, and officinal breynia, was measured. The results were depicted in a graph, showcasing the accuracy levels achieved by the model. From the graph, it can be observed that the MobileNetV2 model demonstrates high accuracy in recognizing most of the leaves belonging to the specified plant species. The average accuracy achieved indicates the model's capability to distinguish between different plant leaves and accurately classify them into their respective categories. The high accuracy achieved by the Plant Reco system validates the effectiveness of the MobileNetV2 model and the robustness of the implemented recognition approach. The accurate identification of medicinal plants is of paramount importance in promoting traditional plant knowledge and facilitating the use of natural remedies for healthcare purposes. Moreover, the user-friendly interface of Plant Reco makes it accessible to a wider audience, encouraging individuals to engage with plant recognition and gain a deeper understanding of the plant world. By providing information on medicinal properties, the system also aids in promoting the sustainable use of medicinal plants and preserving traditional healing practices. Overall, the results and discussion demonstrate the successful implementation of Plant Reco, an automated medicinal plant recognition system, which accurately identifies medicinal plant species and provides valuable information to users. The system's capabilities contribute to enhancing plant knowledge, fostering a connection with nature, and encouraging sustainable healthcare practices (Figs. 4, 5 and 6).

5 Conclusion

In conclusion, the implementation of Plant Reco, an automated medicinal plant recognition system using the MobileNetV2 model, has proven to be highly effective and

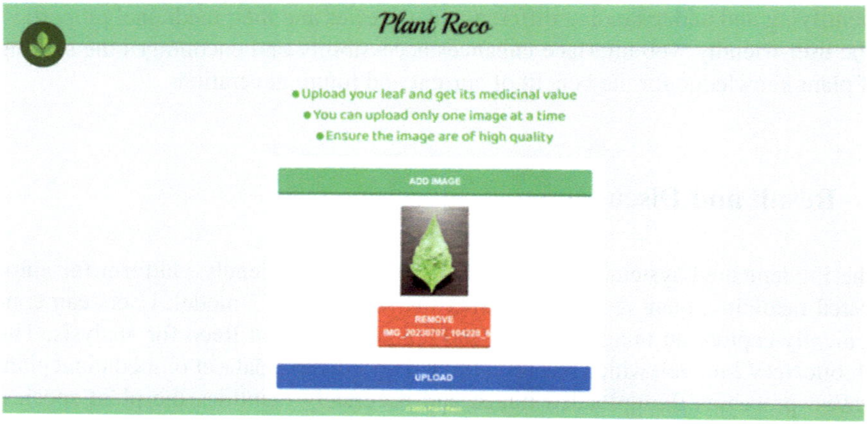

Fig. 4 Image input and data preparation for plant reco

Fig. 5 Recognition output—Plant reco displaying common name, scientific name, brief description, and medicinal usage of recognized plant

valuable. The system offers a user-friendly platform for users to upload leaf images and receive accurate and insightful information about the identified plant species. Through the incorporation of the MobileNetV2 model and feature extraction techniques, Plant Reco successfully recognizes a wide range of medicinal plants based on their leaf features. The high average accuracy achieved for specific plant species, such as bitter bush, black pepper, kokum, and officinal breynia, demonstrates the model's robustness and capability to differentiate between different plant leaves.

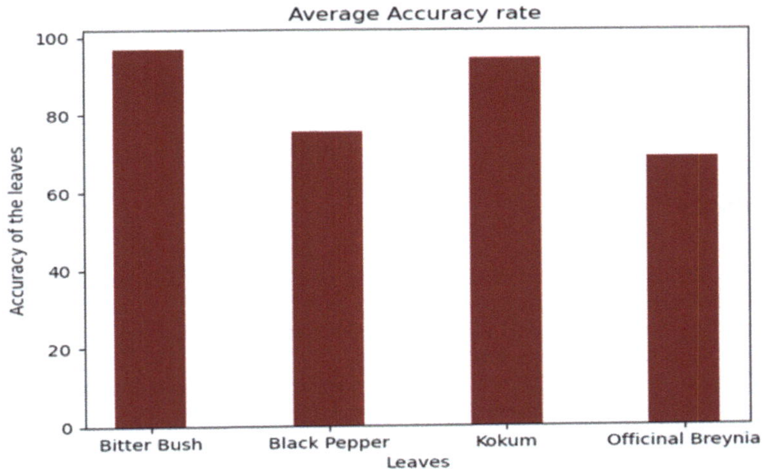

Fig. 6 Average model performance for selected plant species

The system's ability to provide not only the common name and scientific name but also a brief description and medicinal usage of the recognized plant further enhances its utility. Users can gain valuable knowledge about traditional healing practices and the potential health benefits associated with each plant species. The integration of a web interface ensures accessibility and encourages widespread usage of the system, enabling individuals from diverse backgrounds to engage with plant recognition and deepen their understanding of the natural world. Overall, Plant Reco serves as an important tool in preserving traditional plant knowledge, promoting sustainable healthcare practices, and fostering a stronger connection between humans and nature. By empowering users with valuable insights into the vast world of medicinal plants, Plant Reco contributes to the dissemination of crucial plant knowledge to current and future generations. The successful implementation and results of Plant Reco underscore the significance of combining advanced technology, such as the MobileNetV2 model, with traditional wisdom in the field of medicinal plant recognition. As we continue to explore the potential of computer vision and machine learning in healthcare and environmental preservation, Plant Reco stands as a promising example of how these advancements can contribute to a healthier and more harmonious coexistence with nature. Collaborating with botanical gardens, universities, and experts is essential. By involving these stakeholders, the research becomes more robust, credible, and aligned with the goals of plant conservation and education. It also helps establish connections with organizations that have a vested interest in the well-being of plant species and the environment.

Consent for Publication

This is my own work need not require any consent for publication.

References

1. Huixian J (2020) The analysis of plants image recognition based on deep learning and artificial neural network. IEEE Access 8:68828–68841
2. Li L, Zhang S, Wang B (2021) Plant disease detection and classification by deep learning-a review. IEEE Access 9:56683–56698
3. Fei Y, Li Z, Zhu T, Ni C (2023) A lightweight attention-based convolutional neural networks for fresh-cut flower classification. IEEE Access 11:17283–17293
4. Kumar M, Gupta S, Zeng Gao X, Singh A (2019) Plant species recognition using morphological features and adaptive boosting methodology. IEEE Access7: 163912–163918
5. Bakhshipour Amir (2021) Cascading feature filtering and boosting algorithm for plant type classification based on image features. IEEE Access 9:82021–82030
6. Nhan Pham T, Van Tran L, Tan Dao SV (2020) Early disease classification of mango leaves using feed-forward neural network and hybrid metaheuristic feature selection. IEEE Access 8:189960–189973
7. Sun J, Yang Y, He X, Xiaokun W (2020) Northern maize leaf blight detection under complex field environment based on deep learning. IEEE Access 8:33679–33688
8. Prakash Singh A, Sahu P, Chug A, Singh D (2022) A systematic literature review of machine learning techniques deployed in agriculture: a case study of banana crop. IEEE Access 10:87333–87360
9. Kumar Vishnoi V, Kumar K, Kumar B, Mohan S, Ali Khan A (2022) Detection of apple plant diseases using leaf images through convolutional neural network. IEEE Access 11:6594–6609
10. Pratap Singh U, Singh Chouhan S, Jain S, Jain S (2019) Multilayer convolution neural network for the classification of mango leaves infected by anthracnose disease. IEEE Access 7:43721–43729
11. Zhang Z, Gao Q, Liu L, He Y (2023) A high-quality rice leaf disease image data augmentation method based on a dual GAN. IEEE Access 11:21176–21191
12. Yuan Y, Zhi X, Guoqiang L (2021) SPEDCCNN: spatial pyramid-oriented encoder-decoder cascade convolution neural network for crop disease leaf segmentation. IEEE Access 9:14849–14866
13. Barburiceanu S, Meza S, Orza B, Malutan R, Terebes R (2021) Convolutional neural networks for texture feature extraction. Applications to leaf disease classification in precision agriculture. IEEE Access 9:160085–160103
14. Samuel OS (2022) Early cocoa blackpod pathogen prediction with machine learning ensemble algorithm based on climatic parameters. Journal of Information and Organizational Sciences 46(1):1–14
15. Saleem MH, Potgieter J, Mahmood Arif K (2022) A performance-optimized deep learning-based plant disease detection approach for horticultural crops of New Zealand. IEEE Access 10:89798–89822

Mental Health in the Digital Era-NLP Models for Depression and Suicidal Tendency Detection

C. Rohan and V. M. Sapna

Abstract The advent of Natural Language Processing (NLP) as a novel technology, especially with the integration of the BERT (Bidirectional Encoder Representations from Transformers) model, presents a transformative approach in addressing mental health issues in today's digital landscape. BERT's advanced capabilities in understanding the context and nuances of language enable more accurate detection of mental health concerns, including depression and suicidal tendencies, in user-generated text. This exploration presents an in-depth exploration of a cutting-edge NLP-based model designed to detect signs of mental health issues, depression, and suicidal tendencies in user-generated text. The authors emphasize the importance of leveraging the advancements in AI and machine learning for early detection of mental health issues. We introduce a novel approach using NLP models, distinguishing it from the traditional machine-learning models that are commonly employed. The end goal is to facilitate better intervention methods, offering a more comprehensive view of an individual's mental well-being, especially when a vast majority of communication has shifted to digital platforms.

Keywords Mental health · Natural language processing · Suicidal tendencies · Computational linguistics · Contextual language processing

1 Introduction

Mental health, an integral component of overall well-being, has garnered increasing attention, especially in the digital era. The proliferation of digital platforms for communication has opened new channels for understanding and addressing mental health issues. Social media and online forums, where individuals express their

C. Rohan (✉) · V. M. Sapna
PES University, Bengaluru, KA 560085, India
e-mail: chandrashekar.rohans@gmail.com

V. M. Sapna
e-mail: sapnavm@pes.edu

S. M. Thampi et al. (eds.), *Fifth International Conference on Computing and Network Communications*, Lecture Notes in Electrical Engineering 1221,
https://doi.org/10.1007/978-981-97-4711-5_25

emotions and experiences, provide a rich source of data for detecting mental health concerns such as depression and suicidal tendencies.

Depression, a major contributor to the global burden of disease, affects millions worldwide. This underscores the need for effective detection and intervention strategies. Traditional methods of mental health assessment, based primarily on patient-clinician interactions and self-reporting, face limitations in scalability and timeliness, often compounded by stigma associated with mental health conditions.

In response to these challenges, Natural Language Processing (NLP) offers a promising solution. NLP, which lies at the intersection of computer science, artificial intelligence, and linguistics, has seen significant advancements due to breakthroughs in machine learning and deep learning. These technologies enable the nuanced processing and understanding of human language, making NLP a powerful tool for mental health applications.

Our research focuses on developing an NLP-based model for detecting signs of depression and suicidal tendencies in user-generated text. This approach leverages the advancements in AI and machine learning to analyze textual data on digital platforms, identifying patterns indicative of mental distress. As highlighted by Haidt and Sharma et al. [1, 2], and other studies, digital communication provides valuable insights into an individual's mental state, facilitating early intervention.

This paper will explore the unique aspects of our NLP approach, its technical foundations, and the implications of our findings for mental health care in the digital era. We aim to contribute to the evolving landscape of mental health interventions, providing new tools and perspectives for tackling mental health challenges in today's digitally interconnected world.

2 Novelty

In the pursuit of innovating mental health diagnostics, the incorporation of Natural Language Processing (NLP) marks a significant shift from traditional methodologies. Unlike conventional approaches that often rely on overt, explicit expressions in text, NLP enables a deeper, more nuanced analysis. This novel approach recognizes the complexity of human language and emotional expression, particularly in online communications.

This not only enhances the accuracy of detection but also broadens the scope of what can be discerned from digital communications. The adoption of this technology represents a paradigm shift in mental health diagnostics, promising more effective, timely, and empathetic understanding of individuals' mental states, a critical advancement in the realm of digital era healthcare.

2.1 Traditional Versus NLP Approaches

The assessment of suicide risk through online postings has been explored by Saab et al. [3], highlighting the power of computational models in this domain. The combination of ubiquitous sensors and machine learning offers a promising avenue for understanding mental health, as discussed by Castillo et al. [4].

However, traditional methods in the realm of mental health detection through textual data have primarily relied on keyword-based approaches or basic pattern recognition. These methods would identify specific words or phrases commonly associated with mental distress or depressive sentiments, and then categorize the text based on the presence or absence of these indicators. For instance, detecting words such as "sad", "lonely", or "hopeless" might flag a piece of text as potentially indicative of depression.

2.2 Limitations of Traditional Techniques

Lack of Contextual Understanding: Traditional techniques often failed to understand the nuanced and contextual nature of language. A statement like "I'm feeling blue today because of the weather" might be inaccurately flagged due to the presence of the word "blue", even though the sentiment is light-hearted.

Over-Reliance on Keywords: Merely depending on a set of predefined keywords could lead to a significant number of false positives or negatives. Language, especially in the context of mental health, is deeply personal and can vary widely among individuals.

Inability to Detect Implicit Expressions: Often, individuals might express their feelings or distress implicitly, without using overtly negative words. Traditional methods could miss out on these subtle indicators.

2.3 Uniqueness of The NLP Approach

The NLP-based approach, as employed in this research, addresses the aforementioned limitations by offering a more in-depth, contextual understanding of textual data. NLP models, especially advanced ones like BERT, are trained on vast amounts of text, enabling them to grasp the intricate nuances and semantics of language.

Contextual Understanding: BERT, which stands for Bidirectional Encoder Representations from Transformers, is particularly known for its bidirectional understanding of text. This means it doesn't just look at words in isolation but considers the entire sentence or even paragraph to infer meaning.

Reduction in False Positives: With a richer understanding of context, the NLP approach can significantly reduce false positives, as highlighted in the comparison metrics.

Detection of Implicit Expressions: Advanced NLP models can pick up on patterns that might not be immediately obvious, recognizing implicit signs of distress or mental health concerns that traditional methods might overlook.

In essence, the NLP-based model introduced in this research not only offers superior performance metrics but also provides a more comprehensive, nuanced, and human-like understanding of textual data, making it a powerful tool in the early detection of mental health issues in the digital era.

3 Dataset

This section delves into the technical aspects of our dataset, which forms the backbone of our NLP model's effectiveness. We carefully detail the process of selecting, preparing, and structuring our dataset to ensure it's well-suited for identifying mental health indicators like depression and suicidal tendencies. The accuracy and reliability of our model largely depend on the quality and composition of this data. We'll explain how we balanced the dataset and the rationale behind our choice of data sources, providing a clear, straightforward insight into these critical preparatory steps. This focus on dataset integrity is essential for the success of our NLP application in mental health analysis.

3.1 Description and Preparation

The dataset utilized in this research is a collection of 27,977 text entries. A sample of which can be seen in Table 1. These text entries are from a variety of social media platforms mainly Twitter and Reddit. Liu et al. [5] showcased techniques to recognize signs of depression on Twitter, further solidifying the platform's utility in mental health research. Kim et al. [6] pioneered efforts in predicting depression using social media data, emphasizing the potential of such platforms. Platforms like Instagram have also been used to identify markers of depression, as discussed by Reece et al. [7]. Each entry in the dataset corresponds to a message or post from individuals discussing a range of topics, primarily centered around mental health, personal experiences, and related subjects. The dataset is structured with two primary columns:

Text: This column encapsulates the actual content of the message. The text entries vary in length and detail but provide a rich source of information that is crucial for our analysis.

Table 1 A sample table illustrating the shape of the dataset

Text (first 50 characters)	Label
Dear American teens question dutch person heard gu...	0
Nothing look forward life I dont many reasons keep ...	1
Music recommendations im looking expand playlist u...	0
Im done trying feel better the reason im still alive...	1
Worried year old girl subject domestic physical me...	1

Label: The dataset employs a binary labeling system. The label 0 signifies messages that are neutral or non-critical in nature. On the other hand, a label of 1 is used to tag messages that indicate potential mental health distress or concern. This labeling enables us to classify and analyze the severity or criticality of the messages effectively. Sarsam et al. [8] delved into the machine classification of suicide-related communication on platforms like Twitter.

In terms of distribution, the dataset is fairly balanced. Approximately 50.54% of the messages are labeled as 0, while 49.46% carry a label of 1. This balanced distribution is beneficial for our analysis as it prevents any significant bias toward a particular label, ensuring that our models and algorithms operate on a representative sample of data.

3.2 The BERT Model

BERT, which stands for Bidirectional Encoder Representations from Transformers, is a state-of-the-art language representation model (as shown in Fig. 1). Introduced by Kenton et al. [9], BERT has transformed the landscape of natural language processing with its deep bidirectional transformers. It has revolutionized the field of (NLP) due to its superior ability to understand the context of words in a sentence. Instead of looking at words in isolation, BERT examines words concerning their neighbors in both directions, leading to a much richer representation.

Key features of BERT include:

Bidirectionality: Traditional language models, like LSTM or GRU, operate either from left-to-right or right-to-left. In contrast, BERT processes words concerning all other words in the sentence, making it genuinely bidirectional. This feature allows BERT to capture intricate nuances and relationships between words that other models might miss.

Transformers: BERT is built upon the Transformer architecture, which employs self-attention mechanisms to weigh the significance of words in a sentence. This architecture enables BERT to focus more on words that are more relevant in a given context.

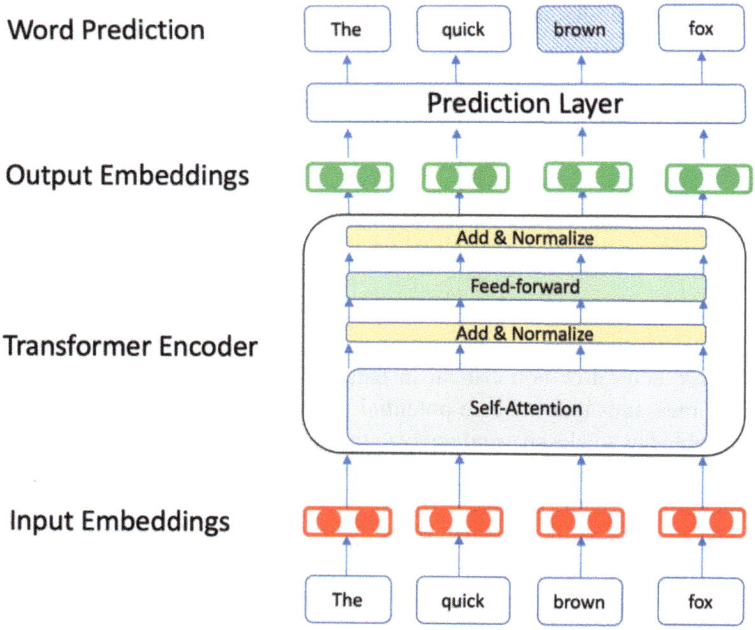

Fig. 1 An illustration of the BERT model

Pre-training on Vast Datasets: BERT is pre-trained on massive amounts of text, including the entire Wikipedia and the BookCorpus dataset. This extensive pre-training allows BERT to develop a comprehensive understanding of language.

Fine-Tuning for Specific Tasks: While BERT is pre-trained on general data, it can be fine-tuned on specific datasets to perform a wide range of tasks, from text classification to question-answering. This adaptability makes BERT incredibly versatile and suitable for diverse NLP challenges. Also, the concept of attention mechanisms, which has been foundational for models like BERT, was detailed by Clark et al. [10].

For our research, BERT offers several advantages. Its deep understanding of context makes it particularly adept at discerning subtle indications of mental health distress in text messages. Furthermore, BERT's capability to capture long-range dependencies in texts helps in analyzing longer messages where crucial information might be spread out.

4 Results and Comparison

In this section, we present the results obtained from our experiments, along with comparisons to benchmark models and previous research.

4.1 Experimental Setup

All experiments were conducted under a consistent environment to ensure fairness in comparison. We employed a train-test split to segregate our dataset, with 80% used for training and the remaining 20% for testing. The models were trained using a consistent set of hyperparameters, and the results were averaged over multiple runs to account for any variability.

4.2 Model Performance

Our primary model, BERT, exhibited promising results. Its capability to understand context and capture intricate relationships in the text was evident from its high accuracy and F1-score (as shown in Fig. 2). The confusion matrix revealed that BERT was particularly adept at minimizing false negatives, which is crucial given the sensitive nature of our task.

In comparison, traditional models like LSTM and GRU performed reasonably but lagged behind BERT in terms of precision and recall. This disparity underscores the importance of bidirectional context understanding, a feature inherent to BERT.

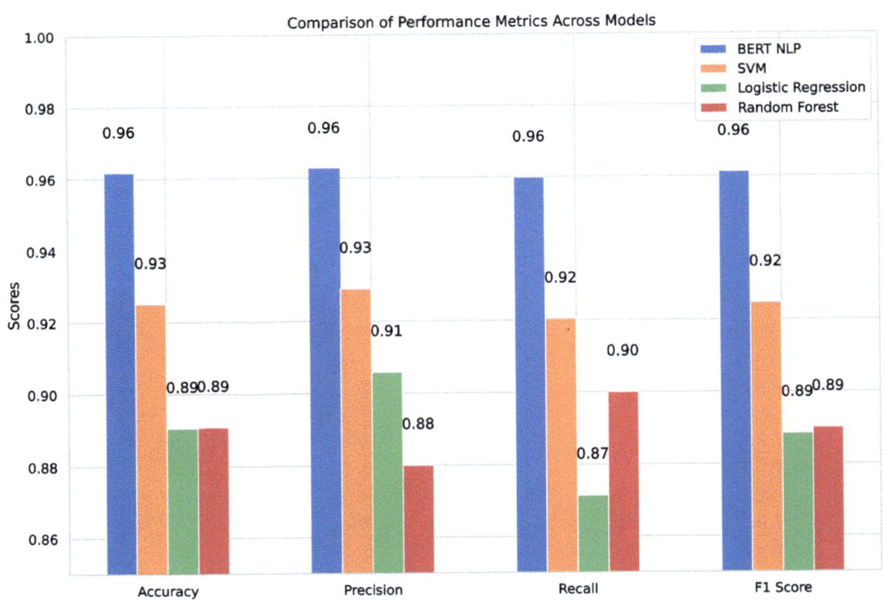

Fig. 2 This comparison plot illustrate the effectiveness of our NLP-Based approach by showcasing several key performance metrics, including accuracy, precision, recall, and F1-score

4.3 Comparison with Existing Work

Compared to previous research in this domain, our approach stands out in its use of advanced NLP models and a balanced dataset. Earlier studies have often shown to have failed to understand the context of the user's text and as a result misinterpret their mental health status. Bayesian techniques, as explored by Ji et al. [11], have also been employed to discover patterns of mental disorders in online forums. While acknowledging that previous research work has provided great insights. The NLP-BERT model seems to have an indisputable edge over the traditional Machine Learning (ML) models.

5 User Interface and Real-Time Application

To make the model accessible and user-friendly, a rudimentary GUI has been designed. Users can input a sample text, which is then analyzed to provide a risk score (refer to Figs. 3 and 4). The model also checks for specific trigger words, flagging them as potential warning signs. This instant feedback can be immensely useful for mental health professionals, allowing them to understand their patients' state rapidly.

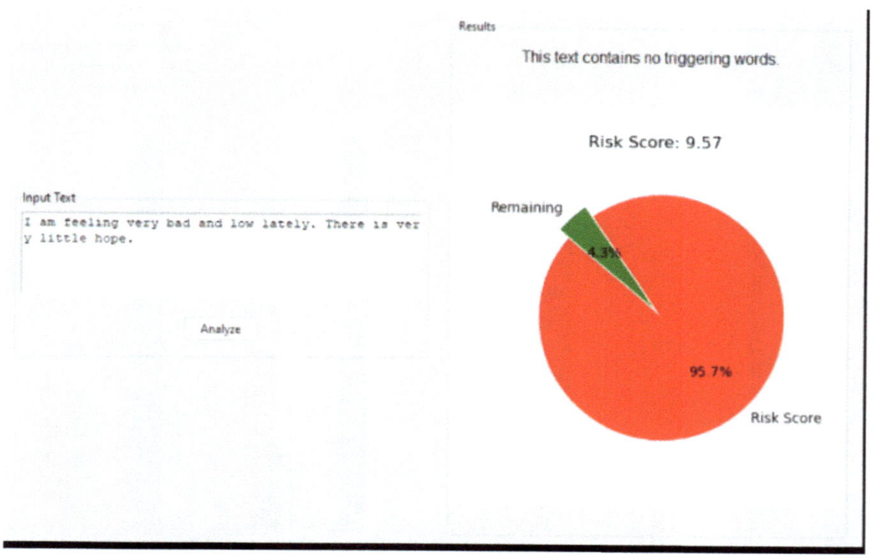

Fig. 3 GUI displaying a text of a user as "At Risk". The model is able to identify that the user's text shows signs of mental health distress

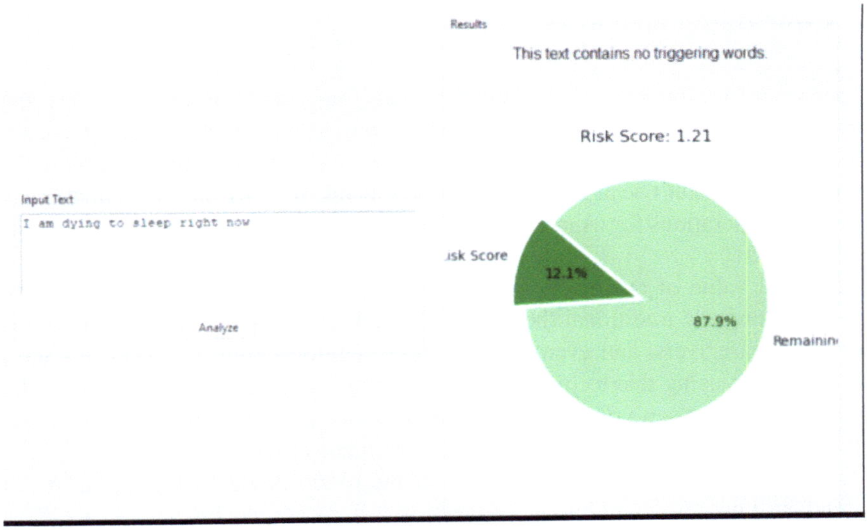

Fig. 4 GUI Displaying text of a user "Not at Risk". The model can clearly discern that even though the word 'dying' is present, when viewed in context, it does not show signs of mental distress

6 Future scope and Potential

The field of mental health diagnosis is always changing and we're about to see some big improvements. This new model is a great example of the exciting changes coming up. As we plan for the future, there are many features we can add to make our current system even better.

One of the key changes is including a person's past messages and when they sent them. This can give us a better understanding when and why a person's feelings might change. By looking at past messages, we can see patterns over long periods of time. This gives us a better picture of someone's emotional changes and helps us notice things that might be easy to miss otherwise.

In addition, there's a lot of data that we haven't used yet which can be very helpful. Forums have been another medium where NLP models, as mentioned by Shrestha et al. [12], have been applied to assess depression and self-harm risks. By using larger datasets, our model can become more versatile. Parapar et al. [13] discussed the importance of early risk prediction on the internet, emphasizing the need for timely interventions. This can lead to offering help sooner and supporting people in a timely way.

7 Concluding Remarks

In our work to improve mental health diagnosis, our new NLP model is a big step forward. However, it's important to see it as a part of a bigger picture in mental health care. Even though our model shows how far we have come in using technology for mental health, it can't be the only solution we rely on. We must also keep in mind the ethical considerations for using social media data as described by Garg and Dhiman et al. [14, 15].

The real value of our model is twofold: it's a sign of new tech advances and it works alongside traditional therapy methods. It can be used in real-time to give therapists, caregivers, and even police important information. However, it's up to us, the people using this technology, to use it wisely. It's important to know what the model can and can't do. While it can give advice and share information, it can't replace the deep understanding and care that humans provide.

In the end, our NLP model shows what the future might look like, but it also reminds us how important human connection is in mental health care. As we move ahead, our model should work alongside human therapists, not replace them, adding to the care and understanding they provide.

References

1. Haidt J, Allen N (2020) Scrutinizing the effects of digital technology on mental health
2. Sharma MK, John N, Sahu M (2020) Influence of social media on mental health: a systematic review. Curr Opin Psychiatry 33(5):467–475
3. Saab MM, Murphy M, Meehan E, Dillon CB, O'Connell S, Hegarty J, Heffernan S, Greaney S, Kilty C, Goodwin J, Hartigan I (2022) Suicide and self-harm risk assessment: a systematic review of prospective research. Arch Suicide Res 26(4):1645–1665
4. Castillo-Sánchez G, Marques G, Dorronzoro E, Rivera-Romero O, Franco-Martín M, De la Torre-Díez I (2020) Suicide risk assessment using machine learning and social networks: a scoping review. J Med Syst 44(12):205
5. Liu D, Feng XL, Ahmed F, Shahid M, Guo J (2022) Detecting and measuring depression on social media using a machine learning approach: systematic review. JMIR Mental Health 9(3):e27244
6. Kim J, Lee J, Park E, Han J (2020) A deep learning model for detecting mental illness from user content on social media. Sci Rep 10(1):11846
7. Reece AG, Danforth CM (2017) Instagram photos reveal predictive markers of depression. EPJ Data Sci 6(1):15
8. Sarsam SM, Al-Samarraie H, Alzahrani AI, Alnumay W, Smith AP (2021) A lexicon-based approach to detecting suicide-related messages on Twitter. Biomed Signal Process Control 65:102355
9. Kenton JDMWC, Toutanova LK (2019) BERT: pre-training of deep bidirectional transformers for language understanding. In Proceedings of NAACL-HLT, vol 1, p 2
10. Clark K, Khandelwal U, Levy O, Manning CD (2019) What does BERT look at? an analysis of BERT's attention. arXiv:1906.04341
11. Ji M, Xie W, Zhao M, Qian X, Chow CY, Lam KY, Yan J, Hao T (2022) Probabilistic prediction of nonadherence to psychiatric disorder medication from mental health forum data: developing and validating Bayesian machine learning classifiers. Comput Intell Neurosci

12. Shrestha A, Serra E, Spezzano F (2020) Multi-modal social and psycho-linguistic embedding via recurrent neural networks to identify depressed users in online forums. Netw Model Anal Health Inf Bioinf 9:1–11
13. Parapar J, Martín-Rodilla P, Losada DE, Crestani F (2023) Overview of eRisk 2023: early risk prediction on the internet. In: International Conference of the Cross-Language Evaluation Forum for European Languages. Springer, pp 294–315
14. Garg M (2023) Mental health analysis in social media posts: a survey. Archiv Comput Methods in Eng 30(3):1819–1842
15. Dhiman DB (2023) Ethical issues and challenges in social media: a current scenario. Available at SSRN 4406610

Lung Cancer Detection Using Convolutional Neural Network Models

Aditi Mishra, Uppari Sai Bhargavi, Sudhanshu Gonge, Rahul Joshi, Vijaykumar Jatti, and Ketan Kotecha

Abstract Lung cancer is one of the world's major causes of death. Lung cancer detection and diagnosis are critical for optimal therapy and patient survival. Deep learning methods for the automatic diagnosis and categorization of lung cancer based on medical pictures have gained popularity in recent years. The purpose of this research is to compare the performance of various algorithms for lung cancer classification using chest CT scan pictures. The collection includes CT scan images of four cancer types: adenocarcinoma, large cell carcinoma, squamous cell carcinoma, and normal. Convolutional neural network (CNN), ResNet50, VGG16, and Inception-v3 are among the algorithms employed in this project. The accuracy of each model in categorizing photos into their appropriate classes is used to evaluate its performance, with ResNet50 having the highest accuracy of 84.76%. The findings of this study could shed light on the efficacy of various deep learning algorithms for lung cancer

A. Mishra (✉) · U. S. Bhargavi · S. Gonge · R. Joshi · K. Kotecha
Department of Computer Science and Engineering, Symbiosis International (Deemed University), Pune, India
e-mail: aditi.mishra.btech2019@sitpune.edu.in

U. S. Bhargavi
e-mail: uppari.bhargavi.btech2019@sitpune.edu.in

S. Gonge
e-mail: sudhanshu.gonge@sitpune.edu.in

R. Joshi
e-mail: rahulj@sitpune.edu.in

K. Kotecha
e-mail: head@scaai.siu.edu.in

V. Jatti
Department of Mechanical Engineering, Symbiosis International (Deemed University), Pune, India
e-mail: vijaykumar.jatti@sitpune.edu.in

A. Mishra · U. S. Bhargavi · S. Gonge · R. Joshi · V. Jatti · K. Kotecha
Symbiosis Institute of Technology, Symbiosis International (Deemed University), Pune, India

S. M. Thampi et al. (eds.), *Fifth International Conference on Computing and Network Communications*, Lecture Notes in Electrical Engineering 1221,
https://doi.org/10.1007/978-981-97-4711-5_26

classification and contribute to the creation of more accurate and efficient lung cancer diagnosis tools.

Keywords Lung cancer · VGG16 · CNN · Inception-v3 · ResNet50

1 Introduction

Lung cancer is the most common and lethal cancer worldwide, accounting for the majority of cancer-related fatalities in both men and women. Lung cancer identification and diagnosis are crucial for effective treatment and improved patient outcomes. Several medical imaging procedures, such as chest CT scans, have established the gold standard for identifying and diagnosing lung cancer over the years. There has been an increase in interest in employing deep learning algorithms to aid in the diagnosis of lung cancer based on medical pictures in recent years. Deep learning algorithms are capable of accurately analyzing and interpreting medical images, and they have demonstrated encouraging results in detecting and categorizing lung cancer from medical imaging.

Some of the most extensively used algorithms for the categorization of lung cancer using medical pictures include convolutional neural networks (CNNs), ResNet50, VGG16, and Inception-v3. To effectively categorize photos into multiple classes, these algorithms use a variety of approaches such as feature extraction, pattern recognition, and deep learning. Despite the potential for deep learning algorithms to improve lung cancer diagnosis and therapy, their effectiveness varies depending on the dataset and algorithm parameters. As a result, a detailed evaluation of these algorithms on a specific dataset is required to assess their usefulness and compare their performance.

The goal of this work is to compare the performance of many deep learning algorithms, including CNN, ResNet50, VGG16, and Inception-v3, for lung cancer classification using chest CT scan pictures. We specifically assess the effectiveness of these algorithms using a dataset of CT scan pictures from four different cancer classes: adenocarcinoma, large cell carcinoma, squamous cell carcinoma, and normal. The performance of these algorithms is then compared to get insight into their usefulness in classifying lung cancer images and lead to the creation of more accurate and efficient lung cancer diagnosis tools.

2 Literature Review

Lam et al. (2016) in [1] provide an overview of lung cancer epidemiology, risk factors, and prevention. They review the evidence for primary and secondary prevention strategies for lung cancer, including smoking cessation, screening, and chemoprevention. Pao et al. (2011) in [2] review new driver mutations in non-small cell lung cancer (NSCLC) and their potential implications for personalized therapy along with discussing the discovery of several key genetic alterations in NSCLC, including mutations in the epidermal growth factor receptor (EGFR), KRAS, and EML4-ALK fusion genes, among others. Li et al. (2018) in [3] explore the potential applications of machine learning (ML) in the prediction of lung cancer. They discuss the challenges associated with developing ML models for lung cancer prediction, such as data quality and quantity issues, and highlight the importance of developing robust and reliable models for clinical decision-making. Li et.al. (2016) in [4] propose the use of deep convolutional neural networks (DCNNs) for lung cancer detection by describing the development and testing of a DCNN model trained on a dataset of CT images, with the aim of distinguishing between lung cancer and non-cancerous nodules. They report promising results, with the DCNN achieving an accuracy of 83.8% in identifying cancerous nodules, outperforming other commonly used machine learning algorithms. Salama et al. (2020) in [5] propose a novel approach for lung cancer detection and classification using the VGG16 convolutional neural network architecture and transfer learning. The authors describe the use of a dataset of CT images, preprocessed using data augmentation techniques, and report an accuracy of 97.22% in classifying lung nodules as either malignant or benign, outperforming other machine learning algorithms. Yao et al. (2021) in [6] propose the use of the Inception-v3 convolutional neural network architecture and transfer learning for prostate cancer detection by describing the use of a dataset of histopathology images, preprocessed using data augmentation techniques, to train and test the model. They report an accuracy of 91.20% in classifying prostate cancer images, outperforming other machine learning algorithms.

A ResNet50 convolutional neural network is proposed by Al-Mahmoud et al. (2021) for classifying lung cancer in [7] which involves preprocessing the lung cancer dataset, followed by training and testing the ResNet50 model, achieving an accuracy of 94.9% using their proposed method, they conclude that the proposed method has the potential to aid in the early detection and diagnosis of lung cancer.

Kim et al. (2017) in [8] provide an overview of targeted therapy for lung cancer, with a focus on immune checkpoint inhibitors (ICIs) where they discuss the various types of ICIs and their mechanisms of action, as well as their efficacy and safety profiles. Lilenbaum et al. (2019) in [9] provide an overview of recent advances in the treatment of lung cancer, i.e., the evolution of treatment options, from traditional chemotherapy to targeted therapies and immunotherapies, and reviews the current standard of care for different types and stages of lung cancer. Along with discussing emerging therapies, such as antibody–drug conjugates and cancer vaccines, and their potential to improve outcomes for lung cancer patients, it is concluded that the field of lung cancer treatment is rapidly evolving, with the development of new therapies and a growing understanding of the molecular mechanisms underlying the disease.

Zhang et al. (2019) in [10] discuss the different screening methods, including low-dose computed tomography (LDCT), chest X-ray, and biomarker testing, and reviews the evidence for their efficacy in detecting lung cancer at an early stage along with discussing the potential harms of screening, such as false-positive results and over-diagnosis, and the need for careful consideration of the risks and benefits. Wang et al.

(2018) in [11] provide an overview of the current state of non-small cell lung cancer (NSCLC) treatment in China. The authors discuss the various treatment options available, including surgery, chemotherapy, targeted therapy, and immunotherapy, and review the recent progress, challenges, and opportunities in each area. Ding et al. (2020) in [12] review the current state and future potential of artificial intelligence (AI) in the diagnosis and treatment of lung cancer including image analysis, natural language processing, and machine learning, and reviews their applications in different stages of lung cancer care, from screening and diagnosis to treatment planning and monitoring.

3 Proposed Methodology

There are various methods that can be used for lung cancer detection and have been discovered. Some of them include:

a. Convolutional Neural Network (CNN)

A convolutional neural network (CNN or convnet) is a machine learning subset. It is one of several types of artificial neural networks utilized for diverse applications and data types. A CNN is a type of network design for deep learning algorithms that are primarily utilized for image recognition and pixel data processing applications (Fig. 1).

The diagram above depicts a typical CNN architecture. The network's first layer is a convolutional layer that applies a series of filters to the input image. Each filter identifies a different aspect of the image, such as borders or corners. The convolution layer's output is then routed via a pooling layer, which decreases the size of the feature maps and improves network efficiency. This convolution and pooling procedure is

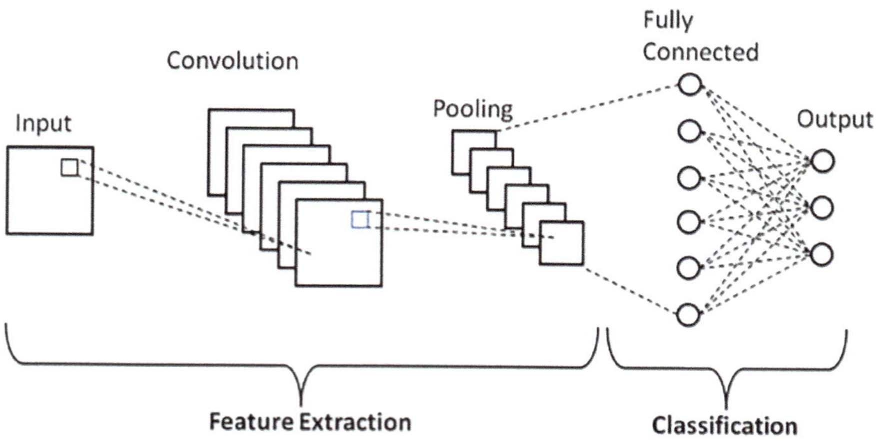

Fig. 1 CNN architecture

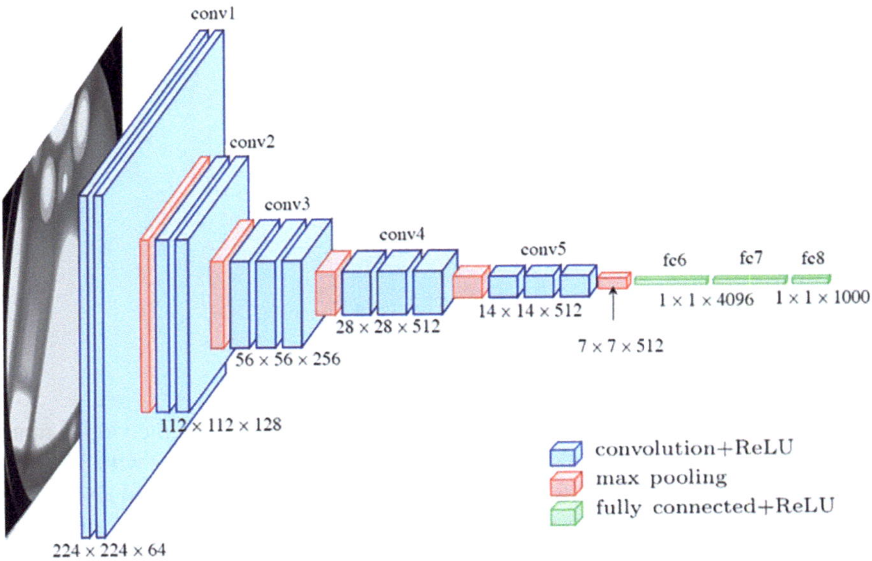

Fig. 2 VGG16 architecture

repeated multiple times, with each layer learning more complicated information. The network's final output is a set of probabilities representing the likelihood of each class.

b. VGG16

VGG16 is a 16-layer convolutional neural network architecture developed by the Visual Geometry Group (VGG) at the University of Oxford. It is notable for its simplicity and obtained state-of-the-art image classification results on the ImageNet dataset in 2014. Several convolutional layers with 3 × 3 filters are followed by max-pooling layers and three fully linked layers in the network architecture (Fig. 2).

The VGG16 architecture is depicted in the diagram above. The network's input is a 224 × 224 RGB image, and the first layer is a convolutional layer with 64 filters and a kernel size of 3 × 3. Following this, numerous more convolutional layers with 3 × 3 filters and max-pooling layers with 2 × 2 kernels are added. The fully linked layers at the network's end each include 4096 neurons and output a probability distribution across the classes.

c. Inception-v3

The diagram above depicts the VGG16 architecture. The input to the network is a 224 × 224 RGB image, and the first layer is a convolutional layer with 64 filters and a 3 × 3 kernel size. Following that, a slew of additional convolutional layers with 3 × 3 filters and max-pooling layers with 2 × 2 kernels are added. At the network's end, the fully linked layers each have 4096 neurons and output a probability distribution across the classes (Fig. 3).

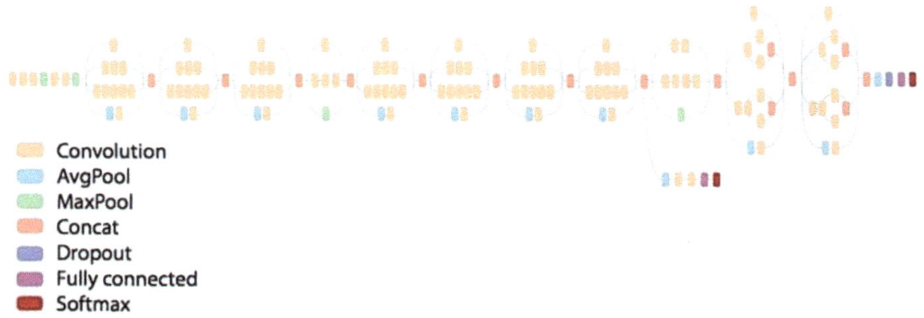

Fig. 3 Inception-v3 architecture

The Inception-v3 architecture is depicted in the diagram above. The network's input is a 299 × 299 RGB image, and the first layer is a convolutional layer with 32 filters and a kernel size of 3 × 3. Following that are several more convolutional layers with varying filter widths and max-pooling layers. The "inception modules" are made up of parallel convolutional layers with varying filter sizes, which are followed by concatenation layers. The network's fully connected layers have 2048 neurons each and produce a probability distribution over the classes.

d. ResNet50

Microsoft launched ResNet50, a deep residual network architecture, in 2015. It is intended to address the issue of vanishing gradients in very deep neural networks. ResNet50 contains 50 layers and includes shortcut connections that skip over several layers during training, allowing gradients to flow more readily. ResNet50 produced cutting-edge results on the ImageNet dataset and has been widely employed in a variety of computer vision tasks (Fig. 4).

The ResNet50 architecture is depicted in the diagram above. The network's input is a 224 × 224 RGB image, and the first layer is a convolutional layer with 64 filters and a kernel size of 7 × 7. Following that are max-pooling layers and numerous additional convolutional layers with varying filter sizes. The shortcut connections are labeled "identity mappings" in the picture, and they allow the gradient to travel directly across the network without going through the intermediate levels. The fully connected layers at the network's end each include 1000 neurons and output a probability distribution over the classes.

The ResNet50 architecture has a number of advantages over other deep neural network architectures. The usage of shortcut links makes it easier to train the network and get higher performance on tough jobs. The architecture is also more efficient than other deep neural network architectures, requiring fewer parameters and taking less time to compute. ResNet50 has been applied to a variety of tasks, including image identification, object detection, and picture segmentation. It's also been developed for use in other fields like speech recognition and natural language processing.

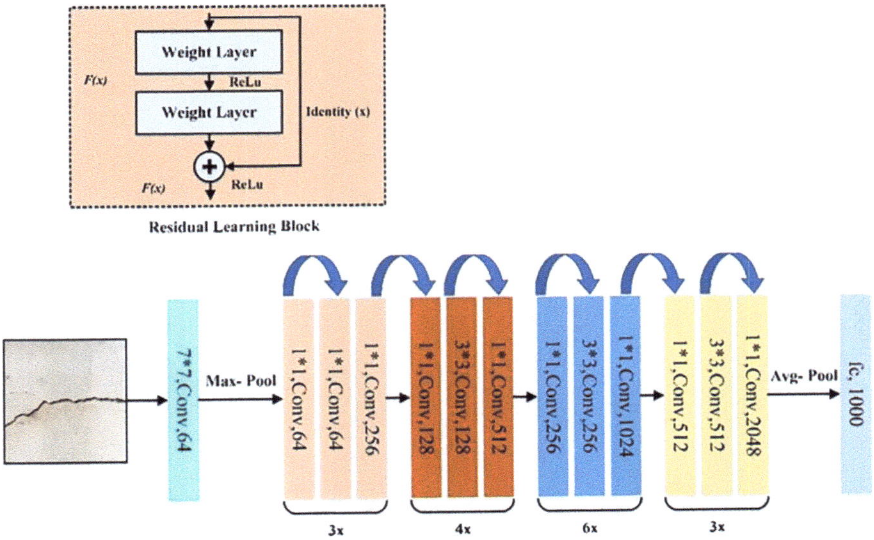

Fig. 4 ResNet50 architecture

4 Implementation

The Keras library and TensorFlow backend are used in the Python programming language to carry out the proposed methodology's implementation. The Kaggle dataset utilized for this study comprises chest CT scan images of individuals with three different forms of lung cancer (adenocarcinoma, large cell carcinoma, and squamous cell carcinoma) as well as normal CT scan images.

Three sets of the dataset—train, test, and validation—are created. Four folders are included in each set, one for each form of lung cancer and one for typical CT scans. There are 613 photos in the train set, 315 in the test set, and 72 in the validation set (Fig. 5).

The graph shows that the number of validation sets in every type is the least and the train dataset has the most number of images.

For the classification challenge, we used convolutional neural networks (CNNs), VGG16, ResNet50, and Inception-v3—four different pre-trained deep learning models. On the provided dataset, the models were improved using transfer learning methods.

The dataset was artificially expanded using image augmentation methods like rotation, zooming, and flipping in order to increase the generalization capacity of the models.

Four convolutional layers made up the CNN model, which was then followed by layers using maximum pooling and two fully connected layers with dropout. The VGG16 model was modified by adding a global average pooling layer, a dense layer with ReLU activation, and a softmax output layer after being first initialized with

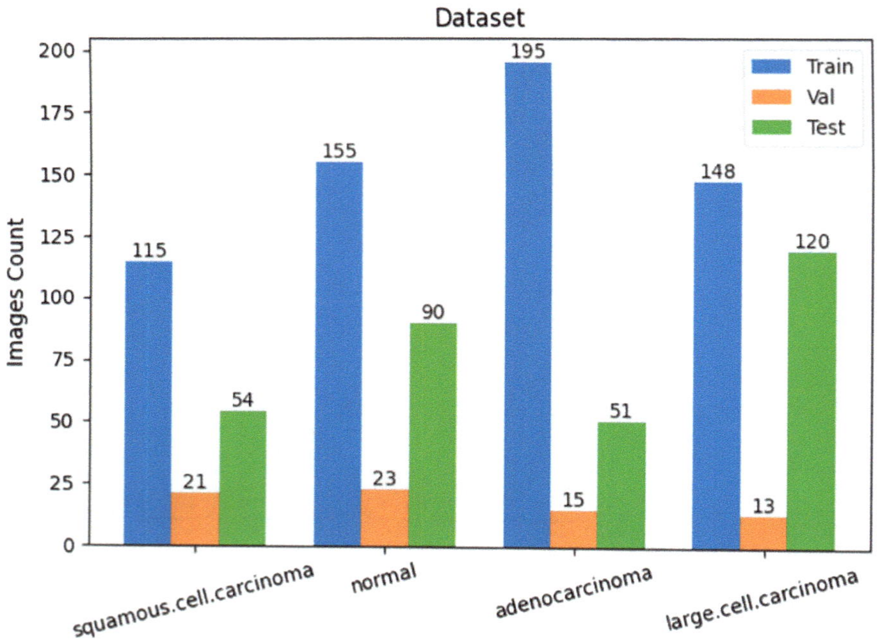

Fig. 5 Graph showing the images count of each type in the dataset

ImageNet weights. By including a global average pooling layer, a dense layer, and a softmax output layer, the ResNet50 model was improved. By including a global average pooling layer, a dense layer, and a softmax output layer, the Inception-v3 model was improved.

With a learning rate of 0.0001 and the categorical cross-entropy loss function as the evaluation metric, all models were trained using the Adam optimizer. Based on the precision of the validation set, the best model was chosen.

5 Results

Our research revealed that ResNet50 had the highest accuracy (84.76%) for classifying lung cancer CT scan images. The accuracy rates for CNN, VGG16, and Inception-v3 were respectively 54.9%, 79.36%, and 81.90%. The performance of many algorithms is thoroughly investigated on the same dataset, which helps researchers and practitioners in the area make decisions on which algorithm to use for a given application.

True Negative **(TN)**: Cases that were accurately labeled as negative (no lung cancer).

***False Positives* (FP)**: Cases that were mistakenly labeled as positives (predicted lung cancer but no lung cancer at all).

***False Negative* (FN)**: Examples of cases that were misclassified as negative (really predicted lung cancer but no lung cancer).

***True Positive* (TP)**: It is an instance where the existence of lung cancer was appropriately categorized as positive. Considering the following to be the case:

The confusion matrix components for each set can be calculated as follows:

For the train set:

TN {train} = \text{Number of instances where Actual train = 0 & Predicted train = 0}.

FP {train} = \text{Number of instances where Actual train = 0 & Predicted train = 1}.

FN {train} = \text{Number of instances where Actual train = 1 & Predicted train = 0}.

TP {train} = \text{Number of instances where Actual train = 1 & Predicted train = 1}.

For the test set:

TN {test} = \text{Number of instances where Actual test = 0 and Predicted test = 0}.

FP{test} = \text{Number of instances where Actual test = 0 and Predicted test = 1}.

FN{test} = \text{Number of instances where Actual test = 1 and Predicted test = 0}.

TP {test} = \text{Number of instances where Actual test = 1 and Predicted test = 1}.

For the validation set:

TN {validation} = \text{Number of instances where Actual validation = 0 and Predicted validation = 0}.

FP {validation} = \text{Number of instances where Actual validation = 0 and Predicted validation = 1}.

FN {validation} = \text{Number of instances where Actual validation = 1 and Predicted validation = 0}.

TP {validation} = \text{Number of instances where Actual validation = 1 and Predicted validation = 1} (Table 1).

For CNN (Figs. 6 and 7):

Table 1 Performance of the different models on the lung cancer classification task

Model	Accuracy (%)
CNN	54.9
VGG16	79.36
ResNet50	84.76
Inception-v3	81.90

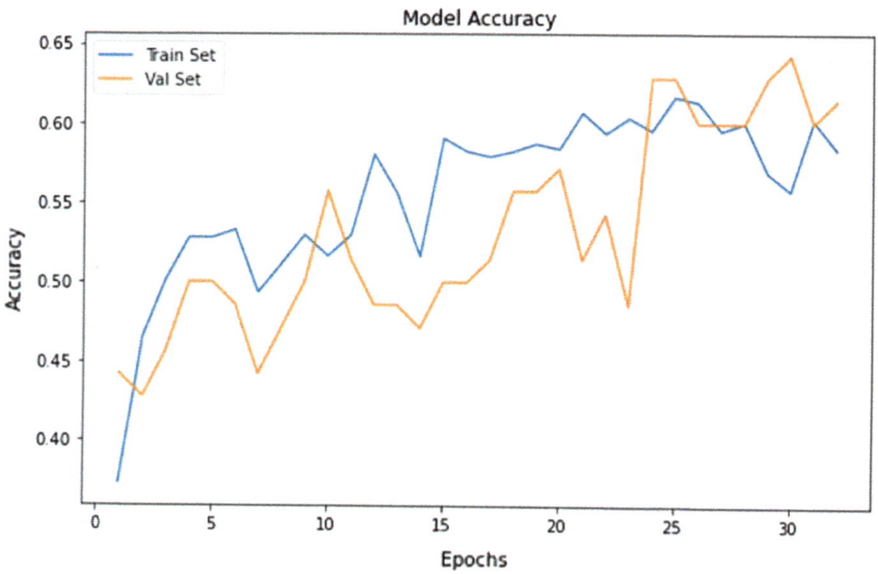

Fig. 6 Accuracy versus epochs graph for model accuracy in CNN

For VGG16 (Figs. 8 and 9):

For Inception-v3 (Figs. 10 and 11):

For ResNet50 (Figs. 12, 13, and 14):

The confusion matrices for each set can be represented as follows, assuming the order is TN, FP, FN, TP:

- **For the training set**: With the values provided: 115, 155, 195, 149

TN_train	FP_train
– – – –	– – – –
FN_train	TP_train

- **For the validation set**: With the values provided: 21, 23, 15, 13

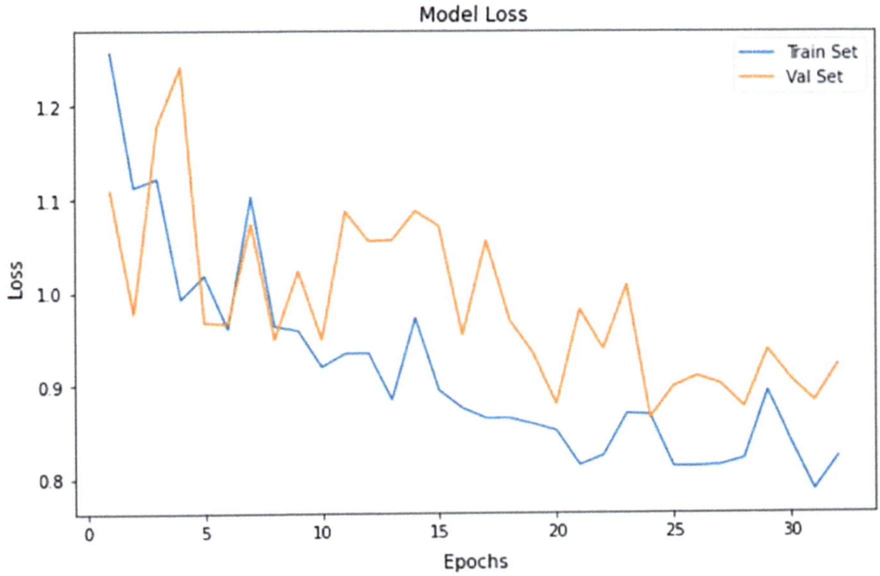

Fig. 7 Loss versus epochs graph for model loss in CNN

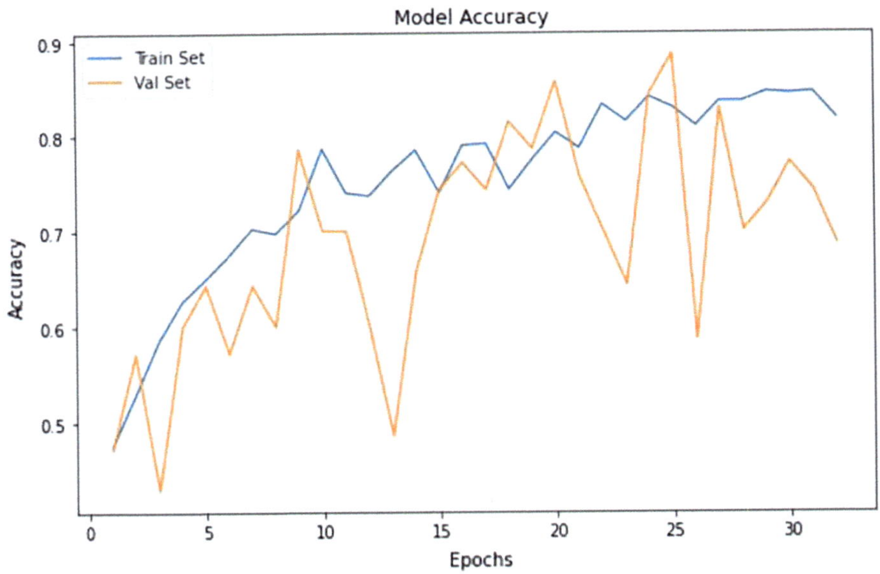

Fig. 8 Accuracy versus epochs graph for model accuracy in VGG16

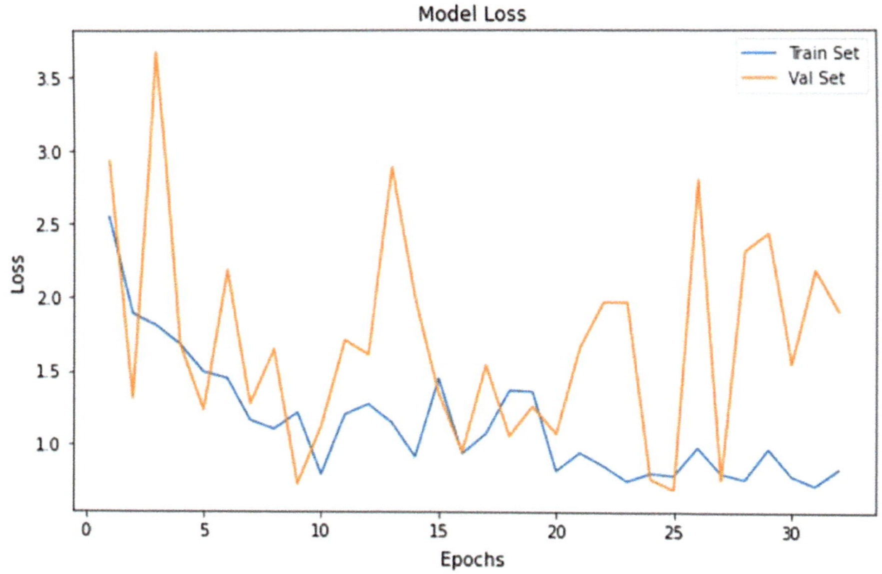

Fig. 9 Loss versus epochs graph for model loss in VGG16

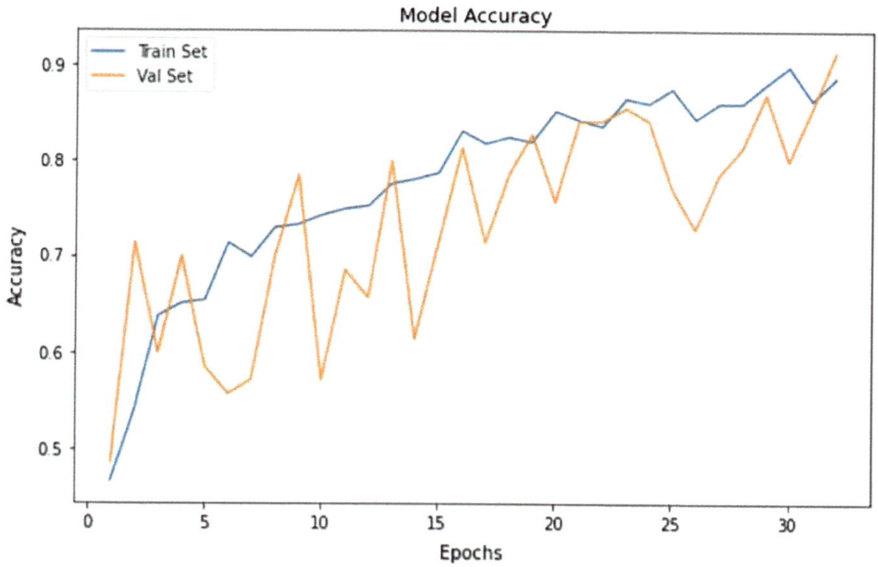

Fig. 10 Accuracy versus epochs graph for model accuracy in Inception-v3

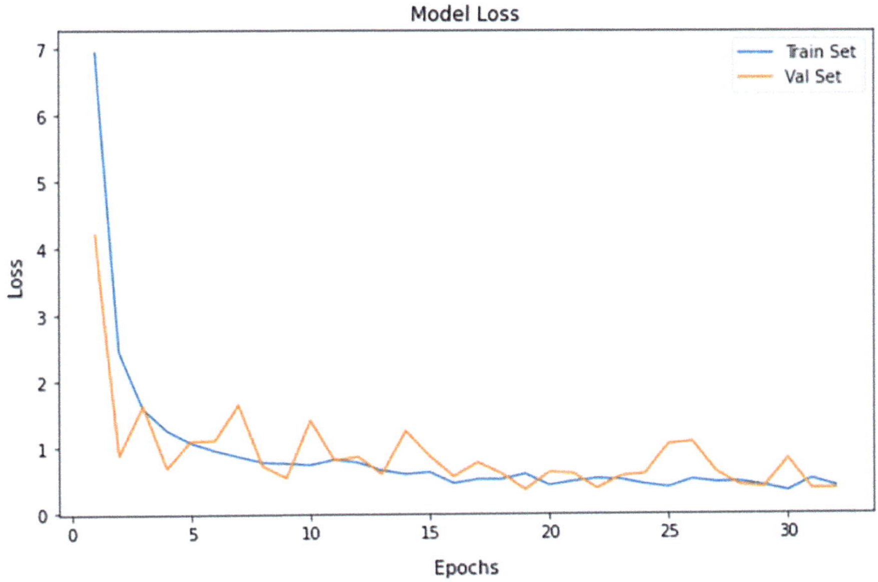

Fig. 11 Loss versus epochs graph for model loss in Inception-v3

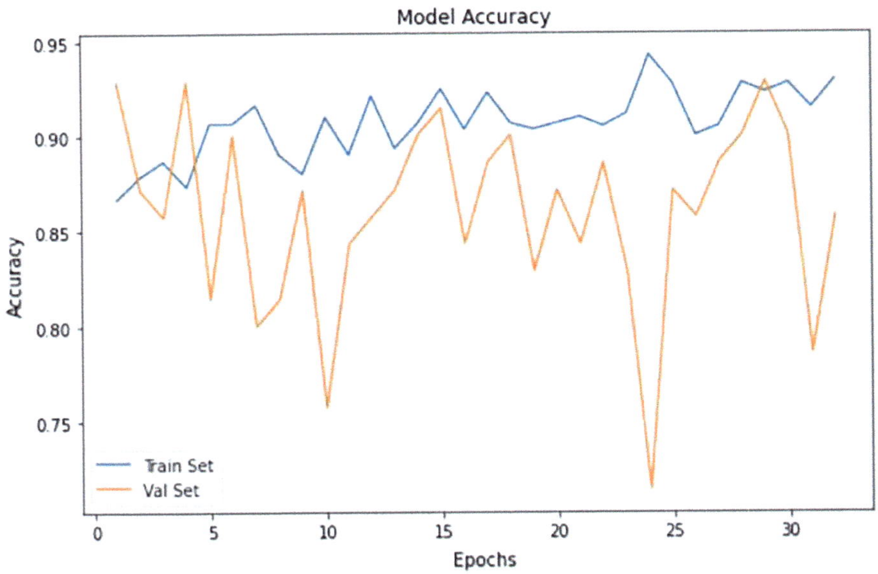

Fig. 12 Accuracy versus epochs graph for model accuracy in ResNet50

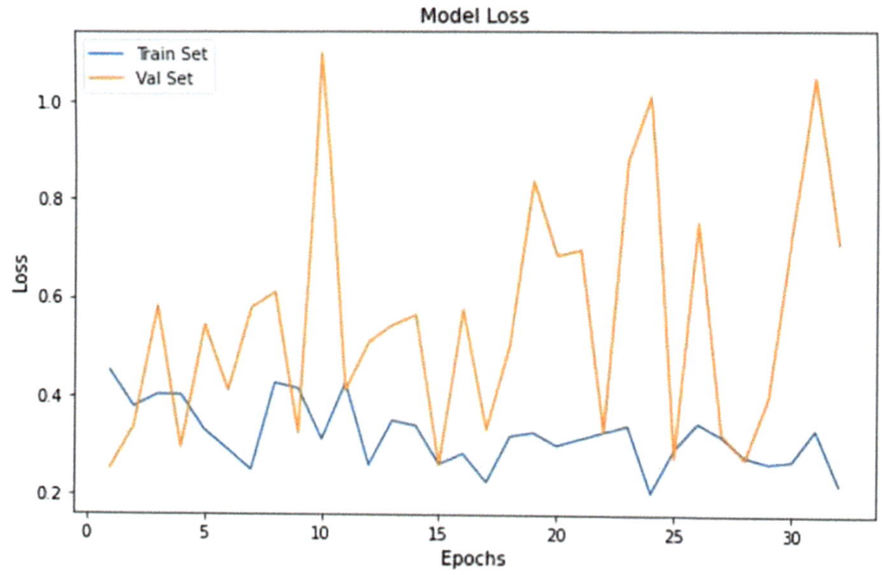

Fig. 13 Loss versus epochs graph for model loss in ResNet50

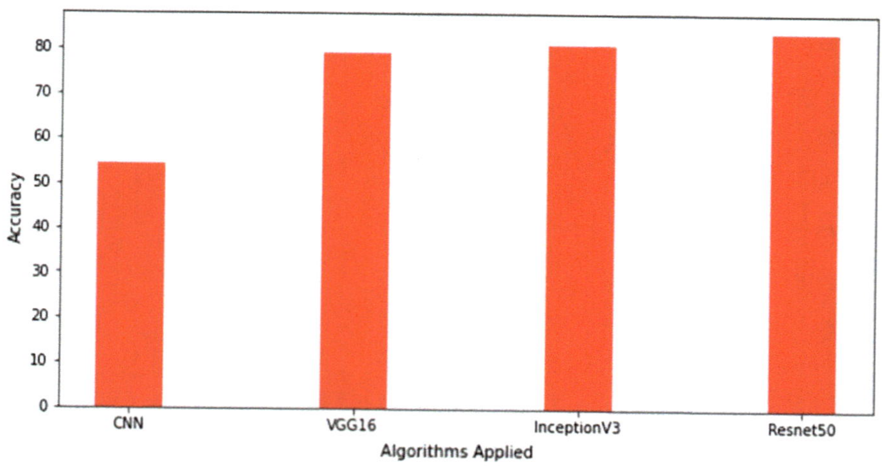

Fig. 14 Graph showing comparison of accuracy in different tested algorithms

TN_validation	FP_validation
−−−−−−−	−−−−−−−
FN_validation	TP_validation

- **For the test set**: With the values provided: 54, 90, 51, 120

TN_test	FP_test
− − −	− − −
FN_test	TP_test

- **Precision**:

  The precision metric is computed by using the following formula for precision:

  Precision = True positives (TP) / (True positives(TP) + False positives(FP)) [15, 16].

  Using the values for the training set:

  Precision train = 149/ 149 + 155 = 149/304 = 0.489.

  For the validation set:

  Precision validation = 13/13 + 23 = 13/36 = 0.361.

  For the test set:

  Precision test = 120/ 120 + 19 = 120/210 = 0.571.

  Hence, around 0.489, 0.361, and 0.571 are the precision values for the training, validation, and test sets, respectively.
- **Recall**:

  The formula for recall is:

  Recall = True Positive (TP) / True Positive (TP) + False Negative (FN).

  Using the values for the training set:

  Recall train = 149/149 + 195 = 149/344 = 0.433.

  For the validation set:

  Recall validation = 13/13 + 15 = 13/28 = 0.464.

  For the test set:

  Recall test = 120/120 + 51 = 120/171 = 0.702.

  Therefore, the estimated recall values for the test, validation, and training sets are 0.702, 0.464, and 0.433, respectively.
- **F1-Score**:

The F1-score is the harmonic mean of precision and recall and is calculated using the following formula:

F1-score = the harmonic mean of the precision and recall.

Using the values for the training set:

F1 train = 2*0.489*0.433 / 0.489 + 0.433 = 2*0.211/ 0.922 = 0.458.

For the validation set:

F1 validation = 2*0.361*0.464 / 0.361 + 0.464 = 2*0.168/ 0.825 = 0.409.

For the test set:

F1 test = 2*0.571*0.702 / 0.571 + 0.702 = 2*0.4/ 1.273 = 0.627.

Table2 shows the $F1$-score values calculated for different models. It is also shown relationship between accuracy and F1-score using graphical representation in Fig. 14 for models used in this research (Fig. 15).

Table 2 F1-score calculated value of the different models on the lung cancer classification task

Model	F1-score in (%)
CNN	45
VGG16	60
ResNet50	72
Inception-v3	68

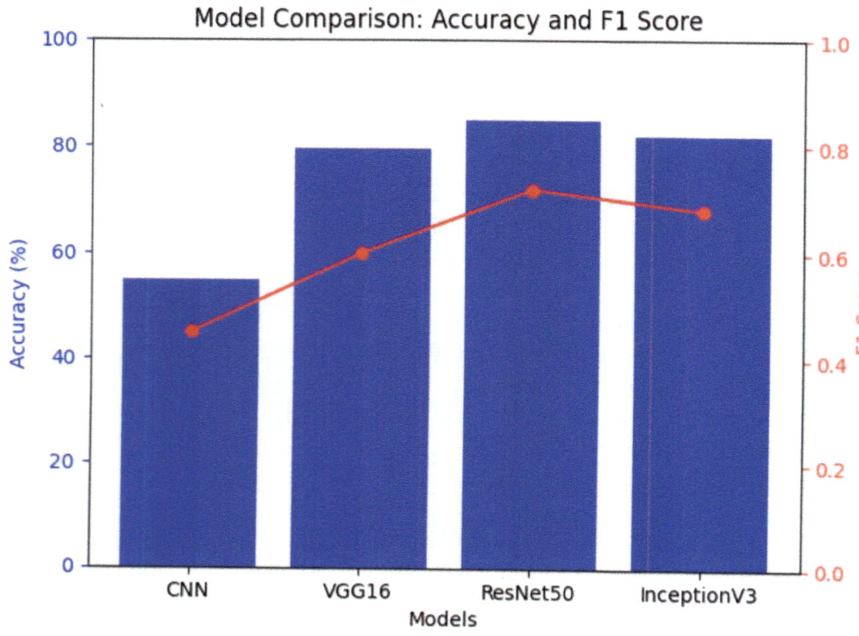

Fig. 15 Graph showing comparison of accuracy and F1-score for different tested algorithms

6 Conclusion

Finally, the article aids in analyzing the effectiveness of several deep learning systems for lung cancer classification utilizing chest CT scan pictures. According to our findings, ResNet50 had the greatest accuracy rate of 84.76%, followed by Inception-v3, VGG16, and CNN. It also explains the F1-score, and accuracy relation for different models is considered for evaluations. Multiple algorithm evaluations on the same dataset can provide insights into their performance and help researchers and practitioners make informed selections about which algorithm to utilize for their unique application. This is especially important in the diagnosis of lung cancer, where early identification and precise diagnosis are vital for effective therapy and patient survival. Our findings emphasize the potential of deep learning algorithms in improving lung cancer diagnosis and treatment. More study is needed, however, to test the efficacy of these algorithms in bigger and more diversified datasets, as well as to build more accurate and efficient lung cancer diagnosis systems.

References

1. Lam WK, White NW (2016) Lung cancer epidemiology, risk factors, and prevention. Lancet Oncol 17(1):e1–e15. https://doi.org/10.1016/S1470-2045(15)00449-8
2. Pao W, Girard N (2011) New driver mutations in non-small-cell lung cancer. Lancet Oncol 12(2):175–180. https://doi.org/10.1016/S1470-2045(10)70087-5
3. Li J, Li Y, Li X (2018) Machine learning applications for the prediction of lung cancer. J Med Syst 42(9):172. https://doi.org/10.1007/s10916-018-1001-6
4. Li H, Wu J, Liu X (2016) Lung cancer detection using deep convolutional neural networks. In: 2016 IEEE international conference on bioinformatics and biomedicine (BIBM), pp 745–748
5. Salama KM, Yaqoub MA, El-Harby AH (2020) A novel approach for lung cancer detection and classification using VGG16 and transfer learning. In: 2020 IEEE Jordan international joint conference on electrical engineering and information technology (JEEIT), pp 456–460
6. Yao C, Xue Q, Yan J (2021) Prostate cancer detection with inception-v3 and transfer learning. In: 2021 IEEE international conference on artificial intelligence and computer applications (ICAICA), pp 208–211
7. Al-Mahmoud AMM, Hussain R (2021) Lung cancer classification using ResNet50 convolutional neural network. In: 2021 IEEE 6th international conference on computing, communication and security (ICCCS), pp 1–6
8. Kim DW, Kim HR (2017) Targeted therapy of lung cancer including immune check point inhibitors. Korean J Intern Med 32(2):199–209. https://doi.org/10.3904/kjim.2016.302
9. Lilenbaum RC (2019) Advances in the treatment of lung cancer. F1000Research, 8, F1000 Faculty Rev-223. https://doi.org/10.12688/f1000research.17375.1
10. Zhang Z, Tang H, Chen X (2019) Early detection of lung cancer: advances and controversies. J Thorac Dis 11(Suppl 4):S461–S468. https://doi.org/10.21037/jtd.2019.01.26
11. Wang X, Yin X, Wang X (2018) Research progress of non-small cell lung cancer treatment in China. J Cancer Res Ther 14(Supplement):S1088–S1093. https://doi.org/10.4103/0973-1482.179203
12. Ding D, Yu H, Liu W (2020) Application of artificial intelligence in the diagnosis and treatment of lung cancer. Front Oncol 10:55. https://doi.org/10.3389/fonc.2020.00055
13. Park J, Kim Y (2018) Early detection of lung cancer: low-dose computed tomography screening in Korea. J Thorac Dis 10(Suppl 33):S3953–S3955. https://doi.org/10.21037/jtd.2018.09.139
14. Zappa C, Mousa SA (2016) Non-small cell lung cancer: current treatment and future advances. Trans Lung Cancer Res 5(3):288–300. https://doi.org/10.21037/tlcr.2016.06.07
15. Pan H, Hou E, Ansari N (2015) M-NOTE: a multi-part ballot based-voting system with clash attack protection. In: 2015 IEEE international conference on communications (ICC), London, pp 7433–7437
16. Pan H, Hou E, Ansari N (2013) Re-note: an e-voting scheme based on ring signature and clash attack protection. In: 2013 IEEE global communications conference (Globecom), IEEE, pp 867–871

Enhancing Cardiovascular Disease Prediction Accuracy Through Optimized Machine Learning and Deep Learning Techniques

S. R. Divyasri and P. S. Sreeja

Abstract Cardiovascular diseases (CVDs) are the ultimate reason for death across the globe and have risen 60% over the last 30 years (Linse et al. in Heliyon 9, 2023) [1]. This study investigates various machine learning (ML) algorithms and deep learning (DL) algorithms for early recognition of heart disease in individuals alongside different techniques to optimize the models and apply them to the heart disease dataset which consists of four databases: Cleveland, Hungary, Switzerland, and Long Beach VA. First, widespread machine learning algorithms like Naive Bayes (NB), Random Forest (RF), Support Vector Machine (SVM), Decision tree (DT), K-nearest neighbors (KNN), and Logistic Regression (LR) algorithms were experimented and evaluated with accuracy, recall, precision, and $f1$-score alongside the Stratified K-fold cross-validation. Furthermore, the same set of models each achieved better results after hyperparameter tuning by automated BayesSearchCV and feature selection with a genetic algorithm. The SVM algorithm outperformed other models by achieving 98.13% as its accuracy. On the same dataset, straightforward deep learning algorithms such as simple neural networks, multilayer perceptron, Gated Recurrent Unit networks, and Convolutional Neural Networks were tested. They achieved less accuracy than Machine learning algorithms, demonstrating that the performance of DL and ML models relies on multifarious aspects, including the complexity of the problem and the quantity of data available for training.

Keywords Cardiovascular diseases · Machine learning · Deep learning · Heart disease dataset · Stratified k-fold · Automated BayesSearchCV · Genetic algorithm

S. R. Divyasri (✉)
Hindustan Institute of Technology and Science, Kelambakkam, Tamil Nadu 603103, India
e-mail: sr.divyasriragu@gmail.com

P. S. Sreeja
Vellore Institute of Technology and Science, Chennai, India
e-mail: sreeja.ramesh.ps@gmail.com

S. M. Thampi et al. (eds.), *Fifth International Conference on Computing and Network Communications*, Lecture Notes in Electrical Engineering 1221,
https://doi.org/10.1007/978-981-97-4711-5_27

1 Introduction

Cardiovascular diseases (CVDs) are a significant global health threat, with 17.9 million people dying from them each year [2]. These diseases are influenced by various factors such as high blood pressure, cholesterol, smoking, obesity, diabetes, a sedentary lifestyle, and family history. Early detection is crucial for preventing complications, improving treatment, reducing mortality, and reducing healthcare costs [3]. By enhancing patient care, making better treatment suggestions, and increasing diagnostic precision, machine learning (ML) has completely transformed the healthcare sector. To find patterns and make predictions, ML algorithms can analyze a lot of medical data, including genomic data, medical images, and electronic health records [4]. The diagnosis of diseases and risk assessment has significantly improved thanks to deep learning, a branch of machine learning.

The challenges in CVD predictions include creating meaningful representations of the features, and identifying pertinent features or risk factors for CVD prediction can be challenging and require domain knowledge. The need for seamless model predictions into clinical practice, adaptation to dynamic patient data, effective feature selection, class imbalance within datasets, the imperative for model interpretability and trust in clinical contexts, variations across different patient demographics and healthcare regions, and variations across different patient demographics and areas are just a few of significant challenges [4].

Hyperparameter tuning, or finding the best configurations for machine learning models, substantially impacts their execution [5]. Automated methods like Bayesian optimization have gained popularity due to their systematic exploration of the hyperparameter space, resulting in increased model efficacy. Feature selection is essential in healthcare applications with many features, as it helps identify relevant variables while reducing noise and overfitting. Effective feature selection improves model interpretability and computational efficiency [6]. Early and accurate detection can lower healthcare costs and improve patient outcomes. Advancements in access to electronic health records and medical imaging data can lead to predictive models for CVD identification. Automated hyperparameter tuning techniques like Bayesian optimization can significantly improve model performance in healthcare applications. In healthcare data analysis, feature selection is crucial to reduce noise and enhance model interpretability. Genetic algorithms can effectively select pertinent features [7].

With the support of deep learning and machine learning algorithms, this research's main goals are to create and validate predictive models for identifying cardiovascular diseases, improve their performance through automated hyperparameter tuning using Bayesian optimization, and increase the efficiency and interpretability of the models by using a genetic algorithm for feature selection. By enhancing early detection of cardiovascular diseases, this study seeks to close the gap between cutting-edge machine learning techniques and significant healthcare challenges. It aims to improve clinical decision support systems, lower healthcare costs, and improve patient outcomes through rigorous experimentation, ultimately promoting disease prevention and better healthcare.

2 Related Works

A review of the literature on numerous methods and models used to forecast heart disease is provided in this section. Recent research by Khan et al. [8] examined the effectiveness of predictive Machine Learning (ML) algorithms [Decision tree (DT), Naive Bayes (NB), Random Forest (RF), Support Vector Machine (SVM), and Logistic Regression (LR)] for patients with CVD using real-time hospital data, where the RF algorithm's prediction accuracy of 85.01% was the highest. In order to analyze and forecast the risk of developing heart disease, Elsedimy et al. [9] proposed a novel quantum-behaved particle swarm optimization Support Vector Machine (QPSO-SVM) model which combines the QPSO algorithm and the SVM classification model using the University of California, Irvine (UCI) Cleveland dataset. The best accuracy for predicting heart disease was obtained by the proposed QPSO-SVM model, which was 96.31%. Similarly, a different study by Chandrasekhar and Peddakrishna [10] aims to enhance heart disease prediction using a soft voting ensemble classifier that combines the predictions of six algorithms, including LR, AdaBoost classifier, RF, K-nearest neighbors (KNNs), NB, and gradient boosting along with GridSearch CV and five-fold cross-validation, resulting in 93.44% accuracy for the Cleveland dataset and 95% for the IEEE Dataport dataset. Recent implementation of Ahamad et al. [11] explored the use of six machine learning algorithms LR, KNN, SVM, DT, RF classifier, and eXtreme Gradient Boosting to predict heart disease. They utilized GridSearch cross-validation to enhance the performance and identify ideal parameters for predicting heart disease using the UCI Kaggle Cleveland (Dataset-I) which achieved 99.03% and comprehensive UCI Kaggle V (Dataset-II) gave 87.91% accuracy.

Reddy et al. [12] focused on extracting and choosing principal components (Principal Component Analysis and Correlation-based Feature Selection) and optimizing hyperparameters of classifiers in a combination of the Cleveland dataset and Statlog dataset obtained from the UCI to detect coronary heart disease using data mining and machine learning techniques (SVM, KNN, J48 DT, AdaBoost M1, Bagging, and Rotation Forest). Saboor et al. [13] used machine learning algorithms like AdaBoost, LR, extra tree, multinomial NB, SVM, linear discriminant analysis, classification and regression tree, RF, and eXtreme Gradient Boosting to achieve high accuracy on datasets for Cleveland heart disease, and SVM got 96.72% while tuning the hyperparameters, which is the best performance. Furthermore, Pathan, et al. [14] examined how feature selection more specifically ANOVA-F test affects the precision of heart disease prediction using machine learning techniques (LR, DT, NB, RF, multilayer perceptron (MLP)) with CVD and Framingham dataset. Recently, Swathy et al. [3] compared and reported the performance metrics, datasets, and tools used for each category of various techniques, including classification, data mining, machine learning, and deep learning, for the prediction and classification of cardiovascular diseases in a review paper.

The methodology presented next in Sect. 3 builds on and broadens the groundwork established by a thorough review of related works, incorporating a variety of deep learning and machine learning methods for the early detection of cardiovascular disease. Addressing the existing gaps in heart disease prediction research, this study aims to investigate the impact of different feature selection techniques on the performance of machine learning algorithms. While previous works [3, 8, 15] have not extensively explored this aspect, the research proposes to assess the efficacy of various optimization algorithms, including genetic algorithms, in enhancing predictive accuracy. Building upon the promising results of the QPSO algorithm [9], this study seeks to broaden the scope of optimization techniques. In contrast to the traditional GridSearch CV method employed in [10], this research intends to leverage more efficient BayesSearchCV hyperparameter tuning approaches that adapt dynamically to the evolving model performance. Furthermore, inspired by recommendations from prior researchers [8–10, 12, 15–18], this paper will explore neural and deep learning models to harness their potential for achieving superior predictive outcomes, thereby contributing to a more comprehensive understanding of cardiovascular disease prediction.

3 Methodology

3.1 Dataset

Heart disease dataset from Kaggle was used for this research purpose. Cleveland, Hungary, Switzerland, and Long Beach VA are the four databases that make up this dataset dated from 1988 along with 13 features (as mentioned in Table 1) excluding the target variable and 1025 instances [19].

These features describe various attributes related to heart health and are used to predict the 'target' variable, which likely represents the presence or absence of heart disease (1 for presence, 0 for absence).

3.2 Methods

Computational Framework and Implementation Details

Python programming was used in Jupyter Notebook v7. 0.6 to create the computational framework for this work. The main libraries used were NumPy for numerical operations, Seaborn and matplotlib for data visualization, data processing done with Pandas, and Scikit-learn for machine learning applications. Data preprocessing,

Table 1 Features of the dataset

Serial No.	Attributes	Description
1	age	The patient's age (in years)
2	sex	The patient's gender (0 means female, 1 means male)
3	cp	Type of chest pain (0—typical angina, 1—atypical angina, 2—non-anginal pain, and 3—asymptomatic)
4	trestbps	Blood pressure at rest (in mm Hg)
5	chol	Serum cholesterol level (in mg/dL)
6	fbs	Fasting blood sugar level greater than 120 mg/dL (true or false: 1 or 0)
7	restecg	Results of the resting electrocardiogram (0 = normal, 1 = abnormal, 2 = indicating possible or certain left ventricular hypertrophy)
8	thalach	Highest heart rate ever (in beats per minute)
9	exang	Angina brought on by exercise (1 = yes, 0 = no)
10	oldpeak	ST depression caused by exercise relative to rest
11	slope	The slope of the ST segment of the peak exercise (0 = flat, 1 = upsloping)
12	ca	Major vessels count (0–3)
13	thal	Types of thalassemia include normal (3), fixed (6), and reversible (7)

which comprised randomization, splitting into training and testing sets using train–test split from Scikit-learn, and standardization using StandardScaler, was the first step in the study. The implementation of a number of machine learning algorithms was done, including the Scikit-learn-available Random Forest Classifier, Logistic Regression, Support Vector Machines (SVM), Neural Networks (MLP Classifier), Decision Tree Classifier, K Neighbors Classifier, and Gaussian Naive Bayes.

Furthermore, sophisticated methods like computing model evaluation metrics, hyperparameter tuning with GridSearchCV and BayesSearchCV from Scikit-optimize, and genetic feature selection using GeneticSelectionCV were integrated. For deep learning, TensorFlow and Keras were used, with layers in the neural network architecture like GRU, Dense, Input, Conv1D, MaxPooling1D, and Flatten. For a rigorous model assessment, the study also used cross-validation procedures with Stratified K-fold and K-fold. This extensive collection of tools, languages, and libraries made sure that the study and assessment of the research's prediction models were done in a thorough and varied manner.

Machine and Deep Learning Algorithm

In order to decipher intricate patterns (involves identifying modest, corresponding variables and their correlations that might contribute to the threat or existence of cardiovascular diseases.) and glean insightful knowledge from data, this research has harnessed the power of machine and deep learning algorithms. This has paved the way for creative solutions and well-informed decision-making as discussed in Sects. 4.1 and 4.2.

A fundamental and popular classification algorithm for binary and multiclass tasks, logistic regression uses the logistic function to model the relationship between dependent binary variables and independent variables. Although its assumption of linear feature-log-odds relationships can limit performance on nonlinear data, its simplicity and interpretability are advantages. Random Forest is a flexible ensemble method that aggregates random subsets of data and features to improve predictive accuracy and reduce overfitting, hence making it robust and efficient for high-dimensional data and complex relationships. Support Vector Machines excel in high-dimensional spaces and handle nonlinear data through kernel functions, but they call for careful hyperparameter selection and seek optimal hyperplanes to separate data classes with the most significant margin. Decision trees split data recursively based on essential features to produce interpretable tree structures, but they are prone to overfitting and can benefit from pruning. While K-nearest neighbors are non-parametric, it is sensitive to K choice and less effective in high dimensions. Based on the Bayes theorem and the conditional independence assumption, Naive Bayes efficiently performs text classification but may need help to capture complex feature dependencies [3, 20, 21].

Perceptrons, also referred to as single-layer neural networks (SNNs), are appropriate for simple linear tasks because they are made up of a single layer of interconnected neurons. Contrarily, Convolutional Neural Networks (CNN) use convolutional layers to extract hierarchical features. Multilayer perceptrons (MLP) are adaptable feedforward networks with a number of hidden layers that can accurately represent complex nonlinear relationships in structured data and hence a popular choice for classification, regression, and recommendation systems. With gating mechanisms for effective learning of long-range dependencies, Gated Recurrent Units (GRUs) are a type of recurrent neural network (RNN) tailored for sequential data. All these above-mentioned algorithms were applied to the given dataset. To ensure robust evaluation, stratified tenfold cross-validation was employed, and the comprehensive results can be found in Tables 2 and 3 [22, 23].

Hyperparameter Tuning and Feature Selection

By using probabilistic modeling to iteratively fine-tune hyperparameters, Bayesian optimization, powered by BayesSearchCV, is known for its ability to decide which hyperparameters to investigate next based on the current understanding of the

Table 2 Results obtained by applying ML models alongside stratified tenfold cross-validation

Model	Accuracy	Precision	Recall	$F1$-score
Logistic regression	84	81	89	85
Random forest	87.21	83	91	87
SVM	93.46	93	94	94
Decision tree	90.33	92	91	94
KNN	86.44	92	93	92
Naïve Bayes	82.53	90	91	90

Table 3 Deep learning algorithms results

Model	Accuracy	Precision	Recall	$F1$-score
SNN	81.95	78	88	83
MLP	79.51	76	86	81
GRU	73.71	72	76	74
CNN	77.56	78	77	77

behavior of the objective function. It's especially helpful for automating the hyper-parameter tuning procedure, which saves time and frequently results in machine learning models that perform better. Parallel to this, the genetic algorithm with cross-validation (GA-CV) uses cross-validation and genetic operators to systematically optimize feature subsets, producing models that are more accurate and reliable [7]. The vast feature space is successfully navigated by GA-CV through intelligent feature combination exploration and adaptation, improving model performance and boosting predictive power in machine learning tasks. When combined, these methods are potent tools for raising the efficiency and performance of machine learning models as depicted in Fig. 1.

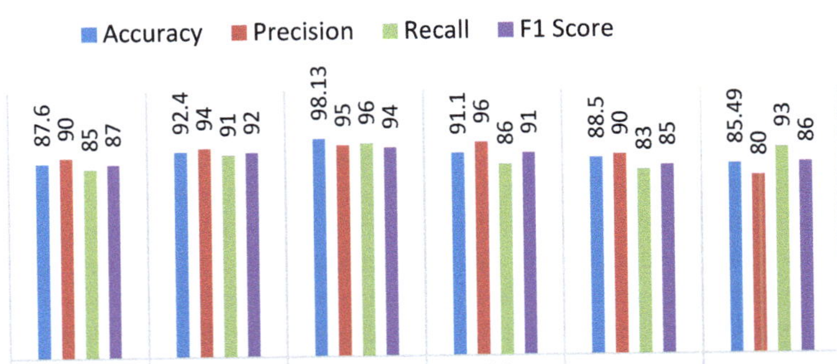

	Logistic Regression	Random Forest	SVM	Decision tree	KNN	Naïve Bayes
Accuracy	87.6	92.4	98.13	91.1	88.5	85.49
Precision	90	94	95	96	90	80
Recall	85	91	96	86	83	93
F1 Score	87	92	94	91	85	86

Fig. 1 Results after hyperparameter tuning by automated BayesSearchCV and feature selection with genetic algorithm

4 Results and Discussions

This section will cover a variety of viewpoints on how well the chosen classification models performed and the results are from training. To determine which models are adequate for the given dataset, first each model's performance individually for both datasets using Stratified tenfold cross-validation was tested. Second, the same dataset was examined with deep learning algorithms. Finally, using an automated BayesSearchCV and feature selection using a genetic algorithm, the performance of the models along with the chosen set of features was assessed to examine the impact of hyperparameter tuning. The accuracy, recall, $F1$-score, and precision evaluation matrices were used to assess the performance of the classifiers.

4.1 Result of Classifiers Using Stratified Tenfold Cross-Validation

In order to predict the CVD, all machine learning models were tested on the dataset in this section using tenfold cross-validation with Stratified K-fold as it came out to be a valuable technique when working with classification problems, especially when the chosen machine learning models wanted to ensure a fair and unbiased evaluation. It helps to obtain more reliable performance estimates and make better-informed decisions about model selection and hyperparameter tuning. The classification outcomes of the ML model for predicting heart disease for the dataset are displayed in Table 2.

With precision of 93%, recall of 94%, and $F1$-score of 94%,SVM produced the highest accuracy reported in Table 2 of 93.46%. It's only sometimes valid that SVM outperforms Random Forest, Logistic Regression, Decision Trees, KNN, and Naive Bayes; the effectiveness of these algorithms varies depending on the dataset's characteristics and the particular problem addressed. For the given data, SVM shows to be less prone to overfitting and less likely to capture noise in the data when it is a small dataset. This outcome also opened the door to the idea that feature engineering, data preprocessing, and hyperparameter tuning can all significantly affect model performance.

4.2 Results of Deep Learning Algorithms

The accuracy of deep learning algorithms shown in Table 3 was lesser than the machine learning algorithms discussed above for the same dataset in Table 2.

These results depicted above give a clear understanding that due to their high complexity, sensitivity to hyperparameters, and requirement for enormous quantities of labeled data, deep learning algorithms can occasionally perform less efficiently than traditional machine learning (ML) algorithms. Deep learning networks frequently require significant computational resources and careful hyperparameter tuning, which might only be available or suitable for some problems. Additionally, they are more challenging to interpret due to their 'black-box' (with a large number of layers and parameters, can be challenging to interpret and understand) nature and more likely to overfit when there is a lack of data. On the other hand, traditional ML models may perform better in situations where feature engineering, quick model iteration, or interpretability are crucial, as they frequently require fewer resources.

4.3 Result of Classifiers After Hyperparameter Tuning by Automated BayesSearchCV and Feature Selection with Genetic Algorithm

BayesSearchCV for hyperparameter tuning and genetic algorithm-based feature selection was applied to all the classifiers as mentioned below. Table 4 lists the best hyperparameter selected for each machine learning algorithm.

Figure 1 shows that SVM with 98.13% of accuracy, 95% of precision, 96% of recall, and 94% of $F1$-score works best after hyperparameter tuning and genetic algorithm feature selection comparatively better than the existing works on same heart disease dataset mentioned in Table 5. Not only SVM but all the ML algorithms have shown a good improvement with automated hyperparameter tuning (BayesSearchCV) which optimizes model settings for better predictive accuracy on a limited healthcare dataset. Genetic algorithm-based feature selection improved each model's efficiency interpretability and reduces overfitting, enhancing the utility of the small healthcare dataset.

This research's findings, when compared to those of other studies, show that combining feature selection with a genetic algorithm and automatic BayesSearchCV hyperparameter adjustment significantly improves the performance of traditional models. The utilization of these methodologies has been demonstrated to be crucial in augmenting the precision of diverse models showcased in this manuscript. For example, without feature selection and hyperparameter tuning, the Support Vector Machine (SVM) gave an accuracy of 93.46%; nevertheless, the same model produced an astounding accuracy of 98.13% when these optimization techniques were applied. On the other hand, various studies have documented the effectiveness of multiple models. In a study by [24], a neural network utilizing Random Forest with a 98% accuracy rate was the best model. Jaiswal et al. [22] used a Convolutional Neural Network (CNN) to attain 94.5% accuracy. A modified version of infinite feature selection and multilayer perceptron in [26] achieved an accuracy of 87.70%, whereas logistic regression in [25] performed an accuracy of 85%. In reference [27], Random Forest

Table 4 Best hyperparameters

Model	Best hyperparameters	Justifications
Logistic regression	• $C = 0.00242$ • max_iter = 100 • penalty = '12' • solver = liblinear	• Lower C value indicates a more regularization • Maximum iteration of 100 is a reasonable value to ensure convergence within a reasonable time • The regularization term penalty = '12' aids in stabilizing the model and reducing overfitting • The 'liblinear' solver can handle both binary and multiclass classifications and is effective for small datasets
Random forest	• max_depth = 5 • max_features = 'sqrt' • min_samples_leaf = 1 • min_samples_split = 2 • n_estimators = 200	• With max_depth set to 5, overfitting is reduced and the trees are kept at a manageable depth • The popular setting of max_features = 'sqrt' contributes to the diversity of the trees • A leaf node can have just one sample if min_samples_leaf is set to 1, which can capture more precise details in the data but may cause overfitting • The tree is encouraged to continue splitting as long as it discovers valid splits with min_samples_split = 2 • Model complexity and performance can be balanced by using 200 estimators, which is big enough to prevent overfitting but not too big
SVM	• $C = 1.0$ • kernel = 'rbf' • gamma = 'scale', • degree = 3	• A reasonable and moderate level of regularization is a C value of 1.0 • The nonlinear decision boundaries in the data can be captured using the Radial Basis Function kernel • Gamma is frequently set to 'scale' because it reduces the impact of individual training samples • When 'degree' is set to 1, a linear kernel is indicated
Decision tree	• criterion = 'gini' • max_depth = 5 • min_samples_leaf = 1 • min_samples_split = 3	• The name 'gini' implies that the model seeks to reduce impurity when splitting • With max_depth = 5, the tree can split relatively shallowly, which can result in a more straightforward model and improved generalization • Using min_samples_leaf = 1, the decision tree can record minute details • min_samples_split = 3 can prevent nodes from splitting when they have too few samples, which can help control the growth of the tree
KNN	• n_neighbors = 8 • weights = 'uniform'	• The model's bias and variance are balanced by the selection of eight neighbors • Because of the 'uniform' weighting system, all nearby data points are equally significant for making predictions
Naïve Bayes	• $C = 0.0304$ • gamma = 0.000465 • kernel = 'linear'	• C should be set to a lower value, such as 0.0304, to help avoid overfitting • A decision boundary that is relatively smooth is suggested by a lower gamma value, such as 0.000465 • The 'linear' kernel is appropriate for tasks where a straightforward linear data separation is effective

Table 5 Outcomes of the existing works

Study	Best model and accuracy
[24]	A neural network using random forest—98%
[22]	CNN—94.5%
[25]	Logistic regression—85%
[26]	Modified variation of infinite feature selection and multilayer perceptron—87.70%
[27]	Random forest—95.61%
[28]	Logistic regression—91.8%

yielded an accuracy of 95.61%, whereas [28]'s Logistic Regression produced an accuracy of 91.8%. The comparative study, which shows competitive and enhanced accuracies in comparison to other methodologies in the field, demonstrates the efficacy of the strategies used in the research.

5 Conclusion

We investigated heart disease detection in this study, an area that has received a lot of attention in recent times. The principal aim was to optimize the performance of traditional models through the utilization of sophisticated methodologies, including genetic algorithm-based feature selection and automated BayesSearchCV hyperparameter tuning. Through our extensive analysis, we were able to demonstrate the critical role that these approaches play in improving the accuracy of different models. One such model is the Support Vector Machine (SVM), which achieved an impressive accuracy of 98.13% when feature selection and hyperparameter tuning were used, as opposed to 93.46% when they weren't. Specifically, our findings demonstrated the complex interactions that exist between the efficacy of the model and variables like the nature of the dataset, the difficulty of the problem, and the volume of training data.

A meticulous investigation of machine learning (ML) models through stratified tenfold cross-validation spotlighted SVM as the standout performer, attaining a precision of 93%, recall of 94%, and an $F1$-score of 94%. The comparison with deep learning algorithms, however, unveiled fascinating insights. While ML models outperformed their deep learning counterparts in terms of accuracy, the latter indicated lower performance due to their sensitivity to hyperparameters, inherent complexities, and demand for substantial labeled data.

Likewise, the detailed tuning of hyperparameters using automated BayesSearchCV and feature selection via a genetic algorithm significantly refined the performance of each ML model. SVM, in particular, emerged as the leader with a staggering accuracy of 98.13%. This not only surpassed the results of existing works but also highlighted the impact of thoughtful data preprocessing, feature engineering, and hyperparameter tuning on model efficacy.

The comparison with the studies mentioned earlier further validated the efficacy of the approach. The methodologies discussed in this paper yielded competitive and improved accuracies compared to other methods, emphasizing the robustness and generalizability of our strategies. In a landscape where heart disease prediction remains a critical healthcare concern, our research contributes valuable insights, demonstrating the transformative potential of advanced techniques in optimizing model performance and paving the way for future advancements in predictive healthcare analytics.

In future, investigating the use of metaheuristic methods and nature-inspired algorithms to optimize the parameters of machine learning classifiers and deep learning techniques may improve the precision of current algorithms for detecting heart disease. To further strengthen the validity of these findings future studies could assess the prevalence of heart disease across different heart disease-related datasets by incorporating additional datasets. Future research can look into sophisticated feature selection methods to find the features that are most useful for predicting cardiovascular disease and should concentrate on putting the knowledge gained from prediction models to use in real-world settings.

References

1. Linse B, Ohlsson M, Stehlik J, et al (2023) A machine learning model for prediction of 30-day primary graft failure after heart transplantation. Heliyon 9. https://doi.org/10.1016/j.heliyon.2023.e14282
2. Biswas N, Ali MM, Rahaman MA, et al (2023) Machine learning-based model to predict heart disease in early stage employing different feature selection techniques. Biomed Res Int 2023. https://doi.org/10.1155/2023/6864343
3. Swathy M, Saruladha K (2022) A comparative study of classification and prediction of Cardio-Vascular Diseases (CVD) using Machine Learning and Deep Learning techniques. ICT Express 8:109–116. https://doi.org/10.1016/j.icte.2021.08.021
4. Venkat V, Abdelhalim H, DeGroat W, et al (2023) Investigating genes associated with heart failure, atrial fibrillation, and other cardiovascular diseases, and predicting disease using machine learning techniques for translational research and precision medicine. Genomics 115. https://doi.org/10.1016/j.ygeno.2023.110584
5. Asif D, Bibi M, Arif MS, Mukheimer A (2023) Enhancing heart disease prediction through ensemble learning techniques with hyperparameter optimization. Algorithms 16:308. https://doi.org/10.3390/a16060308
6. Saidi R, Ncir WB, Essoussi N (2018) Feature selection using genetic algorithm for big data. In: Advances in intelligent systems and computing. Springer, pp 352–361
7. Wutzl B, Leibnitz K, Rattay F, et al (2019) Genetic algorithms for feature selection when classifying severe chronic disorders of consciousness. PLoS One 14. https://doi.org/10.1371/journal.pone.0219683
8. Khan A, Qureshi M, Daniyal M, Tawiah K (2023) A novel study on machine learning algorithm-based cardiovascular disease prediction. Health Soc Care Community 2023:1–10. https://doi.org/10.1155/2023/1406060
9. Elsedimy EI, AboHashish SMM, Algarni F (2023) New cardiovascular disease prediction approach using support vector machine and quantum-behaved particle swarm optimization. Multimed Tools Appl. https://doi.org/10.1007/s11042-023-16194-z

10. Chandrasekhar N, Peddakrishna S (2023) Enhancing heart disease prediction accuracy through machine learning techniques and optimization. Processes 11. https://doi.org/10.3390/pr11041210
11. Ahamad GN, Shafiullah, Fatima H, et al (2023) Influence of optimal hyperparameters on the performance of machine learning algorithms for predicting heart disease. Processes 11. https://doi.org/10.3390/pr11030734
12. Reddy KVV, Elamvazuthi I, Aziz AA, et al (2023) An efficient prediction system for coronary heart disease risk using selected principal components and hyperparameter optimization. Appl Sci (Switzerland) 13. https://doi.org/10.3390/app13010118
13. Saboor A, Usman M, Ali S, et al (2022) A method for improving prediction of human heart disease using machine learning algorithms. Mobile Inf Syst. https://doi.org/10.1155/2022/1410169
14. Pathan MS, Nag A, Pathan MM, Dev S (2022) Analyzing the impact of feature selection on the accuracy of heart disease prediction. Healthc Anal 2. https://doi.org/10.1016/j.health.2022.100060
15. Akella A, Akella S (2021) Machine learning algorithms for predicting coronary artery disease: efforts toward an open source solution. Future Sci OA 7. https://doi.org/10.2144/fsoa-2020-0206
16. Chang V, Bhavani VR, Xu AQ, Hossain MA (2022) An artificial intelligence model for heart disease detection using machine learning algorithms. Healthc Anal 2. https://doi.org/10.1016/j.health.2022.100016
17. Muntasir Nishat M, Faisal F, Jahan Ratul I, et al (2022) A comprehensive investigation of the performances of different machine learning classifiers with SMOTE-ENN oversampling technique and hyperparameter optimization for imbalanced heart failure dataset. Sci Program 2022. https://doi.org/10.1155/2022/3649406
18. Gupta C, Saha A, Reddy NVS, Acharya UD (2022) Cardiac disease prediction using supervised machine learning techniques. J Phys Conf Ser. IOP Publishing Ltd
19. https://www.kaggle.com/datasets/johnsmith88/heart-disease-dataset?select=heart.csv
20. Abuhaija B, Alloubani A, Almatari M, et al (2023) A comprehensive study of machine learning for predicting cardiovascular disease using Weka and SPSS tools. Int J Electr Comput Eng 13:1891–1902. https://doi.org/10.11591/ijece.v13i2.pp1891-1902
21. Tong R, Zhu Z, Ling J (2023) Comparison of linear and non-linear machine learning models for time-dependent readmission or mortality prediction among hospitalized heart failure patients. Heliyon 9. https://doi.org/10.1016/j.heliyon.2023.e16068
22. Jaiswal A, Singh M, Sachdeva N (2023) Empirical analysis of heart disease prediction using deep learning. In: 2023 international conference on advances in computing, communication and applied informatics (ACCAI). IEEE, pp 1–9
23. Bharti R, Khamparia A, Shabaz M, et al (2021) Prediction of heart disease using a combination of machine learning and deep learning. Comput Intell Neurosci 2021. https://doi.org/10.1155/2021/8387680
24. Yadav DC, Pal S, Yadav RK, Pant H (2023) Analyzing risk elements in cardiovascular diseases prediction using neural networks algorithm. In: 2023 10th international conference on computing for sustainable global development (INDIACom). IEEE, pp 262–266
25. Yadav A, Kulpriya S, Upadhyay D (2023) A statistical analysis for heart disease prediction system for next-gen software. In: 2023 International conference on computational intelligence, communication technology and networking (CICTN). IEEE, pp 71–76
26. Modak S, Abdel-Raheem E, Rueda L (2022) Heart disease prediction using adaptive infinite feature selection and deep neural networks. In: 2022 international conference on artificial intelligence in information and communication (ICAIIC). IEEE, pp 235–240
27. Akter B, Shakil R, Rajbongshi A, et al (2022) Utilization of five-distinct dataset to diagnose and predict heart disease: a machine learning approach. In: 2022 13th international conference on computing communication and networking technologies (ICCCNT). IEEE, pp 1–6
28. Hafiz A, Kaur N (2023) Heart disease prediction based on machine learning technique. In: 2023 IEEE 3rd international conference on technology, engineering, management for societal impact using marketing, entrepreneurship and talent (TEMSMET). IEEE, pp 1–7

Design and Development of an e-Health Package for Voice Pathology Detection

Shiksha Rai, S. Rishita, Kandalam Srihari Lasya, and Susmitha Vekkot

Abstract In this paper, a speech-based detection system is developed for voice pathologies like GERD, laryngitis, bulbar palsy, vocal fold leukoplakia, cyst, dysarthria, and aphasia. The e-health set-up comprises three modules for the detection of specific voice disorders leading to degradation in voice quality. Classification of these pathologies is performed using various machine learning algorithms before and after data balancing using SMOTE. The maximum performance is obtained for CatBoost with an accuracy of 56% for the detection of voice disorders before data balancing and 93% post-data balancing. An accuracy of 86.5% is obtained by random forest for detection of dysarthria and an accuracy of 88% is obtained for aphasia using the CatBoost classifier. Further, to make an improvement in the selected features we have customized a generalized feature set for the classification of voice disorders. All the components are effectively incorporated into a user-friendly e-health package, ensuring non-invasive, and precise diagnosis.

Keywords Pathology classification · Random forest · Catboost · Voice disorder detection · Machine learning

1 Introduction

Human communication is the spoken or nonverbal transmission of information between two or more people [1]. Around 1.2% of the world's population is facing some form of voice pathology [2]. Among them, dysarthria is a neuromotor speech disorder marked by impairments in the speech production mechanism. Around 90% of Parkinson's disease (PD) patients are categorized to be with hypokinetic dysarthria [3]. Aphasia is a language disorder related to brain diseases caused due to lesions or damages of the left hemisphere of the brain, thereby affecting speech [4]. The healthcare sector has undergone major changes in the past decade,

S. Rai (✉) · S. Rishita · K. S. Lasya · S. Vekkot
Department of Electronics and Communication Engineering, Amrita School of Engineering, Amrita Vishwa Vidyapeetham, Bengaluru 560035, India
e-mail: v_susmitha@blr.amrita.edu

S. M. Thampi et al. (eds.), *Fifth International Conference on Computing and Network Communications*, Lecture Notes in Electrical Engineering 1221,
https://doi.org/10.1007/978-981-97-4711-5_28

influenced by the COVID-19 pandemic, and the difference in the approach of patients and clinical staff to the treatment process [5].

Voice pathology detection often involves invasive techniques like laryngoscopy, laryngeal electromyography and stroboscopy [6]. Most of these are unpleasant to the patient and often traumatize them. The use of voice signal processing techniques to detect voice pathology is non-invasive and poses minimum discomfort to the patient [4]. Computerized speech processing has proven to be an effective method to deal with diseases that lead to impairment and permanent loss of voice. The motivation for this work arises from the fact that there happens to be a lack of systems which can non-invasively detect pathologies from the voice sample of a patient while maintaining a quick response time and high accuracy. The fact that there are no integrated models which can handle multiple voice pathologies in a common platform is the main motivation for this research. Additionally, identification and utilization of pathology-specific speech features can go a long way in aiding proper diagnosis and treatment. Based on these considerations, this paper proposes a reliable e-health set-up that can detect voice disorders using varied learning algorithms. The research mainly focuses on:

- Detecting different types of voice pathologies (Healthy, Gerd, Laryngitis, Bulbar, Vocal Fold Leukoplakia, and Cyst) using learning algorithms.
- Detecting dysarthria and aphasia using algorithms viz. CatBoost, XGBoost, Support Vector Machine, Random Forest, and Histogram Gradient-based methods.
- Generating a universal custom feature set from the results obtained from the detection of voice disorders, dysarthria and aphasia.
- Integrate all of the preceding modules to produce an e-health package.

Section 2 gives a brief overview of the existing research done on speech processing for voice pathologies. Section 3 encompasses details pertaining to the algorithms experimented with. Section 4 discusses the obtained results, followed by an analysis and discussion. Section 5 concludes the paper by giving future insights.

2 Related Work

Different researchers have used various speech processing techniques for the detection of voice pathologies through methods which include perceptual evaluation and acoustic analysis [7–9]. The algorithms in [10–12] are designed for a reliable detection system that can differentiate between pathological and healthy voices and integrates the model into the mobile app, eliminating the requirement for data transfer and cloud services. The study in [13] identified circumstances as abnormal or healthy using a support vector machine classifier on mel-frequency cepstral coefficients (MFCCs). Vowel /a/ was used for these studies and values of 84.37%, 90.90%, and 80.95%, respectively were the accuracy, specificity, and recall of this model. The study in [13] analyzes the classification of dysarthria severity levels using learning approaches and aural features. Prosodic, phonation and glottal features were used for the analysis. Using SVM, the accuracy was observed to be 60.18% for prosody, 85.87% for articulations, 54.26% for glottal, and 63.9% for phonation. Using random forest classification, the accuracy was observed to be 61.43%, 85.43%, 56.86%, and 68.16% respectively for the above features while deep neural network, yielded an accuracy of 60.98%, 86.77%, 56.50%, and 69.73% for the same.

The research in [14, 15] describe a non-invasive classifier for identifying the type of dysarthria using global statistics of speech variables viz. mean, variance, and standard deviation. In the classification of three categories of dysarthria, an error of 10.5% was achieved using MFCCs and derivatives, mel-filter bank output, perceptual linear prediction and their derivatives, RASTA-PLP, residuals and delta residuals. The study in [16] gives details about a few machine learning algorithms that work well with the data provided by AphasiaBank [17]. SVM and logistic regression perform well with essentially continuous characteristics such as Goodness of Pronunciation (GOP), Dynamic Time Wrapping (DTW), and Duration Measures (DUR). On the other hand, they outperform the phone edit distance (DIST) because their attributes are primarily discrete.

Most of the research indicated above worked with a limited amount of speech samples, and therefore the results cannot be generalized. Nonetheless, the majority of them utilized balanced data for the learning process. We concentrated on working with an imbalanced dataset in this research. Despite the exploration of various dropout rates, the CNN model has consistently found application in various research. The occurrence of overfitting after a few training epochs impedes the validation accuracy from achieving its maximum potential. As a result, we have concentrated this research on more efficient machine learning techniques.

3 Proposed Methodology

This study proposes a generalized feature set for the detection of various voice pathologies and finally concentrates on developing a unified detection method for various voice abnormalities. The proposed architecture is as illustrated in Fig. 1.

3.1 Dataset Description

The databases utilized in this work, and the number of samples drawn from each pathology, are listed in Table 1.

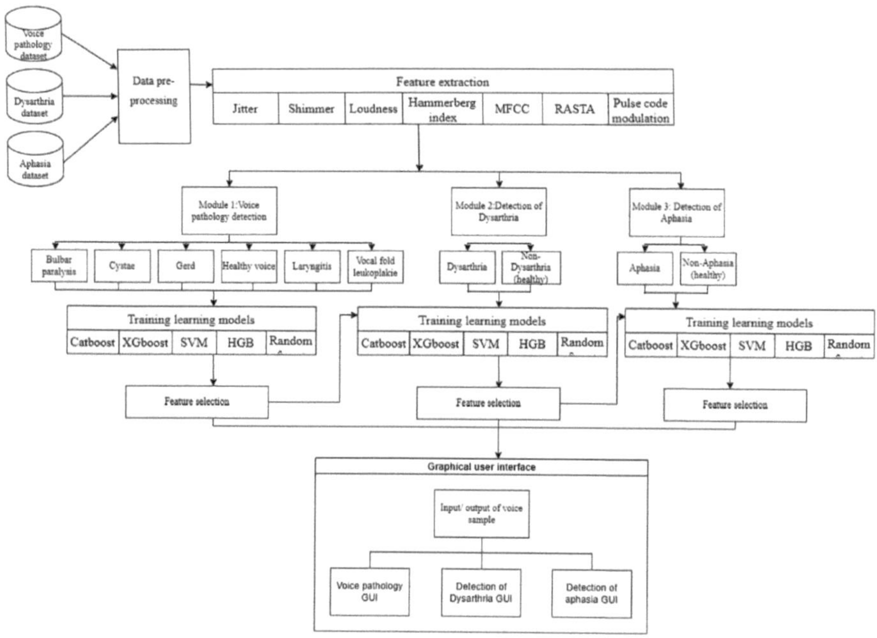

Fig. 1 Proposed architecture of the e-health set-up

Table 1 Data statistics

Module	Dataset	Sample size					
Voice pathology	SVD [18]	Healthy	Laryngitis	Cystae	GERD	Bulbar Palsy	Vocal Fold Leukoplakia
		245	150	84	42	24	147
Dysarthria	TORGO [19]	Healthy			Pathological		
		500			492		
Aphasia	AphasiaBank [17]	54			60		

SVD refers to Saarbuken Voice Dataset [18]

3.2 *Feature Extraction*

OpenSMILE toolkit [20] is used to extract the features. The features are selected by visualizing the extracted data using histograms based on their statistical values. The features were extracted from Emobase, GeMaps, and ComParE2016 feature packages [20].

- Jitter is the cycle-to-cycle variation of pitch [21] as in Eq. (1)

$$\text{Jitter} = \frac{1}{M-1}\sum_{k=1}^{M-1}|F_k - F_{k+1}| \tag{1}$$

F_k represents the period and M is the count of extracted frequency periods
- Shimmer is expressed as the variability of the peak-to-peak amplitude in decibels (B_k) [21] as expressed by Eq. (2).

$$\text{Shimmer} = \frac{1}{M-1}\sum_{k=1}^{M-1}|20\log(B_{(k+1)}/B_k)| \tag{2}$$

- Loudness of a sound is a phenomenon caused by the amplitude of the sound wave [21].
- Hammerberg Index is the difference between the maximum intensity in a lower frequency band versus a higher frequency band [21].
- Rfilt was introduced to support perceptual linear prediction pre-processing [22], as given by Eq. (3).

$$R(k) = \frac{l * \sum(m - (M-1)/2) * k^{-m})}{(1 - \frac{\rho}{y})} \tag{3}$$

The numerator is an odd-order (M) regression filter. The denominator is an integrator with time decay.
- RASTA is used to eliminate any continuous offset and spectral coloring in the voice channel and to smooth out short-term noise changes, by applying a band-pass filter to the energy in each frequency subband [22].
- Zero-Crossing Rate (ZCR) is the rate at which the signal value changes from positive to negative and vice versa as defined by Eq. (4).

$$Z(i) = \frac{1}{2W_L}\sum_{n=1}^{W_L}|\text{sgn}[x_i(n)] - \text{sgn}[x_i(n-1)] \tag{4}$$

- MFCC is a representation of the short-term power spectrum of a signal that is entirely dependent on the linear cosine change of on a non-linear mel scale of frequency [22].

The research for voice pathology detection performed in this work is organized into three modules. As it can be perceived, the SVD dataset is unbalanced. In order to train the classifier, it is utilized to generate a training set that is synthetically nearly class-balanced [23]. Data augmentation refers to the process of artificially expanding a dataset by modifying or transforming the already existing voice data. We used two methods, Synthetic Minority Over-sampling Technique (SMOTE) and data augmentation. The second method uses the initial pre-processed dataset to generate new voice samples by altering speech by certain factors. The techniques that are employed to introduce variations are pitch shifting with parameters of 0.8 and 1.2, time stretching with parameters of − 2 and 2 and noise injection with parameters of 0.1 and 0.3. Training and testing are done on the models using both clean and noise-augmented voice data.

3.3 Feature Analysis and Selection

As illustrated in Fig. 2, the statistical values of jitter differ between healthy and diseased voices. As a result, jitter can be regarded as one of the contributing features to the classification. Similarly, other features were chosen that can help the models perform effectively. These features are fed into machine learning models like Random Forest, Catboost, XgBoost, SVM. This entire technique is carried out with clean data that contains little or no noise in the speech samples. The technique is then performed on the speech samples with noise introduced to obtain more real-time predictions and outcomes. The features are chosen by representing the retrieved data with histograms. The comparative differences between the dysarthric voice and non-dysarthric voice are used to determine which features contribute the most as illustrated in Fig. 2. The features from the previous module were compared with that for dysarthria. The features that do not contribute to either one of the modules were discarded, leaving the feature set with only the most beneficial elements. Rfilt is regarded as a contributing characteristic of the voice pathology module, as represented in Fig. 2d. However, as can be seen from Fig. 2h, Rfilt is not a contributing feature for the dysarthria module as the statistical significance is lesser since dysarthric and non-dysarthric Rilt has similar values. As a result, Rfilt was removed from the customized feature set. The features considered after evaluating the most relevant characteristics for classifying each voice pathology module are listed in Table 2.

3.4 Training Algorithms

Prominently, boosting algorithms were used in this work due to the capability of ensembling the capabilities of various weak classifiers for the task at hand. State-of-the-art classification systems use various deep learning techniques for pathological voice detection. The dataset used in this work consists of 692 samples from all the pathological cases considered in the first module of the e-health package. The idea of the package is to use lightweight models with a reduction in training time and resources. Boosting algorithms like CatBoost and XGBoost are gradient boosting algorithms which are popular for classification and regression tasks and capable of dealing with skewed data. Further details regarding the below algorithms can be found in [24, 25]:

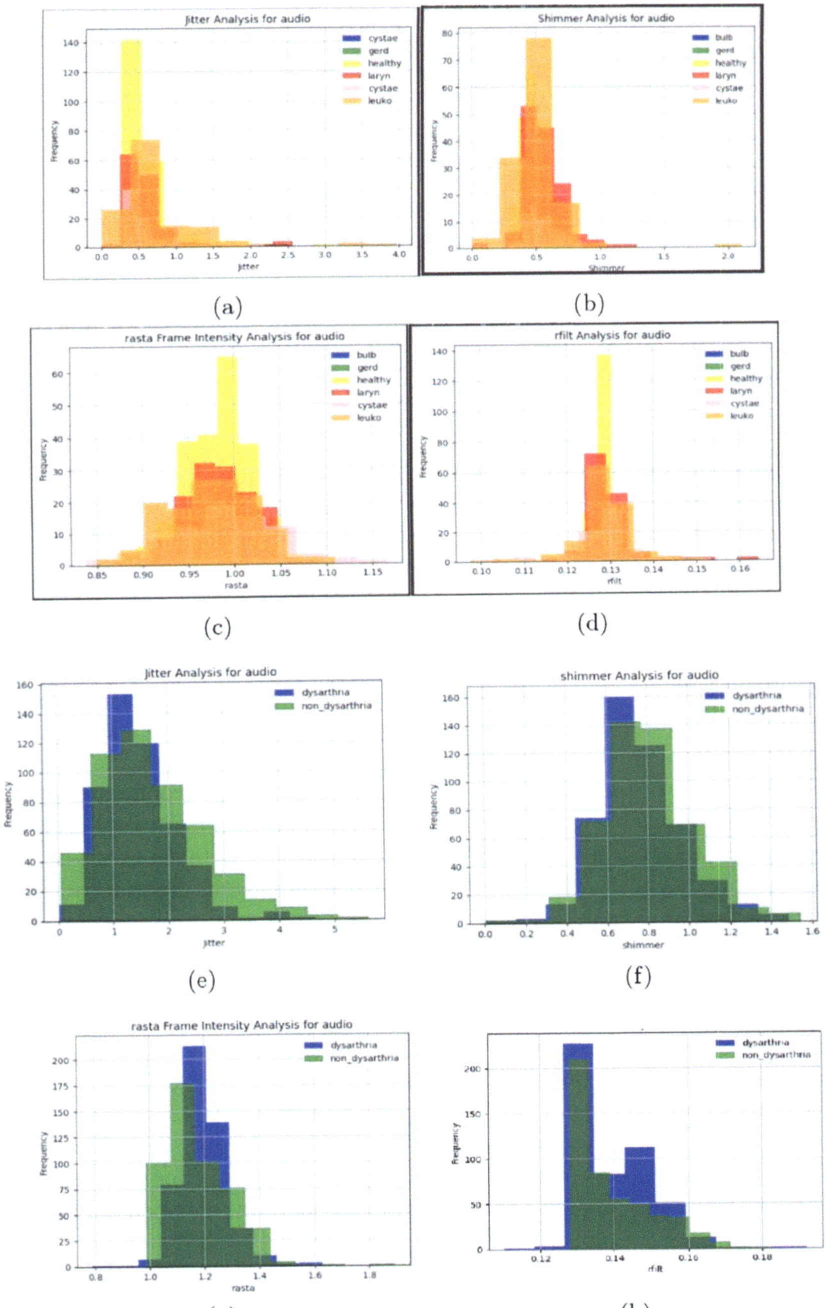

Fig. 2 Analysis of speech features for voice pathology, **a** jitter, **b** shimmer, **c** Rasta, **d** Rfilt, **e** jitter for dysarthria, **f** shimmer for dysarthria, **g** Rasta for dysarthria, **h** Rfilt analysis for dysarthria

Table 2 Optimized feature set after feature selection

Feature set	Features	Voice pathology	Dysarthria	Aphasia
GeMaps	Jitter	✓	✓	✓
(88)	Shimmer	✓	✓	✓
	Loudness	✓		✓
	Hammerberg	✓	✓	✓
	Index			
Compare	PCM	✓	✓	✓
(6373)	RASTA	✓	✓	✓
	RFilt	✓	**X**	**X**
	MFCC	✓	✓	✓

- **Random Forest**—The Random Forest algorithm commences by initially creating multiple decision trees, each trained on a distinct, random subset of the complete training data and a separate random subset of the available features (columns). This strategy enhances the ability to accurately capture the underlying data patterns. This process is repeated for each individual tree in the forest, ultimately yielding a set of decision trees that operate independently of each other.
- **XGBoost**—XGBoost excels in managing missing data, offers flexibility in handling various input formats like sparse data, and efficiently handles large datasets. Additionally, XGBoost integrates early stopping and cross-validation, thereby improving its generalization capabilities.
- **Support Vector Machine(SVM)**—SVM aims to pinpoint the optimal hyperplane serving as a decision boundary by maximizing the margin or the separation distance between distinct classes. The support vectors, which refer to the points nearest to the decision boundary, play a crucial role in determining the characteristics of the hyperplane.
- **CatBoost**—An open-source gradient boosting framework, aims to address the challenges posed by categorical data in machine learning tasks. CatBoost is capable of solving problems related to classification, regression, and ranking with equal effectiveness.
- **Histogram Gradient Boosting (HGB)**—The key idea behind HGB is to reduce the computational complexity of finding the best-split points for continuous features. HGB groups data points into histogram bins and only evaluates potential split points within these bins. This reduces the number of candidate split points and speeds up the training process while still achieving competitive predictive accuracy.

4 Experimental Results and Discussion

One of the main contributions of the paper is the application of various data augmentation techniques to balance the dataset before proceeding to the experimentation. The results are compared based on the skewed and balanced data to understand the difference in scores created by the SMOTE technique incorporated for data balancing. The corresponding results are listed in Fig. 3. CatBoost, XGBoost, Histogram gradient boosting (HGB), SVM, and Random Forest are the learning methods that were employed. The metrics employed for evaluation are benchmark metrics viz. accuracy, precision, recall, and F-score [26, 27]. The term accuracy has been used which is synonymous to the rate of pathology detection in this case. Accuracy is used as it is the standard/benchmark metric in the evaluation of pathology classification. A five-fold cross-validation was conducted with a training-test split of 80:20.

From the above experimental results in Fig. 3, we can draw the conclusion that the learning algorithm CatBoost along with SMOTE is the better combination to use. It provides an accuracy of 93%, precision of 92%, F1-score of 92%, and recall of 93%. In the case of, real-time data with SMOTE, the metrics we obtained are accuracy of 91%, precision of 87%, F1-score of 89%, and a recall of 91%. The investigation shows that the models function more effectively when the data is cleansed using SMOTE. When SMOTE is used, all of the models typically have greater accuracy, precision, F1-score, and recall. This shows that balancing the dataset with SMOTE improves model performance.

Evaluation for dysarthria has also been performed using both clean and real-time data. The results obtained are compared in Fig. 4. The plots that are illustrated in Fig. 4 shows the variations between the two kinds of input. The vertical bars represent results of clean data whereas the solid line marks the results of real-time data. From the results obtained from the dysarthria module, we can observe that random forest classifier gives us the best results with an accuracy of 86.57%, precision of 87%, F1-score of 87% and recall of 87%. The investigation shows that the models perform differently depending on whether the data is clean or real-time. Overall, voice pathological data are handled rather effectively by boosting machine learning algorithms compared to state-of-the-art deep learning [28]. Additionally, in comparison to SVM and AdaBoost on clean data, CatBoost, XGBoost, Random Forest, and HGB produce greater accuracy, precision, F1-score, and recall values. In terms of these metrics on real-time data, XGBoost, Random Forest, and HGB outperform CatBoost, SVM, and AdaBoost. A binary classifier is used for dysarthric voice classification. Training and testing are performed using both clean and noise-augmented voice data to take into account noise and background disturbances when speaking into a smartphone in real-time. The modified feature set as given in Table 2 is fed into machine learning models such as Random Forest, Catboost, and XgBoost for additional training and, eventually, real-time data classification. The revised feature set is input into the same machine learning models for further training and, eventually, real-time data classification.

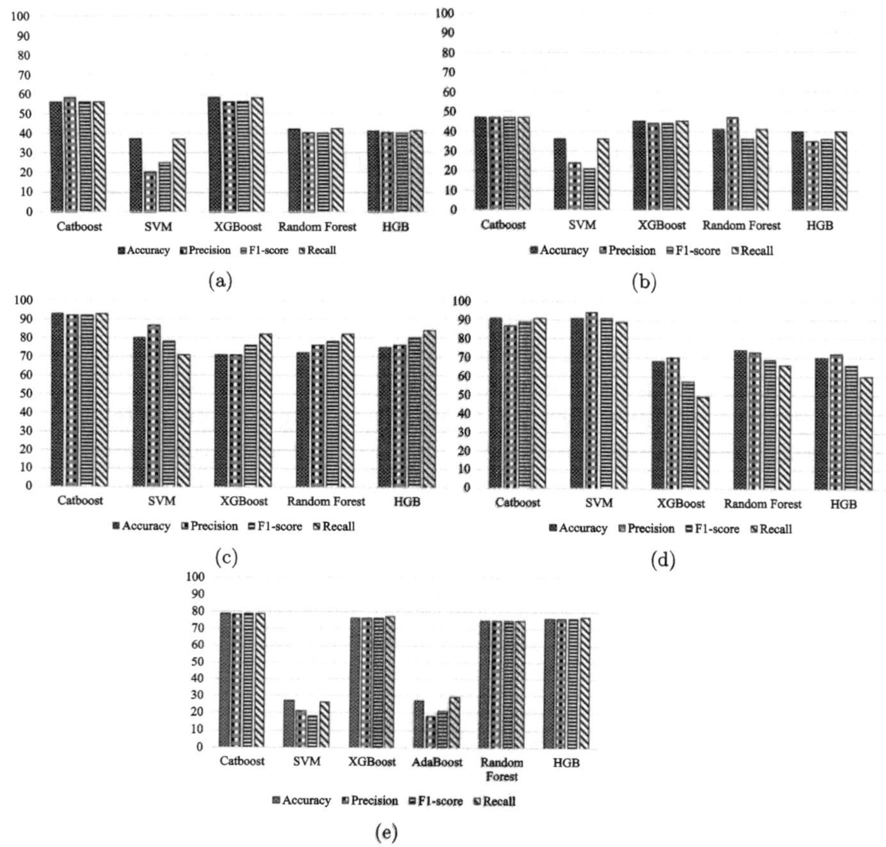

Fig. 3 Evaluation metric plot for detection of voice disorders with **a** clean data before data balancing **b** real-time data before data balancing **c** clean data after data balancing **d** real-time data after using SMOTE **e** real-time data after data augmentation techniques

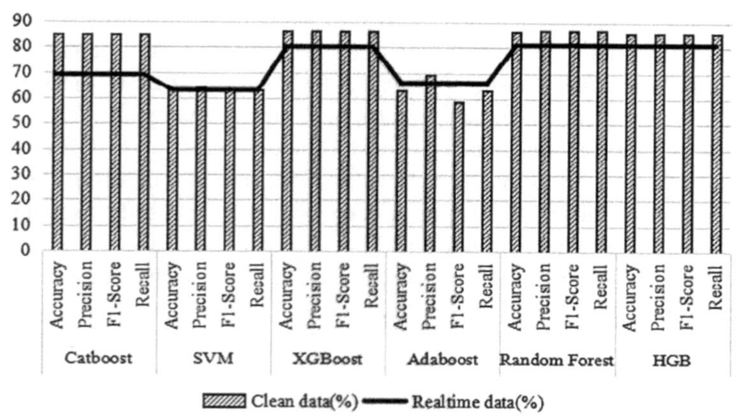

Fig. 4 Comparison of dysarthria detection results using clean data and real-time data

Table 3 Performance evaluation for aphasia module

Model	Clean data				Realtime data			
	Accuracy	Precision	F1-score	Recall	Accuracy	Precision	F1-score	Recall
Catboost	80	89	89	89	88	89	88	89
SVM	51	26	35	51	50	66	44	50
XGBoost	89	80	80	80	76	76	76	76
AdaBoost	63	69	59	63	65.6	66	66	66
Random Forest	86	86	86	86	76.4	77	76	76
HGB	86	87	86	86	76	77	76	76

Table 4 Comparison of proposed model with the state-of-the-art models

Year	Model	Features	Accuracy	Dataset
2019 [7]	SVM	Jitter Shimmer MFCC	84.5	SVD VOice ICar fEDerico II (VOICED)
2020 [26]	DT + ANN	MFCC	87.5	Self-made
2021 [12]	SVM	MFCC	84.37	SVD
2023 [27]	CNN-RNN	ZCR,RMSE, MFCC	87.11	SVD
Proposed	**Boosting**	Jitter, shimmer, MFCCs, loudness, Hammerberg index, RASTA, PCM	**88**	SVD, TORGO, Aphasiabank

The learning algorithms for aphasia detection and results are illustrated in Table 3. From the results obtained from the aphasia module, we can observe that, the Catboost classifier gives us the best results with an accuracy of 80%, precision of 89%, F1-score of 89%, and recall of 89% with clean data. CatBoost, XGBoost, Random Forest, and Histogram Gradient-based handle both types of data quite well. On clean data, CatBoost, XGBoost, Random Forest, and Histogram Gradient-based yield higher accuracy, precision, F1-score, and recall values than SVM and AdaBoost. The customized feature set in Table 2 can be further optimized and used for real-time detection of vocal disorders. A comparative analysis with state-of-the-art models for voice pathology detection, the results of which are depicted in Table 4. From Table 4 the effectiveness of the model in detecting multiple pathologies is at par with the state-of-the-art model for a single task. It is notable none of the prior work encompass an integrated model for voice pathology analysis and detection from speech utilizing a set of common features.

4.1 Integration Into e-Health Package and Sample Questionnaire Development

The three voice pathology modules are integrated into a single e-health platform. Voice pathology detection, dysarthria detection, and aphasia detection are the three modules that are combined into one platform. A sample GUI is developed as part of the work consisting of dialogue boxes for recording the voice and function selection.

Additionally, a questionnaire system was created to analyze the extent of aphasia using some inputs taken from the user. Figure 5 gives a random sample of the questions created to analyze the severity of aphasia from user input. The corresponding result obtained and information given by the system is also depicted in Fig. 5.

Fig. 5 **(a)–(f)** represents the aphasia detection questionnaire created as part of the e-health package and the test results for a random selection

5 Conclusion and Future Scope

From the above experimental results, we can draw the conclusion that for the detection of voice pathology and aphasia, CatBoost classifier with SMOTE works best. For dysarthria, the random forest classifier gives the best metrics. A specialized feature set is generated by removing unnecessary features based on the results of the dysarthria module and retraining the voice pathology module. Furthermore based on the results from aphasia module the features are selected depending on the most contributing feature. Thus, a customized feature set is generated which minimizes the time taken to generate feature sets to aid in the detection of different voice disorders. This will lead to a faster diagnosis in the medical field if applied practically.

In the future, based on the properties of the feature set corresponding to dysarthria, a relation can be derived to the properties from the data corresponding to that of Parkinson's disease. Another improvement which is possible is training the modules with regional accents. This would improve the overall classification over a diverse spectrum. The model can be then further trained with other languages and their respective dialects. This will require extensive effort in the collection of multilingual voice data corresponding to the various pathologies.

References

1. Vekkot S, Gupta D (2022) Fusion of spectral and prosody modelling for multilingual speech emotion conversion. Knowl-Based Syst 242:108360
2. Islam R, Tarique M, Abdel-Raheem E (2020) A survey on signal processing based pathological voice detection techniques. IEEE Access 8:66749–66776
3. Ramachandran VS (2012) Encyclopedia of human behavior, vol 2. Academic Press, pp 177–186
4. Moya-Galé G, Levy ES (2019) Parkinson's disease-associated dysarthria: prevalence, impact and management strategies. Res Rev Parkinsonism 9–16
5. Baudier P, Kondrateva G, Ammi C, Chang V, Schiavone F (2023) Digital transformation of healthcare during the COVID-19 pandemic: patients teleconsultation acceptance and trusting beliefs. Technovation 120:102547
6. Islam R, Abdel-Raheem E, Tarique M (2022) Voice pathology detection using convolutional neural networks with electroglottographic (EGG) and speech signals. Comput Methods Programs Biomed Update 2:100074
7. Verde L, De Pietro G, Alrashoud M, Ghoneim A, Al-Mutib KN, Sannino G (2019) Leveraging artificial intelligence to improve voice disorder identification through the use of a reliable mobile app. IEEE Access 7:124048–124054
8. Chowdary PN, Aravind VS, Vardhan GV, Akshay MS, Aashish MS (2023) A few-shot approach to dysarthric speech intelligibility level classification using transformers. arXiv preprint arXiv:2309.09329
9. Lal GJ, Gopalakrishnan EA, Govind D (2022) A recurrence network approach for characterization and detection of dynamical transitions during human speech production. Circ Syst Signal Process 41(12):6975–6998
10. Cesari U, De Pietro G, Marciano E, Niri C, Sannino G, Verde L (2018) Voice disorder detection via an m-Health system: design and results of a clinical study to evaluate Vox4Health. BioMed Res Int (2018)

11. Raj JR, Jabez J, Srinivasulu SS, Gowri S, Vimali JS (2021) Voice pathology detection based on deep neural network approach. In: IOP conference series: materials science and engineering, vol 1020, no 1. IOP Publishing, p 012001
12. AL-Dhief FT, Latiff NMAA, Baki MM, Malik NNNA, Sabri N, Albadr MAA (2021) Voice pathology detection using support vector machine based on different number of voice signals. In: Proceedings on IEEE Asia-Pacific conference on communications (APCC). IEEE, pp 1–6
13. Joshy AA, Rajan R (2022) Automated dysarthria severity classification: a study on acoustic features and deep learning techniques. IEEE Trans Neural Syst Rehabil Eng 30:1147–1157
14. Mujumdar MV, Kubichek RF (2010) Design of a dysarthria classifier using global statistics of speech features. In: IEEE international conference on acoustics, speech, signal Proceedings on (ICASSP). IEEE, pp 582–585
15. Le D, Licata K, Provost EM (2017) Automatic paraphasia detection from aphasic speech: a preliminary study. In: Proceedings of Interspeech, pp 294–298
16. Rose Y (2022) 32 Managing phonological development data within PhonBank: the Chisasibi child language acquisition study. In: The open handbook of linguistic data management, vol 391
17. Rusu C (2012) Clustering before training large datasets-case study: K-SVD. In: Proceedings of 20th European signal proceedings of conference (EUSIPCO). IEEE, pp 2188–2192
18. Joy NM (2018) Umesh S (2018) Improving acoustic models in TORGO dysarthric speech database. IEEE Trans Neural Syst Rehabil Eng 26(3):637–645
19. Krishna G, Carnahan M, Shamapant S, Surendranath Y, Jain S, Ghosh A, Tran C, Millan JD, Tewfik AH (2021) Brain signals to rescue aphasia, apraxia and dysarthria speech recognition. In: Proceedings of 43rd annual international conference in medicine and biology society (EMBC). IEEE, pp 6008–6014
20. Eyben F, Wöllmer M, Schuller B (2010) Opensmile: the Munich versatile and fast open-source audio feature extractor. In: Proceedings on 18th ACM international conference on multimedia, pp 1459–1462
21. Ancillao A, Galli M, Mignano M, Dellavalle R, Albertini G (2013) Quantitative analysis of pathological female human voice by processing complete sentences recordings. J Laryngol Voice 3(2):46–51
22. Doğdu C, Kessler T, Schneider D, Shadaydeh M, Schweinberger SR (2022) A comparison of machine learning algorithms and feature sets for automatic vocal emotion recognition in speech. Sensors 22(19):7561
23. Flores AC, Icoy RI, Peña CF, Gorro KD (2018) An evaluation of SVM and Naive Bayes with smote on sentiment analysis data set. In: Proceedings of international conference on engineering, applied sciences, and technology (ICEAST). IEEE, pp 1–4
24. Bishop CM, Nasrabadi NM (2006) Pattern recognition and machine learning, vol 4, no 4, p 738. Springer, New York
25. Murphy KP (2012) Machine learning: a probabilistic perspective. MIT Press
26. Milani MM, Ramashini M, Krishani M (202) A real-time application to detect human voice disorders. In: Proceedings of international conference on decision aid sciences and application (DASA). IEEE, pp 979–984
27. Ksibi A et al (2023) Voice pathology detection using a two-level classifier based on combined CNN-RNN architecture. Sustainability 15(4):3204
28. Kumar MR, Vekkot S, Lalitha S, Gupta D, Govindraj VJ, Shaukat K, Alotaibi YA, Zakariah M (2022) Dementia detection from speech using machine learning and deep learning architectures. Sensors 22(23):9311

Computer Vision

Revolutionizing Student Engagement: Real-Time Emotion Detection and Interest Identification in Live Video Streams

M. B. Govind, A. S. Humaid, and G. Malu

Abstract In today's rapidly evolving educational landscape, the analysis of student engagement through the prism of emotion ensures a transformative possibility for educators and learners. We introduce an innovative real-time emotion detection system that transcends conventional methodologies, with a specific emphasis on identifying student interest through emotion analysis in live video streams. This comprehensive approach, utilizing cutting-edge technologies such as data preprocessing, deep learning architecture, exhaustive training, real-time facial detection, emotion inference, and immersive graphical presentation, optimizes the recognition and visualization of emotions. Our methodology harnesses the powerful synergy between a pre-trained VGG19 model, renowned for its precision in feature extraction, and a custom-designed neural network finely tuned for both emotion classification and interest detection. To bolster our system's performance and adaptability, we conducted extensive training and fine-tuning using the FER13 dataset, renowned for its comprehensiveness in emotion recognition spanning a wide range of human expressions. The custom preprocessed images are provided to a pre-trained VGG19 model with a custom neural network. The two dense layers with 256 and 128 units in the network, each accompanied by batch normalization and dropout, also ensure better system accuracy. Extensive data augmentation on the training data allows robust model training, enabling real-time emotion prediction and introducing innovative techniques for identifying student interest, exemplifying a comprehensive approach to emotion recognition and educational technology. Empowered by our meticulously refined model, our system rapidly and accurately predicts emotions, overlaying emotive descriptors onto live video feeds in real-time.

Keywords Student engagement · Emotion detection · Student's interest · Deep learning · VGG19

M. B. Govind · A. S. Humaid · G. Malu (✉)
Digital University Kerala, Technopark Phase IV, Pallipuram, Thiruvananthapuram, Kerala 695317, India
e-mail: malu.g@duk.ac.in

S. M. Thampi et al. (eds.), *Fifth International Conference on Computing and Network Communications*, Lecture Notes in Electrical Engineering 1221,
https://doi.org/10.1007/978-981-97-4711-5_29

1 Introduction

The classroom is a unique environment that fosters direct, one-on-one communication between educators and learners [1]. Online instruction often struggles to replicate the immediate discussions, questions, and concerns addressed, prompt clarification, and active engagement that educators can foster in this physical setting. Teachers can adapt their teaching methods to meet the specific needs of each student by observing and responding to nonverbal cues from their students, such as body language, facial expressions, and emotional states [2]. Technological advancements in recent years, such as artificial intelligence and emotion detection, have opened up new avenues to understand better and capture students' attention and interest in the classroom [3]. Artificial intelligence (AI) systems can analyze students' body language and facial expressions in real-time, enabling the identification of various emotions, including joy, confusion, and boredom, providing instructors with vital insights into their students' engagement and interest levels. In addition to analysis, emotion recognition technology provides immediate feedback to teachers, allowing them to adjust their teaching methods based on their students' emotional responses. For example, if the system detects signs of apathy or uncertainty, the teacher can adapt the pace, content, or teaching style to re-engage the students. This dynamic approach adds flexibility and dynamism to the learning process.

Moreover, recent research on emotion recognition in online learning has underscored the importance of understanding students' emotional experiences during virtual education [4]. This study concentrates on utilizing facial expression recognition models to assess students' emotions during online education, demonstrating the potential of integrating computer simulation perspectives into the learning environment. Artificial intelligence (AI) augments the educational environment by enabling customized learning. The curriculum and content can be customized by technology to fit each learner's interests and emotional states. When a student shows a strong interest in one subject and loses interest in another, the system can dynamically modify the course content to spark their interest again. Educators can utilize this knowledge to encourage students or think of different ways to assess them. Emotion detection involves the recognition of human emotions from visual cues, such as facial expressions. These cues are captured through images, and the subsequent analysis and interpretation hold immense promise in understanding human behavior and response.

In the domain of emotion recognition, machine learning has been a central pillar for automating the classification process, with machine learning algorithms such as Support Vector Machines, k-Nearest Neighbors, Naïve Bayes, and Decision Trees playing pivotal roles in achieving accurate emotion classification. These machine learning techniques offer the advantage of adaptability to various feature sets and data types, making them prominent choices for researchers and practitioners. The primary components consist of face detection, feature extraction, and emotion classification [5]. The literature suggests that feature extraction techniques frequently employed include Bezier curves [6], clustering methods [7], independent component analysis [8], (2D)2FPCA [9, 10], (2D)2 MFPCA [11], principal component analysis [12], local binary patterns [13], and feature-level fusion techniques [14]. Subsequently, these extracted features are provided to classifiers like Support Vector Machines [15], Hidden Markov Models [16], k-Nearest Neighbors [17], Naïve Bayes, and Decision Trees [18] for the task of emotion classification. However, a drawback of conventional systems is their separation of feature extraction and classification phases, which poses a challenge in achieving enhanced system performance [19]. However, it's essential to acknowledge that machine learning-based systems for facial emotion detection come with certain limitations. These limitations encompass difficulties in accommodating variations in lighting conditions, facial expressions, and poses, which can diminish their accuracy in real-world applications. The recognition of intricate and subtle emotions poses a challenge for ML models, as they may not capture the nuanced details necessary for precise emotion classification. Additionally, the demand for a substantial volume of labeled training data can become a bottleneck for ML-driven facial emotion detection, given the resource-intensive nature of data acquisition and annotation. Furthermore, these systems may be prone to biases inherent in the training data, potentially resulting in inaccuracies and disparities in emotion recognition across diverse demographic groups.

To overcome these limitations, deep learning methods have revolutionized the field of machine learning due to their remarkable advances over traditional machine learning techniques. The pivotal advantage of deep learning lies in its ability to harness the power of end-to-end learning, allowing for a more holistic approach to problem-solving [20–22]. Unlike conventional machine learning, which relies on handcrafted feature engineering, deep learning leverages its deep neural networks to automatically extract relevant features, rendering it highly effective in tasks such as segmentation and classification. Additionally, deep learning thrives on large datasets, benefiting from their inherent capacity to enhance model performance. To further bolster dataset sizes, researchers have embraced data augmentation techniques, encompassing translations, normalizations, cropping, noise addition, and scaling, all of which contribute to the effectiveness of deep learning [23, 24]. However, it's important to acknowledge that deep learning, despite its numerous merits, is not without its limitations. One notable drawback is its data hunger, which can make it less practical for applications with limited data availability. Furthermore, deep learning models tend to be computationally intensive, often requiring substantial processing power for training and inference. This can lead to increased time and resource demands, making deep learning less accessible in certain contexts. It is essential to consider these demerits, but the advantages, such as its ability to automatically learn features, the power of large datasets, and the potential for transfer learning, continue to make deep learning a preferred choice in the realm of modern machine learning [25].

Recent advancements in facial expression recognition have underscored the significance of CNN-based methods. For instance, Yingruo Fan and colleagues introduced a novel approach involving a multi-region ensemble CNN. This method extracts features from three distinct facial regions (eyes, nose, and mouth) and subsequently combines them to predict emotions, showcasing promising results on dataset like AFEW 7.0 and RAF-DB [26]. Yingying Wang et al. proposed an innovative emotion recognition method based on an auxiliary model. In this approach, information from three major sub-regions of the face, namely the eyes, nose, and mouth, is combined with the full facial image using a weighted process to maximize information utilization. This model's performance was rigorously evaluated on datasets including CK+, FER2013, SFEW, and JAFFE [27]. In a separate study, Frans Norden and colleagues delved into facial expression recognition, employing architectures such as VGG16 and Resnet50. Their experiments, conducted using the dataset JAFFE and FER2013, revealed that Resnet50 achieved the highest classification accuracy when compared to other state-of-the-art methods [28]. Furthermore, Jyostna Devi Bodapati et al. harnessed VGG16 for feature extraction and a multi-class Support Vector Machine (SVM) for classification, achieving impressive accuracy rates on the CK+ dataset [29]. These recent contributions underscore the increasing importance of CNN-based methods in advancing the field of emotion recognition.

In the current landscape of education and human–computer interaction, our research stands as a critical undertaking with far-reaching implications. In this work, Real-Time Emotion Detection and Interest Identification in Live Video Streams, we delve into the creation of an emotion detection model that leverages the power of VGG19 to accurately classify emotions from facial images from live stream videos. Building upon recent advancements in CNN-based methods, our research contributes to the growing body of work dedicated to improving emotion recognition systems. In contrast to conventional approaches that rely on discrete phases of feature extraction and classification, our methodology embraces the holistic potential of deep learning. Through the utilization of a pre-trained VGG19 architecture, we expedite the development of our model while ensuring its adaptability to the latest innovations in the field. Our research places a specific emphasis on the analysis of student's interests through the application of emotion detection. This emphasis extends to both online and offline learning platforms, making our approach versatile and impactful. By employing this innovative method, we provide a dynamic means of comprehending and engaging with students' emotions, thus making a significant and lasting contribution to the continually evolving domains of education and human–computer interaction. The fusion of emotion detection technology with both online and offline learning platforms holds paramount significance for several compelling reasons.

2 Methodology

The proposed Real-time Emotion Detection and Interest Identification in Live Video Streams utilizes cutting-edge technologies in deep learning. The system is designed to capture the students' emotion and interest during live classes where the speaker might not be able to remotely gauge their attention level. The sequence of procedure, preprocessing, exhaustive training, real-time facial detection, emotion inference, and graphical presentation are the main components of the system that optimize the recognition and visualization of emotions. The pre-trained VGG19 model and a

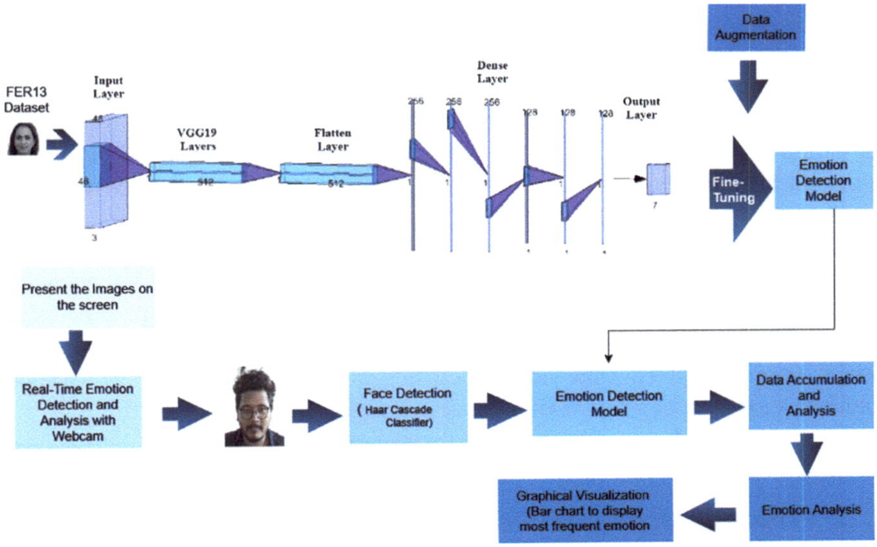

Fig. 1 Proposed system architecture

custom-designed neural network with fine-tuning helped for both emotion classification and interest detection. Training and fine-tuning done using the FER13 dataset. The two dense layers with 256 and 128 units in the network, each accompanied by batch normalization and dropout, are also introduced here. The ImageDataGenerator function in Keras has been used for data augmentation while training, thus enabling real-time emotion prediction and introducing innovative techniques for identifying student interest, exemplifying a comprehensive approach to emotion recognition and educational technology (Fig. 1).

2.1 Data Preparation and Preprocessing

The Facial Expression Recognition 2013 (FER2013) Dataset is a valuable resource within the fields of computer vision and facial expression recognition, consisting of a total of 35,887 grayscale images. These images serve as a foundational dataset for the training and evaluation of facial expression recognition algorithms. The dataset spans seven distinct emotion categories, encompassing anger, disgust, fear, happiness, sadness, surprise, and neutrality. Each image within the dataset is relatively small, typically measuring 48 $\times$ 48 pixels. However, it is important to note that these images were collected from various sources on the internet, leading to inherent variability in image quality and lighting conditions. Consequently, the FER2013 dataset poses challenges to algorithms by presenting diverse facial expressions, poses, and illumination scenarios. Nevertheless, it remains a widely employed benchmark for researchers and machine learning practitioners interested in facial expression recognition.

This data preprocessing process involves converting the pixel values of facial images into a format compatible with deep learning models, thereby ensuring that the data are in an appropriate format for training the model. Subsequently, the dataset is thoughtfully divided into training and testing sets to facilitate evaluation, typically with an 80% training and 10% testing, validation (10%) partition. Furthermore, pixel values are scaled to fall within a range between 0 and 1 to maintain consistent data representation. Additionally, the emotion labels are subjected to one-hot encoding to render them suitable for multi-class classification tasks. This data preprocessing phase is pivotal, as it sets the foundation for the subsequent steps of model architecture design, training, and real-time video processing, all of which collectively contribute to the development of a robust emotion detection system.

2.2 *Proposed Model Architecture*

Our architecture is based on a deep learning model, which is essentially based upon the popular VGG19 architecture created by the University of Oxford's Visual Geometry Group. VGG19, which has 19 layers in its deep structure and is highly skilled at extracting complex hierarchical information from images, has received praise for its ability to perform well in image classification tasks. We enact a rigorous customization procedure to tailor this sturdy architecture to the particular goal of emotion recognition. The first input layer gets the input data ready, and the VGG19 model processes it after that. The foundation is the VGG19 architecture, whose convolutional layers have been painstakingly trained to extract complex visual patterns. The flatten layer reshapes the output for additional processing after the VGG19 layers. The dense1 layer with 256 units is then integrated. Next, a batch_normalization layer is added to improve the model's general stability by normalizing the activations of the preceding layer. One method to avoid overfitting is to deliberately use a 'dropout' layer. The 'dense2' layer with 128 units follows, and to further stabilize the model, a batch_normalization_1 layer comes next. The inclusion of an additional dropout_1 layer strengthens the model's capacity for generalization. Finally, the model is able to forecast the probability of the seven distinct emotional classes thanks to the 'output' layer (dense) with seven units. Furthermore, in order to improve the model's capacity for generalization, we incorporate data augmentation methods, guaranteeing a large and varied training dataset. Overall, these finely tuned layers along with data augmentation methods position our deep learning model as a flexible and potent real-time facial emotion analysis solution. Figure 2 depicts the VGG19 architecture used here.

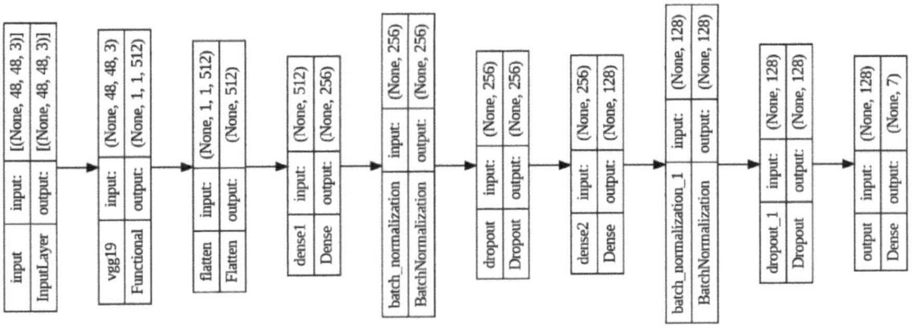

Fig. 2 VGG19 architecture used for emotion recognition

2.3 *Real-Time Emotion Detection and Dynamic Content Presentation*

The dynamic contents including photographs and videos are presented and simultaneously taken student's images to identify student's emotions at that particular instance. Meanwhile, the trained model is transformed into real-time emotion detection mode. It leverages the user's webcam feed, captures video frames using OpenCV, and employs a Haar Cascade classifier to facilitate facial detection. The Haar Cascade face detection algorithm in Eq. 1 leverages the concept of an integral image. This integral image efficiently computes the cumulative sum of pixel values in an input image up to a given point (x, y).

$$I(x, y) = \Sigma\Sigma\ \text{Image}(x', y')\ \text{for}\ x' <= x\ \text{and}\ y' <= \text{y} \tag{1}$$

Haar-like features, which are fundamental to the Haar Cascade approach, are defined by specifying rectangular regions within an image, and their values are calculated using this integral image equation. These features serve as descriptors of image regions and are crucial for object detection. During the detection process, the Haar Cascade algorithm employs a cascade of classifiers, each with its set of Haar-like features and associated thresholds. The computed feature values are compared to these thresholds, aiding in the determination of whether a region contains the object of interest, such as a face. The cascade structure allows for rapid rejection of non-object regions, making the Haar Cascade algorithm efficient for real-time applications and robust object detection in images.

Upon detecting a face, the system extracts the facial region and resizes it to a standardized 48 × 48 pixel format, aligning with the model's input requirements. The pre-trained model, now adept at recognizing emotions, is deployed to predict the emotion conveyed by the resized facial image. The model assigns the highest-confidence emotion label to the detected face. This prediction is then displayed directly on the screen, overlaying the webcam feed, ensuring that users can perceive the emotional analysis in real-time. Additionally, it introduces an innovative feature—dynamic content presentation, where the choice of displayed content dynamically shifts based on the user's detected emotion, fostering a personalized and emotionally responsive interaction This architecture involves data preprocessing to prepare facial expression data, customizing the VGG19-based deep learning model for emotion recognition, and implementing real-time emotion prediction using a webcam feed.

3 Result and Discussions

The Real-Time Emotion Detection and Interest Identification in Live Video Streams uses FER2013 Dataset for training and online live videos for testing the emotions. Figure 3 shows few images of FER2013 dataset. A total of 35,887 grayscale images are used to model the system. The dataset spans seven distinct emotion categories, encompassing anger, disgust, fear, happiness, sadness, surprise, and neutrality. Each image within the dataset is relatively small, typically measuring 48×48 pixels. These images were collected from various sources on the internet, leading to inherent variability in image quality and lighting conditions. This diversity in facial expressions, poses, and illumination scenarios helped to train the system in a better manner.

The three emotion detection models: VGG19-based, CNN, and VGG16, are used in the initial study. These models are subject to evaluation primarily based on their real-time emotion analysis performance as users engage with a series of presented images. The central focus of the evaluation lies in key metrics, namely accuracy and $F1$-scores, which offer a comprehensive understanding of the models' proficiency in accurately identifying and categorizing user-expressed emotions.

During the evaluation process, the VGG19-based model demonstrated an overall accuracy rate of 68%, indicative of its ability to correctly recognize emotions in a substantial proportion of instances. Table 1 shows the accuracy comparison of the

Fig. 3 Various emotions expressed - FER2013 dataset

Table 1 Performance Evaluation of CNN, VGG-16, VGG-19 models

Model	Accuracy
CNN	66
VGG-16	65
VGG-19	68

three models. This analysis proceeds to dissect the performance across individual emotion categories, revealing varying levels of precision and recall. For instance, the 'Happy' emotion exhibited the highest $F1$-score of 0.86, signifying outstanding precision and recall. Conversely, 'Fear' achieved an $F1$-score of 0.54, suggesting potential room for enhancement. Meanwhile, 'Surprise' exhibited robust performance with an $F1$-score of 0.78, whereas 'Angry' and 'Sad' displayed moderate $F1$-scores of 0.54 and 0.53, respectively. 'Disgust' and 'Neutral' yielded $F1$-scores of 0.64 and 0.62, respectively. This meticulous analysis provides a thorough comprehension of the model's performance concerning distinct emotional states.

The fundamental aim of this study revolves around the real-time analysis of emotions as users interact with a sequence of images. The process entails the acquisition of live video footage from the user's webcam, enabling the continuous monitoring of their facial expressions throughout image interactions, and Fig. 4 shows the sample images shown during the live streaming and the expected emotion on seeing these images which are also labeled. Figure 5 showcases the emotion recognized during live streaming and their confidences.

Then, an analysis is performed to depicting the distribution of detected emotions and is graphical plotted as bar charts. The prevailing emotion, most prominently

Fig. 4 Images shown during the live streaming **a** anger, **b** happiness, **c** neutrality, **d** sadness, **e** fear, **f** disgust

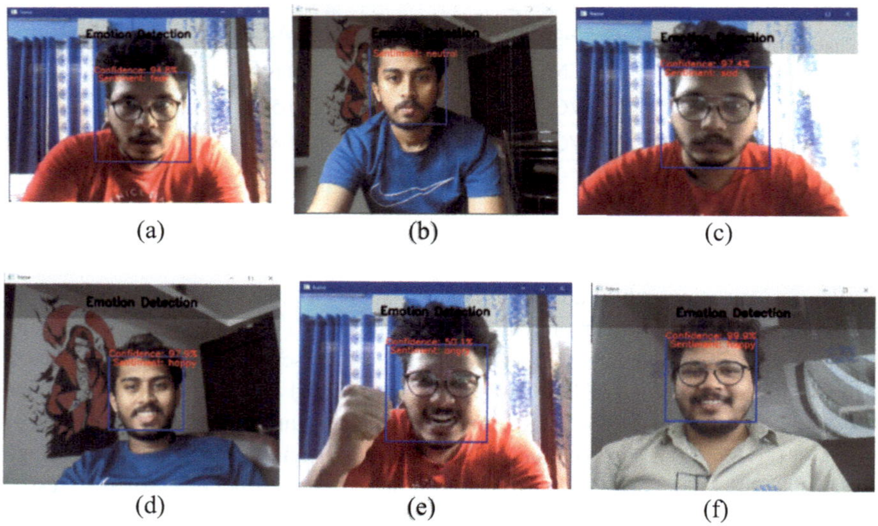

Fig. 5 Emotion recognized during live streaming **a** emotion: fear, confidence: 94.8%, **b** emotion: neutral, confidence: 67.2%, **c** emotion: sad, confidence: 97.4%, **d** emotion: happy, confidence: 97.9%, **e** emotion: angry, confidence: 50.1%, **f** emotion: happy, confidence: 99.9%

expressed, is distinctly labeled within the chart. This real-time emotional analysis serves as a valuable tool, facilitating the evaluation and visualization of users' emotional responses, thereby providing valuable insights into their engagement with the presented images.

In the comparative examination of the three models, the VGG19-based model excelled with the highest accuracy at 68%, closely trailed by the CNN model, which achieved an accuracy rate of 66%. The VGG16 model also demonstrated commendable performance, boasting an accuracy score of 65%. Notably, the VGG19 model exhibited notable enhancements in $F1$-scores across various emotion categories, including 'Angry,' 'Fear,' and 'Sad.' Both the VGG19 and CNN models exhibited exceptional proficiency in recognizing 'Happy' emotions, achieving an impressive $F1$-score of 0.86. These findings underscore the substantial potential of all three models for capturing and categorizing emotions, thus shedding light on their suitability for practical applications. Tables 2 and 3 show the comparison.

The results of this study demonstrate a statistically significant correlation between the variables, providing empirical evidence to support the proposed hypothesis. This

Table 2 Dataset Distribution Metrics

Dataset	Total images	Training	Testing	Angry (0)	Disgust (1)	Fear (2)	Happy (3)	Sad (4)	Surprise (5)	Neutral (6)
FER13 dataset	35,888	28,710	7178	4954	547	5121	8989	6077	4002	6198

Table 3 Model Evaluation Matrices

	Precision	Recall	$F1$-score	Support
Angry	0.59	0.61	0.60	985
Disgust	0.88	0.67	0.76	102
Fear	0.58	0.53	0.55	1043
Happy	0.88	0.83	0.85	1765
Sad	0.47	0.65	0.55	1210
Surprise	0.87	0.67	0.76	795
Neutral	0.64	0.59	0.61	1278
Accuracy			0.68	7178
Macro avg.	0.70	0.65	0.67	7178
Weighted avg.	0.68	0.66	0.67	7178

graphical representation offers a succinct and intuitive overview of the predominant emotional responses recorded during users' interactions with the presented images in real-time. By juxtaposing this visual aid with the textual analysis of the model's performance and the comparative assessment of the three models, readers can swiftly discern patterns and trends in emotional expressions. The bar chart not only complements the narrative but also fosters a deeper comprehension of the research outcomes (Fig. 6). Its inclusion enhances the clarity and impact of the study while facilitating a more accessible and engaging interpretation of the results.

4 Conclusion and Future Work

Our practical solution for real-time facial emotion detection has successfully achieved the primary aim of analyzing the way of interest in students through their emotional responses. It presents a robust framework that amalgamates efficient data preprocessing, a pre-trained VGG19 model, and a rigorous training and evaluation process. This approach not only enables real-time emotion detection but also enhances user engagement by overlaying predicted emotions onto the webcam feed, providing immediate feedback during image presentations. Additionally, the option to save the trained model ensures its adaptability for future projects. Although achieving a commendable accuracy rate of 68%, our solution leaves room for further improvement through the fine-tuning of hyperparameters, refinement of architectural aspects, and enhancements to the user interface. These potential enhancements hold the promise of significantly augmenting the system's versatility and user-friendliness, rendering it an invaluable tool in contexts where understanding the way of interest in students through their emotional responses is of paramount importance, aligning with the core aim of our project.

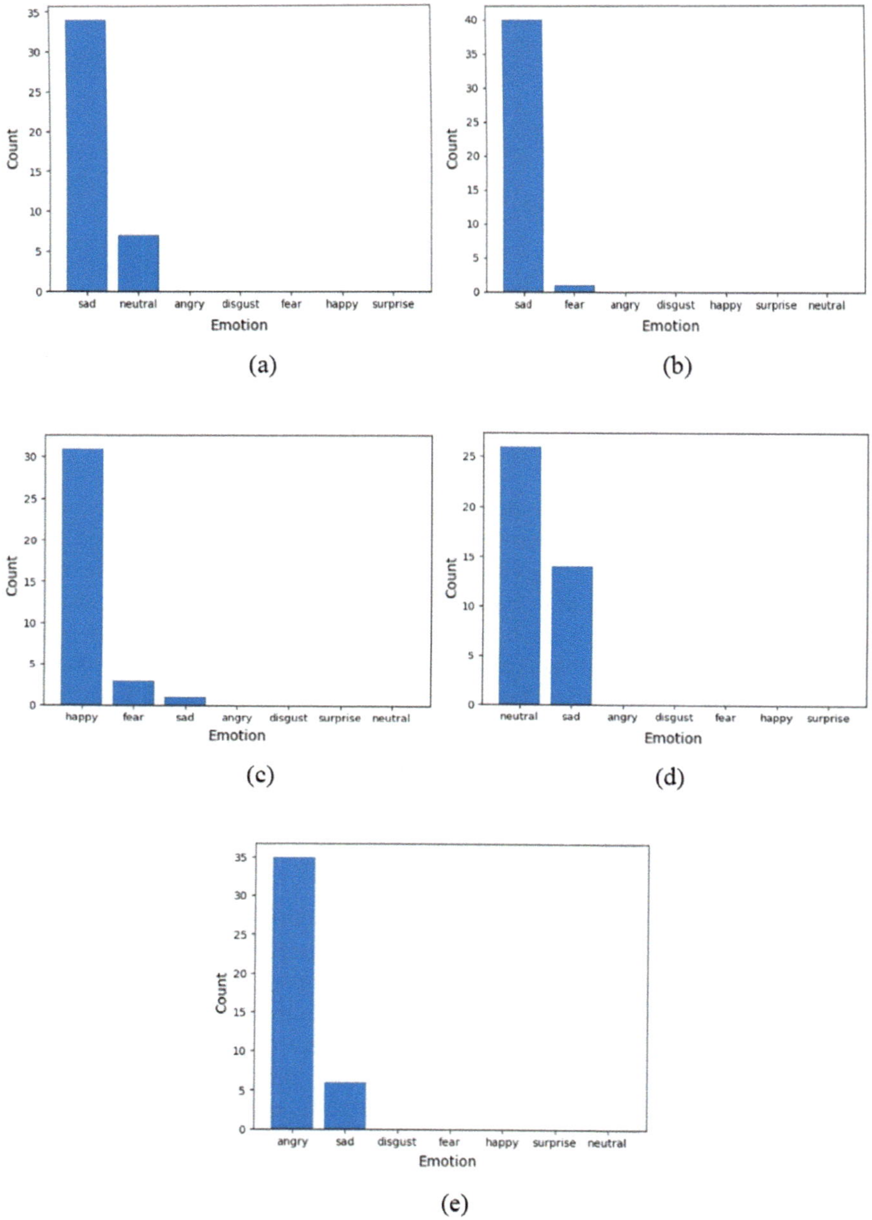

Fig. 6 Results of emotional responses to images, most frequent items displayed during the while showing various images **a** sad and neutral, **b** sad and fear, **c** happy, fear, sad, **d** neutral, sad, **e** angry and sad

We aim to integrate EEG signals to enhance emotion prediction, especially in the context of understanding emotions in autistic children. Developing a specialized facial emotion dataset for autistic children will be a crucial step to train robust emotion detection models. This integration offers the potential for improved emotion recognition, early detection, and personalized learning experiences. Additionally, it has broader implications in human–computer interaction, enabling technology to respond intuitively to users' emotional states. This advancement can significantly benefit autistic children, further our understanding of emotions and neural responses, and contribute to a more empathetic society. The combination of a tailored dataset and an EEG-powered model presents promising prospects for future research in emotion analysis and human–computer interaction.

References

1. Pianta RC, Hamre BK, Allen JP, (2012) Teacher-student relationships and engagement: conceptualizing, measuring, and improving the capacity of classroom interactions, pp 365–386
2. Krithika LB, GGb LP (2016) Student emotion recognition system (SERS) for e-learning improvement based on learner concentration metric. Int Conf Comput Model Secur Procedia Comput Sci 85(2016):767–776 (2016)
3. Imani M, Montazer GA (2019) A survey of emotion recognition methods with emphasis on e-learning environments 147:102423
4. Wang W, Xu K, Niu H, Miao X (2020) Emotion recognition of students based on facial expressions in online education based on the perspective of computer simulation. https://doi.org/10.1155/2020/4065207
5. Tian Y, Kanade T, Cohn JF (2011) Facial expression recognition. In: Handbook of face recognition. Springer, London, pp 487–519
6. Bansal S, Nagar P (2015) Emotion recognition from facial expressions based on Bezier curves. Int J Adv Inf Technol 5(4):5
7. Senthilkumar TK, Rajalingam S, Manimegalai S, and Srinivasan VG (2016) Human facial emotion recognition is achieved through automatic clustering based morphological segmentation and shape/orientation feature analysis. In: 2016 IEEE international conference on computational intelligence and computing research (ICCIC). IEEE, pp 1–5
8. Guo X, Zhang X, Deng C, Wei J (2013) Facial expression recognition is based on independent component analysis. J Multimed 8(4):402–409
9. Wang N, Li Q, Abd El-Latif AA, Peng J, and Niu X (2013) Multibiometrics fusion for identity authentication: dual iris, visible, and thermal face imagery. Int J Secur Appl 7(3):33–44
10. Wang N, Li Q, Abd El-Latif AA, Peng J, Niu X (2013) Two directional two-dimensional modified Fisher principal component analysis: an efficient approach for thermal face verification. J Electron Imaging 22(2):023013
11. Abd El-Latif AA, Hossain MS, Wang N (2019) Score level multibiometrics fusion approach for healthcare. Clust Comput 22(1):2425–2436
12. Mansour AH, Salh GZA, and Alhalemi AS (2014) Facial expression recognition is based on principal component analysis (PCA). arXiv preprint arXiv:1506.01939
13. Shan C, Gong S, and McOwan PW (2009) Facial expression recognition based on local binary patterns: a comprehensive study. Image Vis Comput 27(6):803–816
14. Wang N, Li Q, El-Latif AA, Peng J, Niu X (2013) A novel multibiometric template security scheme for the fusion of dual iris, visible, and thermal face images. J Comput Inf Syst 9(19):1–9
15. Michel P, El Kaliouby R (2005) Facial expression recognition using support vector machines In: Paper presented at 10th international conference on human-computer interaction, Crete, Greece

16. Wang J, Wang S, and Ji Q (2014) Early facial expression recognition using hidden Markov models In: Paper presented at 22nd international conference on pattern recognition. IEEE, pp 4594–4599
17. Thakare PP, Patil PS (2016) Facial expression recognition algorithm based on KNN classifier. Int J Comput Sci Netw 5(6):941
18. Salmam FZ, Madani A, Kissi M (2016) Facial expression recognition using decision trees In: 2016 13th international conference on computer graphics, imaging, and visualization (CGiV). IEEE, pp 125–130
19. Nonis F, Dagnes N, Marcolin F, Vezzetti E (2019) 3D approaches and challenges in facial expression recognition algorithms—a literature review. Appl Sci 9(18):3904
20. Jain N, Nguyen TN, Gupta V, Hemanth DJ (2021) Dental X-ray image classification using deep neural network models. Ann Telecommun
21. Dash R, Nguyen TuN, Cengiz K, Sharma A (2021) FTSVR: finetuned support vector regression model for stock predictions. Neural Comput Appl. https://doi.org/10.1007/s00521-021-05842-w
22. Vu D, Nguyen T, Nguyen TV, Nguyen TN, Massacci F, Phung PH (2019) A convolutional transformation network for malware classification. In: 2019 6th NAFOSTED conference on information and computer science (NICS), pp 234–239
23. Li S, Deng W (2018) Deep facial expression recognition: a survey. arXiv preprint arXiv:1804.08348
24. Pitaloka DA, Wulandari A, Basaruddin T, Liliana DY (2017) Enhancing CNN with preprocessing stage in automatic emotion recognition. Proc Comput Sci 116:523–529
25. Ng HW, Nguyen VD, Vonikakis V, Winkler S (2015) Deep learning for emotion recognition on small datasets using transfer learning. In: Proceedings of the 2015 ACM on international conference on multimodal interaction, pp 443–449
26. Fan Y, Lam JC, Li VO (2018) Multi-region ensemble convolutional neural network for facial expression recognition. In: Proceedings of international conference on artificial neural networks, Rhodes, Greece, pp 84–94
27. Wang Y, Li Y, Song Y, Rong X (2019) Facial expression recognition is based on auxiliary models. Algorithms 12(11):227
28. Norde´n F, von Reis Marlevi F (2019) A comparative analysis of machine learning algorithms in binary facial expression recognition (Dissertation). http://www.diva-portal.org/smash/record.jsf?pid=diva2%3A1329976&dswid=3676
29. Jyostna Devi B, Veeranjaneyulu N (2019) Facial emotion recognition using deep CNN features. Int J Innov Technol Explor Eng (IJITEE) 8(7)

Image Captioning with Multiple Perspectives—A Visual Context-Based Approach

G. Ashwin, V. Chaitanya, Kishan, Rohith, and Priyanka C. Nair

Abstract This study tackles the difficult issues of image captioning while negotiating the complexity of visual data processing. The complexity of visual data and the associated processing requirements make image captioning a daunting task. This research describes an effective method for image captioning that makes advantage of attentional mechanisms. Several models, including Transformer, VGG16, VGG19, Inception, RNN decoders, and Bahdanau's attention mechanism, are contrasted and analyzed to demonstrate the benefits of integrating attention processes. By allowing the model to focus on the appropriate visual component, attention improves the accuracy of the generated captions. Transformer models, in particular, outperform other methods by capturing complicated relationships and delivering accurate output, and they have the highest BLEU Score of 70. The findings emphasize the significance of attention mechanisms and the relevance of selecting a suitable model architecture to maximize picture captioning performance. The variety of potential applications emphasizes the benefits and potential impact of image captioning in a broad range of scenarios. Several use cases are found beneficial, including supporting the visually impaired, improving product descriptions in e-commerce, assisting medical diagnosis, improving image search ability, and allowing effective communication and comprehension of visual information.

Keywords VGG16/19 · RNN decoder · Bahdanau's attention · Transformer · Multi-headed attention · Effnet

---

G. Ashwin · V. Chaitanya · Kishan · Rohith · P. C. Nair (✉)
Department of Computer Science and Engineering, Amrita School of Computing, Amrita Vishwa Vidyapeetham, Bengaluru 560035, India
e-mail: v_priyanka@blr.amrita.edu

S. M. Thampi et al. (eds.), *Fifth International Conference on Computing and Network Communications*, Lecture Notes in Electrical Engineering 1221,
https://doi.org/10.1007/978-981-97-4711-5_30

1 Introduction

Images have an exceptional ability to serve as windows to the outside world, encapsulating and expressing visual data that people may exchange, comprehend, and enjoy. They provide an effective medium for communication as well as the preservation of major events, experiences, and landscapes from our surroundings. Images serve a variety of functions in our life, including recording, exploration, teaching, advertising, and inspiration. They elicit emotions, stimulate memories, and can have a long-lasting impact on our psyche. Despite their great expressive ability, photographs frequently require extra context to successfully convey certain information or concepts. This is where text, the universal language, comes into play. The text allows us to supplement visual information with spoken communication, bridging understanding gaps between varied audiences. Text, when paired with visuals, creates a more rich, informative, and engaging communication experience. This mix of images and text has resulted in the development of numerous multimedia programs and platforms that allow for the seamless integration of visual and textual content. Image captioning, in particular, has emerged as an important topic of study, with machine learning models being built to generate descriptive and insightful captions for photos automatically. The image captioning model is created by a set of machine learning models that predict the features of the image by focusing on various aspects of the image and obtain complete and accurate results. The process of scanning the image to provide a caption that describes the image is central to the believability of the image. Some companies are ready to incorporate the image strategy into their products so that customers can easily get results by clicking on the images. But the requirements of the current model and maintaining the popular image cause problems, lead to incorrect and irrelevant messages. Even with adequate training, some models have difficulty extracting features from images due to poor control of the computational power required for the required features. The aim is to get the results they want quickly and easily just by the click of a button. However, the effectiveness of this strategy is limited by two important problems: First, there is noise in the images. The currently existing models find it difficult to remove noisy images, produce the wrong products, and create irrelevant messages that do not serve the original purpose. Second, image-based processing can require significant computational resources. High computational requirements can make it difficult for models to produce accurate and complete results. Some models have difficulty managing the financial resources required to define the details of the image, even with the appropriate training. The resulting results often lack the necessary precision and are more general.

This study sheds light on the benefits of attention techniques for image captioning, as well as the advantages of using a specific model architecture with their self-attention mechanism. This study also focuses on image captioning utilizing multiple models' attention methods. It emphasizes the significance of meticulous model creation and evaluation in order to attain optimal performance in machine learning activities. Here are some of the endeavors of this paper: The study compares multiple image captioning methods that rely heavily on attention mechanisms. The model's attention mechanism allows it to focus on certain portions of the image as well. Attention mechanisms are critical because they allow models to focus on certain visual regions for accurate and informative captioning. Several research gaps in the field are addressed by the suggested image captioning model. It introduces a balanced attention mechanism for more comprehensive image interpretation, prioritizes computational efficiency to enable real-time applications, emphasizes all parts of the image for richer captions, incorporates discriminative style learning for cross-domain adaptation, and combines resilience and ambiguity to handle complex visual features. These enhancements attempt to increase the model's accuracy, relevance, and emotional expressiveness, making it more useful across a wide range of domains and applications. These improvements contribute to a more accurate, relevant, and emotionally expressive image captioning system, proving efficacy across a wide range of domains and applications. This procedure improves the model's ability to predict subsequent elements. Transformer, Bahdanau's attention mechanism, EffNet, Inception, RNN decoder, VGG16, and VGG19 were among the models tested. The study aims to evaluate the effect of attention and model architecture on performance. It examines how attention enhances results using a VGG16 encoder and an RNN decoder. The study compares the effectiveness of the Transformer model's self-attention mechanism to the VGG16-based approach. The findings emphasize the importance of attention mechanisms and careful model selection in reaching peak performance and precise results in image captioning. In Sect. 1of the document the objectives of the work is elaborated and followed by Sect. 2 that describes the related investigations. Section 3 is methodology which covers model creation, deployment, data gathering, preprocessing, and system architecture. Sections 4 and 5 provides an overview of the findings, contributions, constraints, difficulties, and recommendations for further research.

2 Literature Review

In the research, authors utilize attention mechanisms and parallel LSTMs to elevate the production of emotionally rich captions. As a result of growing interest in emotion-aware image captioning, the introduced system stands out for its distinctive fusion of attention mechanisms and parallel LSTMs, adding valuable insights to the advancements of sentiment-conscious image description [1]. The strategy includes emotional content alongside visual content in an effort to improve existing caption-creation techniques. Based on the results of this work, the model performs better than current methods in both quantitative and qualitative terms. The suggested method aims to create captions that are more interesting and approachable by including sentimental content, especially for social media platforms and people who are visually impaired. It takes a balance of bottom-up and top-down attentional systems to process images effectively [2]. Existing cross-domain image captioning models struggle with large domain shifts, even with attempts to use paired source and unpaired target data. This paper introduces the Style-based Cross-domain Image Captioner (SCIC), addressing the challenge by integrating discriminative style information into the encoder-decoder framework. The model employs a novel "Instruction-based LSTM" to incorporate style instructions, achieving superior performance on MS COCO (source domain) and outperforming prior methods on Oxford-102, CUB-200, and Flickr30k (target domains) with at least a 5% improvement in CIDEr scores [3]. To make the model better it is necessary to focus on enhancing its computational features. One potential improvement is to increase computational efficiency by reducing redundant processing and using parallel processing strategies. The model can analyze data more quickly and can be used for a wide range of applications and domains by increasing computational efficiency. This study [4] introduces a novel bottom-up and top-down attention mechanism for image captioning and visual question answering (VQA). Leveraging faster R-CNN for bottom-up attention and feature weighting by the top-down mechanism, the approach achieves a new state-of-the-art on MS COCO with CIDEr/SPICE/BLEU-4 scores of 117.9, 21.5, and 36.9, respectively. Its broad applicability is demonstrated by securing first place in the 2017 VQA Challenge. In this study [5], attention-based neural encoder-decoder frameworks dominate image captioning, yet their tendency to activate attention for every word generation lacks efficiency, especially for non-visual terms. This paper proposes an adaptive attention model featuring a visual sentinel, dynamically deciding whether to attend to image regions or the sentinel at each step. The model optimally selects when and where to focus on the image, enhancing sequential word generators. The potential of this strategy for subsequent research is demonstrated by experimental results [6]. It integrates ideas, recalls prior ideas, and alterations to create a mental sentence. The intention is to produce subtitles that convey information and ideas while also conveying context, feeling, and prior knowledge. This enables the construction of captivating and compelling subtitles that are appropriate for the task at hand. This work provides a graphical representation of the network prediction [7]. The method considers many types of data, including images and content collected by semantic, geographic, spectral, and other features. This method tries to improve the accuracy and consistency of subtitles created using different methods. The analysis also suggests another idea: use a network transformer. From the captured long-range perspective and image correlation, this change can improve the efficiency of the recording process. Combining visual prediction networks with advanced transformations is a way to create detailed and rich content that will translate all results into the impact of recorded images.

This study evaluates tracking to understand visual perception and overcome the problem of insufficient image data [8]. This work uses a fully convolutional network (FCN) to generate the graph. This study proposes to improve image captioning by combining visual focus (emphasizing key areas of the image) and captioning (emphasizing keywords in the sentence). The model combines these two types of care to create more accurate and important details for the image. The test method demonstrates the understanding of the image and the effectiveness of the written text by combining the differences between visual and verbal data. This study presents an in-depth review of deep neural network image identification techniques based on

CNN-LSTM architecture [9]. It examines the purposes of architecture and how it can be used to create visual and interpretive displays. This study explores training, data preparation procedures, and evaluation steps to identify the potential of the article. However, the study acknowledges the challenge of creating accurate captions for images with ambiguous content or complex ideas. In addition, it is recognized that dealing with blurry or unclear images can complicate the task of creating text. This study reviews various methods for creating signature images and evaluates architectural models for object discovery [10]. It compares different strategies and aims to create an article that focuses on the content or story of the image. However, it emphasizes the importance of checking the consistency and accuracy of information in the registration process, as too much information can lead to inaccurate registration. The study aims to create an article that effectively utilizes the image's content. This study evaluates CNN-RNN, CNN-CNN, and reinforcement-based framework on image captioning [12]. Several evaluation metrics like BLUE, ROUGE, CIDEr, METEOR, and SPICE are analyzed. The above models do not accurately cover all subtle aspects of the image, imbalances in the training and testing image sets impact efficiency. These are the major challenges observed. Image captioning faces challenges due to description variability, mitigated by recent advances using LSTM-powered recurrent neural networks [13]. LSTM units pose complexity and sequential constraints. This paper introduces a convolutional image captioning technique, demonstrating comparable performance to baselines with faster training times on MS COCO. Detailed analysis supports the adoption of convolutional approaches in language generation. The cross modification attention (CMA) module introduced in this paper extends the attention mechanism by adding a hierarchical gating network that modifies both the visual and linguistic features [14]. This modification helps to filter out misleading information from the captions. The deliberation model implemented is optimized using coarse captions as a baseline, and a potential-oriented reward-shaping strategy is employed for reinforcement learning to improve the quality of refinement. Experimental results on Flickr30K and MS COCO datasets show that the proposed CMA-DM approach outperforms single-pass decoding models. In the study [15], a new captioning system called the ruminant framework which consists of an image encoder, base decoder, and ruminant decoder is introduced. Similar to how sentences are refined by humans, word predictions are guided by the output of the basic decoder. An optimization approach based on reinforcement learning is used to train the framework collectively. The performance of the ruminant decoding technique is much better than that of conventional single-pass decoding models, according to experimental findings on MS COCO and Flickr30k. Traditional classification algorithms rely on feature extraction methods, presenting challenges in obtaining unique features. Convolutional neural networks (CNN) show promise in improving classification performance through implicit feature extraction but demand intensive computing resources. To address this [16], a probabilistic ensemble-based classifier is proposed, trained on mutually exclusive classes grouped by hierarchical agglomerative clustering. This model achieves comparable accuracy to traditional ensembles, with faster training times and no dependence on GPUs even for large datasets. Another work [17] explores the classification of images into pleasant and unpleasant categories using deep learning which facilitates quick image discovery. The approach [18] for categorizing news headlines using deep learning is presented in the study. They also include methods such as LSTM, BERT, and word embedding. Open issues include generalizing to many languages and cultural contexts, handling developing language, assuring openness of the model's conclusions, handling headline bias, and scaling to larger datasets. In the research, a technique for classifying tweets' moods using deep learning and word embedding models is introduced [19]. It has a high degree of accuracy and uses explainability modules to make sense of the model's choices. Open issues include handling cultural nuances, eliminating sarcasm and biases, and improving interpretability. The technique also needs to be adjusted to different texts. The research examines how, despite discharge summaries' unstructured character for clinical systems, text mining retrieves relevant information from them [20]. It examines methods, equipment, datasets, and difficulties. Making extraction methods more uniform, privacy-conscious, and adaptive to real-world differences is still a work in progress. This paper focuses on Kannada to English Neural Machine Translation using a unidirectional approach, employing recurrent neural networks (RNN) with a sequence-to-sequence (Seq2Seq) model [21]. The encoder-decoder mechanism, utilizing long short-term memory (LSTM) as the RNN unit, outperforms statistical machine translation (SMT) with an 86.32% accuracy, as measured by Bi-Lingual Evaluation Study (BLEU).

Research papers on image captioning examine several algorithms and tactics to improve the relevancy of the captions. They frequently employ an encoder-decoder architecture, with RNNs like LSTMs decoding captions and CNNs extracting visual data. When creating captions, attention methods are employed to concentrate on important image regions. It is being investigated how to enhance models and performance via reinforcement learning. For better representation learning, transformer networks like BERT and CRFs are taken into consideration. To produce more educative and emotionally compelling captions, these strategies incorporate RNNs, CNNs, attention mechanisms, reinforcement learning, and transformer networks. There are various research gaps in image captioning that must be addressed. To begin, current techniques' unequal attention mechanisms may result in insufficient image interpretation, as they struggle to combine capturing enticing visual aspects with abstract contextual information, potentially leading to biased captions. Second, due to time-consuming computations and redundant processes, inefficient computing in particular models impedes real-time applications and scalability. Third, certain models lack the ability to offer extensive and detailed captions, resulting in captions with insufficient information, particularly when conveying complicated concepts or incorporating crucial context information. Fourth, adapting picture captioning models to new domains is difficult because they may fail to generalize well across varied image contexts and styles, a problem known as the cross-domain adaptation problem. Finally, the inclusion of ambiguous or intricate visual elements can result in inaccurate and contextually inappropriate captions, emphasizing the need for new ways to efficiently handle ambiguity and assure accurate caption production.

Closing these gaps will be critical for improving the efficacy and accuracy of picture captioning systems. The proposed model solves concerns with cross-domain adaptation, inefficient computing, inadequate information in captions, and addressing ambiguity in image content, among other research gaps in image captioning. For deeper and useful captions, it incorporates balanced attention, integrating top-down and bottom-up attentional processes, and capturing both visual aspects and abstract contextual information. To analyze data more quickly and broadly, the model focuses on maximizing computational efficiency by reducing redundant computations and employing parallel processing techniques. Additionally, it adds visual emphasis to all image parts for captions that are more in-depth and provide more precise and significant visual detail. A discriminative style learning module is included in the model to address cross-domain adaptation challenges by modifying the caption style to the target domain and enhancing cross-domain image captioning performance. To deal with ambiguous or complex visual features, the model blends resilience with ambiguity, enhancing the quality and relevance of the output captions. The model extracts diverse contextual information from images using a combination of deep learning and attention techniques to produce high-quality, relevant captions. These improvements aim to improve the generated captions' accuracy, relevance, and emotional expressiveness, making the method more efficient and dependable across variety of domains.

3 Methodology

3.1 Dataset

A popular open-source dataset for picture captioning tasks in computational imaging and natural language processing is the Flickr 8 k dataset [11]. It has about 8,000 pictures, each with five captions, covering different things and events. The collection is composed of numerous text files that provide tokens, annotations, and names for the images. The token txt file connects image file names to their respective captions. The photos are in JPEG format. 6400 training images and 1600 test photos make up the dataset. To provide diversity and randomization, each training instance is applied after being shuffled.

3.2 Preprocessing

Training image names and their corresponding captions are transformed into a data frame format, with each caption and its associated image name occupying a unique row. A vocabulary list is then created, containing all the unique words present in the training caption set. This vocabulary is then processed to remove punctuation marks, numerals, and single-character words. The text data is now refined, eliminating unnecessary elements (noise) that could impede the model's comprehension. This enables the model to focus on the most relevant and informative aspects of the text. Preprocessing vocabulary involves removing punctuation marks, single-character words, and numerical values from a text to improve its understanding. Punctuation removes noise and allows the model to focus on the text's essential content. Single-character words like articles or prepositions reduce noise and increase relevance, allowing the model to focus on more significant aspects. Numeric values, which may represent specific quantities or dates, are removed to focus on the text's content without being distracted by irrelevant information. By removing these elements, the model can better understand the text and its context. The image captioning model is a technique for deep learning whose performance is enhanced by combining batch creation and feature extraction. To begin, the picture file is read, decoded using tf io read_file, and then resized using tf image resize. This guarantees compliance with the expected input size of the model and enhances model performance. The preprocess input function is then used to preprocess the image, normalizing pixel values and applying transformations to make the image conform to the model's specifications. The next step is feature extraction using the transformed image tensor, abbreviated as IMG. For use in processing and analysis in the future, the original image path is kept. After that, the images are separated into batches and processed utilizing matrix operations simultaneously. This method optimizes memory use and training time, enabling the processing of several images at once. For effective use of GPU accelerators and improved efficiency when performing tasks like picture captioning, batches must be formed. Overall, the performance and computational effectiveness of the picture captioning model are improved by the combination of batch creation and feature extraction (Fig. 1)

3.3 Diverse Model Architectures

See Table 1.

3.3.1 VGG16/19 Encoder

The Visual Geometry Group created the VGG16 encoder, a CNN architecture renowned for its efficient feature extraction and picture classification. Following a uniform stride and filters, this uniform architecture has 13 convolutional layers and 3 fully linked layers. The model employs a 2 × 2 window for down sampling and 3 × 3 filters to maintain spatial resolution. The model's consistency provides accurate results, making it a popular pre-trained model in numerous studies. VGG19 CNN architecture is a modified version of the VGG16 encoder and provides a more intricate model for visual feature extraction and classification in computer vision tasks. The VGG19 encoder uses a 3 × 3 filter and a 2 × 2 window with a stride of 2 and consists of 16 convolutional layers and 3 fully linked layers. The representations and feature extractions are improved by the additional layers, making it a popular

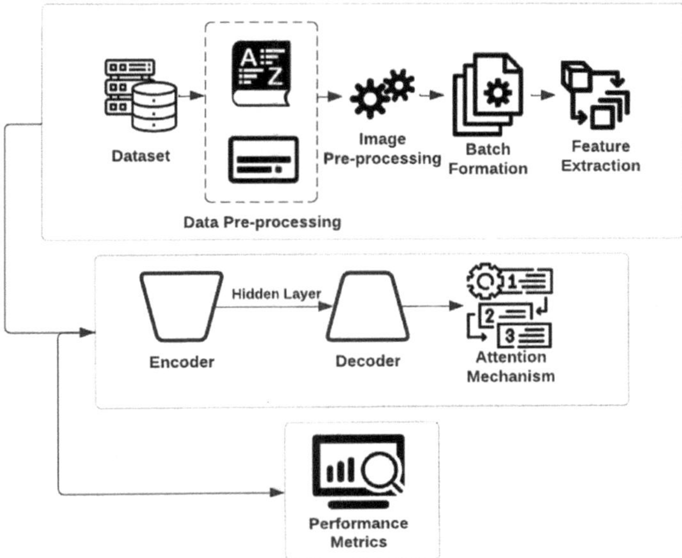

Fig. 1 System design

option for pre-trained models. The VGG19 encoder is frequently used for image feature extraction and classification despite requiring more processing power and complexity than the VGG16 model.

3.3.2 Inception V3

The Inception V3 model is a widely used CNN architecture for image classification and is used as an image encoder in image captions. The model acts as a feature extractor that accepts a preprocessed image as input and removes its classification layers, creating a fixed-length feature vector. This feature vector is a high-level representation of the image and can be used to construct sentences for individual images using transformers or RNNs. The model collects data at different scales which provide unique maps with many variables in a good data collection process. Convolutional layers, pooling layers, and all layers are in 48 layers. The number of maps increases as the spatial size is gradually reduced by using more layers. The model uses a factorization technique to reduce the number of input channels before using larger convolutions to reduce comparison between models to improve efficiency. It also adds the distribution of intermediate levels during learning to solve the vanishing gradient problem making the classification model more comprehensive and provide more accurate signals. Typically, the model is pre-trained on a large image distribution dataset, such as ImageNet. Transfer learning can be facilitated by applying previously learned weights to other tasks or fine-tuning small datasets.

Table. 1 Benefits/drawbacks of the model architectures proposed

Proposed model	Benefits	Drawbacks
VGG16/19 encoder	Captures rich hierarchical features. Pre-trained VGG16/19 models offer knowledge transfer benefits	VGG16/19 down sampling to fixed spatial resolution may discard fine-grained spatial information challenges in resource-constrained or real-time applications
Inception V3	Captures diverse image features. Efficient computation due to its optimized architecture	Challenges in resource-constrained scenarios. Pre-training benefits might be limited compared to simpler architectures
RNN Decoder	Excels understanding sequential context, crucial for coherent language generation	Challenges in capturing long-range dependencies and affecting training efficiency
Bahdanau's attention mechanism	Dynamic focus on different parts of the input sequence, enhancing contextual understanding	The attention mechanism increases computational demands, potentially impacting model training and inference efficiency
Transformer	Enable efficient parallel processing, enhancing training speed	Global attention in Transformers may lead to increased computational overhead in capturing long-range dependencies
Multi-headed attention	Enables the model to focus on different aspects of the input sequence simultaneously, enhancing representation	Adds complexity, potentially requiring more computational resources
EffNet	Effective feature extraction with a balance between depth, width, and resolution	Resource-intensive due to their larger scale, potentially limiting deployment in resource-constrained environments

3.3.3 RNN Decoder

The RNN decoder begins by receiving the encoded image features from the image encoder. The encoded features are often passed through an input embedding layer to map them to a higher-dimensional space that captures semantic meaning and aids in generating more expressive captions. Before generating the first word of the caption, the RNN decoder initializes its hidden state and learns from the encoded image features to incorporate relevant information from the image. This hidden state acts as a memory for the decoder and retains information about previously generated words. The key component, i.e., RNN cell is initialized, a variant of the gated recurrent unit (GRU). The RNN cell takes the input embedding, the previous hidden state, and the attention-weighted image features as inputs. It processes this information and updates the hidden state and generates a word as an output at each time step. Output of the RNN cell is passed through a SoftMax layer to generate a probability distribution based on the vocabulary. The word with the highest probability is selected as the predicted word for that time step. The predicted word is then fed back into the decoder as the input for the next time step, along with the updated hidden state to gradually refine its predictions. The generation of the caption proceeds iteratively, until it reaches an end-of-sequence token appended at the preprocessing state, a predefined maximum length for the caption. Further, the decoder generates a sequence of words, the image captioning system produces descriptive and contextually relevant captions for images.

3.3.4 Bahdanau's Attention Mechanism

It is an attention mechanism that is used in sequence-to-sequence models and addresses the limitations of encoder-decoder models by allowing the decoder to concentrate on different parts of the image dynamically. This model introduces an alignment score, which means the alignment between the decoder's current hidden state and the encoders' hidden states. The attention weights are calculated using a SoftMax function and are sent into the relevant encoder hidden states, which are multiplied across the encoder hidden states and a weighted sum is produced. This weighted sum will be used as the input for the decoder's hidden layers at each step. Using this attention mechanism model can help align the words as per source and translation which results in giving a better and more accurate generation of translations.

3.3.5 Transformer

The Transformer model is a sequence-to-sequence job that doesn't rely on recurrent or convolutional procedures in order to capture long-range dependencies, contextual linkages, and pertinent visual regions. An embedding layer is used to transform the spatial feature maps that the image encoder returns, which reflect the visual information of the image, into a dense vector representation. These embeddings record semantic details on words and their connections to nearby words. The caption input sequence is transferred via the encoder layers of the Transformer, which are made up of feed-forward neural networks and multi-head self-attention processes. The feed-forward neural networks add nonlinearity to change the representations, while the self-attention mechanism enables the model to capture relationships between various parts in the sequence. The image captioning decoder is made up of numerous layers of self-attention and cross-attention techniques. At each time step, the decoder anticipates the following word in the caption sequence by running the output through a SoftMax function. Up until an end-of-sequence token is reached, the operation is repeated.

3.3.6 Multi-headed Attention

It is a transformer-based model which concentrates on multiple parts of the inputs, processes them, and generates a wide range of outputs. In this mechanism, it divides the input into multiple smaller dimension subspaces each called a "head" simultaneously. This mechanism includes three steps. Projection, where the input is sent through multiple subspaces which have unique key values for each head, this step is performed to analyze the different aspects of the image. Attention calculation, which calculates attention values using the dot product of key-value pairs which are then passed through the soft matrix function to get the attention weights. Aggregation, which includes calculation of the weighted sum of values which give context vectors. Then a linear transformation of these context vectors concatenated with each other to produce the final output. The number of heads is a hyper-parameter that can be adjusted as per requirements. Using these multi-heads helps concentrate at different parts of the image and parallel processing of these can help handle complex parameters of the input sequence.

3.3.7 EffNet

The model's remaining layers are employed as feature extractors when the Efficient Net classification head is taken out. The network processes the input image, and the output of one of the intermediary layers is used to represent the image. The RNN decoder or Transformer component uses the retrieved features to create captions based on the content of the images. Initialization of the basic model using EfficientNetB0 includes an input shape defined by (*IMAGE_SIZE, 3). The pre-trained weights of EfficientNetB0 trained on the ImageNet dataset are loaded using the weights = "imagenet" argument. The base model's layers are frozen when base_ model trainable = False is used to stop the weights from changing during training. In order to ensure compatibility with following layers in the model architecture, the output tensor of the basic model is modified to have a constant form regardless of the input size. The keras models are used to generate the CNN_model. Models function, defining the model's input and output layers.

3.4 Attention Plot

An attention plot shows the various input sequence segments that a neural network model concentrates on when producing an output. It is frequently employed in processes like text summarization, image captioning, and machine translation. Attention mechanisms are essential for capturing dependencies between input and output elements in sequence-to-sequence models. The graphic is a grid with the input sequence represented by the x-axis and the output sequence by the y-axis. Each cell represents the attention weight that was given to a particular input position when producing a particular output position. Greater relevance is indicated by higher weights, while less relevance is shown by lower weights. Understanding how the model aligns input and output pieces through the visualization of the attention plot enables debugging and interpretation of the model's behavior. An attention plot, for instance, can be used in machine translation to demonstrate which words or phrases in the source language are given greater consideration when creating words in the target language. This can assist to determine whether the model is paying attention to the right elements of the source sentence while translating.

3.5 BLEU Score

The Bi-Lingual Evaluation Understudy (BLEU) metric is often used to compare the quality of machine translations against reference translations. It computes the similarity of generated and reference translations by counting overlapping word sequences known as n-grams. To capture diverse language features, BLEU examines multiple n values (often 1–4). It penalizes overly short translations and determines an appropriate caption length. To determine the total BLEU score, the scores for individual n-gram precisions are summed using a weighted average. BLEU, on the other hand, has limitations; it may not capture variations in word order or paraphrases and may provide lower scores even for relevant captions. A higher BLEU score suggests greater relevance with training captions, but it does not decide the overall quality of the caption.

4 Results and Discussion

It is observed that Transformer with Effnet generates the most relevant results with respect to the test image captions. The lowest performance across all four BLUE scores is by VGG19 without attention as it is expected to ignore giving weightage to essential words and hence the captions generated are of lower accuracy. Inception with Bahdanau's attention has been an average performer compared to other models. The overall best performer has remained VGG16 with Bahdanau's attention as it maintains consistent high results across all BLUE score variants which indicates that the captions generated are structurally more relevant to the test captions. We have compared BLUE scores of various researches and achieved an improvisation of 87% (Table 2).

Table. 2 Computed BLEU scores generated by different models of test images

BLEU score	VGG19 without attention	Inception with Bahdanau's attention	Transformer with Effnet	VGG16 with Bahdanau's attention
BLEU-1	11.5	46.06	**69.08**	50.01
BLEU-2	8.4	22.36	23.03	36.05
BLEU-3	6.23	10.25	14.08	22.02
BLEU-4	3.45	6.08	4.56	11.7

4.1 User Interface

The user interface model is designed to pick a random image from the test set images (or can be given any image). Once the image is selected, the image is run through four various image captioning models. The resultant captions are displayed, and the user can choose the best suited caption for their use case. Different models focus on different parts of the image to generate a relevant caption, this provides the user with varied perspectives of the same image and does not restrict them with one result. Attention plots visually represent the attention mechanism's focus on input tokens during inference, exposing the model's decision-making process. It is displayed in the user interface to diagnose model behavior, identify biases, and improve model interpretability. This helps the user to align their perspective of the image with that of the model's (Fig. 2)

5 Conclusion and Future Work

Ultimately, attention mechanism-based image captioning is an essential component that improves the model's ability to focus on certain portions of the image, resulting in more accurate and informative captions. A thorough evaluation of potential benefits of many models was carried out, including the Transformer, Bahdanau's attention mechanism, EffNet, Inception, RNN decoder, VGG16, and VGG19. According to the findings, merging attention with a VGG16 encoder and RNN decoder produces better results when trained for longer epochs. However, even when trained for the same number of epochs, the Transformer model beats the others. When compared to the VGG16-based model, the Transformer model's self-attention mechanism is more effective at capturing crucial visual elements. This study also emphasizes the varying efficacy of different model architectures. Model selection and evaluation are critical in reaching peak performance and precise outputs in machine learning activities. Current image captioning models' inability to be explained impacts both their overall effectiveness and user experience negatively. Techniques like attention processes or visual saliency can be used to highlight image regions that have an impact on the caption that is created in order to provide more clarity of execution.

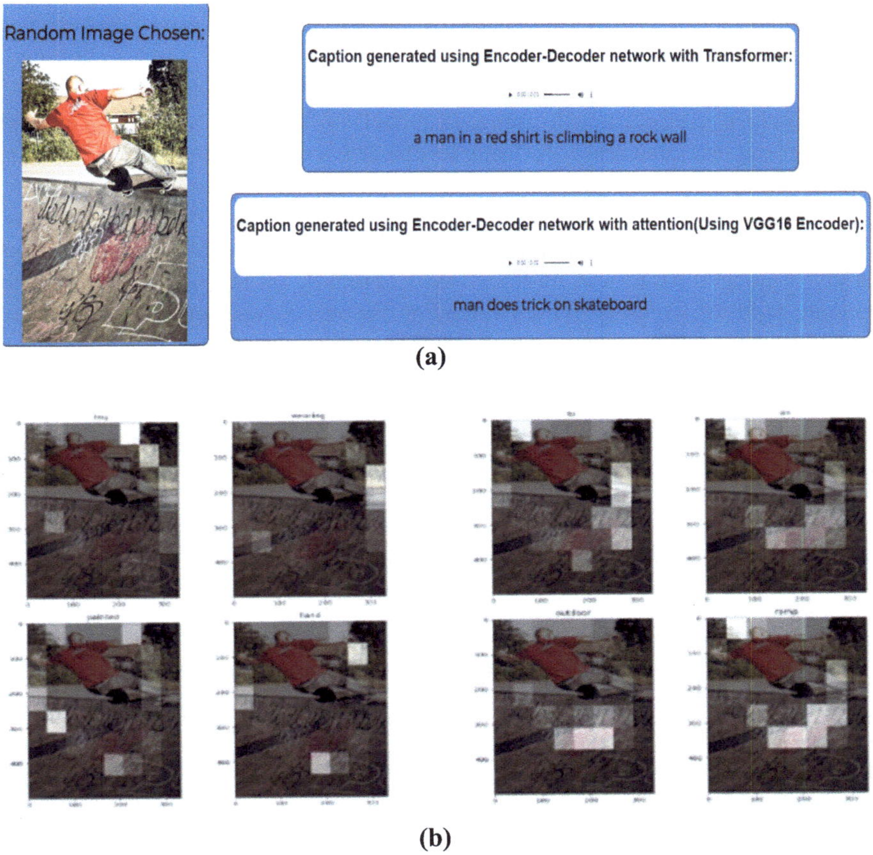

Fig. 2 **a** User interface: generated captions with voice assistant, **b** User interface: attention plot

Generating multiple captions instead of one provides the user the privilege of picking a resultant caption from multiple captions which provides different viewpoints of the same image. It also helps the visually impaired capture much more contextual information than just its plain description. Multimodal comprehension can be incorporated to get beyond the drawback of concentrating just on visual content. It is easier to create more thorough and contextually aware captions when textual context is considered, such as supporting text descriptions or relevant image captions. The focus of current evaluation criteria is human reference overlap; however, they fall short of capturing overall quality, innovation, and relevance. For an in-depth assessment, new metrics assessing novelty, coherence, and alignment with visual and textual signals are required. Accessibility, engagement, and contentment are all increased when voice assistants are incorporated into image captioning. Users can consume information hands-free and have a multimodal experience by turning captions into speech. This cutting-edge UI strategy enhances inclusivity and enables users to concentrate more on visual content.

User evaluation is used to implement a feedback loop to enhance caption quality. With the help of reinforcement learning, the model may be trained to produce captions that better suit user preferences by using feedback as a reward signal. These methods improve picture captioning models by improving explainability, adaptability, contextual comprehension, and evaluation. Also, making use of massive image datasets, like the Flickr30 K dataset, will help the model deal with a very wide and diverse variety of images. Moreover, unsupervised learning offers a powerful approach to generate captions for images, even extending beyond the dataset's boundaries. By employing unsupervised learning techniques, meaningful patterns can be analyzed, and extract features and relationships from the images themselves.

References

1. Li Zhou Ren (2020) Image Emotion Caption Based On Visual Attention Mechanisms. In: 2020 IEEE 6th international conference on computer and communications (ICCC). Chengdu, pp 1456–1460, China https://doi.org/10.1109/ICCC51575.2020.9344900
2. Beddiar Oussalah Tapio (2022) Explainability for medical image captioning. In: 2022 eleventh international conference on image processing theory. Tools and applications (IPTA), Salzburg, Austria. pp 1–6 https://doi.org/10.1109/IPTA54936.2022.9784146
3. Yuan et al (2022) Discriminative style learning for cross-domain image captioning. IEEE Trans Image Process 31:1723–1736. https://doi.org/10.1109/TIP.2022.3145158
4. Anderson P, He X, Buehler C, Teney D, Johnson M, Gould S, Zhang L. Bottom-up and top-down attention for image captioning and visual question answering. In: proceedings of the IEEE conference on computer vision and pattern recognition (CVPR) pp 6077–6086. Salt Lake City UT, USA https://doi.org/10.1109/CVPR.2018.00634
5. Lu J, Xiong C, Parikh D, Socher R. Knowing when to look: adaptive attention via a visual sentinel for image captioning. In: proceedings of the IEEE conference on computer vision and pattern recognition (CVPR) pp 375–383. Honolulu, HI USA https://doi.org/10.1109/CVPR.2017.47
6. Xinxiao Wu, Tong Li (2023) Sentimental visual captioning using multimodal transformer. In: proceedings of the Int J Comput Vis pp 131(4):1–18 https://doi.org/10.1007/s11263-023-01752-7
7. Reshmi Sasibhooshan, Suresh Kumaras, (2023) Image caption generation using visual attention prediction and contextual spatial relation extraction. In: proceedings of the J Big Data 10 18. https://doi.org/10.1186/s40537-023-00693-9
8. Maofu Liu, Lingjun Li, Huijun Hu, Weili Guan, Jing Tian, (2020) Image caption generation with dual attention mechanism. In: proceedings of the IEEE international conference on multimedia and expo (ICME) pp 1–6. London, UK https://doi.org/10.1109/ICME46284.2020.9102722
9. Palak Kabra, Mihir Gharat, Dhiraj Jha, Shailesh Sangle (2020) Image caption generator using deep learning. In: proceedings of the International J Res Appl Sci Eng Technol https://doi.org/10.22214/ijraset.2022.47058
10. Hara Prasad Naik,Prafulla, Laxmipriya (2022) Comprehensive investigation on image caption generation using deep neural networks. In: proceedings of the INFOCOMP: J Comput Sci. 21(1):1–10. 10p
11. Flickr image dataset, Kaggle, https://www.kaggle.com/datasets/hsankesara/flickr-image-dataset.

12. Shuang Liu, Liang Bai, Yanli Hu and Haoran Wang (2018) Image captioning based on deep neural networks. In: proceedings of the 2nd international conference on electronic information technology and computer engineering. Vol 232, pp-01052. https://doi.org/10.1051/matecconf/201823201052
13. Jyoti Aneja, Aditya Deshpande, Alexander Schwing (2017) Convolutional image captioning. In: proceedings of the IEEE conference on computer vision and pattern recognition (CVPR) https://doi.org/10.48550/arXiv.1711.09151
14. Zheng Lian,Yanan Zhang,Haichang Li,Rui Wang,Xiaohui Hu (2023) Cross modification attention-based deliberation model for image captioning. In: proceedings of the IEEE TMM. 53(5):5910-5933
15. Guo Liu Lu (2019) Show, tell, and polish: Ruminant decoding for image captioning. In: proceedings of the IEEE transactions on multimedia pp (99):1–1 https://doi.org/10.1109/TMM.2019.2951226.
16. Neena, Geetha (2017) Image classification using an ensemble-based deep CNN. In: proceedings of the 5th ICACNI. Vol 3, pp 445–456. https://doi.org/10.1007/978-981-10-8633-5_44.
17. Sudarshana, Jyotsna, Amudha (2020) Deep learning model for image classification. In: proceedings of computational vision and bio-inspired computing. Pp 312–320 https://doi.org/10.1007/978-3-030-37218-7_36
18. Madhusmita Khuntia, Deepa Gupta (2023) Indian news headlines classification using word embedding techniques and LSTM model. In: proceedings of procedia international conference on machine learning and data engineering. Vol 218, pp 899- 907 https://doi.org/10.1016/j.procs.2023.01.070.
19. Abubakar, Deepa Gupta and Palaniswamy 2022 Explainable emotion recognition from tweets using deep learning and word embedding models. In: 2022 IEEE 19th india council international conference (INDICON) Kochi India pp. 1-6 https://doi.org/10.1109/INDICON56171.2022.10039878.
20. Priyanka C Nair, Gupta Deepa, Bhagavatula Indira. (2021) A survey of text mining approaches, techniques, and tools on discharge summaries. In: proceedings of advances in computational intelligence and communication technology. Pp 331–348 https://doi.org/10.1007/978-981-15-1275-9_27
21. Gadugoila, Sheshadri, Priyanka Nair and Deepa Gupta (2022) Unsupervised pivot-based neural machine translation for english to kannada. In: 2022 IEEE 19th india council international conference (INDICON) Kochi India, pp. 1-6 https://doi.org/10.1109/INDICON56171.2022.10039732.

Priority Decay Loss-Based Image Inpainting and Outpainting

Aaryan Sharma, Abhishek Singhi, N. Aditya, Alok Kumar, and Jayashree Rangareddy

Abstract The demand for appealing images is expanding in image-driven industries such as advertising, social media, entertainment, healthcare, and historical preservation in modern society. Image inpainting and outpainting techniques are critical in image restoration and enhancement, addressing issues such as degradation and missing areas. Deep learning methods, such as GANs, show promising results in terms of producing aesthetically beautiful and semantically relevant outcomes, but issues such as mode collapse exist, demanding ongoing study for more robust and accurate solutions. To address these challenges, in this research, we introduce a novel image outpainting technique that uses a unique technique called "priority decay loss". In contrast to conventional GANs and outpainting strategies, which rely on one or more fixed loss functions throughout multiple training iterations, our method for outpainting combines dynamic loss prioritizing using two crucial parameters: gamma and alpha. While gamma shows the fundamental priority of a loss, alpha determines the rate of priority decay. By allowing us more control over the training direction and the option to integrate various loss functions, this strategy offers variable training techniques.

Keywords Inpainting · Outpainting · Deep learning · Generative adversarial networks

1 Introduction

In contemporary society, the escalating demand for visually engaging and informative graphics spans various industries. However, a significant challenge arises from the presence of missing or deteriorated elements in medical and forensic images, which hinders comprehensive analysis and interpretation. Recognizing these challenges is essential, and it is imperative to ground our research in a thorough understanding

A. Sharma (✉) · A. Singhi · N. Aditya · A. Kumar · J. Rangareddy
PES University, Bengaluru, India
e-mail: pes1ug20cs003@pesu.pes.edu

S. M. Thampi et al. (eds.), *Fifth International Conference on Computing and Network Communications*, Lecture Notes in Electrical Engineering 1221,
https://doi.org/10.1007/978-981-97-4711-5_31

of existing literature. The exploration of current research in this domain reveals persistent gaps and issues. Factors such as equipment or technical limitations and environmental conditions contribute to the absence or compromise of elements in visual data. These issues, identified through a critical review of literature, underscore the need for innovative approaches to enhance the quality and utility of images in fields like medical and law enforcement. To address these pressing challenges, our research problem proposes a novel approach utilizing advanced image inpainting and outpainting techniques. The identified research problem lies in the inadequacy of current methods to effectively handle missing or compromised elements in visual data, particularly in critical fields where accurate analysis is paramount.

Our novel approach, based on a specially created generative adversarial network (GAN) model, reinforced by advanced deep learning techniques, aims to produce high-quality outcomes. By seamlessly blending visual appeal with important semantic content, our approach represents a departure from conventional methods. Specifically, the avant-garde practice of outpainting goes beyond the confines of the original image, generating new, engrossing content that is both aesthetically pleasing and contextually appropriate. In contrast to conventional GANs and outpainting strategies, which rely on one or more fixed loss functions throughout multiple training iterations, our methodology for outpainting combines dynamic loss prioritizing using two crucial parameters: gamma and alpha. While gamma shows the fundamental priority of a loss, alpha determines the rate of priority decay. By allowing us more control over the training direction and the option to integrate various loss functions, this strategy offers variable training techniques. In a variety of fields, including social media, entertainment, travel, and forensics, this ground-breaking approach has the potential to significantly improve user experiences and results [1]. Our research, rooted in the identified research problem and guided by insights from existing literature, has the power to transform how we engage with images across various domains. This paper delves into the conceptual and technical underpinnings of image inpainting and outpainting, providing a comprehensive exploration of the identified research problem. By examining the underlying technology, philosophical foundations, and significant ramifications of this approach in the field of visual content, our study contributes to fixing flaws and broadening creative horizons.

2 Related Works

Below are a few mentioned pristine research papers that we referred to gain invaluable insights and learnings to head in the right direction. The paper authored by Xiaoming Li, Hengzhi Zhang, Lei Feng, Jing Hu, Rongguo Zhang, and Qiang Qiao presents the EA method, a two-stage image outpainting approach. This technique was created to address issues with produced pictures that contain invalid pixels and the geographical location of useful pixels. The study makes a great contribution to the area of image outpainting by offering creative techniques to improve picture quality by minimizing problems associated with invalid pixels and optimizing the spatial distribution of effective pixels [2]. In this study, Yiwen Xu, Maurice Pagnucco, and Yang Song introduce a unique method for tackling the difficult issue of heterogeneous picture outpainting that dubbed the decoupled high-frequency semantic guidance-based GAN (DHG-GAN). The two-stage method offered in this presented an organized approach to the outpainting work and is intended to improve the quality and diversity of the created pictures. This study addresses the complexities of obtaining numerous plausible outcomes while reconstructing substantial missing portions surrounding a known location [3]. In this work, Dongmin Cha and Daijin Kim describe the dynamic attention map GAN (DAM-GAN), a revolutionary GAN-based model. By concentrating on reducing differences in pixel-level details, this novel strategy helps to improve the overall quality and coherence of created pictures [4]. In this study, Xi Wang, Weixi Cheng, and Wenliang Jia develop a deep learning method based on generative adversarial networks (GANs). The main goal of this approach is to enhance picture production by employing conditional edges as structural priors. Overall, this research presents an advanced GAN-based approach that leverages conditional edges and innovative loss functions to enhance image generation and quality [5].

Advancing further into the dynamic frontier of image processing, our attention turns to pivotal studies that redefine the landscape of image inpainting and outpainting techniques. Geng, Y., Gao, P.,Yang, X., Yan, Y., & Huang, K overcame the challenges of all-sides image outpainting using an encoder decoder system with a multi-view recurrent content transfer module introduced in this paper. The proposed method outperforms current picture inpainting/outpainting techniques and produces visually appealing results for external image outpainting using temporal connection modeling to produce realistic outside areas while maintaining seamless consistency with input regions [6]. Cheng, Y. C., Lin, C. H., Lee, H. Y., Ren, J., Tulyakov, S., & Yang, M. H address the issue of semantically inconsistent image augmentation. The proposed method produces images of higher quality and diversity compared to the current in and outpainting methods. It addresses the issue of repeating structure and texture found in the previous approaches by considering a wide range of latent codes, as well as patch-based generation [7]. Wang, W., Niu, L., Zhang, J., Yang, X., & Zhang, L. introduced a dual-path inpainting network to address the issues with the current inpainting techniques in deep image. The results show that the proposed inpainting method can significantly improve the inpainting results compared to current inpainting approaches. The dual-path network output also produces more realistic results [8]. In this study, Jain, J., Zhou, Y., Yu, N., & Shi, H. developed methodology to create a distinctive inpainting network structure. It combines a coarse-to-fine GAN-based network to enhance structural elements and high-frequency rapid convolutional layers to enhance recurring texture synthesis. The result is an inpainting model that meets the current standards of repeated texture synthesis and structural development, resulting in an impressive visual quality. Numerous studies have demonstrated the efficacy of this technique, as well as the significance of structural and texture elements in influencing image inpainting grid design [9]. The objective of the present study by Xiao, Z., Kreis, K., & Vahdat, A. is to address the dilemma posed by the trade-off between the quality of the sample, the mode coverage of the model, and the sampling speed of the deep generative model.

To address this issue, the authors have developed a method of denoising the diffusion GAN, which eliminates the need for a Gaussian assumption and allows for quicker sampling. This is achieved by replacing the Gaussian with a complex multimodal distribution, resulting in a 2000 times faster sampling speed and a sample quality and sample diversity comparable to the original diffusion model. This is the first time that a denoising GAN has significantly reduced the sampling cost of a diffusion model, making it applicable to real-world applications. Furthermore, this model outperforms conventional GAN models in terms of both mode coverage and sampling diversity [10].

Moreover, the paper by Wang, Z., Zheng, H., He, P., Chen, W., & Zhou, M. presents a novel generalized annular network (GAN) framework, diffusion-GAN. This framework is based on a forward diffusion chain, a timestep-dependent discriminator, and an adaptive diffusion process. The generator is able to learn from the feedback of the discriminator through backpropagation within the adaptive diffusion chain. Compared to traditional GANs, the diffusion-GAN framework produces more accurate images with enhanced stability and data efficiency. Additionally, it outperforms robust GAN baselines across a wide range of datasets. This framework is a potential advancement in GAN training, as its theoretical study of a timestep-dependent discriminator approach confirms its dependable and effective guidance to the generator [11]. Lei Zhang, Kang Liao, Chunyu Lin, and Yao Zhao provide a novel image outpainting method that uses LiDAR depth data to generate accurate RGB scenes in their paper. Their depth-guided outpainting network, which combines depth guidance with multimodal learning modules, outperforms existing methods thanks to a novel constraint approach. Numerous tests using the KITTI and Waymo datasets confirm the superiority of their method in terms of both quantity and quality [12]. The problem of image outpainting—creating varied backdrops from constrained foreground regions—is investigated in the paper by Zhang and Shi. With the use of a cutting-edge regularization technique, they address the mode collapse problem in earlier generative adversarial networks. To improve image quality, they also suggest a feature pyramid discriminator. Their empirical findings demonstrate the model's capacity to generate a variety of high-quality images but at the cost of high computational resources [13]. Moreover, Jiayin Cai and Changlin Li offer a unified framework to solve various picture inpainting conditions in addition to the traditional stroke-type mask. The semantic aware generative adversarial network, or SA-PatchGAN, is enhanced via a method of progressive learning. The model's gradual picture refinement results in better perceptual quality and enhances the discriminator's capacity to adapt to various types of masks [14]. Gang Wei, Jing Guo, Yongzhen K, Kai Wang, Shuai Yang, Nan Sheng developed a three-stage generalized assembly (GAN) model based on human sketching processes to address picture outpainting issues. The EP-net phase predicts missing area edge maps, the IO-Net phase produces outpainting outcomes, and the CP-Net phase predicts lacking area color maps. The proposed ECPIO net was found to be highly successful in Cityscapes, achieving a PSNR score of 27.15, with an SSIM score of 0.78, and was also successful in tests on other publicly available datasets, demonstrating its ability to capture accurate semantic content and structural detail in absent areas [15]. The construction of our network involved extensive analysis, which included a thorough assessment of the strategies outlined in this context.

3 Methodology

3.1 Image Outpainting

Our novel algorithm called outpainting and upscaling network (OupNet) is specifically designed to perform image outpainting. It performs image resizing efficiently by capturing a low resolution 128 × 128 input image and producing a high-resolution 256 × 256 output image. Two important variables are utilized in the context of priority decay losses. The variable gamma shows the fundamental priority of a loss, whereas the variable alpha determines the rate of priority decay. This method offers varied training strategies by giving us more control over the training direction and the ability to combine different loss functions.

As training progresses, gamma is decreased by alpha after each epoch. This indicates that pixel-wise loss eventually gains more significance than adversarial loss. In this learning approach, the model first concentrates more on the adversarial part of the loss before progressively changing its attention to the reconstruction quality (pixel-wise loss). This is usually done with the goal of teaching the model to generate images that are more stable and realistic by focusing on high-level characteristics initially and fine-tuning details later.

The architectural structure is as follows:

Transposition and Convolution Layers: The network starts with an input that accepts an RGB image with a resolution of 128 × 128 (3 color channels). The network comprises a sequence of convolutional layers that rapidly downsample the input image to capture features at different scales. These layers are responsible for lodging hierarchical representations from the input image. Each convolutional layer has specified the values of dimensions as (e.g., 64, 128, 512, 1024, and 2048) and employs a kernel size of (3 × 3). (2, 2) steps are used to reduce spatial boundaries and (1, 1) padding maintains the spatial dimension of the point maps. After the convolutional layers, the network uses a transition state that further processes the features and prepares them for the scaling phase. This state includes a convolution state with a small kernel size (1 × 1) to reduce the number of channels, followed by a convolution transpose property. Convolution transposition layers, also known as deconvolution layers, handle the scaling task. These layers take sampled point clouds and increase their spatial constraints, ultimately producing a final high-resolution image. These layers typically use a (4 × 4) kernel size along with (2, 2) steps to double the spatial boundaries. The padding (1, 1) serves to maintain the specified size of the suitcase.

Functions for LeakyReLU Activation: LeakyReLU activation functions are applied after each convolutional layer to introduce nonlinearity and encourage point knowledge. A negative slope parameter of 0.2 or 0.3 is generally used.

Layers of Batch Normalization: Batch normalization layers are introduced after some of the convolutional layers to stabilize and accelerate training. Batch normalization helps ensure that activations are centered to have approximately zero mean and unit variance, which aids in further effective training.

Tanh Output Layer Activation Function: The final layer of the network is equipped with the Tanh activation function. This is a common choice for the layers of image generation models, as it ensures that the pixel values fall within the range of (−1, 1). The combination of the generator and discriminator in a training setup forms the foundation for the image outpainting process. The generator learns to induce satisfying images, while the discriminator learns to distinguish between real and generated images, performing in a generative adversarial network (GAN) that produces high-quality outpainted images.

To sum up, in the generator's training, the trade-off between adversarial and pixel-wise losses is managed by means of alpha and gamma variables. To generate pictures of greater quality, the generator first emphasizes adversarial loss and then progressively moves its focus to pixel-wise loss. Enhancing the overall quality of the outpainted images and training stability may be achieved by the dynamic modification of loss weights which we have implemented here.

3.2 Image Inpainting

Generator Architecture: The generator is an essential part liable for delivering reasonable and top notch pictures with regard to image inpainting. Working on a 128 × 128 pixel RGB picture with three variety channels, it utilizes an outpouring of convolutional layers to remove mind boggling highlights from the info. These convolutional layers, with a particular number of channels, twofold in limit as the organization advances further into its design (e.g., 64, 128, 256, 512, 4000). LeakyReLU initiation capabilities are applied after each convolutional layer to present nonlinearity, relieving the disappearing slope issue and improving component learning. Cluster standardization layers are decisively consolidated to balance out and speed up preparing by normalizing actuations in every smaller than usual bunch. After the underlying element extraction, a change layer, utilizing a 1 × 1 piece size, is utilized to additionally refine the highlights and set them up for the upscaling stage. The convolutional translate (deconvolution) layers embrace the errand of upscaling the downsampled highlights, multiplying the spatial aspects through layers with a bit size (e.g., 4 × 4) and a step (e.g., 2,2) while keeping up with the ideal result size through cushioning. The last layer is furnished with a Tanh enactment capability, guaranteeing that pixel values fall inside the scope of [−1, 1]. Basically, the generator joins include extraction, clump standardization, and upscaling layers to yield excellent pictures, holding spatial loyalty and visual intelligence.

Discriminator Architecture: The discriminator is responsible for distinguishing between real and generated images. It plays a vital part in the laborious training process. The discriminator network takes images as input, either real images or images generated by the generator. It employs a series of convolutional layers that capture features from the input images. As you move deeper into the network, the number of pollutants generally increases to prize more complex features. LeakyReLU activation function is used after each convolutional subcaste. These activations introduce linearity and enable the network to see the patterns and features in the images. Normalization in the discriminator armature, instance normalization is used after some convolutional layers. This fashion normalizes activations and is frequently salutary for style transfer and image generation tasks. The final affair subcaste consists of a convolutional subcaste with a kernel size of 3×3. It produces a single channel affair that's used to classify the input image as real or generated. In substance, the discriminator armature focuses on point birth and bracket to determine the authenticity of the images (Fig. 1).

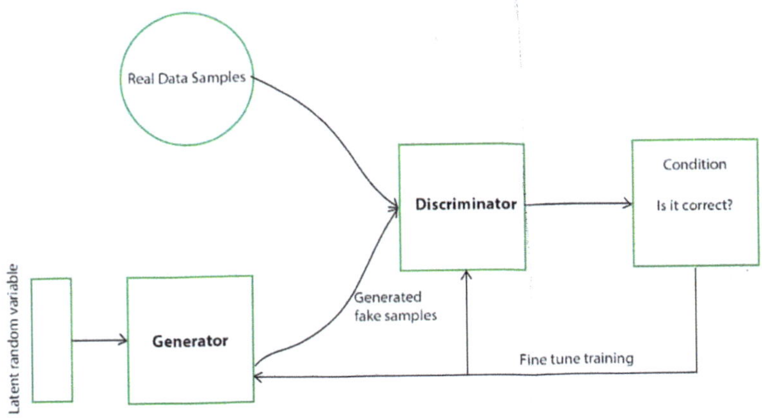

Fig. 1 Generalized architecture [16]

4 Results

4.1 Image Outpainting Examples

The images shown in Fig. 2 demonstrate how the input images have undergone image outpainting. The limits of the input image have been extended on all four sides by outpainting, which has been done on all four sides of the input image.

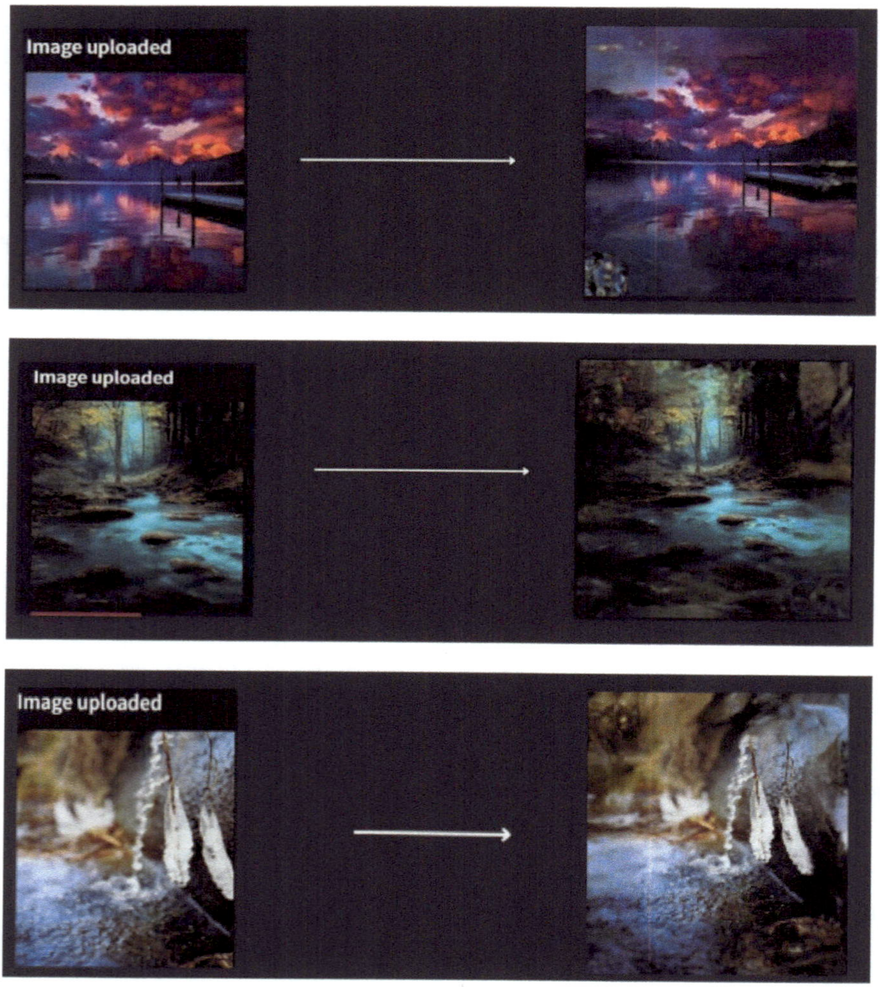

Fig. 2 Outpainting performed on Wikiart and landscape datasets

4.2 Image Inpainting Examples

The images shown in Fig. 3 demonstrate how the GAN's were used to inpaint the input image's masked portion utilizing the context of its surroundings.

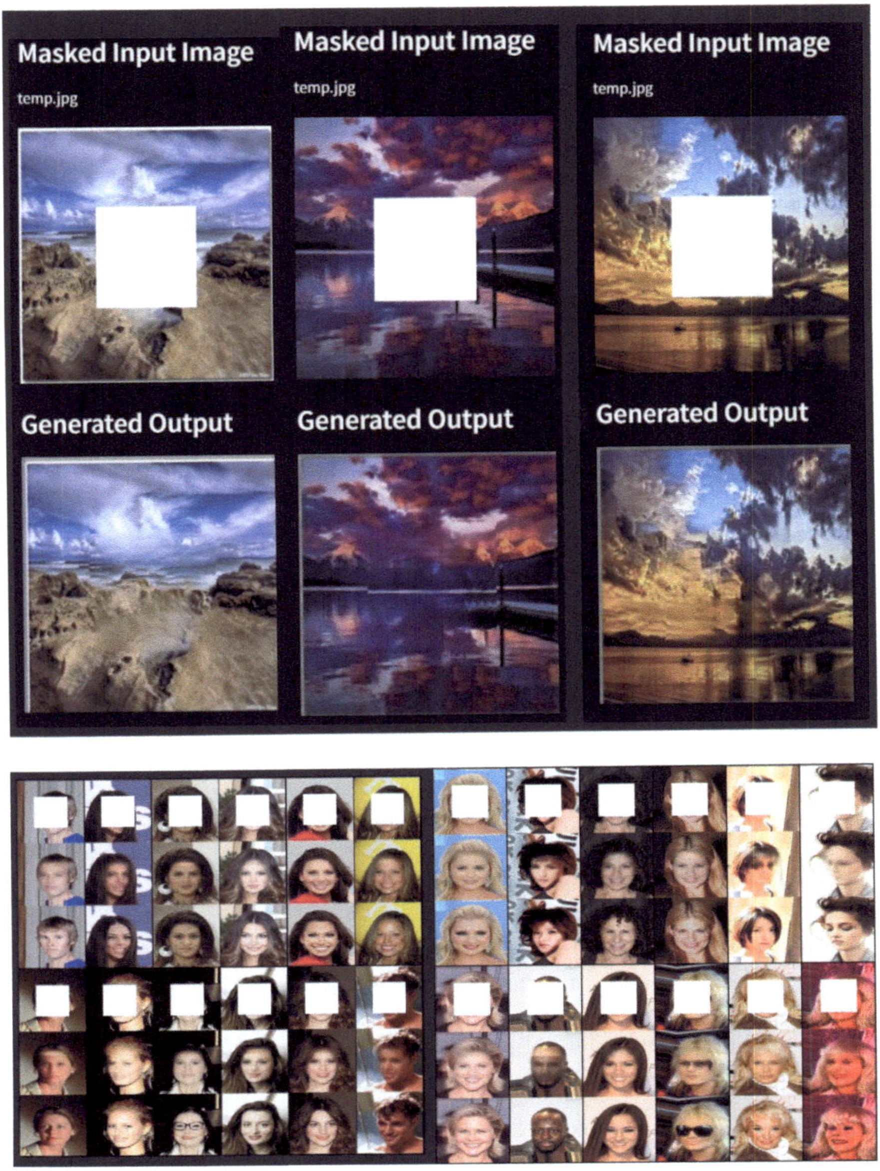

Fig. 3 Image inpainting performed on landscape and celeba datasets

Fig. 4 Generator adversarial loss

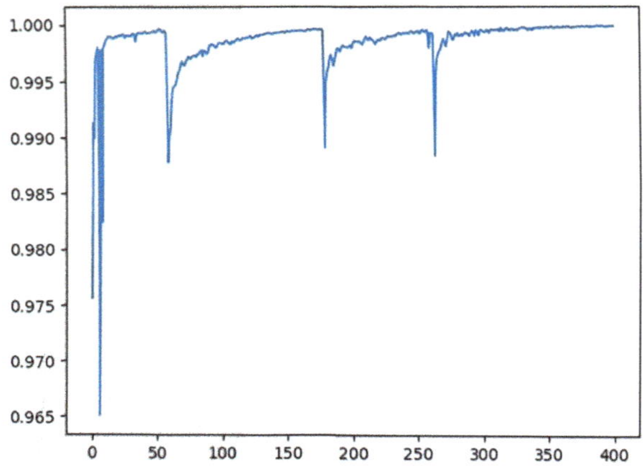

4.3 *Visualizing the Loss Through Graphs*

Outpainting: The generator adversarial loss is 0.999888718128 as seen in Fig. 4.The generator adversarial loss of 0.999888718128 is the generator's performance in an adversary training environment where the generator's goal is to generate real content that can fool the adversary. A high adversary loss indicates that the generator and adversary have a balanced training relationship, the generator is not dominating the adversary, and both elements are likely improving continuously during training. The higher the adversary loss, the better the image quality of the outputs. In other words, the generator is learning to create content that matches the distribution of actual data. It is evident from the significant adversarial loss that the discriminator is successfully taught to separate created samples from real ones. A trained discriminator is essential to driving the generator's advancement.

The generator pixel loss is 0.036329105496406555 as seen in Fig. 5.Low generator pixel loss for image outpainting means that there are fewer outpainting artifacts or distortions. This is very important because it helps to maintain the quality and consistency of the image. A low pixel loss means that the outpainting generator is creating content that matches the ground truth and balancing the training process. The observed high performance, indicated by the low generator pixel loss, stems from several factors. The generator exhibits remarkable fidelity to genuine images and excels in recreating complex details and structures. The low pixel loss indicates that fine-grained features were captured, which is essential for realistic results. The observed higher performance of the model is the result of regularization approaches like dropout and a well-designed architecture with optimal layers and parameters that further reduce overfitting and improve the model's generalization skills (Fig. 6).

Inpainting: The generator pixel loss is 0.1843711 as seen in Fig. 7. The generator pixel loss value refers to the difference between the generated image (the inpainting image) and the actual ground truth image (the ground truth image). A low pixel loss indicates that inpainting algorithms have been able to reconstruct the lost or damaged

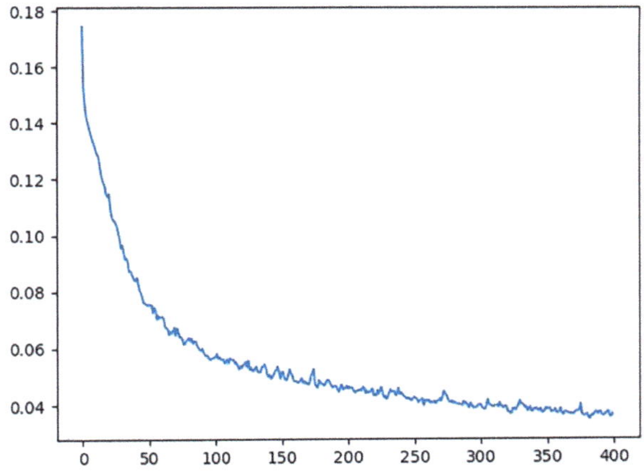

Fig. 5 Generator pixel loss

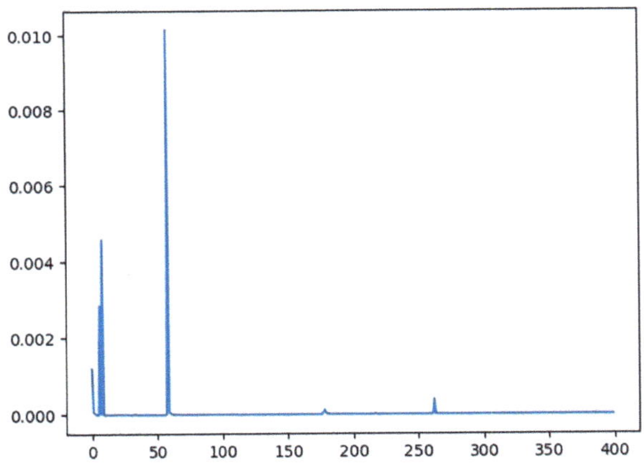

Fig. 6 Discriminator loss

areas of an image with high level of accuracy. This is useful for applications where it is important to retain the visual elements of an original image.

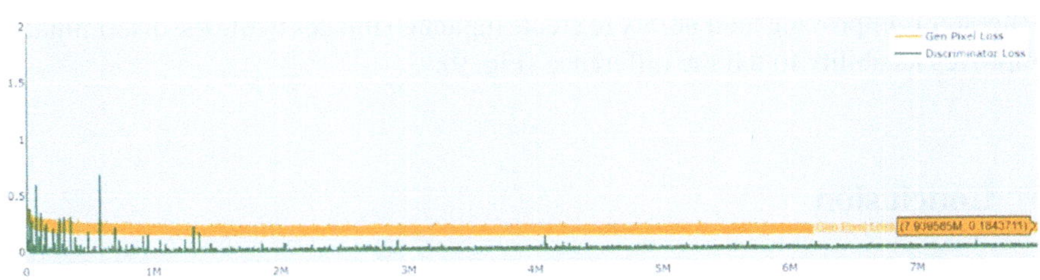

Fig. 7 Generator loss

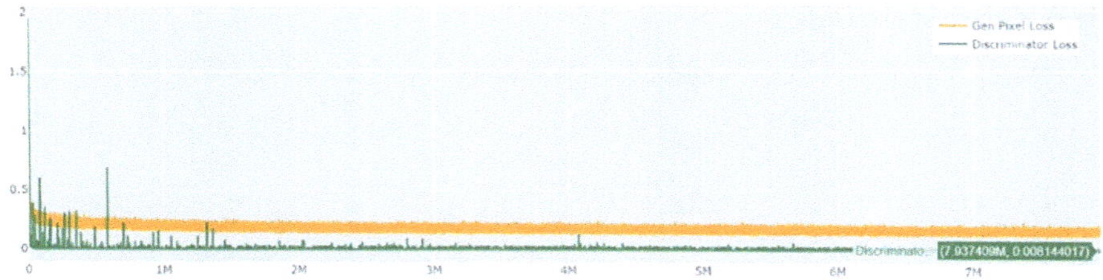

Fig. 8 Discriminator loss

Fig. 9 Generator adversarial loss

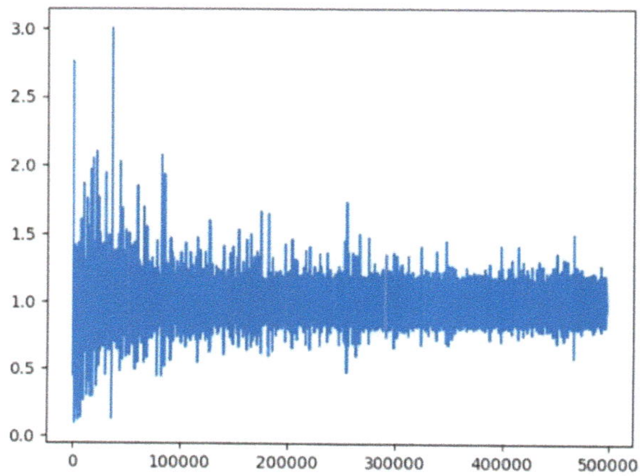

The discriminator loss is 0.008144017 as seen in Fig. 8 The discriminator loss in the context of image inpainting represents how well the discriminator distinguishes between the original image (the original image) and the inpainted image (the inpainted image). The lower the discriminator loss, the better the performance. Low discriminator loss means the discriminator effectively distinguishes between the inpainted image and the original image. Low discriminator loss means that the process of adversarial training converges well. In other words, the generator is competing with the discriminator to create the most realistic inpainted image. The generator is improving their ability to create inpainted images while the discriminator improves its ability to tell the difference (Fig. 9).

5 Conclusion

To sum up, the OuPNet architecture, which successfully combines key elements to enable image outpainting and upscaling, serves as the foundation of our research. This architecture effectively captures the intricate details of an image by integrating convolutional layers for feature extraction. Batch normalization is included to stabilize the network and speed up training, and LeakyReLU activations add the necessary nonlinearity for effective feature recognition. This architecture, which has been designed, addresses the difficulties of outpainting and upscaling in order to produce high-quality, visually consistent images. Importantly, it excels at maintaining image fidelity and spatial information throughout the outpainting process. This model is very well-suited for landscape and sceneric datasets. We have used a powerful GPU to process large datasets.

This may also require efficient data loading methods and storage infrastructure. The size of training datasets affects the scalability of the method. In real-world applications, a large dataset may allow the model to generalize to a larger range of images. Model parallelism is a way to see if a model is too large to fit in the memory of a single GPU. To compute upsampling/downsampling for a segment of the model for each GPU, the model is partitioned among several GPUs. We tested several learning rate schedules/optimizers (like Adam and SGD) and found that with specific problems, these optimizers converged faster or better. It is beneficial to utilize learning rate schedules that change the rate of learning during training. The OuPNet, a key component of our methodology, gives our model the flexibility and performance needed to advance the field of image outpainting, particularly through the creative application of priority decay loss.

References

1. MDPI. https://www.mdpi.com. Last Accessed on 28 Nov 2023
2. Li X, Zhang H, Feng L, Hu J, Zhang R, Qiao Q (2022) Edge-aware image outpainting with attentional generative adversarial networks. IET Image Proc 16(7):1807–1821
3. Xu Y, Pagnucco M, Song Y DHG-GAN (2022) Diverse image outpainting via decoupled high frequency semantics. In: proceedings of the Asian conference on computer vision. pp 3977–3993
4. Cha D, Kim D Dam-gan (2022) Image inpainting using dynamic attention map based on fake texture detection. In: ICASSP 2022–2022 IEEE international conference on acoustics. Speech and signal processing (ICASSP). pp 4883–4887
5. Wang X, Cheng W, Jia W (2022) Structure-guided Image Outpainting. arXiv preprint arXiv: 2212.12326
6. Geng Y, Gao P, Yang X, Yan Y, Huang K (2022) Outpainting natural scenery images by fusing forecasting information. In: J Phys: Conf Ser vol 2278(1):012031. IOP Publishing
7. Cheng YC, Lin CH, Lee HY, Ren J, Tulyakov S, Yang MH (2022) Inout: Diverse image outpainting via GAN inversion. In: Proceedings of the IEEE/CVF conference on computer vision and pattern recognition. pp 11431–11440
8. Wang W, Niu L, Zhang J, Yang X, Zhang L (2022): Dual-path image inpainting with auxiliary GAN inversion. In: Proceedings of the IEEE/CVF conference on computer vision and pattern recognition. pp 11421–11430
9. Jain J, Zhou Y, Yu N, Shi H (2023): Keys to better image inpainting: Structure and texture go hand in hand. In: Proceedings of the IEEE/CVF winter conference on applications of computer vision. pp 208–217
10. Xiao Z, Kreis K, Vahdat A (2021) Tackling the generative learning trilemma with denoising diffusion GANs. arXiv preprint arXiv:2112.07804

11. Wang Z, Zheng H, He P, Chen W, Zhou M (2022) Diffusion-GAN: Training GANs with diffusion. arXiv preprint arXiv:2206.02262
12. Zhang L, Lin C, Liao K, Zhao Y (2023) Towards reliable image outpainting: learning structure-aware multimodal fusion with depth guidance. In: ICASSP 2023–2023 IEEE international conference on acoustics, speech and signal processing (ICASSP) pp 1–5
13. Zhang, L., Wang, J., Shi, J.: Multimodal image outpainting with regularized normalized diversification. In: proceedings of the IEEE/CVF winter conference on applications of computer vision. pp 3433–3442
14. Cai J, Li C, Tao X, Tai YW (2022) Image multi-inpainting via progressive generative adversarial networks. In: Proceedings of the IEEE/CVF conference on computer vision and pattern recognition. pp 978–987
15. Wei G, Guo J, Ke Y, Wang K, Yang S, Sheng N (2023) A three-stage GAN model based on edge and color prediction for image outpainting. Expert Syst Appl 214:119136
16. Generative adversarial network (GAN), https://www.geeksforgeeks.org/generative-adversarial-network-gan. Last Accessed on 28 Nov 2023

Classification of Watermarked and Non-watermark Text in Natural Scene Images: An Xception-Based Approach

Safnaz, Bharathi Pilar, and Shivakumara Palaiahnakote

Abstract In this work, we have explored the Xception architecture for classifying watermarked and non-watermark text in natural scene images. Leveraging depth-wise separable convolutions based on the Inception architecture, the Xception model achieves remarkable accuracy while utilizing fewer parameters. As a convolutional neural network, the Xception model excels in capturing intricate features from images, making it well-suited for discerning subtle differences between watermarked and non-watermarked text. During the training phase, the model requires a substantial dataset comprising both types of images to learn and recognize the distinguishing features effectively. Once trained, the Xception model exhibits high accuracy in classifying new images into their respective categories. Extensive experimentation of publically available standard datasets exhibits the effectiveness of the proposed approach. The performance is evaluated using a comprehensive range of metrics, including accuracy, precision, recall, and $F1$-score, resulting in respective values of 86.5%, 84.76%, 89%, and 86.82%.

Keywords Xception · Transfer learning · Digital watermarking · Natural scene images · Deep learning · Classification · Overfitting · Dropout strategy

Safnaz · B. Pilar (✉)
University College Mangalore, Mangalore, India
e-mail: bharathi.pilar@gmail.com

Safnaz
e-mail: safnaz.ibrahim97@gmail.com

S. Palaiahnakote
University of Malaya, Kuala Lumpur, Malaysia
e-mail: shiva@um.edu.my

S. M. Thampi et al. (eds.), *Fifth International Conference on Computing and Network Communications*, Lecture Notes in Electrical Engineering 1221,
https://doi.org/10.1007/978-981-97-4711-5_32

1 Introduction

The increasing prevalence of digital media has brought about a corresponding surge in the unauthorized distribution of copyrighted material. In order to protect the intellectual property rights of authors, digital watermarking techniques have been developed to embed confidential data into digital documents/images. These watermarks serve to establish the legitimacy and ownership of the media, and it becomes essential to detect and verify the presence of such watermarks. By employing advanced detection algorithms and technologies, the integrity of digital content can be ensured, discouraging unauthorized use and distribution. The use of digital watermarking plays a vital role in safeguarding the rights of content creators and maintaining the integrity of digital media in today's digital age. The watermark embedded in a digital document/image can be retrieved from the multimedia files at a later stage and utilized to establish ownership or provide supporting evidence [1]. Sample images that demonstrate both watermarked and non-watermarked text are given in Fig. 1. A watermarking system comprises two main phases: watermark embedding and watermark detection/extraction. In the embedding process, the watermark is inserted into the host message, which refers to the data that will be encrypted with the watermark. On the other hand, during the extraction procedure, the watermark is removed from the watermarked data to ascertain the ownership of the copyrighted content belonging to the intended user [2]. Despite extensive research efforts over the years, the detection and recognition of watermark text in real-world scenarios still face significant challenges. These challenges can be attributed to three key aspects:

Image Distortions and Degradations: Real-world images often undergo various distortions and degradations, such as compression artifacts, noise, blurring, and geometric transformations. These alterations can significantly impact the visibility and integrity of watermark text, making it harder to accurately detect and recognize watermarks in such images. In unfavorable imaging conditions, the reliability of text,

Fig. 1 Sample images for watermarked and non-watermark text

images, and videos cannot be guaranteed. In uncontrolled scenarios, text instances may experience low resolution and significant distortion due to factors such as inappropriate shooting distance or angle. Blurriness can also occur when the text is out of focus or affected by shaking. Furthermore, inadequate lighting conditions can introduce noise, while highlights or shadows can further distort the watermark text. These imaging challenges make the detection and recognition of watermark text more challenging in real-world situations [3].

Background Complexity and Occlusions: Natural scenes are characterized by unpredictable backgrounds, which can introduce challenges in detecting watermark text. The presence of occlusions caused by external objects or patterns that resemble watermark text (such as brickwork, windows, fences, tree leaves, or traffic signs) can lead to confusion and errors in the detection process. Apart from this, text in natural scenes exhibits a higher degree of diversity and variability compared to text in documents. Watermarked text in natural scenes can vary significantly in terms of language, color, font, size, orientation, and overall appearance. These factors contribute to the complexity of accurately identifying watermark text in natural scenes [4].

Robustness Against Attacks: Watermark detection systems should be resilient against intentional attacks aimed at removing or altering the embedded watermark. Adversarial attacks, such as cropping, scaling, or adding noise, can undermine the effectiveness of watermark detection algorithms, requiring the development of robust and reliable methods to withstand such attacks.

At present, detection of watermarked text in scene images is afflicted with a number of challenged images are shown in Fig. 1. To overcome this issue, we proposed a model that will classify the natural scene images into watermark text or non-watermark text, and it helps in the process of detection, recognition, and removal of the watermark from the image.

The literature reveals the existence of two distinct groups of natural images: those that include watermark text and those that do not. Regardless of the specific size or placement of the watermark text, an image that contains watermark text is classified as a watermark text image. Conversely, an image that does not contain any watermark text is categorized as a non-watermark text image [5]. This paper addresses the classification of watermark text and non-watermark text images in real-world scenarios using Xception architecture. The motivation for exploring the Xception model lies in its effectiveness as a powerful tool for classifying watermarked and non-watermarked text. This technology exhibits immense potential to be harnessed across a diverse array of applications, encompassing critical areas like copyright protection and content authentication, revolutionizing the way we safeguard intellectual property and verify the integrity of digital content.

In this document, we delve into a comprehensive exploration of "Classification of Watermarked and Non-Watermark Text in Natural Scene Images: An Xception-based Approach." To guide you through the intricacies of our analysis, we've organized the content into distinct sections. That's available at the end of Sect. 1. The study is structured as follows: we describe past related studies and works in the next part. The many stages of the proposed solution are presented and thoroughly discussed in

Fig. 2 A few dataset samples of natural scenes with watermark text content under various difficult circumstances **a** text color **b** text size **c** hazy effect **d** cluttered background **e** image with normal text and watermarked text

Sect. refs. 3, "Proposed Methodology." We present the "Experimental Results and Discussion" in Sect. 4. Section 5 concludes the study and outlines the direction for further research (Fig. 2).

2 Related Work

In this section, we highlight notable works that have had a significant impact in the field of image watermarking. Numerous research papers have been published emphasizing the crucial role of watermarks in ensuring data integrity, content authentication, ownership identification, and copyright protection. Van et al. [5] offer a unique approach to classifying handwritten documents online as text or non-text using recurrent neural networks (RNNs) and an enhanced long short-term memory (LSTM) network. The task of labeling a sequence is a process of classifying strokes in a digital ink document into text and non-text groups. Furthermore, the network captures the temporal contextual information of adjacent strokes using a straightforward yet effective approach. The precision of classification is increased by including both local and global contexts.

With regard to managing the problem of separation of text/non-text in the context of freely hand-drawn online documents, Delaye et al. [6] presented a CRF approach in this study. In order to jointly simulate various information sources (local, spatial, and temporal), the categorization accuracy at the stroke level is increased by using a conditional random field.

Sharma et al. [7] developed a novel component-based categorization approach for text frames. The method initially determines possible word clusters from the gradient values derived from the RGB images of a given video frame. The text cluster's Sobel edges are then retrieved and used for additional processing. The proposed approach involves a method for handling text components by first removing any erroneous ones before performing linearity verification. These text elements' linearity is assessed by examining their centroids in a piecewise manner. If the constituent parts of a frame satisfy the specified linearity criterion, it is considered a text frame; otherwise, it is classified as a non-text frame. To showcase its orientation independence, the suggested method has been extensively tested on diverse datasets containing both text and non-text elements in various orientations.

A technique for distinguishing text from non-text ink strokes in digitally uploaded handwritten Japanese manuscripts has been developed using Markov random fields (MRFs) by Zhou et al. [8]. MRFs effectively utilize the spatial relationship between strokes. Support vector machine (SVM) classifiers are trained to categorize both individual strokes and pairs of strokes. The SVM outputs are translated into probabilities to create likelihood clique potentials for MRF. Experimental tests conducted on the TUAT Kondate database demonstrate that the proposed MRF approach outperforms stroke classification based solely on individual strokes and sequence classification based on hidden Markov models (HMMs).

Delaye et al. [9] used the conditional random field (CRF) to focus on integrating contextual information into the classification of text and non-text strokes in online handwritten documents. To achieve this, a CRF methodology was adopted, which combined and integrated various forms of context, such as spatial and temporal interactions. This allowed for comprehensive modeling of contextual information in the categorization process.

Phan et al. [10] created an advanced classifier to categorize individual strokes within a digital ink document as either text or non-text, incorporating both global and local contexts. The global context of the stroke sequence was captured using a neural network architecture, facilitating a holistic comprehension of the entire sequence of strokes.

3 Proposed Methodology

Watermark text can be visually distracting and can affect the overall aesthetics of the image also, Watermark text can be overused, leading to an excessive number of markings on a document or image. The diverse nature of text in the wild, complex imagery background, uneven lighting, and text orientations are the challenges of the watermark text. To overcome those problems, we proposed a model as shown in Fig. 3. This work represents the TensorFlow implementation of the CNN binary classifier using the transfer learning technique (Xception). The proposed approach has three stages: (i) preprocessing, (ii) feature extraction, and (iii) model training. For this experiment, the model utilized a batch size of 16 and an image size of 256 ×

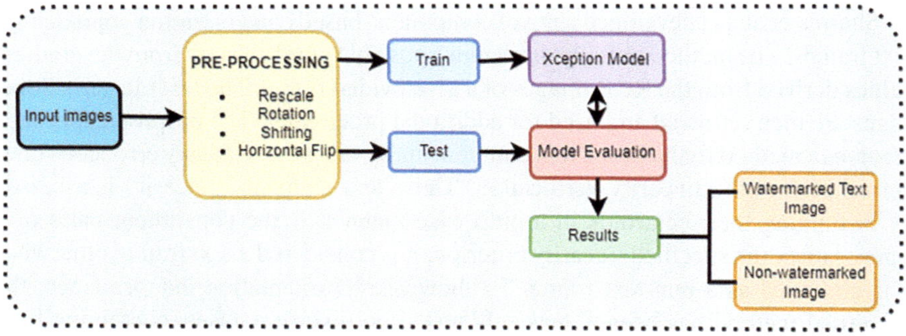

Fig. 3 Architecture of proposed model

256. We employed the ImageDataGenerator class using real-time data augmentation, and it will generate the batches of tensors using image data. It helps to make the model more generalized. The re-scaling factor converts the pixels present in the range [0,255] to a range [0,1] for each input image. Normalization and data augmentation techniques such as rotation, height shift range, width shift range, the range for random zoom, and horizontal flip are used.

The input image is processed in the preprocessing layer to normalize the pixel values and prepare the image for input to the Xception model. The Xception model stands out as a powerful deep convolutional neural network, particularly known for its exceptional performance in image classification tasks. It is constructed using multiple blocks of separable convolution layers and skip connections. To further reduce the spatial dimensions of the feature maps, a global average pooling layer is incorporated into the Xception model. Subsequently, the generated feature vector is passed to a fully connected output layer that produces a single value. This value is then subjected to a sigmoid activation function, resulting in a binary classification output. The output indicates the likelihood that the input image belongs to the positive class.

The input image is fed into the Xception model. It undergoes a series of processing steps within the neural network. The Xception model is designed to handle color images, which are typically represented as the three-dimensional preserve of pixel values, where each pixel contains red, green, and blue color channels. The image goes through multiple blocks of separable convolution layers, which are responsible for extracting features from the image. The convolutional layers apply various filters to the image to detect patterns, shapes, and textures at different spatial scales. Skip connections allow information from early layers to be combined with deeper layers, helping to preserve important features and improving gradient flow during training. After passing through convolutional layers, a global average pooling layer is added. This pooling operation reduces the spatial dimensions of the feature maps and captures the most relevant and significant information from each feature map. The resulting feature vector from the global average pooling layer is fed into a fully connected output layer. This layer computes the final output by performing a weighted sum of the feature vector elements.

To produce a binary classification output, a sigmoid activation function is applied to the value generated by the fully connected output layer. The sigmoid function squashes the output between 0 and 1, representing the probability of the input image belonging to the positive class. Overall, the Xception model processes the input image through a series of convolutional layers and pooling operations, capturing relevant features from the image. The model's architecture and learned parameters allow it to efficiently and accurately classify the input image into the desired classes.

In Fig. 3, we are explaining the various steps involved in watermarked text image classification. The Xception model architecture is shown in Fig. 4 which is used in our framework for the classification task (Fig. 5).

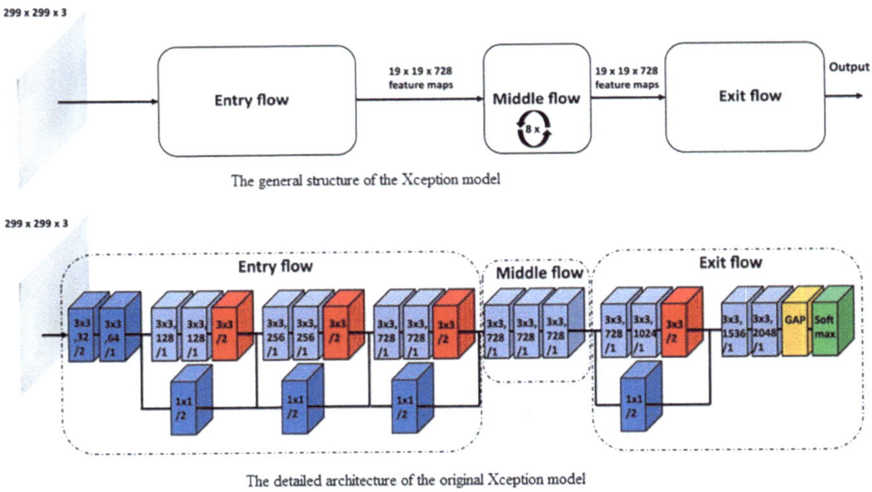

Fig. 4 Architecture of Xception model [11]

Fig. 5 Watermarked and non-watermarked text images

4 Experimental Results and Discussion

The experiments were conducted on the watermarked/not watermarked images dataset from Kaggle, which consists of 12,000 images of both watermarked and non-watermarked images. For our experimental purpose, we have taken only the first 3000 images from both classes. The dataset is divided into training and testing sets, following an 80/20 ratio. Models are pre-trained on the ImageNet dataset. The Xception model is used as a bias for this classifier. The top layer was removed and a simple dense layer with one neuron and sigmoid activation was used instead. The performance of the Xception model can be assessed using various metrics such as accuracy, precision, recall, and F1-score. These metrics offer information on the model's performance and can be used to modify the model's performance. The model achieves an 86.50% accuracy in classifying images containing watermarked or non-watermarked text. Figures 6 and 7 display "accuracy" and "loss" curves during the training process:

4.1 Evaluation Metrics

This section introduces commonly used performance measures in watermarked text classification tasks.

Accuracy: An measure of the model's performance across all classes is accuracy 1. When all classes are equally important, it is helpful. The number of accurate forecasts divided by the total number of predictions is used to compute it.

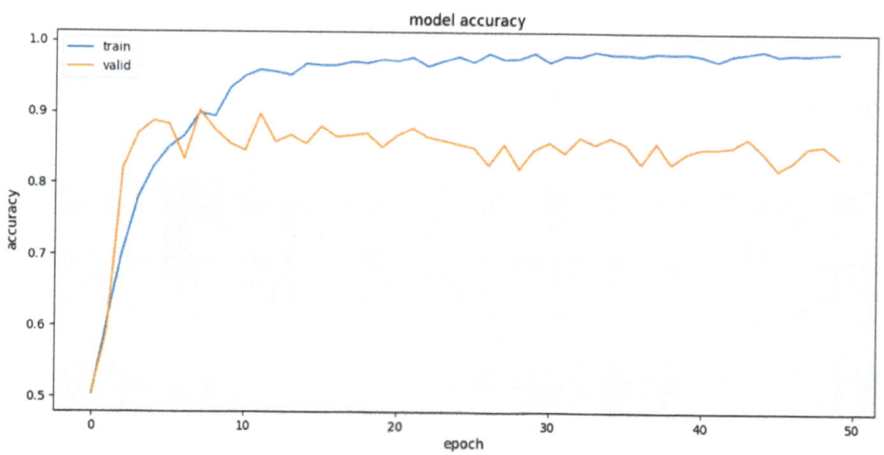

Fig. 6 Model accuracy

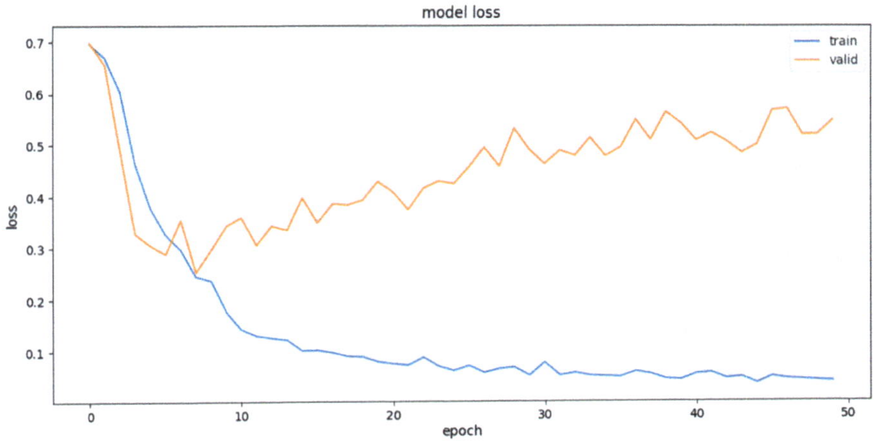

Fig. 7 Model loss

$$\text{Accuracy} = \frac{(\text{True Positive} + \text{True Negative})}{(\text{True Positive} + \text{True Negative} + \text{False Positive} + \text{False Negative})} \tag{1}$$

Precision: The precision 2 is determined by dividing the total number of samples classified as Positive (either properly or erroneously) by the proportion of Positive samples that were correctly identified. The precision gauges how well the model categorizes a sample as positive.

$$\text{Precision} = \frac{\text{True Positive}}{(\text{True Positive} + \text{False Positive})} \tag{2}$$

Recall: The recall 3 is measured as the proportion of Positive samples that were correctly identified as Positive to all Positive samples. The recall gauges how well the model can identify Positive samples. The more positive samples that are identified, the larger the recall.

$$\text{Recall} = \frac{\text{True Positive}}{(\text{True Positive} + \text{False Negative})} \tag{3}$$

***F*1 − Measure**: The harmony between precision and recall is communicated by the harmonic mean of precision and recall 4.

$$F1 - \text{Measure} = 2\frac{(\text{Precision} \times \text{Recall})}{(\text{Precision} \times \text{Recall})} \tag{4}$$

		Predicted Scenario	
Actual Scenario	Positives	True Positive (TP)	False Positive (FP)
	Negatives	False Negative (FN)	True Negative (TN)

Fig. 8 Confusion matrix

Table 1 Results obtained from our model

	Accuracy	Precision	Recall	$F1$-score
Train	99.37	99.25	99.5	99.37
Test	86.5	84.76	89	86.82

Confusion Matrix: Below, confusion metrics are displayed for samples from a given test set that were correctly identified and those that were not (Fig. 8).

As a consequence, the task of classifying the watermarked text and non-watermark text elements from the image has gained recognition. An effective algorithm for watermarked text and non-watermark text separation has been developed, which proficiently distinguishes non-textual content from the textual parts. Figures 6 and 7 illustrate the proposed model's accuracy and loss in order to show their performance. The suggested model performed 50 epochs of training. The model got its maximum accuracy, precision, recall, and F1-score of 86.5%, 84.76%, 89%, and 86.82% respectively as represented in Table 1.

The results are presented in a pictorial form using a confusion matrix as given in figure. This figure depicts the effectiveness of the classification model by comparing the predicted and actual class labels of a given test dataset 9.

5 Conclusion and Future Work

The proposed approach demonstrates the effectiveness of the Xception architecture in classifying watermarked and non-watermark text in natural scene images. Leveraging depth-wise separable convolutions based on the Inception architecture, the Xception model achieves remarkable accuracy while using fewer parameters. As a convolutional neural network, the Xception model excels in capturing intricate features from images, making it well-suited for discerning subtle differences between watermarked and non-watermarked text. During the training phase, the model requires a substantial dataset comprising both types of images to effectively learn and recognize the distinguishing features. Once trained, the Xception model exhibits high accuracy in classifying new images into their respective categories. A thorough evaluation of the model's performance is conducted using a comprehensive range of metrics, including accuracy, precision, recall, and F1-score, resulting in respective values of

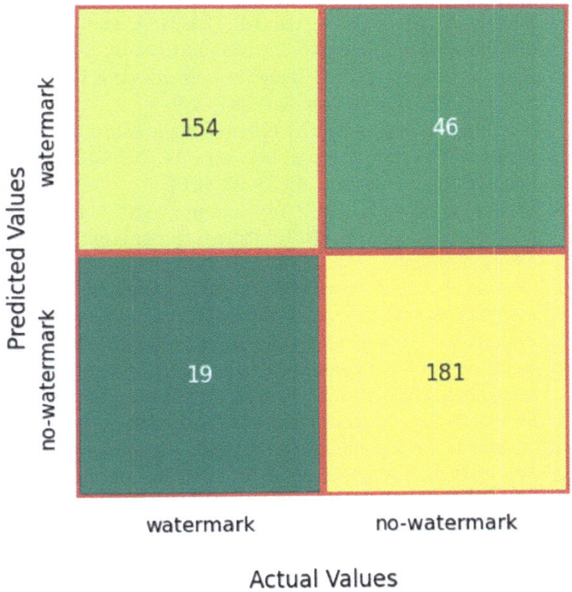

Fig. 9 Confusion matrix

86.5%, 84.76%, 89%, and 86.82%. The experimental results provide clear evidence of the success achieved through the proposed approach (Fig. 9).

In the future, researchers can explore the viability of ensemble methods to enhance watermark image classification. By amalgamating predictions from multiple models like bagging, boosting, or stacking, there is a potential to improve classification accuracy and bolster the robustness of the models and also investigate transfer learning techniques to convey knowledge from diverse domains into watermark classification. For instance, pre-training on relevant tasks such as logo recognition or object detection could potentially yield advantages for watermark classification tasks.

Our system's robustness is crucial; future improvements employ adversarial training and data augmentation with adversarial examples; implement ensemble learning and input validation mechanisms to detect and reject potential attacks; regularly update the model; use regularization techniques; and engage with the security community for ongoing defense against emerging threats. These strategies collectively fortify our Xception model's resilience in classifying watermarked and non-watermarked text amid adversarial challenges.

References

1. Hussein E, Belal MA (2012) Digital watermarking techniques, applications and attacks applied to digital media: a survey. Threshold 5(6)
2. Kadian P, Arora SM, Arora N (2021) Robust digital watermarking techniques for copyright protection of digital data: a survey. Wireless Pers Commun 118:3225–3249

3. Kim Y-W, Moon K-A, Oh I-S (2003) A text watermarking algorithm based on word classification and inter-word space statistics. In: ICDAR, pp 775–779. Citeseer
4. Bian Y, Liang S (2013) Image watermark detection in the wavelet domain using bessel k densities. IET Image Proc 7(4):281–289
5. Van Phan T, Nakagawa M (2014) Text/non-text classification in online handwritten documents with recurrent neural networks. In: 2014 14th International conference on frontiers in handwriting recognition, pp 23–28. IEEE
6. Delaye A, Liu C-L (2012) Text/non-text classification in online handwritten documents with conditional random fields. In: Pattern recognition: Chinese conference, CCPR 2012, Beijing, China, September 24–26, 2012. Proceedings 5, pp 514–521. Springer Berlin Heidelberg
7. Sharma N, Shivakumara P, Pal U, Blumenstein M, Tan CL (2015) Piece-wise linearity based method for text frame classification in video. Pattern Recogn 48(3):862–881
8. Zhou X-D, Liu C-L (2007) Text/non-text ink stroke classification in Japanese hand-writing based on Markov random fields. In: Ninth international conference on document analysis and recognition (ICDAR 2007), vol 1, pp 377–381. IEEE
9. Delaye A, Liu C-L (2014) Contextual text/non-text stroke classification in online handwritten notes with conditional random fields. Pattern Recogn 47(3):959–968
10. Van Phan T, Nakagawa M (2016) Combination of global and local contexts for text/non-text classification in heterogeneous online handwritten documents. Pattern Recogn 51:112–124
11. Dhillon A, Verma GK (2020) Convolutional neural network: a review of models, methodologies and applications to object detection. Prog Artif Intell 9(2):85–112

Object Segmentation in Thermal Images Using YOLOv8

Haider Ali Muften and Ali Retha Hasoon Khayeat

Abstract Thermal images are crucial for object detection in surveillance, security, industrial automation, and vehicular navigation due to their ability to capture heat signatures. However, complexities like low contrast, fluctuating thermal patterns, and lack of thermal-specific datasets necessitate the use of specialized algorithms for precise object identification and location. This research paper delves into the domain of object detection within thermal images, a crucial aspect of applications like surveillance and autonomous systems. Leveraging the YOLOv8 algorithm renowned for real-time and accurate object detection, this study focuses on enhancing detection performance in the challenging realm of thermal imagery. The methodology involves training on a meticulously curated dataset of 1898 thermal images, comprising 1518 for training and 380 for validation. The algorithm's efficacy is demonstrated through a mean average precision (mAP50) of (82%) and map50–90 of (59.3%). These results validate YOLOv8 suitability for thermal imaging applications, propelling advancements in fields reliant on accurate object detection in thermally challenging conditions.

Keywords Yolov8 · Segmentation · CLAHE · Thermal images

1 Introduction

In order to extract two different types of information from the photos, object detection and classification are used. The process of object segmentation is strongly related to the detection. The challenge of extracting complicated, highly organized objects made up of visually distinct sections is one of object segmentation's challenges [1]. The fundamental function of computer vision is image semantic segmentation [2]. Urban scene semantic picture segmentation is essential for autonomous systems [3].

H. A. Muften (✉) · A. R. H. Khayeat
College of Computer Science and Information Technology, University of Kerbala, Computer Science Department, Karbala, Iraq
e-mail: haider.ali@s.uokerbala.edu.iq; ali.r@uokerbala.edu.iq

S. M. Thampi et al. (eds.), *Fifth International Conference on Computing and Network Communications*, Lecture Notes in Electrical Engineering 1221,
https://doi.org/10.1007/978-981-97-4711-5_33

The goal of image semantic segmentation is to give each pixel in an image a semantic label. The fact that an expanding variety of applications, rely on deriving knowledge from pictures highlights the relevance of this high-level job that prepares the way for scene interpretation [1]. Numerous neural networks for semantic segmentation have been suggested in recent years, such as FCN [4], PSPNet [5], UNet [6], etc. The majority of techniques in the literature make use of visible cameras' three-channel RGB pictures. These techniques, however, are unable to deliver the needed performance under challenging environmental circumstances such as low illumination, wet, or foggy because of the constraints of visual imaging. As a result, tasks involving semantic segmentation have been carried out using thermal pictures [7]. One of the solutions that can make up for the shortcomings of ordinary color cameras is a thermal imaging sensor. Thermal sensors are innately resistant to changes in illumination since every item generates infrared radiation that is unaffected by ambient lighting conditions [8].

However, because they rely on visible spectrum pictures, these techniques perform poorly in challenging conditions including foggy, wet, and dark situations. Thermal pictures are collected from thermal radiation, which is steady in any weather and at any time, in contrast to visible spectrum photographs. As a result, thermal pictures have been used in several computer vision applications, such as pedestrian detection [9], visual tracking [10], and person re-identification [11].

2 Related Work

Semantic segmentation refers to associating each pixel of an image with a class label. We review selected semantic segmentation networks in this section.

Sarmiento [12], they suggested using deep learning for segmentation and detection, two methods of capturing pavement distress from 2D RGB photos. On a small dataset of pavement photos, YOLOv4 is utilized for pavement distress detection, and DeepLabv3 is used for pavement distress segmentation. To hasten training, 305 photos were gathered at a 900×600 resolution. The data is split into an 80–20 train-validation split for the 242 training set and 63 for the validation set for both pavement distress detection and segmentation. While YOLOv4 failed to acquire high metric scores, only achieving mAP values of 0.272 on raw photos and 0.286 on pre-processed data, DeepLabv3 was able to reach outcomes of 0.56 in mIoU and 0.58 in the dice coefficient.

Xiong et al. [13], to identify the problematic sections of faulty apples, a semantic segmentation technique based on the BiSeNet V2 deep learning network was developed. The topology of the YOLO V4 network was improved using a model pruning technique. The pruned YOLO V4 network substantially enhanced the detection precision of fault spots in Apple photos. According to the experimental findings, an apple could be classified with an average accuracy of 92.42% and an $F1$ score of 94.31. As part of this study, a collection of 3000 apple photographs was collected. They reduce the size of the supplied pictures to 512×512 pixels. A total of 2400 photos were chosen to make up the training set for the network and 600 images to make up the validation set.

Cao et al. [14], they suggest Swin-Unet, a pure transformer similar to Unet, for medical picture segmentation. Our pure transformer technique offers the greatest performance, as shown by experimental findings on the Synapse dataset, with segmentation accuracy of 79.13% dice-similarity coefficient (DSC) and 21.55% Hausdorff Distance (HD). Similar to the Synapse dataset, the proposed Swin-Unet is trained on the ACDC dataset to perform medical picture segmentation. Swin-Unet is still able to deliver exceptional performance with an accuracy of 90.00% while utilizing the picture data of MR mode as input.

Tang et al. [15], they suggested an automated two-stage liver segmentation technique. Improved faster regions with CNN characteristics (faster R-CNN) are used in the first stage to locate the liver's approximate location. To get the liver contour, the acquired pictures are edited and entered into DeepLab. The suggested method is tested on the MICCAI-Sliver07 and 3Dircadb datasets. The suggested method surpasses cutting-edge approaches in terms of volume overlap error, average surface distance, relative volume difference, and overall score, according to experimental data.

Xiao et al. [16], to solve these problems, they suggested a UNet-like model with a weighted attention mechanism and a skip connection strategy. They use the publicly available benchmark datasets DRIVE and STARE to assess the performance of their technique. 20 training photos and 20 test images are included in the DRIVE dataset. Seven photos in this collection demonstrate early mild diabetic retinopathy. Twenty retinal fundus slides were recorded with a TopCon TRV-50 fundus camera for the STARE dataset. The dataset is split in half between photographs of healthy patients and images of problematic cases, which makes segmentation more difficult. They discover that the accuracy and sensitivity performance of their suggested Weighted Res-UNet can outperform that of the standard UNet model. While the accuracy and sensitivity for the DRIVE data were 0.9655 and 0.7715 respectively, they were 0.9693 and 0.7469 respectively for the STARE data.

Hurtik et al. [17], they introduce Poly-YOLO, an improved and expanded variant of YOLO that includes instance segmentation. Building on the core concepts of YOLOv3, Poly-YOLO eliminates two of its flaws: an excessive number of rewritten labels and an ineffective distribution of anchors. While having just 60% of the trainable parameters of YOLOv3, Poly-YOLO increases map by a relative 40%. Additionally, we provide Poly-YOLO lite, which has fewer options and a lower output resolution. It is three times smaller and twice as quick than YOLOv3, yet it has the same accuracy. Three datasets—Simulator, Cityscapes, and India Driving—are used in the benchmark. Simulator is a 600 × 800 pixel synthetic dataset they created on the internet4 that includes 700 training, 90 validation, and 100 test pictures. A dataset called Cityscapes was recorded while a vehicle was being driven through numerous German cities. The pictures were either taken in the daytime or at night, and they are all 2048 by 1024 pixels in size. The 2474/500/500 image train/valid/test datasets are used. IDD is a dataset that tracks unstructured traffic on Indian roadways and is comparable to Cityscapes. 14,010 photos were used for training, 977 for validation, and 1049 for testing, in that order.

3 Background

3.1 *Contrast Limited Adaptive Histogram Equalization (CLAHE)*

The limits of conventional histogram equalization are solved by CLAHE, an enhanced variant of AHE, or adaptive histogram equalization. In order to preserve crisp field edges, minimize noise while preserving high spatial frequency information, and apply a combination of CLAHE median filtration and edge sharpening, it offers a variety of adaptive contrast limited holographic equalization approaches (CLAHE). The picture can also be subjected to sequential processing, a method that is recorded into an insert macro for recurrent usage [18].

3.2 *You Only Look Once (YOLO)*

A quick and precise object identification paradigm for computer vision is called YOLO. In contrast to other traditional approaches, the YOLO method performs object recognition and placement once on the full image rather than splitting it into smaller portions. The YOLO extracts features from pictures using a convolutional neural network (CNN) architecture [19], and then utilizes convolutional layers to forecast feature positions and kinds. This aids the YOLO's rapid processing and precise object identification. The YOLO model has better accuracy and performance than earlier iterations like YOLOv2 [20], YOLOv3 [21], and YOLOv4 [22]. The YOLO has been used extensively in real-world settings, including auto-driving, object tracking, object identification, and security surveillance, The YOLOv8 is currently the most recent version of YOLO for ultralytics. As a cutting-edge, contemporary model, YOLOv8 [23] builds on the success of earlier iterations by adding new features and improvements to boost functionality, adaptability, and effectiveness. Tasks including detection, segmentation, posture estimation, tracking, and classification are all fully supported by the YOLOv8. Users may employ YOLOv8 features in a variety of applications because of this flexibility [24] (Fig. 1).

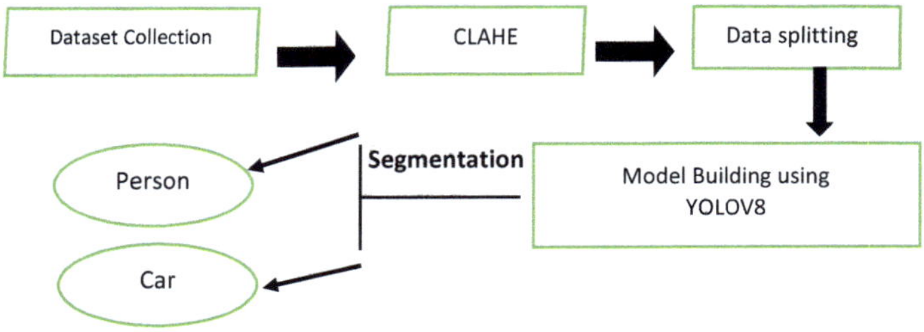

Fig. 1 Diagram block of the suggested deep neural network models is seen in the segmentation model

Table 1 FLIR ADAS classes distribution on train and validation split

	Persons	Cars	Total annotated in images
Train	2353	3504	5857
Validation	593	851	1444
Total of each class	2946	4355	7301

4 Methodology

4.1 Preprocessing

4.1.1 Dataset

FLIR ADAS proposes many images with annotations. These images have been acquired during a large period, 60% by day and 40% by night in various scenes. The images of the FLIR dataset are available under different formats: the raw images directly issued from the sensor and the modified and enhanced ones thanks to specific FLIR algorithms. I used 1898 image and used Ultralytics YOLOv8 to create annotations. Two different classes have been annotated in this dataset: persons and cars (Table 1).

4.1.2 Contrast Limited Adaptive Histogram Equalization (CLAHE)

In order to equalize each region in a certain histogram, this approach breaks a picture into smaller parts called blocks, with a dimension of typically 8×8. It addresses the problem of noise amplification brought on by the employment of HE methods. It was first created to enhance photos' contrast, particularly in low-contrast areas. In contrast, two important components of the CLAHE technique are the block size (BS) and the clipping limit (CL), which govern the quality of the enhanced picture, respectively. The picture does indeed get brighter when CL is increased because of

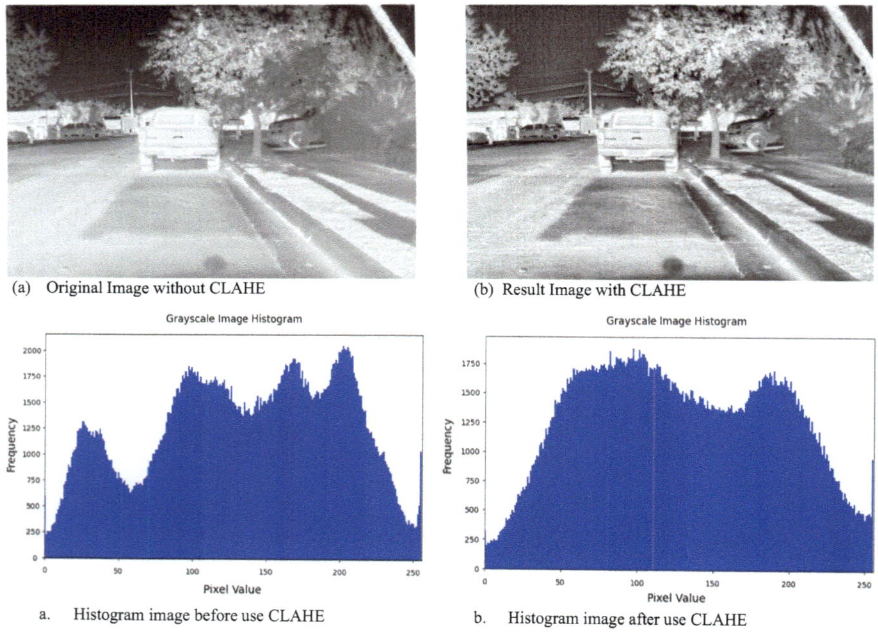

(a) Original Image without CLAHE
(b) Result Image with CLAHE
a. Histogram image before use CLAHE
b. Histogram image after use CLAHE

Fig. 2 Result image use CLAHE method

the input image's extremely low intensity with rising CL, which causes the histogram to flatten.

To demonstrate that the use of CLAHE produced clear picture results in low-light situations, as shown in Fig 2, we conducted trials on several photos with poor illuminations.

4.1.3 YOLOv8

The most recent model in the YOLO family of detection models, noted for their combined detection and segmentation abilities, is called YOLOv8. The architecture consists of a backbone, head, and neck, just like YOLOv5 [25, 26]. It is a top choice for real-time object identification thanks to its redesigned architecture, enhanced convolutional layers (backbone), and more sophisticated detection head. The most recent computer vision methods, including instance segmentation, which enables multiple item recognition in an image or video, are also supported by YOLOv8. The Darknet-53 backbone network, which replaces the older YOLOv7 network [27], is used by the model and is quicker and more accurate. Bounding boxes are predicted by YOLOv8 using an anchor-free detection head. The model's bigger feature map and better convolutional network, which increase its precision and speed, make it more efficient than prior iterations. Pyramid feature networks are another addition to

YOLOv8 that help it recognize objects of various sizes. The model has a user-friendly API that makes it simple to integrate into many applications.

4.2 *Results and Experiments*

4.2.1 Results

In this paper, we conducted experiments using part of the FLIR dataset. The FLIR dataset consists of image samples from 2 (car and person) object categories. With a total of 1898 images, this dataset provides abundant samples for training and evaluation purposes. Additionally, the dataset encompasses multiple viewpoints, which effectively enhances the model's generalization and robustness. The model evaluation metrics are as follows:

Precision: It measures the proportion of correctly detected targets among the ones detected by the model. A higher precision indicates more accurate detection by the model

$$\text{precision} = \text{TP}/(\text{TP} + \text{FP}) \tag{1}$$

In the equation TP represents the number of correctly detected targets. FP represents the number of falsely detected targets [28].

Recall measures the ability of a model to correctly detect targets. A higher recall indicates that the model can better capture the targets.

$$\text{Recall} = \text{TP}/(\text{TP} + \text{FN}) \tag{2}$$

AP: A comprehensive evaluation of the model's performance at different recall levels is by calculating the average precision. It provides an overall assessment of the model's performance at various threshold values.

MAP50: The average precision (AP) at an intersection over union (IOU) threshold of 0.5 is calculated to measure the model's performance in predicting object locations accurately. It quantifies how well the model performs in terms of precision when the predicted bounding boxes have an IOU of at least 0.5 with the ground truth boxes [29].

MAP50-95: The average precision (AP) is calculated by considering the IOU values ranging from 0.5 to 0.95. This comprehensive evaluation takes into account the performance of the object detection model at different IOU thresholds and provides a more comprehensive assessment of its accuracy. It measures the precision of the model's predictions across a range of IOU thresholds, reflecting its ability to accurately localize objects under varying levels of overlap with the ground truth bounding boxes [29].

The **F1 score** is a metric commonly used to evaluate the performance of a binary classification model, such as in object detection where you're determining whether an object is detected correctly or not. It takes into account both precision and recall to provide a balanced measure of the model's accuracy [29].

$$\text{F1 Sore} = 2 * \frac{\text{Precision} * \text{Recall}}{\text{Precision} + \text{Recall}} \tag{3}$$

4.2.2 Model Training Results

The YOLOv8 detector was fine-tuned with images of size 640 × 640 pixels, with a batch size of 16. A stochastic gradient descent, SGD, was used with a learning rate of 0.01. Training used pre-trained weight (yolov8l-seg) and the proposed improved algorithm is trained for 100 epochs. The score on the validation set was calculated for each epoch and the best one was kept at the end of each experiment. The training results of the proposed algorithm are shown in (Fig 3).

Figure 3 shows the PR curves, MAP50 and MAP50-95 scores, and the variation curves for the loss on the training and validation sets for our suggested model (Figs. 4 and 5).

4.2.3 Experiment

Table 2 shows the training metrics of the Yolov8 algorithm on the FLIR dataset. With the use of a pre-trained model and the proposed improvements our approach achieves good results even with 100 training epochs.

We conducted experiments on the FLIR dataset, our proposed improved algorithm with the Yolov8l algorithms. The results of the experiments are shown in (Fig. 6).

5 Conclusions

We have created an object segmentation system that operates in real time and is resilient to varying weather and daylight conditions. In order to do this, we employed YOLOv8, the Yolov8 method, which is currently the state-of-the-art technique with exceptional performance, as the foundation for our modifications. We also suggested CLAHE image processing to enhance and minimize picture noise. Our experimental findings show how well the suggested approach works and how reliable it is at accurately and efficiently segmenting objects in real-world circumstances. Additionally, the outcomes demonstrated that the suggested solution is workable for real-time applications.

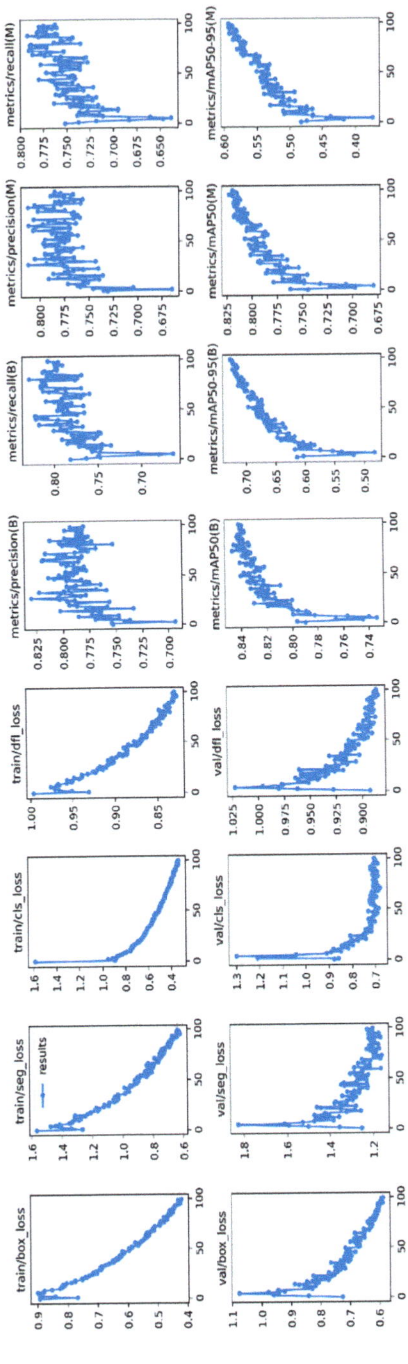

Fig. 3 Model training result

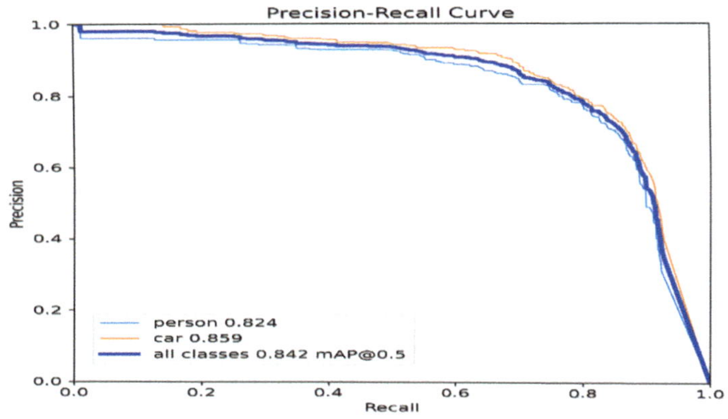

Fig. 4 Depicts the F1 curve of the YOLOv8 algorithm

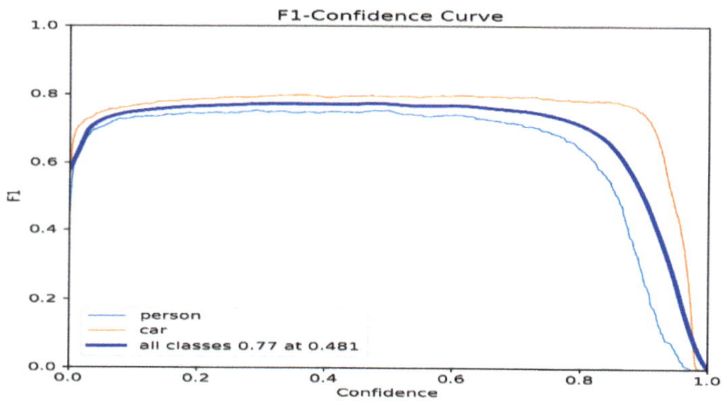

Fig. 5 Depicts the PR curve of the YOLOv8 algorithm

Table 2 Experiments

Algorithm	MAP50	MAP50-95	P	R	F1 score
Yolov8	0.82	0.593	0.785	0.762	0.77

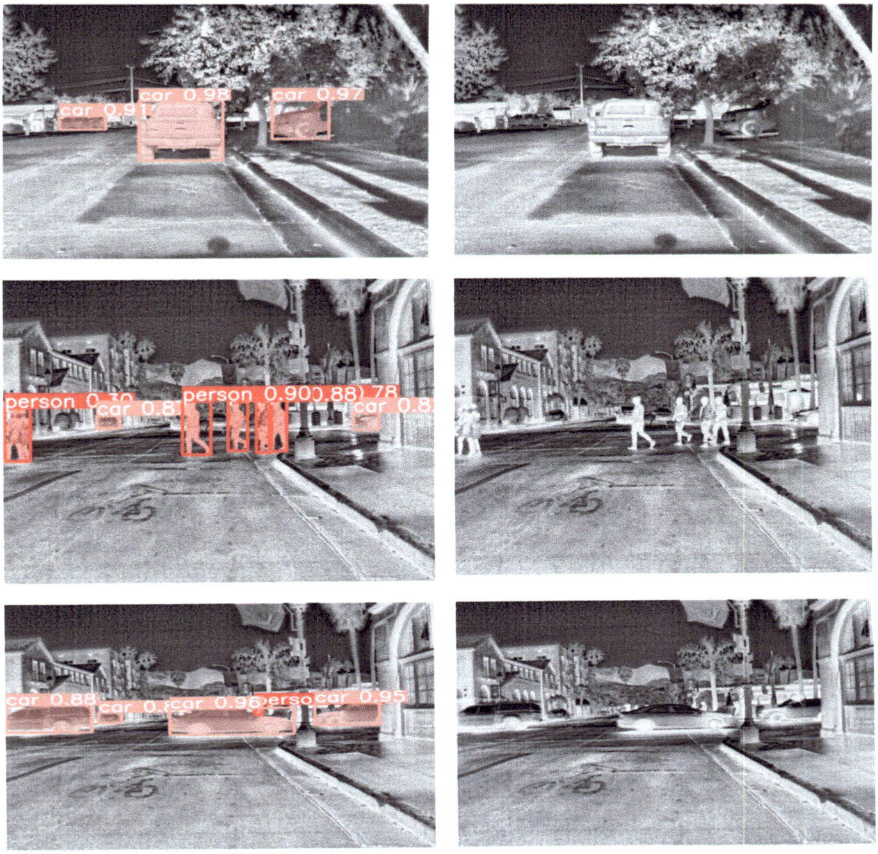

Fig. 6 Chart of target segmentation results

References

1. Li C et al (2020) Segmenting objects in day and night: Edge-conditioned CNN for thermal image semantic segmentation. IEEE Transactions on neural networks and learning systems 32(7):3069–3082
2. Yan R, Yang K, Wang K. (2021) NLFNet: non-local fusion towards generalized multimodal semantic segmentation across RGB-depth, polarization, and thermal images. In: 2021 IEEE international conference on robotics and biomimetics (ROBIO) IEEE
3. Jiao L et al (2019) A survey of deep learning-based object detection. IEEE access 7:128837–128868
4. Sun W, Wang R (2018) Fully convolutional networks for semantic segmentation of very high resolution remotely sensed images combined with DSM. IEEE Geosci Remote Sens Lett 15(3):474–478
5. Zhao H et al. (2017) Pyramid scene parsing network. In: Proceedings of the IEEE conference on computer vision and pattern recognition
6. Zunair H, Hamza AB (2021) Sharp U-Net: Depthwise convolutional network for biomedical image segmentation. Comput Biol Med 136:104699

7. Kütük Z Algan G (2022) Semantic segmentation for thermal images: a comparative survey. In: proceedings of the IEEE/CVF conference on computer vision and pattern recognition
8. Shin U et al. (2021) Unsupervised depth and ego-motion estimation for monocular thermal video using multi-spectral consistency loss. arXiv preprint arXiv:2103.00760
9. Ghose D et al. (2019) Pedestrian detection in thermal images using saliency maps. In: proceedings of the IEEE/CVF conference on computer vision and pattern recognition workshops
10. Kang B et al (2019) Grayscale-thermal tracking via inverse sparse representation-based collaborative encoding. IEEE Trans Image Process 29:3401–3415
11. Wang G.a, et al. (2019) RGB-infrared cross-modality person re-identification via joint pixel and feature alignment. In: Proceedings of the IEEE/CVF international conference on computer vision
12. Sarmiento J.A (2021) Pavement distress detection and segmentation using YOLOv4 and DeepLabv3 on pavements in the Philippines. arXiv preprint arXiv:2103.06467
13. Xiong H, Cai W, Liu Q (2021) MCNet: Multi-level correction network for thermal image semantic segmentation of nighttime driving scene. Infrared Phys Technol 113:103628
14. Cao H, et al. (2022) Swin-unet: Unet-like pure transformer for medical image segmentation. In: European conference on computer vision. Springer
15. Tang W et al (2020) A two-stage approach for automatic liver segmentation with Faster R-CNN and DeepLab. Neural Comput Appl 32:6769–6778
16. Xiao X, et al. (2018) Weighted res-unet for high-quality retina vessel segmentation. In: 2018 9th international conference on information technology in medicine and education (ITME) IEEE
17. Hurtik P et al (2022) Poly-YOLO: higher speed, more precise detection and instance segmentation for YOLOv3. Neural Comput Appl 34(10):8275–8290
18. Musa P, Al Rafi F, Lamsani M (2018) A review: contrast-limited adaptive histogram equalization (CLAHE) methods to help the application of face recognition. In: 2018 third international conference on informatics and computing (ICIC) IEEE.
19. Shin H-C et al (2016) Deep convolutional neural networks for computer-aided detection: CNN architectures, dataset characteristics and transfer learning. IEEE Trans Med Imaging 35(5):1285–1298
20. Huang X et al. (2021) PP-YOLOv2: a practical object detector. ArXiv preprint. arXiv:2104.10419
21. Redmon J. Farhadi A (2018) Yolov3: an incremental improvement. ArXiv preprint. arXiv:1804.02767
22. Bochkovskiy A, Wang CY, Liao HYM, Yolov4 (2020): Optimal speed and accuracy of object detection. ArXiv preprint. arXiv:2004.10934
23. Reis D et al. (2023) Real-time flying object detection with YOLOv8. ArXiv preprint. arXiv:2305.09972
24. Thi-Mai-Phuong Dao NKN, Van-Kien Nguyen (2023) CNN-YOLOv8-Based tomato quality inspection system. A case study in Vietnam. SSRG Int J Electr Electron Eng. pp 31–40
25. Shoman M et al. (2022) A region-based deep learning approach to automated retail checkout. In: Proceedings of the IEEE/CVF conference on computer vision and pattern recognition
26. Aboah A (2021) A vision-based system for traffic anomaly detection using deep learning and decision trees. In: Proceedings of the IEEE/CVF conference on computer vision and pattern recognition.
27. Wang CY, Bochkovskiy A, Liao H.Y.M (2023) YOLOv7: Trainable bag-of-freebies sets new state-of-the-art for real-time object detectors. In: Proceedings of the IEEE/CVF conference on computer vision and pattern recognition
28. Kamal K, Hamid E.Z (2023) A comparison between the VGG16, VGG19 and ResNet50 architecture frameworks for classification of normal and CLAHE processed medical images
29. Zhang L et al (2023) DCF-Yolov8: An improved algorithm for aggregating low-level features to detect agricultural pests and diseases. Agronomy 13(8):2012

A Competent Human Activity Detection Using Integrated Wearable Sensor Data and Machine Learning

Veeramuthu Venkatesh, Pethuru Raj, Roshitha Nedium, Jahnavi Edara, Kalluru Amarnath Reddy, and R. Anushiadevi

Abstract A significant advancement in sensor-based technologies has led to a rapid expansion of applications for the Internet of Things (IoT), which can be used to build any real-time monitoring system. This expansion can be attributed to the significant advancement in sensor-based technologies. Nowadays, an increasing number of elderly people live alone in various locations worldwide, and it is essential to monitor the current state in which their health function or activity is critical. Human activity recognition (HAR) technology based on wearable sensors has many applications, including healthcare, fitness, smart homes, and surveillance. However, despite extensive computational research on HAR, multi-sensor-based activity recognition has numerous open challenges. One of these challenges is recognising complex temporal data, identifying discriminate feature vectors from multi-modal data and reducing data dimensionality. These issues necessitate extensive research. The Internet of Things (IoT) is the foundation for this paper's proposed human activity monitoring model. Using innovative sensor-based technologies, the model is designed to examine the actions of elderly people continuously. This model collects vital information using intelligent sensors or devices powered by the Internet of Things (IoT). Machine learning algorithms analyse this data to identify potential human behaviour risks that can be deduced from their actions. The proposed work's utility is assessed using a dataset made public by UCI. The findings of the experiments, which made use of several different metrics in addition to a parallel coordinate plot (PCP), indicate that the proposed ML-based HAR framework outperforms several state-of-the-art ML techniques and has the potential to achieve a maximum classification accuracy of 96.30%. Following the evaluation of the proposed model, the SVM was found to have committed the best accuracy of 96.30%, which is quite helpful for our objectives. In addition, SVM performed significantly better than every other machine learning algorithm tested for this investigation.

V. Venkatesh · R. Nedium · J. Edara · K. A. Reddy · R. Anushiadevi (✉)
School of Computing, SASTRA Deemed University Tirumalaisamudram, Thanjavur, Tamil Nadu 613401, India
e-mail: anushiadevi@it.sastra.edu

P. Raj
Reliance Jio Cloud Services (JCS), Bangalore 560025, India

S. M. Thampi et al. (eds.), *Fifth International Conference on Computing and Network Communications*, Lecture Notes in Electrical Engineering 1221,
https://doi.org/10.1007/978-981-97-4711-5_34

Keywords Wireless body area network · Internet of Things (IoT) · Wearable sensors · Information fusion · Machine learning (ML)

1 Introduction

Artificial intelligence, big data, 5G, the Internet of Things, and other technologies rapidly evolve, resulting in many new products, technologies, industries, and business models. Wearable technology is just one example of this trend. These technologies are leading to an increase in the number of intelligent sensors, which will have repercussions for human lifestyle and the organisation of society. As a result, new sensor-based technologies have become increasingly popular. This is primarily due to the limitations imposed by space and time, the ease with which a user's privacy can be invaded, excessive energy consumption, and other similar factors. Because of their high processing power, small size, and low cost, sensors have led to the development of wearable sensing devices. Because of their small size, these devices can interact as a part of everyday life. These devices have made detecting human behaviours and emotions possible at any time [1].

The Centres for Disease Control and Prevention (CDC) estimates that approximately 3 million people aged 65 and older go to the emergency rooms of hospitals each year due to injuries sustained in falls. [2]. According to the study's findings, the number of fatalities caused by falls in the United States increased by 33% between 2007 and 2016, which is expected to continue. Furthermore, the report assumed that the death rate from falling would be seven deaths per hour by 2030 [3, 4]. When an elderly person falls, they have a one-in-ten chance of suffering a severe injury, such as a hip fracture, bone breakage, head injury, or injury to the body's deep tissues. People can suffer various social and psychological consequences following an injury and the physical effects of the injury itself. [5]. Furthermore, developing different intelligent technologies has provided an efficient solution to the problems that elderly people who live alone are likely to face. The IoT is one of the most promising technologies currently available, and it has transformed how people live their daily lives intelligently [6].

When all linked devices are connected to the same network, the Internet of Things (IoT) creates an environment where they can communicate. This enables the accurate and timely flow of data needed for decision-making. As a result, the Internet of Things is a viable solution for a wide range of critical applications, including healthcare [7, 8]. Furthermore, the Internet of Things enables the development of many intelligent sensors and monitoring devices that a person can wear to provide early detection and management of their daily activities [9]. However, several challenges must be overcome to build a real-time monitoring system that is dependable and effective at tracking people's day-to-day activities [10]. However, the combination of Internet of Things technology and wearable sensors could be a viable option for predicting and monitoring the primary activities of elderly people.

Wearable technologies and sensors collect data about people's movements and send it to the cloud [11, 12]. The collected data is then used to train powerful machine learning algorithms to make accurate predictions. Wearable sensors can measure several different parameters, including but not limited to the following: heart rate, breathing monitoring, blood oxygen saturation, gesture recognition, acceleration, and inertial monitoring during human movement, body temperature, blood pressure, ion salts (including potassium, sodium, oxides, and calcium), urea, cholesterol, and lactic acid content amongst others. These parameters can now be measured and monitored remotely thanks to microelectronics and communication technology developments. The data can be automatically transmitted to nearby service centres and hospitals. The information contained in this paper may be helpful in the event of an emergency. Wearable sensors are flexible, straightforward to use, and have a longer lifespan and lower price than their desktop counterparts.

A single sensor has been used in a few studies for action recognition, taking into account the mobility and convenience of sensor wearing. On the other hand, a single sensor can only detect motion in its immediate surroundings and has more complex movements; recognition accuracy could be better. In current years, investigators have used a variety of fusion methodologies to determine how well data, features, and classifiers can be used to monitor both activity and health. However, we must carefully consider the number of sensors that will be used because, in real-world applications, using a limited number of sensors can make system recognition faster and more convenient. Furthermore, customers may require assistance carrying multiple sensors simultaneously for wearable sensors.

The number of sensors is essential because of the limitations on personal space and the discomfort that can result. This work proposes an efficient machine learning-based activity identification model using intelligent sensors worn or implanted in people's bodies. This model should be able to make accurate predictions about people's activities. Next, several machine learning algorithms are used to analyse the activity recognition dataset obtained from a smartphone to determine which algorithm will produce more accurate predictions about a person's activities.

The following is a condensed version of the work's most significant contributions:

An introduction is given to several types of sensors as well as the uses for those sensors.

The information fusion strategies of wearable sensors are disassembled and analysed from various perceptions, including multi-modality and multi-location perceptions.

Presented an outline of the more conventional approaches to information fusion and elaborated on the roles that machine learning and deep learning play in information fusion.

Transfer learning and data augmentation are two of the novel approaches to deep learning that have been proposed.

Presented existing problems (limitations) and potential future research directions in deep learning applied to human activity recognition.

In addition, previous research has shed light on the precision of several different fusion techniques by analysing each method's positive and negative aspects. In addition, the following critical issues will be addressed in this section of the paper:

2 Related Work

The primary goal of this section is to present a collection of research studies on the activity recognition model conducted by various researchers. These studies will be shown in the order listed below: In [13] provided a summary of the IoT and the technologies that make it possible, including intelligent sensors, RFID, a variety of communication technologies, and a variety of Internet protocols are all included. The association of smart sensors enables the provision of a wide range of applications that do not require the involvement of a human being. In [14] presented a technique for evaluating the hip joint's therapeutic efficacy in treating children with cerebral palsy created on the human body sensor network (BSN). This technique may be used in the future (CP). This technique employs gravity sensors distributed magnetic angular rate and an algorithm for data fusion from multiple sensors to assess the overall motion function of children with cerebral palsy. In [15] created a system framework that models the upper extremities simultaneously to provide estimations when occlusion is present. To track human joints, this framework combines multiple sensor readings.

Deep CNN-RF is a novel architecture by [16] created to add randomisation to the technique. A CNN and an RF classifier are combined in this new architecture. The simulation showed that hybrid strategies performed significantly better than traditional data mining.

In [17] proposed a new approach to identifying human walking activities based on human walk patterns. The primary goal of this research was to create four distinct "deep hybrid learning" strategies to establish a broad framework for activity recognition and improve its overall effectiveness. In [18] presented a new ensemble model for recognising human activity techniques based on incremental learning that has been processed for continuous learning. This model was created to assist them in identifying human activity. When it comes to personalising in either state, the results of the experiments show that the enhanced ensemble approach outperforms the non-weighted ensemble technique. This is true regardless of whether or not the methods are weighted. The machine learning strategy known as logistic model trees (LMT) was applied to determine human activities in the study [19]. The accuracy of the LMT approach is approximately 94.0%, according to the results of the experiments. In [20], a minimally labelled activity data set was used with a proposed attention-based CNN for the HAR. This study provided a poorly supervised model based on recurrent attention learning during the rigorous data labelling process (RAL). This model avoided the need for additional personnel and computational resources. It was proposed by he and his colleagues [21] that the HAR could benefit from the combination of CNN and LSTM networks. Recent research has suggested the existence of a deep neural network with an accuracy of 95.8%. Convolutional layers and LSTM storage were used in the construction of this network.

3 The Most Common Wearable Devices and Their Primary Applications

The detection of human activity frequently makes use of inertial sensors like gyroscopes, magnetometers, and accelerometers. Wearable accelerometers, for instance, can help improve patient care by measuring gait parameters, detecting falls, and identifying activities of daily living. Patient safety can also be enhanced through the use of recognition activities. A gyroscope, for example, can determine motion direction even when viewed from various angles. This aids in categorising activities as causing no displacement, causing a transitional displacement, or causing a significant displacement. Sitting, standing, and sleeping are examples of zero-displacement activities because they do not require any body movement. Transitional actions help you move from one activity to the next. For example, moving from a seated to a standing position or sitting to a lying down place are transitional activities. Walking, running, and stair climbing are all examples of intense replacement activities. Other types of strenuous physical activity fall into this category as well.

To recognise human actions, physiological sensors such as galvanic skin response (GSR), electroencephalogram (EEG), heat sensors, electrocardiogram (ECG), and electromyography (EMG) are utilised in addition to inertial sensors. These sensors work in tandem with inertial sensors. For example, recognition of hand movements based on EMG data has widespread use in various applications, including robotic exoskeletons, surgical equipment, power prostheses, and computer and automatic video games.

There are several different types of pressure sensors on the market today, such as bending, barometric pressure, force sensing resistors (FSR), and capacitive pressure sensors based on textiles. If the maximum sensing pressure of the sensor is higher than 20 kPa, then it can be put on the knee joint to track human actions like squatting without causing discomfort. If the detection limit of the sensor is as low as 100 pa, it can be used to detect minor physiological activities such as neck vibrations. This is due to the sensor's sensitivity to detect such minute changes. In addition, ultrathin piezoelectric sensors based on piezoelectric nanowires can be implanted on the user's eyelids to measure eye motions when the eye moves to the left or right at varying rates. These sensors can tell if the user is looking up, down, to the left, or to the right. This type of pressure sensor can also function as an eye movement detector.

3.1 *Intelligent Wearable Mobile Devices*

Smartphone technology has advanced to the point where these devices can now function as communication devices and highly effective sensor platforms. A smartphone typically has several sensors, including Bluetooth, Wi-Fi, accelerometers, proximity sensors, barometers, magnetometers, gyroscopes, temperature and humidity sensors, cameras, ambient light sensors, and microphones. Moreover, because smartphones are ubiquitous in our day-to-day lives and don't get in the way of their owners' activities, they make for potentially promising tools for identifying human activity.

A smartwatch is a wearable device that incorporates a powerful computer as well as a variety of sensors. For example, the eWatch platform used had four different types of sensors integrated into it. These sensors included a microphone, a temperature sensor, a light sensor, and a two-axis accelerometer. Recent developments at Facebook Reality Lab have resulted in the development of a magical wristband device with an integrated electromyography sensor. This sensor reads motor nerve signals transmitted from the spinal cord to the wrist and fingers and converts them into digital commands that can be used to control the device. This sensor is intended to pick up on these signals.

Several sensors are utilised for data collection from the Empatica E4 wristband gadget. These sensors, in conjunction with assessing the subject's physical state, sensor fusion techniques, and machine learning (ML) approaches are utilised (such as daily life, running, rest, resistance training, and cycling) in addition to calculating your current energy expenditure and level of physical activity. A central processing unit (CPU), a camera, a small projector, a control device, sensors, and a storage device are typical components of a pair of wearable smart glasses. As a result, numerous applications could be developed for smart glasses, including but not limited to medical and firefighting assistance, travel assistance, etc.

4 Framework for the HAR that Makes Use of Wearable Sensors

Traditionally, methods for detecting human activity used both environmental equipment and ad hoc gadgets worn by the people being studied. RFID, cameras, and acoustic sensors are examples of equipment used in environmental monitoring shown in Fig. 1. You will gather information from various sensors during the "data acquisition" process. Gyroscopes, accelerometers, and compasses are the most prevalent HAR-related operations and categorization sensors. These devices gather information by measuring rotational acceleration. HAR employs many sensors, some more widely used than others. Accelerometers, gyroscopes, and magnetometers are examples of these gyroscopes.

The gyroscope can measure angular velocity in all three dimensions. The unit of measurement for it is degrees per second (degrees/s). A magnetometer is a device

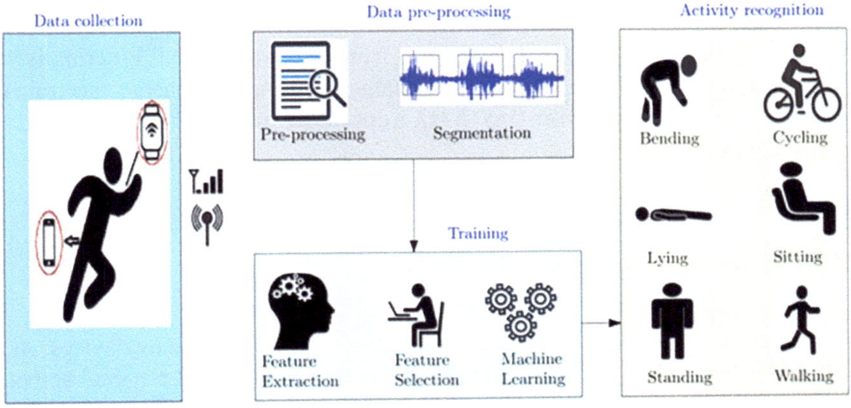

Fig. 1 Human activity recognition using wearable sensors

that can measure the changes that occur in a magnetic field at a specific location over time. The Tesla symbol, represented by the letter T, is used as a unit of measurement and is typically recorded along all three axes. One of the most important considerations during the process's "acquisition" phase is the sample rate, which affects the total number of samples that can be used for the classification phase. The unit of measurement for the sampling rate is the hertz, and the sampling rate is defined as the number of data points that are captured in a single second. For example, a sample rate 50 Hz indicates that data is collected at a rate of 50 different values every second. This parameter is subject to frequent adjustments during the acquisition process.

4.1 Pre-processing—Subsequent Processing

"Data pre-processing" is a critical step in preparing raw data for subsequent processing. Sensor raw data frequently contains artefacts caused by the instruments or measured physical action. To remove these artefacts from the signals, the data must be cleaned. Furthermore, the accelerometer data combines the linear acceleration caused by body movement and gravity. Although gravity represents a bias that can affect classifier performance, it is common to practise removing the gravity component before applying the classifier to the raw signal. Because of the numerous factors discussed in the preceding section, a filtering process is typically used. Filters focus on the frequency component of the movement, and as a result, they are highly efficient devices. The element of the accelerometer signal with the highest frequency is primarily associated with the subjects' activities.

On the other hand, the component of the signal produced by the accelerometer that has the lowest frequency is predominantly associated with the existence of gravity. Consequently, a low-pass filter with a cut-off frequency ranging from 0.1 to 0.5 Hz is often utilised to separate the gravity component from the rest of the signal. After applying a low-pass filter to the signal, the body's acceleration component can be calculated by subtracting the resultant signal from the initial signal. Filtering is yet another method for ridding raw data of any artefacts. Because the energy spectrum of human body motion ranges from 0 to 15 Hz, a cut-off frequency of 15 Hz is adequate for capturing it.

4.2 Pre-processing of Sensor Data

Raw data collected by sensors frequently contains artefacts caused by devices, such as electrical fluctuation or sensor calibration, or by the measured physical activity. However, it could be determined by the instruments themselves. To remove these artefacts from the signals, the data must be cleaned. Furthermore, the accelerometer data combines the linear acceleration caused by body movement and gravity. Although gravity represents a bias that can affect classifier performance, it is common to practise removing the gravity component before applying the classifier to the raw signal. Filters are highly effective devices that work on the frequency component of the movement. A low-pass filter with a cut-off frequency of 0.1–0.5 Hz is commonly utilised to separate the gravity component from the rest of the signal. This helps to effectively isolate the gravity component. After applying a low-pass filter to the signal, the resulting signal is subtracted from the first signal to calculate the element of the body's acceleration corresponding to acceleration.

4.3 The Process of Extracting Features

According to theoretical and empirical research, many algorithms could improve when applied to problem domains with irrelevant and duplicated data. First, extracting features from data requires abstracting each segment into a high-level representation of the same section. This reduces the dimensionality of the data while also identifying the most significant anomalies in the signal.

4.4 Information Extraction and Classification

Regarding activity recognition, data-driven strategies are more effective than model-driven approaches, which are less effective. Artificial intelligence (AI) is a type of data-driven algorithm that has gained popularity recently due to its positive results in data analysis, information extraction, and classification. In addition, machine learning, which includes deep learning approaches, is also included in artificial intelligence algorithms. However, there has yet to be proof of the best classifier, and deciding which classification algorithm to use remains a complex problem for the

HAR community. In addition, the classification algorithm used significantly impacts classification performance.

5 Using a Traditional Teaching Machine to Learn Entails

Machine learning techniques have found widespread application in activity recognition competitions. In addition, methods of increasing sophistication have been developed to deal with the complexity associated with activity recognition tasks.

5.1 Support Vector Machine (SVM)

Is interested in various algorithmic domain transformations. It applies the following concept: it is assumed that the input data x cannot be linearly separated in terms of the classes y existing in the data space. However, it is assumed that there is a higher-dimensional area in which linearity can be achieved. Following the mapping of the data into this space, a hyperplane is generated that is a linear decision surface and is used as a recognition model. The SVM was developed and implemented as a two-class classifier when it was first created. The multi-class support vector machine (SVM) can be computed using one of two ways to this problem: one-versus-one, in which two different classes are classified at the same time, and one-versus-all, in which one type is labelled 0, and the other courses are labelled 1. Both of these approaches are covered in this paper.

5.2 K-*Nearest Neighbours (KNN)*

Is an example of a more general type of method that belongs to the class of instance-based methods. The process of nearest neighbours compares newly created instances to previously stored samples using a distance metric. The closest match determines the class of the newly developed model to an existing example. The latest instance has been assigned the class containing most of the cases of the closest k neighbours. This algorithm belongs to the category of lazy algorithms and is relatively simple to implement. A lazy algorithm's parameters do not require learning while the algorithm is being trained. The number of nearest neighbours, denoted by k, ultimately determines the output of KNN.

5.3 Decision Tree

Algorithm branches, the input values are transferred from the tree's root to the leaves. Decision tree algorithms generate a hierarchical model. The direction of growth from the source to the leaf is a classification rule. It is sometimes necessary to adjust the length of the trees, which necessitates the use of growing and pruning algorithms. Creating a tree from scratch includes determining the split criterion, the halting principle, and the class assignment rule. In HAR, the decision trees J48 and C4.5 are the ones that are utilised the majority of the time.

5.4 Naive Bayes (NB)

Fits into the Bayesian method category. It bases its prediction on a posterior probability estimate, which it obtains by multiplying the likelihood. It uses a conditional probability derived from the training set based on the class and a prior chance. This posterior probability estimation is the foundation for the Bayesian technique to anticipate new instances. The Naive Bayes method assumes that the data are independent of one another based on the class values.

5.5 Adaboost

Is a component of classifier ensembles. Classifier ensembles are any technique combining multiple classifiers into a single system. Two different approaches can be taken to combine classifiers: utilising the same classifiers but modifying the parameter parameters differently (for instance, a random forest with lengths of variable degrees); and using multiple classifiers in conjunction with one another (a few examples include random forest, *k*-nearest neighbours and support vector machines).

5.6 The Collection of Data Was Used in the Development of the Proposed Model

To predict, ML classifiers require data. The primary challenge that Human Activity Recognition scenarios face is the availability of high-quality data. UCI If you're familiar with HAR Dataset, it's used here. Experiments were carried out with thirty willing participants aged 19–48 years. Each participant was instructed to perform six exercises while strapping a smartphone (a Samsung Galaxy S II) to their waist. One of the activities included walking up and down stairs, sitting, standing, and even lying down. Following the generation of the dataset, it was arbitrarily divided into two groups, with 70% of volunteers contributing to the formation of training data and just 30% contributing to the generation of test data.

5.7 *Dataset Attributes*

S. No.	Name of the attributes and their function
1.	The accelerometer acquired the information obtained from the triaxial accelerometer
2.	The value for the Triaxial Angular Velocity is provided via the gyroscope
3.	A feature vector consisting of 561 elements contains frequency and temporal domain variables
4.	The label of the activity provides information about what the action is called
5.	A unique identifier can identify the person being tested in an experiment

6 Experimental Results

This section contains a description and presentation of the experiment's results. In performing performance evaluations, standard metrics such as accuracy, precision, recall, and the F1-score are utilised during the performance evaluation process. Tables 1, 2, 3, 4, 5, 6, 7 and 8 provide an overview of the results of using several machine learning approaches, broken down by individual class categories and presented in their respective columns.

From the above metrics, we can observe that the precision for all the classifiers is highest at Category-6, and the lowest is mostly at Category-5, except for two classifiers. Recall and F1-score have the highest values at Category-6 for all classifiers. The lowest value of Recall is at Category-4, Category-3, and Category-2 based upon the classifiers. F1-score has the lowest value at Category-4 for all classifiers except for KNN at Category-2,3. The support value is highest(537) for Category-6 and lowest(420) for Category-3.

Table 1 Decision tree metrics

Category	Precision	Recall	F1-score	Support
1	0.82	0.93	0.87	496
2	0.83	0.71	0.77	471
3	0.85	0.86	0.85	420
4	0.79	0.73	0.76	491
5	0.76	0.82	0.79	532
6	1.00	1.00	1.00	537

Table 2 KNN metrics

Category	Precision	Recall	F1-score	Support
1	0.68	0.91	0.78	496
2	0.84	0.71	0.77	471
3	0.87	0.69	0.77	420
4	0.82	0.69	0.75	491
5	0.76	0.90	0.82	532
6	1.00	0.95	0.97	537

Table 3 SVC with linear kernel metrics

Category	Precision	Recall	F1-score	Support
1	0.95	1	0.97	496
2	0.98	0.95	0.97	471
3	0.99	0.98	0.98	420
4	0.96	0.89	0.92	491
5	0.91	0.97	0.94	532
6	1.00	1.00	1.00	537

Table 4 SVC with RBF kernel metrics

Category	Precision	Recall	F1-score	Support
1	0.9	1	0.94	496
2	0.94	0.95	0.94	471
3	0.99	0.85	0.92	420
4	0.9	0.86	0.88	491
5	0.88	0.92	0.90	532
6	1.00	1.00	1.00	537

Table 5 Logistic regression metrics

Category	Precision	Recall	F1-score	Support
1	0.94	0.99	0.97	496
2	0.97	0.94	0.96	471
3	0.99	0.97	0.98	420
4	0.98	0.87	0.92	491
5	0.89	0.98	0.93	532
6	1.00	1.00	1.00	537

Table 6 Random forest metrics

Category	Precision	Recall	F1-score	Support
1	0.94	0.99	0.97	496
2	0.97	0.94	0.96	471
3	0.99	0.97	0.98	420
4	0.98	0.87	0.92	491
5	0.89	0.98	0.93	532
6	1.00	1.00	1.00	537

Table 7 CART

Category	Precision	Recall	F1-score	Support
1	0.83	0.94	0.88	496
2	0.84	0.77	0.81	471
3	0.88	0.82	0.85	420
4	0.83	0.75	0.79	491
5	0.79	0.86	0.82	532
6	1.00	1.00	1.00	537

Table 8 Multi-class classification

Classifier	Accuracy	Precision	Recall	F1-score
Decision tree	0.8459	1	1	1
KNN	0.8143	1	0.95	0.97
SVC with linear kernel	0.9633	1	1	1
SVC with RBF kernel	0.9307	1	1	1
Logistic regression	0.9596	1	1	1
Random forest	0.9596	1	1	1
CART	0.8615	1	1	1

6.1 *Performance Analysis*

From the above analysis shown in Fig. 2, we can conclude that the accuracy is highest for the support vector machine with a Linear Kernel of 96.3%. Then random forest and logistic regression have an accuracy of 95.96%. Finally, KNN has the least accuracy at 81.43%. All classifiers have 100% precision. All classifiers have 100% recall, except KNN has 95%. Similarly, all classifiers have 100% F1-score, except KNN has 97%.

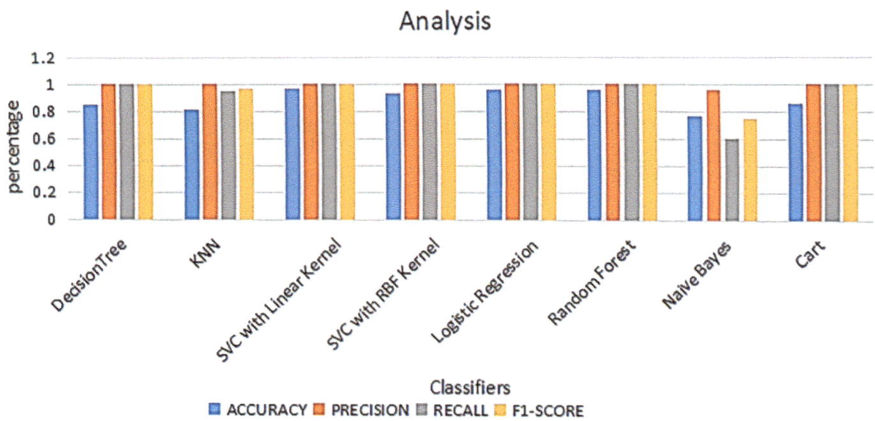

Fig. 2 Performance analysis over various algorithms

7 Conclusion

A wearable device can be close enough to the user to ascertain the user's actions and unique behaviours and cognitive data, which aids in human activity recognition. This can be accomplished by watching the user's movements while the device is attached to the user's body. This can be accomplished by analysing the user's personalised behaviour and cognitive data. A wearable multi-sensor HAR framework is proposed in the scope of this paper. The machine learning methodology serves as the framework's foundation. The testing uses a multi-sensor activity dataset stored in the UCI database. One method for mitigating the effects of overfitting in machine learning algorithms is to use tenfold cross-validation. The confusion matrix and the PCP were used to assess the results of the four distinct classification models (EC, SVM, DT, and KNN). The results of the experiments show that when using the DT algorithm, the proposed approach may achieve an optimal classification accuracy of 99.63%. The key reason for the vast improvement in classifier performance is the methodical approach to data pre-processing and feature extraction technique. It is possible to improve the presented method so that it can be used to recognise more complex everyday events (such as shopping, brushing, and cleaning) on a larger dataset in the future. This would allow more people to use the method. This will enable the way to have a broader range of applications. Furthermore, the algorithm could be deployed on mobile phones, and software that recognises activities for use in healthcare applications could be developed. This would necessitate the development of a straightforward and easy-to use.

References

1. Bota PJ, Wang C, Fred AL, Da Silva HP (2019) A review, current challenges, and future possibilities on emotion recognition using machine learning and physiological signals. IEEE Access 7:140990–141020
2. Tahir A, Taylor W, Taha A, Usman M, Shah SA, Imran MA, Abbasi QH (2022) IoT-based fall detection system for elderly healthcare. In: The Internet of Things for human-centered design: application to elderly healthcare. Springer, Singapore, pp 209–232
3. Nooruddin S, Islam MM, Sharna FA, Alhetari H, Kabir MN (2022) Sensor-based fall detection systems: a review. J Ambient Intell HumIzed Comput 1–17
4. Ren L, Peng Y (2019) Research of fall detection and fall prevention technologies: a systematic review. IEEE Access 7:77702–77722
5. Santos GL, Endo PT, Monteiro KH, Rocha ED, Silva I, Lynn T (2019) Accelerometer-based human fall detection using convolutional neural networks. Sensors 19(7):1644
6. Baig MM, Afifi S, GholamHosseini H, Mirza F (2019) A systematic review of wearable sensors and IoT-based monitoring applications for older adults–a focus on ageing population and independent living. J Med Syst 43:1–11
7. Şengül G, Karakaya M, Misra S, Abayomi-Alli OO, Damaševičius R (2022) Deep learning based fall detection using smartwatches for healthcare applications. Biomed Signal Process Control 71:103242
8. Subhan F, Mirza A, Su'ud MB, Alam MM, Nisar S, Habib U, Iqbal MZ (2023) AI-enabled wearable medical Internet of Things in healthcare system: a survey. Appl Sci 13(3):1394
9. Mobsite S, Alaoui N, Boulmalf M, Ghogho M (2023) Semantic segmentation-based system for fall detection and post-fall posture classification. Eng Appl Artif Intell 117:105616
10. Chan H-L, Ouyang Y, Chen R-S, Lai Y-H, Kuo C-C, Liao G-S, Hsu W-Y, Chang Y-J (2023) Deep neural network for the detections of fall and physical activities using foot pressures and inertial sensing. Sensors 23(1):495
11. Musci M, De Martini D, Blago N, Facchinetti T, Piastra M (2020) Online fall detection using recurrent neural networks on smart wearable devices. IEEE Trans Emerg Top Comput 9(3):1276–1289
12. Torres-Guzman RA, Paulson MR, Avila FR, Maita K, Garcia JP, Forte AJ, Maniaci MJ (2023) Smartphones and threshold-based monitoring methods effectively detect falls remotely: a systematic review. Sensors 23(3):1323
13. Fuqaha A, Guizani M, Aledhari M, Ayash M (2013) Internet of Things: a survey on enabling technologies, protocols and applications. IEEE Commun Surv Tutor 17(4):2347–2352
14. Li J, Wang Z, Qiu S, Zhao H, Wang J, Shi X, Liang B, Fortino G (2021) Multi-body sensor data fusion to evaluate the hippotherapy for motor ability improvement in children with cerebral palsy. Inf. Fusion 70:115–128
15. Chen J-F, Wang C-C, Wu EH-K, Chou C-F (2020) Simultaneous heterogeneous sensor localization, joint tracking, and upper extremity modeling for stroke rehabilitation. IEEE Syst J 14(3):3570–3581
16. Ghate V, Sweetlin Hemalatha C (2021) Hybrid deep learning approaches for smartphone sensorbased human activity recognition. Multimed Tools Appl 80(28/29)
17. Bhaskar Semwal V, Gupta A, Lalwani P (2021) An optimized hybrid deep learning model using ensemble learning approach for human walking activities recognition. J Supercomput 77(11):12256–12279
18. Siirtola P, Röning J (2021) Context-aware incremental learning-based method for personalized human activity recognition. J Ambient Intell HumIzed Comput 12(12):10499–10513
19. Nematallah H, Rajan S, Cretu A (2019) Logistic model tree for human activity recognition using smartphone-based inertial sensors. In: Proceedings of the IEEE sensors conference, Montreal, QC, Canada, 27–30 October 2019, pp 1–4
20. Wang K, He J, Zhang L (2019) Attention-based convolutional neural network for weakly labeled human activities' recognition with wearable sensors. IEEE Sens J 19:7598–7604

21. Deep S, Zheng X (2019) Hybrid model featuring CNN and LSTM architecture for human activity recognition on smartphone sensor data. In: Proceedings of the 20th international conference on parallel and distributed computing, applications and technologies, Gold Coast, Australia, 5–7 December 2019, pp 259–264

Efficiency Unleashed: Reimagining A-to-B Paths in Graphs

K. Sutejas, K. S. Ramalakshmi, and K. S. Srinivas

Abstract Efficiently finding the shortest path in a graph is a fundamental problem with diverse applications across numerous domains, from transportation and network routing to social network analysis and bioinformatics. This paper presents a comprehensive approach to improving the efficiency of shortest path searches in generic graphs. The contributions encompass novel algorithmic enhancements and parallelization techniques to tackle this critical problem. These algorithmic enhancements optimize the traversal of graphs with diverse data types and leverage data encodings to enhance search speed, thus reducing computational overhead and making it feasible to handle large-scale graphs efficiently. To further expedite the shortest path search process, the parallelization technique harnesses the power of modern multi-core processors and distributed computing environments. By efficiently distributing the workload among multiple processing units, this strategy significantly accelerates the state-of-the-art algorithms like Dijkstra's, to explore candidate paths, leading to substantial improvements in query response times. This approach has demonstrated superior performance characteristics in various scenarios and provides an additional layer of optimization. Evaluation of the parallelized approach on benchmark graph datasets has showcased its effectiveness by improving search efficiency and scalability over existing methods like Dijkstra's Algorithm, A* Heuristic Search and the Bellman-Ford Algorithm. By integrating algorithmic enhancements and parallelization, the novel approach opens up new avenues for solving complex shortest path problems more efficiently with limited computational resources. This paper not only contributes to the field of graph algorithms but also provides a valuable toolkit for practitioners seeking to optimize shortest path searches in graph-based data structures.

Keywords Graphs · Shortest paths · Dijkstra's · A* Heuristic search · Bellman-Ford · Parallelization · Encoding · Generic graphs · Optimization

---

K. Sutejas (✉) · K. S. Ramalakshmi
Department of Computer Science and Engineering, PES University, Bengaluru, India
e-mail: sutejaskpes@gmail.com

K. S. Srinivas
Department of Computer Science and Engineering (AIML), PES University, Bengaluru, India
e-mail: srinivasks@pes.edu

S. M. Thampi et al. (eds.), *Fifth International Conference on Computing and Network Communications*, Lecture Notes in Electrical Engineering 1221,
https://doi.org/10.1007/978-981-97-4711-5_35

1 Introduction

Finding the shortest path in a graph is a fundamental and extensively studied challenge in computer science and mathematics, with a wide array of applications in real-world scenarios. From navigation systems guiding travelers through intricate road networks to optimizing data routing in communication networks, from identifying influential nodes in social networks to unraveling complex biological pathways, the efficient computation of the shortest path is at the core of many critical tasks. Consequently, there is an enduring quest to develop algorithms and techniques that can swiftly and accurately identify the shortest path in a graph, regardless of its size, complexity, or data type.

In this era of big data and complex networks, the need for efficient and scalable solutions for shortest path searches has never been more pressing. Traditional algorithms like Dijkstra's and A* have long been cornerstones in this field, providing reliable results in a multitude of applications. However, as datasets grow in size and complexity, the computational demands imposed by these algorithms become increasingly burdensome. Moreover, the diversity of graph types encountered in modern applications, including transportation networks, social networks, biological networks, and more, calls for tailored solutions that can adapt to the specific characteristics of each problem domain.

This paper addresses these challenges by presenting a comprehensive approach to improving the efficiency of shortest path searches in generic graphs. Our approach is multifaceted, encompassing algorithmic enhancements, parallelization techniques, and the integration of state-of-the-art algorithms, all designed to address the unique demands of modern graph-based applications.

This novel approach in this paper introduces algorithmic enhancements that exploit data encodings, leading to faster query response times and efficiently handling a wide range of data types and graph structures.

Recognizing the computational power of modern hardware, the performance of the algorithms is improved through a parallelization framework. By distributing the computation across multiple processing units, the full potential of multi-core processors and distributed computing environments is harnessed, significantly accelerating the exploration of candidate paths.

To complement these enhancements, state-of-the-art shortest path algorithms are incorporated into the framework. These algorithms have demonstrated superior performance characteristics in various settings and augment the efficiency of our approach.

Throughout this paper, an in-depth analysis of this novel approach is provided, evaluating its performance on diverse graph datasets and demonstrating its effectiveness in real-world applications. The experimental results showcase substantial improvements in search efficiency and scalability over existing methods, underscoring the practical relevance of these contributions.

The improvised algorithms represent a significant step forward in the domain of shortest path searches in generic graphs. By combining algorithmic enhancements, parallelization strategies, and the integration of advanced algorithms, it addresses the computational demands of modern applications and paves the way for more efficient and scalable solutions in various domains where efficient path-finding is critical. This paper discusses about the efforts taken until now to improve graph algorithms, three popular shortest path algorithms, improvisations to the same, comparisons between the results and also discusses the scope for further developments.

2 Related Work

At present, several efforts have been taken to analyze and improve the computational efficiency of the shortest path algorithms like Dijkstra's, A* Heuristic Search, and Bellman-Ford.

Fan [1] has proposed an implementation using an improved storage structure using adjacency lists as against adjacency matrix of topological relations between nodes, to eliminate storage redundancy. Additionally, he has reduced the scale of algorithm searching by restricting the search space, thereby minimizing complexity and improving efficiency. However, with these optimizations, the improved algorithm cannot search genuine shortest paths, it can only give an approximate shortest path, compromising the accuracy. Huang [2] introduced an optimization to improve the Dijkstra Algorithm by introducing the concept of an equivalent path and analyzing its influence factor while drawing a comparison with the traditional approach.

Xie [3] approach to shortening Dijkstra's search time by using the heap structure to store the nodes and paths, saved storage space as well. Although the construction of the heap and the heapâŁ™s linear search will waste some time, this approach still proves efficient. Zhu [4], in the intelligent ship path planning use case, improved Dijkstra's algorithm by reducing the redundant points on the original shortest path given by the classical algorithm. A bio-heuristic intelligent optimization algorithm, called the pheromone factor, from ACO, a probabilistic algorithm to find the optimal path, is applied and the results provide a more practical path.

Ju [5] proposed an improvisation over the A* algorithm by considering circular regions around obstacles for local planning as against grid-based A* methods. This led to smoother paths and better obstacle avoidance. Yao [6]'s paper in the context of path planning for virtual human motion implemented an improved A* algorithm by using weighted processing of evaluation function. This reduced the searching steps from 200 to 80 and the searching time from 4.359 s to 2.823 s in the feasible path planning, thus making it more effective and accurate.

Hong [7] introduced optimizations in the A* Star algorithm by implementing a hybrid data structure of the minimum heap and 2D array. This greatly reduced the time complexity of the algorithm. An optimized search strategy that does not check whether the destination has reached in the initial stage of searching for the global optimal path, improved execution efficiency. Zheng [8] has improved the A* algorithm in the use case of AGV Path Planning, by adding the angle evaluation cost function to the cost function of the A-star algorithm, to find the path with the least inflection point. This reduced the search speed and returned an optimal path.

Goldberg [9] proposed an algorithm that achieves the O(nm) worst-case time complexity as the Bellman-Ford algorithm but in superior practice by using topological scan, involving stacks and DFS computation. However, in the acyclic graphs, it runs in the order of O(m) time. Bannister [10] improvised the Bellman-Ford algorithm by randomly permuting the vertices of the graph, and used that permutation to order the vertices within each pass of the algorithm. The paper provides scope to improve the practical runtime results from the experimental observations.

Chen [11] used the shortest path faster algorithm (SPFA) involving queues to store the nodes and reduce redundancy. The improved algorithm resulted in an optimal path by continuous relaxation operation to the new node.

Through an extensive literature review, it has become evident that previous implementations lacked the identification of redundant points, generalization, and the concept of parallelization, thereby compromising the speed, space efficiency, and accuracy of the results delivered. Furthermore, all previous implementations were designed for integer graphs without generalization. Given the limitations observed, this paper proposes a novel implementation that encodes generic nodes into standard integer nodes and introduces parallelization for enhanced algorithm performance with better time complexity and space complexity.

2.1 Background

2.1.1 Dijkstra's Algorithm

Dijkstra's algorithm [12] is a vital tool in graph theory and optimization to find the shortest path in weighted graphs. It is widely used in areas like network routing and logistics. This algorithm works by systematically picking the node with the lowest known distance from the source and updating distances to its neighbors.

Time Complexity: $O(|E| + |V| \log |V|)$, where E is the number of edges and V is number of nodes in the graph.

Space Complexity: $O(|V|)$, where V is number of nodes in the graph.

2.1.2 A* Heuristic Search Algorithm

A* Heuristic Search algorithm is one of the best and most popular techniques used in path-finding and graph traversals. It serves as a powerful tool for solving optimization problems efficiently. A* combines the benefits of both Dijkstra's algorithm and greedy search. Greedy search is a simple and intuitive algorithm that makes a locally optimal choice at each decision point in the graph to find a global optimum.

Algorithm 1 Dijkstra's Algorithm

```
procedure DIJKSTRAALGORITHM(G, s, t)        ▷ G is the graph, s is the source vertex, t is the target vertex
    dist ← array of distances initialized to ∞
    parent ← array of parent nodes initialized to None
    dist[s] ← 0
    Q ← priority queue of vertices, ordered by dist
    ENQUEUE(Q, s)
    while Q is not empty do
        u ← DEQUEUE(Q)
        if u = t then
            return RECONSTRUCTPATH(parent, t)        ▷ Return the shortest path
        end if
        for each neighbor v of u do
            alt ← dist[u] + weight(u, v)
            if alt < dist[v] then
                dist[v] ← alt
                parent[v] ← u
                ENQUEUE(Q, v)
            end if
        end for
    end while
    return []        ▷ No path from s to t exists
end procedure
procedure RECONSTRUCTPATH(parent, t)        ▷ Helper function to reconstruct the path
    path ← []
    while t is not None do
        PREPEND(path, t)
        t ← parent[t]
    end while
    return path
end procedure
```

This algorithm works by considering the cost to reach a particular state as well as an informed heuristic estimate of the remaining cost to reach the goal. This combination of information allows A* to intelligently prioritize exploring paths that are likely to lead to an optimal solution, making it particularly efficient and adaptable for applications such as route planning in GPS systems, robotics path-finding, and puzzle-solving in video games.

$$f(n) = g(n) + h(n) \tag{1}$$

In this Eq. (1), *f(n)* is the evaluation function for a node n, *g(n)* is the cost of the path from the start node to node n and *h(n)* is the heuristic estimate of the cost from node n to the goal node. The A* algorithm selects nodes for expansion based on the lowest *f(n)* value.

Algorithm 2 A* Heuristic Search

```
procedure ASTARSEARCH(G, s, t) ▷ G is the graph, s is the source vertex, t is the target vertex
    Q ← priority queue of vertices, ordered by f          ▷ Priority queue based on the f value
    visited ← set of visited vertices
    g ← array of distances initialized to ∞                ▷ Distance from the source
    h ← array of heuristic values for each vertex to target t
    f ← array of evaluation values initialized to ∞        ▷ f = g + h
    g[s] ← 0
    f[s] ← h[s]
    ENQUEUE(Q, s)
    while Q is not empty do
        u ← DEQUEUE(Q)
        if u = t then
            return RECONSTRUCTPATH(s, t, g)                ▷ Return the shortest path
        end if
        ADD u to visited
        for each neighbor v of u do
            if v is in visited then
                continue                                   ▷ Skip already visited nodes
            end if
            tempG ← g[u] + weight(u, v)
            if tempG < g[v] then
                g[v] ← tempG
                f[v] ← g[v] + h[v]
                ENQUEUE(Q, v)
            end if
        end for
    end while
    return []                                              ▷ No path from s to t exists
end procedure
procedure RECONSTRUCTPATH(s, t, g)                ▷ Helper function to reconstruct the path
    path ← []
    current ← t
    while current ≠ s do
        PREPEND(path, current)
        current ← the neighbor of current with g value that led to current
    end while
    PREPEND(path, s)
    return path
end procedure
```

Time Complexity: $O(b^d)$, where b is the branching factor (the average number of successors per state) and d is the number of nodes in the resulting path.

Space Complexity: $O(b^d)$, where b is the branching factor (the average number of successors per state) and d is the number of nodes in the resulting path.

2.1.3 Bellman-Ford Algorithm

The Bellman-Ford algorithm [13] plays a pivotal role in finding the shortest paths in weighted graphs, even in the presence of negative edge weights-a feature that distinguishes it from some other algorithms like Dijkstra's. Bellman-Ford operates by iteratively relaxing edges in the graph, adjusting distance estimates until they converge to the true shortest distances from a source vertex to all other vertices. This attribute makes it particularly suited for scenarios where negative-weight cycles may exist, as it can detect and report their presence. Various optimizations and adaptations have been proposed to enhance its efficiency, making it a versatile tool for tackling real-world problems involving complex interconnected systems.

Time Complexity: $O(|E| * |V|)$, where E is the number of edges and V is number of nodes in the graph.

Space Complexity: $O(|V|)$, where V is number of nodes in the graph.

3 Dataset

In this study, a widely recognized standard benchmark dataset is used [14]. This dataset comprises directed graphs featuring positive edge weights, generated through a Breadth-First Search (BFS) algorithm initiated from a randomly selected point within the New York City benchmark graphs. These New York City benchmark graphs are part of the 9th DIMACS Implementation Challenge, which was conducted in 2005. This dataset comprises 180 distinct graphs, showcasing a wide range of structural characteristics. These graphs vary in size and complexity, with node counts spanning from 108 to 4882 and edge counts ranging from 290 to 14110.

4 Improvements

Optimization in graphs is crucial for efficiently solving complex real-world problems, such as network routing and resource allocation, by finding the most efficient pathways or solutions. Within the three described algorithms, this paper primarily discusses opportunities for parallelization and highlights the runtime optimizations, to improve performance. In order to handle generic graphs, encoding techniques are done to convert graphs with generic nodes to integer nodes.

Algorithm 3 Bellman-Ford Algorithm

```
procedure BellmanFord(G, s)                         ▷ G is the graph, s is the source vertex
    dist ← array of distances initialized to ∞
    dist[s] ← 0
    parent ← array of parent nodes initialized to None
    for i ← 1 to |V| − 1 do                         ▷ Iterate |V| - 1 times
        for each edge (u, v) in G do
            w ← weight of edge (u, v)
            if dist[u] + w < dist[v] then
                dist[v] ← dist[u] + w
                parent[v] ← u
            end if
        end for
    end for
    for each edge (u, v) in G do                    ▷ Check for negative-weight cycles
        w ← weight of edge (u, v)
        if dist[u] + w < dist[v] then
            return "Negative-weight cycle exists"
        end if
    end for
    paths ← array of shortest paths from s to all vertices   ▷ Initialize empty paths array
    for each vertex v in V do
        path ← ReconstructPath(parent, s, v)
        paths[v] ← path
    end for
    return dist, paths                              ▷ Shortest distances and paths from source s
end procedure
procedure ReconstructPath(parent, s, v)             ▷ Helper function to reconstruct the path
    path ← []
    current ← v
    while current ≠ s do
        Prepend(path, current)
        current ← parent[current]
    end while
    Prepend(path, s)
    return path
end procedure
```

4.1 *Improved Dijkstra's Algorithm*

With reference to the classical Dijkstra's Algorithm 1 for all source-all destinations, it is noticed that in line 7, the 'while loop' which performs the traversals can be parallelized. The 'for loop' at line 12 which checks the neighbors, computes new distance, and updates the distance array and parent array, can be parallelized by making line 14, the 'if statement' critical. The path reconstruction for all destinations can also be parallelized and this algorithm can be called for all sources parallely. However, the experimental results show that the 'while loop' in line 7 cannot be parallelized due to its interference with the dequeue operation, giving inaccurate results. The pseudo-code of the improved algorithm with these optimizations is given by Algorithm 4. In order to obtain all source-all destinations, the call for the algorithm is parallelized to all sources.

4.1.1 Mathematical Formulation

Using adjacency list, with only 1 core,
Time for visiting all vertices: $O(E + V)$, where E is the number of edges and V is number of nodes in the graph
Time for processing one vertex: $O(\log V)$
Time for path construction: $O(V^2)$
Time Complexity of Dijkstra's Algorithm with Path reconstruction: $O((V + E)\log E + V^2)$

Using adjacency list, for 'p' cores,
Time for visiting all vertices: $O(\frac{V+E}{p})$, where E is the number of edges and V is number of nodes in the graph
Time for Path Reconstruction: $O(\frac{V^2}{p})$
Time Complexity of Parallelized Dijkstra's Algorithm with Path reconstruction: $O(\frac{(V+E)\log E}{p} + \frac{V^2}{p})$

4.2 *Improved A* Heuristic Search*

With reference to the classical A* Heuristic Search Algorithm 2 for a single source-single destination, the 'while loop' in line 10 performing traversal and the 'for loop' in line 16 which checks the neighbors, computes new distance, heuristic and the evaluation value and updates the distance array and parent array, can be parallelized while keeping the 'if statement' in line 21 in the critical section. The experimental results show that the 'while loop' cannot be parallelized due to its interference with the dequeue operation. The pseudo-code of the improved A* heuristic Search can be found at Algorithm 5.

4.2.1 Mathematical Formulation

Using adjacency list, with only 1 core,
Time for visiting all vertices: $O(E + V)$, where E is the number of edges and V is number of nodes in the graph
Heuristic Time Complexity: $O(\phi)$

Algorithm 4 Parallelized Dijkstra's Algorithm

```
procedure ParallelizedDijkstra(source)                  ▷ source is the source node index
    n ← number of nodes in the graph
    Initialize paths as an empty array of arrays to store paths
    Initialize distance as an array of size n filled with ∞
    distance[source] ← 0.0
    Initialize parent as an array of size n filled with −1
    Initialize visited as an array of size n filled with false
    Initialize a priority queue pq as a min-heap of pairs: (distance, vertex)
    pq.push({0.0, source})
    while not pq is empty do
        u ← top vertex from pq                  ▷ Critical section to access the top element
        if visited[u] then
            continue
        end if
        visited[u] ← true
        for each neighbor in adjList[u] do                          ▷ Parallelized loop
            v ← neighbor.first
            weight ← neighbor.second
            if distance[u] + weight < distance[v] then
                Critical Section:
                distance[v] ← distance[u] + weight
                parent[v] ← u
                pq.push({distance[v], v})
            end if
        end for
    end while
    Path Reconstruction
    for i from 0 to n − 1 do                                        ▷ Parallelized loop
        if distance[i] = ∞ then
            paths[i] ← empty array
            continue
        end if
        Initialize an empty array path
        current ← i
        while current ≠ −1 do
            Insert current at the beginning of path
            current ← parent[current]
        end while
        paths[i] ← path
    end for
    return paths
end procedure
```

Algorithm 5 Parallelized A* Heuristic Search

```
procedure ParallelizedAStar(start, goal, heuristic)   ▷ start and goal are node indices, heuristic is a heuristic function
    n ← number of nodes in the graph
    Initialize distances as an array of size n filled with ∞
    distances[start] ← 0.0
    Initialize a priority queue Q as a min-heap of pairs: (−distance, vertex)
    Q.push({0, start})
    Initialize an array parent of size n filled with −1
    while not Q is empty do
        (current, distance) ← Q.top()        ▷ Critical section to access the top element
        Q.pop()
        if distance > distances[current] then
            continue
        end if
        for each neighbor in adjList[current] do        ▷ Parallelized loop
            next ← neighbor.first
            weight ← neighbor.second
            newDistance ← distance + weight
            if newDistance < distances[next] then
                Critical Section:
                distances[next] ← newDistance
                parent[next] ← current
                Q.push({−(newDistance + heuristic(dec[current], dec[next])), next})
            end if
        end for
    end while
    Initialize an empty array path
    current ← goal
    while current ≠ −1 do
        Insert current at the beginning of path
        current ← parent[current]
    end while
    return path
end procedure
```

Time for processing one vertex: $O(\log V)$
Time Complexity of A* Heuristic Search Algorithm: $O((V + E) * \phi * \log V)$

Using adjacency list, for 'p' cores,
Time for visiting all vertices: $O(\frac{V + E}{p})$, where E is the number of edges and V is number of nodes in the graph
Time Complexity of Parallelized A* Heuristic Search: $O(\frac{(V + E) * \phi * \log V}{p})$

4.3 Improved Bellman-Ford Algorithm

With reference to the classical Bellman-Ford Algorithm 3 for single source-all destinations, the 'for loop' in line 5 iterates (V-1) times, and the 'for loop' in line 6 which traverses through the edges to remove the negative weights, can be parallelized, by making the 'if statement' in line 8 critical. The 'for loop' in line 14 which checks for negative cycles in the graph can also be parallelized. The path reconstruction for all destinations done by the 'for loop' in line 21 can also be parallelized. Experimentally, the parallelizations resulted in better performance, time complexity and space complexity. The pseudo-code of the improved Bellman-Ford Algorithm 6.

4.3.1 Mathematical Formulation

Using adjacency list, with only 1 core,
Time for traversing all edges (v–1) times: $O(E * V)$, where E is the number of edges and V is number of nodes in the graph
Time for Path Reconstruction: $O(V^2)$
Time Complexity of Bellman-Ford Algorithm: $O((V * E) + V^2)$

Using adjacency list, for 'p' cores,
Time for traversing all edges (v-1) times: $O(\frac{V * E}{p})$, where E is the number of edges and V is number of nodes in the graph
Time for Path Reconstruction: $O(\frac{V^2}{p})$ **Time Complexity of Parallelized Bellman-Ford Algorithm:** $O(\frac{(V * E)}{p} + \frac{V^2}{p})$

5 Empirical Results and Discussion

From the dataset that consisting of directed graphs with positive edge weights that were grown by breadth-first search from a random point in the New York City, 180 to 4882 nodes were chosen to test the enhanced algorithm's performance. The programs were tested on Intel i7 9th generation system and produced the following results for each of the algorithm. The results were drawn in comparison to the top graph libraries currently in use, namely C++ based Boost graph Library and Python-based NetworkX and iGraph.

The mentioned tables draw comparisons between the libraries based on their algorithmic and complete runtime. Algorithmic runtime is the time taken to run the algorithm, given the constructed graph, and complete runtime is the total time taken for the execution of the program from the construction of the graph to the end of the algorithm.

Algorithm 6 Parallelized Bellman-Ford Algorithm

```
procedure ParallelizedBellmanFord(source)          ▷ source is the source node index
    n ← number of nodes in the graph
    Initialize paths as an empty array of arrays to store paths
    Initialize distance as an array of size n filled with ∞
    Initialize parent as an array of size n filled with −1
    distance[source] ← 0.0
    for i from 0 to n − 1 do
        for each node u from 0 to n − 1 do                       ▷ Parallelized loop
            for each neighbor in adjList[u] do                   ▷ Parallelized loop
                v ← neighbor.first
                weight ← neighbor.second
                if distance[u] + weight < distance[v] then
                    Critical Section:
                    distance[v] ← distance[u] + weight
                    parent[v] ← u
                end if
            end for
        end for
    end for
    Check for negative cycles:
    local_negative_cycle ← false
    for each node u from 0 to n − 1 do                           ▷ Parallelized loop
        for each neighbor in adjList[u] do                       ▷ Parallelized loop
            v ← neighbor.first
            weight ← neighbor.second
            if distance[u] + weight < distance[v] then
                local_negative_cycle ← true
            end if
        end for
    end for
    if local_negative_cycle then                ▷ Critical section for error reporting
        throw runtime error("Negative cycle detected.")
    end if
    Path Reconstruction:
    for i from 0 to n − 1 do                                     ▷ Parallelized loop
        if distance[i] = ∞ then
            paths[i] ← empty array
            continue
        end if
        Initialize an empty array path
        current ← i
        while current ≠ −1 do
            Insert current at the beginning of path
            current ← parent[current]
        end while
        paths[i] ← path
    end for
    return paths
end procedure
```

Table 1 Algorithm and complete runtime in (μs) for A* Heuristic search

		New method		BGL		NetworkX		iGraph	
Nodes	Edges	AR	CR	AR	CR	AR	CR	AR	CR
108	290	0	959	570	2337	1000	46,989	998	335,000
136	368	0	1002	744	2349	1007	2036	999	336,000
171	470	0	1956	1080	3257	1041	3034	1000	337,001
214	584	0	1992	1266	4056	1998	3998	1000	383,010
268	726	0	2004	1307	4612	2000	4000	1001	313,004
335	910	0	2016	1631	5197	2001	14,000	1001	287,900
419	1140	0	2999	1969	6197	2033	8040	1002	264,000
524	1406	0	3998	2233	6574	3002	6991	1004	313,001
655	1758	0	4996	2563	38,402	3010	8009	1009	335,966
819	2226	0	5959	2859	39,538	4002	18,993	1031	306,042
1024	2824	960	8984	3272	28,104	4993	15,990	1035	286,045
1280	3548	999	9999	3579	12,757	5001	15,998	1040	288,051
1600	4442	1000	12,961	4951	21,097	6999	20,000	1042	344,040
2000	5596	1972	19,002	6675	24,215	8000	27,001	2000	317,984
2500	6946	1999	22,001	8532	28,555	10,008	44,010	2002	284,999
3125	8646	2960	29,957	15,688	34,547	13,007	44,996	2002	292,000
3906	10,736	3000	30,999	20,280	46,679	19,988	59,999	3000	342,004
4882	14,110	4000	42,966	26,990	184,239	26,999	75,031	3002	367,999

It can be inferred from Tables 1, 2 and 3 that as the cost of computation increases with the increase in the graph size, the parallelized method performs significantly better and is greatly optimized in comparison to the traditional approaches implemented as a part of the graph libraries.

6 Conclusion and Future Scope

This paper highlights the importance of optimizing graph theory algorithms, particularly the shortest path search, showcasing their relevance in solving real-world challenges like routing and resource management through reduced computational complexity and enhanced practicality. This paper also provides a solid foundation for extending its implementation to state-of-the-art algorithms like Johnson's algorithm, while also opening the door to research opportunities in other graph-related areas, including Minimum Spanning Trees and Centrality Measures.

The research primarily focuses on two key enhancements: utilizing hashmap encoding to convert generic graphs into integer graphs and implementing parallelization techniques in algorithms like Dijkstra's, A* Heuristic Search, and Bellman-Ford Algorithms. The empirical results provide strong evidence that these implementations can significantly boost the efficiency of these algorithms.

Table 2 Algorithm and complete runtime in (μs) for Bellman-Ford's Algorithm

		New method		BGL		NetworkX		iGraph	
Nodes	Edges	AR	CR	AR	CR	AR	CR	AR	CR
108	290	0	988	570	2337	23,012	24,011	4004	149,994
136	368	985	1959	744	2349	30,958	31,994	5001	144,001
171	470	1001	3013	1080	3257	37,009	39,010	6999	158,960
214	584	1034	4049	1266	4056	45,001	47,043	10,998	141,998
268	726	2000	5000	1307	4612	77,013	79,006	18,991	166,002
335	910	2995	4995	1631	5197	125,965	128,000	27,000	152,999
419	1140	3988	6988	1969	6197	198,008	201,001	42,008	216,994
524	1406	6988	11,998	2233	6574	323,994	328,002	65,000	209,000
655	1758	6999	11,998	2563	38,402	510,975	515,002	139,990	276,998
819	2226	9995	15,991	2859	39,538	1,227,957	1,235,991	167,953	335,967
1024	2824	11,965	20,000	3272	28,104	2,061,041	2,070,035	267,988	424,998
1280	3548	19,049	33,055	3579	12,757	2,319,994	2,331,006	395,038	546,005
1600	4442	25,958	38,984	4951	21,097	3,591,965	3,604,954	617,999	770,993
2000	5596	35,990	50,958	6675	24,215	5,440,000	5,455,005	987,958	1,147,958
2500	6946	49,999	67,944	8532	28,555	11,524,000	11,541,998	1,663,007	1,824,005
3125	8646	77,007	100,979	15,688	34,547	15,080,665	15,104,522	2,414,970	2,575,978
3906	10,736	109,999	140,030	20,280	46,679	22,564,001	22,591,039	4,016,006	4,188,005
4882	14,110	141,964	178,958	26,990	184,239	45,273,714	45,325,712	6,616,001	6,791,002

Table 3 Algorithm and complete runtime in (*milli* s) for Dijkstra's Algorithm

		New method		BGL		NetworkX		iGraph	
108	290	69	70	42	44	41	42	529	658
136	368	134	136	56	366	60	72	742	883
171	470	227	228	97	117	84	93	1453	1592
214	584	381	383	150	174	149	158	2816	2979
268	726	871	875	234	237	249	259	5833	5985
335	910	1284	1288	438	442	330	340	10,359	10,508
419	1140	2243	2246	558	563	507	521	19,696	19,841
524	1406	3903	3907	857	862	901	913	39,890	40,041
655	1758	6778	6785	1349	1423	1431	1445	78,523	78,705
819	2226	11,831	11,841	2100	2328	2380	2395	156,813	156,949
1024	2824	20,548	20,561	3625	3688	4430	4448	275,239	275,386
1280	3548	34,726	34,735	5210	5252	6810	6830	497,456	497,598
1600	4442	61,412	61,425	8452	8496	10,441	10,462	1,144,850	1,144,993
2000	5596	108,312	108,329	12,119	12,256	17,500	17,527	1,144,993	1,956,648

References

1. Fan D, Shi P (2010) Improvement of dijkstra's algorithm and its application in route planning. In: 2010 Seventh international conference on fuzzy systems and knowledge discovery, vol 4. IEEE, pp 1901–1904
2. Huang Y (2018) Research on the improvement of dijkstra algorithm in the shortest path calculation. In: 2017 4th International conference on machinery, materials and computer (MACMC 2017). Atlantis Press, pp 745–749
3. Xie D, Zhu H, Yan L, Yuan S, Zhang J (2012) An improved Dijkstra algorithm in GIS application. In: World automation congress 2012. IEEE, pp 167–169
4. Zhu Z, Li L, Wu W, Jiao Y (2021) Application of improved Dijkstra algorithm in intelligent ship path planning. In: 2021 33rd Chinese control and decision conference (CCDC). IEEE, pp 4926–4931
5. Ju C, Luo Q, Yan X (2020) Path planning using an improved a-star algorithm. In: 2020 11th international conference on prognostics and system health management (PHM-2020 Jinan). IEEE, pp 23–26
6. Yao J, Lin C, Xie X, Wang AJ, Hung C-C (2010) Path planning for virtual human motion using improved A* star algorithm. In: 2010 Seventh international conference on information technology: new generations. IEEE, pp 1154–1158
7. Hong Z, Sun P, Tong X, Pan H, Zhou R, Zhang Y, Han Y, Wang J, Yang S, Xu L (2021) Improved a-star algorithm for long-distance off-road path planning using terrain data map. ISPRS Int J Geo-Inf 10(11):785
8. Zheng T, Xu Y, Zheng D (2019) AGV path planning based on improved a-star algorithm. In: 2019 IEEE 3rd advanced information management, communicates, electronic and automation control conference (IMCEC). IEEE, pp 1534–1538
9. Busato F, Bombieri N (2015) An efficient implementation of the bellman-ford algorithm for Kepler GPU architectures. IEEE Trans Parallel Distrib Syst 27(8):2222–2233
10. Bannister MJ, Eppstein D. Randomized speedup of the Bellman-Ford algorithm, pp 41–47
11. Chen H, Suh H-J (2012) An improved bellman-ford algorithm based on SPFA. J Korea Inst Electron Commun Sci 7(4):721–726
12. Dijkstra EW (1959) A note on two problems in connexion with graphs. Numer Math 1(1):269–271
13. Walden D (2008) The bellman-ford algorithm and distributed bellman-ford
14. Planken LRL, Weerdt M, Krogt RPJR (2011) Benchmark instances for experiments in "Computing all-pairs shortest paths by leveraging low treewidth—New York (Figure 6). TU Delft—Delft University of Technology; Faculty EEMCS; Algorithmics group (2011). https://doi.org/10.4121/UUID:4B6D7F56-F09B-4507-8E1A-3078D6E0266A . https://data.4tu.nl/articles/_/12693512/1

Natural Language Processing

Improving Document-Based Question Answering with Multi-modal Learning: Batch Size and Token Modality Experiments

K. Raagulbharatwaj, Janaki Meena Murugan, and Jeganathan Lakshmanan

Abstract Token modalities underpin the model's ability to effectively fuse and process diverse data types, essential for comprehensive understanding and robust performance over various multi-modal learning tasks. Batch size holds paramount importance in deep learning research, as it profoundly affects GPU resource utilization efficiency and training stability, significantly shaping model generalization and convergence dynamics. In this paper, we investigate the role of batch size and token modalities on document-understanding by fine-tuning LayoutLMv3 on a small subset of the DocVQA. To highlight the importance of spatial layout on an extremely challenging task like Document-based VQA we conducted an ablation study where we only fed the text tokens and compared the model's performance with its multi-modal baseline. Additionally, we evaluated the significance of batch size on learning and generalizing in computationally constrained environments by training our model with varying sample size and batch size, and the results are presented. Our experimental results demonstrate an average improvement of approximately 38% when utilizing multi-modal learning approaches over relying solely on text-based approaches, especially in scenarios where training data is limited. When training the model with different sample sizes and batch sizes, we observed that employing a larger batch size effectively alleviates over-fitting, resulting in improvements of up to 16%. As a recommendation, we advise selecting an optimal batch size that maximizes GPU memory utilization while preserving training stability.

Keywords LayoutLMv3 · Multi-modal learning · Spatial layout · Document visual question answering (DocVQA)

This work is funded by MeitY (Government of India), Sanction number L-14011/4/2022-HRD.

K. Raagulbharatwaj (✉) · J. M. Murugan · J. Lakshmanan
Vellore Institute of Technology, Chennai, Tamil Nadu, India
e-mail: raagulbharatwaj@gmail.com

S. M. Thampi et al. (eds.), *Fifth International Conference on Computing and Network Communications*, Lecture Notes in Electrical Engineering 1221,
https://doi.org/10.1007/978-981-97-4711-5_36

1 Introduction

Natural Language Processing (NLP) is an interdisciplinary field that seeks to enable machines to comprehend, interpret, and generate human language, making it a pivotal area of research in the global AI community. The development of NLP has been heavily influenced by groundbreaking works from prominent researchers across the globe. "Word2Vec" by Mikolov et al. [1] introduced efficient word embeddings, laying the foundation for capturing semantic relationships in large text corpora. Additionally, attention mechanisms, as presented in "Neural Machine Translation by Jointly Learning to Align and Translate" by Bahdanau et al. [2], revolutionized sequence-to-sequence learning by enabling models to focus on relevant parts of the input during decoding. The Transformer architecture, proposed in "Attention Is All You Need" by Vaswani et al. [3], has surpassed traditional recurrent neural networks (RNNs) in handling long-range dependencies and capturing global context. Pre-trained language models, such as BERT by Devlin et al. [4], have further elevated NLP performance by leveraging unsupervised pre-training on extensive text data. Contextualized word embeddings, like ELMo by Peters et al. [5] and GPT by Radford et al. [6], have also made notable contributions to NLP by capturing context-specific word meanings. These seminal works exemplify the progress in NLP and have significantly influenced the direction of research, fostering global collaborations to address new challenges and further the application of NLP in real-world scenarios. As NLP continues to evolve, the insights from these influential works serve as valuable guidance for researchers seeking to push the boundaries of language understanding and generation.

In parallel with the influential works that have driven the progress of NLP, researchers have explored a diverse range of NLP tasks to achieve deeper language understanding and real-world applications. These tasks encompass sentiment analysis [7], machine translation [3], named entity recognition [8], text classification [9], and question answering [10]. Addressing the limitations of traditional NLP models and leveraging information from various modalities, multi-modal learning has emerged as a promising direction. By combining information from multiple sources, such as text, images, speech, and other modalities, multi-modal learning aims to gain a more comprehensive understanding of data. With its potential to interpret complex input and exhibit more human-like cognitive abilities, multi-modal learning has been widely explored in tasks such as visual question answering [11], image captioning [12], and multi-modal sentiment analysis [13]. Notably, the development of multi-modal transformers, as introduced in "VisualBERT: A Simple and Performant Baseline for Vision and Language" by Li et al. [14], has significantly advanced the field. Multi-modal transformers effectively process input from diverse sources, creating a unified representation that captures both linguistic and visual information. This progress opens up exciting possibilities for developing more sophisticated AI systems, bridging the gap between language and perception, and unlocking new frontiers in language understanding and generation.

Visual Question Answering (VQA) is an intriguing application of multi-modal learning, combining computer vision and natural language processing to enable machines to answer questions about images, leveraging advancements in NLP from works like Mikolov et al. [1], Vaswani et al. [3], and Devlin et al. [4], which have elevated NLP performance through unsupervised pre-training and contextualized word embeddings. While VQA focuses on generic images, "Document-based Visual Question Answering" (DocVQA) presents new challenges due to document complexities, including layouts and varying text sizes. Overcoming the limitations of traditional NLP models, DocVQA requires efficient methods for aligning document regions with questions and generating accurate answers. To address these challenges, researchers have introduced benchmark datasets like DocVQA [15] and innovative approaches such as [16], a layout-aware pre-training method. As NLP and computer vision evolve, DocVQA exploration drives deeper document-understanding and multi-modal integration, opening new frontiers in language understanding and generation, with the insights from VQA guiding this promising direction.

Layout models have emerged as powerful tools in the domain of document analysis, demonstrating remarkable capabilities in comprehending complex documents. These models, such as LayoutLM and its successors like LayoutLMv2 and LayoutLMv3, leverage a combination of computer vision and natural language processing (NLP) techniques to effectively capture both textual content and spatial layout information. By incorporating spatially aware self-attention mechanisms and global-enhanced layout encoding, layout models can comprehend intricate document structures, varying font sizes, and the hierarchical organization of content [17, 18]. This enables them to accurately identify relevant document regions in response to questions and generate contextually appropriate answers in tasks like Document-based Visual Question Answering (DocVQA). Notably, their layout-aware pre-training approach eliminates the need for pre-trained CNN or Faster R-CNN backbones for visual feature extraction, resulting in a more streamlined architecture [18]. As a result of these advancements, layout models have showcased their generality and superiority in various document-centric tasks, from document classification to information extraction, thereby making them highly promising tools in advancing the field of Document AI in real-world scenarios.

When fine-tuning a large model such as LayoutLMv3 for a downstream task like Document-based Visual Question Answering (DocVQA), several challenges are likely to emerge. The initial and prevalent issue encountered is a compromise in performance, primarily attributable to the inadequacy of available training data. Another significant hurdle that arises during training is the frequent occurrence of GPU memory exhaustion, often stemming from the use of a substantial batch size. Batch size also determines the training stability of the model especially when trained on small datasets. Mitigating the latter problem is crucial, as it directly impacts the model's training. While reducing the batch size might seem like a solution, it raises concerns about potential impacts on the model's generalization abilities. Therefore, it is imperative to carefully consider this trade-off during the training process. This research endeavors to offer solutions to the aforementioned challenges through our proposed approach.

2 Proposed Methodology

In order to address the aforementioned issue, we have undertaken two distinct studies aimed at elucidating optimal strategies for achieving peak performance when training an large model such as LayoutLMv3 within resource-constrained environments. Study-1 is an ablation study that trains the model first with text and layout tokens then with text tokens alone. Study-2 analyzes how well the model generalizes by training it with different sample sizes and batch sizes. The execution of these experiments entails a systematic progression through the following stages, which are detailed as follows. The first stage is common for both the experiments.

2.1 Data Sampling and Preprocessing

The initial phase involves the acquisition and preprocessing of the dataset. For this study, we have used the dataset introduced by Mathew et al. in 2020, known as DocVQA dataset [15]. This dataset was established as part of the broader DocVQA project and encompasses a substantial compilation of varied document images accompanied by associated interrogatives. The dataset is thoughtfully designed. Its contents encompass a diverse array of document images, including tables, forms, paragraphs, and more.

1. The first stage entails the extraction of the training split from the DocVQA dataset, which is initially in JSON format. This extracted data is then loaded into memory. Each entry in the dataset comprises the following attributes: Question ID, Question content, Image data, Document ID, UCSF Document ID, UCSF Page Number, Answer content, and Data Split indication.
2. Subsequently, extraneous details are eliminated from each sample, retaining only pertinent information crucial to the task Question, Image, and Answer.
3. In the subsequent phase, the dataset is stratified into six equitably sized folds, with each fold containing 1000 documents. Out of these, 5000 documents are allocated for training, while the remaining 1000 are designated for validation. Due to computational limitations, the data is processed in parallel.
4. The document images within each fold are then input into the LayoutLMv3-ImageProcessor, a crucial step in generating layout tokens. This processor also performs Optical Character Recognition (OCR) to extract the words within the document and establishes bounding boxes for each word. Each of the six folds is individually preserved as HuggingFace data shards, facilitating subsequent processing.
5. The training shards from all five folds are amalgamated, and a sub-word-finder module is employed. This sub-word-finder identifies matches between the listed answers in each sample and the words present within the document. When a match is identified, the start and end positions of the matching words are recorded.

Fig. 1 Data preprocessing steps as a flowchart

6. Having completed the tagging of answer positions, the words and questions are presented to the LayoutLMv3Tokenizer for tokenization, a pivotal step in the sequence.
7. Finally, the dataset is organized into batches and primed for training. The validation data is fashioned from the designated validation fold (Fig. 1).

2.2 *Study-1: Ablation Study on Token Modalities*

In the realm of document-understanding, the availability of comprehensive training data stands out as a central limitation. The scarcity of adequately labeled datasets poses a significant challenge to the development and performance of models. Addressing this challenge necessitates a multi-modal learning approach that harnesses the synergistic relationship between layout and text tokens. In this context,

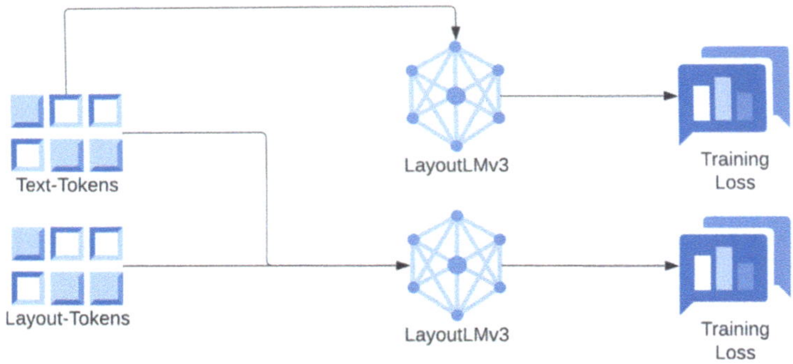

Fig. 2 Experimental design of the Ablation study

layout tokens encapsulate the spatial organization of document elements, while text tokens capture the semantic nuances of textual content. The amalgamation of these two modalities equips algorithms with the ability to discern hierarchical structures, contextual intricacies, and linguistic subtleties, providing the model with additional information to learn from (Fig. 2).

To test the significance of this additional mode of data we embark on a pivotal stride in the form of an ablation study, thoughtfully crafted to underscore the importance of multi-modal learning in mitigating data limitations. This empirical investigation unfolds as follows: Initially, the model is trained with access to both text and layout tokens, serving as the baseline for performance evaluation. A comparative analysis is then conducted, contrasting this baseline with the model's performance when exclusively utilizing text tokens.

2.3 *Study-2: Study on the Impact of Batch Size on Generalization*

The primary objective of this study was to meticulously examine the influence of batch size on the performance of the model, with a specific focus on how computational limitations come into play. One critical aspect we investigated was how higher batch sizes could potentially lead to GPU RAM constraints, while lower batch sizes might impact the model's ability to generalize effectively.

We utilized two distinct sets of samples: one encompassing 3000 documents and the other comprising 5000 documents. We conducted experiments using varying batch sizes, specifically 2, 10, and 12 for the 3000-document sample, and 2 and 10 for the 5000-document sample. A comprehensive overview of the batch sizes employed in conjunction with the respective sample sizes is presented in Table 1. We conducted training for LayoutLMv3 using the datasets mentioned above, employing the batch sizes specified in Table 1. We assessed the model's performance primarily based on validation loss (Fig. 3).

Table 1 Batch sizes and sample sizes used for training

Sample size	Batch sizes
3000	2, 10, 12
5000	2, 10

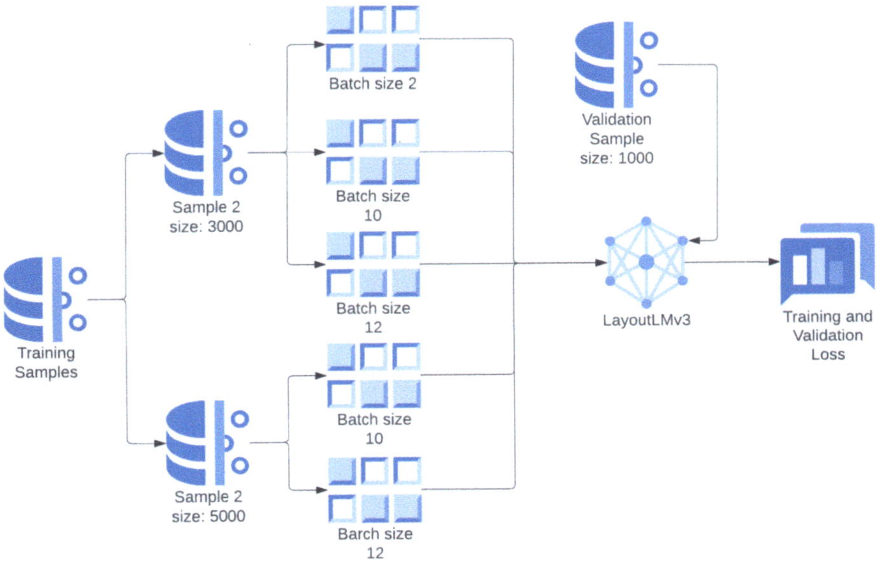

Fig. 3 Experimental design of study-2

3 Results

3.1 Results of Study-1: Ablation Study on Token Modalities

The ablation study aimed to assess the impact of incorporating both text and layout information in the Visual Question Answering (VQA) task on document images compared to using text alone. Two models were employed for comparison: the Text-only model, relying solely on textual information extracted from document images, and the multi-modal model, which integrated both text and layout features. The training loss of the two models is presented in Table 2 (Fig. 4).

Table 2 Training loss comparison: Text alone versus text + layout

Training steps	Text	Text + Layout
1000	6.362206	2.257500
2000	4.367871	1.372100
3000	4.430867	0.728500
4000	3.514406	0.557400
5000	3.969666	0.339200
6000	3.674368	0.363000
7000	3.333303	0.322200
8000	2.762063	0.154200
9000	2.216875	0.061400
10,000	1.63429	0.069000

Fig. 4 Figure depicting the training loss of the models over 10,000 training steps. Blue curve represents the training loss of the text only model and the orange curve represents the training loss of the multi-modal model

3.2 *Results of Study-2: Study on the Impact of Batch Size on Generalization*

Study-2 was designed to tackle the challenges related to batch size constraints during model training. The model underwent training across various batch size and dataset size combinations, as previously outlined, over 10 epochs. Subsequently, validation was conducted using the validation shard, and the outcomes are detailed below. Tables 3 and 4 provide a comparison of both training and validation loss over various combinations of sample size and batch size (Figs. 5, 6, 7, 8, , 9 and , 10).

Table 3 Training and validation loss comparison of the model with training sample size 3000

(a) Batch Size 12

Step	Training loss	Validation loss
500	3.523	2.211
1000	1.971	1.975
1500	1.433	1.796
2000	1.11	1.72
2500	0.9273	1.768

(b) Batch Size 10

Step	Training loss	Validation loss
500	2.58	2.438
1000	1.65	1.749
1500	1.279	1.656
2000	0.9875	1.655
2500	0.8016	1.776
3000	0.7659	1.779

(c) Batch Size 2

Step	Training loss	Validation loss
1000	3.201	2.557
2000	2.18	2.366
3000	1.877	1.572
4000	1.545	2.017
5000	1.258	1.908
6000	1.261	2.199
7000	1.12	2.542
8000	0.8804	2.946
9000	0.9508	2.973
10000	0.6725	2.874
11000	0.578	3.059
12000	0.6263	3.445
13000	0.4535	3.351
14000	0.3564	3.38
15000	0.3495	3.513

Table 4 Training and validation loss comparison of the model with sample size 5000

(a) Batch Size 2

Step	Training loss	Validation loss
1000	3.619	2.425
5000	1.596	1.455
9000	1.127	2.048
13000	0.7465	2.396
17000	0.4987	2.475
21000	0.2692	2.936
25000	0.2282	3.031

(b) Batch Size 10

Step	Training loss	Validation loss
500	3.419	2.211
1000	1.967	1.513
1500	1.43	1.514
2000	1.125	1.571
2500	0.9175	1.468
3000	0.7552	1.448
3500	0.6436	1.466
4000	0.5515	1.57
4500	0.4746	1.704
5000	0.4109	1.745

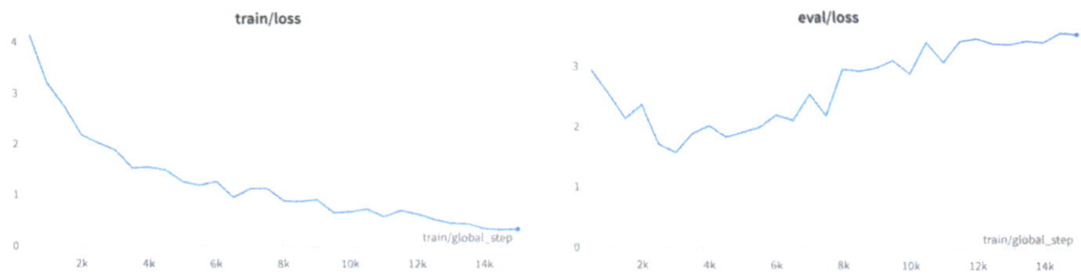

(a) Train loss (b) validation loss

Fig. 5 Plot depicting training and evaluation loss versus training steps (sample size: 3000, batch-size: 2)

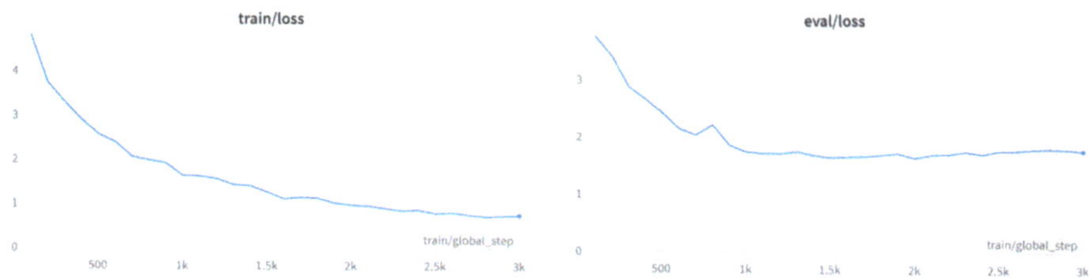

(a) Train loss (b) validation loss

Fig. 6 Plot depicting training and evaluation loss versus training steps (sample size: 3000, batch-size: 10)

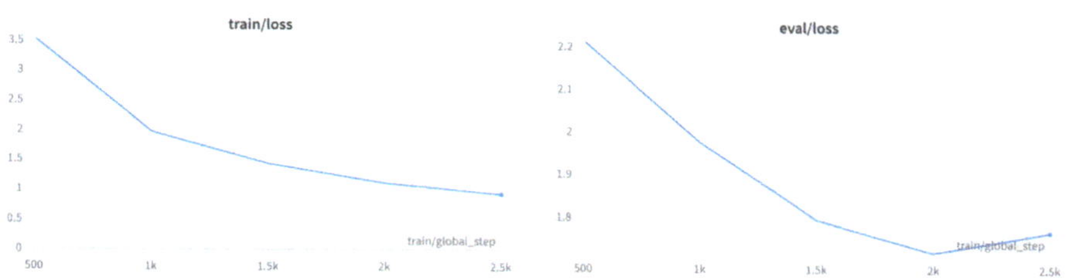

(a) Train loss (b) validation loss

Fig. 7 Plot depicting training and evaluation loss versus training steps (sample size: 3000, batch-size: 12)

4 Inferences

4.1 Inference from Study-1

1. The study provided strong evidence that incorporating multiple modalities consistently improved model performance across the entire training process. This improvement was evident through significantly reduced training loss values, indicating that the multi-modal approach enhanced the model's overall accuracy.

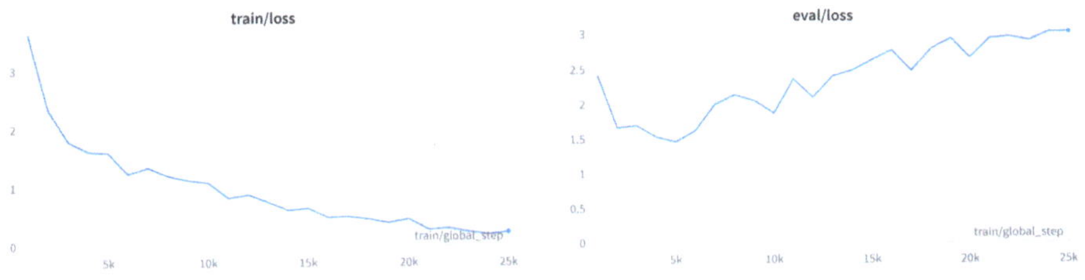

(a) Train loss (b) validation loss

Fig. 8 Plot depicting training and evaluation loss versus training steps (sample size: 5000, batch-size: 2)

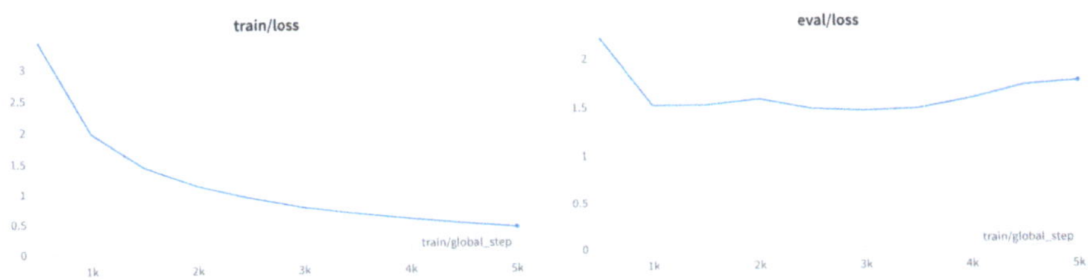

(a) Train loss (b) validation loss

Fig. 9 Plot depicting training and evaluation loss versus training steps (sample size: 5000, batch-size: 10)

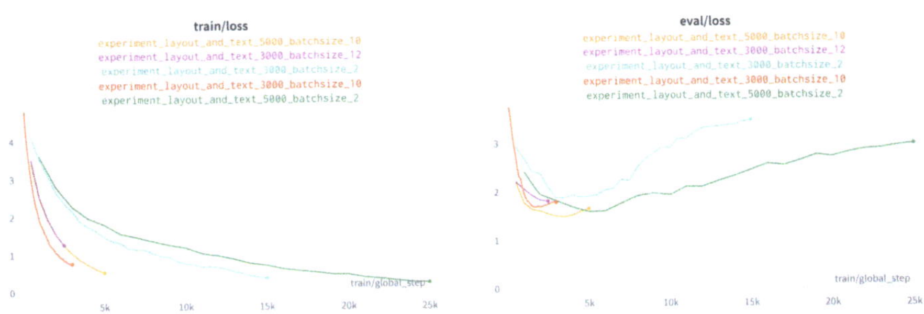

(a) Train loss (b) validation loss

Fig. 10 Plot depicting comparison of training and evaluation loss versus training steps across various batch sizes

2. Inference-1 directly stems from the trend observed in the model's training loss. At each training step, the Text + Layout model exhibited substantially lower training loss values, indicating improved performance and accuracy. This performance superiority was evident from the initial training steps to the later stages, where the gap in training loss between the two models remained consistently significant. The same can be realized via a graphical representation of the results.

3. These findings carry broad implications for various tasks constrained by limited training data. By integrating multiple modalities, models can effectively address data scarcity issues, leading to improved performance and generalization across different domains and complexities. This underscores the importance of multi-modal information processing, emphasizing the need to consider multiple sources of information, thus enhancing model capabilities in a wide range of applications facing data limitations.

4.2 *Inference from Study-2*

1. The examination of the 3000-document sample yields valuable insights into the model's response to varying batch sizes during training. It's worth noting that a batch size of 10 consistently showcased superior performance, consistently yielding the lowest validation loss at multiple training stages. In a closely following manner, the batch size of 12 also demonstrated competitive results, further affirming the efficacy of larger batch sizes when dealing with a smaller dataset.
2. Conversely, the batch size of 2 did not yield competitive performance, likely due to insufficient exposure to training examples per iteration. While the smaller batch size might lead to faster convergence, it also resulted in quicker over-fitting, indicating that the model failed to learn effectively from limited data instances.
3. The results observed for the 5000-document sample are consistent with the patterns identified in the 3000-document sample, providing additional support for the idea that larger batch sizes contribute to improved generalization and help mitigate over-fitting.
4. As previously emphasized, it is imperative for researchers to approach the use of larger batch sizes with careful consideration of the associated trade-offs, which include heightened demands on GPU memory and longer training duration. Conversely, smaller batch sizes may show faster convergence but are vulnerable to over-fitting, which can impede the model's ability to generalize. The results above provide direct evidence of this specific trade-off.

5 Conclusion

In the realm of fine-tuning large models such as LayoutLMv3 for an intricate task like Document-based Visual Question Answering (DocVQA), we have navigated through a terrain riddled with challenges. Chief among these challenges is the persistent hurdle of limited training data, which inevitably leads to a performance compromise. This compromise stems from the inherent difficulty of the task and highlights the need for innovative solutions to bridge the gap between model capability and available data resources.Simultaneously, we grappled with the complex issue of GPU memory exhaustion, which frequently reared its head due to the adoption of larger batch sizes, a strategy commonly used to mitigate this and expedite training is the use of smaller batch sizes. However, this solution introduced a new conundrum: the risk of compromising model generalization. Reducing the batch size seemed to alleviate memory constraints but raised concerns about the model's ability to generalize effectively, potentially jeopardizing overall performance.

In response to these multifaceted challenges, our research has taken a step forward in addressing them. By incorporating multiple modalities, such as text and layout information, we've successfully improved model performance across training phases, as evidenced by significantly reduced training loss values. Additionally, our exploration of batch size variations has given us an idea of how to identify an optimal batch size for training, effectively addressing GPU memory constraints while preserving the model's robust generalization capabilities.

References

1. Mikolov T, Chen K, Corrado G, Dean J (2013) Efficient estimation of word representations in vector space. ArXiv. /abs/1301.3781
2. Bahdanau D, Cho K, Bengio Y (2014) Neural machine translation by jointly learning to align and translate. ArXiv. /abs/1409.0473
3. Vaswani A, Shazeer N, Parmar N, Uszkoreit J, Jones L, Gomez AN, Kaiser L, Polosukhin I (2017) Attention is all you need. ArXiv. /abs/1706.03762
4. Devlin J, Chang M, Lee K, Toutanova K (2018) BERT: pre-training of deep bidirectional transformers for language understanding. ArXiv. /abs/1810.04805
5. Peters ME, Neumann M, Iyyer M, Gardner M, Clark C, Lee K, Zettlemoyer L (2018) Deep contextualized word representations. ArXiv. /abs/1802.05365
6. Radford A, Narasimhan K, Salimans T, Sutskever I (2018) Improving language understanding by generative pre-training
7. Pang B, Lee L (2008) Opinion mining and sentiment analysis. Found Trends Inf Retr 2:1–135. https://doi.org/10.1561/1500000011
8. Sang EF, De Meulder F (2003) Introduction to the CoNLL-2003 shared task: language-independent named entity recognition. ArXiv. /abs/cs/0306050
9. Severyn A, Moschitti A (2015) Learning to rank short text pairs with convolutional deep neural networks, pp 373–382. 10.1145/2766462.2767738
10. Rajpurkar P, Zhang J, Lopyrev K, Liang P (2016) SQuAD: 100,000+ questions for machine comprehension of text. ArXiv. /abs/1606.05250
11. Agrawal A, Lu J, Antol S, Mitchell M, Zitnick CL, Batra D, Parikh D (2015) VQA: visual question answering. ArXiv. /abs/1505.00468
12. Xu K, Ba J, Kiros R, Cho K, Courville A, Salakhutdinov R, Zemel R, Bengio Y (2015) Show, attend and tell: neural image caption generation with visual attention. ArXiv. /abs/1502.03044
13. You Q, Jin H, Wang Z, Fang C, Luo J (2016) Image captioning with semantic attention. In: Proceedings of the IEEE conference on computer vision and pattern recognition (CVPR), pp 4651–4659
14. Li LH, Yatskar M, Yin D, Hsieh C, Chang K (2019) VisualBERT: a simple and performant baseline for vision and language. ArXiv. /abs/1908.03557
15. Mathew M, Karatzas D, Jawahar CV (2020) DocVQA: a dataset for VQA on documents. ArXiv. /abs/2007.00398
16. Xu Y, Li M, Cui L, Huang S, Wei F, Zhou M (2019) LayoutLM: pre-training of text and layout for document image understanding. ArXiv. https://doi.org/10.1145/3394486.3403172

17. Xu Y, Xu Y, Lv T, Cui L, Wei F, Wang G, Lu Y, Florencio D, Zhang C, Che W, Zhang M, Zhou L (2020) LayoutLMv2: multi-modal pre-training for visually-rich document understanding. ArXiv. /abs/2012.14740
18. Huang Y, Lv T, Cui L, Lu Y, Wei F (2022) LayoutLMv3: pre-training for document AI with unified text and image masking. ArXiv. /abs/2204.08387
19. LeCun Y, Bottou L, Bengio Y, Haffner P (1998) Gradient-based learning applied to document recognition. Proc IEEE 86:2278–2324. https://doi.org/10.1109/5.726791
20. Simonyan K, Zisserman A (2014) Very deep convolutional networks for large-scale image recognition. arXiv preprint arXiv:1409.1556
21. Hastie T, Tibshirani R, Friedman J (2009) The elements of statistical learning: data mining, inference, and prediction, 2nd edn. Springer
22. Heaton J (2017) Ian goodfellow, yoshua bengio, and aaron courville: deep learning. MIT Press
23. Caruana R (1997) Multitask learning. Mach Learn 28. https://doi.org/10.1023/A:1007379606734
24. Ruder S (2017) An overview of multi-task learning in deep neural networks. arXiv preprint arXiv:1706.05098
25. Anderson P, Fernando B (2018) Bottom-up and top-down attention for image captioning and VQA. ArXiv. /abs/1707.07998
26. Zhang C, Luo C, Lu J, Liu A, Bai B, Bai K, Xu Z (2020) Read, attend, and exclude: multi-choice reading comprehension by mimicking human reasoning process. In: Proceedings of the 43rd international ACM SIGIR conference on research and development in information retrieval

Emotional Attention in Text Search for Deeper Context, a Case for Analysing Review Helpfulness

Amit Oberoi and Brejesh Lall

Abstract In this paper, we present an approach for utilising emotions and sentiments for enhancing the context handling capability of Natural Language Processing algorithms. We base our work on the premise that emotions of a person get reflected in the textual content generated by the individual. We thus conceptualise that by making use of existing tools for sentiment analysis and emotion extraction, we can improve capability of machines to understand humans better. We study the effect of sentiment patterns within an utterance and associated emotions on the perceived helpfulness of Amazon reviews. We propose an approach involving use of emotional information as an incremental add-on to transformer architecture for enhanced application-specific textual analytics. In this work we have experimented with various options of enriching BERT embedding with emotional information present in Amazon product reviews. Our results indicate that adding emotional patterns to BERT embedding yields improved performance in predicting helpfulness of a review, thus demonstrating that emotions within text can be utilised in a deliberate and scientific manner for better understanding by machines. The emotion augmented embedding yielded 1.85% better accuracy, 0.5% better precision, 6% better Recall and 3% better F1 Score than original BERT embedding. Our approach does not need fine tuning of BERT which is resource intensive, and can be generalised to an application specific use case.

Keywords Natural language processing · Emotional patterns · Sentiment analysis · BERT · Language representation · Embedding

---

A. Oberoi (✉) · B. Lall
Indian Institute of Technology, New Delhi, India
e-mail: amit.oberoi@ee.iitd.ac.in

B. Lall
e-mail: brejeshi@iitd.ac.in

S. M. Thampi et al. (eds.), *Fifth International Conference on Computing and Network Communications*, Lecture Notes in Electrical Engineering 1221,
https://doi.org/10.1007/978-981-97-4711-5_37

1 Introduction

Natural Language Processing based applications are being increasingly used across the globe for human machine interaction, analysis of voluminous textual data, generative tasks and automation functions. With rapid advances in artificial intelligence based techniques, it is only a matter of time when applications will be able to understand human emotions and deeper meanings behind those emotions. The fundamental challenge in recognising intent from text can be addressed if contextual information embedded in human generated content can be processed effectively by machines. One such aspect is emotional variation and tone, part of which is covered under the umbrella of sentiment analysis. Sentiments within a piece of text can be considered to be a sum of emotions or strong feelings which are being experienced by a person. Thus, if we can analyse these in depth, we may be in a better position to assess the meaning of what a human is trying to say. Further, if we are able to classify specific emotions from text, such as anger, joy, sadness, etc., we would have the opportunity to enhance contextual understanding. With the adoption of transformer based architecture which was first introduced by Vasnavai [21], the ability of machines to comprehensively make use of the attention mechanism to understand context behind words and even sentences has increased significantly over the years. Further, representations such as BERT (Bidirectional Encoder Representations from Transformers) which was first introduced by Delvi, Chang, Lee and Toutanova [7], have made algorithms more powerful in performing effective language tasks. However, transformer based representations have their own limitations which include heavy computational resources for training and fine tuning as mentioned by Chen [6], as also dependence upon word-sense contextual embedding. Hence, for text applications which rely on outcomes associated with feelings, pre-trained BERT models need to be fine-tuned. Further, the embeddings do not capture entire latent space in language, and in order to do that we would theoretically need larger and larger models which can represent large dimension vectors making them computationally inefficient. Therefore, for certain application-specific requirements involving interpretation of deeper feelings, we need to modify our approach and work out representations which can give attention to emotions and sentiments. We thus propose a conceptual framework to incorporate information related with feelings present in text, for refining transformer-based embedding. In this paper we have used Amazon product reviews to assess the applicability of our conceptual framework, by ascertaining improved prediction capability of BERT embedding concatenated with emotional embedding. Our results are encouraging and indicate scope for undertaking further exploratory work in this field. The remaining part of this paper is organised as follows. Conceptual Framework is given at Sect. 2. Related work covering aspects covered by people in similar fields is given at Sect. 3. Approach Adopted by us during this study including Experiments and Results are given at Sect. 4. Limitations and suggestions for future work have been covered in Sect. 5. Finally we have concluded at Sect. 6.

2 Conceptual Framework

2.1 *Sentiment Pattern*

Sentiment within a sentence can be thought to be an amalgamated sum of emotions giving out the overall flavour of the feelings associated with the text. For the sentence x the sentiment $S(x)$ can be depicted as an assessment of positive or negative feelings embedded within the text. A particular utterance or a review r will have one or more sentences. If a review has l sentences the entire review will have a sequence of sentiment scores given by $\boldsymbol{r} = [S_1, S_2, \ldots, S_l]$. If we can discern the sentiment associated with a sentence accurately enough to classify it as either positive or negative, then the polarity of a sentence within a review can from a pattern. Thus, we will have $\psi(\boldsymbol{r}_k) = [p, n, p, p, \ldots, n]$ as the pattern for the kth review. The flow of emotions in an utterance can be thought to be related with the emotional state of the originator. The pattern is indicative of the overall flow of feelings in the review.

2.2 *Emotional Intensity*

If we can extract emotions from text and quantify their intensity then we have a emotional intensity vector $\boldsymbol{e}(r_k) = [\bar{e}_{1k}, \bar{e}_{2k}, \ldots, \bar{e}_{Nk}]$ for the kth review, where N is the number of emotions and $\bar{e}_{pk}$ is the mean intensity of emotion e_p in the kth review. $E = \{e_1, e_2 \ldots e_N\}$ is a set comprising of N emotions that we are able to classify a text document into and find its emotional intensity. For example, using the NRC Lexicon [14] we can classify and extract emotional intensities of eight different emotions namely, trust, anticipation, joy, sadness, surprise, anger, fear and disgust, and, two sentiments which are positive and negative.

2.3 *Emotional Feature Set*

Considering the above, for a given review r_k we have two emotion-specific features relating to the feelings of the person who has generated the content. One is the sentiment pattern $\psi(\boldsymbol{r}_k)$ and other is the mean emotional intensity $\boldsymbol{e}(r_k)$. Analysing $\psi(r_k)$ and $\boldsymbol{e}(r_k)$ together can yield insightful information about aspects related with feelings of the originator, which can influence the overall perception of the person reading the text. Thus, if machines can generalise these inputs in some form, and use these for application-specific scenarios, the contextual understanding capability of machines can be enhanced.

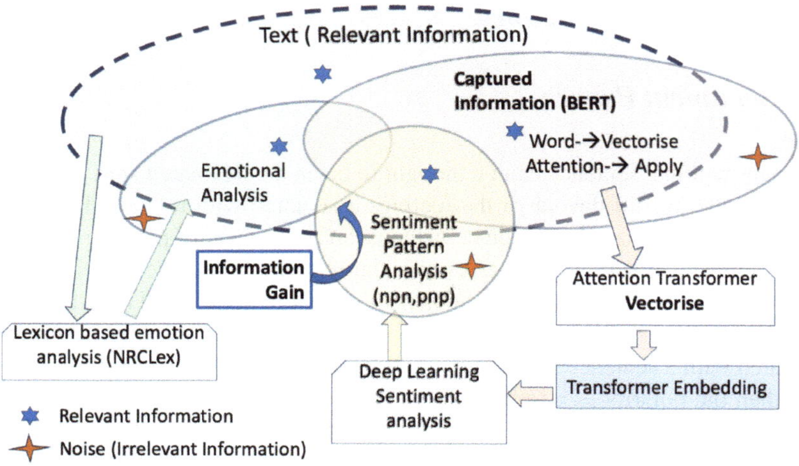

Fig. 1 Conceptual depiction of information gain achieved by utilising emotions and associated patterns within text—emotional patterns $\psi(r_k)$ and mean emotional intensity $e(r_k)$ extracted from text can give derivable insights into context for application-specific cases

2.4 *Embedding Emotional Information in Text*

Conceptual depiction of Information Gain given in Fig. 1 relies on the presumption that transformer based architecture such as BERT can capture significant part of information from text by using its attention mechanism. The same can be utilised for extracting sentiments, either through a pre-trained model or after application-specific training. Sentiment polarity patterns give an indicative insight into some additional information part of which cannot be originally captured by BERT. In addition, by employing lexicon based approach, we can extract emotional intensity of each review, which can further yield emotion-specific information. Part of relevant information which cannot be captured by transformer architecture (in this case BERT) is therefore captured by the emotional and sentiment pattern analysis, leading to an overall information gain. We can appreciate that there is still a large part which remains inaccessible to machines, and which is the motivation for improving NLP algorithms. Also, the transformer based sentiment analysis and lexicon based emotional assessment have their own elements of irrelevant data which is noise, as shown in the Fig. 1.

2.5 *Sentiment Pattern for Helpfulness of Review*

In order to corroborate our conceptual framework, we used Amazon Review dataset wherein one of the features, i.e. 'helpful', is dependant upon the flow of feelings within text. If we were to extract an emotional pattern using sentiment analysis, and calculate the polarity of each sentence within the review, the patterns generated would be of the type 'npn', or 'pnp' or 'nppn', or 'pnnp' etc. Here 'p' denotes positive sentiment and 'n' denoted a negative sentiment. When a person writes a product review, three situations can emerge.

1. The person is in a balanced frame of mind and writes a holistic review covering all aspects of the product.
2. The person is agitated or excessively frustrated with the product.
3. The person is excessively happy and writes all positive aspects.

These aspects can be thought to emerge as sentiment patterns. The context and usefulness of a review can be viewed from two perspectives.

1. **Buyer's perspective**. A buyer will find a review to be helpful if it assists him to make a decision to purchase or otherwise. Any review therefore which gives a clear 'yes' or 'no' would be viewed as more helpful by the buyer. Such reviews need to be clear and concise and non-ambiguous.
2. **Seller's perspective**. A seller would like to search for key indicators which can give him true feedback about the product. Therefore, balanced and maturely written reviews encompassing both good as well as bad or undesirable aspects would be most helpful for the seller. While analysing product reviews, we find that those reviews which were able to convey relevant and meaningful information about the product were the ones which were balanced and contained both positive as well as negative aspects.

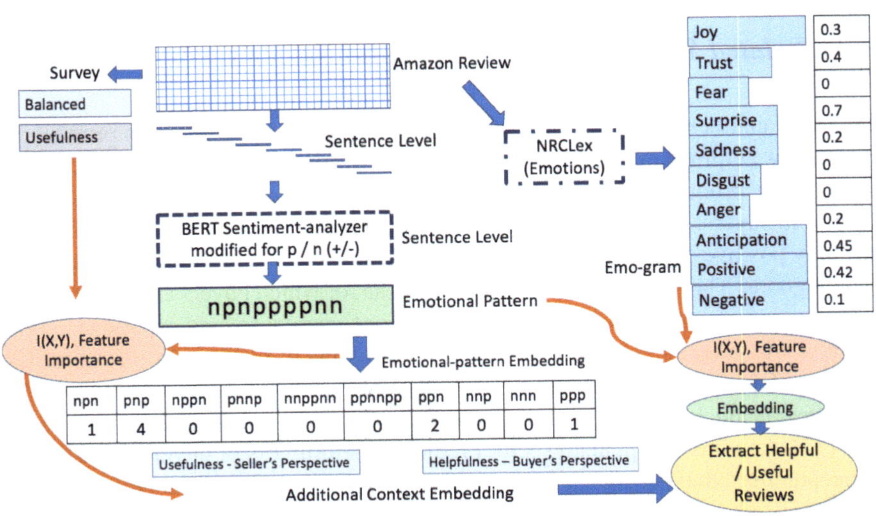

Fig. 2 Analysing sentiment patterns and emotional intensity for extracting additional information—an example of a review depicted as 1400002001 for pattern and corresponding emotional intensities; we analyse feature importance scores and mutual information $I(X, Y)$ for arriving at weights for patterns and emotion embedding; conceptually, we should be able to predict helpful and useful reviews

We can thus conjecture that the flow of sentiment polarity within a review combined with extracted emotions can give us insights into whether the review is helpful for buyer or seller or both. These two aspects can be combined intelligently after analysing their effect, to give a prediction algorithm as depicted at Fig. 2.

3 Related Work

3.1 Sentiment Analysis and Emotions from Text

A holistic survey of Sentiment Analysis has been covered by Birjali and Maroune in [5] wherein aspect, sentence and document level analysis of sentiments have been talked about. Various approaches for Emotion and Sentiment Analysis have been covered in [16]. Although sentiment analysis amalgamates all emotions within textual data, there are situations wherein we would require to extract specific emotions. An overview of emotion detection (ED) has been covered by Acheampong in [1, 2]. For our work we selected the NRC Emotion Lexicon [14] which has evolved through crowd-sourcing [13]. This lexicon makes use of an extensive vocabulary to assess a piece of text and evaluate the intensity of eight emotions namely, anger, fear, sadness, disgust, trust, joy, surprise and anticipation. It also gives out the intensity score for two sentiments, which are positive and negative.

3.2 Related Work in Predicting Helpful Reviews

Mudambi and Schuff [15] attempted to establish a relationship between product rating, review helpfulness and product type. They claimed that product type, specifically whether the product is a search or experience good, is important to make a review helpful to buyers. Also, they found reviews which are moderate are more helpful than strongly positive or negative (extreme) reviews for experience goods, but not for the search goods. They defined search goods as those for which buyers are able to obtain adequate information on quality of product prior to purchase, and experience goods as those which are sampled or purchased for evaluating product quality. Further, they also found that length of reviews generally improve the helpfulness of the review, but this effect is much more for search goods than experience goods. Thomas, Sinha and Sen [17] examined script analysis for predicting helpfulness of buyer reviews. They conceptualised that people share a cognitive script for helpful reviews in a particular domain. Bilal and Alamazroi [4] have proposed fine tuning of BERT base model for classifying helpfulness of product reviews. They found that fine-tuned BERT models performed better than others on the Yelp Open dataset. The accuracy achieved was approximately 0.707 on the fine-tuned BERT model with sequence length of 320. Martin and Pu in [10] attempted to propose emotions as a predictor for helpfulness

Table 1 Complimentary Cumulative Conditional Probability $P(h/\text{pattern}_{\text{freq}}) = P(h = 1/\text{freq} >= F)$ calculated on balanced dataset; h denotes helpfulness of a review assessed by a buyer; note that NA implies there are no reviews with pattern frequency higher than F

Freq (f)	$P(h/npn)$	$P(h/pnp)$	$P(h/nppn)$	$P(h/pnnp)$	$P(h/nnppnn)$	$P(h/ppnnpp)$	$P(h/ppn)$	$P(h/nnp)$	$P(h/nnn)$	$P(h/ppp)$	$P(h/\text{total})$
0	0.500	0.500	0.500	0.500	0.50	0.500	0.500	0.500	0.500	0.500	0.500
1	0.618	0.606	0.692	0.643	0.69	0.690	0.603	0.618	0.645	0.590	0.590
2	0.713	0.733	0.823	0.768	1.00	0.857	0.721	0.712	0.753	0.699	0.687
3	0.767	0.789	0.769	0.842	NA	NA	0.750	0.750	0.806	0.731	0.698
4	0.793	0.775	0.857	0.889	NA	NA	0.875	0.842	0.750	0.886	0.735
5	0.786	0.850	1.000	0.667	NA	NA	0.903	0.850	0.750	0.900	0.789
6	1.000	0.909	1.000	1.000	NA	NA	0.889	0.846	1.000	1.000	0.814
7	1.000	1.000	NA	NA	NA	NA	1.000	0.889	NA	1.000	0.816
8	NA	1.000	NA	NA	NA	NA	1.000	1.000	NA	1.000	0.850
9	NA	NA	NA	NA	NA	NA	NA	1.000	NA	NA	0.931

Table 2 Complimentary Cumulative Conditional Probability $P(h/\text{emotion}) = P(h = 1/P(e_i) >= X)$ calculated on balanced dataset; h denotes helpfulness of a review assessed by a buyer; e_i is the emotion under consideration ; anticp - anticipation; note that NA implies there are no reviews with emotional intensity $P(e_i)$ beyond the particular value X

$P(e_i)$	$P(h/\text{anger})$	$P(h/\text{fear})$	$P(h/\text{trust})$	$P(h/\text{surprise})$	$P(h/\text{positive})$	$P(h/\text{negative})$	$P(h/\text{sadness})$	$P(h/\text{disgust})$	$P(h/\text{joy})$	$P(h/\text{anticp})$
0	0.500	0.500	0.500	0.500	0.500	0.500	0.500	0.500	0.500	0.500
0.06	0.527	0.538	0.515	0.554	0.527	0.507	0.526	0.492	0.549	0.519
0.12	0.382	0.395	0.494	0.521	0.538	0.474	0.424	0.328	0.533	0.477
0.18	0.348	0.340	0.444	0.517	0.552	0.427	0.407	0.226	0.534	0.409
0.24	0.435	0.395	0.349	0.444	0.549	0.320	0.375	0.294	0.479	0.332
0.3	0.444	0.176	0.283	0.333	0.540	0.272	0.308	0.333	0.411	0.308
0.36	0.667	0.200	0.227	0.500	0.537	0.394	0.500	NA	0.462	0.327
0.42	0.500	0.200	0.189	0.500	0.503	0.317	0.444	NA	0.571	0.317
0.48	0.500	0.200	0.185	0.500	0.462	0.342	0.444	NA	0.500	0.324

of a review, by demonstrating better performance by using a feature vector of 20 emotions which were extracted using the GALC lexicon [20]. Lu, Tasparas [9] have attempted to associate helpfulness of a review with the social context of the reviewer. There is however no study analysing sentiment patterns and emotions in the manner proposed in this paper.

4 Approach Adopted, Experiments and Results

4.1 Data Set

We used the Amazon Book Review dataset available at [11] and even at [18]. The dataset contains information about helpful votes in addition to the review and rating of books. Helpful criteria in the data set is included as total number of votes endorsing the review as helpful along with the total number of votes overall. For the purpose of our analysis, we used a threshold of 50% as the classification threshold to decide whether a review was helpful or not. In case more than 50% of the people (amongst those who voted) found the review to be helpful we labelled the helpfulness feature as +1, otherwise is was labelled as 0. This data was thereafter utilised for analysing relationship between helpfulness and emotional and sentiment pattern information.

4.2 Sentiment Pattern Vector

Our goal in this work is to explore performance improvement by enriching our representations with sentiment and emotional information. To keep sentiment analysis simple, we used the BERT pre-trained sentiment analyser. We selected Twitter-RoBERTa which is a variant of the RoBERTa (A Robustly Optimised BERT Pre-training Approach) model that has been specifically trained on Twitter data. We used the Cardiff NLP, Twitter-RoBERTa-base-sentiment-latest model [3, 8] which is a pre-trained RoBERTa-based model fine-tuned to perform sentiment analysis on text data, especially Twitter text. This tool originally classifies a sentence into positive, negative and neutral, based on maximum predicted probability score. On examination, we found that with three classes the patterns were complex. Hence we decided to use an initial exploratory approach of incorporating only 'p' (positive) and 'n' (negative) classification for each sentence by modifying the output to look only at max probability between 'p' and 'n' . We thereafter evaluated sentiment patters present within a review and identified ten different significant patterns as depicted at Table 3 and also mapped at Fig. 2 (the frequency of occurrence of respective pattern has been mapped against the pattern vector).

Table 3 Weights derived from feature importance scores for embedding

Emotional intensity			Sentiment pattern		
Emotion	Weight	Replication	Pattern	Weight	Replication
Positive	0.158	79	ppn	0.158	111.0
Trust	0.128	64	ppp	0.152	106.0
Anticipation	0.114	57	pnp	0.150	105.0
Joy	0.110	55	nnp	0.147	103.0
Negative	0.110	55	npn	0.117	82.0
Sadness	0.088	44	nnn	0.104	73.0
Surprise	0.084	42	pnnp	0.072	51.0
Fear	0.079	40	nppn	0.063	44.0
Anger	0.066	33	ppnnpp	0.021	14.0
Disgust	0.064	31	nnppnn	0.016	11.0
	Total	500		Total	500

4.3 Emotional Vector

The emotional score for each review was calculated using a lexicon based tool National Research Council Canada (NRC) Lex [12] which is based on the concept given at [14]. The NRC lexicon has approximately 27,000 words and includes WordNet synonyms from the NLTK Library. The probability of occurrence of eight different emotions was extracted as depicted at Fig. 2 and the resulting emotional vector which also included two sentiments positive and negative, was utilised for enriching the BERT embedding (Fig. 3).

4.4 BERT Embedding

We utilised BERT base which has 12 layers and 768 dimensional encoding per token in the hidden layers. We utilised output of the last hidden layer which gives a 768 dimension vector for each token. Since each token in the transformer embedding has contextual information with respect to every other token, we used the starting CLS token as the representational embedding for the entire review. Note that we assumed each review to be limited to 512 tokens, which is the capability of our BERT Base. Incidentally, all our reviews were well within this length. Also, BERT incorporates padding the tokens for this length. Over iterative runs, we found that concatenating BERT CLS with mean-pool of all tokens gave better results for representing entire review as depicted in Table 4. Thus, we had a 1×1536 dimension BERT generated vector representing one review.

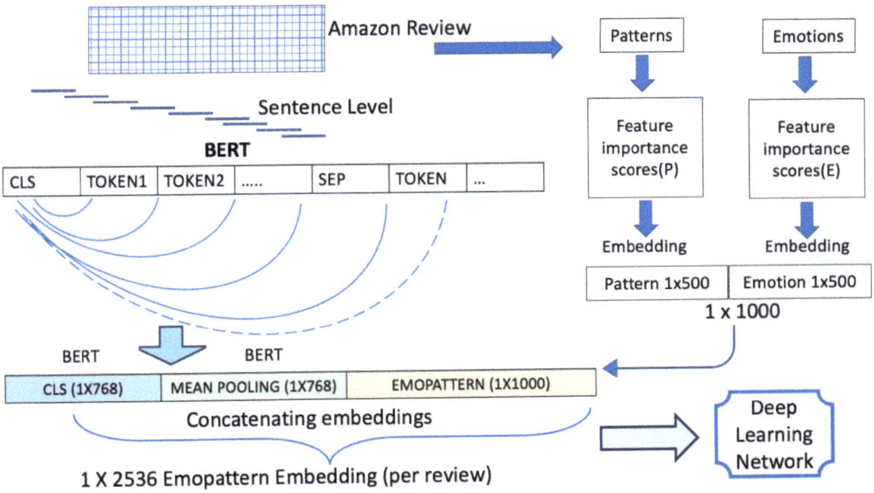

Fig. 3 Enhanced BERT embedding—we utilise feature importance scores for weights; concatenating emotional pattern (sentiment patterns and emotional intensity); we get a 1×2536 dimension vector for each review

4.5 *Enriched Embedding*

We used mutual information scores and also feature importance calculated through random forest regression for earmarking significant features of sentiment patterns and emotions, which were most relevant for predicting helpfulness of a review, as shown at Fig. 4a and b. In addition, we analysed conditional probabilities of occurrence of helpfulness given patterns and also that of helpfulness given emotional intensity as depicted at Tables 1 and 2. The mutual information as as well as the conditional probability is indicative of information in patterns and emotions contributing to helpful assessment of a review. With increase in frequency of patterns present within a review, we found an increasing probability of finding the review helpful. Also, conditional probability of helpfulness given emotions indicates a weak but discernible dependence. We needed to retain context captured by BERT and also ensure that we add on to this information an optimal weight of sentiment patterns and emotions to effectively achieve improved performance. In doing so, we have attempted to search for deeper context. We finally homed on to a vector length of 500 each for patterns and emotions. We used feature replication with each feature replicated as per the weight of feature importance, as depicted at Table 3. We thus had a vector of size 1×1000 which we concatenated with the BERT 1×1536 vector after standardisation, giving a resultant vector of dimension 1×2536 for each review as shown at Fig. 3.

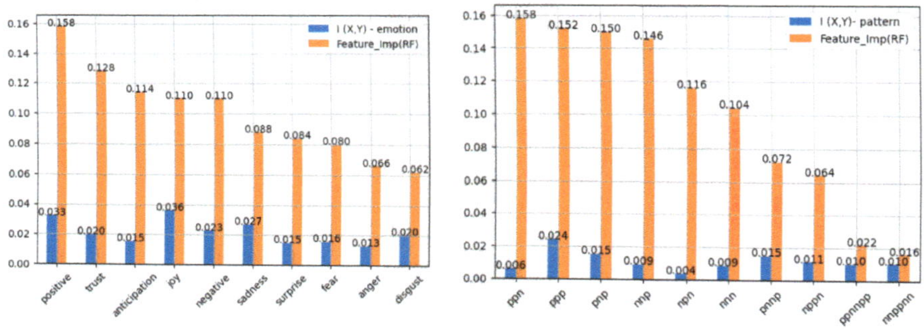

(a) Emotions (b) Patterns

Fig. 4 Mutual information $I(X, Y)$ and feature importance score obtained through random forest regression—after iterative trials, we used the feature importance scores as weights for feature replication (1×500)

4.6 Deep Learning Model and Training

We utilised a seven layer deep learning model with Adam optimiser, with a learning rate of 0.001 and a batch size of 48. For assessing performance, we used a balanced sample of the Amazon Book reviews dataset having train, test, validation split of 60:30:10. The training loss curves for various embedding options are given at Fig. 5. We limited the number of samples to 3600 (balanced, with 1800 each for helpful and unhelpful reviews) and we found that the deep learning model was able to converge within 30–50 epochs with a training size of 2160 samples. We carried out a 5 fold cross validation to evaluate and tune model hyper-parameters.

4.7 Results Achieved

Performance of all embedding options in predicting helpfulness of reviews has been compared at Table 4 and Fig. 6. We found that there was an overall performance improvement of 1.85% in accuracy, 0.5% in precision, 6% in Recall and 3% in the overall F1 Score than original BERT 1536 embedding for predicting helpful reviews. There was no fine tuning needed for BERT base model. We also found that pre-trained BERT sentiment analyzer as well as NRC Lex tools for extracting sentiment patterns and emotional intensity were efficient and utilised minimal resources. Also, we found the model to generalise well on the entire dataset.

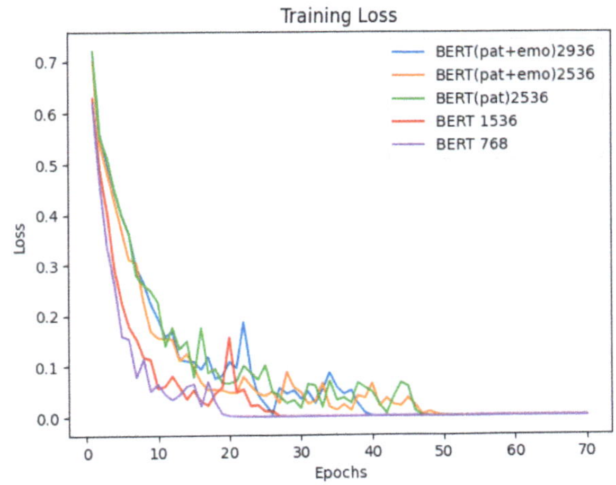

Fig. 5 Training loss while training a 7-layer deep learning model

5 Limitations and Future Work

Our main emphasis in this work has been to explore whether sentiment patterns and emotions extracted from text can yield greater insights and thus be utilised for application-specific NLP tasks. Owing to time and resource constraints, we utilised simplified versions of pre-trained BERT sentiment analyzer, and the lexicon based NRCLex. More refined approaches for accurate assessment of emotions and sentiments can be adopted in the future for a better and more enriched embedding. Also, multiple options of emotional embedding with BERT can be explored for further improving performance. Since emotions are very personal and text based perceptions vary as per culture, future studies can include impact of cultural and language variations. Although, fine tuning BERT takes considerable time and resources, a comparison of performance of fine fine-tuned BERT model for predicting helpful reviews vis-a-vis the approach given in this paper can be undertaken after refining the approach. Also, we have talked about helpful reviews from both buyer's as well as seller's perspective. However, in this work, we have addressed only the former. A detailed analysis after carrying out diligent survey by including sellers can be undertaken to study the impact of looking at reviews from seller's perspective.

6 Conclusion

In this study, we have explored utilisation of emotional information hidden within text for enhancing overall performance of application-specific NLP models. We have demonstrated that emotions can help extract meaningful context, and have proposed a framework which utilises existing sentiment patterns along with emotional intensity for application-specific tasks. The advantage of this approach is that there is

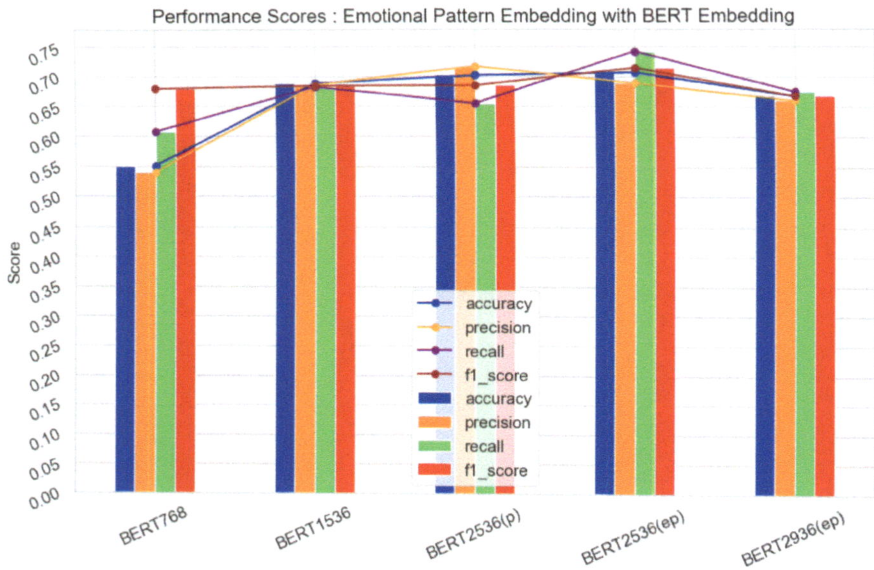

Fig. 6 Performance Scores for various embedding options—we find that BERT 2536(ep) utilising BERT CLS + BERT Mean-pool + Emotion embedding (500) + Sentiment Pattern embedding (500) has better performance (1.85% better accuracy, 0.5% better precision, 6% better Recall and 3% better F1 Score than original BERT 1536 embedding)

Table 4 Performance of BERT embedding and emotionally enriched BERT embedding—the BERT 2536(ep) embedding having BERT CLS(786) + BERT mean-pool (786) + patterns (500) + emotions (500) was found to be the best

Embedding	Structure	Accuracy	Precision	Recall	F1-score
BERT 768	CLS(768)	0.5500	0.5400	0.6075	0.6793
BERT 1536	CLS(768) + Mean-Pool(768)	0.6885	0.6859	0.6825	0.6842
BERT 2536(p)	BERT1536 + Patt(1000)	0.7021	0.7178	0.6550	0.6849
BERT 2536(ep)	BERT1536 + Patt(500) + Emo(500)	0.7071	0.6891	0.7425	0.7148
BERT 2936(ep)	BERT1536 + Patt(700) + Emo(700)	0.6675	0.6602	0.6750	0.6675

Patt pattern, *Emo* emotions

no requirement of fine-tuning complex transformers and also that it has potential for gaining useful insights into latent information present in textual content. We observed performance enhancement of over 3% improvement in F1 Score as compared to traditional attention based approach using BERT, depicted at Fig. 6 and Table 4. This would be promising in specific use case scenarios wherein we want to emphasise on outcomes which are governed by the feelings of a person expressed within text. We have thus been able to successfully propose an approach wherein we add 'feeling' to 'meaning' thereby augmenting the existing options of embedding textual information. Lastly, for future work we propose further exploration and use of similar approaches to extract and utilise emotions within a body of text.

7 Declarations

- Funding—The has been no external funding for this research work.

- Conflict of interest/Competing interests—The authors declare that there is no conflict of interest.
- Ethics approval—Not Applicable.
- Consent to participate—Not Applicable.
- Consent for publication—The authors have no objection to publishing this article.
- Availability of data and materials—available at [11, 18].
- Code availability—available on author's public GitHub repository [19].
- Authors' contributions—conceptualisation, data curation, coding and experimentation, writing- original draft, final manuscript preparation—A.O; guidance, editing and review—B.L.

References

1. Acheampong FA, Nunoo-Mensah H, Chen W (2021) Recognizing emotions from texts using an ensemble of transformer-based language models. In: 2021 18th International computer conference on wavelet active media technology and information processing (ICCWAMTIP), pp 161–164. https://doi.org/10.1109/ICCWAMTIP53232.2021.9674102, ISSN: 2576-8964
2. Acheampong FA, Wenyu C, Nunoo-Mensah H (2020) Text-based emotion detection: advances, challenges, and opportunities. Eng Rep 2(7):e12189
3. Barbieri F, Camacho-Collados J, Neves L, Espinosa-Anke L (2020) TweetEval: unified benchmark and comparative evaluation for Tweet classification. http://arxiv.org/abs/2010.12421, arXiv:2010.12421 [cs]
4. Bilal M, Almazroi AA (2022) Effectiveness of fine-tuned BERT model in classification of helpful and unhelpful online customer reviews. Electron Comm Res. https://doi.org/10.1007/s10660-022-09560-w
5. Bird S (2006) NLTK: the natural language toolkit. In: Proceedings of the COLING/ACL 2006 interactive presentation sessions. Association for Computational Linguistics, Sydney, Australia, pp 69–72. https://doi.org/10.3115/1225403.1225421, https://aclanthology.org/P06-4018

6. Chen X, Cheng Y, Wang S, Gan Z, Wang Z, Liu J (2021) EarlyBERT: efficient BERT training via early-bird lottery tickets. http://arxiv.org/abs/2101.00063, arXiv:2101.00063 [cs]
7. Devlin J, Chang MW, Lee K, Toutanova K (2019) BERT: pre-training of deep bidirectional transformers for language understanding. https://doi.org/10.48550/arXiv.1810.04805, http://arxiv.org/abs/1810.04805, arXiv:1810.04805 [cs]
8. Face H (2023) Cardiff NLP Twitter-RoBERTa-base-sentiment-latest. Hugging Face. https://huggingface.co/cardiffnlp/twitter-roberta-base-sentiment-latest
9. Lu Y, Tsaparas P, Ntoulas A, Polanyi L (2010) Exploiting social context for review quality prediction. In: Proceedings of the 19th international conference on world wide web. ACM, Raleigh North Carolina USA, pp 691–700. https://doi.org/10.1145/1772690.1772761, https://dl.acm.org/doi/10.1145/1772690.1772761
10. Martin L, Pu P (2014) Prediction of helpful reviews using emotions extraction. Proc AAAI Conf Artif Intell 28(1). https://doi.org/10.1609/aaai.v28i1.8937, https://ojs.aaai.org/index.php/AAAI/article/view/8937
11. McAuley J. Amazon review data at Stanford University. https://snap.stanford.edu/data/amazon/productGraph/
12. Metalcorebear: NRCLex: an affect generator based on TextBlob and the NRC affect lexicon. Note that lexicon license is for research purposes only. https://github.com/metalcorebear/NRCLex
13. Mohammad SM, Turney PD (2013) Crowdsourcing a word-emotion association lexicon (2013). http://arxiv.org/abs/1308.6297, arXiv:1308.6297 [cs]
14. Mohammad SM, Turney PD (2013) NRC emotion lexicon. Research, National Research Council of Canada
15. Mudambi SM, Schuff D (2010) Research note: what makes a helpful online review? A study of customer reviews on Amazon.com. MIS Quarterly 34(1):185–200. https://doi.org/10.2307/20721420, https://www.jstor.org/stable/20721420 (publisher: Management Information Systems Research Center, University of Minnesota)
16. Nandwani P, Verma R (2021) A review on sentiment analysis and emotion detection from text. Soc Netw Anal Min 11(1):81. https://doi.org/10.1007/s13278-021-00776-6
17. Ngo-Ye TL, Sinha AP, Sen A (2017) Predicting the helpfulness of online reviews using a scripts-enriched text regression model. Exp Syst Appl 71:98–110. https://doi.org/10.1016/j.eswa.2016.11.029
18. Ni J. Amazon review data at UC San Diego Jacobs School of Engineering. https://cseweb.ucsd.edu/~jmcauley/datasets/amazon_v2/
19. Oberoi A (2023) Amitobe-code/emotional_embedding. https://github.com/amitobe-code/emotional_embedding, original-date: 2023-10-23T05:37:54Z
20. Scherer KR (Dec2005) What are emotions? And how can they be measured? Soc Sci Inf 44(4):695–729. https://doi.org/10.1177/0539018405058216, https://doi.org/10.1177/0539018405058216, publisher: SAGE Publications Ltd
21. Vaswani A, Shazeer N, Parmar N, Uszkoreit J, Jones L, Gomez AN, Kaiser L, Polosukhin I (2023) Attention is all you need (2023). https://doi.org/10.48550/arXiv.1706.03762, http://arxiv.org/abs/1706.03762, arXiv:1706.03762 [cs]

Mitigating Overfitting in Bimodal Data Integration with GLOW for Large and Sparse Datasets

Wentao Yu, Dorothea Kolossa, and Robert Nickel

Abstract The end-to-end training of neural networks with multimodal data poses challenges beyond those observed for the training with unimodal data. The difficulty lies frequently in a network's capacity to overfit and generalize at different rates in each modality, resulting networks may thus perform worse than networks trained on unimodal data alone. In the present study, we introduce a late fusion methodology designed for the bimodal binary classification task under the assumption that the logit spaces of both modalities can be modeled by a joint Gaussian distribution. On the one hand, it is suitable when ample bimodal data is available, avoiding the effect of *catastrophic* fusion, where unimodal models may perform better than the bimodal systems. On the other hand, the proposed technique is particularly well suited for situations in which significant amounts of unimodal data are available to train separate networks for each modality, but where only a limited amount of joint, bimodal data is available to train the fusion mechanism. The proposed method is assessed over the PAN 2018 gender identification dataset, which involves the identification of an authors' (binary) gender label through their tweets and posted images. The cross-validation results demonstrate that our proposed hybrid model-based method exhibits superior performance and robustness compared to competing state-of-the-art neural-network-based approaches.

Keywords Overfitting and generalizing · Joint Gaussian model · Bimodal

https://www.tu.berlin/en/mtec/about.

W. Yu (✉) · D. Kolossa
Electronic Systems of Medical Engineering, TU Berlin, Berlin, Germany
e-mail: wentao.yu@tu-berlin.de

D. Kolossa
e-mail: dorothea.kolossa@tu-berlin.de

R. Nickel
Electrical and Computer Engineering Bucknell University, Lewisburg, USA
e-mail: robert.nickel@bucknell.edu

S. M. Thampi et al. (eds.), *Fifth International Conference on Computing and Network Communications*, Lecture Notes in Electrical Engineering 1221,
https://doi.org/10.1007/978-981-97-4711-5_38

1 Introduction

Multimodal fusion has been extensively explored and discussed in the literature [1–3]. The predominant fusion strategies typically rely on deep learning approaches, where multimodal models are oftentimes constructed by combining multiple unimodal models. These deep-learning-based multimodal models have shown significant performance improvements compared to their unimodal counterparts in various tasks [2–4], as multimodal learning has generally more access to information than unimodal learning.

However, deep-learning-based multimodal models have also encountered unique challenges, suggesting that they may not always be universally more effective than their unimodal counterparts. One of the many challenges in end-to-end multimodal learning is that during the training procedure, the network may either overfit or generalize at different rates for each of the modalities [1]. The performance of the resulting network will thus be suboptimal. The issue becomes even more pronounced when only a limited amount of multimodal training data is available. In addition, many conventional neural-network-based fusion approaches suffer from a time-consuming optimization process. In [4], for example, a 15-day training duration was required, with NVIDIA's Volta-based DGX-1 multi-GPU system with 7 T V100 GPUs, each having 32GB memory, to train a bimodal model using the training set of an audio-visual speech recognition corpus. Extending a trained multimodal model to incorporate additional modalities thus presents a substantial challenge, if it necessitates a retraining of the entire multimodal model. Furthermore, large trainable parameter counts tend to exacerbate overfitting.

Hence, exploring fusion algorithms that efficiently combine different modalities with reduced model complexity is valuable. For that purpose, this study focuses on a scenario where only the *posterior probabilities* and their *ground-truth labels* of each unimodal model are available for integration experiments. For this purpose, we propose the Gaussian Logits glObal stream Weighting (GLOW) method, utilizing posterior probabilities and ground-truth labels from the training set to estimate a global stream weight. This weight is then used to combine the unimodal logits from the test set to generate bimodal posterior probabilities. We investigate the issue of overfitting by comparing the proposed fusion strategy with several bimodal baseline models on differently sized subsets of the training data. When training samples are limited, the neural-network-based baseline models, such as dynamic stream weighting and representation fusion, exhibit overfitting, while the proposed fusion algorithm demonstrates, on average, a superior performance compared to the baseline models. Thus, our proposed model has practical implications, as it can help to significantly reduce the cost of multimodal data annotation. Moreover, the approach requires optimizing only a few parameters, significantly reducing the optimization duration and parameter count compared to neural-network-based fusion approaches. This reduction may be of great value for online multimodal learning tasks. The code is available on GitHub.[1]

---

[1] https://github.com/wentaoxandry/GLOW.git.

The subsequent sections of this paper are structured as follows: Sect. 2 provides the essential background by introducing the unimodal and bimodal baseline models. In Sect. 3, the proposed algorithm is discussed in detail. The experimental setup, presented in Sects. 4 and 5, comprehensively analyzes the performance of all unimodal and bimodal models. Finally, Sect. 6 summarizes the strategies, draws conclusions, and briefly outlines future work.

2 Baseline Models

The proposed algorithm is evaluated using the PAN 2018 author profiling dataset [5]. The dataset consists of a collection of tweets and images posted by individuals who self-identify[2] as either male (with class label $c = 0$) or female (with class label $c = 1$). The goal of the task is to predict the gender of an individual based on their posted collection of tweets and images. In the following, we provide a concise overview of, both, the unimodal and bimodal baseline models we employed for the task. In our notation we generally use bold uppercase letters to denote either matrices or multidimensional arrays and we use bold lowercase letters to denote vectors. Furthermore, we use $\sigma(x)$ to denote the logistic sigmoid function, i.e., $\sigma(x) = \frac{1}{1+e^{-x}}$ and $\sigma^{-1}(x) = \ln\left(\frac{x}{1-x}\right)$. The vectorized version $\sigma(\mathbf{x})$ is assumed to operate in an element-by-element fashion.

In this study, we adopt the BERT model [6] as our ***text model***, owing to its remarkable performance in various text classification tasks. As the PAN 2018 dataset was collected from the Twitter platform, we employ the `BERTweet-large` model in our experiments. The BERTweet model contains a built-in text pre-processing scheme, eliminating the need for additional manual text pre-processing steps in our experiments. We employ the text segmentation strategy outlined in [7] by aggregating all tweets of an author into a single document, which is then segmented into multiple sub-documents, each with a predetermined word length L. To maintain text continuity, the segmentation introduces an overlap of length O between adjacent sub-documents. The text classification is performed on these segmented sub-documents, each represented by a token vector $\mathbf{t}_{k,m}$, where k denotes the author index and m refers to the mth sub-document. The collection of M sub-documents of author k are denoted as $\mathbf{T}_k = [\mathbf{t}_{k,0}, \mathbf{t}_{k,1}, \cdots, \mathbf{t}_{k,M}]$, with identical ground-truth labels $c_{k,m} = c_k \in \{0, 1\}$. The BERTweet model is fine-tuned on our training data and generates posterior probabilities $p(\hat{c}_{k,m}|\mathbf{t}_{k,m})$ for class labels $c_{k,m}$ as well as classification token (CLS) embeddings $\mathbf{r}^t_{k,m} \in \mathbb{R}^{1024\times 1}$. Ultimately, the class prediction $\hat{c}_k$ is obtained through hard voting over all sub-document predictions $\hat{c}_{k,m}$, with $m \in \{0, 1, \cdots, M-1\}$.

The ***image model*** employed in this study is based on a pre-trained vision transformer model (`ViT-large`, [8]). For each author the dataset contains ten

[2] Future work should certainly also consider individuals who do not necessarily self-identify within a binary gender regime. Such a task would require techniques, however, that go beyond the two-class methods considered in this paper.

images they posted on Twitter. These images are resized to 256 × 256 while preserving the RGB channels. The input image set for author k is represented as $\mathbf{I}_k = [\mathbf{I}_{k,0}, \mathbf{I}_{k,1}, \mathbf{I}_{k,2}, \cdots, \mathbf{I}_{k,9}]$, where $\mathbf{I}_{k,n} \in \mathbb{R}^{256\times256\times3}$ for $n \in \{0, 1, \cdots, 9\}$. The ViT model is then utilized to extract image representations $\mathbf{r}^i_{k,n} \in \mathbb{R}^{1000\times1}$ with $\mathbf{r}^i_{k,n} = \text{ViT}(\mathbf{I}_{k,n})$. The author-level image classification representation of the k-th author $\mathbf{r}^i_k$ is obtained by averaging all image representations $\mathbf{r}^i_{k,n}$. Finally, the posterior probability $p(\hat{c}_k|\mathbf{I}_k)$ for class label c_k is computed by passing $\mathbf{r}^i_k$ through a feed-forward classifier. All ViT model parameters are fine-tuned during the training stage.

As mentioned earlier, to emphasize our focus on the fusion strategy, this work primarily considers scenarios where only the posterior probabilities of the unimodal model and their corresponding ground-truth labels are available for fusion. Consequently, the parameters of the unimodal models in the subsequent bimodal baseline models remain frozen.

Stream weighting enjoys a sound information theoretic justification [9]. Numerous studies have employed the stream weighting strategy, yielding remarkable performance improvements compared to their unimodal counterparts [10–12]. The stream weighting approach integrates multiple modalities by a linear combination of the posterior probabilities of each unimodal model with appropriately chosen weights. To incorporate the text segmentation strategy into the bimodal model, all the bimodal models discussed in this study combine the mth text sub-document prediction of author k with the corresponding image prediction of author k, resulting in the m-th bimodal prediction for author k. Subsequently, the final bimodal prediction for author k is derived through a hard-voting process.

The ***fixed stream weighting*** (FSW) method utilizes a grid search, where a fixed stream weight λ is chosen from a search grid between 0 to 1 with a step size of 0.01 on the training set. Using λ, the bimodal posterior probability is obtained through

$$p_f(\hat{c}_{k,m}|\mathbf{o}_{k,m}) = \lambda \cdot p(\hat{c}_{k,m}|\mathbf{t}_{k,m}) + (1-\lambda) \cdot p(\hat{c}_k|\mathbf{I}_k), \tag{1}$$

with $\mathbf{o}_{k,m}$ denoting the combined observation information of the text and image features. The optimal fixed stream weight λ^* is determined by maximizing the accuracy on the training set.

In contrast, ***dynamic stream weighting*** (DSW) is implemented through

$$l(\hat{c}_{k,m}|\mathbf{o}_{k,m}) = \lambda_{k,m}[0] \cdot l(\hat{c}_{k,m}|\mathbf{t}_{k,m}) + \lambda_{k,m}[1] \cdot l(\hat{c}_k|\mathbf{I}_k) \tag{2}$$

with $\lambda_{k,m}[0], \lambda_{k,m}[1] \in [0, 1]$ representing the weights for the first and second modality and where $l(\cdot)$ denotes the logits, as provided by the respective unimodal networks. The weights are estimated using a feed-forward weight estimator (FC) and a sigmoid function $\sigma(\cdot)$

$$[\lambda_{k,m}[0], \lambda_{k,m}[1]]^T = \sigma(\text{FC}([(\mathbf{r}^t_{k,m})^T, (\mathbf{r}^i_k)^T]^T)). \tag{3}$$

The bimodal posterior probability $p_d(\hat{c}_{k,m}|\mathbf{o}_{k,m})$ is obtained by applying a softmax function to the estimated logits $l(\hat{c}_{k,m}|\mathbf{o}_{k,m})$.

This work also considers ***representation fusion*** (RF) as a baseline. Representation fusion aims to integrate high-level bimodal features. The representation fusion baseline obtains the bimodal posterior probabilities by passing the concatenated text and image model representations, namely $\mathbf{r}^t_{k,m}$ and $\mathbf{r}^i_k$, through a feed-forward network. Again, the text representation of the m-th sub-document of author k is concatenated with its corresponding image representation of author k to yield

$$[\, p_r(\hat{c}_{k,m} = 0|\mathbf{o}_{k,m}),\, p_r(\hat{c}_{k,m} = 1|\mathbf{o}_{k,m})\,]^T = \text{softmax}(\text{FC}([(\mathbf{r}^t_{k,m})^T,\ (\mathbf{r}^i_k)^T]^T)). \quad (4)$$

The bimodal prediction of author k is then, again, estimated through hard voting.

As a final baseline, we adopt the ***stream-wise max pooling*** (SMP) fusion strategy used in the winning model of the PAN 2018 author profiling challenge [3]. Here, the fused matrix $\mathbf{G}_{k,m} \in \mathbb{R}^{1024\times 1000}$ is defined as the outer product of the text representation and the image representation: $\mathbf{G}_{k,m} = \mathbf{r}^t_{k,m} \times (\mathbf{r}^i_k)^T$. Text pooling features $\tilde{\mathbf{r}}^t_{k,m} \in \mathbb{R}^{1024\times 1}$ are obtained by performing column-wise max pooling on the fused matrix $\mathbf{G}_{k,m}$. Image pooling features $\tilde{\mathbf{r}}^i_k \in \mathbb{R}^{1000\times 1}$ are obtained via row-wise max pooling on $\mathbf{G}_{k,m}$. The final bimodal posterior probability is then obtained through Eq. (4) with text and image pooling features as inputs.

3 System Overview

With the new method that we are proposing in this section we are building on foundational concepts from Bayesian estimation, fusion, and decision theory [13, 14], in particular logit and probit models [15, 16]. Consider a bimodal binary classification task, i.e., a task where we fuse information about two classes from two discriminators. For simplicity, we use p_1 to denote the posterior probability estimate of classifier 1 for class $c = 0$ and we use p_2 to denote the posterior probability estimate of classifier 2 for class $c = 0$. The commensurate posterior probabilities for class $c = 1$ become $(1 - p_1)$ and $(1 - p_2)$ respectively. Relating this notation back to our discussion from Sect. 2, and to our experiments described in Sect. 4, we have $p_1 = p(\hat{c}_{k,m} = 0|\mathbf{t}_{k,m})$ and $p_2 = p(\hat{c}_k = 0|\mathbf{I}_k)$. By defining

$$\beta_1 = \sigma^{-1}(p_1) \quad \text{and} \quad \beta_2 = \sigma^{-1}(p_2) \quad (5)$$

we can relate the information contained in p_1 and p_2 back into the respective logit-space. In the conventional fixed stream weighting approach, we estimate an optimal constant coefficient λ for the mixing of the probability values. In our proposed ***Gaussian Logits glObal stream Weighting*** (GLOW) approach, we conduct an optimization by, alternatively, estimating a weight α in the β-space with

$$\beta_\alpha = \alpha \cdot \beta_1 + (1 - \alpha) \cdot \beta_2 \quad \text{and} \quad \alpha \in [0, 1], \quad (6)$$

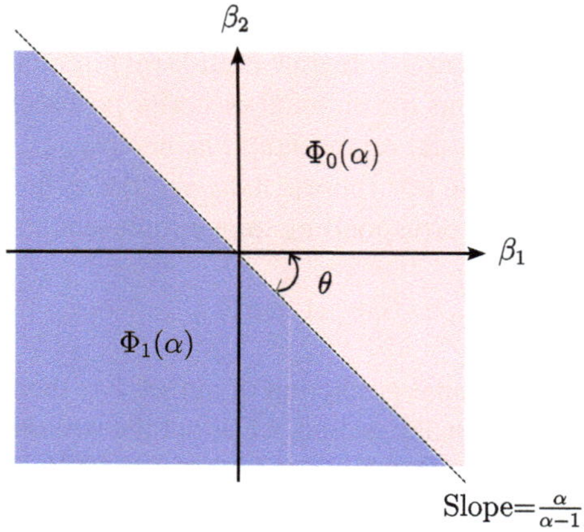

Fig. 1 Joint distribution space of β_1 and β_2

where β_1 and β_2 represent the logits obtained from classifiers 1 and 2 respectively after Eq. (5). Given β_α, we decide on class $c = 0$ if $\beta_\alpha \geq 0$ and we decide on class $c = 1$ if $\beta_\alpha < 0$, which is commensurate with the *fused probability* $p_\alpha = \sigma(\beta_\alpha)$ for class $c = 0$ being greater or smaller than $\frac{1}{2}$ respectively. We can calculate the accuracy $\mathrm{A}(\alpha)$, i.e., the probability of correctly classified predictions, of the proposed scheme as a function of α as follows:

$$\mathrm{A}(\alpha) = P[\beta_\alpha \geq 0|c = 0] \cdot P[c = 0] + P[\beta_\alpha < 0|c = 1] \cdot P[c = 1]. \tag{7}$$

Figure 1 depicts the commensurate decision regions $\Phi_0(\alpha)$ and $\Phi_1(\alpha)$ in $[\beta_1, \beta_2]$-space. Considering the conditions $\beta_\alpha \gtrless 0$, we can easily obtain a line with slope $\frac{\alpha}{\alpha-1}$ to separate the two regions $\Phi_0(\alpha)$ (light red area) and $\Phi_1(\alpha)$ (dark blue area): Samples in $\Phi_0(\alpha)$ are classified as class $c = 0$, and samples in $\Phi_1(\alpha)$ are classified as class $c = 1$. Based on the illustration in Fig. 1, the conditional probabilities from Eq. (7) can be calculated with

$$\begin{aligned} P[s_c \cdot \beta_\alpha > 0|c] &= P[s_c \cdot (\alpha \cdot \beta_1 + (1-\alpha) \cdot \beta_2) > 0|c] \\ &= \int\limits_{\Phi_c(\alpha)} f(\beta_1, \beta_2|c)\, d\beta_1 d\beta_2 \quad \text{for} \quad c \in \{0, 1\}, \end{aligned} \tag{8}$$

in which the $f(\beta_1, \beta_2|c)$ functions denote the respective conditional probability density functions in $[\beta_1, \beta_2]$-space and $s_c = +1$ for $c = 0$ and $s_c = -1$ for $c = 1$.

Calculating the above integrals in closed form would, in general, be a daunting task. However, both β_1 and β_2 are logit values and most neural networks estimate logit values through a linear combination of a sizeable number of hidden state variables. We may, thus, appeal to the central limit theorem to argue that β_1 and β_2 are reasonably well modeled by a Gaussian probability distribution [3]:

[3] Here, the central limit theorem does technically not support the assumption of joint Gaussianity in this case, but we assume joint Gaussianity for practical reasons.

$$f(\beta_1, \beta_2|c) = \frac{1}{2\pi} \cdot \frac{1}{\sqrt{\det(\mathbf{Q}_c)}} \cdot \exp[-\frac{1}{2}(\boldsymbol{\beta} - \boldsymbol{\mu}_c)^T \mathbf{Q}_c^{-1}(\boldsymbol{\beta} - \boldsymbol{\mu}_c)] \tag{9}$$

with $\boldsymbol{\beta} = \begin{bmatrix} \beta_1 \\ \beta_2 \end{bmatrix}$, $\boldsymbol{\mu}_c = \begin{bmatrix} \mu_1^c \\ \mu_2^c \end{bmatrix}$, and $\mathbf{Q}_c = \begin{bmatrix} (\sigma_1^c)^2 & \sigma_1^c \sigma_2^c \varrho^c \\ \sigma_1^c \sigma_2^c \varrho^c & (\sigma_2^c)^2 \end{bmatrix}$ for $c \in \{0, 1\}$.

The mean vector $\boldsymbol{\mu}_c$ and the covariance matrix $\mathbf{Q}_c$ can be estimated from the posterior probabilities p_1 and p_2 of the samples that belong to class c after they have been transformed to the $[\beta_1, \beta_2]$-space through Eq. (5). To simplify the integration over regions $\Phi_0(\alpha)$ and $\Phi_1(\alpha)$ we introduce a change of variables in Eq. (8) that rotates the coordinate system from $\boldsymbol{\beta}$ to $\boldsymbol{\beta}'$:

$$\boldsymbol{\beta}' = \begin{bmatrix} \beta_1{}' \\ \beta_2{}' \end{bmatrix} = \mathbf{R} \begin{bmatrix} \beta_1 \\ \beta_2 \end{bmatrix} = \mathbf{R}\boldsymbol{\beta} \quad \text{with} \quad \mathbf{R} = \begin{bmatrix} \cos\theta & -\sin\theta \\ \sin\theta & \cos\theta \end{bmatrix}. \tag{10}$$

Matrix $\mathbf{R}$ rotates the coordinate system counterclockwise by an angle of θ. When we choose $\tan(\theta) = \frac{\alpha}{1-\alpha}$ then the integrals in Eq. (8) can be rewritten as

$$\int\limits_{\Phi_0(\alpha)} f(\beta_1, \beta_2|c = 0) d\beta_1 d\beta_2 = \int\limits_0^{\infty} \int\limits_{\infty}^{-\infty} f'(\beta_1{}', \beta_2{}'|c = 0) d\beta_1{}' d\beta_2{}', \tag{11}$$

and

$$\int\limits_{\Phi_1(\alpha)} f(\beta_1, \beta_2|c = 1) d\beta_1 d\beta_2 = \int\limits_{-\infty}^{0} \int\limits_{\infty}^{-\infty} f'(\beta_1{}', \beta_2{}'|c = 1) d\beta_1{}' d\beta_2{}'. \tag{12}$$

The inner integrals of Eqs. (11) and (12) evaluate as a marginal of a Gaussian density and the outer integrals simply become integrals over a one-dimensional Gaussian density. In summary, we obtain

$$P[\beta_\alpha > 0|c = 0] = \frac{1}{2} + \frac{1}{2}\text{erf}\left(\frac{\alpha\mu_1^0 + (1-\alpha)\mu_2^0}{\sqrt{2}\sqrt{\alpha^2(\sigma_1^0)^2 + (1-\alpha)^2(\sigma_2^0)^2 + 2\alpha(1-\alpha)\sigma_1^0\sigma_2^0\varrho^0}}\right), \tag{13}$$

where μ_1^0, μ_2^0, σ_1^0, σ_2^0 and ϱ^0 can be obtained from the mean $\boldsymbol{\mu}_0$ and covariance $\mathbf{Q}_0$ of the set of all training samples $\boldsymbol{\beta}$ that belong to class $c = 0$, and

$$P[\beta_\alpha < 0|c = 1] = \frac{1}{2} - \frac{1}{2}\text{erf}\left(\frac{\alpha\mu_1^1 + (1-\alpha)\mu_2^1}{\sqrt{2}\sqrt{\alpha^2(\sigma_1^1)^2 + (1-\alpha)^2(\sigma_2^1)^2 + 2\alpha(1-\alpha)\sigma_1^1\sigma_2^1\varrho^1}}\right), \tag{14}$$

where μ_1^1, μ_2^1, σ_1^1, σ_2^1 and ϱ^1 can be estimated from the mean $\boldsymbol{\mu}_1$ and covariance $\mathbf{Q}_1$ of the set of all training samples $\boldsymbol{\beta}$ that belong to class $c = 1$.

The fixed optimal weight α^* is found during training by maximizing the accuracy $\mathrm{A}(\alpha)$ from Eq. (7). For our experiments we assume $P[c = 0] = P[c = 1]$. Thus, α^* can be calculated through:

$$\begin{aligned} \alpha^* &= \arg\max_{\alpha} \mathrm{A}(\alpha) = \arg\max_{\alpha} \left(P[\beta_\alpha > 0|c = 0] + P[\beta_\alpha < 0|c = 1]\right) \\ &= \arg\max_{\alpha} \left(\operatorname{erf}\left(\frac{\alpha\mu_1^0 + (1-\alpha)\mu_2^0}{\sqrt{2}\sqrt{\alpha^2(\sigma_1^0)^2 + (1-\alpha)^2(\sigma_2^0)^2 + 2\alpha(1-\alpha)\sigma_1^0\sigma_2^0\varrho^0}} \right) \right. \\ &\left. -\operatorname{erf}\left(\frac{\alpha\mu_1^1 + (1-\alpha)\mu_2^1}{\sqrt{2}\sqrt{\alpha^2(\sigma_1^1)^2 + (1-\alpha)^2(\sigma_2^1)^2 + 2\alpha(1-\alpha)\sigma_1^1\sigma_2^1\varrho^1}} \right) \right). \end{aligned} \tag{15}$$

Finally, after we obtained the fixed optimal α^* during training, we can calculate the fused prediction probability $p_g(\hat{c}_{k,m}|\mathbf{o}_{k,m})$ as the result of our GLOW approach from:

$$p_g(\hat{c}_{k,m} = 0|\mathbf{o}_{k,m}) = \sigma(\alpha^* \cdot \beta_1 + (1 - \alpha^*) \cdot \beta_2) \tag{16}$$

$$\text{and} \quad p_g(\hat{c}_{k,m} = 1|\mathbf{o}_{k,m}) = 1 - p_g(\hat{c}_{k,m} = 0|\mathbf{o}_{k,m}). \tag{17}$$

Similarly to the process used for our baseline systems, the bimodal prediction of author k in our experiments is, again, estimated through hard voting on the individual decisions for $\hat{c}_{k,m}$ over all m.

To assess the efficacy of our proposed model, we conducted an additional experiment, referred to as the ***Naive Logit Space Optimization*** (NLSO) approach. This approach closely resembles the GLOW method, wherein posterior probabilities for each modality are transformed into the β-space using Eq. (5), followed by the optimization of an α parameter for fusion within the β-space, as defined in Eq. (16). The distinction lies in the fact that GLOW estimates the probability of correctly classified predictions by assuming that the samples from each modality in β-space conform to a joint Gaussian distribution, whereas NLSO directly computes the accuracy on the training set.

In the NLSO optimization phase, fusion in β-space is conducted through Eq. (6). The bimodal probability of class $c = 0$ is computed as $p_n = \sigma(\beta_\alpha)$ and the probability of class $c = 1$ becomes $(1 - p_n)$, respectively. Consequently, the accuracy $\mathrm{A}_{\mathrm{NLSO}}(\alpha)$ can be calculated based on the predicted bimodal probabilities p_n and the ground-truth labels. The optimal α^*_{NLSO} is determined through a grid search using

$$\alpha^*_{\mathrm{NLSO}} = \arg\max_{\alpha} \mathrm{A}_{\mathrm{NLSO}}(\alpha) \quad \text{on the training set.} \tag{18}$$

4 Experimental Setup

This study focuses on author gender identification using the PAN 2018 dataset. The dataset is categorized based on the language of the tweets, namely English, Arabic, and Spanish. For this work, only the English language subset is considered. The English dataset consists of a training set with 3000 author samples and a test set with 1900 author samples. Each author sample includes 100 tweets and 10 posted images. The text model is optimized using the text re-segmentation strategy, where the sub-documents are segmented with a word length of $L = 500$ and an overlapping length of $O = 128$.

All neural networks in this study were trained using the PyTorch library.[4] To prevent overfitting, early stopping was employed if the evaluation accuracy did not improve for four consecutive epochs. For the text and image models, the AdamW optimizer was used to optimize the parameters of the BERT model. The learning rate was scheduled using the `get_linear_schedule_with_warmup` from the transformer model, with a warm-up period of four epochs and an upper limit of 2e-5. The bimodal baseline models were optimized with the SGD optimizer, and the learning rate was scheduled using `ReduceLROnPlateau`, reducing it by a factor of 0.1 if the evaluation accuracy did not improve. The training of all neural networks was performed on a single GPU.

5 Results

Table 1 presents the accuracies of our ***unimodal*** models compared to the best-performing unimodal models from the PAN 2018 author profiling challenge. The text-only model in this study, incorporating the document re-segmentation strategy, demonstrates superior performance compared to the winning text-only model in the PAN 2018 challenge [17]. The winning text-only model employed n-gram term frequency-inverse document frequency (tf-idf), applied a truncated singular-value decomposition (SVD) for feature dimension reduction and utilized a linear Support Vector Machine (SVM) for classification. On the other hand, the image-only model in this work performs slightly worse than the top-performing image-only model in the PAN 2018 challenge [3], which utilized a pre-trained VGG-16 model for extracting high-level image representations. The author-level representation was obtained by averaging 10 image representations for each author. We were, unfortunately, unable to replicate the performance achieved by the winning image-only model, for which no source code was available. As a result, the subsequent experiments in this study were conducted utilizing the image-only model as detailed in Sect. 2.

For our ***bimodal*** experiments, Fig. 2a and b show scatter plots of $[\beta_1, \beta_2]$-space samples for the training and testing sets, respectively (as theoretically depicted in Fig. 1), to visually demonstrate the proposed GLOW approach. In these figures, the

[4] https://github.com/pytorch/pytorch.

Table 1 Unimodal model accuracy compared with the winning model of the PAN 2018 challenge

	Our model	PAN 2018 top unimodal model
Text-only	**0.8452**	0.8221
Image-only	0.8057	**0.8163**

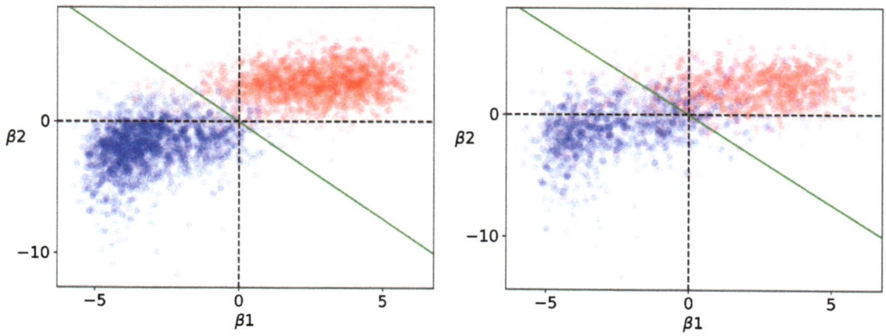

(a) Training set (b) Testing set

Fig. 2 $\boldsymbol{\beta}$-space sample distribution

red points represent samples annotated as class 0 ($\Phi_0(\alpha)$ in Fig. 1), while the blue points represent class 1 ($\Phi_1(\alpha)$ in Fig. 1). The dashed lines indicate the β_1- and β_2-axis, while the oblique line represents the estimated slope obtained through $\frac{\alpha^*}{\alpha^*-1}$. Figure 2a illustrates that the distribution of training samples in $[\beta_1, \beta_2]$-space is effectively separated using the estimated decision boundary. Figure 2b demonstrates that the test samples in $[\beta_1, \beta_2]$-space are, equally, appropriately separated into their respective classes by the identical decision boundary. It can, furthermore, be seen in Fig. 2a and b that our assumption of joint Gaussianity is not strictly satisfied. The proposed algorithm, nevertheless, performs well, as is evident from the accuracies reported in Table 2.

Table 2 shows the results of our experimental performance comparison between the proposed GLOW approach, the NLSO approach, the baseline methods described in Sect. 2 (FSW, SMP, RF, and DSW), as well as the result reported for the winning bimodal model from the PAN 2018 challenge, listed as PAN [3]. The methods were compared with respect to decision accuracy (Acc.), the number of trainable parameters (#params), and the expected calibration error (ECE) [18]. ECE is calculated with

$$\text{ECE} = \sum_{m=1}^{M} \frac{|B_m|}{n} \left|\text{acc}(B_m) - \text{conf}(B_m)\right|, \tag{19}$$

where n denotes the total number of samples, with M signifying the number of bins employed. The set B_m characterizes the indices of samples whose predictive confidence aligns with the m-th bin interval, where $m \in \{1, \cdots, M\}$. The functions acc(·)

Table 2 The accuracy (Acc.), the number of trainable parameters of the fusion model (#params), the expected calibration error (ECE), the time duration (in seconds) required for the training (T_t) and evaluation (T_e) processes across all examined models

Model	Acc.	#params	ECE	T_t	T_e	Model	Acc.	#params (k)	ECE	T_t	T_e
GLOW	0.8789	**1**	**3.36**	**0.058**	**0.002**	RF	**0.8794**	2027	6.55	241	3
NLSO	0.8726	**1**	3.92	0.331	**0.002**	DSW	0.8731	2027	4.82	144	3
FSW	0.8684	**1**	5.88	120	3	PAN [3]	0.8584	–	–	–	–
SMP	0.8789	2027k	8.02	241	3						

GLOW: proposed Gaussian logits global stream weighting; *FSW*: fixed stream-weighting; *NLSO*: naive logit space optimization; *SMP*: stream-wise max pooling; *RF*: representation fusion; *DSW*: dynamic stream-weighting; *PAN* [3]: the winning bimodal model of the PAN 2018 challenge

and conf(·) denote the accuracy and the average prediction confidence associated with the set B_m, respectively.

It is interesting to note that all of the accuracy results were roughly within the same ballpark. The method with the lowest accuracy was the winning bimodal model from the PAN 2018 challenge (PAN [3]). A factor that might explain the poorer performance is, that the unimodal models employed in PAN [3] were not the same as the ones we used for all the other models (see Table 1). We therefore decided to 'replicate' the PAN [3] approach, but with our unimodal models as discussed in Sect. 2. The resulting SMP approach performed on par with our other methods. The second poorest performance with respect to accuracy was found for the FSW method. This is not surprising since many studies have highlighted the superiority of dynamic techniques of stream weighting over fixed stream weighting, especially for the described DSW approach in [19, 20]. The accuracies for the techniques, GLOW, NLSO, SMP, RF, and DSW were all very similar to each other, with the proposed GLOW method coming in a close second after the RF method. The differences between those methods, however, are so small that it is not even clear if they are statistically significant.

A much more important metric for the benefit of the proposed GLOW method, when trained on large size data sets, is *not* provided by accuracy measures, but by its performance with respect to the ECE. The calibration error measures roughly how much a model's *estimate* for a class probability deviates from the *actual* probability in the given scenario. In the expected calibration error the GLOW method outperforms all other methods by a significant margin, with the SMP method coming in last. Again, the good performance of GLOW in ECE is commensurate with the fact that it has been derived directly from underlying probabilistic principles.

In addition, it is valuable to consider the number of trainable parameters required for each fusion model. This work considers a scenario in which only the posterior probabilities, ground-truth labels, and/or the respective modality representation vectors ($\mathbf{r}^t$ and $\mathbf{r}^i$) of the unimodal data are available for bimodal training. Therefore, we only calculate the number of trainable parameters involved in the bimodal fusion models, excluding the unimodal parameters from consideration. In the proposed GLOW model, optimization is performed for a single parameter, α, which contrasts with the neural-network-based baseline models (DSW, RF, and SMP) that require the training of over 2027 k parameters for a classifier or weight estimator. Another advantage of the proposed GLOW model is that it estimates a single weight from the training set in a single step. This differs from neural-network-based models such as dynamic stream weighting or representation fusion, which require multiple epochs to optimize the fusion model, see Table 2 (T_t column). Therefore, the proposed approach further accelerates optimization time. In terms of inference, fixed stream weighting methods offer an advantage compared to neural network-based models, requiring less computational resources during inference, as illustrated in Table 2 (T_e column). The proposed model necessitates only the computation of the mean value and covariance matrix in betaspace, both of which can be estimated recursively with new incoming data and label pairs. As such, the optimal mixture coefficient α^* can be updated in an adaptive fashion, making the method suitable for online learning tasks as well.

Lastly, we investigated the NLSO, FSW, and RF model's propensity for overfitting by comparing the performance of different models with varying sizes of bimodal training data, while keeping the pre-trained unimodal models fixed. The sub-training sets used in this investigation comprise 16, 32, 64, 128, and 256 author samples, respectively. These sub-training sets are selected randomly from the overall training set. To comprehensively evaluate each model's robustness, the sub-training sets are independently selected ten times for each specific sample size. By doing so, we can gauge a model's performance across various training set sizes.

Figure 3 compares the performance of two global stream weighting approaches, NLSO and GLOW. It is observed that the proposed GLOW model exhibits greater robustness and outperforms the NLSO model in average on the sub-training sets. To further shed light on the performance difference, we visualized a sample distribution in $[\beta_1, \beta_2]$-space, utilizing the GLOW and NLSO approaches to classify the author samples into their pre-defined classes in Fig. 4a and b. Figure 4a illustrates the sub-training set comprising 256 author samples in $[\beta_1, \beta_2]$-space, where different decision boundaries separate the two data groups. Notably, the GLOW decision boundary demonstrates superior separation of the data samples.

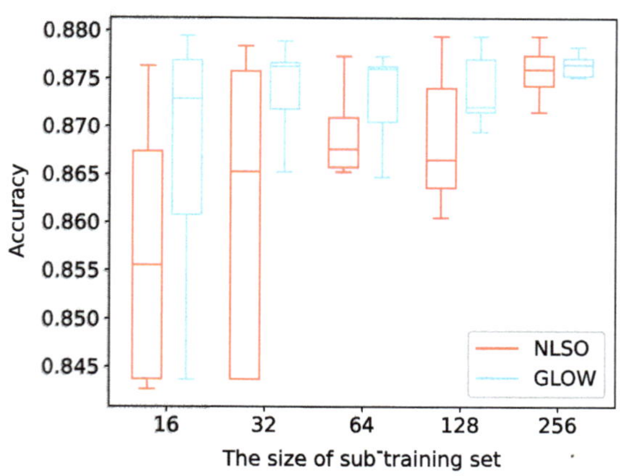

Fig. 3 Accuracy of NLSO and GLOW across different training set sizes

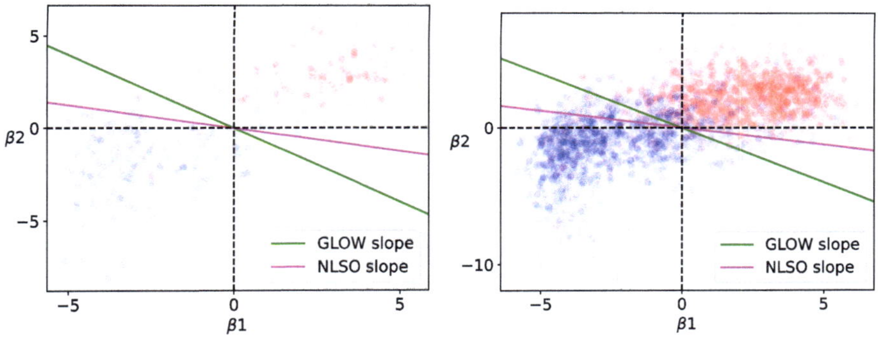

(a) Sub-training set in β-space (b) Test set β-space samples

Fig. 4 Decision boundary (slope) difference between GLOW and NLSO for subset training with 256 author samples only

Fig. 5 Accuracy of FSW and GLOW across different training set sizes

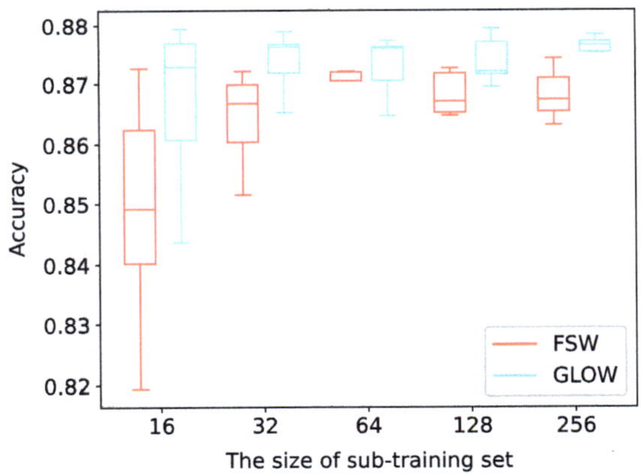

Figure 4b presents the test set author samples with the same decision boundaries as depicted in Fig. 4a, reaffirming the superior performance of the GLOW approach compared to the NLSO method. We attribute the improvement to the fact that GLOW estimates the distribution through the mean vector and covariance matrix of the samples in the $[\beta_1, \beta_2]$-space under the joint Gaussianity assumption, thereby introducing a type of regularization. Thus, even in scenarios with limited training data, the estimated distribution can still effectively represent the data distribution, supporting generalization of the fusion model. In contrast, NLSO and FSW estimate weights solely based on the accuracy of the training samples, without the assumption of a distribution model, resulting in a loss of generalization capability. The same reason can explain why GLOW is more robust than FSW when only 16 or 32 training samples are available, as shown in Fig. 5.

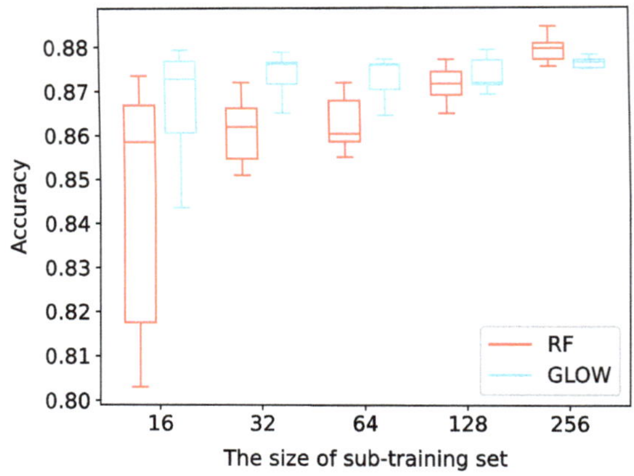

Fig. 6 Accuracy of RF and GLOW across different training set sizes

Figure 6 presents the performance comparison between the proposed GLOW model and the RF model across different sub-training sizes. In scenarios where a limited number of training samples are available, the GLOW model clearly demonstrates superior performance in terms of both accuracy and robustness when compared to the RF model.

6 Conclusion and Discussion

This study introduces a novel fusion strategy to overcome the challenges posed by overfitting and generalization in neural-network-based bimodal models. The proposed approach shows promising results, whether the available bimodal data is limited or abundant, as long as the unimodal constituent models are trained on a substantial amount of data. Our new approach, termed Gaussian Logits glObal stream Weighting (GLOW), achieves this by estimating a constant stream weight from the (potentially limited) bimodal training set. During inference, the estimated stream weight is then used for modality integration in the logit space.

When compared to other state-of-the-art bimodal stream weighting baseline models that were trained on the entire training set, the proposed GLOW model achieves superior performance in terms of a small expected calibration error while exhibiting reduced complexity during the learning process and a comparable accuracy to the state-of-the-art systems. Our cross-validation results, furthermore, consistently show the superior performance of the GLOW model in terms of accuracy and robustness when compared to the neural-network-based baseline fusion models trained on small sub-training sets. The GLOW algorithm thus demonstrates its efficacy in mitigating overfitting issues when trained on limited bimodal data, establishing the GLOW model as a robust solution for bimodal application scenarios, where joint data is scarce.

In our current work, we specifically focus on bimodal binary classification. The next objective is to extend the GLOW framework to multimodal binary classification tasks, as well as to bimodal multi-class classification tasks. Results on these scenarios, however, will have to be deferred to upcoming papers.

References

1. Wang W, Tran D, Feiszli M (2020) What makes training multi-modal classification networks hard? In: IEEE/CVF conference on computer vision and pattern recognition (CVPR)
2. Yu W, Zeiler S, Kolossa D (2021) Fusing information streams in end-to-end audio-visual speech recognition. In: IEEE international conference on acoustics, speech and signal processing. IEEE, pp 3430–3434
3. Takahashi T, Tahara T, Nagatani K, Miura Y, Taniguchi T, Ohkuma T (2018) Text and image synergy with feature cross technique for gender identification. In: Working notes papers of the CLEF, vol 10
4. Yu W, Zeiler S, Kolossa D (2021) Large-vocabulary audio-visual speech recognition in noisy environments. In: IEEE 23rd international workshop on multimedia signal processing. IEEE, pp 1–6 (2021)
5. Rangel F, Rosso P, Gómez MM, Potthast M, Stein B (2018) Overview of the 6th author profiling task at PAN 2018: multimodal gender identification in twitter. In: Working notes papers of the CLEF, pp 1–38
6. Devlin J, Chang M-W, Lee K, Toutanova K (2018) Bert: pre-training of deep bidirectional transformers for language understanding. arXiv preprint arXiv:1810.04805
7. Yu W, Boenninghoff B, Kolossa D (2022) BERT-based ironic authors profiling
8. Dosovitskiy A, Beyer L, Kolesnikov A, Weissenborn D, Zhai X, Unterthiner T, Dehghani M, Minderer M, Heigold G, Gelly S, et al. (2020) An image is worth 16x16 words: transformers for image recognition at scale. arXiv preprint arXiv:2010.11929
9. Kulhavý R, Kraus FJ (1996) On duality of exponential and linear forgetting. IFAC Proc 29(1):5340–5345
10. Gurban M, Thiran J-P, Drugman T, Dutoit T (2008) Dynamic modality weighting for multi-stream HMMs in audio-visual speech recognition. In: Proceeding of the 10th international conference on multimodal interfaces, pp 237–240
11. Heckmann M, Berthommier F, Kroschel K (2002) Noise adaptive stream weighting in audio-visual speech recognition. EURASIP J Adv Sig Proc 1–14:2002
12. Nefian Ara V, Liang L, Pi X, Liu X, Murphy K (2002) Dynamic Bayesian networks for audio-visual speech recognition. EURASIP J Adv Sig Proc 1–15:2002
13. Christopher M (2006) Bishop. Springer, Pattern recognition and machine learning. Information Science and Statistics Series
14. Koliander G, El-Laham Y, Djuric PM, Hlawatsch F (2022) Fusion of probability density functions. Proc IEEE 110(4)
15. Aldrich JH, Nelson FD, Scott Adler E (1984) Linear probability, logit, and probit models. Sage Publications, Inc
16. Lu YG, Mao, Li C (2021) Few-shot learning for named entity recognition based on BERT and two-level model fusion. Data Intell 3(4):568–577
17. Daneshvar S, Inkpen D (2018) Gender identification in Twitter using N-grams and LSA. In: Proceeding of the 9th international conference of the CLEF association. CEUR-WS
18. Naeini MP, Cooper G, Hauskrecht M (2015) Obtaining well calibrated probabilities using Bayesian binning. In: Proceedings of the AAAI conference on artificial intelligence, vol 29

19. Bellegarda JR (2009) A dynamic cost weighting framework for unit selection text-to-speech synthesis. IEEE Trans Audio, Speech, Lang Proc 18(6):1455–1463
20. Meutzner H, Ma N, Nickel R, Schymura C, Kolossa D (2017) Improving audio-visual speech recognition using deep neural networks with dynamic stream reliability estimates. In: IEEE international conference on acoustics, speech and signal processing. IEEE, pp 5320–5324

Title-Based Topic Modeling on E-learning Web Content Titles Using BERTopic Model

T. B. Lalitha and P. S. Sreeja

Abstract In today's information-rich digital age, the volume of web content available to users has become overwhelming, making it challenging for individuals to find relevant and personalized content. Recommendation systems have emerged as a transformative solution, catering to individual users by offering customized suggestions aligned with their unique interests. This research explores a novel approach that utilizes topic modeling techniques on web content titles for recommendation purposes. Topic modeling, a subfield of natural language processing (NLP) is utilized to automatically identify latent topics within a large corpus of text. The proposed work begins by collecting a diverse dataset of web content titles across the domains. It employs a combination of other state-of-the-art topic modeling algorithms like BERTopic modeling and statistical model to uncover the underlying topics in the titles. By leveraging this approach on web content titles, aim to extract meaningful themes and categorize the content efficiently. Then preprocess the data to remove irrelevant information, ensuring that the subsequent topic modeling process yields accurate and meaningful results. This approach not only expedites the recommendation process but also conserves computational resource. Once the topics are identified, associate them with appropriate metadata, such as user preferences, and content types. This step forms the foundation of our content-based recommendation approach. Then maps the user's interests to the most relevant topics, enabling us to present a tailored list of web content titles. By recommending content based on underlying themes rather than just keywords, this approach surpasses traditional methods, ensuring more accurate and diverse suggestions for users. The results demonstrate the system's ability to provide highly personalized recommendations, enhancing user satisfaction and engagement. By delving into the semantic structure of content rather than relying

T. B. Lalitha (✉)
Department of Computer Applications, Hindustan Institute of Technology and Science, Chennai, India
e-mail: lalitha.srm@gmail.com

P. S. Sreeja
Department of Computer Applications, Vellore Institute of Technology, Chennai, India

S. M. Thampi et al. (eds.), *Fifth International Conference on Computing and Network Communications*, Lecture Notes in Electrical Engineering 1221,
https://doi.org/10.1007/978-981-97-4711-5_39

solely on metadata, offers a dynamic and effective solution to tackle information overload and deliver truly personalized web content recommendations that are tailored to the users' interest.

Keywords Topic modeling · Recommendation systems · Personalized content · BERTopic modeling · Natural language processing (NLP) · Web content

1 Introduction

In the vast realm of textual information that flooding our information-driven world, overwhelming amount of textual data is generated and consumed daily. Whether its news articles, academic research papers, social media posts, customer reviews, or product descriptions, the sheer volume of text can make it challenging to extract meaningful descriptions efficiently. The need to efficiently organize, categorize, and retrieve this information becomes increasingly crucial. One powerful tool that has emerged as a beacon of light in the sea of text to address this challenge is topic modeling.

Topic modeling is an approach within the field of natural language processing (NLP) and machine learning, designed to unveil and extract the central subjects or concepts that pervade throughout a compilation of documents. It is popular unsupervised learning method used for text analysis and information retrieval. The goal of topic modeling is to identify topics within a corpus of text without prior knowledge of what those topics might be. It provides a way to organize and understand large volumes of textual data by automatically grouping related documents and assigning topics to each document based on words and patterns it contains [1]. It is particularly useful for understanding large volumes of unstructured textual data and extracting meaningful insights extensively supports decision-making and understanding.

1.1 The Significance of Topic Modeling in Web Content Titles

Document titles serve as gateways to the content they encapsulate, offering a concise representation of the core theme or subject matter. They play a vital role in capturing the essence of the document and guiding readers' expectations. However, as the volume of documents grows, it becomes increasingly difficult to manually categorize and organize them effectively. This is where title-based topic modeling steps in as a valuable tool. Topic modeling in web mining refers to the application of topic modeling techniques to analyze and extract meaningful topics from web-related data, including web pages, documents, URLs, and other online content [2]. Web mining involves discovering patterns and knowledge from the large data available on the internet. Topic modeling enhances web mining by providing a structured way to understand and to unearth hidden patterns and themes within document titles, enabling efficient retrieval, organization, and analysis [3]. By automatically identifying the underlying themes and topics within titles, can streamline the process of document management and provide users with more relevant and meaningful access to information.

1.2 Topic Modeling Techniques

In the domains of statistics and natural language processing, a topic model represents a statistical technique designed to unveil the underlying 'topics' present within a corpus of documents. Topic modeling techniques are algorithms and methods used to uncover latent thematic structures within a collection of documents [4]. These techniques help in identifying and extracting hidden topics or themes that frequently occur across the documents. Here are some prominent topic modeling techniques:

- **Latent Dirichlet Allocation (LDA)** [5, 6]: LDA is one of the most widely used and foundational topic modeling algorithms. This method operates under the assumption that every document comprises a blend of a few distinct topics, with each topic being characterized by a distribution across words. LDA aims to reverse-engineer this process to discover the underlying topics and their associated word distributions.
- **Non-negative Matrix Factorization (NMF)** [7, 8]: NMF is a matrix factorization technique that decomposes the term-document matrix into two lower-dimensional matrices, one representing topics and the other representing document-topic associations. It ensures that all matrix elements are non-negative, which is particularly useful for interpreting the resulting topics.
- **Latent Semantic Analysis (LSA)** [7]: LSA uses singular value decomposition (SVD) to reduce the dimensions of the term-document matrix and extract latent semantic relationships between words and documents. While not explicitly designed for topic modeling, LSA can be used for discovering related concepts.
- **Probabilistic Latent Semantic Analysis (pLSA)** [7]: Similar to LDA, pLSA is a probabilistic variant of LSA. It models the generation of a document as a probabilistic mixture of topics and aims to capture the joint distribution of words and documents.
- **Correlated Topic Model (CTM)** [9]: CTM extends LDA by allowing topics to be correlated. This can better capture the relationships between topics and potentially improve the quality of topics extracted.
- **Hierarchical Dirichlet Process (HDP)** [7]: HDP is an extension of LDA that infers the number of topics from the data. It allows for an arbitrary number of topics to be generated, making it more flexible than traditional LDA.
- **Dynamic Topic Models (DTM)**: DTM extends topic modeling to account for temporal evolution in document collections. It models how topics change over time, which is useful for analyzing topics in datasets with a temporal dimension.

- **Word Embedding-based Techniques**: Techniques like Word2Vec, Doc2Vec, Top2Vec, and FastText can be used to learn distributed representations of words and documents. These embeddings can then be clustered or analyzed to extract topics.
  - *Word2Vec* [10]: word2vec serves as a tool for crafting distributed numerical vectors that encapsulate the essence of words, transforming textual content into a format that encapsulates semantic meaning and intricate word connections.
  - Doc2Vec: Doc2Vec vectors symbolize the theme and inherent significance of a document. This approach extends the principles of word2vec, encompassing documents within a generalized vector representation.
  - Top2Vecv [11]: Top2Vec stands as an innovative algorithm that serves the dual purpose of topic modeling and semantic search, proficiently identifying inherent topics within textual data and creating interconnected embeddings for topics, documents, and words in a unified manner.
  - FastText [12]: FastText, a freely available library developed by Facebook AI Research (FAIR), facilitates the acquisition of word embeddings and classifications, enabling the creation of both unsupervised and supervised learning algorithms to derive vector representations for words.
- **Deep Learning-based Approaches**: Variational Autoencoders (VAEs) and Generative Adversarial Networks (GANs) have been applied to topic modeling, often in conjunction with other techniques, to learn complex topic structures.
- **Online Variational Bayes (Online LDA)**: Online LDA is a variant of LDA that allows for incremental updates as new documents arrive, making it suitable for real-time applications.
- **BERT-based Approaches** [13]: Most recent approaches involve using pre-trained transformer-based models like BERT to extract topics from text. By fine-tuning Bert on a specific topic modeling task, it's possible to obtain accurate and context-aware topic representations.

Each of these techniques has its strengths, weaknesses, and specific use cases. The choice of technique depends on the nature of the data, the desired level of interpretability, and the specific goals of the analysis. Implementing and experimenting with these techniques can lead to valuable insights and a better understanding of the underlying topics within a document collection.

1.3 Applications and Benefits

The fundamental objective of topic modeling involves the automated identification and extraction of the predominant subjects within a collection of documents, without requiring manual labeling or categorization. Its applications encompass a range of objectives, including:

- **Content Understanding**: Topic modeling helps to gain a deeper understanding of the content and themes present in a collection of documents. It can reveal hidden patterns, trends, and relationships that might not be immediately obvious.
- **Document Organization and Categorization**: By assigning documents to different topics, topic modeling can assist in organizing and categorizing a large corpus of text. This can be particularly useful for information retrieval and content organization.
- **Information Retrieval** [14]: Topic modeling can enhance information retrieval systems by associating documents with relevant topics. This makes it easier for users to find specific information within a large dataset.
- **Recommendation Systems** [15]: Topic modeling can be used to recommend similar or related content to users based on their preferences or the topics they are interested in.
- **Content Summarization**: It can aid in summarizing large volumes of text by focusing on the main topics and discarding less relevant information.
- **Trend Analysis** [8]: It can be used to understand the evolution of topics over time can be useful for tracking the trends and shifts in public interest.
- **Sentiment Analysis** [13]: Combining topic modeling with sentiment analysis can provide deeper insights into the sentiments associated with different topics.

1.4 Limitations and Considerations

While title-based topic modeling offers numerous advantages, it is not without challenges. Determining the optimal number of topics remains a subjective task, and topics might not always be accurately represented due to the brevity of titles. Furthermore, titles might be creatively crafted or intentionally misleading, affecting the accuracy of topic extraction. The subjective determination of the optimal number of topics and the need for human interpretation to label the topics are some potential hurdles. They have limitations like:

- **Subjectivity in Interpretation**: Topics are often labeled and interpreted by humans, introducing a level of subjectivity.
- **Choosing Optimal Parameters**: Determining the number of topics or other model parameters can be challenging and might require experimentation.
- **Lack of Context**: Topic models might struggle with capturing nuanced contextual information in certain cases.
- **Dependency on Data Quality**: The quality and preprocessing of the input data significantly impact the reliability of topic modeling outcomes.

By taking advantage of topic modeling algorithms, this work aims to automatically extract the underlying themes and topics from the titles of web articles, blog posts, and other online resources. The rest of this paper addresses different facets of the study, which includes Sect 2 as background overview and literature survey, Sect 3 as methodologies for the proposed work, Sect 4 as results and discussion and finally followed by conclusion. This enables a clear presentation of the problem, the techniques employed, the findings and their implications.

2 Background and Related Works

The proliferation of web content has led to an information overload, necessitating innovative approaches to navigate and derive insights from this digital landscape. The titles of web pages serve as a condensed representation of the content they encapsulate, providing users with a glimpse into the themes and subjects covered. Analyzing these titles enables efficient content categorization, search result refinement, and trend identification. Traditional methods of organizing web content rely on keywords, metadata and manual categorization. These approaches of content analysis and keyword-based indexing fall short in capturing the nuances and complexities present in web titles. Topic modeling, a data-driven and automated method that offer a solution by automatically uncovering hidden patterns and topics, thereby enhancing information retrieval and content management. As technology continues to evolve, further research and innovation in this area promise to revolutionize information retrieval, content categorization, and trend analysis, contributing various work under this scope. In the subsequent part, we will delve into several relevant studies concerning the field of topic modeling in detail.

The principal objective of the research presented by Allahyari et al. [6] is to leverage and integrate the semantic knowledge graph of concepts in an ontology (specifically DBpedia) with probabilistic topic models (such as LDA) to improve the quality of topic labeling. The aim is to automatically generate interpretable and semantically relevant labels for topics discovered in text corpora. The two specific corpora which are BAWE corpus and the Reuters news set corpus are used in the evaluation and comparison of different topic modeling techniques. The research proposes a knowledge-based topic modeling approach, KB-LDA, that incorporates ontological concepts as a distribution over words for each topic, allowing for more extensive and specific topic descriptions.

Wahyudi et al. [16] proposed a study to analyze and model topics in various forms of text data, such as social media messages, user reviews, and news headlines. The study utilizes the Latent Dirichlet Allocation (LDA) algorithm, which helps identify hidden topics within a corpus of text. The corpus used was 106,209 news headlines collected from online news media website 'https://news.detik.com', from March to October 2020. The researchers aim to determine the best number of topics and evaluate the interpretability of the generated topics. The study also focuses on understanding the concerns and complaints expressed by users in different contexts, such as online stores, public complaint sites, and news media during the COVID-19 pandemic in Indonesia.

The study represented by Weisser et al. [17] is to evaluate and compare the performance of different topic models, namely Latent Dirichlet Allocation (LDA), Gibbs Sampling Dirichlet Mixture Model (GSDMM), and Generalized Polya Urn Model

(GPM). The study proposes a new evaluation method called Pseudo-Document Simulation, which involves simulating pseudo-documents and comparing the resulting topic models with the true document topic prevalence values. The dataset used in this context refers to the collection of tweets or text data from twitter related to COVID-19. The study aims to provide an alternative to conventional evaluation methods and address their shortcomings.

Latif et al. [7] proposed works objective is to analyze and evaluate different topic modeling techniques for Urdu text tweets. The document discusses the use of unsupervised machine learning techniques to auto-discover the subject of a document from large amounts of text. The work focuses on conducting experiments with various topic modeling techniques like LDA, NMF, LSA, pLSA, HDP on the dataset of 0.8 million Urdu tweets. It also explores the challenges and preprocessing efforts required for handling Urdu tex, which is low-resource language with different syntax and morphological structure compared to English. It highlights the limitations of topic modeling evaluation and provides insights into the performance of various topic modeling algorithms. The document aims to establish a baseline for advanced techniques and compare them to conventional models in order to better understand their effectiveness in handling short-text documents.

This study proposed by Athukorala et al. [8] is to give an accurate topic modeling algorithm for short texts, specifically addressing the challenges of higher dimensionality and extreme sparseness in short-text representation. The goal is to develop a model that can effectively identify meaningful topic clusters in real-world short-text data, using datasets as Twitter COVID-19 Tweets Data Set, Organic Food Data Set, Twitter data set from posts related to cancer, health, sports domain, and improve the accuracy and topic coherence compared to existing methods. The study aims to leverage neighborhood-based information and non-negative matrix factorization (NMF) techniques to compensate for information loss and enhance the performance of topic modeling on short texts.

Alhaj et al. [18] proposed a novel text classification model called Optimal Configuration Determination for Arabic text Classification (OCATC). The OCATC method aims to improve the performance of Arabic text classification by finding the optimal solution (configuration) from a large search space. It utilizes the Particle Swarm Optimization (PSO) algorithm to select the best architecture from a set of classifiers and feature selection methods with an optimal number of features. The CNN dataset, BBC dataset, Alj-News dataset, Alarabiya dataset, 1000 docs from different newspaper websites were used to extract and convert features from textual documents into numerical vectors using the TF-IDF approach. Comprehensive experiments were undertaken to assess the performance of OCATC, and the findings showcase its supremacy over individual classifiers and other cutting-edge techniques.

Uthirapathy et al. [13] discussed two important techniques in natural language processing: Latent Dirichlet Allocation (LDA) topic modeling and the Bidirectional Encoder Representation from Transformers (BERT) classification model. It explains how LDA identifies topics in documents and maps words to those topics, while BERT is a deep learning method used for sentiment analysis. The document also mentions the use of these techniques in analyzing social media data related to climate change by adopting the Twitter climate change sentiment dataset available on Kaggle. Overall, the document provides insights into the application of LDA and BERT in text mining and sentiment analysis.

Nabiilah et al. [19] proposed a study to detect toxic comments on social media in Indonesia using a pre-trained model. The researchers aimed to develop a multi-label classification model that could accurately identify comments containing toxic sentences. The user comments from various social media platform such as Instagram, Twitter, and Kaukus were collected to frame a toxic comment dataset. The goal was to evaluate the performance of different pre-trained models, such as IndoBERT, Multilingual BERT (MBERT), and Indo RoBERTa Small, and determine the optimal model for toxic comment analysis. The study aimed to achieve a high F1 Score, with the IndoBERT model achieving the best result of 0.8897.

Abuzayed et al. [20] presented a study to experiment with BERTopic, a topic modeling technique, using different pre-trained Arabic language models as embeddings. The goal is to compare the results of BERTopic with two popular topic modeling techniques, LDA and NMF. The study aims to evaluate the performance of BERTopic in terms of topic modeling quality, specifically using the Normalized Pointwise Mutual Information (NPMI) measure. The overall goal is to determine if BERTopic can provide better results compared to LDA and NMF in the context of Arabic topic modeling.

Hutama et al. [21] proposed a study is to develop a classification system for hoax news in low-resource languages, specifically Indonesian. The goal is to improve the performance of the classification system by using a deep learning approach that combines a pre-trained multilingual model (mBERT, XLM-R) with topic distribution using the BERTopic model. The study aims to compare the proposed method with other models and evaluate its performance in terms of accuracy, precision, recall, and F1 score. The ultimate goal is to provide an effective and efficient method for automatically detecting hoax news in Indonesian language news articles.

Alhaj et al. [10] presented a work to improve the classification of cognitive distortions (CDs) in Arabic content on Twitter. CDs are biased perspectives and beliefs that can negatively impact mental health. The goal is to develop a machine learning-based approach that enriches the representation of short texts by defining latent topics within tweets. This enrichment step uses unsupervised learning techniques, specifically a transformer-based topic modeling algorithm called BERTopic. BERTopic utilizes a pre-trained language model (AraBERT) to provide a topic distribution. The aim is to enhance the classifier's ability to distinguish between different CD categories and improve the overall classification performance. The performance of the approach is compared with classification using word2vec features. The results show that the proposed approach improves the classification results, with DT and bagging classifiers showing the most improvement.

Uncovska et al. [22] proposed a study to scrutinize the correlation between the prescription and reimbursement status of mHealth apps and the user experience. The study aims to answer specific research questions related to the user experience, such as the relationship between prescription and reimbursement status and user ratings, as well as the differences in user experiences between app groups. The study utilizes a mixed-methods approach, combining quantitative analysis of user star ratings and qualitative analysis of written reviews using sentiment analysis and topic modeling techniques. The study aims to provide insights into the user experience of mHealth apps and contribute to the development of effective and patient-centered digital medical products.

Rajendran et al. [15] proposed a study to enhance the user-based approaches in recommender systems by incorporating user browsing history and topic modeling. The researchers aimed to personalize the user experience and address the cold-start problem, where new users have limited information available. They proposed an approach that combines Wikipedia data and user browsing history to extract user preferences and improve the accuracy of recommendations. The study evaluated the performance of different approaches in terms of prediction and classification accuracy in the movie and restaurant domains. The results showed that the proposed enhanced User-User Collaborative Filtering (CF) approach that incorporates LDA-based topic modeling outperformed other methods, indicating its effectiveness in improving recommendation accuracy.

In our exploration of topic modeling, have observed that the predominant focus within this field encompasses sentiment analysis [19] and trend analysis [8]. These studies often involve mining textual data from various social media platforms [2, 3, 7, 8, 13, 17–19], subsequently utilizing the extracted content to classify topics and create relevant mappings. However, the survey has indicated that there is a relatively sparse concentration on a specific aspect: the utilization of web content organizing and categorization to map topics for the purpose of generating recommendations.

With this understanding in mind, the current endeavor will be directed toward delving into this lesser-explored area. The goal is to comprehensively investigate and address the mapping of topics using web content titles, with the ultimate objective of achieving more pertinent and valuable results in the field of recommendations.

3 Methodology of Title-Based Topic Modeling

The proposed model gives a novel approach for topic modeling for web content titles. This approach strives to take the unique characteristics of web content titles such as exact use of the word, context- dependence and linguistic patterns, to enhance the accuracy and relevance of the topics. By harnessing state-of-the-art techniques, title-based topic model aims to include BERTopic model which is introduced by (Maarten Grootendost, 2020), [23], and statistical models to address the robust solution for capturing the essence of web content titles and find appropriate topics and recommend to the users.

The workflow model for the title-based topic modeling is given in Fig. 1.

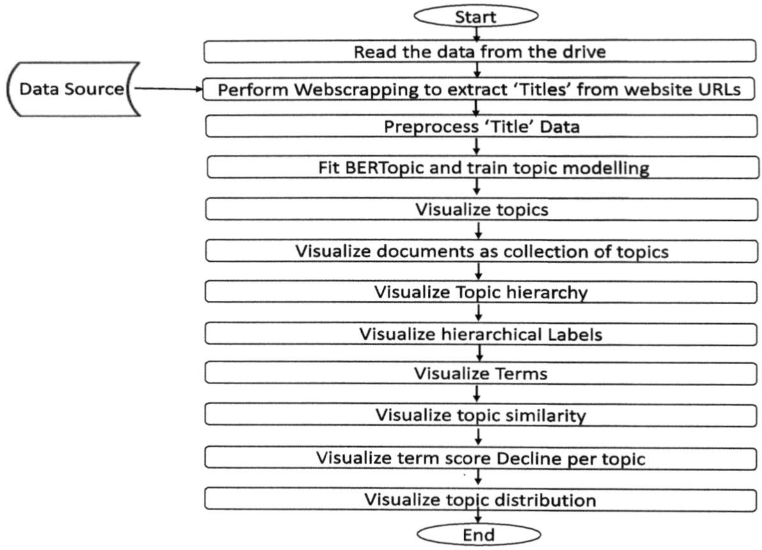

Fig. 1 Workflow of Title-based topic modeling

3.1 Dataset Depiction

The dataset collection process involves web scraping where it uses web content mining techniques to obtain the required web content documents from the search engine sites. Web scraping is an automated tool used to extract large amounts of accurate data from web pages which uses bots to access websites directly and retrieve contents. This process includes steps like sending input keywords to the Google search engine, scraping URLs and page content and extracting desired featured data. The collected data is further processed and stored in a structured format for analysis and classification. The dataset is acquired through this web scraping process and stored in an Excel spreadsheet [24]. It includes details such as the number of data, Google rank list, URL, Keyword, keyword density, and total number of words. Then, the dataset is stored in the google drive which is a cloud storage service for an access friendly and convenient environment. Figure 2 shows the exploratory dataset for this work.

The proposed work reads the data from the google drive which enhances the flexibility without the constraints of physical proximity and suitable for quick access to the files. It helps in automatically saves the changes made allowing collaborative editing streamlined.

	Page listing number on google	Website	keywords	word densities	word count
0	3	https://www.informit.com/articles/article.aspx...	Coordinationbetweenthreads-wait'	0.00	3353
1	4	https://www.programmerall.com/article/5083760149/	Coordinationbetweenthreads-wait'	1.02	1917
2	5	https://www.programmerall.com/article/5619303223/	Coordinationbetweenthreads-wait'	2.32	1606
3	6	https://www.tutorialspoint.com/importance-of-w...	Coordinationbetweenthreads-wait'	4.48	565
4	7	https://stackoverflow.com/questions/37026/java...	Coordinationbetweenthreads-wait'	2.83	9494
...	...	...	...	...	...
1560	4	http://coddingbuddy.com/article/55934002/read-...	writingtoafileusingvariousAPIsinJava	9.15	2332
1561	5	https://www.c-sharpcorner.com/article/file-cla...	writingtoafileusingvariousAPIsinJava	6.82	2131
1562	6	https://www.informit.com/articles/article.aspx...	writingtoafileusingvariousAPIsinJava	0.00	3246
1563	8	https://www.computerworld.com/article/2076078/...	writingtoafileusingvariousAPIsinJava	0.00	4069
1564	9	https://www.webucator.com/article/how-to-handl...	writingtoafileusingvariousAPIsinJava	7.98	915

1565 rows × 5 columns

Fig. 2 Exploratory dataset

3.2 *Extracting Titles from URLs*

The title of a web page serves as a digital introduction, a glimpse into the content that lies within. Users rely on titles to access the relevance of a web page before navigating further. Extracting titles from URLs provides an efficient and structured way to access and present the crucial piece of information to the users. The process of extracting titles from URLs is also achieved through web scraping. This technique involves fetching and parsing the HTML content of a web page. Key steps in title extraction include:

- *HTTP Request*: Sending an HTTP request to the URL to fetch the HTML content of the page.
- *HTML Parsing*: Utilizing parsing libraries like BeautifulSoup in Python to navigate and extract specific elements from the HTML structure.
- *Title Tag Retrieval*: Locating the '< title >' tag, a standard HTML element that typically contains the webpage's title, and extracting the text within it.

From the dataset, the data samples were restricted by selecting the initial 1100 rows, while the remaining rows were excluded by dropping them. The website URLs are extracted from the respective column of dataset, and individual row labels or indices are subsequently assigned to each row sample. This helps in accessing specific row, performing arithmetic or logical operation involving index, merging or joining

```
Website
https://www.informit.com/articles/article.aspx?p=31682&seqNum=5                    0
https://www.programmerall.com/article/5083760149/                                  1
https://www.programmerall.com/article/5619303223/                                  2
https://www.tutorialspoint.com/importance-of-wait-notify-and-notifyall-methods-in-java  3
https://stackoverflow.com/questions/37026/java-notify-vs-notifyall-all-over-again  4
                                                                                 ...
https://opensource.com/article/20/5/functional-java                             1095
https://opensource.com/article/20/1/javastream                                  1096
https://www.webucator.com/article/how-to-handle-java-files-with-streams/        1097
https://www.c-sharpcorner.com/article/java-8-stream-api/                        1098
https://programmerall.com/article/9693985079/                                   1099
Length: 1100, dtype: int64
```

Fig. 3 Website URLs data

DataFrames based or their index. Figure 3 shows the data samples of website URLs and its corresponding indices.

The extraction of website titles will be facilitated through the utilization of the Beautifulsoup() Python library. The process of extracting the titles from URL uses Beautifulsoup to request and fetch the titles for every sample.

Methodology:

Input: webpage URLs from the dataset.

- Import the necessary Modules.
- Create the requests instance and provide the URL.
- Utilize the requests object within the Beautifulsoup() function.
- Implement the 'title' tag to locate and extract all instances of the specified tag ('title').

Output: Titles of the webpage URLs.

The extracted titles are appended to the dataset respective samples. Figure 4 shows the appended dataset:

During the process of title extraction from URLs, encountered a few challenges such as variability in Title Placement, dynamic content, linguistic ambiguity and ethical and legal Aspects. Most common in this, titles may not always be found within the '< title>' tag, causing the extraction as null value. Scraping contents from the licensed websites causes to contradict the principles of ethical guidelines and rules, and occurring in forbidden error. So, most of the sites that are chosen to extract values were open unlicensed websites which are freely accessible. Even though few of them had some issues in fetching the content title causing null values and errors. Analyzing such rows, they are subsequently removed to obtain a refined dataset for further approach.

3.3 Data Preprocessing

The raw extracted titles often require preprocessing to transform them into a structures, clean, and usable format. In this data preprocessing procedure, is to clean and

	Page listing number on google	Website	Title	keywords	word densities	word count
0	3	https://www.informit.com/articles/article.aspx...	Cooperation between threads I Concurrency in J...	Coordinationbetweenthreads-wait"	0.00	3353
1	4	https://www.programmerall.com/article/5083760149/	403 Forbidden	Coordinationbetweenthreads-wait"	1.02	1917
2	5	https://www.programmerall.com/article/5619303223/	403 Forbidden	Coordinationbetweenthreads-wait"	2.32	1606
3	6	https://www.tutorialspoint.com/importance-of-w...	Importance of wait() notify() and notifyAll() ...	Coordinationbetweenthreads-wait"	4.48	565
4	7	https://stackoverflow.com/questions/37026/java...	multithreading - Java: notify() vs. notifyAll(...	Coordinationbetweenthreads-wait"	2.83	9494
...	...	...	...	...	...	...
1095	2	https://opensource.com/article/20/5/functional...	Why use Java Streams instead of loops I Openso...	Javastream	0.85	3374
1096	3	https://opensource.com/article/20/1/javastream	Data streaming and functional programming in J...	Javastream	1.39	4864
1097	4	https://www.webucator.com/article/how-to-handl...	How to Handle Java Files with Streams I Webucator	Javastream	6.30	915
1098	Screenshot	s://www.c-sharpcorner.com/article/java-8-s...	Java 8 - Stream API	Javastream	3.00	1987
[illegible]	7	[illegible]	403 Forbidden	Javastream	6.98	1027

Fig. 4 Dataset with appended title column

refine the data in the 'title' column, ensuring that it contains relevant and usable values for further analysis. The initial step involves focusing on the 'title' column of the dataframe. This column is singled out for processing, as we are particularly interested in the content it holds. Next, perform filtering on the rows within the 'title' column. This filtering is done based on specific values: '403 forbidden', 'loading…', and null values. Rows containing these values are identified. Once the rows with the specified values are identified, they are dropped from the dataframe. This removal process ensures that only rows with relevant content remain, effectively eliminating any irrelevant or redundant data points. Then the dataframe is left with a cleaner set of data samples, containing only valuable information that is pertinent to the analysis. The data type of the 'title' column is then converted from its initial format (usually 'object') to strings. This conversion aims to transform the data from a format that is primarily readable by machines to one that is more human-readable and interpretable. Then applied TF-IDF for vectorization of the titles, that leverages word frequencies to assess the significance of words in relation to a specific document. The 'title' column is further processed to present its contents in the form of a list. This arrangement makes it easier to work with the data, particularly when dealing with multiple titles. The cumulative effect of these steps is a dataset Fig. 5, that has been significantly enhanced in terms of cleanliness and usefulness for subsequent processing.

By filtering out irrelevant or problematic data points, converting data types, and presenting data in a list format, the dataset is prepared for further analysis or application without the distraction of irrelevant or messy information.

Fig. 5 Cleansed dataset

	Title
0	Cooperation between threads I Concurrency in J...
3	Importance of wait() notify() and notifyAll() ...
4	multithreading - Java: notify() vs. notifyAll(...
5	java - How can the wait() and notify() methods...
6	Java 101: Understanding Java threads, Part 3: ...
...	...
1094	Java 8 - An Introductory article to Java Strea...
1095	Why use Java Streams instead of loops I Openso...
1096	Data streaming and functional programming in J...
1097	How to Handle Java Files with Streams I Webucator
1098	Java 8 - Stream API

862 rows × 1 columns

4 Results of Title-Based Topic Modeling

The process of title-based topic modeling follows a structure similar to traditional text-based topic modeling. The main difference lies in the input data: rather than analyzing the content of documents, the focus is shifted to the patterns and keywords present in titles. Title-based topic modeling, a dynamic technique, utilize the potential of BERTopic modeling, which leverages contextual embeddings from transformer-based language models like BERT (Bidirectional Encoder Representations and Transformers) and statistical model for comparative analysis.

4.1 Applying BERTopic Model

Interactive Title-based topic modeling employing BERTopic model, represents an advanced and pioneering approach in extracting profound insights from the succinct yet powerful medium of web content titles. By integrating contextual embeddings from BERT with advanced topic modeling, this methodology illuminates clustering documents contents that exhibit not only lexical but also semantic similarities. This gives a big advantage over the traditional topic modeling techniques like Latent Dirichlet Allocation (LDA) [11] employs a probabilistic methodology, while BERTopic leverages transformers (BERT embeddings) and class-based TF-IDF [23] to establish compact clusters [20].

Leveraging a BERT-based language model, contextual embeddings are generated for the preprocessed titles. BERT embeddings encapsulate the semantic context of words, enabling a deeper understanding of text than the traditional topic models [25]. Where the reduced BERT embeddings are employed to unveil title-based topics. Dimensionality reduction techniques, such as UMAP [20] or t-SNE, are applied to reduce the high-dimensional BERT embeddings while retaining the essence of semantic relationships between titles.

The BERTopic model is initialized with a designated number of topics, clustering titles based on their semantic proximity within the reduced-dimensional space. It uses UMAP and HDBSCAN algorithms [10, 26] for unfolding the structural reference and hierarchical clustering. Uniform Manifold Approximation and Projection for Dimension Reduction (UMAP) algorithm, builds a representation of the data in a high-dimensional graph, followed by the creation of a low-dimensional graph that is optimized to closely mirror the structural attributes of the original high-dimensional graph. Here UMAP preserves semantic information within the text data via umap_model parameter. To visualize the generated topics, involves embedding the c-TF-IDf representation of topics into a two-dimensional space using UMAP. Subsequently, these two dimensions are visualized using Plotly library, allowing for creation of an interactive view. This visualization technique provides a concise and interactive representation of the topics, enabling a more in more intuitive and efficient comprehension of the topic distribution. Figure 6 shows the 2D representation of the topics there in the titles which is considered as document, using Plotly interactive graph. When the slider is utilized for topic selection, which subsequently illuminates with pertinent information about the chosen topic, encompassing its size and associated terms.

Employing the previously described technique, we can gain an overview of topic relationships through visualization. Nonetheless, for a more detailed examination, a more nuanced strategy might be sought. This entails visualizing the documents within the topics to assess their correct assignment or coherence. To facilitate this, the function recalculates document embeddings and reduces them to a two-dimensional space, optimizing visualization. It's important to note that this procedure can be computationally intensive. Figure 7 ensures the representation of document visualization. Where the Slider hovers over the topic, then the related all titles can be visualized one by one. This procedure involves clustering documents based on their shared vocabulary, grouping together those documents that employ similar words, much like it clusters words that appear in a similar context across documents. The

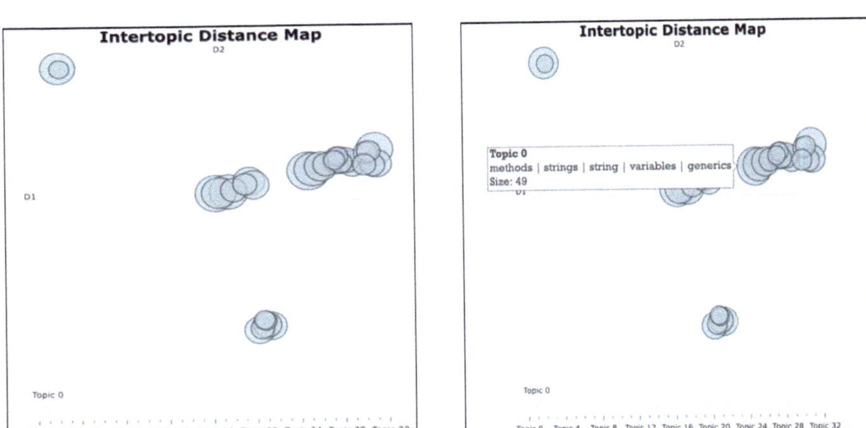

2D representation of Topics using ploty interactive graph

Fig. 6 Visualize topics

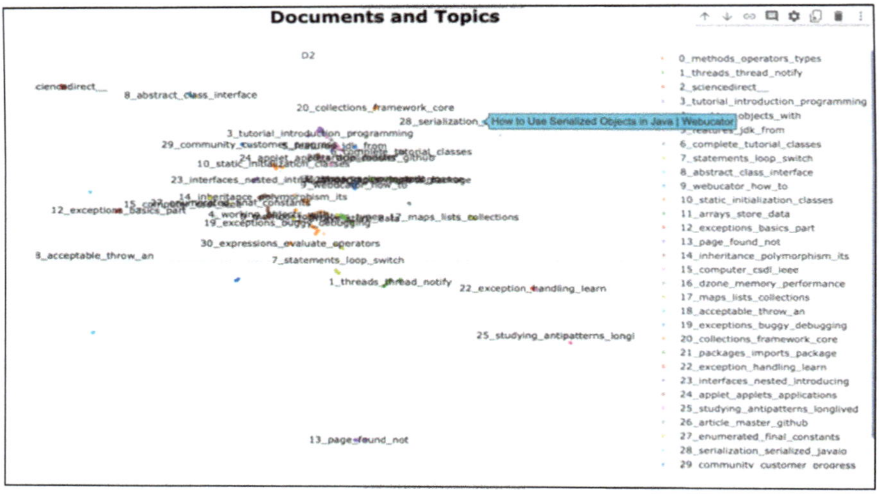

Fig. 7 Visualize documents

top-scoring keywords within each cluster then assume the role of the main title for that particular topic. Currently, the topic labels are generated through an automated process that selects the top 3 words within a topic and combines them using the underscore (_) separator to form the label. This approach streamlines the categorization and labeling of topics within the dataset, facilitating a more organized and comprehensible structure for analysis.

The generated topics can undergo a hierarchical reduction, potentially revealing a hierarchical structure. To explore this, the scipy.cluster.hierarchy module can be employed to create clusters and visualize their interrelationships. This approach assists in determining a suitable value for nr_topics, aiding in the reduction of the number of topics that have been generated. By delving into the hierarchical organization of topics, this process contributes to informed decision-making regarding the optimal number of topics for analysis. Figure 8 shows the visualization of topic hierarchy.

While visualizing the hierarchical structure provides insights into the arrangement, gaining an understanding of how topic representations evolve when merging topics can be equally valuable. To achieve this, the initial step involves computing the representations of hierarchical topics. To render these outcomes in a visual format, it suffices to provide the obtained hierarchical_topics as input to the.visualize_hierarchy function. Hovering over the black circles on the visualization grants access to topic representations specific to that hierarchy level. This facilitates a comprehensive grasp of the implications stemming from merging particular topics. While some mergers might seem coherent, others might not align as logically. Furthermore, this visualization aids in identifying sub-topics nestled within broader thematic categories, offering deeper insights into the hierarchical structure. Figure 9 shows the

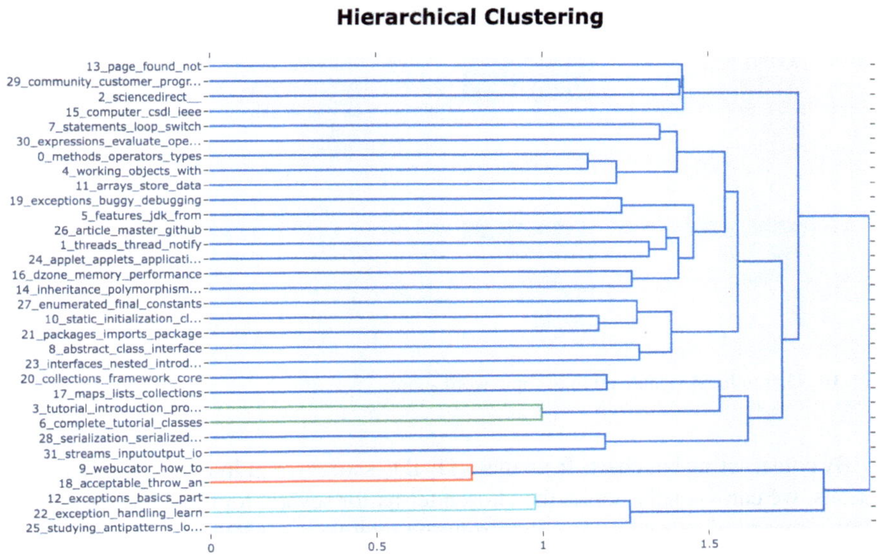

Fig. 8 Visualize topic hierarchy

hierarchical labels of the topics and using slider, can visualize the sub-topics within that.

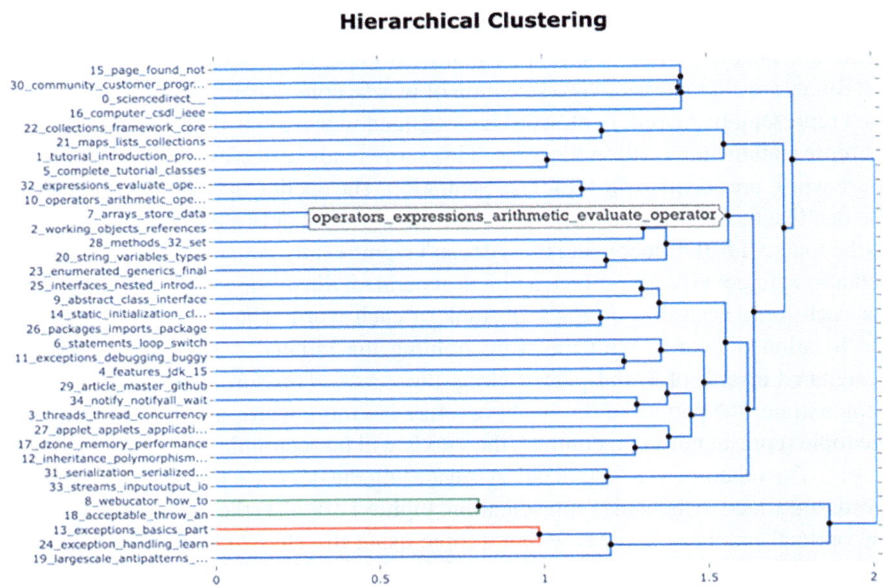

Fig. 9 Hierarchical labels

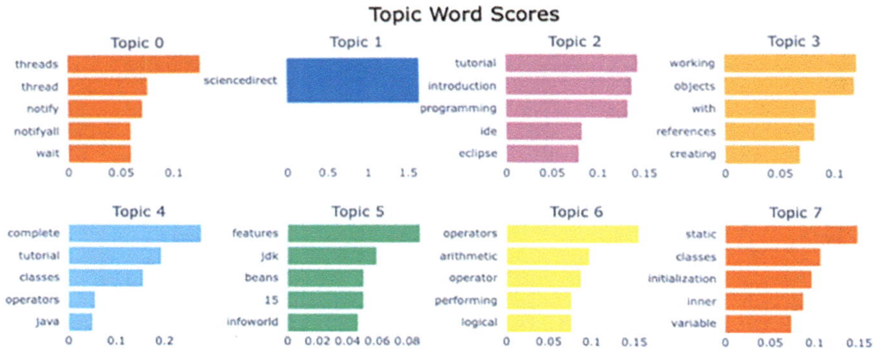

Fig. 10 Hierarchical visualization of topic word score

By constructing bar charts from the c-TF-IDF scores for individual topic representations, we can visually depict the chosen terms for several topics. These bar charts offer a means of extracting insights from the relative c-TF-IDF scores, both across and within topics. Additionally, this visualization facilitates effortless comparisons between different topic representations, enabling a comprehensive understanding of the relationships and significance of the selected terms within each topic. Figure 10 illustrates the hierarchical visualization of topics word score in a sample bar chart.

After generating topic embeddings using both c-TF-IDF and embeddings, a similarity matrix can be created by employing cosine similarities based on these topic embeddings [23]. The resultant matrix will provide insights into the degree of similarity between various topics. To visualize this relationship. Figure 11 shows the generated heatmap of topic similarity matrix as depicted below.

Topics are characterized by a selection of words, commencing with the most relevant representative word. Each word is associated with a c-TF-IDF score, signifying its representativeness within the topic. This score tends to decrease incrementally as more words are added to the topic representation. The decline in scores is attributed to the fact that the addition of words beyond a certain point only marginally contributes to the total c-TF-IDF score, and thus, doesn't significantly enhance the topic's representation. To get visualize effect, a plot is generated, illustrating the c-TF-IDF scores for each topic according to the term rank of each word. The term 'rank' signifies the location of words, with the word holding the inflated c-TF-IDF score being designated a rank of 1 and placed along the *x*-axis. This visualization effectively demonstrates the pattern of diminishing returns as more words are incorporated into the topic representation. In contrast, the y-axis will be accumulated by the c-TF-IDF scores. The outcome is a visualization illustrating the decrease in c-TF-IDF score as words are added to the topic representation. Figure 12 depicts the process of selecting the optimal number of words within a topic using the elbow method and by using the logarithmic scale on *y*-axis to enhance the clarity of individual topics.

To generate the topic-document probability matrix, can activate the calculate_probabilities = True option when employing an HDBSCAN model [26]. In scenarios

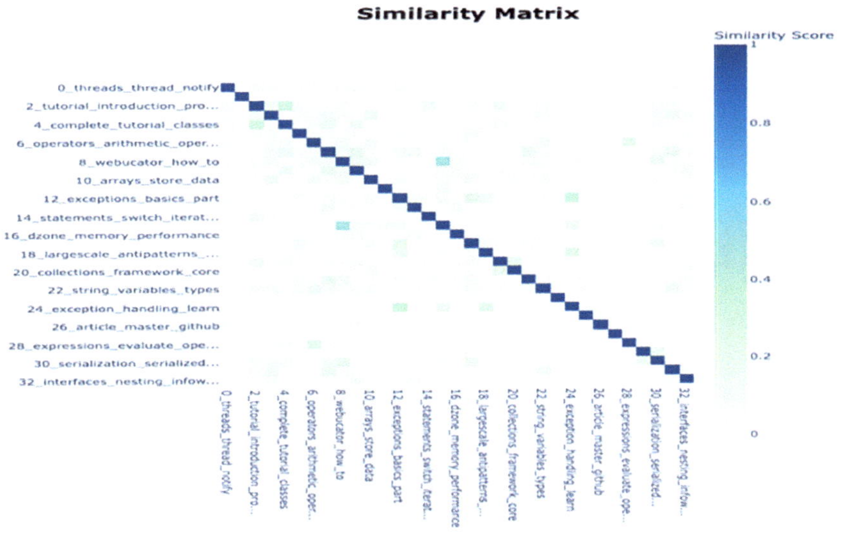

Fig. 11 Visualize topic similarity

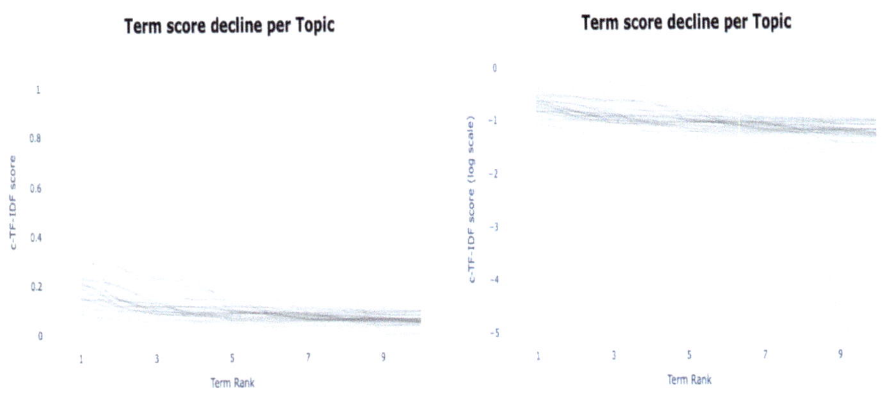

Fig. 12 Visualize term score decline

where a non-HDBSCAN model is utilized, the topic distributions can be approximated subsequent to model training. Subsequently, the probs or topic_distr variable can be passed to.visualize_distribution, enabling the visualization of either probability or topic distributions. While topic distribution offers valuable insights, it might be beneficial to explore how individual tokens contribute to specific topics. To achieve this, initial calculations of topic distributions on a token level are required, followed by visualization to provide a detailed overview of token-topic relationships. Figure 13 shows the visual probability distribution of topics and a sample topic distribution for the index 100 as below.

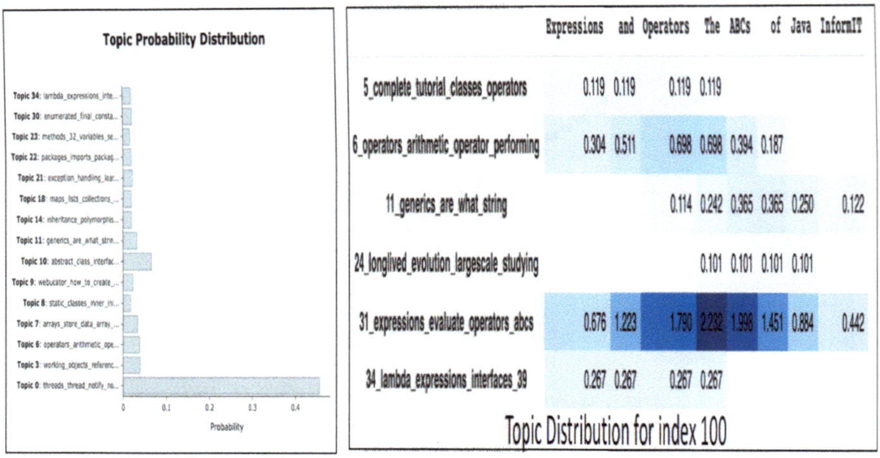

Fig. 13 Visualize topic probability distribution

This topic distribution value refers to the probability distribution of topics assigned to specific title of index 100 within the corpus. Each title is associated with a probability distribution over all the topics in the BERTopic model, indicating the likelihood that the title belongs to each topic. These probabilities provide insights into the degree of relevance or association of a title with different topics. A higher probability indicates a stronger assurance of association with a particular topic as compared to other topics. The final classified topics are used to derive impartial classifications and metrics from textual, qualitative data, lending a quantitative analysis of structured textual data in the recommendation process. This transition facilitates the integration of qualitative information into quantitative research, particularly in the context of hypothesis testing and empirical investigations for the recommenders. This bridging of the qualitative-quantitative divide empowers researchers to draw meaningful insights from diverse data sources, enriching the research process and broadening the scope of empirical inquiry.

5 Conclusion

In conclusion, this research presents a novel approach to recommendation systems that utilizes title-based topic modeling on e-learning web content titles using the BERTopic model. By automatically identifying latent topics within a large corpus of text, we aim to extract meaningful themes and categorize content efficiently, expediting the recommendation process and conserving computational resources. The proposed methodology involves a series of steps, including data collection through web scraping, data preprocessing, and topic modeling using the BERTopic algorithm. It also presents several visualizations to help interpret the results, including topic similarity matrices, hierarchical topic labels, and topic distributions. The results demonstrate that title-based topic modeling can effectively identify latent topics within e-learning web content titles. Additionally, our visualizations provide valuable insights into the underlying themes and topics within the corpus which helps in furthering this study.

However, it also acknowledges several limitations of this approach, including the subjective nature of topic labeling and the potential for misleading or creatively crafted titles. Also note that the quality and preprocessing of the input data significantly impact the reliability of topic modeling outcomes. Overall, this work contributes to the growing body of research on recommendation systems and topic modeling in e-learning. We believe that our approach has the potential to improve the efficiency and accuracy of recommendation systems, ultimately benefiting learners and educators alike. Future research could explore the use of more advanced natural language processing techniques or incorporate additional data sources to further improve the accuracy and reliability of topic modeling outcomes.

References

1. Li X, Lei L (2021) A bibliometric analysis of topic modelling studies (2000–2017). J Inf Sci 47(2):161–175. https://doi.org/10.1177/0165551519877049
2. Wang Q, Qian Y, Song R et al (2013) Mining subtopics from text fragments for a web query. Inf Retrieval 16:484–503. https://doi.org/10.1007/s10791-013-9221-8
3. Daud A, Muhammad F (2012) Group topic modeling for academic knowledge discovery. Appl Intell 36:870–886. https://doi.org/10.1007/s10489-011-0302-3
4. Blei DM, Lafferty JD (2009) Topic models. Text Min ClIfication Clust Appl 10(71):34
5. Wang T, Liu Y 92017) JSEA: a program comprehension tool adopting LDA-based topic modeling. Int J Adv Comput Sci Appl (IJACSA) 8(3). https://doi.org/10.14569/IJACSA.2017.080359
6. Allahyari M, Pouriyeh S, Kochut K, Arabnia HR (2017) A knowledge-based topic modeling approach for automatic topic labeling. Int J Adv Comput Sci Appl 8(9):335
7. Latif S, Shafait F, Latif R (2021) Analyzing LDA and NMF topic models for Urdu Tweets via automatic labeling. IEEE Access 9:127531–127547
8. Athukorala S, Mohotti W (2022) An effective short-text topic modelling with neighbourhood assistance-driven NMF in Twitter. Soc Netw Anal Min 12:89. https://doi.org/10.1007/s13278-022-00898-5
9. Daenekindt S, Huisman J (2020) Mapping the scattered field of research on higher education. A correlated topic model of 17,000 articles, 1991–2018. High Educ 80(3):571–587
10. Alhaj F, Al-Haj A, Sharieh A, Jabri R (2022) Improving Arabic cognitive distortion classification in Twitter using BERTopic. Int J Adv Comput Sci Appl 13(1):854–860. https://doi.org/10.14569/IJACSA.2022.0130199
11. Egger R, Yu J (2022) A topic modeling comparison between lda, nmf, top2vec, and bertopic to demystify twitter posts. Front Sociol 7:886498
12. Joulin A, Grave E, Bojanowski P, Douze M, Jégou H, Mikolov T (2016) Fasttext. zip: compressing text classification models. arXiv preprint arXiv:1612.03651
13. Uthirapathy SE, Sandanam D (2023) Topic modelling and opinion analysis on climate change twitter data using LDA And BERT Model. Procedia Comput Sci 218:908–917. https://doi.org/10.1016/j.procs.2023.01.071

14. Tong Z, Zhang H (2016) A text mining research based on LDA topic modelling. In: International conference on computer science, engineering and information technology, pp 201–210
15. Rajendran DPD, Sundarraj RP (2021) Using topic models with browsing history in hybrid collaborative filtering recommender system: experiments with user ratings. Int J Inf Manag Data Insights 1(2):100027. https://doi.org/10.1016/j.jjimei.2021.10002
16. Wahyudi MDR, Fatwanto A, Kiftiyani U, Wonoseto MG (2021) Topic modeling of online media news titles during COVID-19 emergency response in Indonesia using the latent dirichlet allocation (LDA) algorithm. Telematika 14(2):101–111
17. Weisser C, Gerloff C, Thielmann A et al (2023) Pseudo-document simulation for comparing LDA, GSDMM and GPM topic models on short and sparse text using Twitter data. Comput Stat 38:647–674. https://doi.org/10.1007/s00180-022-01246-z
18. Alhaj YA, Dahou A, Al-qaness MA, Abualigah L, Abbasi AA, Almaweri NAO, Elaziz MA, Damaševičius R (2022) A novel text classification technique using improved particle swarm optimization: a case study of Arabic language. Future Internet 14(7):194. https://doi.org/10.3390/fi14070194
19. Nabiilah GZ, Prasetyo SY, Izdihar ZN, Girsang AS (2023) BERT base model for toxic comment analysis on Indonesian social media. Procedia Comput Sci 216:714–721. https://doi.org/10.1016/j.procs.2022.12.188
20. Abuzayed A, Al-Khalifa H (2021) BERT for Arabic topic modeling: an experimental study on BERTopic technique. Procedia Comput Sci 189:191–194. https://doi.org/10.1016/j.procs.2021.05.096
21. Hutama LB, Suhartono D (2022) Indonesian hoax news classification with multilingual transformer model and BERTopic. Informatica 46(8). https://doi.org/10.3389/fsoc.2022.886498
22. Uncovska M, Freitag B, Meister S et al (2023) Rating analysis and BERTopic modeling of consumer versus regulated mHealth app reviews in Germany. NPJ Digit Med 6:115. https://doi.org/10.1038/s41746-023-00862-3
23. Grootendorst M (2022) BERTopic: neural topic modeling with a class-based TF-IDF procedure. arXiv preprint arXiv:2203.05794
24. Lalitha TB, Sreeja PS (2023) Potential web content identification and classification system using NLP and machine learning techniques. Int J Eng Trends Technol 71(4):403–415. https://doi.org/10.14445/22315381/IJETT-V71I4P235
25. Alcoforado A, Ferraz TP, Gerber R, Bustos E, Oliveira AS, Veloso BM, Siqueira FL, Costa AHR (2022) ZeroBERTo: leveraging zero-shot text classification by topic modeling. In: International conference on computational processing of the Portuguese Language. Springer International Publishing, Cham, pp 125–136. https://doi.org/10.1007/978-3-030-98305-5_12
26. de Groot M, Aliannejadi M, Haas MR (2022) Experiments on generalizability of BERTopic on multi-domain short text. arXiv preprint arXiv:2212.08459

Knowledge Aware Artificial Intelligence Models for Text Generation in Energy Industry

Athira Puthanveetil Madathil, Malathi Murugesan, Ali Youssef, Lorenzo Paladini, and Giacomo Veneri

Abstract Artificial Intelligence (AI) has been revolutionizing various sectors with its remarkable applications. Generative AI, in particular, has gained significant traction in both industrial and research domains. In this work, we focus on leveraging this technology for an industrial use case, specifically for generating high-quality data (i.e., text data) for service request reports in manufacturing units. Addressing safety concerns is crucial in energy production sites. Field engineers raise several issues, including product safety impact, environmental concerns, health and safety, and quality issues, as well as fleet implications. The goal of our application is to classify these concerns into either unsafe or safe operating conditions, allowing the system to trigger warnings and enable engineers to prioritize and address them effectively. To implement this application successfully, a substantial amount of data is required. However, in real-world scenarios, available data is often insufficient. In this paper, we have utilized several text generation models for our application and compared their performance based on the quality of the generated text. We have successfully generated samples that closely resemble the original text using large language models. Notably, we achieved an $\approx 30\%$ improvement in text classification task performance by employing a large language model (GPT2, Falcon, and LLama-2).

---

http://bakerhughes.com.

A. P. Madathil (✉) · M. Murugesan
Baker Hughes, Neon Building West Tower, Bangalore, India
e-mail: Athira.PuthanveetilMadathil@bakerhughes.com

M. Murugesan
e-mail: MALATHI.MURUGESAN@bakerhughes.com

A. Youssef · L. Paladini · G. Veneri
Baker Hughes, Via Felice Matteucci 2, Firenze, Italy
e-mail: ali.youssef@bakerhughes.com

L. Paladini
e-mail: Lorenzo.Paladini@bakerhughes.com

G. Veneri
e-mail: giacomo.veneri@bakerhughes.com

S. M. Thampi et al. (eds.), *Fifth International Conference on Computing and Network Communications*, Lecture Notes in Electrical Engineering 1221,
https://doi.org/10.1007/978-981-97-4711-5_40

Keywords Deep learning · Industrial use case · NLP · Text generation · Text classification

1 Introduction

In the energy transformation industry, the company's knowledge base enhancement, quality improvement of outcomes, defect-free manufacturing, and customer satisfaction in product delivery are critical aspects. The accumulation of knowledge requires substantial effort in analyzing data from various sources that are presented in full-service equivalent reports by human operators at the sites. Handling incoming service requests and addressing the most pressing concerns assigned by domain experts necessitates significant time and efforts investments. Meanwhile, AI techniques have demonstrated their potential in digitalizing the industry by utilizing continuously generated, processed, and analyzed data and segregating the concerns into various types of reports.

The main challenge that industries face is the lack of available data. To address the issue of unbalanced data and the unavailability of labeled samples, researchers have employed various data augmentation strategies. However, this approach is often unsuccessful due to the poor quality of the generated data that does not resemble human-written text. To overcome this challenge, we have applied language models that generate text similar to that written by human operators [34]. We investigated various state-of-the-art generative models [8] and compared their performances on our industrial use case.

We structure the paper as Sects. 1 and 2 introduce the paper and related works, Sect. 3 describes the industrial problem statement (see Sect. 3.1), the proposed approach approach (see Sect. 3.2) and the data set used in our experiments (see Sect. 3.3) together with details of the setup of the experiments and chose models, Sect. 4 provides the results obtained by measuring the perplexity of the generated text and the improvement of text classification performance by using the proposed approach for text augmentation, and last Sect. 5 reports conclusions and future works.

2 Related Works

Data Augmentation is a popular technique widely used in machine learning to increase the size of training data set by generating new examples from the existing ones. In the context of Natural Language Processing(NLP),research has been done [26] to increase the number of samples [29]. It can improve the performance of models by addressing issues such as class imbalance, data scarcity, and improving model generalization [29]. Recently, researchers have widely used the pretrained language models such as BERT [5], GPT [19], and RoBERTa [12], Falcon [15], Llama [28] achieving state-of-the-art results on various NLP tasks. These models are trained on large amounts of text data and can be fine-tuned on specific tasks with limited amounts of labeled data. For example, the authors in Ref. [11] show that their approach improve the performance of the model on text classification tasks by using BERT to generate new sentences.

The model GPT was used in Ref. [4] to rephrase each sentence in the training samples into multiple similar samples. The augmented samples can then be used in downstream model training. The approach in this paper [19] uses GPT2 model to produce long sequences and perform comparative analysis. The decoder-based transformers [20] have the capability to produce high quality human-like text,without human supervision. The approach used in this paper, [27] produces domain-specific keywords and then progressively refine them into complete paragraphs in consecutive stages, but it lacks controllability. Diffusion models have displayed enormous success in domains such as audio [10]. and images [22, 24]. Due to the discrete nature of language [2], it is not largely used in text generation tasks. So more research is going on in this area so that diffusion models for text also can be used widely. In Ref. [17], a decoding method was proposed to relax the discrete language model outputs to continuous variables and back propagate gradient information from the right context.

Text generation techniques using pretrained language models has shown promising results in improving the performance of NLP models on different tasks. These models can generate new examples that are similar or semantically related to the input sentence; therefore it can address issues such as class imbalance and data scarcity. Large Language Models(LLMs) have shown great potential as AI assistants, demonstrating good capabilities in complex reasoning tasks across various fields. Pretrained LLMs like BLOOM [25], LLaMa-2 [28], and Falcon [1] have been publicly released, making them available on a wide range of seen and unseen NLP tasks such as summarisation, question answering [30] and creative text generation [23, 28].

3 Problem Statement

3.1 Service Requests Application

Within the energy transformation domain, the capacity to expeditiously identify and rectify problems, coupled with a robust continuous improvement framework, is of paramount importance, as it effectively prevents non-conformity and diminishes the likelihood of quality-related challenges.

Field engineers from the site raise numerous concerns, encompassing product safety implications, environmental impacts, health and safety considerations, and quality-related issues, as well as fleet effects. To effectively address these concerns, it is essential to meticulously examine the full range of cases collected through reporting tools and to uncover concealed critical issues periodically raised by human operators

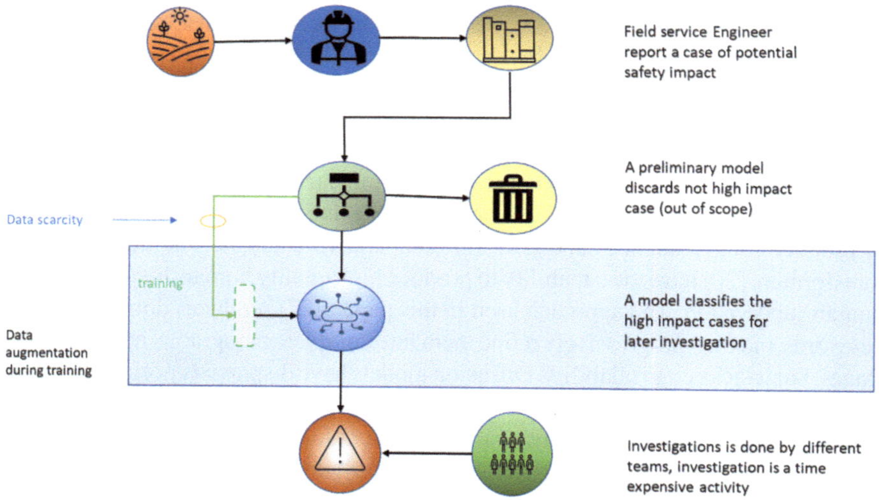

Fig. 1 Workflow of the use case: Field service engineer submits a case for investigation; a preliminary machine learning model (out of scope) rejects low impact cases; a second model classifies high impact cases for later investigation. Due to data scarcity we propose to augment the data set for training and to produce better outcomes

(field service engineers). In this endeavor, we strive consistently to guarantee that our digital products conform to specifications mandated by regulators, and that all processes are devoid of defects. By implementing a zero-defect escape approach, we ensure that our products are flawless before shipment to the customer. This is crucial in our industry, as it contributes to reduced rework, fosters customer loyalty, and promotes continuous improvement.

The Fig. 1 shows the workflow of industrial use-case: field service engineer submit a case for investigation; a preliminary machine learning model (out of scope) rejects low impact cases; a second model classifies high impact cases for later investigation. Due to data scarcity and different ways of reporting cases and the subjectivity of operators and investigators, it is very difficult to develop an efficient classifier; in this work we compare different data augmentation techniques to demonstrate their effectiveness even in industrial cases.

3.2 *The Proposed Approach*

Text generation involves creating content that closely resembles human-written text. Researchers often employ AI techniques to address data scarcity issues and develop robust models [18]. Therefore, it is important to develop efficient systems that not only augment the database but also obtain high-quality data while minimizing computational resource consumption [14]. In this work, we implemented and assessed various knowledge-aware language generation models, specifically designed to generate open-ended text from the energy domain. We investigated the use of state-of-the-art AI models for text generation in the field of engineering, with a focus on evaluating their performance in generating high-quality text [33]. To do this, we conducted a study using the perplexity as a metric to compare the performance of various generative models on text generated by field engineers [13]. We also employed a simple machine learning classifier CatBoost, to assess the quality of the generated text, and evaluated its performance in a classification task using different texts generated by the models.

Indeed, perplexity is a well-known method for evaluating the quality of text, however we preferred to simulate the final use case with a CatBoost model as it is more similar to the production environment. We chose a simplified model to avoid dependency between text generation task and classification task.

In particular, we created our baseline based on Bidirectional Encoder Representation of Transformer architecture and Masked Language Model (MLM)as a baseline generative model [32] and compared the performance of different models on text generation. In addition to evaluating the performance of MLM, we also investigated the efficacy of a transformer-based generative model GPT2 [20], which has been shown to achieve great results in various NLP tasks. Furthermore, we investigated the use of two recent large language models,Falcon 7B [16] and LLAMA-2 7B [28], which have shown promising results in our restricted domain. Finally, to assess the quality of the generated text, we employed a straightforward classification model to prove that the generated text can improve the classification performance. Our results indicate that the models we investigated have the potential to generate a good quality text that can be used in practical applications. Specifically, we found that the classifier was able to accurately distinguish between the predefined classes of the text. This suggests that the models we evaluated are capable of generating text that is not only coherent, but also contextually appropriate and semantically meaningful. Our findings suggest that these models have the potential to generate high-quality text that can be used in real-world applications.

3.3 Experimental Setup

Data Set. We used data set consists of 5000 unlabeled text descriptions, which represent quality issues and service requests written in the form of reports. These descriptions were collected from various sources by field engineers and are intended to be evaluated by domain experts. The data set contains approximately 27000 sentence tokens in total (The data is proprietary and is not open-sourced or available for public to conduct experiments). We have fine tuned the GPT2 language models on 80% of samples and evaluate on 20% of samples; while, we used the pretrained Bert MLM, LLAMA-2 and Falcon for the other text generation experiments. All experiments were conducted on NVIDIA DGX-A100 machine.

Table 1 Test set

Class name	Description	Test size
EHS impact	Potential issue of health safety	28
Fleet impact	Potential issue of safety due to installation	64
Product impact	Potential issue of product safety	23

Table 2 Tokens generated by the Bert MLM for the masked words

Text sequence	Masked original words	Generated-words BertMLM
1	*Tripped, enclosure, leak, flame*	*Tripped*, package, get, *flame*
2	*Discharge, solved, replace, type*	*Discharge, solved*, increase, *type*

To evaluate the effectiveness of the generated text, we utilized a simple Boosting classifier with a Bag of words encoder. We implemented the classifier using the CatBoost library [6] and applied the following hyper parameters: 500 iterations, learning rate of 0.1, and early stopping on the training set. We opted for a simpler model to compare the results of the augmentation techniques and avoid introducing bias due to the classifier and encoder. The data set for the classification task consisted of 381 labeled text into 3 classes (EHS Impact, Fleet Impact and Product Impact). We divided it into 266 for training and 115 to perform the base line using stratified sampling (70% for training and 30% for validation), see Table 1. Then, we use each generative model to augment the number of training data set to 1064 for each classification experiment. Then We used those 1064 generated text samples to calculate the perplexity of each model.

We would like to remind that the main problem is the scarcity of data; in fact (and fortunately) not many safety cases have been reported on the last 5 years.

Bert Masked Language Model. We experimented with Masked Language Model (MLM), as a base case to formulate a system that predicts a word hidden intentionally in a sentence [7]. We utilized the Bert model for this task as it is efficient in understanding the meaning of surrounding text and deriving the right context around the masked words. The task is challenging as the words can have different meanings in various places and the context would not be the same. The model needs to understand multiple representations of words.

Since Bert has the capacity for learning the bidirectional representations of words, we applied this model to find the words hidden in sentences [5] (see Table 2).

The inputs to the Bert model are text sequences with 15 percent of the words in the sequences masked. Based on the contextual knowledge of the unmasked words in the text sequence, the Bert MLM predicts the original values of the masked words. To make the prediction more accurate, model learns the bidirectional representation of the sentences.

The prediction of Bert MLM for text sequences given below are coined in Table 2.

This case is being created to document the issue and provide investigation follow up power service now references FERT for EEAV the GTGc turbine power generator [MASK] *due to flame detection inside the* [MASK] *while being test run on diesel fuel as part of commissioning of the liquid fuel system on investigation it was found that the liquid fuel manifold had a small crack in a weld which allowed diesel to* [MASK] *inside the enclosure a small quantity of this diesel was in contact with the hot chamber and ignited flashed, causing the* [MASK] *detection the flame detection cleared immediately.*

Knocking at [MASK] *side of the driven pumps of GV and DLGL compressors DLGL the issue has been* [MASK] *by increasing the oil level in low tank to increase the retention time and so the time GG the solution is to* [MASK] *the driven pump suction pipe from to diameter in the aim to have oil flow speed in our* [MASK] *criteria and so improve modification of the suction pipe.*

GPT2. Decoder-based models use a Transformer decoder in their architecture (also typically stacked) and are great for future prediction, which makes them suitable for text generation [19]. The GPT models [20] are decoder based transformers which produces high quality human-like text, when trained on large corpora, and are good at language generation tasks. Since GPT2 was trained with causal language modeling, it is powerful in predicting tokens in a sequence and generating syntactically coherent text [31].

The following sentences have been generated by the GPT2 model as open-ended text:

The current value of low-key trip is bar g the timer pulse duration was just minutes which in our experience make high pressure with red led pulse every so often and in general hp goes offline before minutes and the delay till key recovery rate reaches threshold at critical values was approximately second for ltc.

The following are reported defective from one motor, we cannot send additional information to see if the replacement motor damaged, and the inspection completed since previous inspection results are as before and need be in field services engineers works for further action.

Falcon. The Falcon model is a casual decoder-only model train more on 1 trillion tokens. We have experimented with Falcon-7B-Instruct model, which has been fine tuned on instructions and conversational data. Since the architecture of Falcon is optimized for inference with multi-query and flash attention, it is able to generate high quality data, according to human instructions.
The query prompting used to generate new data is:

```
"[INST]
Question: Summarize the given text in less than
200 words without putting the query into the answer
Context: {text}[\INST]"
```

The following sentence has been generated by the Falcon (7B) model as open-ended text:

There appears to be a discrepancy in the check of the valve selection process of valves and LPG (Liquid Propane Gas) valves, specifically for the 5 units mentioned. As per the email correspondence mentioned, a bar is required to fix the current issue. The attached materials and explanation can be used to better understand the issue and solution. Please proceed with procurement of goods for the affected units.

LLama-2. We used LLama-2 with 7 billions parameters (7B); we preferred the 7B parameters versus 13B and 70B because more compliant with our production environment. The same prompting of Falcon (see Sect. 3.3) was applied.

Below an example of the generated text by LLama-2 (7B):

Lifting of compressor spare bundle from horizontal to vertical was attempted, but the bundle lost balance during lifting. The lifting details provided in document were not sufficient, and a dedicated drawing with modifications required for the bundle container is code 8080-678. Additional materials necessary to perform the operation on site include a rectangular beam.

4 Results

In this section, we present the results of our experiments, focusing on the capability of the language models in generating coherent and high-quality text. We begin by evaluating the models' performance in generating text similar to the original field engineers' inputs. We then evaluate the quality of the generated text by comparing its performance on text classification tasks. Our evaluation include:

- *Perplexity* to measure how well the language models can generate text that is similar to the original input in terms of syntax, semantics, and coherence. Lower perplexity values indicate better performance.
- *F1-score* to evaluate the text classification task, where higher $F1$-scores indicate better performance.

4.1 Perplexity

Perplexity is used as a metric to compare the performance of our generative models [13]. The use of perplexity as a metric is justified because the aim was to measure the degree of uncertainty [3] in the generated tokens over a large language sequence, since the models should be aware of the context of the surrounding words.

The definition of perplexity is:

$$\mathrm{PP}(W) = 2^{H(W)} : \; H(W) \text{ is the entropy} \tag{1}$$

Perplexity measures how well the model predicts the probability distribution of a test data set. A lower perplexity value indicates better performance [21]. The perplexity results including the parameter size of various models are summarized in Table 3.

Table 3 Performance comparison of various generative models

Generative models	Parameter size	Perplexity
Bert MLM	110M	23.35
GPT2 pretrained	117M	9.8
LLama-2	7B	5.5
Falcon	7B	9.00
GPT2 fine tuned	117M	40.9

The perplexity of the Llama-2, GPT2 pretrained and Falcon is very good (< 10), but in general we consider perplexity < 50 as a good indication that model can generate text that is similar to the original input in terms of syntax, semantics, and coherence [9]. So, we concluded that our evaluation revealed that all models were capable of producing satisfactory text for our specific application.

4.2 Classification

To evaluate performance of the quality of the generated text, we used a classifier and a standard text encoder in order to classify both the original data set and the augmented data set. The classification report metric has been generated to evaluate the performance of the classifier. The below report also shows averaging the total true positives, false negatives, and false positives.

The false positive rate is calculated as FP/FP+TN, where FP is the number of false positives and TN is the number of true negatives (FP+TN being the total number of negatives). It's the probability that a false alarm will be raised: that a positive result will be given when the true value is negative.

The baseline experiment consists of using the boosting classifier with the original data set in order to evaluate the performance of the classifier without applying text augmentation. The train data set size was composed of 266 text descriptions. We applied text augmentation on the original data set using Bert MLM, GPT2, LLAMA-2 and Falcon models in order to generate a new balanced data set with a 3 times bigger size than the original. We use the boosting classifier again with the new augmented text data as a training data set in order to evaluate the the classification performance; the test set was defined at the beginning and remained the same in all experiments (see Table 1). The performance results including the classification report and the confusion matrices are summarized in Tables 5 and 4.

Comparing the performance results across all the three tables, we can see that Llama-2 and Falcon performance are quite similar and we achieved 20% of improvement in the classification task. We report that Bert MLM improved the performance of $\approx 24\%$, and GPT2 model fine tuned outperformed the pretrained models.

Table 4 Confusion matrices: original data set (**a**) and training set text augmented with MLM BERT uncased, GPT2, LLAMA-2 (7B), Falcon (7B) (**b–e**) and fine tuned GPT2 (**f**)

	EHS impact	Fleet impact	Product impact
a. *Confusion matrix for original data set with no data augmentation of training set*			
EHS impact	4	21	3
Fleet impact	2	61	1
Product impact	6	14	3
b. *Confusion matrix for original data set with GPT2 text augmentation during training*			
EHS impact	23	0	5
Fleet impact	17	30	17
Product impact	0	0	23
c. *Confusion matrix for original data set with LLama-2 (7B) text augmentation during training*			
EHS impact	11	13	4
Fleet impact	0	64	0
Product impact	0	8	15
d. *Confusion matrix for original data set with Falcon (7B) text augmentation during training*			
EHS impact	13	11	4
Fleet impact	0	64	0
Product impact	0	9	14
e. *Confusion matrix for original data set with Bert MLM text augmentation during training*			
EHS impact	17	7	4
Fleet impact	1	63	0
Product impact	0	7	16
f. *Confusion matrix for original data set with fine tuned GPT2-based text augmentation during training*			
EHS impact	19	5	4
Fleet impact	1	63	0
Product impact	0	4	19

5 Conclusion

In industrial applications, data scarcity poses problems that would possibly require the necessity of a sophisticated system to enhance the size of the data set to improve the efficiency of our application. We leveraged different generative AI models to generate text similar to the reports written by the field engineers at the energy manufacturing sites.

We demonstrated and proved through a series of experiments that by utilizing language models, that brings out high quality text and thus mitigate the data sparsity problems. The knowledge aware generative models are very efficient in understanding the context by creating the data which is on par with human generated text,thus improved the overall quality and quantity of the database.

Table 5 Classification report: original data set (**a**) and training set text augmented with MLM BERT uncased, GPT2, LLAMA-2 (7B), Falcon (7B) (**b–e**) and fine tuned GPT2 (**f**)

Class	Precision	Recall	*f*1-score	Support
a. *Classification report for original data set with no data augmentation of training set*				
EHS impact	0.33	0.14	0.20	28
Fleet impact	0.64	0.95	0.76	64
Product impact	0.43	0.13	0.20	23
Accuracy			**0.59**	115
b. *Classification report for original data set with GPT2 text augmentation during training*				
EHS impact	0.57	0.82	0.68	28
Fleet impact	1.0	0.47	0.64	64
Product impact	0.51	1.0	0.68	23
Accuracy			**0.66**	115
c. *Classification report for original data set with LLama-2 (7B) text augmentation during training*				
EHS impact	1.00	0.39	0.56	28
Fleet impact	0.75	1.00	0.86	64
Product impact	0.79	0.65	0.71	23
Accuracy			**0.78**	115
d. *Classification report for original data set with Falcon (7B) text augmentation during training*				
EHS impact	1.00	0.46	0.63	28
Fleet impact	0.76	1.00	0.86	64
Product impact	0.78	0.61	0.68	23
Accuracy			**0.79**	115
e. *Classification report for original data set with Bert MLM text augmentation during training*				
EHS impact	0.94	0.61	0.74	28
Fleet impact	0.82	0.98	0.89	64
Product impact	0.80	0.70	0.74	23
Accuracy			**0.83**	115
f. *Classification report for original data set with fine tuned GPT2-based text augmentation during training*				
EHS impact	0.95	0.68	0.79	28
Fleet impact	0.88	0.98	0.93	64
Product impact	0.83	0.83	0.83	23
Accuracy			**0.88**	115

Text augmentation improved the accuracy from 59% to 78–88%

Firstly, we verified the perplexity of the models: all the experiments reported reasonable perplexity score; therefore, we can conclude that models were able to generate good quality texts. Finally, we applied a classifier for the industrial use case. Our experiments showed that LLAMA-2 and Falcon models produced excellent text for our purpose, but surprisingly the best generative model was still GPT2 fine tuned on our custom data set. Indeed, in a technical domain with very specific acronyms and abbreviations; the language varies between countries and specialists. Hence, the fine tuned model can understand the "slang" better and improve the text generation. The assumption was confirmed by the improvement in performance between GPT2 pre-trained vs GPT2 fine tuned.

Further, there are various methods where we can demystify the black box of generative models by numerous XAI methods such as intrinsic explainability, attention-based, saliency methods, and extrinsic explainability etc. By these techniques we can explain how a model generated a specific text and the reasoning behind it. We are planning for an elaborate work on the explainability of generative models in the near future. We plan to enhance our work by optimizing the existing Llama2 and falcon Large Language Models (LLMs) by way of quantization. Quantizing the LLMs reduces the precision of model parameters thus saves memory and helps in the deployment of these small models on various devices including edge devices making the inference simpler.

This work is not intended as a benchmark between models but we want to report a real use case to suggest the use of generative models to increase the training data set.

References

1. Almazrouei E, Alobeidli H, Alshamsi A, Cappelli A, Cojocaru R, Debbah M, Goffinet E, Heslow D, Launay J, Malartic Q, Noune B, Pannier B, Penedo G (2023) Falcon-40B: an open large language model with state-of-the-art performance
2. Austin J, Johnson DDNW, Ho J, Tarlow D, van den Berg R (2021) Structured denoising diffusion models in discrete state-spaces
3. Cover Thomas M, Thomas Joy A (2006) Elements of information theory (Wiley series in telecommunications and signal processing). Wiley-Interscience, USA
4. Dai H, Liu Z, Liao W, Huang X, Wu Z, Zhao L, Liu W, Liu N, Li S, Zhu D et al (2023) Chataug: leveraging chatgpt for text data augmentation. arXiv preprint arXiv:2302.13007
5. Devlin J, Chang M-W, Lee K, Toutanova K (2019) Bert: pre-training of deep bidirectional transformers for language understanding. ArXiv, abs/1810.04805
6. Dorogush AV, Ershov V, Gulin A (2018) Catboost: gradient boosting with categorical features support
7. Ghazvininejad M, Levy O, Liu Y, Zettlemoyer L (2019) Mask-predict: parallel decoding of conditional masked language models. In: Conference on empirical methods in natural language processing
8. Iqbal T, Qureshi S (2022) The survey: text generation models in deep learning. J King Saud Univer—Comput Inf Sci 34(6, Part A):2515–2528
9. Rafal J, Vinyals O, Shazeer N, Wu Y (2016) Exploring the limits of language modeling, Mike Schuster
10. Kong Z, Ping W, Huang J, Zhao K, Catanzaro B (2021) Diffwave: a versatile diffusion model for audio synthesis. In: International conference on learning representations
11. Kumar V, Choudhary A, Cho E (2020) Data augmentation using pre-trained transformer models. arXiv preprint arXiv:2003.02245
12. Liu Y, Ott M, Goyal N, Du J, Joshi M, Chen D, Levy O, Lewis M, Zettlemoyer L, Stoyanov V (2019) Roberta: a robustly optimized bert pretraining approach. arXiv preprint arXiv:1907.11692
13. Miaschi A, Brunato D, Dell'Orletta F, Venturi G (2021) What makes my model perplexed? a linguistic investigation on neural language models perplexity. In: Proceedings of deep learning

inside out (DeeLIO): the 2nd workshop on knowledge extraction and integration for deep learning architectures, Online. Association for Computational Linguistics, pp 40–47
14. Mishra A, Laha A, Sankaranarayanan K, Jain P, Krishnan S (2019) Storytelling from structured data and knowledge graphs: an NLG perspective, pp 43–48
15. Penedo G, Malartic Q, Hesslow D, Cojocaru R, Cappelli A, Alobeidli H, Pannier B, Almazrouei E, Launay J (2023) Outperforming curated corpora with web data, and web data only, the refinedweb dataset for falcon LLM
16. Penedo G, Malartic Q, Hesslow D, Cojocaru R, Cappelli A, Alobeidli H, Pannier B, Almazrouei E, Launay J (2023) The RefinedWeb dataset for Falcon LLM: outperforming curated corpora with web data, and web data only. arXiv preprint arXiv:2306.01116
17. Qin L, Welleck S, Khashabi D, Choi Y (2022) COLD decoding: energy-based constrained text generation with langevin dynamics. In: Oh AH, Agarwal A, Belgrave D, Cho K (eds) Advances in neural information processing systems
18. Qiu X, Sun T, Yige X, Shao Y, Dai N, Huang X (2020) Pre-trained models for natural language processing: a survey. Sci China Technol Sci 63:1872–1897
19. Qu Y, Liu P, Song W, Liu L, Cheng M (2020) A text generation and prediction system: pre-training on new corpora using BERT and GPT-2. In: 2020 IEEE 10th international conference on electronics information and emergency communication (ICEIEC), pp 323–326
20. Radford A, Jeffrey W, Child R, Luan D, Amodei D, Sutskever Ilya et al (2019) Language models are unsupervised multitask learners. OpenAI blog 1(8):9
21. Reddy S (2023) Evaluating large language models for use in healthcare: a framework for translational value assessment. Inf Med Unlocked 41:101304
22. Rombach R, Blattmann A, Lorenz D, Esser P, Ommer B (2021) High-resolution image synthesis with latent diffusion models. 2022 IEEE/CVF conference on computer vision and pattern recognition (CVPR), pp 10674–10685
23. Rozière B, Gehring J, Gloeckle F, Sootla S, Gat I, Tan XE, Adi Y, Liu J, Remez T, Rapin J et al (2023) Code llama: open foundation models for code. arXiv preprint arXiv:2308.12950
24. Saharia C, Chan W, Saxena S, Li L, Whang J, Denton EL, Ghasemipour SKS, Ayan BK, Mahdavi SS, Lopes RG, Salimans T, Ho J, Fleet DJ, Norouzi M (2022) Photorealistic text-to-image diffusion models with deep language understanding. ArXiv, abs/2205.11487
25. Scao TL, Fan A, Akiki C, Pavlick E, Ilić S, Hesslow D, Castagné R, Luccioni AS, Yvon F, Gallé M et al (2022) Bloom: a 176b-parameter open-access multilingual language model. arXiv preprint arXiv:2211.05100
26. Shorten C, Khoshgoftaar TN, Furht B (2021) Text data augmentation for deep learning. J Big Data 8:1–34
27. Tan B, Yang Z, Al-Shedivat M, Xing E, Hu Z (2020) Progressive generation of long text. CoRR, abs/2006.15720
28. Touvron H, Martin L, Stone K, Albert P, Almahairi A, Babaei Y, Bashlykov N, Batra S, Bhargava P, Bhosale S, Bikel D, Blecher L, Ferrer CC, Chen M, Cucurull G, Esiobu D, Fernandes J, Fu J, Fu W, Fuller B, Gao C, Goswami V, Goyal N, Hartshorn A, Hosseini S, Hou R, Inan H, Kardas M, Kerkez V, Khabsa M, Kloumann I, Korenev A, Koura PS, Lachaux M-A, Lavril T, Lee J, Liskovich D, Lu Y, Mao Y, Martinet X, Mihaylov T, Mishra P, Molybog I, Nie Y, Poulton A, Reizenstein J, Rungta R, Saladi K, Schelten A, Silva R, Smith E-M, Subramanian R, Tan XE, Tang B, Taylor R, Williams A, Kuan JX, Xu P, Yan Z, Zarov I, Zhang Y, Fan A, Kambadur M, Narang S, Rodriguez A, Stojnic R, Edunov S, Scialom T (2023) Llama 2: open foundation and fine-tuned chat models
29. Wei J, Zou K (2019) Eda: Easy data augmentation techniques for boosting performance on text classification tasks. arXiv preprint arXiv:1901.11196
30. Wu C, Zhang X, Zhang Y, Wang Y, Xie W (2023) Pmc-llama: further finetuning llama on medical papers. arXiv preprint arXiv:2304.14454
31. Wu Z, Michel G, Chris B, Yizhe Z, Xiang G, Chris Q, Rik K-K, Jianfeng G, Hannaneh H, Mari O, Bill D (2021) A controllable model of grounded response generation. In: Proceedings of the AAAI conference on artificial intelligence 35(16):14085–14093

32. Yamaguchi A, Chrysostomou G, Margatina K, Aletras N (2021) Frustratingly simple pretraining alternatives to masked language modeling
33. Zhang H, Song H, Li S, Zhou M, Song D (2022) A survey of controllable text generation using transformer-based pre-trained language models. ArXiv, abs/2201.05337
34. Zhang L, Sun J-T (2009) Text generation. Springer US, Boston, MA, pp 3048–3051

Exploring Cost-Effective Machine Learning Classifiers for Text Data: A Literature Review

Abbas Saliimi Lokman, Mohamed Ariff Ameedeen, and Ngahzaifa Ab. Ghani

Abstract The escalating size of Machine Learning (ML) models in Natural Language Processing (NLP) systems presents a challenge for researchers with limited computing resources. This assertion is substantiated by the current research landscape, where major industrial corporations dominate the State-of-The-Art (SoTA) NLP-ML model leaderboard. As substantial ML models entail high computing costs, an economical approach within this research domain is crucial. This paper aims to provide a literature review as a foundational analysis toward exploring the right direction in designing cost-effective or economical NLP-ML models, specifically for text data classification problem. Before delving into a domain-specific NLP-ML review, an introduction to ML algorithms is provided. This introductory section discusses an overview of the most common ML algorithms in recent years, along with the problem domains these algorithms attempt to address, with classification being one of those challenges. The review section comprises three interconnected yet independent subsections: ML for classification, ML for NLP, and the economics of NLP-ML models. Synthesizing information from these review subsections, the subsequent sections present a summary and a future direction. While the summary aims to provide a cohesive conclusion to the entire review, the future direction section includes an example case with a proposed methodology, illustrating where the economical ML model might be implemented. Collectively, these final sections provide a justified direction on how to design an economical NLP-ML classifier, with the hope that it becomes a basic guide for researchers interested in pursuing this field.

Keywords Machine learning · Classification · Cost-effective · Economical · Natural language processing · Neural network

---

http://www.umpsa.edu.my.

A. S. Lokman (✉) · M. A. Ameedeen · N. Ab. Ghani
Faculty of Computing, Universiti Malaysia Pahang Al-Sultan Abdullah,
26600 Pekan, Pahang, Malaysia
e-mail: abbas@umpsa.edu.my

S. M. Thampi et al. (eds.), *Fifth International Conference on Computing and Network Communications*, Lecture Notes in Electrical Engineering 1221,
https://doi.org/10.1007/978-981-97-4711-5_41

1 Introduction

In the ever-expanding landscape of Machine Learning (ML) and Natural Language Processing (NLP), the pursuit of cost-effective or economical solutions has become paramount. To explore this direction, a fundamental understanding of ML system architecture and algorithms is crucial. At its core, Machine Learning is a computerized system that learns patterns from data. Unlike traditional computer systems, where logic is manually encoded into algorithms, an ML system's logic is automatically derived from patterns in the provided (training) data. This counter-intuitive process allows an ML system to comprehend complex logic from patterns in large datasets-an accomplishment impossible to achieve through manual encoding by a human programmer in a traditional computer system.

In general, the most common Machine Learning (ML) algorithms to date can be categorized into two *learning/training* categories: Supervised and Unsupervised. At a fundamental level, Supervised learning is an ML algorithm that learns patterns in data from given examples, while Unsupervised learning is an ML algorithm that learns patterns by discovering them in data autonomously, without given examples. Within these two categories, various ML algorithms have been proposed. To visualize these algorithms in their respective categories, Fig. 1 showcases the tree structure for the most common algorithms in both Supervised and Unsupervised categories.

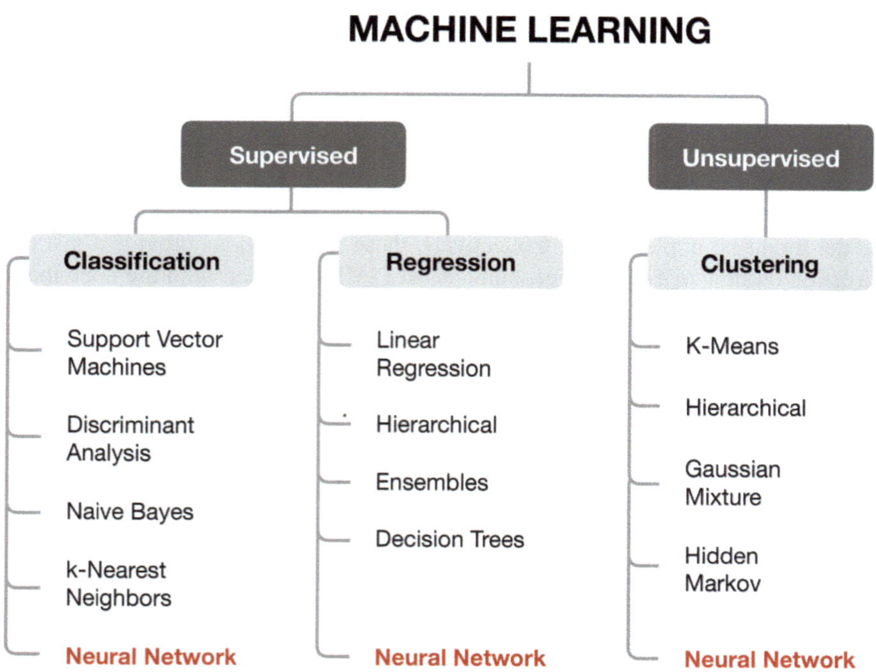

Fig. 1 The most common machine learning algorithms in recent years [1]

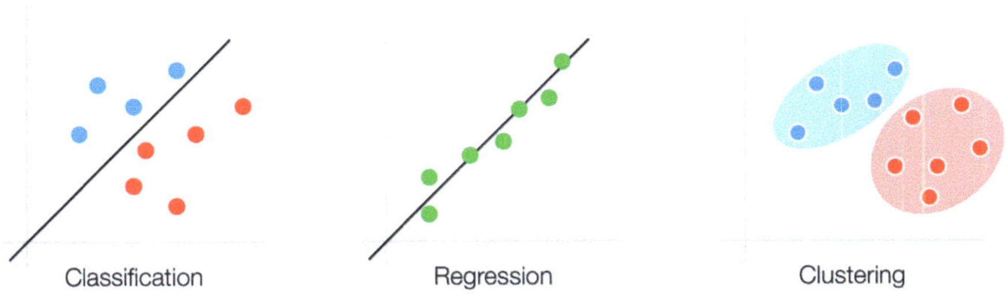

Fig. 2 Simple visualization of classification, regression and clustering

Referring to Fig. 1, ML algorithms can be further divided into three subcategories: Classification, Regression, and Clustering. These subcategories essentially represent the problems for which the algorithms are designed to solve. In the realm of Classification, the algorithm addresses the task of classifying input into one of two classes/labels (binary classification) or more than two classes/labels (multiclass classification). An example of a classification problem is identifying whether an input image depicts a cat or a dog. In Regression, the algorithm is tasked with predicting continuous output based on given input features. An example of a regression problem is predicting the value of a house based on factors such as the year of construction, location, and size. Lastly, in Clustering, the algorithm learns to categorize input into a certain number of groups based on input features. An example of a clustering problem is dividing a group of people into two subgroups based on their family, education, and work background. Figure 2 visually represents classification, regression, and clustering in their simplest forms.

Referring to Fig. 1, the most dominant algorithm in the modern ML field is the Neural Network (NN). The capability and versatility to address all three main ML problems (classification, regression, and clustering) are perhaps the reasons for its prominence. While other algorithms are relatively rigid and lack expandability, NN is highly flexible and extensible. This is exemplified by its ability to incorporate various algorithms with different capabilities and architectures into a cohesive and functional system, all while maintaining its core architecture as the fundamental ML mechanism for that particular system. To illustrate this flexibility, Fig. 3 visualizes various vanilla and expanded NN algorithms under both Supervised and Unsupervised categories, as well as the intersection between them.

This paper aims to present a literature review that ultimately supports the goal of designing economical or cost-effective ML classifier for text data. Broadly speaking, this review sits at the intersection of ML algorithms, the classification problem, and the domain of Natural Language Processing (NLP)—a computational process tailored for text data. The preceding introductory paragraphs have presented a brief discussion of the literature surrounding Machine Learning (ML) and a review of the most common and dominant ML algorithms to date, with the aim of addressing the first domain, which is the domain of ML algorithms. Moving forward, the next section will provide a review of ML usage in the classification domain, covering the

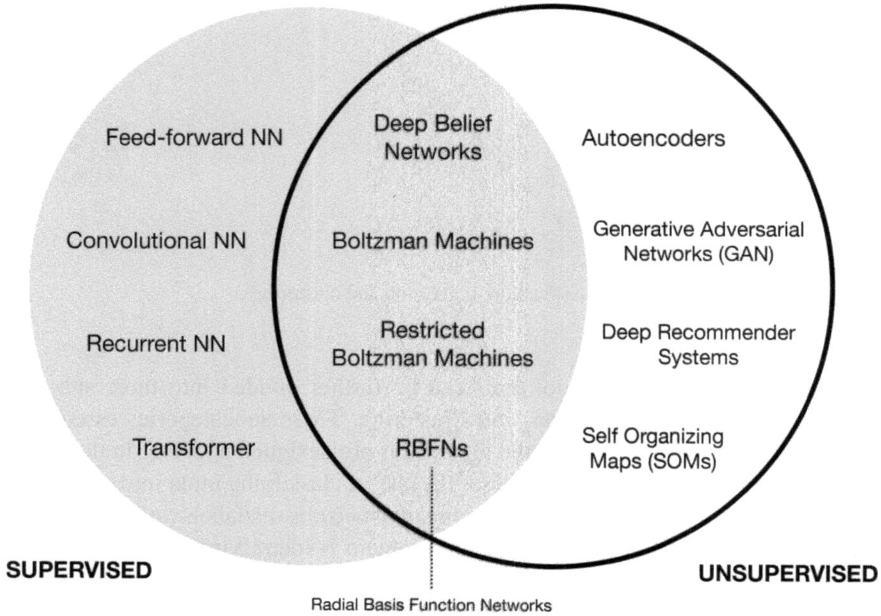

Fig. 3 Supervised and unsupervised neural network models

second domain (the classification problem), and the subsequent sections will delve into a review regarding ML models in the NLP field to address the third domain. Intersecting all domains, the final review of this paper will discuss the economic merits of ML models within the NLP domain before proposing a justifiable direction on how to design an economical NLP-ML classifier model. Overall, the following are contributions of this paper:

- A literature review focusing on ML algorithms for the classification problem in the NLP domain.
- An examination of the economic merits of current NLP-ML models.
- A justifiable direction on how to design an economical NLP-ML classifier.

2 The Review

This section is divided into three subsections that will serve as independent yet interconnected reviews surrounding economical ML classifiers for text data. These subsections are ML for classification, ML for NLP, and the economics of NLP-ML models.

2.1 ML for Classification

As mentioned in the introductory section, classification is the task of categorizing items into one of two classes/labels (binary classification) or into more than two classes/labels (multiclass classification). Referring back to Fig. 1, the five ML algorithms listed under the classification subcategory are Support Vector Machines (SVM), Discriminant Analysis (DA), Naïve Bayes (NB), k-Nearest Neighbors (kNN) and Neural Network (NN). Among all five algorithms, NN and SVM are perhaps the most commonly used algorithms in recent years justified by the number of publications within past 5 years (2019–2023). To back up this claim, the following Table 1 outlines the result counts for Google Scholar search on those five ML algorithms.

Based on Table 1, further exploration within classification domain is conducted toward SVM and NN while neglecting DA, NB, and kNN algorithms due to lack of implementation in recent years (implying disadvantageous over SVM and NN). As mentioned earlier, classification problem can be binary or multiclass and given its general paradigm, multiclass classification is naturally harder to solve. Extending Table 1 result, Table 2 outlines the result counts for Google Scholar search on SVM and NN multiclass classification implementation.

Comparing Tables 1 and 2, multiclass classification can be seen as harder to solve as publication counts for multiclass against all classification problem is 17,500 per 228,000 for SVM and 17,200 per 276,000 for NN. Although both are able to perform multiclass classification process, SVM and NN are doing it differently [2–4]. For SVM, multiclass classification is not natively supported. This is because the nature of SVM is binary classification, that is producing a hyperplane to divide datapoints into two classes. To perform multiclass classification, SVM is utilizing its binary classification process after breaking down the multiclass classification problem into several binary classification problems (a binary classifier per each pair of classes).

Table 1 Google Scholar search results for publication year 2019–2023

Search keywords	Result counts
Neural network classification	276,000
Support vector machine classification	228,000
Discriminant analysis classification	17,800
Naive bayes classification	17,400
k-nearest neighbors classification	18,100

Table 2 Google Scholar search results for publication year 2019–2023

Search keywords	Result counts
Support vector machine multiclass classification	17,500
Neural network multiclass classification	17,200

Although this approach has resulted in shorter training time (because datapoints are narrowed down per each binary classification), inference time is becoming longer due to hierarchy of classification processes (compared to NN).

In contrast to SVM, multiclass classification is natively supported by NN. By making n number of classes as NN output nodes, multiclass classification can be performed natively using softmax function. Softmax is a generalized logistic function toward multiple dimensions. For logistic function, output will either be 1 or 0 therefore can be implemented for binary classification. Generalizing it, softmax function produces a sum of 1 output (e.g., 0.2, 0.3 and 0.5 is a sum of 1) therefore can be implemented for multiclass classification because of its larger dimension. In NN implementation, a sum is a total value of all combined final layer outputs. To solve multiclass classification problem, a highest value produced by one of final layer's node will be denoted as the selected class. For example, outputs for class A, B and C are 0.2, 0.3 and 0.5 respectively. In this example, class C will be denoted as selected class due to its highest value. Related to previous SVM discussion regarding training and inference time, NN will take longer time for training (because each datapoints are being compared to all classes in every iteration), but shorter time for inference (because NN algorithm will only run the whole classification process once, compared to multiple times for SVM algorithm (multiple binary classifier)).

2.2 *ML for NLP*

A Natural Language Processing (NLP) model is a computerized system designed to process and manipulate human natural language data. Typically built upon various mathematical functions, the objective of an NLP model is to replicate human capabilities in handling natural language. These capabilities can be broadly categorized into two major groups: (i) the ability to interpret the semantics of language (understanding the language), and (ii) the ability to perform specific language tasks such as question-answering, language translation, and text classification (performing downstream NLP tasks). An NLP model cannot perform downstream tasks without first understanding the language. To enable an NLP model to comprehend human natural language, words need to be represented by mathematical values that computers can interpret. Furthermore, for a computer to 'meaningfully' interpret human natural language, these mathematical values must represent the word's semantic value and not just its syntactic value (interpreting the meaning, not just the syntax's patterns). In the NLP field, representing a word using a semantically meaningful mathematical value is termed Word Representation (WR). Overall, there are two types of WR methods: static representation (generated from word embedding methods such as word2vec [5] and GloVe [6]) and dynamic representation (generated from contextualized embedding methods such as BERT [7] and ELMo [8]).

Emerging methods in generating contextualized WR are through the usage of Machine Learning (ML) algorithm (particularly Deep Artificial Neural Network or Deep Learning). Known as Deep Neural Language Model (DNLM) or simply Large

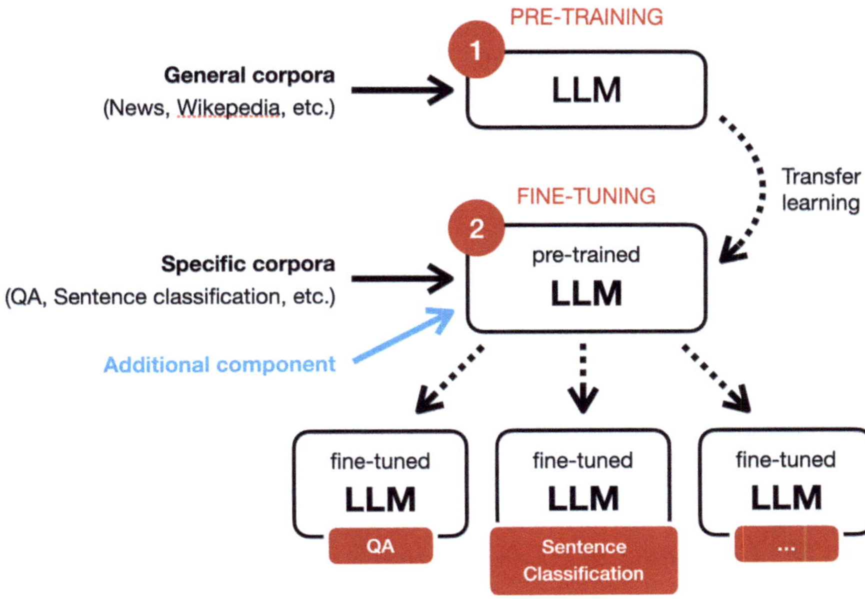

Fig. 4 LLM transfer learning process

Language Model (LLM), these models are usually versatile where they are not only capable of understanding language, but also capable of using it in various NLP downstream tasks [9–11]. To acquire these capabilities, a model is usually trained in two stages: Pre-training for language understanding (using general corpora such as news article and Wikipedia page) and fine- tuning for language usage (using downstream task dataset such as Question-Answering and sentence classification). In architectural perspective, pre-training is done to the original model structure while fine-tuning requires modification and/or additional component. Figure 4 generally illustrates the pre-training and fine-tuning process for LLM that later resulted in ready-to-use NLP model for downstream task.

To further simulate the pre-training and fine-tuning processes, let's assume the downstream task "Classification of English film review: Positive or negative comment". For this problem, LLM will first be trained using general English corpora (unsupervised pre-training using super big datasets), then a binary classification component will be added into the LLM and the whole model will be re-trained again (with previous trained weights) using English film review dataset (supervised fine-tuning using smaller dataset). The English film review dataset will contain collection of film review text, paired with associated label/class (positive or negative review). After both pre-training and fine-tuning processes are completed, the NLP model is ready to classify a new film review comment as being positive or negative, thus solving the task "Classification of English film review: Positive or negative comment".

On previous example, NLP model for downstream task requires the training of LLM model on task-specific dataset (the fine-tuning process). This practice, despite its versatility (re-using pre-trained LLM for various downstream tasks a.k.a. *transfer learning*), it is also expensive in economic point of view. This is because fine-tuning process will need to go through the whole LLM parameters even though the needed task (patterns to be learned) is simple (e.g., binary classification). To grasp the understanding of natural language and producing contextualized WR, LLM needs to have millions to billions of parameters. These needed parameters are what makes LLM so expensive to train [12]. Although needed for language understanding, these parameters are not entirely needed for NLP downstream tasks. This is because a pre-trained LLM is already capable of producing contextualized WR that can be used as input for smaller ML model (given the right dataset and sometimes additional component). For this reason, fine-tuning LLM for simple downstream task can be define as not economical.

2.3 The Economics of NLP-ML Models

The economics of NLP models can be measured using its parameter size. Although the architectural complexity between each model is different, parameter size/counts is the biggest contributor toward the amount of computing power needed to manage each model (training and inference). As per Sharir's method [12], cost (in USD) for model training is calculated per its parameter counts. Table 3 showcases the reported style of Sharir's method.

From Table 3, it shows that the cost to train BERT [7] increases rapidly per the model's size (fully loaded is a multiple training runs with hyperparameter tuning process). As a background literature, BERT is a LLM that is based upon the Transformer architecture [13]. Despite the fact that Transformer accumulates a lot of parameters hence meaning more expensive in nature, it is still considered as a very capable ML architecture, particularly in producing contextualized representation of text. This capability can be demonstrated by Transformer implementation in recent notable LLM showcases in Table 4.

Table 4 outlines the size (parameter counts), base architecture and developer of some notable recent LLM, sorted from smallest to largest in size, that can also be translated as cheapest to most expensive in cost. From Tables 3 and 4, it can be

Table 3 The cost to train different size of Bert models [12]

Single run cost (in USD)	Fully loaded cost (in USD)	Parameter counts
2,500	50,000	110 million
10,000	200,000	340 million
80,000	1,600,000	1.5 billion

Table 4 Notable decent LLM size, architecture and developer

Model name	Parameter counts	Base architecture	Developer
ELMo [8]	94 million	LSTM	AllenNLP
BERT [7]	340 million	Transformer	Google
XLNET [14]	340 million	Transformer	Google Brain + CMU
RoBERTa [15]	355 million	Transformer	Facebook
GPT-2 [16]	1.5 billion	Transformer	OpenAI
Megatron-lm [17]	8.3 billion	Transformer	Nvidia
T5 [18]	11 billion	Transformer	Google
Turing-NLG [19]	17 billion	Transformer	Microsoft
LLaMa [20]	65 billion	Transformer	Meta AI
GPT-3 [21]	175 billion	Transformer	OpenAI
PaLM [22]	540 billion	Transformer	Google

concluded that LLM is super expensive to train. This claim can be further justified by *Developer* column in Table 4 in which big technological corporation names are presented instead of universities or small companies. This indicates that only organizations with access to extensive computing resources can train LLM. To put things into perspective, it takes 4 d to train BERT (Large) with 340 million parameters on 16 GB of training data using 64 TPU chips (or approximately 1 day with 280 T V100 GPU), 1 day to train Roberta with 355 million parameters on 160 GB of training data using 1024 T V100 GPU, and 2.5 d to train XLNET with 340 million parameters on 113 GB of training data using 512 TPU chips. In US dollar perspective, training BERT model with 16 GB data can cost anywhere between $50k and $1.6 m (depending on the chosen model size and training procedure) and training T5 model for a single run will cost more than $1.3 m [12].

Looking at economical points of view, ML model architecture performance is a trade-off between Accuracy, Economy and Complexity. As illustrated in Fig. 5, ML model economical score can be divided into four segments: most economical, most expensive, and two average segments. By plotting the Complexity (mainly the parameter counts) and Accuracy (score to certain task) values, an ML model economical score segment can be determined. Although such is the general case, accuracy performance is strictly related to what problem the model intends to solve. If the problem scale is small, increasing model's complexity and parameter counts might not produce better accuracy but only making the model less economical. As such, depending on the problem at hand, one should always start with simple (less parameter counts) model first before venturing into more complex and less economical model.

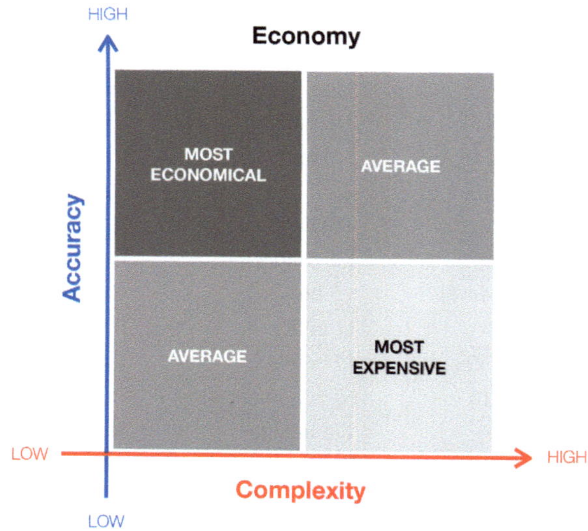

Fig. 5 ML model economical score segments

3 Summary of Review

As mentioned previously in ML for NLP section, versatility is the current direction for emerging NLP models. LLM transfer learning through pre-training and fine-tuning is currently the go-to method for solving various NLP downstream tasks. Although pre-training of DLNM requires extensive computing resources (as presented in previous section), fine-tuning is more economically accessible. This is because the required dataset for fine-tuning is much smaller than what is required for pre-training (megabyte for fine-tuning versus gigabyte for pre-training). Nevertheless, fine-tuning is still relatively expensive due to the LLM size. As also previously mentioned in prior section, fine-tuning process still needs to go through the whole LLM parameters albeit small dataset input and un-complex pattern to be learned. Consequently, performing specific text classification or other downstream tasks through the fine-tuning process remains costly, as indicated by the model size metric. Given this realization, it is logical to utilize the pre-trained LLM solely for producing Word Representations (WR) and then substitute the fine-tuning process with a smaller ML model for economic purposes, with WR serving as input to the smaller ML model. When focusing on the downstream task of NLP classification process, a question arises: *What ML model is the most economical for solving the classification problem of text data?*

From previous ML for Classification subsection, it can be summarized that Neural Network (NN) is the most prevalent algorithm used for classification problem. And referring to previous Fig. 3, there are four supervised NN models/architectures that can be used for classification task. Those models are Feed-forward NN, Recurrent NN, Convolutional NN and Transformer. In its core, recurrent, convolutional and transformer are all implementing the feed-forward NN algorithm. Being the vanilla

Table 5 Supervised ANN model's components comparison

Architecture	Components
Feed-forward	Input, hidden and output layers (vanilla)
Recurrent	Vanilla + recurrent mechanism on hidden layers (base RNN)
	Vanilla + recurrent + memory cell (LSTM)
	Vanilla + recurrent + different memory cell (GRU)
Convolutional	Vanilla + kernel + convolutional cell
Transformer	Vanilla (multiple) + positional encoding + attention mechanism

or original form of NN, feed-forward is basically the simplest form of NN model. Comparing these four models' components, feed-forward is the most economical model in the sense that other models are more complex with additional components and more parameters. Table 5 compares all four discussed supervised ANN models with regards to their implemented components.

In technical view, word representation (WR) or vector representation of word is a series of real numbers in multidimensional vector space. WR is stored in a computer as a single or multidimensional array value, hence can become input to smaller ML model provided that the model can accept parallel values as its first level input. Based on this premise and comparison of all four candidates in Table 5, it is justifiable to claim that feed-forward is the answer to the question *What ML model is the most economical to solve the classification problem of text data?*. Going a step back to the root of ML algorithm hierarchy, classification is a problem under supervised ML domain. Be it binary or multiclass, NN algorithm can be considered the best choice to solve the classification problem given its popularity in recent research publications. Finally, in regards to economical merit, feed-forward NN is proven to be the most economical model given its straight forward and simple architecture that is also able to accept parallel values of multidimensional WR as its first level input. To appreciate the surrounding algorithms and methods that leads to selecting feed-forward NN as the answer to the posed question, Fig. 6 visualizes the literature path taken toward finalizing the selection process.

4 Future Direction

Collectively, the preceding sections have illustrated that the most cost-effective or economical ML algorithm for developing a text classification model is the feed-forward NN. As highlighted in the previous *Summary of Review* section, this economical feed-forward NN model can replace the computationally-expensive LLM fine-tuning process for the downstream NLP task. In this alternative approach, a pre-trained LLM serves as the Word Representation (WR) generator for this more streamlined model. To visually compare the current State-of-The-Art (SoTA) methodology

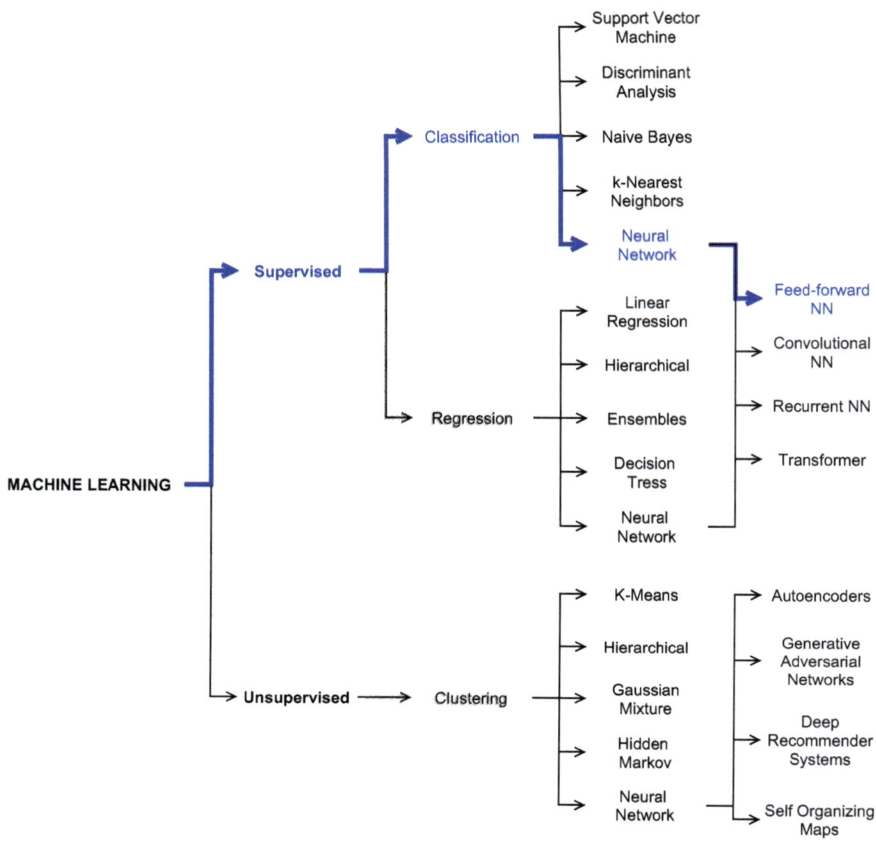

Fig. 6 Literature path toward selecting feed-forward NN

(utilizing LLM for both pre-training and fine-tuning processes) with the proposed economical methodology, Fig. 7 illustrates the flow and components of the proposed approach.

Referring to Fig. 7, the proposed methodology substitutes the *Additional component* in the State-of-the-Art (SoTA) methodology with the *External weight* component. In the SoTA methodology, the *Additional component* typically represents an extra layer in the LLM architecture designed to address downstream NLP tasks such as question-answering, language translation, and text classification. In the proposed methodology, the *External weight* acts as this component by determining the significance or insignificance of the Word Representation (WR) using an external weightage value. For instance, in addressing a text classification problem, the *External weight* value can be the factor that enhances the importance of certain words in the given document, thereby elevating the input WR to assist the base feed-forward neural network in performing the classification. An example method for generating such an *External weight* value is provided by Lokman et al. [23].

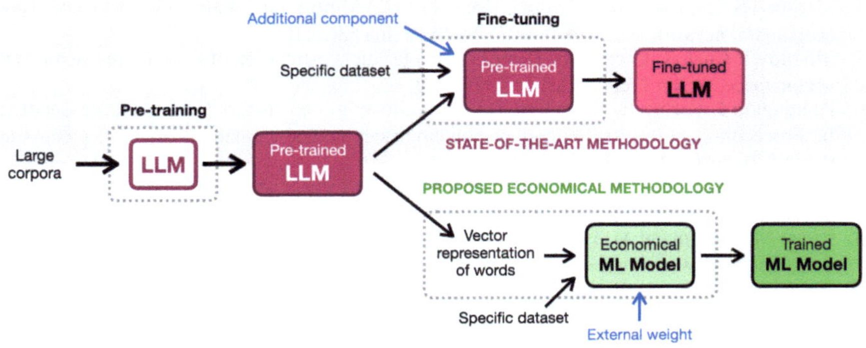

Fig. 7 Proposed economical methodology for NLP downstream tasks

5 Conclusion

This paper provides a well-justified approach for designing an economical Natural Language Processing (NLP)—Machine Learning (ML) classifier, derived from a selective literature review encompassing NLP and ML. The direction is summarized through an examination of three key subsections within the NLP-ML domain which are: ML for Classification, ML for NLP, and the economics of NLP-ML models. Additionally, the paper incorporates an introductory review of ML and its prevailing algorithms in recent years. In essence, the overarching goal of this paper is to lay the groundwork for analyzing the optimal path in creating an economical NLP-ML model. This exploration is particularly significant due to the prevailing trend of escalating NLP-ML model sizes, which poses a challenge for researchers with constrained computing resources. Ultimately, the paper aspires to encourage further research in this direction, with the fundamental objective of establishing economical ML models as the new standard in a predominantly non-economical ML field.

Acknowledgements The authors would like to thank Universiti Malaysia Pahang Al-Sultan Abdullah for providing financial support under Internal Research Grant RDU220315.

References

1. Duc TL, Leiva RG, Casari P, Östberg PO (2019) Machine learning methods for reliable resource provisioning in edge-cloud computing: a survey. ACM Comput Surv (CSUR) 52(5):1–39
2. Tiwari P, Upadhyay D, Pant B, Mohd N (2022) Multiclass classification of disease using CNN and SVM of medical imaging. In: International conference on advances in computing and data sciences. Springer International Publishing, Cham, pp 88–99
3. Vidyarthi A, Agarwal R, Gupta D, Sharma R, Draheim D, Tiwari P (2022) Machine learning assisted methodology for multiclass classification of malignant brain tumors. IEEE Access 10:50624–50640

4. Maqsood S, Damaševičius R, Maskeliūnas R (2022) Multi-modal brain tumor detection using deep neural network and multiclass SVM. Medicina 58(8):1090
5. Mikolov T, Chen K, Corrado G, Dean J (2013) Efficient estimation of word representations in vector space, ArXiv Preprint ArXiv:1301.3781
6. Pennington J, Socher R, Manning CD (2014) Glove: global vectors for word representation. In: Proceedings of the 2014 conference on empirical methods in natural language processing (EMNLP), pp 1532–1543
7. Devlin J, Chang MW, Lee K, Toutanova K (2018) Bert: pre-training of deep bidirectional transformers for language understanding. ArXiv Preprint ArXiv:1810.04805
8. Peters ME, Neumann M, Iyyer M, Gardner M, Clark C, Lee K, Zettlemoyer L (2018) Deep contextualized word representations. In: Proceedings of the 2018 conference of the North American chapter of the association for computational linguistics: human language technologies, vol 1 (Long Papers). New Orleans, Louisiana. Association for Computational Linguistics, pp 2227–2237
9. Shaikh S, Daudpota SM, Imran AS, Kastrati Z (2021) Towards improved classification accuracy on highly imbalanced text dataset using deep neural language models. Appl Sci 11(2):869
10. Roshanzamir A, Aghajan H, Soleymani M (2021) Baghshah, transformer-based deep neural network language models for Alzheimer's disease risk assessment from targeted speech. BMC Med Inf Decision Mak 21:1–14
11. Zhao WX, Zhou K, Li J, Tang T, Wang X, Hou Y, Wen JR et al (2023) A survey of large language models. arXiv preprint arXiv:2303.18223
12. Sharir O, Peleg B, Shoham Y (2020) The cost of training nlp models: a concise overview. ArXiv Preprint ArXiv:2004.08900
13. Vaswani A, Brain G, Shazeer N, Parmar N, Uszkoreit J, Jones L, Gomez AN, Kaiser L, Polosukhin I (2017) Attention is all you need. Adv Neural Inf Proc Syst 30
14. Yang Z, Dai Z, Yang Y, Carbonell J, Salakhutdinov R, Le QV (2019) XLNet: generalized autoregressive pretraining for language understanding. In: Proceedings of the 33rd international conference on neural information processing systems, pp 5753–5763
15. Liu Y, Ott M, Goyal N, Du J, Joshi M, Chen D, Levy O, Lewis M, Zettlemoyer L, Stoyanov V, Allen PG (2019) Roberta: a robustly optimized bert pretraining approach. ArXiv Preprint ArXiv:1907.11692
16. Radford A, Wu J, Child R, Luan D, Amodei D, Sutskever I (2019) Language models are unsupervised multitask learners. OpenAI Blog 1(8). https://openai.com/blog/better-language-models
17. Shoeybi M, Patwary M, Puri R, LeGresley P, Casper J, Catanzaro B (2019) Megatron-lm: training multi-billion parameter language models using model parallelism. ArXiv Preprint ArXiv:1909.08053
18. Raffel C, Shazeer N, Roberts A, Lee K, Narang S, Matena M, Zhou Y, Li W, Liu PJ (2020) Exploring the limits of transfer learning with a unified text-to-text transformer. J Mach Learn Res 21(140):1–67
19. Rosset C (2020) Turing-NLG: a 17-billion-parameter language model by Microsoft. Microsoft Res. https://www.microsoft.com/en-us/research/blog/turing-nlg-a-17-billion-parameter-language-model-by-microsoft
20. Touvron H, Lavril T, Izacard G, Martinet X, Lachaux MA, Lacroix T, Rozière B, Goyal N, Hambro E, Azhar F, Rodriguez A (2023) Llama: open and efficient foundation language models. arXiv preprint arXiv:2302.13971
21. Brown TB, Mann B, Ryder N, Subbiah M, Kaplan J, Dhariwal P, Neelakantan A, Shyam P, Sastry G, Askell A, Agarwal S, Herbert-Voss A, Krueger G, Henighan T, Child R, Ramesh A, Ziegler DM, Wu J, Winter C, Amodei D (2020) Language models are few-shot learners. Adv Neural Inf Proc Syst 33(NeurIPS 2020):1877–1901
22. Chowdhery A, Narang S, Devlin J, Bosma M, Mishra G, Roberts A, Barham P, Chung HW, Sutton C, Gehrmann S, Schuh P (2023) Palm: scaling language modeling with pathways. J Mach Learn Res 24(240):1–113
23. Lokman AS, Ameedeen MA, Ghani NA (2023) General words representation method for modern language model. J Telecommun Electron Comput Eng (JTEC) 15(1):1–5

Alex: A Virtual Assistant for Natural Interaction in Mixed Reality Environment

Shafina Abd Karim Ishigaki, Ajune Wanis Ismail, Nur Ameerah Abdul Halim, and Md. Sah Hj Salam

Abstract This study introduces "Alex," a virtual assistant designed for seamless and natural interactions within a mixed reality (MR) environment. Leveraging advancements in voice command and speech recognition technologies, Alex incorporates a speech-to-text (STT) system and natural language processing (NLP) to transcribe and respond to user commands. The integration of a virtual assistant avatar enhances MR interactions, reducing the time needed for users to instantiate objects and enhancing collaborative telepresence experiences. The chapter presents a comprehensive list of voice commands and details testing in an MR application, providing an overview of the implementation of natural voice commands in MR. The study concludes with suggestions for future research, emphasizing the ongoing exploration and refinement of natural voice command applications in the MR landscape.

Keywords Virtual assistant · Mixed reality · Natural language processing · Speech-to-text · Voice commands · Speech recognition

---

S. A. K. Ishigaki (✉) · A. W. Ismail · N. A. A. Halim
Mixed and Virtual Reality Research Lab, ViCubeLab, Faculty of Computing, Universiti Teknologi Malaysia, 81310 Johor Bahru, Malaysia
e-mail: shafina@graduate.utm.my

A. W. Ismail
e-mail: ajune@utm.my

N. A. A. Halim
e-mail: nurameerah@graduate.utm.my

Md. S. H. Salam
Faculty of Computing, Universiti Teknologi Malaysia, 81310 Johor Bahru, Malaysia
e-mail: sah@utm.my

S. M. Thampi et al. (eds.), *Fifth International Conference on Computing and Network Communications*, Lecture Notes in Electrical Engineering 1221,
https://doi.org/10.1007/978-981-97-4711-5_42

1 Introduction

Mixed reality (MR) is a new type of real-world experience that is enhanced by a layer of computer graphics-based interaction tied to a specific activity [1]. MR seamlessly superimposes two-dimensional (2D) and three-dimensional (3D) objects, including audio files, films, and textual material, onto the real world. An MR interface allows the user to see the real world through a portable or head-mounted device (HMD) that overlays graphics on the surroundings. In MR, user interaction must be as intuitive and robust as possible, with the user interacting directly with the object [2]. The MR system has been described as a hybrid of real and virtual elements in a real environment that operates interactively and in real time. MR allows users to experience the real world as a more integrated and better environment when combined with augmented data.

In recent years, significant progress has been made in the field of personal virtual assistants, owing primarily to underlying technologies such as virtual (VR) and augmented reality (AR). A virtual agent, on the other hand, is a computer-generated avatar designed to help humans communicate with digital environments. A "chatbot" is a piece of software that can carry on text or voice conversations in place of a human service provider. Chatbots become more capable of mimicking human conversation as artificial intelligence (AI) advances, eventually replacing human workers [3, 4]. The related works to this study are depicted in Fig. 1. The interaction between the virtual assistant providing a hint and the student selecting a response is depicted in Fig. 1a. In Fig. 1b, a virtual assistant in augmented reality demonstrates the ancient leather tanning process step by step. Meanwhile, Fig. 1c depicts the Mini-Me, an adaptive avatar designed to improve MR remote collaboration.

It wasn't until 2011 that the first modern computerized voice assistant, "Siri," was released [8]. Various companies, including Microsoft with Cortana and Amazon with Alexa, have incorporated spoken dialogue systems into their product development [9]. Notably, early voice assistants such as Siri and Alexa lacked a visual representation or avatar [10], they solely relied on voice interaction. In response to this limitation, our initial work introduced a virtual assistant avatar designed to enhance the voice assistant experience in AR. Figure 2 illustrates the preliminary outcomes of this project, showcasing the virtual avatar's capability to respond to queries and

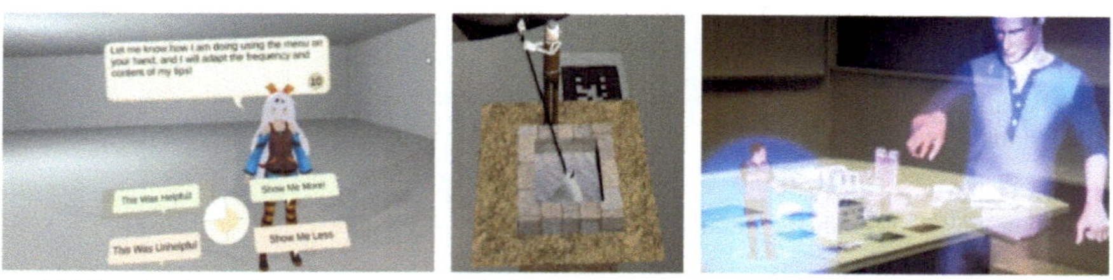

(a) Virtual assistant in VR [5] (b) Virtual assistant in AR [6] (c) MiniMe in MR[7][7]

Fig. 1 Related works. **a** Virtual assistant in VR [5]. **b** Virtual assistant in AR [6]. **c** Mini-Me in MR [7]

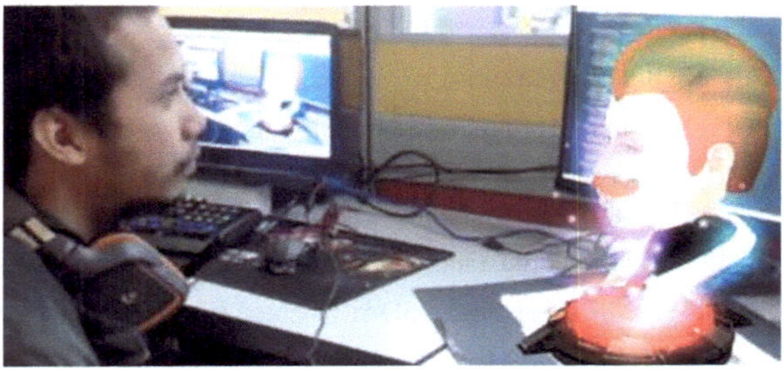

Fig. 2 Virtual assistant in AR application

execute commands, thereby bridging the gap between voice interaction and visual representation.

Voice command and recognition technologies represent significant innovations in the twenty-first century [8]. Their applicability in MR offers a distinct advantage for facilitating user interaction [9], providing alternative interfaces that eliminate the need for physical contact or button pressing to execute specific functions. With their widespread accessibility, speech technologies can be effectively employed in diverse yet constrained scenarios [8]. For instance, the HoloLearn application developed by [9] for Microsoft HoloLens aims to enhance the self-sufficiency of individuals with cognitive disabilities by assisting in acquiring fundamental everyday life skills. Their study revealed that integrating a speech-based virtual assistant into an MR environment improves user performance during activities. Marques et al. [10] conducted a study integrating MR and a virtual assistant for asynchronous remote guidance in industrial scenarios, creating a highly adaptable system supporting basic tasks based on instructions from prior sessions. Furthermore, studies by [11, 12] have explored the use of voice assistants in MR and VR, demonstrating a positive reception among users for expediting decision-making processes, particularly in shopping contexts.

Natural language processing (NLP) facilitates an immersive experience by means of speech recognition. NLP facilitates enhanced and more significant interactions and relationships with avatars, as well as other digital characters or assets [13]. NLP enables users to utilize their vocal commands to navigate within the MR environment. The advent of voice command and voice recognition technology has emerged as a noteworthy achievement in the twenty-first century, facilitating human interaction with various devices and applications via speech recognition capabilities. The utilization of voice input is a method that enables users to effectively communicate their intentions [14]. This technology facilitates the ability of virtual assistants [15] to effectively and reliably process human voices and provide valuable and beneficial services. This paper elucidates the process of integrating voice commands into MR systems and explores the evolution of the virtual assistant known as Alex.

Table 1 shows the related works on a diverse exploration of the applications and implications of MR environments, each focusing on distinct domains and functionalities. Notably, the work in 2022 by [16] delves into the educational realm, employing mixed-reality simulation for teacher preparation in reading. In contrast, [9] from 2020 concentrates on leveraging speech-based virtual assistants within MR applications

Table 1 Related works virtual assistant in MR

Year	Related works
2022	Using mixed reality simulation in teacher preparation in reading [16]
2020	Speech-based virtual assistants in mixed reality applications for people with cognitive disabilities [9]
2020	A virtual assistant for natural interactions in museums [15]
2022	A MR scene with a 3D augmented avatar during autonomous driving [17]
2019	HoloLearn: Mixed reality experiences for persons with neurodevelopmental disorders [18]
2022	Question answering with BERT: Designing a 3D virtual avatar for cultural heritage exploration [19]
2018	Mini-Me: An adaptive avatar for mixed reality remote collaboration [7]
2018	Exploring virtual agents for augmented reality [20]

to support individuals with cognitive disabilities, showcasing the potential for inclusive technology. The study in 2020 by [15] extends the scope to cultural and educational settings, presenting a virtual assistant tailored for natural interactions within museums. Moving into automotive contexts, [17] introduces a MR scene with a 3D augmented avatar during autonomous driving in 2022. Shifting the focus to neurodevelopmental disorders, [18] in 2019 explores HoloLearn, emphasizing mixed reality experiences tailored for persons with specific cognitive needs. In cultural heritage exploration, [19] explores question answering with BERT, integrating a 3D virtual avatar in 2022. Earlier works, such as [7] in 2018, introduces "Mini-me," an adaptive avatar designed for mixed reality remote collaboration, while [20] investigates virtual agents in augmented reality in the same year. This comparative overview highlights the breadth of applications within MR research, spanning education, accessibility, cultural heritage, collaboration, and automotive contexts.

2 Proposed System

This section presents the proposed system that enables the user of the HMD to issue voice commands to a virtual assistant named Alex in MR. The virtual assistant avatar will possess the capability to receive voice input data, analyze the data, carry out the requested action, and provide auditory feedback in response to the commands. The process will entail the transformation of spoken language into written text (speech-to-text or STT), the conversion of written text into spoken language (text-to-speech or TTS), and the establishment of a MR environment. Therefore, the proposed method will be segmented into three distinct processes. The initial step involves utilizing voice commands, subsequently interacting with an avatar virtual assistant and the MR application.

2.1 Speech Interaction Module

Speech recognition is a strategy that enhances the functionality of virtual assistants. Currently, there is a wide array of languages accessible to assist individuals, and this number is steadily increasing. By incorporating voice commands, natural interaction obviates the need for users to rely on gestures to physically touch and manipulate virtual objects. The user simply needs to focus on the virtual item and then verbally articulate their instruction. The ability to utilize voice commands instead of navigating a complex user interface would greatly facilitate users in navigating through layered menus with a single instruction. Users can optimize efficiency and expedite the interaction process by utilizing voice commands. Therefore, in this study, we utilize a straightforward vocal instruction to incorporate the object into the mixed reality scene. Figure 3 depicts the speech interaction module, where the user utilizes a HMD to issue commands and communicate with the MR application. This module utilizes a microphone to capture voice commands.

In order to execute the voice commands, it is necessary to perform STT conversion to recognize the command. Once the command has been recognized, it will generate the corresponding text. The instruction will be executed based on the textual output. Figure 3 illustrates the STT conversion process. As indicated by Fig. 4, it is necessary to analyze the voice prior to transmitting it to the speech recognition decoder. The speech recognition decoder utilizes the acoustic model and language model to analyze various alignments of potential pronunciations and word sequences in order to determine the most likely one.

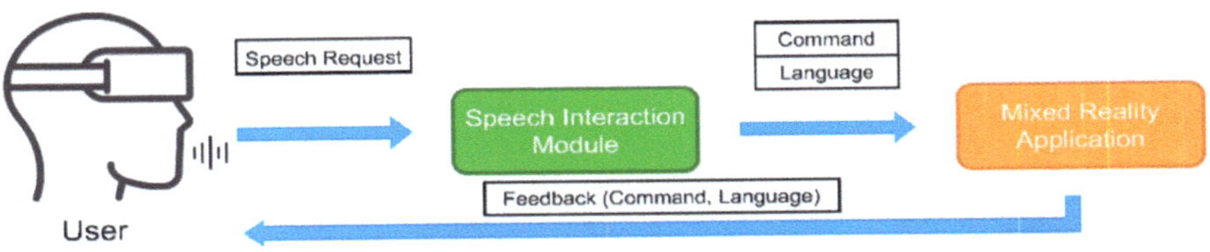

Fig. 3 Speech interaction module in MR application [21]

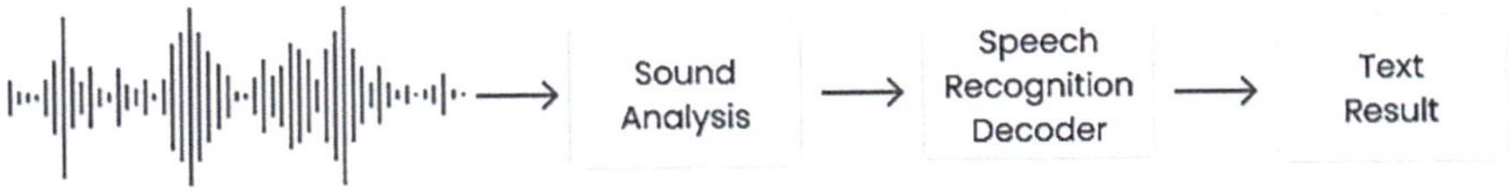

Fig. 4 Process of the STT conversion

2.2 *Virtual Assistant*

Following the conversion of STT the virtual assistant will then carry out the command based on the textual output data obtained from the speech. For instance, if the user instructs the virtual assistant, Alex, to place a building on the table in MR, Alex will carry out the command and provide a response to the user upon completion. The voice commands in this study are limited to either the provided database or a guided conversation.

The system's avatar is named Alex. Alex is a male character that has been animated using the Unity software. The virtual assistant avatar in MR will project audio as feedback to the voice commands, providing information to the user. The study employs Microsoft Azure Cognitive, a speech service, to transform text into synthesized speech that closely resembles human speech. Speech synthesis is a term used interchangeably with TTS. The procedure for converting TTS utilizing the Microsoft Azure Cognitive service. According to Fig. 5, the text file is executed using a C# script in the Unity platform to generate the audio. Microsoft Azure Cognitive has facilitated the process by offering a speech system development kit (SDK) specifically designed for the unity platform, making it significantly easier. Provided that the internet connection remains operational, the SDK can retrieve the script and make use of the available services. The script will reproduce the audio from the SDK.

A flowchart depicting the process of a voice-guided virtual personal assistant is shown in Fig. 6. Once the user enters the MR environment, Alex remains in standby mode until the user says, "Hi Alex." Subsequently, the virtual assistant has commenced. The guided dialogue has been initialized and the voice conversation has commenced. The guided conversation will provide instruction and solicit feedback for confirmation. Then the task has been executed in accordance with the instructions.

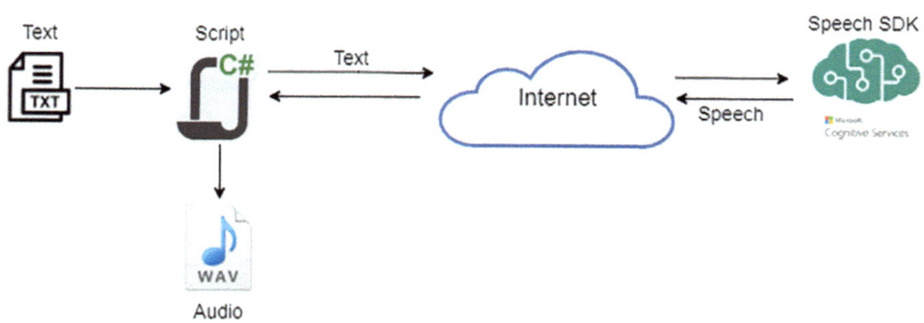

Fig. 5 Process of TTS conversion using a speech SDK

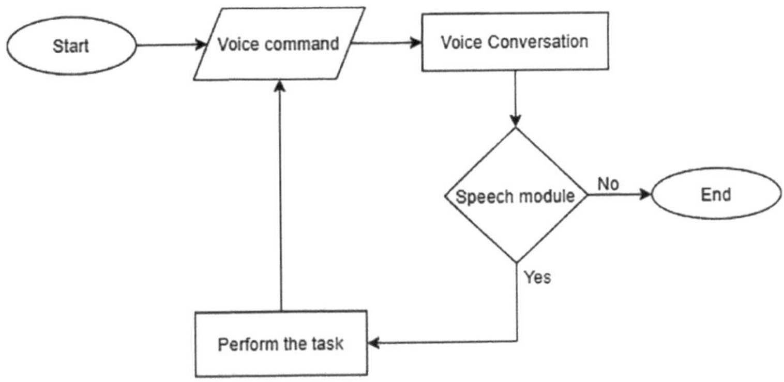

Fig. 6 Guided dialogue for Alex

2.3 *MR Environment*

VR devices are inherently interconnected with the HMD. Various HMD devices offer distinct levels of immersion. Several models of HMD are available, including Oculus Go, Quest, and Rift. HMD effectively separates individuals from their physical surroundings by fully immersing them in a simulated reality.

With regard to Fig. 7, prior to initiating the application, the user must establish the boundary of the workspace area. The VR user must consistently stay within the boundaries of the play area. It is advisable to remove any obstacles in the area to prevent users from getting injured upon collision with them. The application cannot run due to the failure to establish the play area boundary. The HMD device is connected to a personal computer (PC) for setup. The HMD provides the user with an extremely engaging VR environment, but it can potentially disconnect the user from the real world. This poses a difficulty in communicating with viewers who are not using the headset and in sharing the interactive experience. The weight of the Oculus Quest 2 is lighter than that of the Oculus Quest, with a weight of 503 grams compared to 571 grams. The device includes a solitary fast-switch LCD screen with a resolution of 1832×1920 pixels per eye and a refresh rate of up to 120 Hz. Additionally, it possesses six degree of freedom (6DOF) input capabilities.

The virtual assistant, named Alex, is depicted in MR as illustrated in Fig. 8. Alex's audio output is only accessible to the user wearing the HMD as Alex's primary function is to provide assistance and engage in interaction with the user. Alex will receive user voice commands and execute them to construct the urban city.

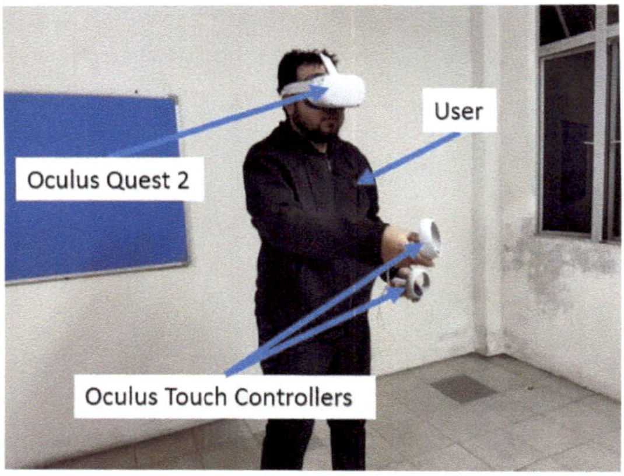

Fig. 7 MR device setting

Fig. 8 MR environment and Alex avatar

3 Test Application

This section details the outcomes of our proposed technique for voice commands and virtual assistant integration within an MR environment, leveraging the Alex avatar assistant to enhance interaction efficiency. Utilizing voice commands, Alex demonstrates the capability to add or generate 3D objects on the workbench tabletop in response to specific sequences. To initiate interaction with Alex, users employ the greeting "Hello Alex" prompting the awakening of Alex's avatar. Once initiated, users can issue a variety of eight commands, each rigorously tested within the MR environment.

As illustrated in Fig. 9a, the voice command to add a platform was successfully executed, resulting in the placement of a black platform on the workbench. Upon issuing the command "Alex, create a platform," Alex promptly responds with "The platform has been created," confirming the immediate addition of the platform. Moving to Fig. 9b, the user instructs Alex to incorporate a petrol station and subsequently construct a building adjacent to it. Responding dynamically to the command "Create a building across the petrol station," Alex promptly adds two structures to the mixed reality scene. In Fig. 9c, the user further directs Alex to accompany them to a residence across from the building.

The virtual assistant seamlessly follows user voice commands, as presented in Table 2, operating in a sequential manner to expedite the process and eliminate the

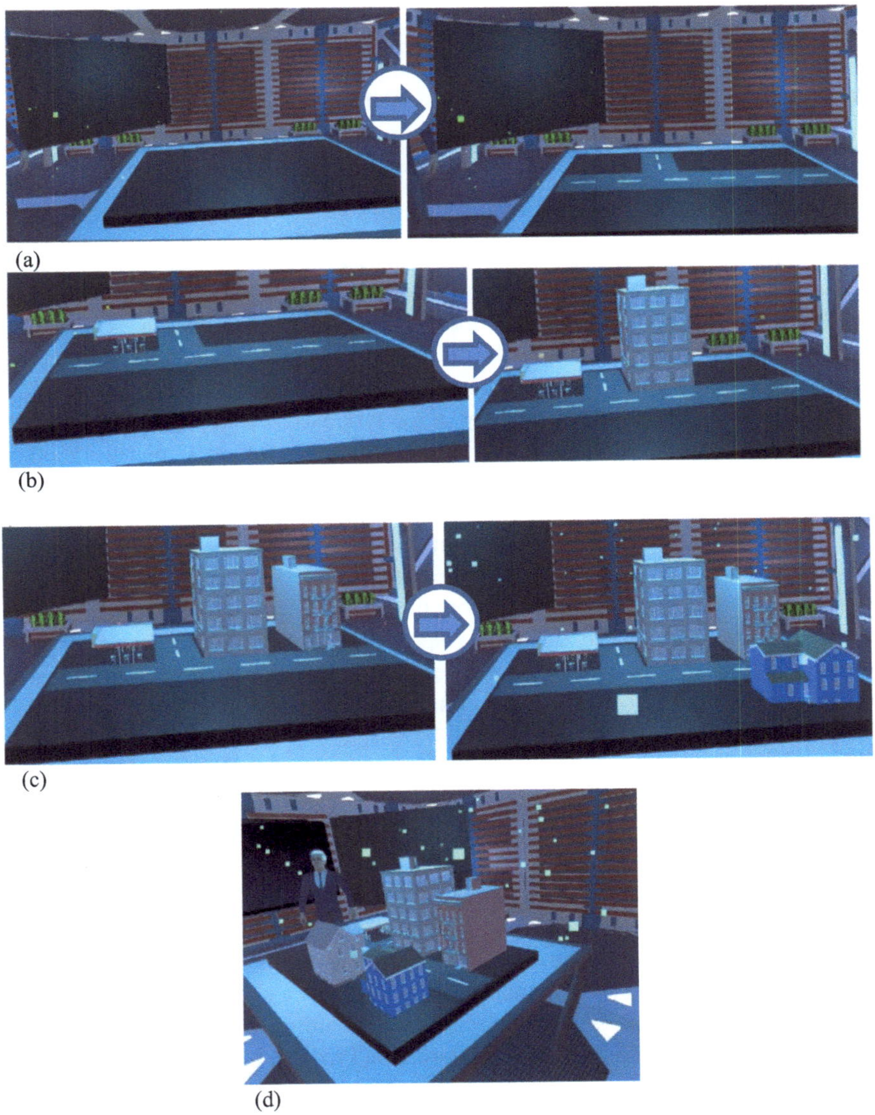

Fig. 9 Test application workflow. **a** Creating platform, the user says "*Alex creates a platform.*" **b** Adding a gas station and a building, uses a command "Add a gas station" and "*Create a building across the gas station.*" **c** Creating a house by "creates a house in front of the building." **d** Completed scene

Table 2 Voice command with response from the virtual assistant

Voice commands	Virtual assistant Alex's feedback
Alex creates a platform	Platform has created
Alex creates an inverted *T* junction	Inverted *T* junction created
Add a gas station	Gas station added
Add the building across the gas station	Ok
Add another building	No problem
Creates a house in front of the building	Done
Create another house	Is it okay now?
Yes, thank you Alex	Ok. You're welcome

need for users to manually navigate the user interface panel. This approach enhances interaction by allowing users to instantiate objects with a simple vocal command, avoiding the reliance on gestures or tools. Alex's immediate responses following object creation further contribute to the fluidity of the interaction process, showcasing the potential of voice-activated commands in mixed reality environments.

Our results demonstrate the efficacy of the proposed technique, showcasing the seamless integration of voice commands and a virtual assistant, particularly embodied by the Alex avatar, within the MR environment. The versatile execution of eight tested voice commands, ranging from the addition of platforms to the construction of complex scenes, highlights the adaptability and responsiveness of the system. The simplicity of initiating interaction with the greeting "Hello Alex" and the avoidance of manual interface navigation underscore the user-friendly nature of our approach. Overall, the immediate and dynamic responses of the virtual assistant to voice commands pave the way for more intuitive and efficient user interactions in MR settings, opening new possibilities for enhanced engagement and immersion.

4 Conclusion

The objective of this study is to investigate and implement voice commands in order to engage in conversations with the virtual assistant in MR. The user can interact with the mixed reality system by directly using their voice to manipulate objects. The local user initiates the voice command, which prompts a response from the virtual assistant named Alex in the MR environment. The utilization of voice commands enhances reliability and expedites the interaction process, eliminating the need for extensive physical movement. This feature will enhance the user's immersion and concentration in the scene by minimizing distractions caused by additional gestures or movements. In [22], researchers conducted a practical demonstration of using actual hand gestures in MR along with speech to enhance the speed and efficiency of interaction. Our local user utilizes a virtual assistant to generate the object and modify its texture using voice commands.

In future work, our focus will involve a comprehensive evaluation of the voice command system's performance in the MR environment. We plan to assess latency, measuring the response time between user-initiated commands and the virtual assistant's feedback, aiming to optimize efficiency and enhance user immersion. An additional enhancement for our prototype would be to incorporate more 6DOF object manipulation, encompassing translation, scaling, and rotation. Integrating speech and natural gestures [23, 24] for seamless interaction is a promising area for future research. Additionally, it is recommended to improve the virtual assistant avatar by training it with artificial intelligence in order to expand the database of commands and responses. The constraint that arises is the use of language slang or informal diction when formal English is necessary. This study implemented sequential voice commands in the form of guided conversations. In the future, we will transition to an AI platform that utilizes NLP to improve voice command functionality.

Acknowledgements We would like to extend our gratitude to the Ministry of Higher Education under Fundamental Research Grant Scheme, UTM, FRGS/1/2023/ICT10/UTM/02/2 which has funded this research. We are most grateful to the Mixed and Virtual Reality Laboratory (mivielab) and the ViCubeLab at Universiti Teknologi Malaysia (UTM) for providing the technical support and facilities.

References

1. Azuma RT (1997) A survey of augmented reality. Presence Teleoperators Virtual Environ 6:355–385
2. Harborth D (2017) A systematic literature review on augmented reality augmented reality in information systems research: a systematic literature review
3. Adam M, Wessel M, Benlian A (2021) AI-based chatbots in customer service and their effects on user compliance. https://doi.org/10.1007/s12525-020-00414-7
4. Luo X, Tong S, Fang Z, Qu Z (2019) Frontiers: machines vs. humans: the impact of artificial intelligence Chatbot disclosure on customer purchases. 38, 937–947, https://doi.org/10.1287/mksc.2019.1192
5. Sloan H, Zhao R, Aqlan F, Yang H, Zhu R (2022) Adaptive virtual assistant for virtual reality-based remote learning. In: 2022 ASEE annual conference & exposition
6. Maçães G, Pimenta W, Carvalho E (2011) Using augmented reality virtual assistants to teach the traditional leather tanning process. In: 6th Iberian conference on information systems and technologies (CISTI 2011). IEEE pp 1–7
7. Piumsomboon T, Lee GA, Hart JD, Ens B, Lindeman RW, Thomas BH, Billinghurst M (2018) Mini-me: an adaptive avatar for mixed reality remote collaboration. https://doi.org/10.1145/3173574.3173620
8. Joachimczak M, Liu J (2017) Real-time mixed-reality telepresence via 3D reconstruction with Hololens and commodity depth sensors. https://doi.org/10.1145/3136755.3143031
9. Vona F, Torelli E, Beccaluva E, Garzotto F (2020) Exploring the potential of speech-based virtual assistants in mixed reality applications for people with cognitive disabilities. ACM Int Conf Proceed Ser. https://doi.org/10.1145/3399715.3399845

10. Marques B, Silva S, Teixeira A, Alves J, Dias P, Santos BS (2023) Towards asynchronous mixed reality remote guidance supported by a virtual assistant: proposal of a conceptual model. In: 2023 IEEE Conference on virtual reality and 3D user interfaces abstracts and workshops (VRW). IEEE pp 370–374
11. Jain S, Schweiss T, Bender S, Werth D (2021) Omnichannel retail customer experience with mixed-reality shopping assistant systems. In: Advances in visual computing: 16th international symposium, ISVC 2021, Virtual event, October 4–6, 2021, Proceedings, Part I. Springer. pp 504–517
12. Morotti E, Donatiello L, Marfia G (2020) Fostering fashion retail experiences through virtual reality and voice assistants. Proc. 2020 IEEE Conf. Virtual Real. 3D User Interfaces, VRW 2020. pp 338–342. https://doi.org/10.1109/VRW50115.2020.00074
13. Das P, Acharjee K, Das P, Prasad V (2015) Voice recognition system: speech-to-text. J Appl Fundam Sci 1:191
14. Hirschberg J, Manning CD (2015) Advances in natural language processing. Science 80(349):261–266. https://doi.org/10.1126/SCIENCE.AAA8685
15. Duguleană M, Briciu VA, Duduman IA, Machidon OM (2020) A virtual assistant for natural interactions in museums. Sustain 12:6958. https://doi.org/10.3390/SU12176958
16. Allen AA, Stecker PM (2023) Using mixed-reality simulation in teacher preparation in reading. Interv. School Clinic 59(1):59–65. https://doi.org/10.1177/10534512221130077
17. Voinea GD, Gîrbacia F, Postelnicu CC, Duguleana M, Antonya C, Soica A, Stănescu RC (2022) Study of social presence while interacting in metaverse with an augmented avatar during autonomous driving. Appl Sci 12:11804. https://doi.org/10.3390/APP122211804
18. Pietro Crovari Dott Francesco Vona D (2023) HoloLearn : mixed reality experiences for persons with neurodevelopmental disorders
19. Farella M, Chiazzese G, Lo Bosco G (2022) Question answering with BERT: designing a 3D virtual avatar for cultural heritage exploration. MELECON 2022—IEEE Mediterr Electrotech Conf Proc. pp 770–774. https://doi.org/10.1109/MELECON53508.2022.9843028
20. Wang I, Smith J, Ruiz J (2019) Exploring virtual agents for augmented reality. Conf Hum Factors Comput Syst Proc. https://doi.org/10.1145/3290605.3300511
21. Siyaev A, Jo GS (2021) Towards aircraft maintenance metaverse using speech interactions with virtual objects in mixed reality. Sensors 21:2066. https://doi.org/10.3390/S21062066
22. Karim Ishigaki SA, Ismail AW, Kamruldzaman MQ (2022) MR-MEET: mixed reality collaborative interface for HMD and handheld users. 2022 IEEE Glob. Conf. Comput. Power Commun. Technol. GlobConPT 2022. https://doi.org/10.1109/GLOBCONPT57482.2022.9938307
23. Ismail AW, Iman MA (2023) Implementation of natural hand gestures in holograms for 3D object manipulation. Virtual Real Intell Hardw 5:439–450
24. Aladin MYF, Ismail AW (2019) Designing user interaction using gesture and speech for mixed reality interface. Int J Innov Comput. 9, https://doi.org/10.11113/ijic.v9n2.243

Education Informatics

Transforming Education: The Impact of Generative AI

Elezabeth Mathew and Hesham El-Sayed

Abstract Generative AI has the capacity to transform the educational system. This study examines the advantages and disadvantages of using Generative AI, for example, ChatGPT in the classroom, and the hype around the use of generative artificial intelligence. The study includes a taxonomy of Generative AI, use cases in industries, an analysis of the reason for hype, the effect on academia, and ethical considerations. The study uses a short survey to find the perception of students in using ChatGPT, Generative AI, to define the desired results, choose the right level of automation, ensure ethical considerations are taken into account, and assess effectiveness. For teachers and students, the impact of Generative AI on teaching and learning and ethical concerns are vital.

Keywords Generative artificial intelligence · Academia · Hype cycle · Ethics · ChatGPT · Teaching and learning · Proactive approach

1 Introduction

1.1 Generative Artificial Intelligence: GAI

Generative artificial intelligence (GAI) is an artificial intelligence (AI) that can create new content instead of expert systems that can only analyze or act on existing data. Modern generative artificial intelligence consists of a discriminator or transformer model trained on a corpus or dataset that can map the input data into a latent high-dimensional space and a generator model that can generate stochastic

E. Mathew (✉) · H. El-Sayed
College of Information Technology, United Arab Emirates University, Al Ain, United Arab Emirates
e-mail: elezabeth.mathew@uaeu.ac.ae

H. El-Sayed
e-mail: helsayed@uaeu.ac.ae

S. M. Thampi et al. (eds.), *Fifth International Conference on Computing and Network Communications*, Lecture Notes in Electrical Engineering 1221,
https://doi.org/10.1007/978-981-97-4711-5_43

behavior, creating novel content in every new trial even from the same prompts as an input, performing unsupervised, semi-supervised, or supervised learning, depending on the particular methodology [1]. Generative artificial intelligence models differ from predictive machine learning systems in producing new material rather than only addressing classification or regression problems through discrimination. These models, in particular, have the ability to differentiate between information and produce information from altered input information, or prompts [2]. According to the Forbes [23] survey, 43% of college students claim to have used AI tools like ChatGPT, and 50% claim to have used those tools to complete assignments or take examinations. According to this statistic, 1 in 5 college students admit to using AI to accomplish their coursework. The majority of students who have used AI apps stated that they did so for fun, out of curiosity, or for personal projects.

Given that the most common generative AI models are transforming various industries, especially higher education, it is essential to present a state-of-the-art review of them [3, 4]. In this way, generative AI models won't take the place of people but rather improve our material by serving as a source of inspiration for writers or enhancing academic content. The document is structured in the manner shown below in order to provide information for a professional working in any industry that could benefit from these models. First, a taxonomy of the most popular generative models to have emerged on the market is given. The next section includes finding the effect on academia and a proactive approach toward managing it and finally the future direction and conclusion.

1.2 Generative AI Model Taxonomy

An effort to classify the current generative AI models into a taxonomy whose categories describe the key mappings between each multimedia input and output kind of data is drawn before going into details of effect on academia. The taxonomy is depicted in Table 1.

Reinforcement learning and generative models have advanced together, improving control and optimizing the created material [5]. Generative models have improved controllability and few-shot learning, allowing them to produce content from a small sample size or under particular circumstances [6]. Researchers had been investigating techniques for scaling up training and strategies to improve large-scale generative models' effectiveness and environmental friendliness. In-depth generative models were being created by researchers and businesses for various industries, including healthcare, the arts, and finance [7]. The ethical issues and safety considerations related to generative AI, such as bias, abuse, and possible injury, are receiving increased attention from the AI community [8].

Table 1 Taxonomy of popular generative AI

Format		AI Models				
Text-To-Image	DALL-E 2	Stable diffusion	Muse	Imagen	Style-GAN 2	CLIP
Text-To-Video	Phenaki	Soundify				
Text-To-Audio	Audio LM	Jukebox	Whisper			
Text-To-3d	DreamFusion	Magic3D				
Image-To-Text	Flamingo	Visual GPT	CLIP			
Text-To-Text	Chat GPT 4	LaMDa	PEER	Speech from brain	BERT and transformers	T5
Text-To-Code	Codex	Alpha code				
Text-To-Science	Galactica	Minerva				
Other models	AlphaTensor	GATO	Human motion diffusion model			

1.3 Generative AI Use Cases Across Multiple Industries

Numerous industries have found uses for generative AI, which has changed how companies run and engage with customers. Here is a breakdown of several cutting-edge generative AI use cases across many industries, as shown in Table 2.

These are only a few instances; generative AI is constantly developing, with new applications appearing in response to demands from specific industries. The

Table 2 Generative AI use cases across many industries

Industries		Usages	
Health care	Medical imaging	Drug discovery	Electronic Health records (HER)
Entertainment	Content creation	Video game design	
Finance	Algorithmic trading	Fraud detection	
Retail	personalized Recommendations	Virtual Try-On	
Automotive	Autonomous vehicles	Design and prototyping	
Marketing	Content generation	A/B testing	
Education	Tutoring and personalized learning	Language learning	
Legal	Document generation	Predictive analysis	
Energy	Energy optimization	Weather forecasting	
AI research	Model development		

technology is anticipated to have a transformative effect across a wide range of areas as it develops and becomes more widely available. However, as generative AI is more thoroughly incorporated into numerous industries, ethical considerations, such as bias mitigation and responsible AI use, will become more crucial. Nevertheless, GAI is famous in most industries [15].

High-resolution medical images produced by generative AI models can help with diagnosis and treatment planning [21]. Music, art, and literature are produced using generative AI, which generates original works based on particular genres or techniques [22]. To help stop financial fraud, generative models examine transaction data and produce anomaly alerts [22]. In order to train autonomous car systems, generative AI creates synthetic data which helps to simulate driving situations [22]. AI creates tailored educational materials and customizes coursework to meet the needs of each learner [22]. These are some fields where generative AI is constantly transforming its capability and improvising on a daily basis.

1.4 Hype Cycle of Generative AI

The research and consultancy company Gartner coined the term "Hype Cycle" to describe how expectations and adoption of emerging technologies evolve over time. The hype cycle of Generative AI is shown in Fig. 1 on its up-to-date course based on current trends in the industry. The capability of generative AI in producing impressive outputs, creative applications, content generation, automation, research advancement, personalization, etc., has made significant strides and hence the hype overall. The hype cycle of AI is shown in Fig. 1, adapted from Gartner (2023).

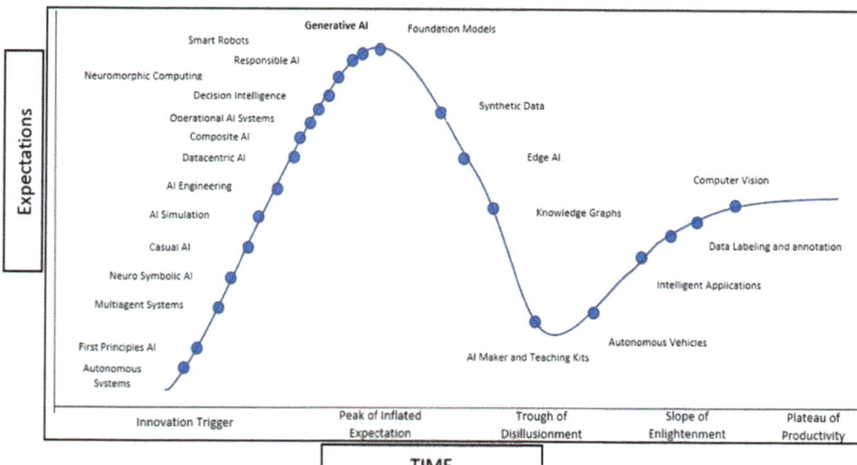

Fig. 1 Hype cycle of artificial intelligence, 2023, adapted from Gartner (2023)

1.5 *Phases of Hype Cycle*

There are multiple distinct phases in the hype cycle [9]:

An innovation or new technology is born during the Innovation trigger phase. It frequently begins with a discovery or breakthrough that may not yet be well-known or understood. Researchers and early adopters are interested in these advances. During the peak of the inflated stage, the technology receives a lot of attention, and anticipations are frequently raised. The technology is the subject of much excitement and is anticipated to transform the game. However, at this time, neither the practical uses nor the restrictions are fully comprehended or understood. Reality sets in after the peak of exaggerated expectations. Frequently, the technology falls short of unrealistic expectations, which causes disillusionment and suspicion. Setbacks failed projects, or the recognition of the difficulties with the technology may all occur during the trough of the disillusionment phase. Lessons from the disillusionment phase are applied throughout the slope of the enlightenment phase. As the technology develops, a more precise grasp of its advantages and disadvantages becomes apparent. Expectations become more reasonable as successful use cases and best practices emerge. The technology reaches a stable and valuable condition during the plateau of the productivity stage of the hype cycle. It has been widely embraced and incorporated into various uses and businesses. The usefulness of the technology is well known, and it has become an accepted element of the technical environment. As depicted in the figure, Generative AI is at its peak currently. It still needs to undergo other phases to reach maturity, the plateau of productivity.

2 Effect on Academic Integrity

Many academics have experienced a significant deal of concern due to the publication of ChatGPT, and the commentary that has followed. This anxiety is driven partly by the need to maintain academic integrity in university classrooms. The need to uphold academic integrity in university courses is one factor that fuels this fear [16]. The author says these technologies enable "a way of avoiding learning" and "basically high-tech plagiarism." We contend that dismissing the potential of adopting such technologies would be a mistake by pointing to academic honesty. But this technology has indeed dramatically altered both students' learning outcomes and their classroom experience.

IT Mindfulness is a useful framework for this investigation that can be found in the IS research. According to [18] definition of IT mindfulness, there are four components:

(1) Orientation to the present, (2) awareness of distinction, (3) openness to novelty, and (4) awareness of diverse perspectives. Instructors can use this theoretically constructed concept to encourage pupils to investigate technological tools. For instance, a teacher may encourage students to use ChatGPT to compose an essay on a subject pertinent to the course. Then, the teachers would ask for a review of this essay from the students. The exploring process should be aided by analyzing

the essay's strengths and shortcomings [19]. It is crucial to educate pupils about the ethical ramifications of technology use as well as the possibilities and limitations of such use. According to research, a person's faith in technology "depends on what the technology can do for the person, such as being reliable on providing timely help quickly." We have to understand that generative AI is just one among a number of technologies that have changed how college students learn in the classroom. Technologies like calculators, email, Google search, and statistical software also caused disruption and change in the past.

2.1 Short Survey

A short survey was conducted to find students' perceptions of ChatGPT usage using *jotform*. The questionnaire is attached with appendix A. The questionnaire consisted of 20 questions with both open-ended and closed-ended queries. The sample consists of 11 students in one of the Higher Education Institutes of the Middle East. Though sample collected is small because the survey is done on Graduate students, and it is a strength at the same time as most of them are ages 25–45. The survey could be reliable as it is conducted among adult students. The demographical and employment details of the participants are shown in Fig. 2.

Students were given a questionnaire with quantitative and open-ended questions for qualitative analysis. 100% of students support the use of ChatGPT. The survey showed that the students use it primarily for project help or rephrasing sentences. The students believe ChatGPT-generated English structure is better than their language formatting as non-native speakers. The students think the content in ChatGPT is somewhat reliable, but at the same time, the user must have an idea of the topic. The responses show that using ChatGPT is very easy and they use it at least once a week. "I believe ChatGPT should be used to enhance knowledge rather than foster[ing] laziness," wrote another respondent. The reason for using ChatGPT includes but is not limited to the quality, quantity, and English grammar itself. It is worth mentioning that few students specified about the authenticity of the content as a concern.

Fig. 2 Demography and employment status of participants in percentage

2.2 Ethical Considerations

Responsible AI development and deployment are becoming more important. Given how generative AI interacts with pedagogy, privacy, equity, and other issues, using it in education raises a number of ethical issues. The application of generative AI in education, also known as "educative AI," [8] presents the following ethical challenges:

- Bias and Fairness: Like any AI systems, generative AI models are susceptible to biases inherited from training data. This may result in the creation of biased information, including suggestions or writing prompts. Stereotypes and disparities in schooling can be strengthened through bias [10].
- Quality Control: It is essential to guarantee the accuracy and quality of material produced by AI. Learning outcomes may suffer if instructional materials produced by AI contain errors or false information [10].
- Plagiarism and Academic Integrity: It is possible to generate essays, papers, or assignments using generative AI. This raises questions regarding academic honesty and plagiarism because students might submit works made by AI without giving proper credit [11].
- Privacy Concerns: Privacy issues may arise when student data is gathered and processed in order to tailor educational content. It's critical to safeguard kids' privacy, particularly when utilizing AI to assess their academic achievement and behavior [11].
- Accessibility: Students with disabilities might not always be able to access AI-generated information. Institutions of higher learning must make sure that materials produced by AI are accessible and fit a range of student needs [11].
- Social and Moral development: The overuse of AI in teaching and tutoring could result in a decline in the human element of learning. The learning process requires interaction with peers and teachers, which is crucial [10].
- Equity and Access: Internet and technological access are not available to every student equally. If AI in education is not deployed in a way that considers students' access to technology, it may worsen already-existing disparities [11].
- Depersonalization: Although AI can customize learning experiences, there is a risk of depersonalization, where students lose interest in the learning process because too much emphasis is placed on AI-driven interactions [12].
- Teacher Autonomy: The job of instructors may come under scrutiny as a result of the usage of AI in education. Ethics demands that teachers' autonomy and knowledge be protected and that AI should serve and complement them rather than displace them [13, 14].
- Transparency and Explainability: Explaining how AI algorithms produce specific educational recommendations or content may be difficult. Lack of transparency can undermine confidence and make it challenging for teachers and pupils to comprehend decisions made by AI [12, 13].
- Data Security: It's crucial to protect the data that AI systems use. Educational institutions must safeguard student information from theft or misuse, which could have serious ethical and legal repercussions [12, 14].

- Accountability: It might be challenging to determine who is to blame when AI-generated content has biases or inaccuracies. It is essential to establish distinct lines of responsibility [15].

Planning carefully, being transparent, having strong data governance standards, and continuously evaluating AI systems in education are all necessary to address these ethical issues. To make sure that the application of generative AI in education is in line with moral values and students' interests, it is crucial for educational institutions, policymakers, and AI developers to collaborate. Along with the use of generative AI, ethical standards for institutions and publications also need to change. Some magazines have already offered guidance to authors.

2.3 Proactive Approach by Academia: Solution

The academic sector is currently undergoing a digital change [20]. An essential lesson from using ChatGPT while drafting a brief essay is that the tool showed superior writing abilities in English than one can do as a non-native speaker. Having this encounter brings to light a number of significant elements. First, asking intriguing questions requires human intelligence. Second, GAI can only respond with information based on previously recorded information used in its training. Thirdly, creating the knowledge that may be used to train GAI involves human intelligence. And finally, since GAI is currently unable to research, it still requires human intelligence. Since English is not their first language, many researchers, including myself, have already utilized Grammarly, spell checks, and professional editing to improve their writing. With the introduction of ChatGPT, language boundaries are no longer an obstacle to research or its dissemination. The ability to write well in English is no longer the deciding element in how widely our research findings are shared [16]. Hence, writing text/reports may not be the most important skill to focus on when assessing our pupils' intellectual ability.

The resistance to this technology is consistent with user resistance to IT, which has been well-documented for years, and it is evidence that academia is beginning to undergo a digital transition. It's more critical than ever to teach our pupils how to pose probing questions and come up with thoughtful responses when it comes to college essays/reports. Studies examining how ChatGPT and/or other generative AI applications affect student learning outcomes, like academic success, motivation, and engagement, are equally crucial [8]. This is the intellectual foundation that will help society and students alike. Hence, the grading criteria for each component needs a transformation depending on the possibility of using generative AI. For example, if there is a project report, usually the report weighs 70%, and the presentation with viva weighs 30%. It shall be changed to 35% for the report, 35% for the presentation, and 30% for Q&A or something similar. Because students wouldn't be as driven or engaged in generating something that requires critical thought if they believe the grade is just based on the report they wrote.

3 The Future of Education and Generative AI

Students, teachers, and researchers should know we are all participating in this technological revolution. However, there is a need to tighten regulations or enforce them more. It is a time to emphasize the significance of independent thought by utilizing this innovation to facilitate learning and understanding [20]. It's our capacity for independent thought that makes us better people. The marathon is long, and the real prize isn't a higher essay/report score but a better quality of life for everyone. GAI may undermine creativity and critical thinking if it performs all or many tasks for students and lecturers. Therefore, the evaluation criteria for students' ability to present and critically evaluate the report they generated should be more heavily weighed than the report itself.

Exploring how generative AI will impact knowledge transfer and acquisition for people, teams, and organizations will require research. Looking for ways to make GAI and other technologies more transparent and less biased is crucial. Regulations, moral principles, and ethical rules regarding generative AI, such as ChatGPT, have not yet been developed. The current work opens up new possibilities for investigating how these institutions change. For example, the research could focus on "How can the academic community better respond to disruptive, developing technologies that could endanger current methods of research, teaching, and learning?" or "How can the academic community extend the debate and improve its comprehension of ChatGPT and other conversational tools driven by AI?" Incorporating generative AI into educational processes also requires considering human teachers' contribution. Although generative AI has the potential to support and improve teaching and learning, it is vital to take into account how it may alter the role of teachers and the skills required for them to be successful in the classroom [8].

4 Conclusion

Without a question, Generative AI, represented by ChatGPT, is one of the most revolutionary AI tools which is at its peak to be created in recent years. For organizations, societies, and people as a whole, it offers both major opportunities and problems. While generative AI can significantly increase productivity in various contexts, these technologies also pose several practical issues involved in moral, ethical, and even political aspects. For instance, it is generally known that ChatGPT in the education sector has difficulties/poses a dilemma as generative AI presently lacks well-established rules and moral standards. A significant practical obstacle is that authorities cannot hold offenders accountable through legal action when such technologies are willfully abused or misused. Therefore, it is essential to pass new legislation to control these technologies. Due to their worldwide character, international cooperation is also required to fully reap the advantages of technologies like ChatGPT. It is crucial to approach ChatGPT integration into educational practices responsibly, ethically, and with the support of a solid theoretical framework. ChatGPT and other generative AI, also known as "educative AI," can offer students highly personalized and interactive learning experiences. By doing this, it will be possible for educational institutions to benefit from new developments and stay up with the evolving state of education.

Appendix

1. Questionnaire used for survey

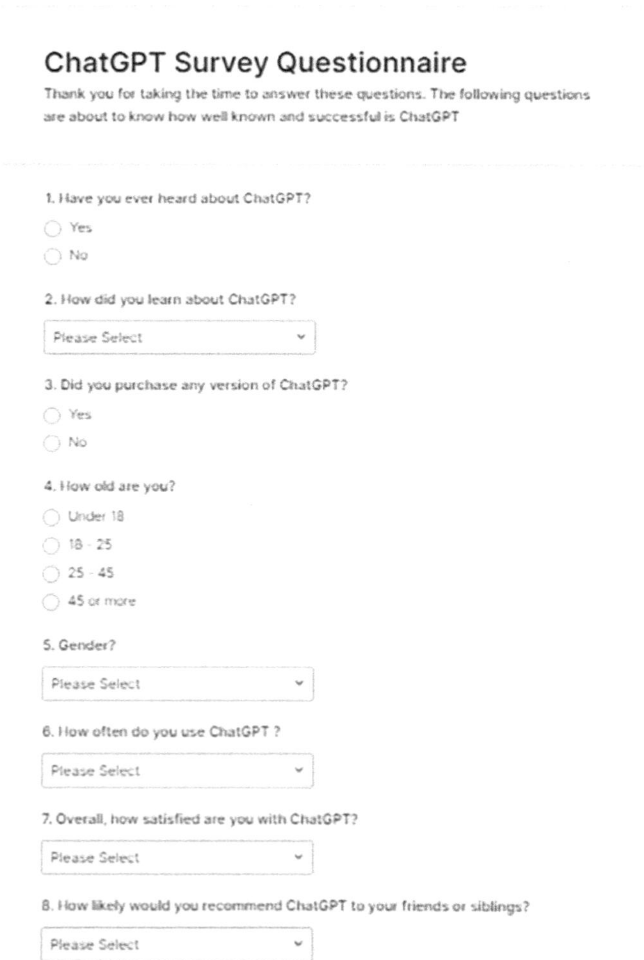

ChatGPT Survey Questionnaire

Thank you for taking the time to answer these questions. The following questions are about to know how well known and successful is ChatGPT

1. Have you ever heard about ChatGPT?

- Yes
- No

2. How did you learn about ChatGPT?

Please Select

3. Did you purchase any version of ChatGPT?

- Yes
- No

4. How old are you?

- Under 18
- 18 - 25
- 25 - 45
- 45 or more

5. Gender?

Please Select

6. How often do you use ChatGPT ?

Please Select

7. Overall, how satisfied are you with ChatGPT?

Please Select

8. How likely would you recommend ChatGPT to your friends or siblings?

Please Select

9. What particular aspect(s) of ChatGPT do you like?

10. What particular aspect(s) of ChatGPT do you dislike?

11. What factors motivated you to purchase ChatGPT?

Please Select

13. What is the main reason you use ChatGPT?

Please Select

14. How easy it to use ChatGPT?

Please Select

15. How often do you use ChatGPT?

Please Select

16. Nationality

- ◯ UAE National
- ◯ Non-UAE national

17. Do you think ChatGPT is reliable?

- ◯ Completely reliable
- ◯ Somewhat reliable
- ◯ Neither nor
- ◯ Not reliable

18. How satisfied were you with the available options in ChatGPT?

Please Select

19. Please rank what you look for in order of importance when selecting content from ChatGPT?

	1 - Not important	2	3 - Maybe	4	5 - Very important
English	◯	◯	◯	◯	◯
Quantity (word count)	◯	◯	◯	◯	◯
Quality of content	◯	◯	◯	◯	◯
Grammar	◯	◯	◯	◯	◯
Better than what you wrote.	◯	◯	◯	◯	◯

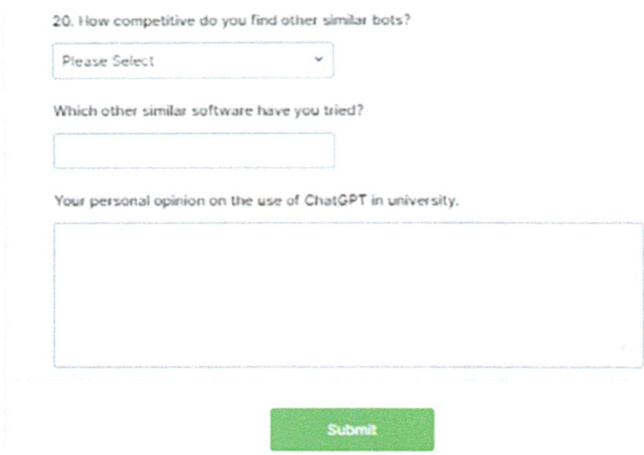

References

1. Murphy KP (2022) Probabilistic machine learning: an introduction. MIT press
2. Creswell A, White T, Dumoulin V, Arulkumaran K, Sengupta B, Bharath AA (2018) Generative adversarial networks: an overview. IEEE Signal Process Mag 35(1):53–65
3. Kandlhofer M, Steinbauer G, Hirschmugl-Gaisch S, Huber P (2016) Artificial intelligence and computer science in education: from kindergarten to university. In: 2016 IEEE frontiers in education conference (FIE). IEEE, pp 1–9
4. Susnjak T (2022) Chatgpt: the end of online exam integrity? arXiv preprint arXiv:2212.09292
5. Tirinzoni A, Poiani R, Restelli M (2020) Sequential transfer in reinforcement learning with a generative model. Proceedings of the 37th international conference on machine learning, In: Proceedings of machine learning research 119:9481–9492. Available from https://proceedings.mlr.press/v119/tirinzoni20a.html
6. Garcia X, Bansal Y, Cherry C, Foster G, Krikun M, Johnson M, Firat O (2023) The unreasonable effectiveness of few-shot learning for machine translation. Proceedings of the 40th international conference on machine learning, In: Proceedings of machine learning research 202:10867–10878. Available from https://proceedings.mlr.press/v202/garcia23a.html
7. Hataya R, Bao H, Arai H (2022) Will large-scale generative models corrupt future datasets? arXiv preprint arXiv:2211.08095
8. Su (苏嘉红) J, Yang (杨伟鹏) W (2023) Unlocking the power of ChatGPT: a frame-work for applying generative AI in education. ECNU Rev Educat 6(3):355–366. https://doi.org/10.1177/20965311231168423
9. https://www.gartner.com/en/research/methodologies/gartner-hype-cycle
10. Wach K, Duong C, Ejdys J, Kazlauskaitė R, Korzynski P, Mazurek G, Paliszkiewicz J, Ziemba E (2023) The dark side of generative artificial intelligence: a critical analysis of controversies and risks of ChatGPT. Entrep Bus Econ Rev 11(2):7–30. https://doi.org/10.15678/EBER.2023.110201
11. Liu M, Ren Y, Nyagoga LM, Stonier F, Wu Z, Yu L (2023) Future of education in the era of generative artificial intelligence: consensus among Chinese scholars on applications of ChatGPT in schools. Futur Educ Res
12. Kshetri N (2023) The economics of generative artificial intelligence in the academic industry. Computer 56(8):77–83. https://doi.org/10.1109/MC.2023.3278089
13. Christudas BCL, Kirubakaran E, Jeba Thangaiah PR (2018) An evolutionary approach for personalization of content delivery in e-learning systems based on learner behavior forcing

compatibility of learning materials. Telemat Inform 35(3):520–533, ISSN 0736–5853, https://doi.org/10.1016/j.tele.2017.02.004
14. Bozkurt A, Sharma RC (2023) Challenging the status quo and exploring the new boundaries in the age of algorithms: reimagining the role of generative AI in distance education and online learning. Asian J Distance Educ 18(1). Retrieved from https://asianjde.com/ojs/index.php/AsianJDE/article/view/714
15. Zhong H, Chang J, Yang Z, Arachchige PCM, Pathmabandu C, Xue M (2023) Copyright protection and accountability of generative AI: attack, watermarking and attribution, WWW '23 companion: companion proceedings of the ACM web conference , pp 94–98. https://doi.org/10.1145/3543873.3587321
16. Dwivedi YK, Kshetri N, Hughes L, Slade EL, Jeyaraj A, Kar AK, Wright R (2023) So what if ChatGPT wrote it? Multidisciplinary perspectives on opportunities, challenges and implications of generative conversational AI for research, practice and policy. Int J Inf Manage 71:102642
17. Cotton DR, Cotton PA, Shipway JR (2023) Chatting and cheating. Ensuring academic integrity in the era of ChatGPT. edarix.org
18. Jensen ML, Dinger M, Wright RT, Thatcher JB (2017) Training to mitigate phishing attacks using mindfulness techniques. J Manag Inf Syst 34(2):597–626
19. Mollick ER, Mollick L (2022) New modes of learning enabled by AI chatbots: three methods and assignments. Available at: https://doi.org/10.2139/ssrn.4300783Pflügner
20. Zhao L, Mao H, Compton BJ, Peng J, Fu G, Fang F, Heyman GD, Lee K (2022) Academic dishonesty and its relations to peer cheating and culture: a meta-analysis of the perceived peer cheating effect. Educ Res Rev 36:100455. ISSN 1747-938X. https://doi.org/10.1016/j.edurev.2022.100455
21. Szolovits P, Patil RS, Schwartz WB. Artificial intelligence in medical diagnosis. https://doi.org/10.7326/0003-4819-108-1-80
22. Konen W (2019) General board game playing for education and research in generic AI game learning. In: 2019 IEEE conference on games (CoG), London, UK, 2019, pp 1–8. https://doi.org/10.1109/CIG.2019.8848070
23. Umbrello S, De Bellis AF. Book: artificial intelligence safety and security, 1st Edition, First Published 2018, Imprint chapman and hall/CRC Page 15 eBook ISBN 9781351251389. https://www.forbes.com/sites/michaeltnietzel/2023/03/20/more-than-half-of-college-stu-dents-believe-using-chatgpt-to-complete-assignments-is-cheating/?sh=71ac9b0618f9

Developing and Deploying Open-Access Web-Based Educational Tools for Teaching Data Preprocessing and Classification in Bioinformatics

Bhavya Ottappurakkal, Dhanush Kumar, Asha Vijayan, Rakhi Radhamani, Krishnashree Achuthan, and Shyam Diwakar

Abstract Machine learning and data mining techniques have become indispensable in the field of bioinformatics with the advancements in big data analysis such as omics, biomedical imaging, and signal processing. Toward developing a pedagogical solution for data preprocessing, this paper explores design architecture of easily accessible, free-cost novel open-source web-based virtual laboratory to simplify the data preprocessing and visualization steps for big data analysis. A user-friendly interface combines computational capabilities with HTML and JavaScript as the underlying technology. This virtual laboratory architecture was designed to be compatible for access using mobile phones, tablets, and desktop computers having an active internet connection, making the education resource usage independent of end user dependencies such as specific operating systems, installed apps or other not so easily available applications. The software implementation was created to allow student learners to preprocess and classify data, incorporating techniques like variable type conversion, normalization, standardization. With this web-based virtual laboratory, large datasets can also be preprocessed and classified that may help as education resources for students, data analysts and researchers from different domains.

Keywords Virtual laboratory · Data preprocessing and classification · Machine learning · Higher education

B. Ottappurakkal · D. Kumar · A. Vijayan · R. Radhamani · S. Diwakar (✉)
Amrita Mind Brain Center, Amrita Vishwa Vidyapeetham, Amritapuri Campus, Clappana P.O., Kollam, Kerala 690525, India
e-mail: shyam@amrita.edu

K. Achuthan
Center for Cybersecurity Systems and Networks, Amrita Vishwa Vidyapeetham, Amritapuri Campus, Clappana P.O., Kollam, Kerala 690525, India

S. Diwakar
Department of Electronics and Communication Engineering, School of Engineering, Amrita Vishwa Vidyapeetham, Amritapuri Campus, Clappana P.O., Kollam, Kerala 690525, India

S. M. Thampi et al. (eds.), *Fifth International Conference on Computing and Network Communications*, Lecture Notes in Electrical Engineering 1221,
https://doi.org/10.1007/978-981-97-4711-5_44

1 Introduction

Across the EdTech sector, data mining and machine learning techniques are widely used within bioinformatics to analyze and extract knowledge from biological databases [1–4]. The techniques have also been used in classrooms to pedagogically explore and analyze biological data, and they have been applied in various areas such as preprocessing, classification, clustering, and feature selection. Data mining approaches also investigate disease mechanisms, discover biomarkers, and build predictive models for the prognosis, early detection, and treatment of cancer [5, 6]. In bioinformatics classrooms, teaching and learning practical methods in data mining has some skill-related challenges that includes learners to be familiar in basic statistical thinking in order to derive meaningful inferences [7]. Another challenge is the complexity of the data in bioinformatics, which includes large amounts of data called "big data," unobserved information, and noisy measurements [8–10]. To overcome these challenges, it is imperative to prioritize training in bioinformatics, particularly in the realm of data mining. Furthermore, learning materials need to be designed to promote better conceptual understanding [11].

From a pedagogical perspective, for a student learner or an educator, data preprocessing needs to be explained in terms of tasks such as data management, modeling, feature extraction, and feature selection [12]. Preprocessing is essential to handle the complexity of information, integrate heterogeneous data sources, and ensure data quality [13]. It helps in transforming raw data into a suitable format for analysis, reducing noise, and improving the efficiency and accuracy of subsequent data mining and analysis tasks. The Weka machine learning workbench is a popular tool that provides a comprehensive environment for automatic classification, regression, clustering, and feature selection in bioinformatics research [14]. Over the years, several tools have been developed to address the challenges and complexities associated with preprocessing raw data. Weka, KNIME, and RapidMiner are popular open-source platforms offering comprehensive suites of preprocessing tools [4, 15]. These tools provide users with a range of functionalities for data cleaning, transformation, feature selection, and integration. Additionally, Python libraries such as Pandas, Scikit-learn, and NumPy offer extensive functionality for data preprocessing [16].

However, students face several challenges when using data preprocessing and classification tools in many interdisciplinary classrooms including a knowledge gap in interpreting and utilizing captured data, the selection of appropriate techniques and tools affecting data quality, handling text preprocessing complexities, etc. [17, 18]. To overcome these challenges, a web-based virtual laboratory tool was implemented that simplified and democratized the data preprocessing process and classification, making it accessible to a wide range of student users. The virtual laboratory (VL) is a flexible and user-friendly method that allows students to carry out practical-based experiments in a virtual space [19, 20]. The resources in VL are categorized as pedagogical, human, and technical factors [21]. By leveraging HTML and JavaScript as foundational technologies, VLs enable the creation of web-based virtual environments accessible through standard browsers on any platform [19, 22, 23]. This accessibility, coupled with its cost-effectiveness, fosters collaboration and facilitates the sharing of equipment and resources [24, 25]. Moreover, VLs offer learning tools and data visualization capabilities that enhance students' learning experiences. A key advantage with VLs is their provision of simulations, virtual experiments, and interactive activities that allow students to learn practical skills and knowledge in a secure and cost-effective manner [26–31]. Within the virtual environment, students can engage in hands-on experiments, familiarize themselves with laboratory safety protocols, and analyze experimental data [32]. VLs also provide opportunities for students to interact with measuring tools, compare actual and virtual outcomes, and receive feedback from teachers [33]. Additionally, VLs can be integrated with multimedia content and educational video games to improve knowledge acquisition and develop cognitive abilities [34]. These learning tools and data visualization capabilities make virtual laboratories effective and practical resources for enhancing student learning outcomes in various subjects, including physics and biology [35–37]. Additionally, VLs have been used in the field of biotechnology to virtualize wet-lab techniques and experiments, resulting in improved student performance and enhanced interaction-based lab education.

Web-based tools provide a useful and approachable platform for social media monitoring and allow users to interact with and assess social media data without the need for sophisticated programming skills. For businesses, marketers, and researchers, these tools can significantly increase the usefulness and efficiency of social media analytics. [38]. In the field of e-commerce personalization, web-based solutions for data preprocessing and classification are essential for customizing the online purchasing experience for customers. These solutions gather real-time data about product views, purchases, and user interactions by connecting to e-commerce sites [39].

While data preprocessing and classification is widely acknowledged as an essential part of data mining, current conventional teaching approaches and tools often fall short when it comes to usability, accessibility, and performance. A problem addressed in this study is to overcome the lack of a comprehensive web-based virtual lab tool that integrates multiple preprocessing, classification techniques and offers a user-friendly interface freely for students and instructors. To improve current teaching methods for data preparation and classification modeling, this study explores the development and evaluation of a web-based utility created for both students and educators.

This paper focuses on the design, development and deployment of an open-source student community-centric virtual laboratory that is freely available online for teaching and learning data preprocessing and classification methods to students from interdisciplinary backgrounds such as in biotechnology and bioinformatics. The design features and components of the web-based tool were aimed at improving accessibility, scalability, usability, and collaboration among student–teacher community facilitating peer to peer interaction in higher education level.

2 Implementation of a Virtual Laboratory Platform for Data Preprocessing and Classification in Augmenting Bioinformatics Education

The key components in developing our web-based online educational resource for data preprocessing, classification included a user-friendly graphical user interface (GUI) with data preprocessing and classification steps built in through various modules. Data preprocessing involved four main steps, i.e., data cleaning, formatting, transformation, and reduction (Fig. 1). Users could provide raw data as input and the output data obtained is a cleaned, formatted, transformed, and reduced file. Raw data may contain missing, redundant, or incomplete data and the output file can be used for classification, prediction, or clustering. The lab is freely available at https://virtual-labs.github.io/exp-weka-tool-au/.

Data cleaning involved identifying and addressing incomplete, incorrect, inaccurate, or irrelevant elements within the dataset. The data formatting module involves converting or rearranging data to a format for further analysis. The data transformation module employed various functions to convert data to a range of values using methods such as normalization techniques. Data reduction was to reduce the data to get only relevant data using different sampling techniques, feature selection, or extraction techniques. Various steps and algorithms were implemented for developing a user-interactive platform for virtualizing the data mining experiment with perceived usefulness by learner groups without instructor support.

2.1 Identification of a Dataset

The initial process for a learner involved uploading a widely recognized attribute-relation file format (ARFF) containing the dataset. For example, in the common weather dataset (see Fig. 2) the user could upload ARFF format which included three main sections: metadata, attribute definitions, and data instances, which was identified by keywords such as "@relation," "@attribute," and "@data." Firstly, the "@relation" section served to declare the relationship or name of the dataset under consideration, effectively providing a clear identification (e.g., "weather" as the dataset name). Secondly, the "@attribute" section defined different attributes (e.g., "outlook," "temperature," etc.) and the corresponding class label (e.g., "True," "False," "Yes," "No," etc.). Finally, the "@data" section specified the data segment in the file, which encapsulates the actual instances or data points (for instance: sunny, 85, 85, FALSE, no).

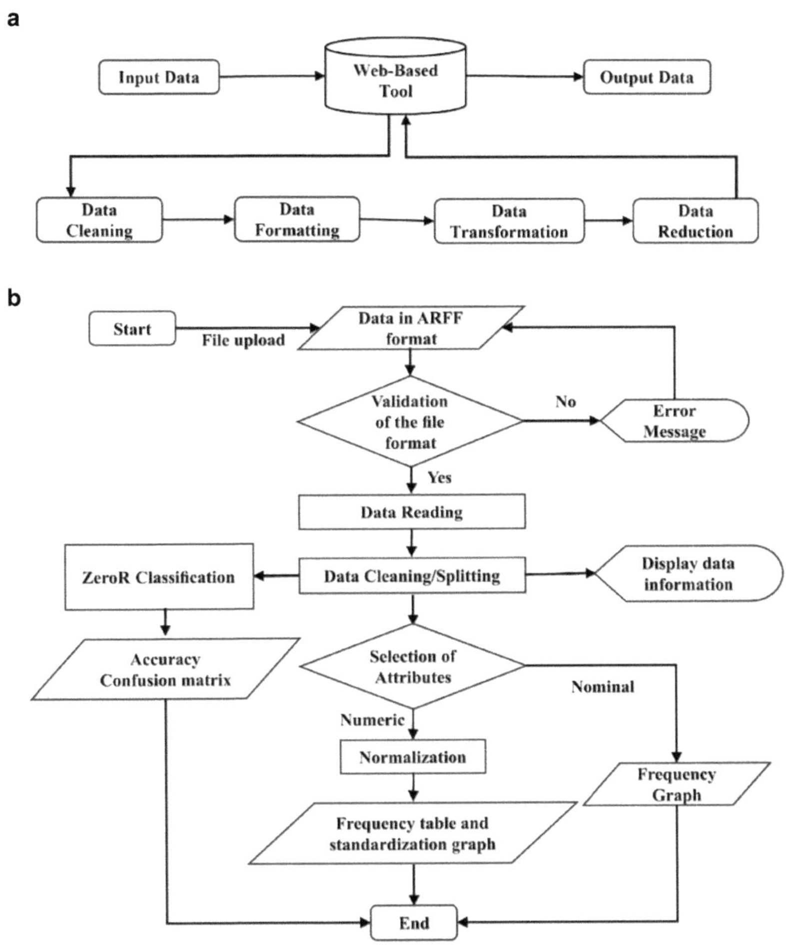

Fig. 1 Pipeline for data preprocessing and classification in machine learning included in bioinformatics module, **a** overview of the steps involved in data preprocessing, **b** steps involved in implementing user-interactive virtual laboratory web-based tool for learning machine learning applications

2.2 *Data Reading*

In the virtual lab platform, once the ARFF data file was selected, the subsequent step was to extract the valuable information stored within the file for subsequent processing and analysis. The learner may use the web platform to perform a data reading procedure that involves parsing a file to precisely extract information, and identifying and retrieval of attribute definitions, which describe the distinct characteristics or features of the data. The process also ensured the learner to understand

```
@relation weather

@attribute outlook {sunny, overcast, rainy}
@attribute temperature real
@attribute humidity real
@attribute windy {TRUE, FALSE}
@attribute play {yes, no}

@data
sunny,85,85,FALSE,no
sunny,80,90,TRUE,no
overcast,83,86,FALSE,yes
rainy,70,96,FALSE,yes
rainy,68,80,FALSE,yes
rainy,65,70,TRUE,no
overcast,64,65,TRUE,yes
sunny,72,95,FALSE,no
sunny,69,70,FALSE,yes
rainy,75,80,FALSE,yes
sunny,75,70,TRUE,yes
overcast,72,90,TRUE,yes
overcast,81,75,FALSE,yes
rainy,71,91,TRUE,no
```

Fig. 2 Weather dataset in ARFF file format, an example for executing virtual laboratory simulation

attribute definitions that encapsulate essential details such as attribute names, data types (numeric, nominal, string, etc.), and the potential range of attribute values.

2.3 *Validation of File Format*

As an implementation break point, a user could perform file format validation as a crucial role in ensuring the integrity and compatibility of data. In the case of web-based tools, file format validation typically began by performing a straightforward check of the file extension verifying whether the file extension matches the expected format. For simplification of learning process, in the case of an ARFF file, the tool ensures that the file ends with the extension ".arff".

After validating the extension, the tool verified whether the ARFF file's content adheres to a specific pattern that defines a valid ARFF file header structure. If the content did not confirm to the pattern, an alert box was displayed, and the execution of the code was terminated. This validation process ensures that student learners know that the data file begins with a line starting with "@RELATION" followed by one or more whitespace characters and additional content. It allowed for zero or more lines starting with "@ATTRIBUTE" and their corresponding content. Finally, it optionally allowed for a line starting with "@DATA."

If the regular expression used for pattern matching fails to find a match in the ARFF data string, the user gets notification in the GUI that the file content was invalid. In such cases, an alert box was displayed to the user, providing information about the encountered issue, and ceasing further code execution facilitating explicit interaction by the users. This module acted as a fundamental validation mechanism for virtual laboratory experiments, ensuring that the learner correctly identifies the provided ARFF file content adhered to the expected header structure.

2.4 Data Cleaning/Splitting

The goal was to ensure the students availing the virtual learning experience understood that the data quality and integrity can be altered by replacing, modifying, or removing problematic data points. Extracting information by splitting the file into individual lines and iterating over each line was defined as the data splitting process. The learners used the web-based platform to interpret relevant details that were retrieved from the ARFF file by carefully examining each row in the ARFF file. With this, the pedagogical flow allows learners to interpret data cleaning as involving identifying and addressing incomplete, incorrect, inaccurate, or irrelevant elements within the dataset.

Handling Missing Value. The tool enables identification of missing values as in ARFF files indicated by a question mark "?" or the string "NaN." The learner is expected to learn to understand how missing values significantly impact the accuracy of the results, making the process of handling them relevant. In the web-based learning tool, missing values are removed or replaced with a default value that supports removing instances in the dataset.

Handling Duplicate Data. The web-based tool allows learners to compare the records in the dataset to identify duplicates that were carried out by comparing all the attribute values or a subset of attributes that uniquely identify a record. The objective was to analyze rows with identical values in the selected columns by the user community.

Cleaning Attribute Values. As in many data mining platforms, if there were specific attributes or instances that need to be excluded from the dataset, that could be filtered out based on predefined criteria. For instance, attributes with a high percentage of missing values or instances with outliers can be removed from the dataset to ensure data quality.

2.5 Data Transformation

When a specific attribute was chosen by a user, the virtual laboratory tool was set to automatically generate a data frequency graph. The graph visually represented the frequency distribution of values within that attribute helping users to quickly understand the distribution pattern and identify potential outliers or anomalies. The generated graph helps student learners to explore the data, gain insights, and make data-driven decisions based on the frequency distribution of the selected attribute. The developed web-based tool enables data transformation through the process of normalization using min–max and Z-score normalization methods.

2.6 Data Reduction

A filter-based feature extraction approach specifically targets the conversion of numeric features to nominal ones. A numeric-to-nominal filter, an unsupervised technique, was used to convert numeric features into nominal ones as they help to group similar data, thus reducing the complexity of data and helping in better visualization.

3 Steps to Execute the Virtual Laboratory-Based Educational Tool for Machine Learning and Data Analysis

This section discusses the graphical user interface (GUI) of the implemented virtual laboratory web-based tool for data preprocessing and classification. The GUI was designed with HTML and modules were developed using JavaScript. The initial crucial step in data preprocessing involved uploading the required dataset in the ARFF format. Once the data was uploaded, through user interactions and process flow, it undergoes a series of processes such as data validation and data cleaning. Subsequently, the tool allowed to visualize details extracted from the uploaded dataset including the relation, sum of weights, number of instances, and attributes of the dataset (Fig. 3a). This visualization aided the learners to understand the dataset, enabling informed decision-making for subsequent analysis.

3.1 Selection of Attributes

The tool identified and listed all the attributes in the dataset (for example outlook, temperature, humidity, windy, and play in Fig. 3a). To facilitate attribute selection, the tool provided a checkbox within the virtual lab platform. Users can conveniently choose the desired attribute from the provided list. Upon selection, the interface displayed relevant information about the attribute, including its name (e.g., Outlook), type (e.g., nominal), and the number of distinct values it possesses (e.g., 3). If the selected attribute is nominal, the tool displays the frequency of each feature of all the

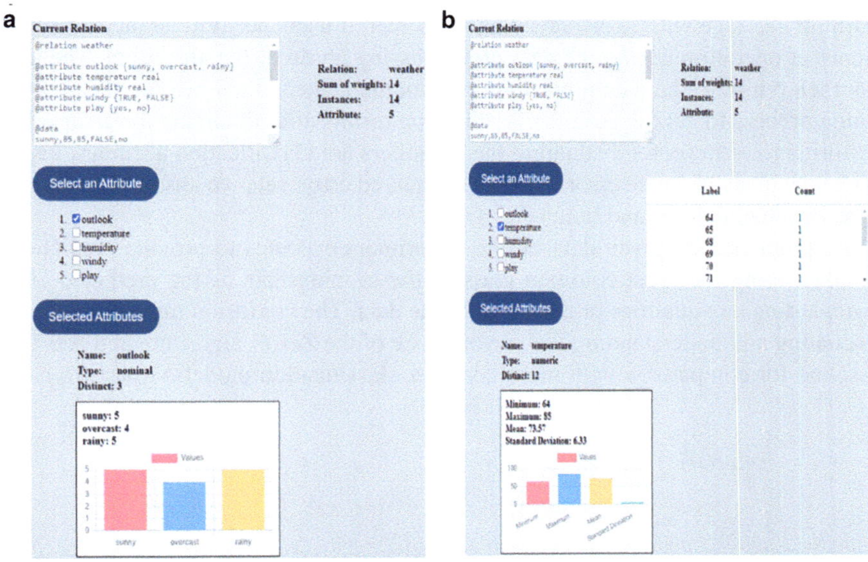

Fig. 3 Visualization of graphical user interface for data preprocessing, **a** outline of the attribute selection procedure and frequency graph, **b** overview of numeric attribute selection and display of the statistical characteristics and frequency table with respect to the given dataset

instances (e.g., sunny = 5; overcast = 4; and rainy = 5) in the dataset and visualizes it as a bar plot.

This visual representation allowed interdisciplinary learners, especially from non-computing backgrounds, to comprehend the distribution and prevalence of different attribute values. On the other hand, if the attribute was numeric (e.g., temperature), the tool was designed to display the minimum, maximum, mean, and standard deviation of the instance values and bar plot along with name, type, and number of distinct values (Fig. 3b). This was aimed to enable users to deduce statistical characteristics and trends within the dataset. A user could select an attribute at a time to provide more information. The virtual laboratory also allowed users to reset the entire data preprocessing using the "reset" button.

3.2 *Classification*

The classification module includes the simplest classification algorithm ZeroR, a naive classification methodology used as a starting point for classification tasks and provides a baseline for evaluating more complicated classifiers. The algorithm manages by identifying the most frequently occurring class label within the training dataset and makes categorization straightforward so that the most frequent result is picked as the forecast. Additionally, in the context of a data classification module, a warning message will be displayed on the screen if users perform an incomplete or incorrect operation during the data preprocessing phase of the algorithm. The user can then return to the data preprocessing module and can then execute the classification process to get the desired classification results after finalizing the errors. The performance criteria for evaluating the classifiers are classification accuracy, kappa statistics, mean absolute error, root mean squared error, relative absolute error, root relative squared error, and time (Fig. 4).

As output of the given dataset, the experiment executes to provide predictions by identifying the most common class in the training data as the predicted class disregarding any qualities or attributes of the data. The confusion matrix is used for evaluating and understanding the performance of the ZeroR algorithm and acts as a baseline for comparison with more complex classification models. Apart from the

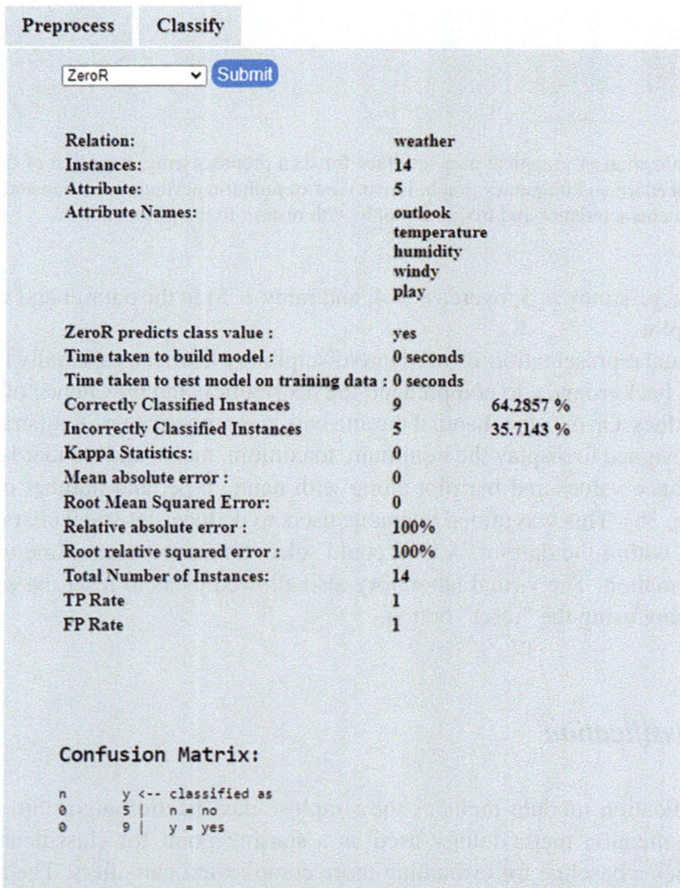

Fig. 4 Graphical user interface illustrates statistical outcomes using ZeroR classification in machine learning

simulation interface, the platform provided theoretical and procedure sections of the experiment as reading materials and pre-test and post-test questions as evaluation strategies to enhance learning pedagogy [40].

ZeroR serves as a teaching tool in educational settings by presenting baseline models and emphasizing the value of comparing against them to show the flaws in overly simplest methods. It serves as a basic benchmark by which more complex algorithms can be evaluated.

4 Discussion

The paper explores implementation of a virtual laboratory web-based tool for learning data preprocessing and classification in a bioinformatics classroom for offering valuable insights for effective data analysis, categorization and organizing objects. The goal of developing data preprocessing and classification as a virtual laboratory experiment was to provide a user-friendly interface and a seamless computing environment for students to learn techniques with and without teachers and at a convenient time and place with accessibility over the internet. Adopting an open-source freely available model, the virtual laboratory emphasizes the importance of creating an intuitive and accessible learning environment that supports student engagement and exploration. The utilization of a web-based open-access virtual lab facilitates the need to provide students with accessible resources to enhance their hands-on learning experience. Studies also support online laboratories offer flexibility in accessing educational materials and allow students to practice data preprocessing concepts in a simulated environment.

During the development of this web-based tool, a structured four-step approach was adopted. The student learners can repeat and perform data cleaning, formatting, transformation, and data reduction as a sequential framework for understanding and applying data preprocessing techniques. When compared to other data preprocessing tools for text mining, the implemented web-based tool simplifies the data preparation process by greatly minimizing the time and effort required for routine data preprocessing steps. The tool will allow users to quickly proceed to the main task of data analysis and generate datasets that were ready to be loaded into data mining workbenches for further analysis and tasks such as classification and feature prediction. This work can also be extended to complex and big data sources. Through the design, we hypothesized that student learning may have to involve the importance of data validation, data cleaning techniques, data transformation through normalization, as in regular classroom courses. The web-based tool helps students to understand the importance of generating and working with small datasets, applying different data scaling methods and interpreting their impact on data analysis. The data cleaning techniques can be extended in real life and streaming data to equip students with the skills to address common data quality issues and prepare datasets for further analysis. An important feature of our virtual laboratory is the user-friendly interface for attribute selection that permits students to interpret data points within larger datasets and gain insights through data frequency graphs. Through this process the web-based educational technology solution promotes active exploration and analysis, enabling students to make informed decisions based on statistical characteristics and trends within the dataset. Overall, the newly developed virtual laboratory serves as a skilling resource for biological data analysis and biotechnology and bioinformatics students, equipping them with the basic skills to preprocess and analyze real-life laboratory datasets effectively.

5 Conclusion and Future Work

This study showcases the design considerations for a virtual laboratory web-based tool for data preprocessing and classification, designed to support students in understanding and practicing various steps involved in data cleaning, formatting, transformation, and data reduction and ZeroR classification. The development of this virtual lab tool utilized popular web technologies such as HTML and JavaScript, ensuring its accessibility, ease of learning, flexibility, interactivity, and cost-effectiveness for educational purposes. In our implementation, users could easily upload datasets, explore dataset details, select attributes, and access comprehensive information about the selected attributes, facilitating the process of making informed decisions during subsequent data analysis processes. Further enhancements can and need to be made to expand the purpose and scope of data analysis. This includes our ongoing processes to extend functionality, incorporating additional preprocessing techniques, and integrating it with algorithms for classification or clustering and platforms to target the virtual laboratory as an interactive textbook for those within and beyond classrooms. Limitations of this study include the pedagogical surveys to evaluate the effectiveness of a data analysis-based virtual laboratory in decision-making skills, and to link the data-driven approach with real-world situations.

Acknowledgements This work derives direction and ideas from the Chancellor of Amrita Vishwa Vidyapeetham, Sri Mata Amritanandamayi Devi. This study was supported by the Virtual Laboratory Project Phase III extension of NMEICT, the Ministry of Education, the Government of India, and partially by Embracing the World Research for a Cause Initiative.

References

1. Dua S, Chowriappa P (2012) Data mining for bioinformatics. CRC Press
2. Lan K, Wang D, Fong S, Liu L, Wong KKL, Dey N (2018) A survey of data mining and deep learning in bioinformatics. J Med Syst 42:1–20
3. Dumancas GG, Adrianto I, Bello G, Dozmorov M (2017) Current developments in machine learning techniques in biological data mining, Bioinform Biol Insights, SAGE Publications Sage UK: London, England, vol 11. 1177932216687545

4. Magana AJ, Taleyarkhan M, Alvarado DR, Kane M, Springer J, Clase K (2014) A survey of scholarly literature describing the field of bioinformatics education and bioinformatics educational research. CBE Life Sci Educ 13(4):607–623. https://doi.org/10.1187/cbe.13-10-0193
5. Tsai MF, Yu SS (2015) Data mining for bioinformatics: design with oversampling and performance evaluation. J Med Biol Eng 35:775–782
6. Janani SR, Subramanian R, Karthik S, Vimalarani C (2023) Healthcare monitoring using machine learning based data analytics, Int J Comput Commun Control, Vol 18, https://doi.org/10.15837/ijccc.2023.1.4973
7. Ranganathan S (2005) Bioinformatics education—perspectives and challenges. PLoS Comput Biol 1(6):e52
8. Kuonen D (2003) Challenges in bioinformatics for statistical data miners. Bull Swiss Stat Soc 46:10–17
9. Singh et al (2013) Towards human-computer synergetic analysis of large-scale biological data, in BMC bioinformatics, 1–18
10. Li Y, Chen L (2014) Big biological data: challenges and opportunities. Genomics Proteomics Bioinform 12(5):187
11. Kleinschmit AJ et al (2021) Community development, implementation, and assessment of a NIBLSE bioinformatics sequence similarity learning resource. PLoS ONE 16(9):e0257404
12. Altman RB (2003) Complexities of managing biomedical information. OMICS 7(1):127–129
13. Zomaya AY, Elloumi M (2013) Biological knowledge discovery handbook: preprocessing, mining and postprocessing of biological data. John Wiley & Sons
14. Frank E, Hall M, Trigg L, Holmes G, Witten IH (2004) Data mining in bioinformatics using Weka. Bioinformatics 20(15):2479–2481
15. Naik A, Samant L (2016) Correlation review of classification algorithm using data mining tool: WEKA, Rapidminer, Tanagra, Orange and Knime. Proced Comput Sci 85:662–668
16. Stančin I, Jović A (2019) An overview and comparison of free Python libraries for data mining and big data analysis, In: 2019 42nd International convention on information and communication technology, electronics and microelectronics (MIPRO), 977–982
17. Garc'ia S, Luengo J, Herrera F (2015) Data preprocessing in data mining, Springer. Vol 72
18. Kathuria A, Gupta A, Singla RK (2021) A review of tools and techniques for preprocessing of textual data. Adv Intell Syst Comput 1227:407–422. https://doi.org/10.1007/978-981-15-6876-3_31
19. Achuthan K et al (2011) The VALUE @ amrita virtual labs project: using web technology to provide virtual laboratory access to students, In: Proceedings—2011 IEEE global humanitarian technology conference, GHTC, 117–121. https://doi.org/10.1109/GHTC.2011.79
20. Diwakar S, Radhamani R, Nizar N, Kumar D, Nair B, Achuthan K (2019) Using learning theory for assessing effectiveness of laboratory education delivered via a web-based platform: Proceedings of the 15th international conference on remote engineering and virtual instrumentation, In: Lecture notes in networks and systems, 639–648. https://doi.org/10.1007/978-3-319-95678-7_70
21. Francis SP, Kanikkolil V, Achuthan K (2016) Learning curve analysis for virtual laboratory experimentation, In: 2016 International conference on advances in computing, communications and informatics, ICACCI 2016, Institute of electrical and electronics engineers Inc., Nov. 1073–1078. https://doi.org/10.1109/ICACCI.2016.7732187
22. Prieto-Blázquez J, Arnedo-Moreno J, Herrera-Joancomart J (2008) An integrated structure for a virtual networking laboratory IEEE. Trans Ind Electron 55(6):2334–2342
23. Stark E, Bistak P, Kozak S, Kucera E (2017) Virtual laboratory based on Node. JS technology, In: 2017 21st International conference on process control (PC), 386–391
24. Achuthan K, Nedungadi P, Kolil VK, Diwakar S, Raman R (2020) Innovation adoption and diffusion of virtual laboratories. Int J Online Biomed Eng 16(9):4–25. https://doi.org/10.3991/ijoe.v16i09.11685

25. Diwakar S et al (2014) Usage and diffusion of biotechnology virtual labs for enhancing university education in India's urban and rural areas, In: E-Learning as a socio-cultural system: a multidimensional analysis, IGI Global, 63–83. https://doi.org/10.4018/978-1-4666-6154-7.ch004
26. Bose R (2013) Virtual labs project: a paradigm shift in internet-based remote experimentation. IEEE Access 1:718–725. https://doi.org/10.1109/ACCESS.2013.2286202
27. Achuthan K, Murali SS (2015) A comparative study of educational laboratories from cost & learning effectiveness perspective, In: Advances in intelligent systems and computing, Springer Verlag, 143–153. https://doi.org/10.1007/978-3-319-18473-9_15
28. Seifan M, Robertson N, Berenjian A (2020) Use of virtual learning to increase key laboratory skills and essential non-cognitive characteristics. Educ Chem Eng 33:66–75. https://doi.org/10.1016/j.ece.2020.07.006
29. August SE, Hammers ML, Murphy DB, Neyer A, Gueye P, Thames RQ (2015) Virtual engineering sciences learning lab: Giving STEM education a second life. IEEE Trans Learn Technol 9(1):18–30
30. Diwakar S et al (2014) Complementing neurophysiology education for developing countries via cost-effective virtual labs: case studies and classroom scenarios, [Online]. Available: http://amrita.edu/virtuallabs
31. Kolil VK, Muthupalani S, Achuthan K (2020) Virtual experimental platforms in chemistry laboratory education and its impact on experimental self-efficacy, Int J Educ Technol Higher Educ, 17, 1 https://doi.org/10.1186/s41239-020-00204-3
32. Achuthan K, Raghavan D, Shankar B, Francis SP, Kolil VK (2021) Impact of remote experimentation, interactivity and platform effectiveness on laboratory learning outcomes, Int J Educ Technol Higher Educ, 18, 1, https://doi.org/10.1186/s41239-021-00272-z
33. Achuthan K, Brahmanandan S, Bose LS (2015) Cognitive load management in multimedia enhanced interactive virtual laboratories. Adv Intell Syst Comput 320:143–155. https://doi.org/10.1007/978-3-319-11218-3_15
34. Papaconstantinou M, Kilkenny D, Garside C, Ju W, Najafi H, Harrison L (2020) Virtual lab integration in undergraduate courses: insights from course design and implementation. Can J Learn Technol 46(3)
35. Abramov V et al (2017) Virtual biotechnological lab development. Bionanoscience 7(2):363–365. https://doi.org/10.1007/s12668-016-0368-9
36. Achuthan K, Kolil VK, Diwakar S (2018) Using virtual laboratories in chemistry classrooms as interactive tools towards modifying alternate conceptions in molecular symmetry. Educ Inf Technol (Dordr) 23(6):2499–2515. https://doi.org/10.1007/s10639-018-9727-1
37. Radhamani R, Kumar D, Nizar N, Achuthan K, Nair B, Diwakar S (2021) What virtual laboratory usage tells us about laboratory skill education pre—and post-COVID-19: Focus on usage, behavior, intention and adoption, Educ Inf Technol (Dordr), 26, https://doi.org/10.1007/s10639-021-10583-3
38. Karyukin V et al (2022) On the development of an information system for monitoring user opinion and its role for the public. J Big Data 9(1). https://doi.org/10.1186/s40537-022-00660-w
39. Petprasit W, Jaiyen S (2015) E-commerce web page classification based on automatic content extraction, In: 2015 12th International joint conference on computer science and software engineering (JCSSE), 74–77. https://doi.org/10.1109/JCSSE.2015.7219773
40. Radhamani R et al (2018) Virtual laboratories in biotechnology are significant educational informatics tools, 1547–1551. https://doi.org/10.1109/ICACCI.2018.8554596

Enhancing Online Learning: A Pose and Emotion-Based Approach for Student Attention Monitoring

Sarath Krishna Chingapurathu, L. Jeganathan, and M. Janaki Meena

Abstract With the increasing popularity and preference of online classes, teachers are at a disadvantage of losing face-time with students and are faced with the challenge of monitoring their students' attention levels remotely. This research, Visio-Focus, presents a non-invasive student monitoring system using deep learning and computer vision to measure and analyze the attention levels of students. It has two main components namely Pose Estimation and Emotion Recognition which processes the captured facial images of the student through the webcam at 15 frames per second to measure the attention level. The Pose Estimation model uses pure computer vision and the Emotion Recognition model uses a modified ResNet deep convolutional network that could classify emotions portrayed into 6 different labels at about 70% accuracy. This research proposes a novel and reliable approach to measuring attention levels, which will have a significant impact on the education field. The analytical survey of attention levels after each lecture will enable educators to tailor their teaching approach to specific students, resulting in better academic development.

Keywords Computer vision · Attention · Open-CV · CNN · Transfer learning · Pose estimation

---

Supported by MeitY (Government of India).

S. K. Chingapurathu (✉) · L. Jeganathan · M. Janaki Meena
School of Computer Science and Engineering, Vellore Institute of Technology, Chennai, TN 600127, India
e-mail: sarathkrishna.chinga2020@vitstudent.ac.in

L. Jeganathan
e-mail: jeganathan.l@vit.ac.in

M. Janaki Meena
e-mail: janakimeena.m@vit.ac.in

S. M. Thampi et al. (eds.), *Fifth International Conference on Computing and Network Communications*, Lecture Notes in Electrical Engineering 1221,
https://doi.org/10.1007/978-981-97-4711-5_45

1 Introduction

The traditional model of education, where students attend classes in person, has been challenged by the increasing popularity of online education. While online education offers several benefits such as flexibility and convenience, it also presents challenges for educators, one of which is the inability to monitor the attention levels of students during a lecture. This challenge has led to the emergence of a new research domain, student attention detection, involving computer vision, NLP, and deep learning. However, the writing requires further elaboration to clearly articulate the research problem and contributions.

Computer vision and deep learning have the potential to develop systems that can detect and monitor the attention levels of students during online classes. These technologies involve the use of cameras and sensors to capture and analyze data such as facial expressions, eye movements, and body posture. In this context, the integration of eye-tracking technology could offer more precise insights into students' focus. This aspect should be acknowledged for a comprehensive understanding of attention detection systems. By analyzing this data, educators can gain insights into the engagement and attention levels of students during a lecture. The development of attention detection systems using computer vision and deep learning has the potential to revolutionize online education by enabling educators to better monitor and engage with students in a virtual environment. However, it is crucial to address privacy concerns related to the continuous monitoring of students and ensure ethical considerations are implemented in the deployment of these systems. As the technology continues to evolve, it is expected that attention detection systems will become more accurate and will be widely adopted in the education industry. Additionally, further research and development are needed to enhance the real-time capabilities of these systems, ensuring immediate feedback for educators to adjust their teaching strategies dynamically. Another challenge that educators face in online education is the inability of the teachers to gauge the understanding level of the students. Attention detection systems can help address this challenge by providing educators with data that can be used to personalize the learning experience for each student. Furthermore, attention detection systems can be used to identify and intervene when a student is struggling with a particular topic. To maximize the benefits, it is essential to integrate machine learning algorithms that can analyze historical data to predict students' comprehension levels and provide timely interventions. For example, if a student's attention levels drop during a particular part of a lecture, the teacher can intervene by providing additional explanations or resources to help the student understand the material. This can help improve the overall learning outcomes and reduce the risk of students falling behind. Additionally, collaboration with psychologists and educational experts is vital to ensure that interventions are not only timely but also tailored to individual learning styles and needs. Finally, attention detection systems can provide valuable insights into the effectiveness of teaching methods and materials. By analyzing the attention levels of students during different parts of a lecture or using different teaching methods, educators can gain insights into what works toward helping the student learn better and gain much insights from the lecture. However, it is crucial to consider diverse learning styles and preferences among students to ensure that the conclusions drawn from attention data are applicable across a broad spectrum of learners. This can help improve the quality of education and drive innovation in teaching methods and materials. Moreover, ongoing feedback loops between educators and developers should be established to iteratively enhance attention detection systems based on real-world classroom experiences. Having understood the importance of measuring the attention levels of students, this paper attempts to measure the attention level of students in an online class through computer vision and deep transfer learning techniques. The integration of Pose Estimation and Emotion Recognition models signifies a multifaceted approach that could provide a more comprehensive understanding of attention dynamics in online classes.

2 Literature Review

In this section, we review the relevant literature on attention detection. The literature survey also highlights the limitations of existing systems and the need for a comprehensive and non-invasive approach to monitoring student attention.

Gao et al. [1] employed an invasive approach to wearable glasses with cameras for eye tracking. Their deep learning-based approach achieved over 90% accuracy in detecting changes in attention level based on the gaze tracking algorithm, which led to its application in various domains. Lemme et al. [2] surveyed various invasive and non-invasive techniques for accurate eye localization in real-world settings, offering insights into improving eye-tracking technologies. Santoso and Hardjono [3] explored non-invasive automatic attention detection using eye tracking for online learning, demonstrating the relevance of eye-tracking technology in education. Hussein et al. [4] proposed a non-invasive approach for analyzing student attention using eye-tracking technology. Their method classified student attention into three levels based on pupil position, which showed promising results for real-time feedback.

Fernandes et al. [5] predicted student attention using whole-body features. Even though invasive, by analyzing various body positions, they achieved over 80% accuracy in attention prediction, offering the potential for tailored support in e-learning systems. Li et al. [6] proposed a recognizing attention based on facial expressions using a non-invasive convolutional neural network. Their method accurately detected changes in attention level, showcasing the potential of neural networks in attention analysis paving the way for the use of deep learning algorithms in attention study. Zhou et al. [7] detected attention levels in e-learning through facial expression recognition and eye movement tracking, highlighting non-invasive multi-modal approaches for engagement assessment. Kamarudin and Akmeliawati [8] yet again surveyed attention detection techniques in classrooms, providing an overview of various methodologies, non-invasive and invasive, employed for attention analysis of recent times. Alberico et al. [9] studied attentional engagement in multimedia learning, highlighting the effects of attention on recall and cognitive load during learning activities. It was invasive due to the need for EEG sensors for brain activity analysis.

Wang et al. [10] developed a non-invasive student attention monitoring and feedback system-based solely on machine learning in relation to the emotions displayed. Their system showcased the potential of machine learning techniques in attention analysis. Diaz et al. [11] used non-invasive facial Pose Estimation to measure student attention in a traditional classroom setting. Their approach achieved an accuracy of over 85% in detecting attention changes, providing insights for improving engagement. Cai et al. [12] explored non-invasive real-time attention tracking in online education via computer vision. Their study demonstrated accurate detection of attention changes, highlighting opportunities for personalized feedback through their custom dataset. Sharma and Kaur [13] combined facial expressions and skin conductance to analyze student attention. Their study provided insights into multi-modal approaches for attention estimation, though being invasive because of the need for skin sensors and other electrodes. Karadayi et al. [14] employed non-invasive sentiment analysis in speech tracks encountered in the classroom to monitor attention and engagement in a classroom. The proposed method accurately detected changes in attention level and engagement, showcasing the potential for enhancing teaching strategies.

Overall, these studies reflect the potential of diverse computer vision techniques in analyzing student attention. Each approach contributes valuable insights into monitoring and enhancing student engagement, paving the way for improved educational strategies. However, all the above papers did not have a specified scoring mechanism to measure the attention level but rather a classifier separate for each model. Moreover, there is no substantial research done on creating an ensemble-based model that would consist of two or more models that would contribute quintessential information toward gauging the attention level exhibited by the student. With a plan to propose a more comprehensive system, we unify the two models, namely a Pose Estimation module and an Emotion Recognition module that would contribute information on the pose and emotion displayed throughout the session. A composite score is generated from the data collected through these models and the student is graded based on the score he/she receives.

3 Methodology

The aim of this research paper is to develop a non-invasive attention monitoring system to compute the attention level. This system, VisioFocus, is composed of two main modules namely a Pose Estimation system and an Emotion Recognition system. The overall architecture of the system would be as follows (Fig. 1).

3.1 Pose Estimation System

The system leverages a sophisticated Pose Estimation module, grounded in the principles of computer vision. This module plays a pivotal role in calculating the precise

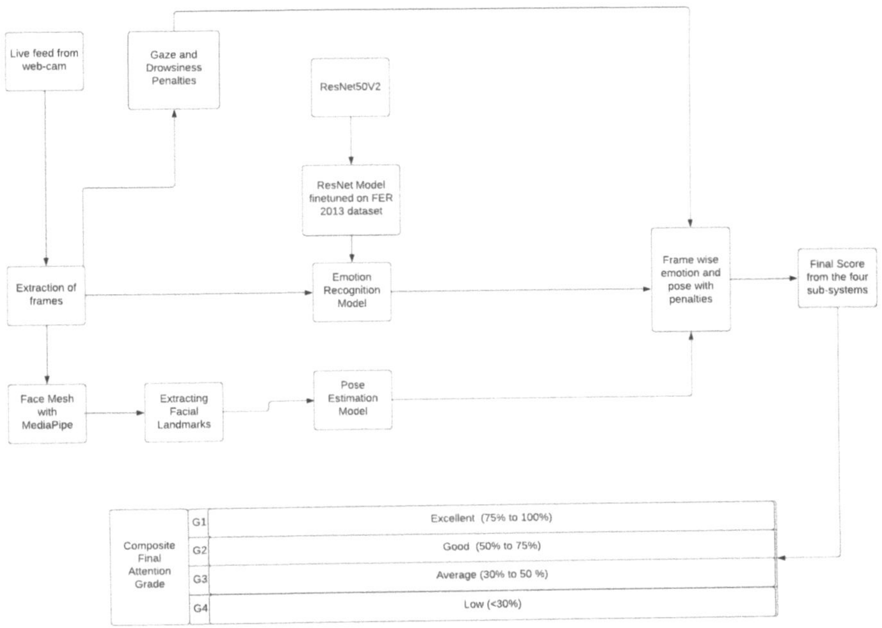

Fig. 1 Architecture for VisioFocus

projection angle of the user's nose, providing us with valuable insights into the direction of the user's head. By analyzing these subtle yet informative cues, we gain a deeper understanding of the user's posture and head orientation during the monitoring session.

In the subsequent algorithm, we delve into the intricacies of our head Pose Estimation technique. The algorithm operates by computing the yaw and pitch of the user's head, i.e., angles of rotation along the X and Y axes. At the start, the MediaPipe package is initialized which is used to create the Haar Cascade Classifier, used to detect facial landmarks. Once the relevant 2D and 3D landmarks are extracted, the system initializes the camera and distortion matrices for utilization in the SolvePnP package. SolvePnP is a critical computer vision function employed to solve the Perspective-n-Point (PnP) problem, allowing the accurate calculation of an object's three-dimensional pose using its corresponding two-dimensional image coordinates. The function returns two matrices, the rotational and translational matrices. The angles are then processed to match the coordinate axes to find the pitch and yaw angles along the y and x axes, respectively. These angles are the key to categorizing the user's head position into distinct classes, including 'Head at Left,' 'Head at Right,' 'Head Down,' 'Head Up,' or 'Head Straight'. Each classification has a threshold unique to it, exceeding which it would be classified as such. We don't attach any scientific merit to these thresholds as we discovered these values experimentally through trial and error. Relevant pose scoring metrics for each class are described in the algorithm (Figs. 2, 3 and 4).

Algorithm 1 Pose Estimation Algorithm

Initialize video capture from the default camera
Initialize the FaceMesh model with confidence thresholds
Initialize the Face Cascade Classifier for face detection
while WebCam video capture is open **do**
    Read a frame from the camera
    Detect faces in the frame using the Face Haar Cascade Frontal Face classifier
    Process the frame with the FaceMesh model
    Initialize variables for 3D face landmarks and 2D face landmarks
    if no face landmarks are detected **then**
        $text \leftarrow$ "Face not available"
        $PoseMetric \leftarrow 0.2$
    else
        for each 3D and 2D face landmark **do**
            Extract relevant landmarks
            Convert the 2D and 3D landmarks to a NumPy array
            Set the camera matrix as follows :

$$\text{cam\_matrix} = \begin{bmatrix} f_x & 0 & \frac{w}{2} \\ 0 & f_y & \frac{h}{2} \\ 0 & 0 & 1 \end{bmatrix}$$

            Where:
            - f_x is the focal length in the horizontal axis.
            - f_y is the focal length in the vertical axis.
            - w is the width of the image frame.
            - h is the height of the image frame.
            Set the distortion parameters

$$\text{dist\_matrix} = \begin{bmatrix} 0 \\ 0 \\ 0 \\ 0 \\ 0 \end{bmatrix}$$

            Setting them to zeros initially assuming that there is no distortion at the beginning
            Use solvePnP to compute the rotation and translation vectors
            Extract the rotational matrix and angles. The angles are then processed from the rotational matrix and then the pose is computed as follows
            if $x < -10$ **then**
                $text \leftarrow$ "Head at Left"
                $PoseMetric \leftarrow 0.7$
            else if $x > 10$ **then**
                $text \leftarrow$ "Head at Right"
                $PoseMetric \leftarrow 0.7$
            else if $y < -10$ **then**
                $text \leftarrow$ "Head Down"
                $PoseMetric \leftarrow 0.9$
            else if $y > 10$ **then**
                $text \leftarrow$ "Head Up"
                $PoseMetric \leftarrow 0.6$
            else
                $text \leftarrow$ "Head straight"
                $PoseMetric \leftarrow 1.0$
            end if
            Display head pose information on the frame with projection from the tip of the nose
        end for
    end if
    Display the text and use the Pose Metric while calculating the weighted total
end while

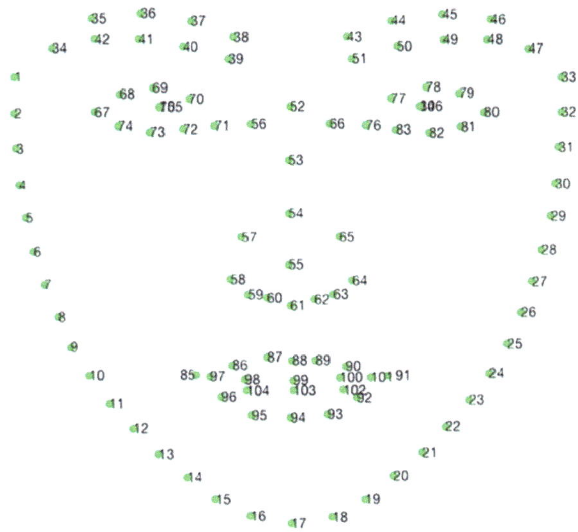

Fig. 2 The 106—landmark mesh used in the Haar cascade classifier [15]

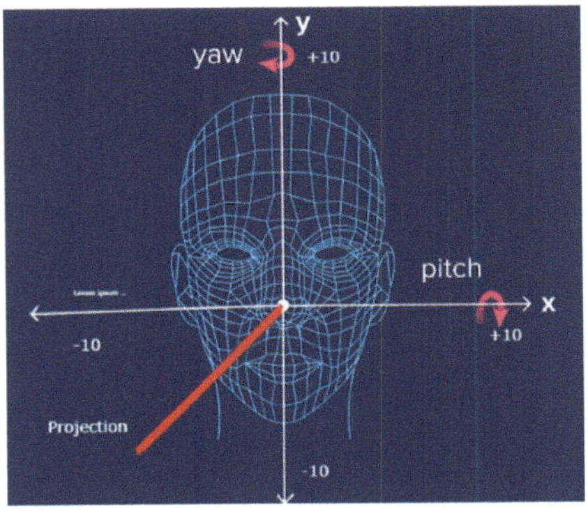

Fig. 3 The projection angles, as used in the above-mentioned algorithm

3.2 *Emotion Recognition System*

The Emotion Recognition model trained with the FER 2013 dataset was developed using a comprehensive methodology that involved several crucial steps, including data preprocessing, model selection, model training, and evaluation.

Dataset: The dataset used in this study is the Facial Expression Recognition 2013 (FER2013) dataset, which is a publicly available dataset consisting of 35,887 grayscale images of faces with seven different emotion labels—Angry, Happy, Neutral, Sad, Surprised, Fear, and Disgust. For our purpose of Emotion Recognition during online classes, we take only 5 classes out of this dataset namely Happy, Surprised, Neutral, Sad, and Angry because the other two emotions will not provide any required information about the user's emotional state during the class. The required

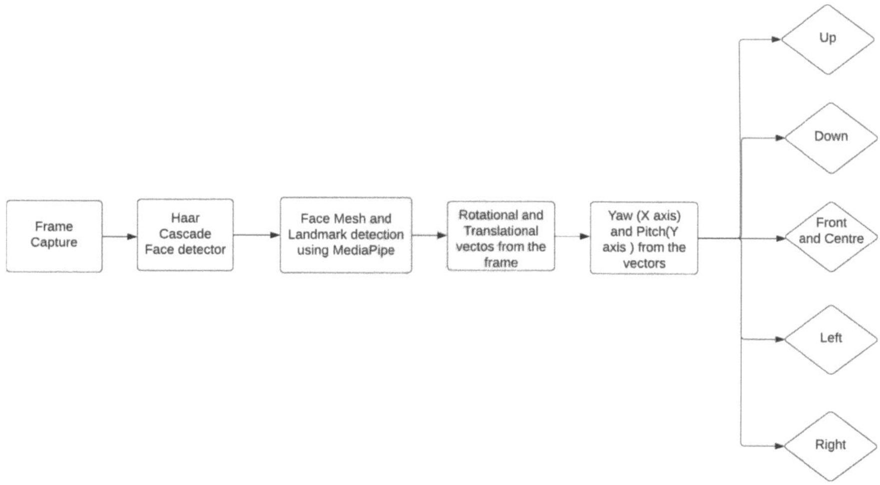

Fig. 4 Block diagram of the pose estimation module

annotated images were collected and the dataset is split into three parts—a training set of 25,000 images, a validation set of 5,000 images, and a test set of 5,887 images.

Data Pre-processing: Before training the models, the images were pre-processed to enhance the quality of the images and to extract the facial features. Each image was resized to a fixed size of 224 × 224 pixels, converted to grayscale, and then normalized to have zero mean and unit variance. To extract the facial features, the pre-trained Haar Cascade Classifier was used to detect faces in the images. The faces were then aligned using facial landmark detection to ensure that the eyes, nose, and mouth were in the same position across all images. The aligned faces were then cropped and resized to 48 × 48 pixels. Then they are normalized to the pixel values between 0 and 1, and converted the labels to one-hot encoded vectors for ease of input.

Model Selection and Training: We first created a custom CNN model for the Emotion Recognition task. The CNN had 11 layers, including 5 convolutional layers with batch normalization, two pooling layers, and one fully connected layer. The input image is first passed through a convolution layer with 32 filters of size 3 × 3. The resulting feature maps are then passed through a ReLU activation function and a max-pooling layer of size 2 × 2. This process is repeated with 4 other convolutional layers with 64 filters of size 3 × 3, followed by another ReLU activation and max-pooling layer of size 2 × 2. Three more convolution layers are used with 128, 256, and 512 filters. The output is then flattened and passed through a fully connected layer of 1024 units with a ReLU activation function, followed by a final output layer of 6 units with a softmax activation function. The model was trained using the categorical cross-entropy loss function and the Adam optimizer with a learning rate of 0.001. We

Fig. 5 Architecture of ENet

Table 1 Comparison of different models after training for 50 epochs

Model name	Training accuracy (%)	Validation accuracy (%)
ENet	83.30	64.91
ResNet50V2	84.45	69.35
InceptionV3	74	65.23
VGG16	85.67	63.64

named this model ENet and we trained it on the FER 2013 dataset. On training and cross-validation, we obtained a training accuracy of 84% and a validation accuracy of 65.32%. Since the performance metrics weren't up to the mark for the system, we turned to transfer learning techniques for better accuracy and precision (Fig. 5).

For transfer learning, we used three pre-trained models: ResNet50v2, VGG16, and InceptionV3. ResNet50v2 is a deep residual network that has shown excellent performance in a variety of computer vision tasks. It consists of 50 layers and has skip connections that allow for the efficient training of deep networks. VGG16 is a classic architecture that consists of 16 layers, including several convolutional and fully connected layers. InceptionV3 is a more recent architecture that uses a combination of convolutional layers with varying kernel sizes and pooling layers to extract features. By leveraging these pre-trained models, we were able to take advantage of their learned features and adapt them to our specific task of Emotion Recognition. The use of pre-trained models allowed us to significantly reduce the time and resources required for training our models from scratch, while still achieving high accuracy in our predictions. Each of the four models was trained on the training set and evaluated their performance on the validation set. The categorical cross-entropy loss function and the Adam optimizer were used to train the models. To prevent overfitting, early stopping with patience of 10 was employed in each model, which was trained for 50 epochs at a batch size of 64. The learning rate was initially set to 0.001 and was reduced by a factor of 10 after 10 and 20 epochs to improve convergence. All the trained models were evaluated using accuracy, precision, recall, and AUC-ROC as the performance metrics.

Based on the results obtained, we can see that the ResNet50V2 model performed the best, achieving a training accuracy of 84.45% and a validation accuracy of 69.35%. Even though the other models performed well in training, they lacked accu-

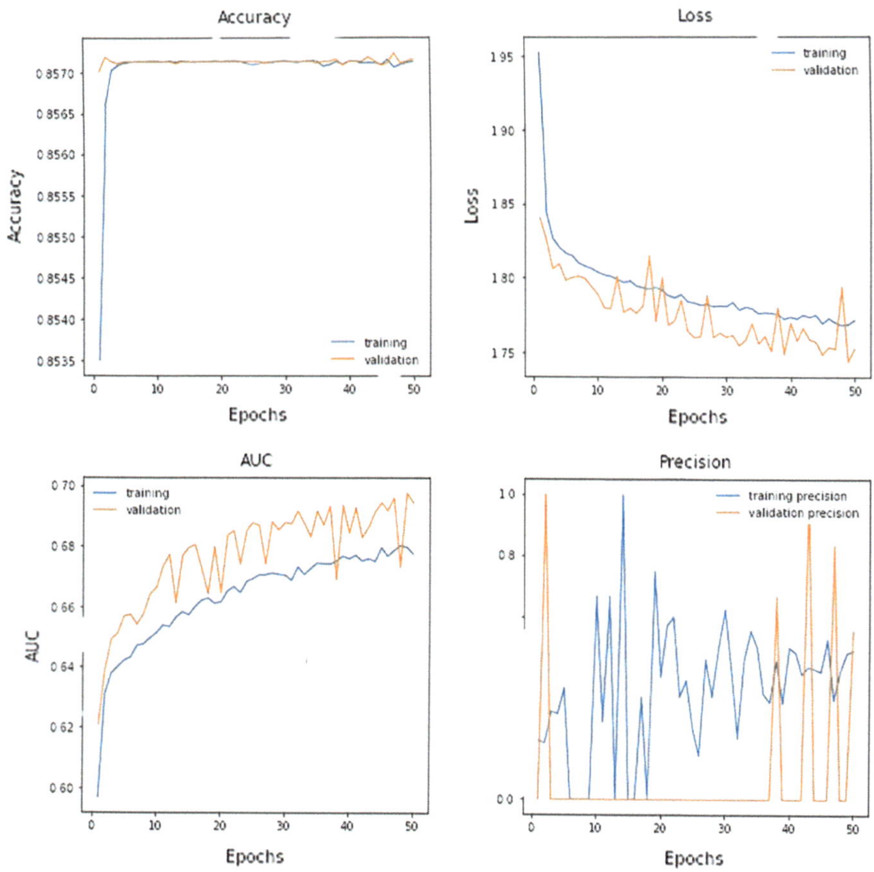

Fig. 6 Training metrics for ResNet50V2

racy in validation and testing as seen from the Table 1. Hence we go forward with the ResNet50V2 model for the Emotion Recognition model (Figs. 6 and 7).

3.3 *Ensemble of Both the Models*

We propose the ensemble system that amalgamates the outcomes of the individual Pose Estimation and Emotion Recognition models to derive a comprehensive assessment of the user's cognitive engagement and attention levels during interactive sessions. Each of these models contributes unique insights: the Pose Estimation system evaluates the user's body posture and the Emotion Recognition system gauges their emotional state. By combining these distinct metrics, a holistic representation of the user's cognitive and emotional states is achieved.

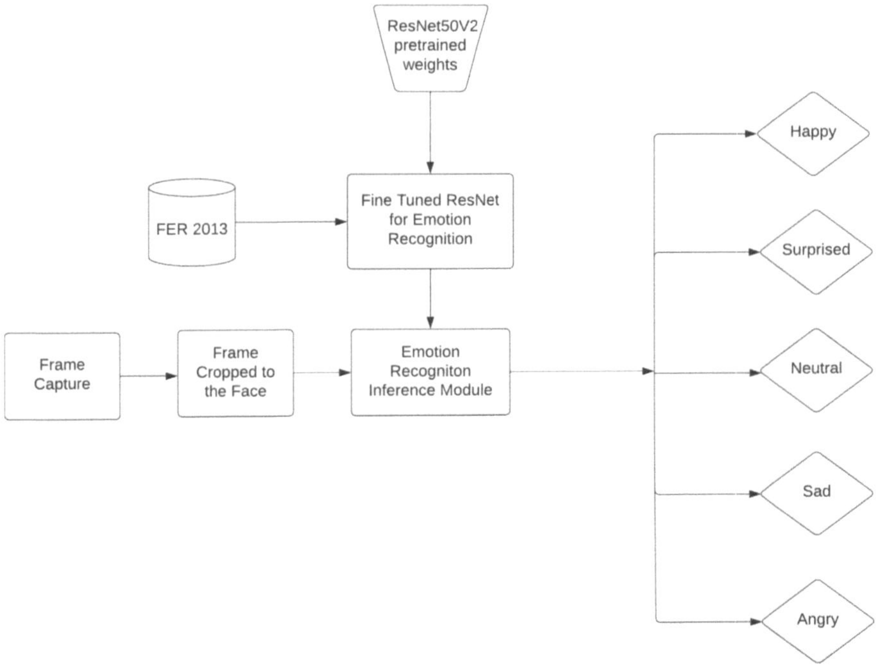

Fig. 7 Block diagram of emotion recognition system

Pose Estimation Score (PES) : This is the first level that represents a person who is paying attention and is in a positive emotional state. This could be an indication that they are engaged and interested in the content being presented. There are 5 fundamental poses up, down, left, right, and looking straight forward. Each of these poses would have a multiplying factor for each second there are in this pose to get a weighted score for this whole system. Looking straight at the screen would indicate the highest attention pose and thus it would gain the highest score. Second comes looking down, when even though the student might be attentive and taking notes, a little bit of penalty is incurred. Rest every pose is classified as inattentive. The equation given below is the formula for PES and the table shows the multiplying factors for each pose:

$$\text{PES} = \frac{1}{T_{\text{total}}} \sum_{i}^{\text{poses}} (T_i \cdot \text{PoseMetric}_i) \tag{1}$$

- Where poses = {left, right, up, down, straight forward}
- Pose Metric considered as depicted in the Algorithm 1

Table 2 Emotion attention metrics

Emotion	Metric
Happy	1.0
Surprised	1.0
Neutral	1.0
Angry	0.7
Sad	0.5

Emotion Analysis Score (EAS): The emotion analysis score EAS is a metric that quantifies the emotional states exhibited by individuals during a monitoring session. It captures the emotional dynamics of users by evaluating their facial expressions and corresponding emotions using the ResNet50V2 model. The score is computed by summing the weighted contributions of emotional states over discrete time intervals. Each interval's duration (T_i) is multiplied by a specific emotion metric (EmotionMetric_i) that corresponds to the emotional intensity of the detected expressions. Considering the vast array of emotions shown by a student in the class, we score them based on how much the emotion correlates to their understanding. For instance, a happy or surprised face would depict a good understanding and would get the complete score for the time spent happy. When sad or angry emotions are shown, it would lead to the inference that the student might not be attentive. Hence those emotions would only get a partial score. The equation given below is the formula for EAS and the Table 2 shows the Emotion Metric for each emotion displayed:

$$\text{EAS} = \frac{1}{T_{\text{total}}} \sum_{i}^{\text{emotions}} (T_i \cdot \text{EmotionMetric}_i) \tag{2}$$

- Where emotions = {happy, surprised, neutral, angry, sad}.

3.4 Final Grading of the Ensemble Score

The Composite Attention Score is the composition of the individual scores from the Pose Estimation and Emotion Recognition modules. This integrated metric, brought together by weighted average, combines all the weighted scores from the two modules and offers a comprehensive assessment of the user's attention during online sessions. The Composite Attention score can be calculated as follows:

$$\text{Composite Attention Score} = \frac{100 \times (\text{PES} + \text{EAS})}{2} \tag{3}$$

From the Composite Attention Score, we can create a grading system to universally grade the attention exhibited by the user into four grades. The following insights can be gathered from the grades with their corresponding interpretations:

Grade 1: Excellent Attention (75–100%):

- Users exhibit high attention cues, such as happy and surprised expressions, while maintaining a direct gaze.
- Computationally, the user would receive high PES and EAS thus leading to a higher final composite score.
- Implies optimal focus and responsiveness during the session.

Grade 2: Good Attention (50–75%):

- Users display moderate attention, occasionally showing neutral or slightly negative emotions, while still keeping a centered gaze.
- Shows a slight reduction in either PES or EAS, indicating that he might be physically or mentally disturbed enough to show the deflection.
- Indicates adequate engagement with a room for improvement.

Grade 3: Average Attention (25–50%):

- Indicates adequate engagement with room for improvement.
- Users demonstrate average to below average attention levels, with varying emotional responses and gaze behavior. The scores exhibited would be below average in both PES and EAS thus leading to an average or below composite score.
- Suggests the need for interventions to enhance attention and maintain a consistent focus.

Grade 4: Low Attention (< 25%):

- Users exhibit minimal to no attention cues, reflecting limited engagement and responsiveness.
- Substantial reduction in both PES and EAS could be seen which resulted in a poor final composite score.
- Requires immediate attention and intervention to address the lack of focus and enhance participation.

4 Results

In our research, we aimed to develop a system that could accurately quantify the level of attention of students in a classroom setting. To achieve this, we utilized a combination of two attributes—pose and emotion—to provide a more comprehensive and reliable measure of attention. Overall, our research provides a promising approach for quantifying attention levels in real-time using multiple attributes and can potentially have applications in various fields, including education, healthcare, and psychology.

5 Limitations, Future Work, and Conclusion

Thus, we have proposed a comprehensive mechanism for gauging the attention of students in classrooms. Despite the promising results of our research, we still have not validated our work with real-time subjects in the setting of a remote classroom, We plan to validate the same and explore various other sources of data for quantifying student attention. Apart from validation, there are still some limitations that need to be addressed. For example, the system's performance may vary depending on the individual's ethnicity, age, and gender, and further research is needed to determine the system's generalization across different populations. Additionally, the system's accuracy may be affected by environmental factors such as lighting and background noise, and further research is needed to determine how these factors affect the system's performance. In light of these limitations, our research serves as a perfect starting point for further investigations in this area of multi-modal attention analysis.

In conclusion, our research provides a promising approach for quantifying attention levels in real-time using multiple attributes. Our method has proven to be reliable and effective, demonstrating its potential to be used as a valuable tool for monitoring and improving attention in various diverse domains.

Acknowledgements This work is funded by the Ministry of Electronics and Information Technology, Government of India. Sanction number L-14011/4/2022-HRD.

References

1. Gao X, Cao Y, Jiang J, Liu W (2015) Eye tracking in the wild. In: Proceedings of the IEEE conference on computer vision and pattern recognition, pp 2555–2564
2. Lemme A, Kinner L, Bulling A (2016) Accurate eye localization in real-world settings: a comparison of state-of-the-art algorithms. In: Proceedings of the 18th ACM international conference on multimodal interaction, pp 3–10
3. Santoso A, Hardjono B (2017) Automatic attention detection using eye tracking technology for online learning. Int J Inf Educ Technol 7(5):330–334
4. Hussein M, Mohamed A, Alkanhal R (2017) Analyzing student attention in classroom using eye-tracking technology. J Educ Technol Dev Exchange (JETDE) 10(1):1–14
5. Fernandes RA, Silva CB, Santos AL (2018) A method for predicting student attention using whole bodily features. In: 2018 IEEE frontiers in education conference (FIE). IEEE, pp 1–8
6. Li X, Li Q, Li L (2018) Recognizing student attention with facial expressions using convolutional neural network. In: 2018 IEEE international conference on information and automation (ICIA). IEEE, pp 1910–1914
7. Zhou Y, Gao H, Chen Y (2018) Attention-level detection in e-learning based on facial expression recognition and eye movement tracking. Multimedia Tools Appl 77(3):3463–3485
8. Kamarudin N, Akmeliawati R (2018) A review of attention detection in the classroom. J Eng Appl Sci 13(4):888–893
9. Alberico C, Bozza G, Giglio F (2019) Attentional engagement in multimedia learning: effects on recall and cognitive load. Comput Human Behav 92:442–449
10. Wang X, Chen L, Wu X (2019) Student attention monitoring and feedback system based on machine learning. IEEE Access 7:158931–158940

11. Diaz G, Carro RM, Martínez-Mirón EA (2020) Measuring student attention in a traditional classroom using face detection. Sensors 20(8):2365
12. Cai X, Zhan Z, Yu L, Zhang X, Liu Y (2021) Real-time attention span tracking in online education via computer vision. IEEE Access 9:13165–13176
13. Sharma S, Kaur P (2021) Analyzing Attention Span of Students using Facial Expressions and Skin Conductance. Int J Innov Technol Explor Eng 10(8):123–126
14. Karadayi J, Dogan R, Turhan MA (2022) An attention monitoring and engagement detection system for the classroom. Interact Learn Environ 1–19
15. Liu Y, Shen H, Si Y, Wang X, Zhu X, Shi H, Hong Z, Guo H, Guo Z, Chen Y, Li Bi, Xi T, Yu J, Xie H, Xie G, Li M, Lu Q, Wang Z, Lai S, Wei X (2019) Grand challenge of 106-point facial landmark localization, pp 613–616. https://doi.org/10.1109/ICMEW.2019.00117
16. Ouellette P, Casanova MF (2018) Utilizing an eye-tracking system to evaluate attention to written words during a free-viewing task. Front Psychol 9:1216
17. Gao X-S, Hou X-R, Tang J, Cheng H-F (2003) Complete solution classification for the perspective-three-point problem. Pattern Anal Mach Intell, IEEE Trans 25(8):930–943

Learning in the Digital Space: A Review of Assessment Approaches in Education Metaverse

Nur Ameerah Abdul Halim and Ajune Wanis Ismail

Abstract The advent of the education metaverse has transformed the landscape of learning and teaching, offering immersive and dynamic. The Education Metaverse represents a transformative paradigm shift in the field of education, leveraging immersive technologies to provide dynamic and interactive learning experiences. This paper explores various aspects of the education Metaverse, including conceptual frameworks, pedagogical approaches, and assessment approaches, drawing insights from recent research works in this emerging field. The findings reveal a growing interest in immersive and personalized learning, augmented by innovative assessment methodologies that go beyond traditional testing. The paper concludes by highlighting key areas for future research and development, emphasizing the need for continued exploration of this promising educational frontier experiences for learners.

Keywords Education metaverse · Digital learning · Online learning · Assessment · Assessment strategies

1 Introduction

The rapid evolution of technology has given rise to a new era in education, where virtual worlds and digital environments have become integral to the learning experience [1]. While Covid-19 pandemic declared in 2020, our people now have to adapt living in a community where face-to-face interactions were significantly limited [2, 3] causing distance learning to be a significant activity during pandemic through online platform such as Zoom or Microsoft Teams. However, it primarily relies on two-dimensional (2D) display, having lacks of immersion and engagement as in

N. A. Abdul Halim (✉) · A. W. Ismail
Mixed and Virtual Environment Research Lab (mivielab), ViCubeLab, Faculty of Computing, Universiti Teknologi Malaysia, Johor Bahru, Johor, Malaysia
e-mail: nurameerah@graduate.utm.my

A. W. Ismail
e-mail: ajune@utm.my

S. M. Thampi et al. (eds.), *Fifth International Conference on Computing and Network Communications*, Lecture Notes in Electrical Engineering 1221,
https://doi.org/10.1007/978-981-97-4711-5_46

physical classroom which leads to students struggling to stay focused during remote lecturers. In this context, the concept of the "education metaverse" has emerged as a transformative force, reshape traditional educational paradigms. It provides opportunities for students from diverse backgrounds, including those in rural and urban areas, to engage in collaborative learning activities, interact with digital tutors, and create and innovate within a simulated virtual environment [4]. The education metaverse represents a convergence of interactive technologies such as virtual reality (VR), augmented reality (AR), mixed reality (MR), and immersive digital environments that provide students and educators with novel and interactive spaces for teaching and learning [5, 6].

Learning in the metaverse gives students the flexibility of studying anytime and anywhere, tailoring their learning experience to their needs. It also uses blockchain technology to prevent academic misconduct [1]. The metaverse revolutionizes the conventional teaching model, shifting it from a static to a dynamic approach, surpassing other online teaching platforms when it comes to fostering student-centered, collaborative learning [7]. This shift has not only redefined the medium but also prompted a reevaluation of how we assess and measure learning outcomes. Therefore, in this paper, we aim to present a comprehensive exploration of assessment methodologies within the education metaverse. As educators harness the power of data analytics and behavioral insights to craft tailored assessments, they empower learners to take charge of their education, aligning the metaverse's potential for innovation with meaningful educational outcomes.

With this paper, we aim to contribute to a deeper understanding of how educators can leverage the metaverse to inspire, engage, and assess students in ways previously unimaginable in traditional education settings. Therefore, this paper is organized as follows. Section 2 discusses on the background of this study. In the next section, we study on the assessment on education, and lastly is the discussion and future work.

2 Background

The term Metaverse has emerged as a transformative concept in the realm of digital technology and education. Coined initially in science fiction literature "Snow Crash", the metaverse has evolved into a dynamic and immersive virtual space that blurs the boundaries between the physical and digital worlds [8]. The term metaverse is a combination of the word 'meta', a Greek word that means "beyond" and the word 'verse' which refer to the root of universe [9]. Within the metaverse, individuals can interact, collaborate, and engage in a wide array of activities [10], from social interactions to business transactions and, notably, educational endeavors [2, 4]. Hwang and Chien [11] define the metaverse as an emerging paradigm within the next-generation internet, with the objective of constructing a completely immersive, hyper-spatiotemporal, and self-sustaining virtual shared environment where individuals can engage in work, recreation, and social interactions. The development of the metaverse facilitates everyday human communication and interaction through the internet [12].

In the years following, the metaverse concept found more vivid depictions in science fiction movies like "Ready Player One," "Lucy," and "The Matrix". Despite that, in 2007, a metaverse roadmap was proposed as a fusion of both virtually-enhanced physical reality and physical-persisted virtual space by a metaverse research organization, Acceleration Studies Foundation (ASF) [2, 13]. According to Zhang et al. [14], there are four scenarios which were categorized in the metaverse roadmap including AR, lifelogging, virtual worlds, and mirror worlds. Utilizing technologies and devices like AR and VR, users crafted avatars to represent themselves and engage in various activities within a virtual world.

In the context of the educational metaverse, the configurations of AR, VR, and MR systems play a crucial role in enhancing learning experiences. AR superimposed the virtual elements into the physical environment, typically utilizing AR glasses or smartphones with cameras and sensors [15, 16]. VR immerses learners in a fully simulated environment through headsets and hand controllers, requiring powerful computers for optimal performance [17]. Meanwhile, MR integrates both real-world and virtual elements, often implemented through MR headsets with spatial mapping capabilities and gesture recognition [18, 19]. These configurations enable users to craft avatars, fostering interactive and engaging educational activities within virtual worlds, aligning with the metaverse roadmap's categories of AR, virtual worlds, and mirror worlds.

Min and Cai [20] attribute three distinguishing characteristics to the Metaverse that set it apart from conventional VR or AR: "shared," "persistent," and "decentralized." While "shared" and "decentralized" are terms commonly understood, the concept of "persistent" is somewhat abstract. Some interpretations define "persistent" as the Metaverse's continued existence irrespective of time, space, or the user's presence [20]. Furthermore, the integration of AI technology is essential to ensure that the Metaverse operates in accordance with the rules established by its creator. Consequently, an AR or VR system could be considered a component of the Metaverse, facilitating the presentation of virtual content.

2.1 *Education in Metaverse*

In education, the Metaverse is poised to revolutionize the learning paradigm by expanding opportunities beyond physical classrooms, echoing a decade-long argument among educators [21]. This concept represents a paradigm shift in how we perceive and harness technology for learning, offering unprecedented opportunities for innovative pedagogical approaches and redefining the traditional classroom experience. The formal education system encompasses primary or elementary, secondary, and tertiary or higher education levels. As asserted by Birabil and Ogeh [22], higher or tertiary education holds the highest position among these levels as it surpasses the secondary or high school tier in terms of educational attainment.

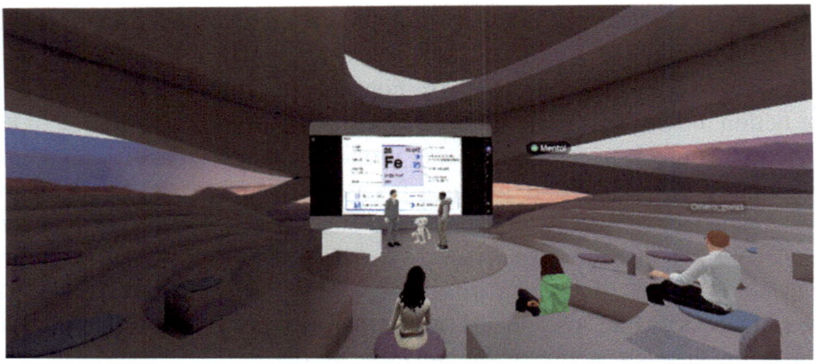

Fig. 1 Example view of education being conducted in metaverse environment

According to [23], online and virtual learning offer the advantage of delivering educational content while minimizing engagement risks, all without the constraints of time or location. Hence, numerous countries have ventured into distinct approaches and techniques to encourage educational reform and innovation through the metaverse. Wang et al. presents an education metaverse ecosystem in guiding applications that focusing on education in metaverse [24]. Meanwhile, Mystakidis [25] introduced a metaverse course as a strategy to enhance learning performance by maintaining learners' interest and active engagement in online learning. His findings suggest that metaverse-based content has the potential to stimulate learner interest. Furthermore, the metaverse offers the facilitation of the social interactions among learners' avatars within the virtual environment and their real-world counterparts. Consequently, it is reasonable to infer that the integration of the metaverse and education will likely witness a substantial increase in research and studies in the coming years.

Education within a metaverse has the potential to enrich student experiences, ensuring the availability of diverse learning opportunities rooted in genuine human interaction (see Fig. 1). Kress's multimodal learning theory [26] posits that the greater the engagement of human senses in the learning process, the more effectively students comprehend and retain information. Facilitating multimodality plays a pivotal role in education by promoting an understanding of the myriad ways people communicate and express themselves. By utilizing different modes, such as visual, auditory, linguistic, gestural, and spatial, students can engage in learning through a multitude of avenues, cultivating a versatile and adaptive learning style [27].

2.2 *Learner Engagement in Education Metaverse Space*

Learner engagement is a critical aspect of education in the metaverse. It is depicted by a significant focus on engaging and captivating activities. A common approach to defining engagement is to characterize it as the quality of the learner's experience across four dimensions: emotional states (such as interest and enjoyment), observable behavioral states (indicating active involvement in tasks), cognitive engagement (reflecting concentration), and volitional states (related to future actions or agency) [28]. It goes beyond traditional forms of engagement due to the immersive and interactive nature of this digital environment. To illustrate the complexities of student engagement effectively, it is most beneficial to conduct an analysis and offer a real-life case example. The dialogue within an assignment serves as a tangible representation of the affective aspects of engagement.

In the educational metaverse, learners establish virtual identities through digital mapping, mirroring their physical, cultural, psychological, and spiritual aspects, thereby creating a comprehensive representation within the virtual realm. Avatars representing students can interact with one another, fostering communication and teamwork. Collaborative projects, discussions, and group activities are seamlessly facilitated, creating an environment that mirrors real-world interactions. In the study by Han et al. [29], they highlight the interaction and learner engagement as a receiver, collaborator, constructor between learners and avatars, learner engagement with human–machine. In the education metaverse, students are transported into immersive and interactive digital environments that stimulate their senses and encourage active participation. Immersion in a three-dimensional (3D) or VR setting allows learners to explore subjects in ways not possible in a traditional classroom [30]. For instance, in a history lesson about ancient civilizations, students can virtually visit historical sites, interact with simulations of historical figures, and witness events as if they were there [5]. The metaverse supports multimodal learning, catering to various learning styles and preferences. Students can engage with content through visual, auditory, linguistic, gestural, and spatial modes, providing a rich and diverse learning experience.

3 Assessment in Education

Assessment in education is a multifaceted process that plays a crucial role in evaluating students' understanding, progress, and the effectiveness of instructional methods. It involves the systematic collection and interpretation of data to inform teaching and learning practices. This section discusses the assessment strategies which has adapted to diverse learning environments, including traditional physical classrooms, online learning platforms, and the cutting-edge Education Metaverse.

3.1 Assessment in Physical Learning

Assessment is a broad term that refers to the systematic gathering of information to inform decisions about students, curricula, programs, schools, and educational policies [31]. When we speak of "evaluating a student's proficiency," it signifies the process of gathering data to determine the extent to which the student has successfully attained the desired learning outcomes. Traditionally, assessment methods have long been established and widely practiced for physical learning. These methods encompass a range of techniques aimed at measuring students' understanding, progress, and competencies. In this section we highlight several key approaches to assessment in physical learning environments, including formative assessment [32], summative assessment [33], authentic assessment [34] and observation & participation [35].

Formative assessment is an ongoing and interactive process that provides feedback to both students and educators during the learning journey [32]. It is often used to inform instruction and help students understand their strengths and areas for improvement. Formative assessment methods may include quizzes, class discussions, peer assessments, and in-class activities. Meanwhile summative assessment, in contrast, is typically conducted at the end of a learning period [33] which commonly include final exams, standardized tests, and term papers. These assessments are used to assign grades and measure the attainment of learning objectives.

Authentic assessment aims to evaluate students' abilities to apply knowledge and skills in real-world contexts. This approach often involves project-based assignments, simulations, or performance assessments that mirror tasks encountered in professional settings. Other than that, in physical learning settings, direct observation of students' engagement and participation can be a valuable assessment tool [36]. Educators can assess students' contributions to class discussions, group projects, and practical demonstrations to gauge their involvement and understanding.

3.2 Assessment in Online Learning

Online education has become quite popular in recent years with the advancements of computer sciences and the Internet. Online education mainly deals with synchronous and asynchronous teaching–learning systems. Synchronous systems enable teachers, students, or professionals to interact in real-time or at the same time in a digital virtual space. A few common synchronous platforms are Microsoft Teams, Zoom, Webex, and Skype. On the other hand, in asynchronous systems, the participants can participate according to their own time and schedule.

Moore [37] initially delineated the teaching methodologies employed in distance education, defining it as a domain distinct from learning behavior. Within this context, communication between educators and learners is facilitated through various mediums such as print, electronics, and machinery. A specific subset of distance education is internet-based online classes, which grant students considerable autonomy in terms of choice, control over learning schedules and environments, access to educational materials, and task execution. Moreover, international students residing in their home countries and participating in online courses gain the advantage of convenient access to authentic learning materials and textbooks, as well as an opportunity to experience the academic culture and ambiance of a university. The term "effectiveness" in education circles refers to the outcome of carefully planned online learning experiences. The increasing prevalence of online classes, accelerated by the challenges of COVID-19, has led to a significant body of literature exploring the efficacy of online education.

As in traditional learning, both formative and summative assessment are also suitable to be implemented in online learning. Swan [38] performed analysis of 73 online courses, various assessment methods were identified, including discussions, written assignments, projects, quizzes/tests, and group work. Notably, nearly 75% of the courses utilized online discussions as a graded component, while approximately 50% incorporated written assignments and quizzes/tests. Meanwhile, Arend [39] conducted a similar study with 60 courses, uncovering comparable assessment methods such as online discussions, exams, written assignments, experimental or problem assignments, quizzes, journals, projects, and presentations. In alignment with Swan's findings, a substantial proportion of these courses featured online discussions as graded activities (83%), along with quizzes/tests (83%) and written assignments (63%).

With the increasing prevalence of online learning platforms and the emergence of the metaverse, assessment methodologies have adapted to the digital landscape. One of the advantages of online learning assessment is its flexibility. Students can often choose when and where they complete assessments, accommodating various learning styles and schedules. Additionally, online assessments can be adaptive, tailoring questions to individual learners' abilities, which enhances the accuracy of assessment results and allows for a more personalized learning experience.

3.3 Assessment in Education Metaverse

Assessment within the education metaverse represents a pivotal aspect of this transformative educational paradigm. The education metaverse introduces a dynamic and adaptable assessment landscape that goes beyond traditional testing, embracing the principles of authenticity, personalization, and real-world application. Within the Education Metaverse, assessment takes on various innovative forms, moving far beyond the confines of pen-and-paper exams. VR simulations, AR scenarios, and immersive experiences provide new avenues for assessing students' skills, knowledge, and competencies.

Table 1 reveals a growing interest in conceptual and theoretical frameworks for assessment within the Education Metaverse. These frameworks serve as the foundational structures upon which assessment methodologies are built. Notably, Chen et al. [40] proposed a 5-layer framework that emphasizes the layers of physical, data, computational, interaction, and application within the Metaverse. This conceptual framework sets the stage for understanding the intricate layers of this digital learning environment, which can have a profound impact on assessment strategies. Other studies, such as Wang et al.'s [41] work on immersive learning and Mitra's [42] exploration of blended and personalized learning, underline the importance of conceptualizing how the Metaverse can be harnessed for assessment purposes.

Table 1 Summary of related works on assessment approach for education metaverse

Year	Authors	Proposed framework	Pedagogical approach	Assessment approach	Design criteria
2023	Chen et al. [40]	Conceptual	Immersive and real-time learning	– Interactive quizzes – Gamifying content	– Virtual Avatars – Virtual Learning Resources – Virtual Teaching Scenes
2023	Shu and Gu [47]	Theoretical	Smart, immersive collaborative learning	– AI-based assessment – High-order thinking projects – Assessment subjects are diversified, including teachers, peers, and AI technology – Data visualization is utilized on the results	– Smart pedagogy – AI-enabled supporting service – Data-driven teaching design – Multimodal Course Resources
2023	Joshi and Pramod [44]	Theoretical	immersive and adaptive personalized real-time learning ecosystem	– Formative/ summative assessments – Progress capturing, evaluation, assessment, certification, remote interaction & monitoring, submissions and demonstrations through virtual labs, intermittent quizzes – Automated assessment tools (e.g. CourseMarker, DOMjudge, and Formative)	– Facilitating a fully immersive, virtualized and globalized learning environment – AI/ML driven academic administration – Cross Interactive Space
2023	Mitra [42]	Conceptual	Blended learning, personalized learning, Virtual learning, Collaborative learning	– Tests/Quizzes – Automated Assessments	– Effective professional training and development – Simulation potentials of the metaverse – Innovative mentoring platform

(continued)

Table 1 (continued)

Year	Authors	Proposed framework	Pedagogical approach	Assessment approach	Design criteria
2023	Wu et al. [43]	Conceptual	Language learning, interactive learning, Collaborative learning, game-based language learning, task-based	– Language learning activities – Language learning games – Immersive tasks	– Onion-like architecture: 5-layers (Transdiciplinarity, Pedagogy, Technology, Actor and Learning (central focus))
2023	Wang et al. [41]	Conceptual	Immersive learning	– Preserved data in blockchain to be assessed by the teachers – AI analyzes and assesses these data from multiple dimensions to realize the diversification of student performance evaluation	– Technical supports and theoretical basis – Education courses to include element listening, speaking, and interpreting
2023	Roy et al. [49]	Theoretical	Immersive learning	– Evaluate students' performance with observation on how students solve problems or complete projects	– Interactive capabilities – Customized user experience – Independent and dependent variable

(continued)

Table 1 (continued)

Year	Authors	Proposed framework	Pedagogical approach	Assessment approach	Design criteria
2023	Yuan et al. [50]	Theoretical	Immersive learning	– Learning tasks which include speech repetition, knowledge explanation, game, stage dressing speech – Reflection (Re-hearing system recorded speeches, peer reflection exchange and discussion and observed class performance reflected) – Allow group discussion, peer evaluation and communication within the group – Analyzed the learning experience of students based on questionnaires and semi-structured interviews	– Adopted ICARE model – Introduction – Generally introduce content – C—Connect or content – Assist to organize new knowledge – A—Application – Applying learned knowledge – R—Reflection – Allowing learners to reflect – E—Extension – Extending the learning experience
2023	Onecha et al. [5]	Conceptual	Challenge-based learning (CBL), team-based learning, flipped classrooms, project-based learning, reflective learning	– Formative and summative assessment – Industry –community projects, videos of real cases, case studies, gamification activities, interdisciplinary activities, problem-solving activities, storytelling, real material practices, hands-on activities, role play	– 1st phase: Obtain data on the usefulness of the propose environment with questionnaires to the student – 2nd phase: Obtain data on the improvement gained in learning and understanding during virtual visits (Questionnaire)

(continued)

Table 1 (continued)

Year	Authors	Proposed framework	Pedagogical approach	Assessment approach	Design criteria
2022	Fung et al. [48]	Conceptual	Personalized learning	– Online Assessments (formative and summative online assessment methods) – Online quizzes – Built-in continuous feedback. Multiple-choice questions. Simulated clinical skills. Clinical examinations using real-time communication technologyv – True/false, fill in the blank, fill in multiple blanks, and essay question – Speed grader & Plagiarism check software	– Comparison of online assessment with traditional assessment (benefits, concept and framework, methods, and difficulties) – Highlights the common formative and summative online assessment methods – Assessment Feedback – Formulated 6 components need to be considerate of for the online assessment
2022	Zhang et al. [14]	Conceptual	Immersive, virtual experiment learning, blended learning, personalized learning	– Combine with formative and summative data – (example: quizzes, discussion group) – Pay more attention to learners' growth	– Involves collection of technologies (technological infrastructure) – Include learning scene, resources, logging, analysis and authentication
2022	Wang et al. [24]	Theoretical	Immersive learning	– Hands-on practice	– Consists of four major hubs: – Instructional design and performance technology hub – Knowledge hub – Research and technology hub – Talent and training hub

(continued)

Table 1 (continued)

Year	Authors	Proposed framework	Pedagogical approach	Assessment approach	Design criteria
2022	Dahan et al. [52]	Theoretical	Immersive learning	– Enable students to experience real-life example and scenarios in the metaverse – This paper does not include assessment for the student in their proposed Framework	– ELEM consists of a user, devices, and metaverse components – Metaverse special technologies
2022	Wang et al. [53]	Conceptual	Gamified Learning and Task-based Modules, Collaborative learning, Learner-driven Activities	– This paper did mention that their platform includes the features of learning assessment in the metaverse for the courses, however this paper did not discuss on the assessment in detail	– Learning assessment in the Metaverse – Interaction with presentations – Teaching experience with augmentation that benefits from visualizing knowledge and from 3D virtual entities
2022	Jovanović et al. [46]	Conceptual	Gamified & Collaborative Learning	– MicroLessons feature - quest-based challenges, problem-solving puzzles – Problem-solving activities	– Web Platform as tools for teachers – Access Control: utilized multi-factor identity verification – Intelligent Agents and MicroLessons – VLE User Interface

(continued)

Table 1 (continued)

Year	Authors	Proposed framework	Pedagogical approach	Assessment approach	Design criteria
2021	Ng et al. [45]	Conceptual	Personalize learning	– AI Tutors	– Specifically, for English education metaverse – Two stage stochastic integer programming (SIP) – First—minimize the reservation cost – Next, to minimize the on-demand cost
2021	Demir [54]	Conceptual	Adaptive and personalize learning	– Assign task for students to write a blog – Assign quizzes – Problem solving questions (algebra)	– New or improved learning/teaching approaches – Essential/ transforming technologies – Enriching and supporting technologies – Identified the required technologies
2020	Wang [55]	Theoretical	Virtual learning, inquiry, collaborative learning	– Five feedback model	– Users – External environment – Virtual learning environment only – Including avatars
2020	Suzuki et al. [56]	Conceptual	Collaborative learning	– Virtual training and questionnaire – Discussion within group	– Remote sharing system – Possibility to simulating various real-life activities

Table 1 also provides an overview of various assessment approaches proposed in recent research related to the Education Metaverse. These approaches encompass a wide range of pedagogical and technological innovations designed to evaluate and measure learning outcomes in immersive digital environments. Below, we discuss some key themes and trends emerging from the table.

Immersive and Interactive Assessments

A common thread among several studies is the emphasis on immersive and interactive assessment methods. The study by [40, 41, 43], all highlight the importance of using immersive experiences and interactive activities like quizzes, games, and language learning tasks to assess students within the Education Metaverse. This approach aligns with the idea that assessment should reflect the dynamic and engaging nature of immersive learning environments. However, Chen et al. [40] focuses their research on the framework design criteria which are virtual avatars, virtual learning resources, and virtual teaching scene, while Wang et al. [41] addresses both technical supports and theoretical basis in their proposed framework. Meanwhile, Wu et al. [43] adopted an onion-like architecture of 5-layers with learning as central focus.

Personalization and Adaptive Learning

Several studies, including Joshi and Pramod [44], Zhang et al. [14], and Ng et al. [45], emphasize the significance of personalization and adaptive learning in the Education Metaverse. These approaches involve tailoring assessments to individual learners' needs, providing continuous feedback, and adapting the learning path based on performance. Adaptive assessments aim to optimize the learning experience and ensure that students receive appropriate challenges and support.

Gamification and Collaborative Learning

Gamified elements and collaborative learning are prominent features in the assessment approaches proposed by Mitra [42], Onecha et al. [5], and Jovanović et al. [46]. Gamification introduces elements like badges and game-based activities to motivate and engage students in the assessment process. Collaborative learning encourages teamwork, problem-solving, and real-world application of knowledge, often assessed through projects, case studies, and interdisciplinary activities.

Innovative Assessment Technologies

The integration of advanced technologies in assessment is evident in several studies. For instance, Shu and Gu [47] advocate for AI-based assessments and high-order thinking projects, while Fung et al. [48] discuss the use of automated assessment tools, plagiarism checks, and real-time communication technologies. These technologies not only streamline the assessment process but also offer opportunities for data-driven insights into student performance. Shu and Gu [47] has also proposed for a smart teaching environment of the education model in metaverse which allow seamless connection between offline, online and formal or informal learning.

Reflective and Real-Life Assessments

Some studies, such as Roy et al. [49] and Yuan et al. [50], propose assessments that involve reflection, observation, and real-life application. These assessments go beyond traditional testing methods and focus on evaluating how students solve problems, perform tasks, and engage in real-world scenarios, mirroring the experiential nature of the Education Metaverse.

Diverse Pedagogical Approaches

Another noteworthy aspect is the diversity of pedagogical approaches within the Education Metaverse. Researchers explore blended learning, personalized learning, team-based learning, flipped classrooms, and more, highlighting the adaptability of the Metaverse to various educational contexts. Immersive learning explores the phenomenon of learning when a student becomes deeply mentally engaged in a state characterized by technology, storytelling, and challenging experiences, to the extent that their awareness of the physical world is diminished.

In the study by Felemban et al. [51], they focus on the challenges of assessing learning outcomes in immersive environments, and proposes a virtual observation model that maps observing learners in classrooms with observing and assessing students in these virtual environments.

4 Discussion and Future Work

In this paper, we discussed the surrounding of the education Metaverse which has a promising landscape of innovative pedagogical approaches, immersive learning experiences, and evolving assessment methods. As we reflect on the findings presented in the previous sections, it becomes evident that the education Metaverse represents a paradigm shift in education, offering new possibilities for both learners and educators. The integration of VR and AR technologies, along with the development of conceptual frameworks, has paved the way for immersive and interactive learning experiences that transcend traditional classroom boundaries. One of the key takeaways from this exploration is the need for continued research and development in this emerging field. While the education Metaverse shows immense potential, it also presents unique challenges, for examples on how to design a teaching module with suitable assessment approaches to enhanced the learner's capability in understanding and progressing through learning in the educational metaverse space. Addressing ethical concerns, data privacy, and refining learning environment design are crucial issues for the Metaverse's educational potential.

Other than that, assessment within the education Metaverse is still in its early stages, and future work should explore advanced assessment methodologies that harness the capabilities of immersive technologies. This includes the development of AI-driven assessment tools, real-time feedback mechanisms, and competency-based assessments that align with the dynamic nature of virtual learning environments. Therefore, future work should concentrate on providing comprehensive training and professional development opportunities for teachers to effectively navigate and utilize these immersive learning environments. This includes pedagogical training, technical skills development, and the integration of Metaverse tools into the curriculum.

In conclusion, the education Metaverse represents a groundbreaking evolution in education, offering learners unprecedented opportunities for immersive and interactive learning experiences. However, realizing its full potential requires ongoing research and development, addressing ethical considerations, and ensuring accessibility and inclusivity for all. By embracing these challenges and opportunities, we can shape the future of education and provide learners with dynamic, engaging, and personalized learning experiences in the Education Metaverse.

Acknowledgements We would like to extend our gratitude to the Ministry of Higher Education under Fundamental Research Grant Scheme, UTM, FRGS/1/2023/ICT10/UTM/02/2 which has funded this research. We are most grateful to the Mixed and Virtual Reality Laboratory (mivielab) and the ViCubeLab at Universiti Teknologi Malaysia (UTM) for providing the technical support and facilities.

References

1. Lin H, Wan S, Gan W, Chen J, Chao HC (2022) Metaverse in education: vision, opportunities, and challenges. In: Proceedings—2022 IEEE international conference on big data. Big Data 2022
2. Kye B, Han N, Kim E, Park Y, Jo S (2021) Educational applications of metaverse: possibilities and limitations
3. Lee A, Park W, Park H (2022) Exploring experience and applicability of metaverse based medical humanities course. J Hum Soc Sci 13
4. Zhai XS, Chu XY, Chen M, Shen J, Lou FL (2023) Can Edu-metaverse reshape virtual teaching community (VTC) to promote educational equity? An exploratory study. IEEE Trans Learn Technol
5. Onecha B, Cornadó C, Morros J, Pons O (2023) New approach to design and assess metaverse environments for improving learning processes in higher education: the case of architectural construction and rehabilitation. Buildings 13. https://doi.org/10.3390/buildings13051340
6. Bharath G, Rupa C, Karthik M, Leela Bhavani A, Manish Chowdary V (2021) Revelation of geospatial information using augmented reality. In: 2021 International conference on wireless communications, signal processing and networking. WiSPNET 2021
7. Díaz JEM (2020) Virtual world as a complement to hybrid and mobile learning. Int J Emerg Technol Learn 15
8. Stephenson N (2003) Snow crash: a novel. Spectra
9. Villalonga-Gómez C, Ortega-Fernández E, Borau-Boira E (2023) Fifteen years of metaverse in higher education: a systematic literature review. IEEE Trans Learn Technol
10. Kim S (2020) Metaverse: digital world, world of emerging items. Hwaseong PlanB Design 376
11. Hwang GJ, Chien SY (2022) Definition, roles, and potential research issues of the metaverse in education: an artificial intelligence perspective. Comput Educ Artif Intell 3
12. Akour IA, Al-Maroof RS, Alfaisal R, Salloum SA (2022) A conceptual framework for determining metaverse adoption in higher institutions of gulf area: an empirical study using hybrid SEM-ANN approach. Comput Educ Artif Intell 3
13. Smart J, Cascio J, Paffendorf J (2022) Co-Authors accelerating.org metaverseroadmap.org Lead Reviewers Futuring and Innovation Center
14. Zhang X, Chen Y, Hu L, Wang Y (2022) The metaverse in education: definition, framework, features, potential applications, challenges, and future research topics. Front Psychol
15. Abdul Halim NA, Ismail AW (2021) Designing ray-pointing using real hand and touch-based in handheld augmented reality for object selection. Int J Innov Comput 11
16. Rupa C, Srivastava G, Ganji B, Tatiparthi SP, Maddala K, Koppu S, Chun-Wei Lin J (2022) Medicine drug name detection based object recognition using augmented reality. Front Public Health 10
17. Lee H, Woo D, Yu S (2022) Virtual reality metaverse system supplementing remote education methods: based on aircraft maintenance simulation. Appl Sci 12
18. Mohamad Yahya Fekri A, Ajune Wanis I (2019) A review on multimodal interaction in mixed reality environment. In: IOP conference series: materials science and engineering

19. Muhammad Nur Affendy N, Ajune Wanis I (2019) A review on collaborative learning environment across virtual and augmented reality technology. In: IOP conference series: materials science and engineering
20. Min T, Cai W (2022) Portrait of decentralized application users: an overview based on large-scale Ethereum data. CCF Trans Pervasive Comput Interact 4
21. Nunan D, Richards JC, Benson P (2016) Language learning beyond the classroom. ELT J 70
22. Birabil DR, Tete S (2020) EDUCATION in Nigeria: challenges and way forward. Int J Acad Res Reflect 8
23. Pellas N, Dengel A, Christopoulos A (2020) A scoping review of immersive virtual reality in stem education. IEEE Trans Learn Technol
24. Wang M, Yu H, Bell Z, Chu X (2022) Constructing an Edu-metaverse ecosystem: a new and innovative framework. IEEE Trans Learn Technol 15
25. Mystakidis S (2022) Metaverse. Encyclopedia 2:486–497
26. Kress GR, Van Leeuwen T (2001) Multimodal discourse: the modes and media of contemporary communication. (No Title)
27. Kress G (2005) Design and transformation: new theories of meaning. In: Multiliteracies: lit learning. Routledge, pp 149–157
28. Sinatra GM, Heddy BC, Lombardi D (2015) The challenges of defining and measuring student engagement in science. Educ Psychol
29. Han Z, Tu Y, Huang C (2023) A framework for constructing a technology-enhanced education metaverse: learner engagement with human–machine collaboration. IEEE Trans Learn Technol
30. Nor'a MNA, Ismail AW (2019) Integrating virtual reality and augmented reality in a collaborative user interface. Int J Innov Comput 9
31. Brookhart SM, Nitko AJ (2018) Educational assessment of students. Hum Mov Sci 53
32. Wood DF (2018) Formative assessment: assessment for learning. In: Understanding medical education: evidence, theory, and practice. https://doi.org/10.1002/9781119373780.ch25
33. Dolin J, Black P, Harlen W, Tiberghien A (2018) Exploring relations between formative and summative assessment. In: Contributions from science education research
34. Ortega DP, Minchala OE (2017) Assessing students in an authentic and ongoing manner in the english classroom. Theory Pract Lang Stud 7. https://doi.org/10.17507/tpls.0703.01
35. Rahman MM (2018) Exploring teachers practices of classroom assessment in secondary science classes in Bangladesh. J Educ Learn 7. https://doi.org/10.5539/jel.v7n4p274
36. Lekwa AJ, Reddy LA, Shernoff ES (2019) Measuring teacher practices and student academic engagement: a convergent validity study. Sch Psychol Q 34
37. Moore MG (1972) Learner autonomy: the second dimension of independent learning. Convergence 5
38. Swan K (2001) Virtual interaction: design factors affecting student satisfaction and perceived learning in asynchronous online courses. Distance Educ 22
39. Arend BD (2019) Course assessment practices and student learning strategies in online courses. Online Learn 11. https://doi.org/10.24059/olj.v11i4.1712
40. Chen X, Zhong Z, Wu D (2023) Metaverse for education: technical framework and design criteria. IEEE Trans Learn Technol. https://doi.org/10.1109/TLT.2023.3276760
41. Wang Q, Tang L, Wang Y (2023) Potential applications of the metaverse in higher English education. Open J Soc Sci 11. https://doi.org/10.4236/jss.2023.111031
42. Mitra S (2023) Metaverse: a potential virtual-physical ecosystem for innovative blended education and training. J Metaverse 3. https://doi.org/10.57019/jmv.1168056
43. Wu JG, Zhang D, Lee SM (2023) Into the brave new metaverse: envisaging future language teaching and learning. IEEE Trans Learn Technol
44. Joshi S, Pramod PJ (2023) A collaborative metaverse based A-La-Carte framework for tertiary education (CO-MATE). Heliyon 9. https://doi.org/10.1016/j.heliyon.2023.e13424.
45. Ng WC, Yang Bryan Lim W, Ng JS, Xiong Z, Niyato D, Miao C (2022) Unified resource allocation framework for the edge intelligence-enabled metaverse. In: IEEE international conference on communications

46. Jovanović A, Milosavljević A (2022) VoRtex metaverse platform for gamified collaborative learning. Electronics 11. https://doi.org/10.3390/electronics11030317
47. Shu X, Gu X (2023) An empirical study of a smart education model enabled by the Edu-Metaverse to enhance better learning outcomes for students. Systems 11
48. Fung CY, Su SI, Perry EJ, Garcia MB (2022) Development of a socioeconomic inclusive assessment framework for online learning in higher education. In: Socioeconomic inclusion during an Era of online education
49. Roy R, Babakerkhell MD, Mukherjee S, Pal D, Funilkul S (2023) Development of a Framework for metaverse in education: a systematic literature review approach. IEEE Access 11
50. Yuan J, Liu Y, Han X, Li A, Zhao L (2023) Educational metaverse: an exploration and practice of VR wisdom teaching model in Chinese Open University English course. Interact Technol Smart Educ 20(3):403–421. Emerald Publishing Limited
51. Felemban S, Gardner M, Callaghan V, Pena-Rios A (2017) Towards observing and assessing collaborative learning activities in immersive environments. In: Communications in computer and information science
52. Dahan NA, Al-Razgan M, Al-Laith A, Alsoufi MA, Al-Asaly MS, Alfakih T (2022) Metaverse framework: a case study on E-learning environment (ELEM). Electronics 11
53. Wang Y, Lee LH, Braud T, Hui P (2022) Re-shaping Post-COVID-19 teaching and learning: a blueprint of virtual-physical blended classrooms in the metaverse era. In: Proceedings—2022 IEEE 42nd international conference on distributed computing systems workshops. ICDCSW 2022
54. Demir KA (2021) Smart education framework. Smart Learn Environ 8
55. Wang J (2020) The design of 3D-virtual learning environments in the view of system theory. In: 4th International conference on culture, education and economic development of modern society (ICCESE 2020) https://doi.org/10.2991/assehr.k.200316.154
56. Suzuki SN, Kanematsu H, Barry DM, Ogawa N, Yajima K, Nakahira KT, Shirai T, Kawaguchi M, Kobayashi T, Yoshitake M (2020) Virtual experiments in metaverse and their applications to collaborative projects: the framework and its significance. Procedia Comput Sci 176:2125–2132. Elsevier

Prediction of Student Performance in Online Classes During COVID-19 Pandemic

N. Sudharshana, J. Naren, and D. Sumathi

Abstract Despite global disruptions caused by the COVID-19 pandemic, education sector adapted and continued to function online. Undergraduate courses happened online. Variations in students' performance emerged due to differing approaches in incorporating online education into routine life amidst challenging conditions at home and in society during the pandemic. It was necessary to find out whether online education aided academics. A solution based on a machine learning is proposed and implemented in the paper to forecast an individual's performance using data gathered from a user survey among university students regarding their perceptions of online education. The user survey comprised questions about factors that could affect academic performance. A diverse range of algorithms are meticulously evaluated to identify one that delivers precise predictions of student performance, leading to an effective restructure of educational landscapes to replicate the immersive and engaging experience of in-person classes, fostering a conducive environment for students to wholeheartedly focus on academic pursuits.

Keywords Student academic performance prediction · Data mining · Online education · Social media · Pandemic

1 Introduction

During the COVID-19 pandemic, educational institutions adopted an online mode of teaching to impart education to not lose out on days. However, it became difficult to assess whether a student performed well or not as there was no face-to-face interaction. A majority of the students were distracted and did not pay full attention to online

---

N. Sudharshana
Chennai, India

J. Naren (✉) · D. Sumathi
Department of Computer Science and Engineering, Alliance College of Engineering and Design, Alliance University, Bengaluru, India
e-mail: naren.jeeva3@gmail.com

S. M. Thampi et al. (eds.), *Fifth International Conference on Computing and Network Communications*, Lecture Notes in Electrical Engineering 1221,
https://doi.org/10.1007/978-981-97-4711-5_47

classes due to non-interactive sessions. However, with machine learning and data analytics, data was collected from students and performance was predicted, aiding university staff in teaching better and conducting additional sessions to improve students' performance. Thus, a need to predict students' performance has arisen to improve overall experience of online classes.

In the paper, a methodology to address the critical knowledge gap surrounding student performance in online learning environments has been proposed. The dataset is based on a questionnaire, going beyond traditional methods that solely rely on historic academic records, providing a holistic understanding of student experience, nature of residence, type of device used to attend classes, nature of residence, type of social media users if any, effectiveness of practical classes and online cultural conducted if any, and overall experience of online classes factors in the expanded scope, determining academic success in digital realm. Educators and administrators gain valuable insights from this data collection strategy. Features [1] like intermediate marks, B. Tech performance till third semester, etc., were considered and machine learning algorithms like Naïve Bayes, decision trees, K-means, and hierarchical clustering were used for prediction. Durán Acevedo [6] and [8] also considered galvanic skin response, socioeconomic factors, and others. In [7] students' poor performance based on the attributes was predicted using Naïve Bayes, SVM, KNN, decision tree, and random forest algorithms. The management and staff can use this information to conduct more interactive sessions and change teaching techniques to improve student experience. K-nearest neighbors, decision tree, random forest and voting lassifier are used to predict student performance. The data-driven approach points to the area of improvement and helps tailor interventions to address specific needs of students.

2 Literature Survey

Uday Kumar et al. [1] introduced a system for predicting student performance using data from K L University between 2013 and 2017, applied clustering techniques, including K-means and hierarchical clustering, based on parameters like intermediate marks and B. Tech performance. Decision trees and Naïve Bayes were also utilized. The study concluded that K-means had higher accuracy than hierarchical clustering, and decision trees outperformed Naïve Bayes.

Aoulad Ali [2] proposed classification models to predict MOOC dropout rates, employing algorithms such as KNN, SVM, logistic regression, Ada boost, and random forest. Testing on two semesters' course data revealed random forest and Ada boost as the most accurate predictors.

Habib et al. [3] presented a framework for exploring challenges in online learning at Assumption University of Thailand during the COVID-19 outbreak. Survey data indicated that technology-related issues were less common than those linked to instructors' attitudes and aptitudes, with the most critical issues being (1) instructor's attitude, (2) instructor's aptitude, and (3) technology issues.

Zhang et al. [4] proposed a student performance prediction model incorporating data collection, learning behavior analytics, algorithm model building, and prediction. Using an optimized logistic regression algorithm, students' behavior was analyzed and performance was accurately predicted. However, the model lacked consideration for subjective factors, leaving room for future improvements.

Boddeti Sravani et al. [5] implemented a machine learning model predicting students' academic performance based on a dataset with features like age, gender, academic records, and medical history. Linear regression achieved highly accurate results, highlighting the significant influence of students' backgrounds on performance.

Durán Acevedo et al. [6] developed a methodology for academic stress detection in engineering students, utilizing an electronic nose system and galvanic skin response. Classification methods like linear discriminant analysis, K-nearest neighbors, and support vector machine achieved high success rates.

Rao et al. [7] used six classification algorithms to detect factors for poor student performance. Logistic regression achieved 97.12% accuracy on test data, providing insights into reasons for failure and suggesting improvements through machine learning.

Rao et al. [8] analyzed student data using Naïve Bayes and ensemble learners to predict performance. An improved random forest algorithm enhanced accuracy, focusing on identifying at-risk students based on factors like gender, family size, and parents' status.

Khalfan al-Mayahi et al. [9] created a machine learning model predicting students' educational performance using previous data and performance in the last stage of school, achieving 87% accuracy.

Gopal et al. [10] proposed a framework identifying factors affecting students' satisfaction and performance in online classes during the COVID–19 pandemic. Factors like quality of instructor, course design, prompt feedback, and student expectations impacted satisfaction and performance.

Yaseen et al. [11] proposed a framework analyzing online learning's impact on education during the COVID-19 pandemic. Increased absenteeism, dropouts, and increased workload for both students and faculties were observed.

Alshabandar et al. [12] designed predictive models for students' assessment grades and final performance in MOOCs. Student engagement with digital material significantly impacted success.

Keshtkar et al. [13] developed a model predicting negative learning outcomes early in the semester using logistic model tree, aiding in identifying students at risk of failure in offline courses.

Sriram et al. [14] proposed a framework exploring the effects of personality features on students' future career positions in addition to academic performance. Deep classifiers successfully classified data, suggesting practical applications for career guidance.

Ramanujam et al. [15] conducted a study on hybrid recommender systems with deep learning, revealing higher performance compared to collaborative and content-based models.

Elakia et al. [16] proposed a data mining-based system predicting student details, including career options and the possibility of violence, enabling early intervention.

Ganesan et al. [17] proposed a hybrid recommendation system using big data and deep learning to provide accurate learning resources for online learners, emphasizing the development of learner profiles.

Naren et al. [18] developed a collaborative recommendation system for graduate students to choose electives, relieving the dilemma of selecting courses each semester.

Gayathiri et al. [19] proposed a framework quantifying the relationship between academic performance and Internet usage behavior using machine learning algorithms. The work aimed at improving management skills in education.

Reddy et al. [20] assessed student performance through data mining and ontology-based applications, providing feedback on factors like confidence level and time management.

Ashokkumar et al. [21] presented an overview of data mining and machine learning techniques for predicting student performance, emphasizing the early identification of slow learners.

Ganesan et al. [22] explored the impact of personality features on students' careers and academic performance, highlighting the role of technology in helping students realize their potential.

Li et al. [23] investigated middle school students' attitudes toward home-based online education during COVID-19, identifying influential variables such as time spent on homework and family relationships.

Goundar et al. [24] used classification algorithms to predict student success in a computing science course, with random forest outperforming traditional decision tree algorithms.

Azzali et al. [25] studied student engagement and exam completion speed and its effect on performance before and during the COVID-19 pandemic, finding both disadvantages and advantages.

Lux et al. [26] examined the impact of online classes on accounting students' engagement and satisfaction, identifying variables related to engagement and suggesting practical applications for support and flexibility.

2.1 Visualization of Features

See Figs. 1, 2, 3, 4, 5, 6, 7, 8, 9, 10 and 11.

3 Methodology

The methodology followed in this paper is given as a flowchart. As stated initially, to gain insights into student perceptions of online classes, a survey was conducted among undergraduate students. The data collected was cleaned to obtain correct data

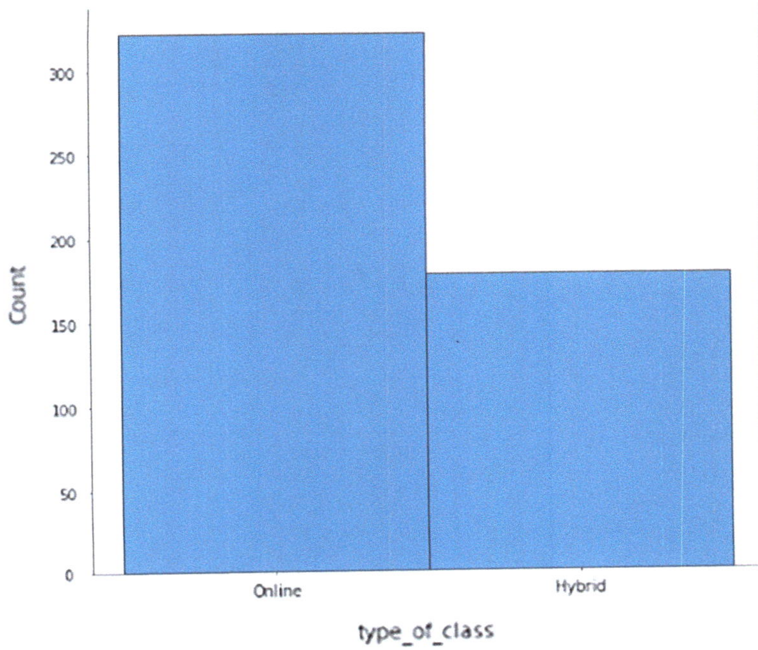

Fig. 1 Type_of_class

Fig. 2 Device_used

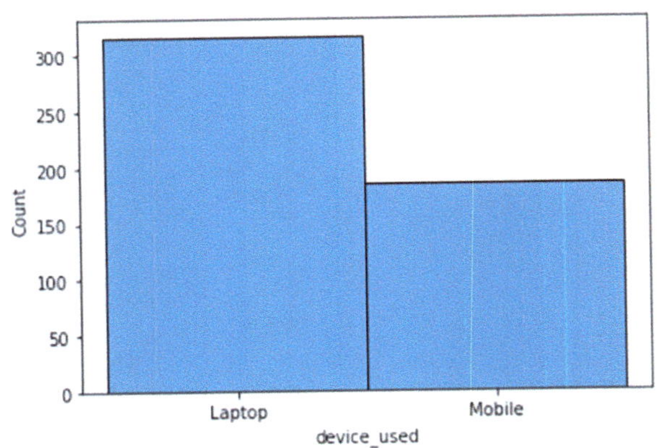

to represent factors in form of class and numerical-based data, which helped to find dependent and independent factors for model training. Next, the model was trained by classification algorithms. The results were evaluated based on accuracy generated by models. Algorithm with highest accuracy was considered most effective. If the results were not satisfactory, data cleaning was performed to look for adopting better approaches in cleaning data and the subsequent steps were followed as before (Fig. 12).

Fig. 3 Nature_of_residence

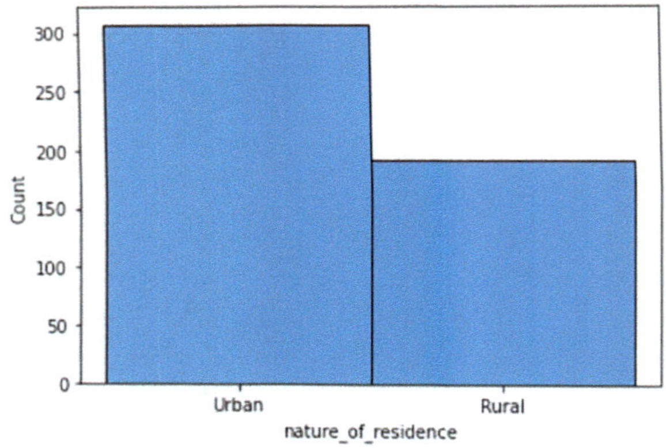

Fig. 4 Difficulty_level

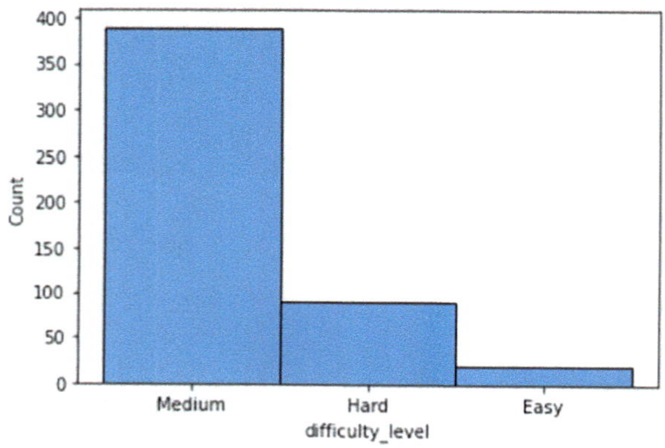

Fig. 5 Social_media_usage

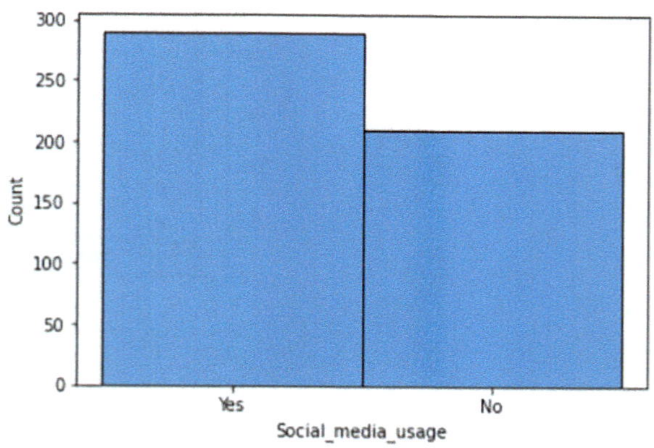

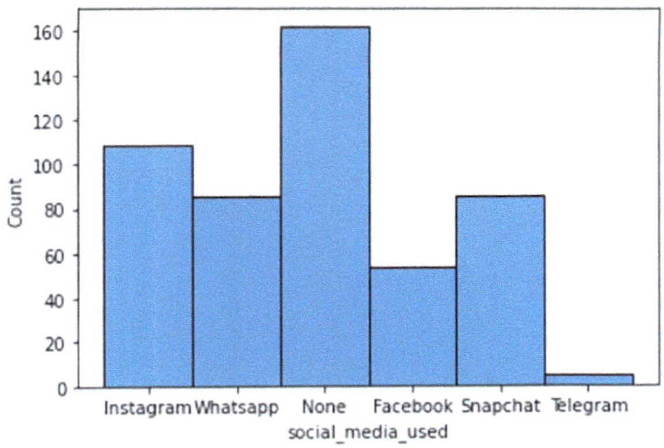

Fig. 6 Social_media_used

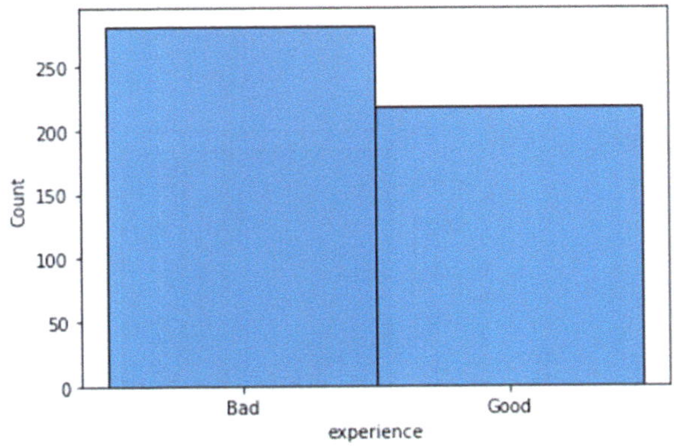

Fig. 7 Experience

3.1 Data Collection

The dataset for this problem statement was prepared from survey conducted among college students, comprised of features such as device used for online education—laptop or mobile, type of class sessions—fully online or hybrid mode, nature of student's residence—urban or rural area, difficulty level of subjects, difficulties faced in online classes, social media usage during class hours, type of social media used, type of lecturers' interaction in class, whether practical classes are conducted or not, effectiveness of such practical classes, online cultural events conducted or not and overall experience of online education. The features were collected and converted to a row-columnar labeled dataset. Conversion to machine-understandable datasets was done in the process of data cleaning.

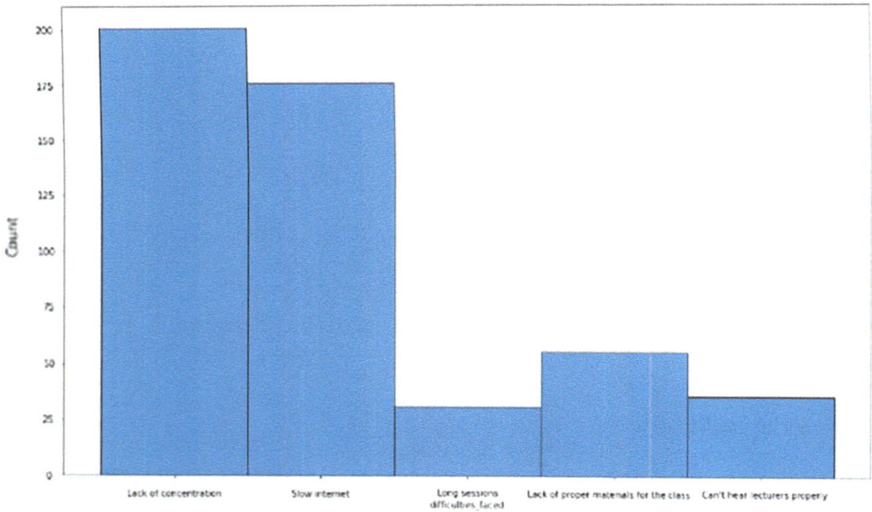

Fig. 8 Difficulties_faced

Fig. 9 Practical_classes_conducted

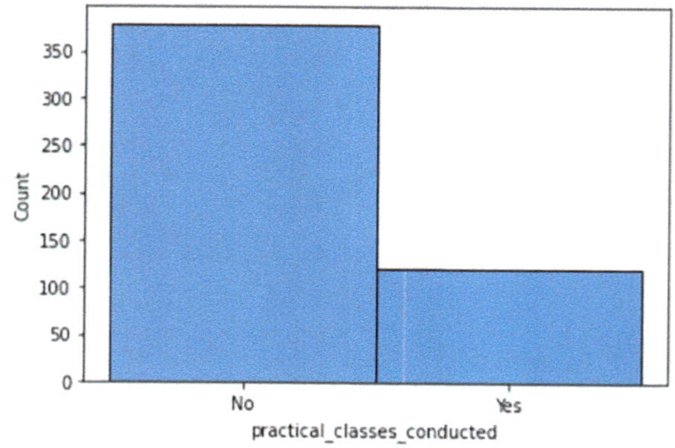

Fig. 10 Practical_class_effectiveness

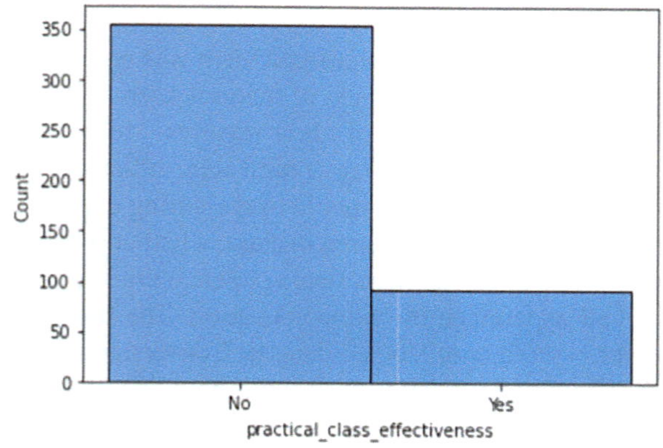

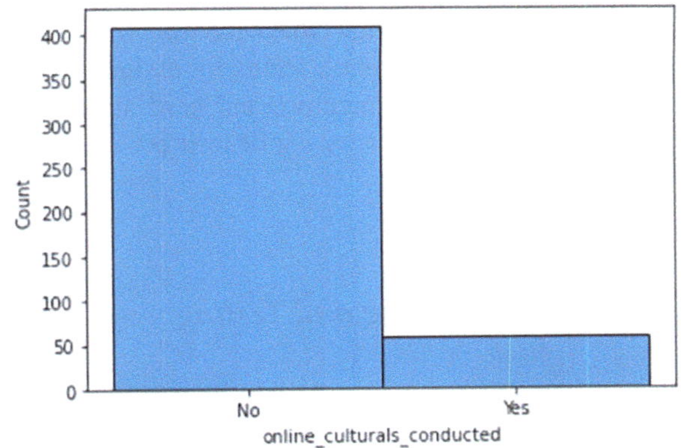

Fig. 11 Online_culturals_conducted

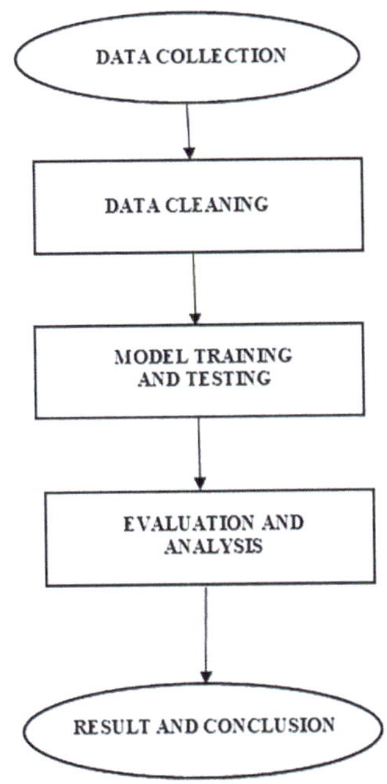

Fig. 12 Architectural framework

3.2 Data Cleaning

Next, the dataset was checked for null values and it was found that attributes 'practical_class_effectiveness' and 'online_culturals_conducted' contained null values. Univariate analysis in which only one variable is analyzed with data visualization techniques was performed. Barplots were plotted for each variable with the seaborn library from Python and null values were filled using information obtained from barplot analysis. The dataset obtained was in categorical form, so it was converted into numerical form through label encoding. In label encoding, the categorical values were converted into numeric forms by assigning numbers in increasing order for e.x for attributes easy, medium, and hard are assigned 0, 1, and 2, respectively. Subsequently the dataset was split into 80–20 ratios as training and testing sets, respectively.

3.3 Model Training and Testing

Commonly used algorithms for student performance prediction:

Commonly employed algorithms for predicting student performance include K-nearest neighbors (KNN), decision trees, random forests, and voting classifiers.

K-Nearest Neighbors (KNN). K-nearest neighbors is a nonparametric method utilized for both classification and regression. In KNN, an object's classification is determined by a majority vote from its k-nearest neighbors. This algorithm is sensitive to the local structure of data and employs distance metrics like Euclidean distance, Manhattan distance, Hamming distance, and Minkowski distance. KNN finds applications in tasks such as recommendation systems and image classification.

Decision Tree: A decision tree is a supervised learning algorithm used for classification and regression. It recursively partitions data into subsets based on input feature values. At each node, a feature optimally splitting data is selected by the algorithm. Leaf nodes represent the decision or prediction, be it a class label for classification or an output value for regression. However, decision trees can be prone to overfitting, mitigated by techniques like pruning or limiting the tree's depth. It is a white box model, providing transparency into the decision-making process.

Random Forest: Random forest is an ensemble machine learning algorithm for classification and regression. It builds multiple decision trees and averages predictions. The ensemble technique, known as bagging, involves training trees on different subsets of data and features, reducing overfitting. Feature randomness is employed, selecting random feature subsets for each split in each tree. Random forest effectively addresses overfitting concerns.

Voting Classifier: A voting classifier is an ensemble technique combining predictions from multiple base classifiers to enhance overall accuracy. Diverse base classifiers are typically employed, and the voting classifier can operate in hard or soft voting modes. Hard voting considers majority decision, while soft voting uses probability or weighted average of base classifier decisions. The 'VotingClassifier' from 'sklearn.ensemble' in Python's Scikit-learn library is a notable implementation, compatible with ensemble methods like bagging and boosting.

3.4 Evaluation and Analysis

Definitions

F1 Score: $F1$ score, or F-measure, gauges a test's accuracy, calculated from precision and recall. Precision is true positive results divided by all positive results, and recall is true positive results divided by all samples that should have been identified as positive. The $F1$ score is the harmonic mean of precision and recall.

$$F1 \text{ score} = \frac{2 \times (\text{Precision} \times \text{Recall})}{(\text{Precision} + \text{Recall})}$$

Accuracy: The metric measures correctly identified cases, often used when all classes are equally important. Accuracy prioritizes true positives and true negatives. In contrast, the $F1$ score is preferred when false negatives and false positives (Table 1 and 2).

$$\text{Accuracy} = \frac{(\text{TP} + \text{TN})}{(\text{TP} + \text{TN} + \text{FP} + \text{FN})}$$

Based on training and testing done, it is apparent that KNN algorithm was most effective and accuracy wise, for use in student performance prediction tasks where the features being used are similar to those used in this study. When it comes to $F1$ score, the random forest algorithm seems to be the best candidate and therefore it would make sense to use this algorithm if we aim to optimize $F1$ score over the accuracy while predicting student performance.

Table 1 Accuracy scores for different algorithms

Sl. no.	Algorithm name	Accuracy obtained (%)
1	K-nearest neighbor	80
2	Decision tree	97
3	Random forest	78
4	Voting classifier	79

Table 2 $F1$ scores for different algorithms

Sl. no.	Algorithm name	$F1$ score (%)
1	K-nearest neighbor	78.96
2	Decision tree	75.44
3	Random forest	94.44
4	Voting classifier	78.46

4 Mathematical Model

The meanings of the involved notations are listed out as follows:

$$C\text{—means}$$

$$i\text{—means}$$

$$j\text{—means}$$

This mathematical model is proposed for maximizing students' performance in online classes as discussed below:

$$\text{Max}(Ci, Xi, Yi, Zi, Ki, Pi, Ei) \geq 0$$

$$Ci\text{—}i\text{th of the taken class } Ci = \begin{cases} 0, & \text{online} \\ 1, & \text{offline} \end{cases}$$

$$Xi\text{—}i\text{th of the nature } Xi = \begin{cases} 0, & \text{urban} \\ 1, & \text{rural} \end{cases}$$

$$Yi\text{—}i\text{th of the difficulty level } Yi = \begin{cases} 0, & \text{easy} \\ 1, & \text{med} \\ 2, & \text{hard} \end{cases}$$

$$Zi\text{—}i\text{th of the social media } Zi = \begin{cases} 0, & \text{No} \\ 1, & \text{Yes} \end{cases}$$

$$Ki\text{—}i\text{th of the taken class } Ki = \begin{cases} 1, & \text{Yes} \\ 0, & \text{No} \end{cases}$$

$$Pi\text{—}i\text{th of the Practical\_Class\_Conducted } Pi = \begin{cases} 0, & \text{not - conducted} \\ 1, & \text{conducted} \end{cases}$$

$$Ei\text{—}i\text{th effectiveness } Ci = \begin{cases} 1, & \text{Yes} \\ 0, & \text{No} \end{cases}$$

Metrics to be considered: offline class, practical class, and social media class.

(Offline, Rural, Hard, Social Media Usage, Good Experience, Practical Conducted, Experience).

$$S1(1, 1, 2, 1, 1, 1, 1)$$
$$S2(1, 1, 1, 1, 0, 0, 1)$$
$$S3(1, 1, 0, 1, 1, 0, 1)$$
$$S4(1, 1, 1, 1, 0, 1, 0)$$
$$S5(1, 1, 2, 0, 1, 1, 1)$$
$$S6(1, 1, 1, 1, 0, 1, 1)$$
$$S7(1, 0, 0, 1, 1, 1, 1)$$
$$S8(1, 0, 0, 0, 1, 1, 1)$$
$$S9(0, 0, 0, 1, 1, 1, 1)$$
$$S10(0, 0, 1, 0, 1, 0, 0)$$

5 Conclusion

The usefulness of ML in building models predicting student performance accurately largely comes down to algorithm of choice. In this paper, several models such as KNN, random forests, decision trees, etc., were used to build models to predict student performance, and as expected varying results were obtained for accuracy and other measures of performance. The algorithm is selected based on data being worked upon. Some algorithms work best with some forms of data and not so much with other formats. The current work will potentially aid academic institutions like schools and colleges to predict in whether a student is about to pass or fail based on input data and thus help them take corrective measures and change teaching methodologies to enhance learner experience. The work can also be transformed to a web-based application to make access easier for institutions. In future research, improving data quality must be focused upon thereby improving accuracy of models, and hence helping understand student performance better and assisting institutions in deciding right about teaching techniques educating students for optimal performance.

References

1. Institute of Electrical and Electronics Engineers, & Hindusthan Institute of Technology. (n.d.). In: Proceedings of the international conference on electronics and sustainable communication systems (ICESC 2020), pp 02–04, July 2020
2. Aoulad Ali H, Mohamed C, Abdelhamid B, el Alami T (2021) Prediction MOOCs for the student by using machine learning methods. In: Proceedings—11th international conference on virtual campus. JICV 2021. https://doi.org/10.1109/JICV53222.2021.9600310
3. Habib S, Parthornratt T (2021) Students' perception of online classes and exams held during COVID-19 pandemic: the engineering faculty's experience at Assumption University. In: 6th International STEM education conference. ISTEM-Ed 2021. https://doi.org/10.1109/iSTEM-Ed52129.2021.9625135

4. Zhang W, Huang X, Wang S, Shu J, Liu H, Chen H (2017) Student performance prediction via online learning behavior analytics. In: Proceedings—2017 international symposium on educational technology. ISET 2017, pp 153–157. https://doi.org/10.1109/ISET.2017.43
5. Jain College of Engineering, Institute of Electrical and Electronics Engineers. Bangalore Section., & Institute of Electrical and Electronics Engineers. (n.d.). In: 2020 International conference for emerging technology (INCET), Belgaum, India. 5–7 Jun 2020
6. Durán Acevedo CM, Carrillo Gómez JK, Albarracín Rojas CA (2021) Academic stress detection on university students during the COVID-19 outbreak by using an electronic nose and the galvanic skin response. Biomed Signal Process Control 68. https://doi.org/10.1016/j.bspc.2021.102756
7. Rao BS, Jagadeesh Sai K, Anirudh LK, Sravani CH (2020) Student academic performance prediction using machine learning classification algorithms. Int J Adv Sci Technol 29(5):9314–9322
8. AISSMS Institute of Information Technology, & Institute of Electrical and Electronics Engineers. (n.d.). In: 2020 International conference on emerging smart computing and informatics (ESCI), AISSMS Institute of Information Technology, Pune, India. 12–14 Mar 2020
9. Institute of Electrical and Electronics Engineers. (n.d.). ICUMT 2020 : the 12th International congress on ultra modern telecommunications and control systems : 5–7 October
10. Gopal R, Singh V, Aggarwal A (2021) Impact of online classes on the satisfaction and performance of students during the pandemic period of COVID-19. Educ Inf Technol 26(6):6923–6947. https://doi.org/10.1007/s10639-021-10523-1
11. Yaseen H, Alsoud AR, Nofal M, Abdeljaber O, Al-Adwan AS (2021) The effects of online learning on students' performance: a comparison between UK and Jordanian Universities. Int J Emerg Technol Learn 16(20):4–18. https://doi.org/10.3991/ijet.v16i20.24131
12. Alshabandar R, Hussain A, Keight R, Khan W (2020) Students performance prediction in online courses using machine learning algorithms. In: Proceedings of the international joint conference on neural networks. https://doi.org/10.1109/IJCNN48605.2020.9207196
13. Keshtkar F, Cowart J, Crutcher A (2016) Predicting risk of failure in online learning platforms using machine learning algorithms for modeling students' academic performance
14. Sriram S, Naren J (2018) Bharati Vidyapeeth's Institute of Computer Applications and Management (BVICAM)
15. Ramanujam SS, Sivaneshwar P, Naren J, Madhumitha S, Vithya G (n.d.) A study on hybrid recommender system with deep learning and deployment in big data
16. Elakia G, Aarthi NJ (2014) Application of data mining in educational database for predicting behavioural patterns of the students. Int J Eng Technol. https://www.researchgate.net/publication/264458763
17. Ganesan V (2019) A novel Architecture for the development of an e-learning hybrid recommender system with deep learning techniques in big data environment some of the authors of this publication are also working on these related projects: quantum sensor image capturing View project. In: 3rd International conference on innovative computing and communication (ICICC-2020). https://www.researchgate.net/publication/338525987
18. Naren J, Banu MZ, Lohavani S (2020) Recommendation system for students' course selection. In: Somani AK, Shekhawat RS, Mundra A, Srivastava S, Verma VK (eds) Smart systems and IoT: innovations in computing. Springer Singapore, pp 825–834
19. Naren J, Ganesan V, Gayathiri P, Dhivyhalakshmi KK, Ramalingam P (2021) A machine learning model for predicting academic performance of students through internet usage. In: Mallick PK, Bhoi AK, Chae G-S, Kalita K (eds) Advances in electronics, communication, and computing. Springer Nature Singapore, pp 315–324
20. Naren J, Ganesan V, Reddy B (2019) Predicting academic performance in students by an analytical study on big data machine learning techniques. Int J Psychosoc Rehabil 23:371–376. https://doi.org/10.37200/IJPR/V23I1/PR190247
21. Naren J, Ashokkumar K, Raghavendran V (2016) Semantic feedback on student's performance with data mining techniques: State of the art survey. In: 2016 3rd International conference on computing for sustainable global development (INDIACom), pp 1605–1609

22. Ganesan V (2019) Deep learning 5 some of the authors of this publication are also working on these related projects: quantum sensor image capturing. In: 3rd International conference on innovative computing and communication (ICICC-2020). Int J Psychosoc Rehabil 23. https://www.researchgate.net/publication/338526152
23. Li Y, Chen X, Liu S, Wang L (2022) Exploring middle school students' attitudes and satisfaction about home-based online education during the COVID-19 epidemic and the influential variables. Educ Stud 58(2):177–199. https://doi.org/10.1080/00131946.2021.2010080
24. Goundar S, Deb A, Lal G, Naseem M (2022) Using online student interactions to predict performance in a first-year computing science course. Technol Pedagog Educ 31(4):451–469. https://doi.org/10.1080/1475939X.2021.2021977
25. Azzali S, Mazza T, Tibiletti V (2022) Student engagement and performance: evidence from the first wave of COVID-19 in Italy. Account Educ, pp 1–22. https://doi.org/10.1080/09639284.2022.2081813
26. Lux G, Callimaci A, Caron M-A, Fortin A, Smaili N (2022) COVID-19 and emergency online and distance accounting courses: a student perspective of engagement and satisfaction. Account Educ, pp 1–35. https://doi.org/10.1080/09639284.2022.2039729

MIX
Papier aus verantwortungsvollen Quellen
Paper from responsible sources
FSC® C105338

If you have any concerns about our products,
you can contact us on
ProductSafety@springernature.com

In case Publisher is established outside the EU,
the EU authorized representative is:
Springer Nature Customer Service Center GmbH
Europaplatz 3, 69115 Heidelberg, Germany

Printed by Libri Plureos GmbH
in Hamburg, Germany